Principles of Environmental Sampling

Second Edition

Principles of Environmental Sampling

Second Edition

Lawrence H. Keith, Editor

ACS Professional Reference Book

AMERICAN CHEMICAL SOCIETY
WASHINGTON, DC

Library of Congress Cataloging-in-Publication Data

Principles of environmental sampling / Lawrence H. Keith, editor.—2nd ed.

p. cm.—(ACS professional reference book)

Includes bibliographical references and index.

ISBN 0-8412-3152-4

1. Environmental sampling.

I. Keith, Lawrence H., 1938– . II. American Chemical Society. III. Series.

GE45.S25P75 1996
628.5´0287—dc20 96-9386
CIP

PRINTED IN THE UNITED STATES OF AMERICA

About the Editor

Lawrence H. Keith is a Corporate Fellow at Radian International LLC, in Austin, TX. At Radian he helps develop and market high-purity analytical reference materials and also helps manage technical contracts. A pioneer in environmental sampling and analysis, method development, handling of hazardous chemicals, and the preparation of environmental analytical standards, he has published many books and technical articles on these subjects. His recent publications have involved electronic books and expert systems as well as the traditional printed medium. Recently he helped organize the successful international conference Pacifichem '95, and he is the co-organizer from the American Chemical Society (ACS) Division of Environmental Chemistry, Inc., for the joint ACS–U.S. Environmental Protection Agency annual Waste Testing and Quality Assurance Symposium. He has also organized many symposia for the ACS and has served as chair, secretary, alternate councilor, and program chair for the Division of Environmental Chemistry, Inc. Keith is the current editor of the division's newsletter, *EnvirofACS*, and also serves as chair of the Subcommittee on Environmental Monitoring and Analysis for the ACS Committee on Environmental Improvement. He teaches ACS short courses on environmental sampling and analysis and has lectured throughout the United States and abroad on topics involving obtaining reliable environmental data. In addition, Keith also serves on academic, industrial, and government advisory boards.

Contents

Quality Assurance and Quality Control

Sampling Waters

Sampling Air

Sampling Biota

Sampling Solids and Hazardous Wastes

Glossary and Indexes

Contributors

Albert, Richard *page 653*
Center for Food Safety and Applied Nutrition
U.S. Food and Drug Administration
Washington, DC 20204
Current address: 25 Lennon Court
Unit 26
South Boston, MA 02107

Anderson-Sprecher, Richard *page 203*
Statistics Department
University of Wyoming
Laramie, WY 82071

Baldocchi, D. D. *page 599*
Atmospheric Turbulence and Diffusion Division
National Oceanic and Atmospheric Administration
Oak Ridge, TN 37831

Barcelona, Michael J. *pages 41, 693*
University of Michigan
Department of Civil and Environmental Engineering
181 EWRE Building
Ann Arbor, MI 48109–2125

Barnard, Thomas E. *page 169*
Metcalf & Eddy International, Inc.
30 Harvard Mill Square
Wakefield, MA 01880

Bayne, C. K. *page 277*
Computer Science and Mathematics Division
Oak Ridge National Laboratory
Oak Ridge, TN 3781–6418

Bertoni, Malcolm J. *page 111*
Environmental Research Planning Department
Research Triangle Institute
1615 M Street, N.W., Suite 740
Washington, DC 20036

Black, S. C. *page 139*
Analytical Services Section
Bechtel Nevada Corporation
P.O. Box 98521, NLV082
Las Vegas, NV 89193

Bollinger, Mark *page 267*
ENSECO/Rocky Mountain Analytical Laboratory
4955 Yarrow Street
Arvada, CO 80002

Bone, Larry I. *page 737*
Dow Chemical Company
2030 Building
Midland, MI 48674

Borgman, Leon E. *pages 203, 753*
Statistics Department and Department of Geology and Geophysics
University of Wyoming
Laramie, WY 82071

Bourke, John B. *page 671*
Analytical Laboratories
New York State Agricultural Experiment Station
Cornell University
Geneva, NY 14456

Bryant, Mark A. *page 393*
Harding Lawson Associates
5580 Havana Street, Suite 5A
Denver, CO 80239

Callaway, Owen *page 267*
ENSECO/Rocky Mountain Analytical Laboratory
4955 Yarrow Street
Arvada, CO 80002

Carlberg, Kathy *page 267*
ENSECO/Rocky Mountain Analytical Laboratory
4955 Yarrow Street
Arvada, CO 80002

Carter, Kevin R. *page 727*
EnSys Environmental Products
P.O. Box 14063
Research Triangle Park, NC 27709

Chow, Judith C. *page 539*
Desert Research Institute
University and Community College System of Nevada
5625 Fox Avenue
Reno, NV 89506

Claff, Roger E. *page 693*
American Petroleum Institute
1220 L Street, NW
Washington, DC 20005

Comeau, Joseph *page 693*
Inchcape/Aquatec Testing Services
55 South Park Drive
Colchester, VT 05446

Cowgill, U. M. *page 317*
Mammalian and Environmental Toxicology
Dow Chemical Company
Midland, MI 48674

Dick, Elie M. *page 237*
Mintra, Inc.
9297 Parkside Draw
Woodbury, MN 55125-7507

Divita, Frank, Jr. *page 539*
Desert Research Institute
University and Community College System of Nevada
5625 Fox Avenue
Reno, NV 89506

Edwards, Larry O. *page 471*
GeoMatrix Consultants, Inc.
180 Blue Ravine, Suite C
Folsom, CA 95630

Edwards, P. G. *page 3*
Radian International LLC
P.O. Box 201088
Austin, TX 78720–1088

Eng, Leslie *page 693*
Trillium, Inc.
28 Grace's Drive
Coatesville, PA 19320

Feenstra, Stan *page 693*
Applied Groundwater Research Ltd.
The Pentagon Building, Suite 207
2550 Argentia Road
Mississauga, ON, L5N 5R1
Canada

Flatman, George T. *pages 203, 753, 779*
U.S. Environmental Protection Agency
National Exposure Research Laboratory
Characterization Research Division
P.O. Box 93478
Las Vegas, NV 89193–3478

Floyd, M. P. *page 337*
State of California Department of Water Resources, Central District
3251 S Street
Sacramento, CA 95816

Garner, Forest C. *page 679*
Lockheed Engineering and Management Services Company
P.O. Box 15027
Las Vegas, NV 89109

Gerow, Ken *page 753*
Department of Zoology and Physiology and Statistics Department
University of Wyoming
Laramie, WY 82071

Gordon, Sydney M. *page 401*
Battelle Memorial Institute
505 King Avenue
Columbus, OH 43201

Hagen, Donald F. *page 297*
3M Corporate Research Laboratories
3M Center, Building 201–1W–29
St. Paul, MN 55144–1000

Hicks, B. B. page 599
Atmospheric Turbulence and Diffusion Division
National Oceanic and Atmospheric Administration
Oak Ridge, TN 37831
Current address: National Oceanic and Atmospheric Administration
Air Resources Laboratories
1315 East–West Highway
Room 3151
Silver Spring, MD 20910

Holcombe, L. J. page 361
Radian International LLC
1093 Commerce Park Drive
Oak Ridge, TN 37830

Horwitz, William page 653
Center for Food Safety and Applied Nutrition
U.S. Food and Drug Administration
Washington, DC 20204

Inn, Kenneth G. W. page 185
National Institute of Standards and Technology
Building 245, Room C229
Gaithersburg, MD 20899

Jackson, Larry P. page 743
Western Research Institute
Laramie, WY 82071

Keith, Lawrence H. page 3
Radian International LLC
P.O. Box 201088
Austin, TX 78720–1088

Kent, Robert T. page 377
IT Corporation
2499-B Capital of Texas Highway
Austin, TX 78746

Kern, John W. page 203
Western Ecosystems Technology, Inc.
1402 South Greeley Highway
Cheyenne, WY 82007

Klodowski, Harry F., Jr. page 63
Law Offices of Harry F. Klodowski, Jr.
Grant Building, Suite 3321
330 Grant Street
Pittsburgh, PA 15219–2202

Klopp, Kris *page 693*
Wisconsin Department of Natural Resources
P.O. Box 7921
Madison, WI 53707

Koerner, C. E. *page 155*
Radian International LLC
2455 Horsepen Road
Herndon, VA 20171

Kulkarni, Shrikant V. *page 111*
Center for Environmental Measurements and Quality Assurance
Research Triangle Institute
Research Triangle Park, NC 27709

Lewis, David L. *pages 3, 85*
Radian International LLC
P.O. Box 201088
Austin, TX 78720–1088

Lewis, Robert G. *page 401*
National Exposure Research Laboratory (MD–77)
U.S. Environmental Protection Agency
Research Triangle Park, NC 27711

Liggett, Walter S. *page 185*
National Institute of Standards and Technology
Building 820, Room 353
Gaithersburg, MD 20899

Malley, Michael J. *page 393*
Harding Lawson Associates
5580 Havana Street, Suite 5A
Denver, CO 80239
Current address: 9800 Richmond Avenue
Suite 200
Houston, TX 77042

Markell, Craig *page 297*
New Products Department
3M Industrial and Consumer Sector
3M Center, Building 209–1W–24
St. Paul, MN 55144–1000

Maskarinec, Michael P. *pages 277, 693*
Chemical and Analytical Sciences Division
P.O. Box 2008
Oak Ridge National Laboratory
Oak Ridge, TN 37831–6120

Meyers, T. P. *page 599*
Atmospheric Turbulence and Diffusion Division
National Oceanic and Atmospheric Administration
Oak Ridge, TN 37831

Minnich, Marty *page 693*
Lockheed Environmental System and and Technology Company
Environmental Science and Technology Division
980 Kelly Johnson Drive
Las Vegas, NV 89119

Moore, Michael B. *page 693*
ENSR Consulting and Engineering
Acton, MA 01720
Current address: The Johnson Company, Inc.
100 State Street
Montpelier, VT 05602

Nesbitt, Kevin J. *page 727*
EnSys Environmental Products
P.O. Box 14063
Research Triangle Park, NC 27709

Newburn, Lorance H. *page 225*
Environmental Division
Isco, Inc.
531 Westgate Boulevard
Lincoln, NE 68528–1586

Norris, James E. *page 259*
ERM-Southeast, Inc.
1110 Montlimar Drive
Suite 560
Mobile, AL 36609

Parr, Jerry *pages 267, 693*
ENSECO/Rocky Mountain Analytical Laboratory
4955 Yarrow Street
Arvada, CO 80002

Patton, G. L. *page 3*
Radian International LLC
1801 Broadway, Suite 1300
Denver, CO 80202

Payne, Katherine E. *page 377*
IT Corporation
2499-B Capital of Texas Highway
Austin, TX 78746

Pilgrim, Mary J. *page 693*
Alaska Department of Environmental Conservation
Juneau Environmental Analysis Laboratory
10107 Bentwood Place
Juneau, AK 99801–8552

Rogalla, J. A. *page 621*
Radian International LLC
10389 Old Placerville Road
Sacramento, CA 95827

Rose, Candice *page 693*
Argonne National Laboratory
9700 South Cass Avenue
Argonne, IL 60439

Schulte, Robert M. *page 693*
Delaware Department of Natural Resources and Environmental Control
715 Grantham Lane
New Castle, DE 19720–4801

Siegrist, Robert *page 693*
Oak Ridge National Laboratory
P.O. Box 2008
Oak Ridge, TN 37831–6038

Smith, James S. *pages 393, 693*
Trillium, Inc.
28 Grace's Drive
Coatesville, PA 19320

Spittler, Terry D. *page 671*
Analytical Laboratories
New York State Agricultural Experiment Station
Cornell University
Geneva, NY 14456

Stapanian, Martin A. *page 679*
Lockheed Engineering and Management Services Company
P.O. Box 15027
Las Vegas, NV 89109

Steele, David P. *page 393*
Walter B. Satterthwaite Associates, Inc.
720 North Five Points Road
West Chester, PA 19380

Stinchfield, Matt
CET Environmental Services, Inc.
1010 East Palmdale, Suite 101
Tucson, AZ 85714

Tanner, Roger L.
Environmental Chemistry Division
Department of Applied Science
Brookhaven National Laboratory
Upton, NY 11973
Current address: Tennessee Valley Authority
Environmental Research Center
Chemical Engineering Building 2A
Muscle Shoals, AL 35662

Taylor, John K.
National Institute of Standards and Technology
Gaithersburg, MD 20899

Urban, Michael J.
Envirotech Research, Inc.
777 New Durham Road
Edison, NJ 08817

Watson, John G.
Desert Research Institute
University and Community College System of Nevada
5625 Fox Avenue
Reno, NV 89506

Williams, C. Herndon
Radian Corporation
8501 North Mopac Boulevard
Austin, TX 78759

Williams, Llewellyn R.
Quality Assurance and Method Development Division
Environmental Monitoring Systems Laboratory
U.S. Environmental Protection Agency
944 East Harmon Avenue
Las Vegas, NV 89114

Winegar, Eric D.
Air Toxics Limited
180 Blue Ravine Road, Suite B
Folsom, CA 95630

Yfantis, Angelo A. *page 779*
Computer Science Department
University of Nevada–Las Vegas
Las Vegas, NV 89154

Young, Susan J. *page 671*
U.S. Food and Drug Administration
Washington, DC 20204

Preface

The goal of this book, like its predecessor, is to help ensure consideration of the many variables and special techniques that are needed to plan and execute sampling activities that will provide representative environmental samples for analysis. Failure to consider all of the important factors will result in samples that do not represent the site from which they were taken. Analysis of such unrepresentative samples then results in data that may lead to wrong conclusions (which can be very expensive). The logic is simple: If the right kinds of samples are not collected from the right areas at a site and then preserved, prepared, and analyzed correctly, wrong answers will be obtained. They may be precise and accurate answers, but they will be wrong in that they will not represent the condition of the site with respect to the absence, presence, or representative concentrations of the pollutants of interest.

There are special techniques and devices for sampling different matrices (e.g., air, water, soil, sediment, vegetation, and animals) and even different techniques and devices for sampling varying types of the same matrix (e.g., surface waters, groundwater, and drinking water). There are also different sampling and preservation techniques for different types of analytes within the same matrix (e.g., metals, volatile organic compounds, semivolatile organic compounds, cyanides, and ammonia in water samples).

In addition to matrix complications, the greatest source of variability in most environmental samples is the heterogeneous distribution in time or space of the analytes of interest (pollutants). For example, a soil or vegetation sample taken at one spot and a related sample from a nearby spot may differ in concentration of the pollutants of interest by 100% or more. Additionally, when a sample of air or water at a certain spot is compared with another sample taken from the same spot the next day or hour, or even within a few minutes, the concentrations of the pollutants may be found to have changed by 100% or more between samplings.

Unfortunately, the problems do not end here. After samples that are most representative of an environmental site have been taken, they must be correctly preserved, stored, and prepared for analysis (unless they are analyzed on the spot in the field). During this time they can be contaminated or some of the analytes of interest may be lost through volatilization, hydrolysis, or reactions with other chemicals. Different types of pollutants have varying degrees of propensity to undergo these types of losses. In addition, various matrices have great variability with respect to their complexity and thus to potentially interfering compounds (e.g., clean drinking water may contain free chlorine that can react with phenols, and complex industrial wastewaters may contain acids, bases, or organic compounds that can react with each other or promote hydrolysis).

It is obvious that if the environmental samples that are submitted for analysis are not representative of the site from which they came, then it does not matter how accurate and precise the laboratory measurements on those samples are; they still will not provide an accurate estimate of the state of pollution at that environmental site. Despite this common sense logic, most quality assurance–quality control (QA–QC) sampling and analysis plans still focus on laboratory sources of error and variability. It is not uncommon for sampling sources of error (which include the heterogeneous distribution of pollutants in time or space) to be several hundred percent larger than the analytical sources of error. At the same time, laboratory QC focuses on maintaining accuracy (bias) and precision around a 5–10% range with a 95% confidence level. (What is wrong with this picture?)

A special feature of this second edition is the inclusion of a postcard for ordering free software that helps in estimating how many environmental samples need to be collected to meet desired levels of confidence that (1) any analytes of interest that may be present at a site will not be missed or (2) randomly collected samples are of sufficient number so that the analytes of interest will be within a desired tolerance level of their average true concentration. This software will also help in calculating how many QC samples need to be analyzed to provide a selected level of confidence for not exceeding a desired rate of false positives or false negatives (i.e., for not exceeding a desired rate of wrong answers). Parameters can be adjusted and answers recalculated easily to estimate how much lower the confidence level will be if fewer samples are collected and analyzed. These answers translate into informed decisions instead of "educated guesses" and thus provide a great advance over how we have been making most of these decisions until now.

No wonder obtaining samples that are representative of the environment from which they came is such a difficult process. In 1988 the American Chemical Society Committee on Environmental Improvement cosponsored a symposium about environmental sampling principles from which the first edition of this book was published. In 1994 the same committee cosponsored a second symposium. Manuscripts from it were selected for peer review and combined with others from the first edition to form this second edition. The new material represents 60% of the book; 40% of the chapters from the first edition were retained because they are still excellent resources of current technology for which we have no updates. An advantage of retaining selected chapters from the first edition is that it makes the second edition more complete; in other words, people who do not have the first edition will not have to find a copy of it to supplement the information in this edition. As in the first edition, the chapters are grouped into sections governed by matrix types plus QC and QA topics.

I hope this book will be as useful to people as the popular first edition was. I and the other authors have tried to provide information in a form that does not just present facts. Rather, we have tried to supplement information with advice based on our many years of experience. This format seemed more useful because it conveys knowledge instead of just information. And knowledge is what we need for environmental sampling and analysis to be meaningful—we are all inundated with a sea of facts and information. The trick is to separate meaningful environmental data from the mass of useless or less useful data and, through knowledge, to apply it to make our world a cleaner, better place in which to live. Regressing to previous levels of technology is not an acceptable option; we only move forward, never backward, with technology—especially when it affects our standard of living.

LAWRENCE H. KEITH
February 7, 1996

Planning and Sample Design

Chapter 1

Determining What Kinds of Samples and How Many Samples To Analyze

Lawrence H. Keith, G. L. Patton, David L. Lewis, and P. G. Edwards

Statistical considerations, in conjunction with data quality objectives (DQOs), are used to estimate how many samples are required for individual needs. A free Microsoft Windows-based program (DQO-PRO) in the form of a simple calculator interface makes estimating numbers of samples, error tolerance levels, and confidence levels easily understood and quickly calculated. It also allows quick assessments of changes to DQOs. When DQO-PRO *is used in conjunction with the commercial program* Practical QC, *answers can quickly be obtained to the basic questions of how many and what kinds of quality control (QC) and environmental samples are needed. These answers are explained within the context of four basic principles of environmental sampling and analysis.*

The most frequently asked questions involving environmental sampling and analysis are (1) What kinds of samples are needed? and (2) How many of them are needed? The answers to these questions lie at the very heart of every environmental sampling and analysis project. Once these questions are answered, the planning for the rest of a project can commence and the answers to many more questions can be obtained. These additional questions involve the intricate details of sampling and analysis, such as:

- How will the results of the sampling and analysis project be used?
- Where should the samples be collected?
- What containers should be used to collect them?
- How should the samples be preserved?
- What preparations are necessary for sample analysis?

3152–4/96/0003$19.50/0 © 1996 American Chemical Society

- How should the samples be analyzed (e.g., what types of instruments, detectors, and analytical methods should be used)?
- How should the data be evaluated and presented?

The rest of the chapters in this book are dedicated to these many intricate details. They contain the advice and wisdom of many of our most experienced scientists. This chapter, however, focuses on how to determine what kinds of samples are needed and how many of them are needed.

For the question of how many samples are needed, answers are derived using statistical equations and the researcher's specific requirements, or *data quality objectives* (DQOs). However, although the equations have been known for years, they are not frequently available in a convenient form for use by chemists, project managers, samplers, regulators, and others who would use this information more often if they could understand it and if it were in an easily used form. We have prepared a sophisticated, but easy-to-use, series of computer programs called *DQO-PRO* that instantly calculate the number of samples needed on the basis of an individual's DQOs. Programmed using Windows Visual Basic, *DQO-PRO* comprises a trilogy of computer programs (*Enviro-Calc, HotSpot-Calc,* and *Success-Calc*) to calculate the number of samples needed to meet specific DQOs. *DQO-PRO* has a user interface like a common calculator and is accessed using Microsoft Windows.[1]

DQO-PRO provides answers for three objectives: (1) determining the rate at which an event occurs, (2) determining an estimate of an average within a tolerable error, and (3) determining the sampling grid necessary to detect localized points of contamination or *hot spots. DQO-PRO* facilitates understanding the significance of DQOs by showing the relationships between numbers of samples and DQO parameters such as (1) confidence to detect a specific rate of false-positive or false-negative conclusions, (2) tolerable error in an estimate of an average analyte concentration, and (3) probability of failing to detect a hot spot versus sampling grid size. The user has only to type in his or her requirements, and the calculator instantly provides the answers. Guidance for using the calculator is provided in hypertext help files. Users access the help text with a click of the mouse button.

Four Principles of Environmental Sampling and Analysis

The most simplistic view of an environmental investigation includes two basic kinds of samples: (1) samples from an environmental site that are used to assess the kind and level of pollution at that site (i.e., the analytes and their concentrations), and (2)

---

[1]To obtain *DQO-PRO,* please consult the postcard contained in this book or the Web site at http://www.instantref.com/inst-ref.htm. An IBM-compatible computer with Windows 3.1 or higher and a 3.5-in., 1.4-megabyte hard drive are required to install and run *DQO-PRO.* The program is free. It will also be able to be downloaded from the ACS Division of Environmental Chemistry Web site in late 1996.

quality control (QC) samples that are used to assess the kinds and amounts of *bias* and *imprecision* in the measurement system used to analyze the environmental samples. Depending on the kinds of QC samples selected for analysis, the data from the measurement system may reflect bias and imprecision from the laboratory, the field (including sampling), or both.

Viewed this way, the two primary questions posed in the introduction diverge into four basic questions:

1. What kinds of environmental samples are needed?
2. How many environmental samples are needed?
3. What kinds of QC samples are needed?
4. How many QC samples are needed?

Four principles of environmental sampling and analysis have been formulated, one for each of the four basic questions. These principles are as follows:

1. *Environmental samples must be representative of the portion of the environment being investigated.* This principle is critical to determining what kinds of environmental samples are needed.
2. *Procedures for sampling and analysis influence each other, and so plans for sampling and analysis are codependent.* The minimum numbers of samples needed may be directly influenced by, among other things, the objectives to be resolved and the choice of analytical methods, which affect the measurement system used to analyze the samples.
3. *QC samples must be representative of the environmental samples being analyzed.* This principle is critical to determining what kinds of QC samples are needed because the QC samples must provide potentials for error that are similar to those of the environmental samples.
4. *QC samples are used to provide an assessment of the kinds and amounts of bias and imprecision in data from analysis of the environmental samples.* The minimum numbers of QC samples needed are influenced by the rates of measurement errors possible (e.g., rates of *false positives* and *false negatives*) and the *confidence level* that is considered acceptable in correctly detecting this potential for error.

The First Principle

Environmental samples must be representative of the portion of the environment being investigated. Almost an infinite number of samples and measurements could be collected at any environmental site. Collectively, those data would be statistically referred to as the population at a site. The purpose of environmental sampling and analysis is to assess a small, but informative, portion of a population and then draw an inference about that population from the data gathered. For example, if an analyte such as mercury were identified and measured at detectable concentrations in samples of soil and river water from an environmental site, then it might be inferred that remaining portions of soil and river water also probably contain mercury at similar concentrations. If only soil samples had been collected and analyzed, then, in the absence of special knowledge about the site, it could not be inferred that the river water samples would also probably be contaminated with mercury. In the second case, soil samples would not be representative of the river water. Thus, samples of more than one *matrix* may be needed to provide an adequate assessment of an environmental site.

Matrix Considerations

Primary matrices include water, soils, biota, air, solid wastes, and liquid wastes. However, each of these primary matrices includes many different kinds of samples. For example, water samples may include those from lakes, rivers, underground sources, rain, industrial wastewaters, estuaries, and seas. Likewise, each of the other primary types of matrices has multiple subdivisions (e.g., sand and clay; plants and animals; indoor and outdoor air; ashes and sludge; and liquids that are more dense than water or less dense than water). In addition, combinations of some matrices can be sampled, such as aqueous sediments at the bottom of a lake or stream, solid particulate matter in air, and liquid aerosols in air.

Other Considerations

Most of the many different types of matrices require specialized equipment to obtain samples that will be representative of the parent (remaining) environmental population. In addition, the analytes of the investigation themselves will influence the type of sampling equipment needed as well as the procedures taken to preserve the samples if they are to be sent to a laboratory for analysis at some later time. For example, heavy metals; cyanides; ammonia; volatile, semivolatile, and nonvolatile organic materials; bromine-containing compounds; and cyclic chlorinated hydrocarbons require different types of containers, pH conditions, and protection from heat, light, and air as part of their preservation.

These many and varied considerations for the large variety of analytes in many different matrices make the collection of environmental samples difficult. Failure at this point (which is essentially the start of the chain of events that includes sampling, sample preparation, analysis, data evaluation, and reporting) means that all remaining time and money spent to collect, analyze, evaluate, and report the results will be wasted or at least compromised. In other words, if the environmental samples do not represent the portion of the environment being investigated to resolve a specific objective, then their analysis is usually pointless.

For example, what good would it do for a laboratory to report highly accurate and precise results on the concentration of polychlorinated biphenyls (PCBs) in samples of water from a lake if the objective was to determine the distribution of PCBs in the lake system and the majority of the PCBs were adsorbed to aqueous sediments? The data would not provide an accurate representation of the contamination of the lake system with PCBs, but it may provide an accurate representation of the PCB concentrations in just the lake water in the areas sampled.

As another example, why should one bother to analyze for volatile organic compounds (VOCs) such as benzene, xylene, or toluene in soil if the current accepted method of sampling and preservation results in large losses through volatilization? The chapter by Smith et. al. (*1*) discusses some examples of this kind of bias. If large errors (bias) occur before samples are introduced into an analytical instrument, then it does not matter that a laboratory can produce accurate and precise analytical results with respect to calibration standards; the data still do not represent the environmental site.

Sampling: The Weak Link

The activities associated with collecting and preserving environmental samples are generally acknowledged to be the weakest link in the chain of activities that comprise environmental sampling and analysis. Sampling activities are often the weakest link because they usually contribute the largest amount of errors in conclusions about the "true" condition of the site being investigated. Figure 1 illustrates the relative sources of variability in the chain of typical sampling and analysis events. The smallest sources of overall variability (analysis and data handling) are typically where most quality assessment (QA) or QC efforts are focused because steps involving laboratory analysis and data handling are the most easily controlled and assessed (i.e., we know how to do that well). However, laboratory and data handling variability usually accounts for the smaller sources of variability and incorrect conclusions associated with environmental sampling and analysis. Thus, the upper activities in the chain of events in Figure 1 also must be assessed if conclusions about the true conditions of contamination at a site are to be meaningful. Unless this assessment is done, we continue to fool ourselves about the conclusions drawn from environmental analytical data, and all of the data validation, laboratory certifications, and analytical reference standard certifications in the world will not make a difference.

Novices sometimes assume that, because highly sophisticated instrumentation such as that found in laboratories is not used, sampling must be relatively easy. Thus, inexperienced, low-paid technicians may be assigned the tasks of collecting environmental samples. The folly of this false effort to reduce the costs of environmental sampling and analysis should be apparent. A better way to reduce costs and protect an investigation from drawing wrong conclusions would be simply not to collect the samples; the conclusions might be just as good and would be much cheaper.

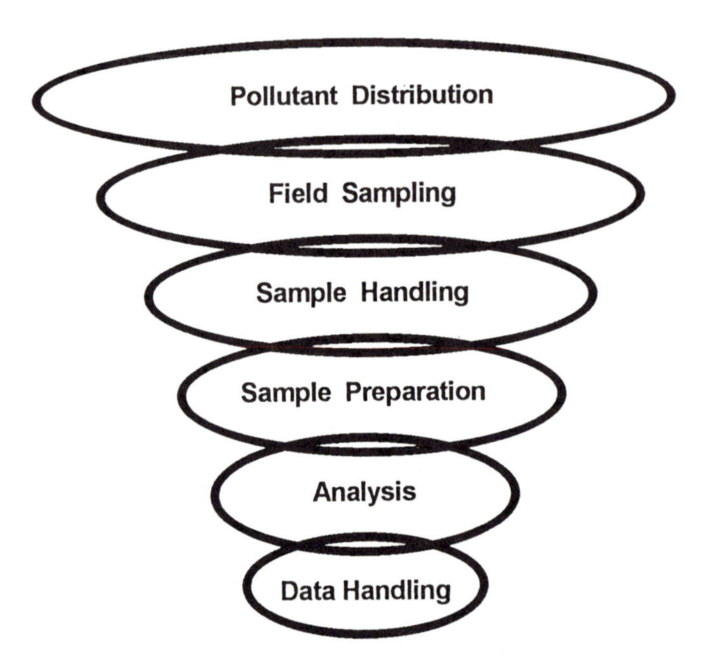

Figure 1. Sources of variability in the sampling and analysis chain of events.

Many of the other chapters in this book are devoted to this first principle because of its basic underlying importance. In fact, the other three principles are meaningless unless this first one is followed.

The Second Principle

Procedures for sampling and analysis influence each other, and so plans for sampling and analysis are codependent. In the preceding discussion, ways in which sampling can affect conclusions drawn from the analysis of environmental samples were illustrated. This relationship is direct and usually very clear. However, the relationship in the opposite direction can also influence the number of environmental samples needed. The analytical methods selected may influence not only the number of environmental samples that are needed but also the amounts of samples (i.e., their volume or weight) and the way they are preserved if they are not analyzed immediately in the field. Both the amount of a sample to be collected and the way in which samples are to be preserved (e.g., pH; temperature; and protection from light, heat, or air) will influence the sampling procedures. Thus, the influence of analytical procedures on sampling are also important. For this reason, plans for sampling and analysis are usually codependent. When sampling and analytical plans are developed independently of each other, the probability of obtaining data that will meet the objectives of the investigation may be significantly lowered.

Defining "Good" Data

Because almost every environmental sampling and analysis effort involves different, very specific objectives, an almost infinite number of considerations could be made unless a narrow focus is placed on the purpose of performing a particular sampling and analysis effort. That focus is what determines whether the resulting data are "good". "Good" is a strictly relative term. It defines whether the data obtained can be used with the specified level of confidence to make an inference about the environmental population from which samples were collected and analyzed.

What is "good" for one purpose may be totally inadequate for another. Attributes that are commonly used to assess the goodness of data for a particular use include:

- the level of confidence in the correct identification of the analyte(s);
- the concentration levels at which the analyte(s) can be identified or quantified with some specified level of confidence; and
- the degree to which the environmental samples represent the remaining unsampled population at an environmental site.

DQOs

The definition of a goal or objective of an environmental investigation is a primary component of the first step in developing DQOs (2). DQOs are statements that define the confidence required in conclusions drawn from data produced by a particular sampling and analysis project. Data cannot be evaluated as good or not good until the use of that data has been clearly defined.

Developing DQOs is a structured way to plan data collection and analysis efforts. It was developed by the U.S. Environmental Protection Agency (EPA) Quality Assurance Management Staff (QAMS) to (1) help define specific questions that an environmental project is intended to answer, (2) identify the decisions that will be made when using the resulting data, and (3) define the allowable risk of decision errors in specific and quantifiable terms. The quantifiable terms directly influence the numbers of environmental samples that are needed.

The DQO process comprises seven steps:

1. State the problem to be resolved;
2. Identify the decision(s) that must be made;
3. Identify all the inputs needed to make the decision(s);
4. Define the study boundaries (e.g., in space, time, and analytes);
5. Develop a decision rule for each decision;
6. Specify limits on decision errors; and
7. Optimize the design for collecting data.

This process results in qualitative and quantitative statements that specify study objectives, define the types and amounts of data needed, define the statistical populations that the data represent, and specify tolerable errors. *Tolerable errors* are those specified by the investigator (e.g., that the average concentration of an analyte be within ±10% of the true value). The tolerable errors that are chosen for closeness to true values of average analyte concentrations will affect the numbers of environmental samples that will be needed; the smaller the tolerable errors that are chosen, the larger will be the numbers of samples needed.

The tolerable errors that are chosen for acceptable rates of false positives and false negatives in the data likewise affect the numbers of QC samples that will be needed. These are discussed under the fourth principle.

Calculating Numbers of Environmental Samples Needed

The numbers of environmental samples needed for many projects can be calculated using two sets of equations. The first set of equations may be used when the purpose of environmental sampling is to estimate the average concentration of target analyte(s) within the study boundaries (specified in step 4 of the DQO process). The second set of equations may be used when the purpose of environmental sampling is to determine if a localized point of contamination (a hot spot) exists within the study boundaries.

Numbers of Samples Based on Average Pollutant Concentration. Equation 1 has been relatively widely used since its inclusion in a discussion on how to estimate numbers of environmental samples in a 1983 ACS journal article titled "Principles of Environmental Analysis" (3). Equation 1 makes the important assumption that the pollutants are distributed over the entire study area. The objective of the sampling and analysis effort is assumed to be to determine the average concentration of the pollutants within the study boundaries, which are referred to as a sampling site. Correctly defining sampling site boundaries is thus very important when using eq 1. The resulting characterization is for the site as a whole:

$$n = (z\sigma_p/E)^2 \tag{1}$$

where n is the number of samples calculated, z is the value of the standard normal variant (1.96 with a 95% confidence level), σ_p is the standard deviation for the sample population (obtained from the method), and E is the tolerable error in the estimate of the mean for the characteristic of interest (an arbitrary value of the amount of error one is willing to accept in the data).

Equation 1, however, does not consider the number of samples that one has available in order to estimate the variability. Therefore, a similar approach was developed with eqs 2 and 3 (4). These equations allow an iterative determination of the number of samples needed to meet the objectives for tolerable error and the confidence in estimating a mean concentration. This iteration is necessary because the t

distribution is the most appropriate for determining the number of samples when variability is estimated from sampling data. However, because one rarely has historical data to estimate the variability in a process or at an investigative site [the coefficient of variation (CV) or standard deviation (SD) of a method is usually used to estimate variability], the Z distribution must be used initially to estimate a number of samples. The number of samples calculated via the Z distribution is then introduced into the equation using t (retaining all other parameters as specified for the Z), and the required n is calculated again. This procedure should be iterated until the n stabilizes. We have used this iterative process in the *Enviro-Calc* computer program (which is a part of the *DQO-PRO* program). We assumed for our purposes that three iterations were sufficient. One fallacy of this assumption is that as the variability in the population approaches the error tolerated by the user, the formula will fluctuate wildly for several iterations. As a general rule, if the variability estimate is >75% of the tolerable error, then the estimated number of samples may not be accurate. The user should evaluate the need for such low error (relative to the variability) and consult a statistician. However, the use of eqs 2 and 3 should provide estimates of the numbers of samples needed that should be more accurate than those using eq 1.

Two sets of equations implemented in *Enviro-Calc* are presented below. The first set is to be used when the variability estimate is presented in absolute terms using the SD. The second set of equations is used when the variability is presented in relative terms using the CV [relative standard deviation (RSD)].

The equations for determining the number of samples needed to estimate a mean concentration when variability is estimated in absolute terms using the SD are:

$$\text{first pass equation: } n = [Z(1 - \alpha/2)(\text{SD}/E)]^2 \quad (2)$$

$$\text{second pass equation: } n = [t(1 - \alpha/2)(\text{SD}/E)]^2 \quad (3)$$

The equations for determining the number of samples to estimate a mean concentration when variability is estimated in relative terms using the coefficient of variation (CV or RSD) are:

$$\text{first pass equation: } n = \{Z(1 - \alpha/2)[\text{CV}/E(r)]\}^2 \quad (4)$$

$$\text{second pass equation: } n = \{t(1 - \alpha/2)[\text{CV}/E(r)]\}^2 \quad (5)$$

where Z is the standard normal deviate from the Z distribution using α for a two-tailed distribution; SD is the standard deviation for a sample set; E is the amount of error tolerable in the estimate of the average in absolute terms (e.g., 4 μg/L); $E(r)$ is the amount of error tolerable in the estimate of the average in relative terms (e.g., 5%); α is type I error (rate of false positives, e.g., 5%), but β (type II error; false negatives) may also be substituted in the equation; and t is the t statistic.

These equations have absolute and relative applications:

- For absolute applications, the standard deviation in eqs 2 and 3 is used, and the allowable error is stated in absolute terms. The standard deviation and the allowable error must be in the same terms.
- For relative applications, the coefficient of variation in eqs 4 and 5 is used, and the allowable error is stated in relative terms (i.e., as percentage).

Although simple random sampling plans are often used in environmental investigations, the assumption that the measurements follow a normal distribution is less certain. Therefore, unless one has previous information indicating that the assumption of normality is reasonable, the number of samples estimated by *EnviroCalc* should be considered to be sufficient to gather preliminary information about investigative media. Additional sampling may be required in a second or third phase after initial data have been analyzed and the underlying assumptions have been tested.

Defining the Sampling Site. For example, one might consider a physical site that contains areas of contaminated soil through which a stream flows that empties into a lake. Within the physical site a number of sampling sites may be identified during the DQO process. They could include the following:

- soils in one or more topographical areas;
- air over one or more portions of the physical site;
- vegetation in selected areas of the physical site;
- water from one or more areas of the stream;
- water from one or more areas of the lake;
- water from one or more groundwater wells in the area;
- aqueous sediment from one or more areas of the stream; and
- aqueous sediment from one or more areas of the lake.

Each of these specific physical areas could become individual sites for collecting samples if, during the DQO process, they were determined to be necessary. Different sampling equipment would be needed for different matrices, and usually the concentration of pollutant(s) would need to be determined for each area site or at least for each matrix. During the DQO process it would be decided which matrices and which physical areas of the site would need to be sampled. During that process it would also be decided whether any or all of the data from the area sites could be combined. For example, should

- all of the water sample data be combined,
- all of the groundwater sample data be combined,

- all of the stream sites data be combined, or
- data from each water area site be evaluated individually?

The temporal aspects of sampling at a selected site must also be defined. For example, will the site be sampled just one time or multiple times (e.g., daily, weekly, or monthly, and for how long)? These decisions must be made before the number of samples needed can be calculated.

Selecting the Analytical Method. Once the decisions have been made as to what and how to sample, then a decision must be made as to how the samples will be analyzed. This decision is necessary because the RSD or the SD for the analytical method that will be used must either be known or estimated. This information is necessary because analytical methods often significantly differ in the amount of variability they introduce in the data. Analytical variability is usually method-, analyte-, and matrix-specific. Analytical variability is also concentration-specific and usually increases as the analyte concentration level decreases.

Most environmental analytical methods accommodate multiple pollutants. The number of pollutants covered by a particular method may range from half a dozen analytes to several hundred. Exceptions include some methods for elements (involving atomic absorption spectrophotometers) or for analytes such as cyanide and ammonia. In practical terms one need only select the analyte of interest that has the largest SD or RSD listed with a method because the calculation of how many samples will be needed is related to the variability (i.e., SD or RSD) in the data that the method is expected to produce. Therefore, by selecting the analyte of interest with the largest variability in a method (highest SD or RSD), the DQOs of the other analytes analyzed by that method should also be met because, when calculated, they will require fewer numbers of samples (when all other inputs to the equation remain unchanged).

One must pay particular attention to the analytes of interest. When a method covers additional analytes, they should not be considered in these calculations because they could result in larger numbers of samples being collected than are needed if their SDs or RSDs are higher than those of the pollutants of interest.

The numbers of samples to be collected at a site are thus in part related to the method that will be used. When multiple methods will be used (to analyze all the pollutants of interest) and the samples will be subdivided (to accommodate multiple sample preparations for analysis), the number of samples to be collected will depend on the method with the largest variability (highest SD or RSD).

The values for SDs or RSDs may be obtained from each of the methods, or they may conveniently be found in one or more publications that contain short method summaries. Three sources of method summaries that contain pertinent information needed to make decisions concerning which, among several methods, would be best for analysis of samples for a particular purpose are listed here. Each contains SDs or RSDs, detection levels, instrumentation required, interferences, sample collection and preservation information, a brief description of the method and the matrices for which it is appropriate, and reference sources. The purpose of these publications is to help people quickly and easily find the information needed to calculate how many samples will need to be collected and to decide which method is best for meeting their specific DQOs. These three publications contain similar information in different formats to accommodate most people's needs. They are:

1. *Instant EPA's Best Methods* (5), published in Microsoft Windows format. This publication contains more than 1500 analyte–method summaries with the powerful and intuitive Windows searching engine. Data are presented in lists ordered two different ways: (1) analytes by method, by which one selects a method and sees all the analytes it covers; and (2) method by analyte, by which one selects an analyte and sees all the methods in the publication that will cover it. In addition to internal hyperlinks to, for example, abbreviations and definitions, this electronic publication is externally hyperlinked using a lightning bolt "hot key" to other publications in the Professional PC References/Windows Series. Links to the other publications are by chemical name and by method number.
2. *EPA's Sampling and Analysis Methods Database (2nd ed.)* (6), published in Microsoft DOS format. This publication also contains over 1500 analyte–method summaries, many of them the same as in *Instant EPA's Best Methods* (5), with the same easy searching engine used with the popular first edition. It is menu-driven and contains more than twice the number of analyte–method summaries as the first edition.
3. *Compendium of EPA's Sampling and Analysis Methods (2nd ed.)* (7), published in a book format. This book contains the same 1500 analyte–method summaries included in *EPA's Sampling and Analysis Methods Database (2nd ed.)*. They are published as a bound book, complete with an index, and they provide a complimentary and convenient source for people who do not have or do not want to use a computer. Although some may find this book easier to read than a computer screen, it lacks the searching capability of the two electronic publications.

A caution must be noted here. *Calculations of estimated numbers of samples needed that are based only on analytical method SDs or RSDs will result in underestimated numbers of samples because only the analytical variances are considered.* Earlier discussions pointed out that variances from the sampling operations are usually larger than those from the analytical operations. Thus, to obtain more accurate estimates of numbers of samples needed, SDs or RSDs from overall sampling and analysis operations should be used. These are not available from EPA methods and must be obtained experimentally or estimated. A general guideline that may be helpful when overall sampling and analysis variances are not available is to triple the analytical method SDs or RSDs; this method is a poor substitute for the real data, and it risks under- or overestimating numbers of samples needed. Therefore, using only analytical method variances ensures the underestimation of the numbers of samples needed to meet one's objectives.

Selecting a Tolerable Error. Another component of eqs 1–5 is the amount of tolerable error that one determines to be acceptable in a decision based on the data produced from using an analytical method. If the amount of tolerable error is lower, then the number of samples that need to be collected is larger. When one uses absolute units of variability (SD), the input must also be in absolute units. When one uses relative units of variability (%RSD), the input must also be in relative units (i.e., percentages). Tolerable error is a subjective number that the researcher arbitrarily selects on the basis of specific DQOs.

Selecting a Confidence Level. The final input to the equations is the minimum acceptable level of confidence in the statistical estimate. In calculating the estimated average concentration, one must enter a desired confidence level, which is a relative number expressed as a percentage. Like the tolerable error, the confidence level is a strictly subjective value that is selected consistent with an individual's DQOs. Typical confidence levels range from 80 to 99%, with 95% being the most commonly used. For a given calculation, if the desired confidence level is increased, then the number of samples required will become larger.

***Using* Enviro-Calc.** All of this information—the SD or RSD from a method or from an estimate, the maximum tolerable acceptable error in the data, and the minimum percent confidence level in the data—may be entered into the equations and calculated by hand or with a calculator. However, these calculations are cumbersome and time-consuming, especially for comparisons of the numbers of samples required when considering multiple methods, multiple analytes, various amounts of tolerable error, or various levels of percent confidence in the data.

Enviro-Calc is a sophisticated but easy-to-use computer program that instantly calculates the number of samples needed based on eqs 2–5. It has the familiar user interface of a calculator (Figure 2). Programmed using Windows Visual Basic, it is the first of a trilogy of computer programs used to help calculate the number of samples needed to meet specific objectives in a sampling and analysis project.

Figure 2 illustrates the ease with which *Enviro-Calc* is used. Using the calculator in its RSD configuration and data in Table I, pentachlorophenol at 100 μg/L in water using EPA Method 1625 has a RSD of 21%. If the maximum tolerable analytical error that will be accepted in an estimate of the average concentration of an analyte is ±10%, and a 95% confidence is desired, then 20 samples will need to be analyzed. The information in Table I was obtained from *Instant EPA's Best Methods* (5).

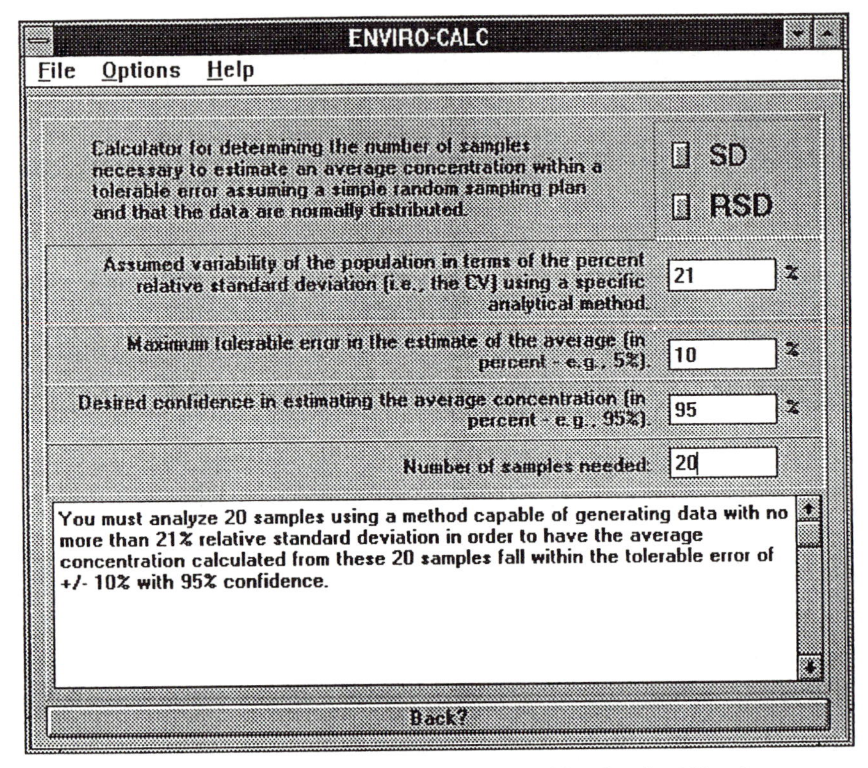

Figure 2. An example using Enviro-Calc *with pentachlorophenol at 100 μg/L.*

Table I. Precision as RSD for Pentachlorophenol in Water at 100 μg/L for Four EPA Methods

EPA Method Number	*RSD (%)*
1625	21
625	34
8270	34
8040	24

NOTE: RSD is relative standard deviation.

However, the great utility of this and the other *DQO-PRO* programs lies in their ability to help make rapid assessments of changes in project objectives. Often initial project objectives (and DQOs based on them) cannot be met because of budget constraints, lack of time, inability of the method to perform at its published levels because of matrix effects or interferences, or many other problems. Now these programs allow quick assessment (either before or after a project's samples have been analyzed) of the consequences of a project that did not meet all of the desired objectives that were selected. With *Enviro-Calc* one can easily determine the consequences to either the maximum tolerable error or the confidence level if fewer samples were analyzed. For example, if only 10 samples were able to be analyzed for pentachlorophenol at 100 μg/L using EPA Method 1625, then:

- at the same 95% confidence level the maximum analytical tolerable error would increase from 10 to 16%, or
- at the same 10% analytical tolerable error the confidence level would decrease from 95 to 92%.

Likewise, if one wanted to know the minimum number of samples required to increase the desired confidence level from 95 to 99% with an analytical tolerable error of ±10% and pentachlorophenol at 100 μg/L using EPA Method 1625, one could instantly discover that the answer is 34 samples.

Enviro-Calc can also be used to compare the numbers of samples required for specific DQOs for various methods. For example, from the data in Table I, analyzing for pentachlorophenol at 100 μg/L with a ±10% tolerable error and a 95% confidence limit would require 47 samples using EPA Method 625, 25 samples using EPA Method 8040, and 47 samples using EPA Method 8270.

This kind of information arms one to readily assess objectives and modify them as needed to meet time and budget limitations. In addition, when reports of analytical data are evaluated, if the minimum numbers of samples are not available to achieve stated objectives, then the approximate confidence levels that the data have for a desired analytical tolerable error in the data can quickly be calculated. This capability allows rapid assessment of whether the data will be useful for a specific need.

Numbers of Samples Based on Sampling Grids. The many different types of sampling protocols (e.g., stratified sampling, or random sampling at sites judgmentally chosen) are all combinations or variations of three basic types of sampling (*8*): random sampling, systematic sampling, and judgmental sampling.

Systematic sampling typically involves placing a grid over a map, or marking the sampling site, and then collecting samples within each grid area or at every node where the lines cross. Various shaped grids are used. The most common shapes are triangles, squares, or rectangles. Systematic sampling is usually used with contaminated land-based sites. It may be used to collect samples of soil from a landfill, to drill wells to collect samples of groundwater, or to collect aqueous sediments from the bottom of a lake or the sea.

Systematic sampling is the second statistical treatment to be considered in this chapter (random sampling was covered already with *Enviro-Calc*). Different equations are used and a different calculator interface was formatted for systematic sampling. The calculator is named *HotSpot-Calc*.

***Using* HotSpot-Calc.** *HotSpot-Calc* is designed to determine the grid spacing needed to detect the presence of a single hot spot of a specified size and shape with a specified probability of missing the hot spot. This calculator is based on the following key assumptions:

- The hot spot is circular or elliptical (moderately elliptical or long elliptical).
- Sample measurements are collected on square, rectangular, or triangular grids.
- The definition of a hot spot is clear and agreed to by all decision makers.
- There are no misclassification errors (i.e., there are no false-positive or false-negative measurement errors).

This last assumption is the most often overlooked and requires careful consideration of the QA program and its design in order to prevent misclassification errors. Ways in which false positives and false negatives (caused by the measurement process) can be estimated and reduced are discussed under the fourth principle of environmental sampling and analysis.

The objectives of hot spot sampling are fundamentally different from the objectives of the *Enviro-Calc* sampling models. Whereas the *Enviro-Calc* model focuses on estimating the site-wide average concentration or the percentage of an area contaminated, the primary objective of hot spot sampling is to determine if localized areas of contamination exist. These localized areas of contamination may be due to spills, leaks, buried waste, or any number of other events by which contamination might be confined to a relatively small area. A single site might have multiple hot spots of different origins.

Basically, hot spot sampling involves performing a systematic search of a site for hot spots of a certain specified shape and area. The search is conducted by sampling every area or every point using a two-dimensional grid. If a three-dimensional grid is desired, then multiple layers of the two-dimensional grid may be used for rough estimates. The probability of finding a hot spot is determined as a function of the specified size and shape of the hot spot, the pattern of the grid, and the relationship between the size of the hot spots and the grid spacing. For example, if a square grid is used to search for circular hot spots of radius r, then the probability of locating a hot spot, if one exists, is 100% when the distance between grid points is r. Obviously, this probability decreases as the grid spacing increases relative to hot spot size.

Assumptions. The methods discussed for *HotSpot-Calc* are based on those described by Gilbert (9), which make the following assumptions:

1. A hot spot may be a surface area, or a volume at any depth below the surface (i.e., at a particular soil horizon), but the surface projection of the hot spot is assumed to be circular or elliptical in shape.
2. Samples are collected on a two-dimensional grid of a specified pattern.

3. The distance between grid points is large relative to the projected surface area of the sample that is actually removed for analysis.
4. The criteria for defining a hot spot are unambiguous with respect to the measurement method and the concentration considered "hot", and no classification errors occur in applying these criteria.

The concept of false-positive errors is slightly different for hot spot sampling than for the other two sampling models (*Enviro-Calc* and *Success-Calc*) that are described in this chapter. The reason is that for both of the other models, a decision is based on an estimate of the mean or an estimate of the frequency with which some characteristic occurs; however, in the hot spot model, a decision is made for each individual sample result.

For *Enviro-Calc* and *Success-Calc,* false-positive and false-negative errors are incorrect conclusions (i.e., the risk of concluding that the average concentration or the average percent with some characteristic is higher or lower than the criteria applied to the decision rule the models use). A false-positive conclusion error is made if, by chance alone, a sample set overrepresents the higher concentrations at a site and underrepresents the lower concentrations. In the hot spot model, it is not possible to make this type of false-positive conclusion error. The only way to make a false-positive error in the hot spot model is to have a classification error. A classification error occurs when a particular sample is classified as hot when it actually is not (or vice versa). Thus, a false-positive error can only be attributed to the analytical process, and not to the sampling process. It is impossible with the hot spot design to "overrepresent" an area in the same sense as it is possible with the other two models, because the parameter that is used to make the decision is an individual result, and not an average. The last assumption in the preceding list of assumptions for hot spot sampling is that there are no classification errors. Thus, this assumption is the same as saying that the assumed false-positive rate is zero.

Although it is not possible to have a false-positive error attributable to sampling, it is possible to have false-negative errors. Just as with the other two models, one potential source of false-negative error is chance underrepresentation of higher concentration sampling points, which causes the hot spots to be missed. The risk of this type of false-negative error is controlled by grid size and the number of samples collected. As with the other two models, it is also possible to have false-negative errors as a result of analytical results that underestimate the actual concentration in a particular sample. However, this type of false negative is excluded (by the assumption of no classification errors) from the sample size calculations discussed in this section.

Sampling Design. Hot spot sampling uses a systematic sampling design. Typically, sampling is conducted on either a square, rectangular, or a triangular grid. The sampling design for hot spot sampling depends on four factors:

1. the size of the hot spot, in terms of the length of the long axis of the ellipse (where L is equal to one-half of the length of the long axis or the radius of a circle);
2. the shape (S) of the hot spot, in terms of the ratio of the long axis of the ellipse to the short axis of the ellipse;
3. the acceptable risk (β) of failing to detect a hot spot that is present; and
4. the grid pattern, in terms of grid spacing (G) and geometry (square or triangular).

When these four factors are considered, a few general principles can be used to guide the development of a hot spot sampling plan. First, for any hot spot, $0 < S \leq 1$, and for a circular hot spot, $S = 1$. For a given grid spacing and design, the probability of locating a circular hot spot of a given radius L is higher than for an elliptical hot spot having the same value for L. Also, if the ellipse is "skinnier" (i.e., has a longer horizontal axis), then the hot spot is more difficult to locate. In other words, hot spots become more difficult to locate as the shape factor S gets smaller. The other general principle is that the probability of locating a hot spot of a given size and shape using a triangular grid with a grid spacing G is higher than when using a square grid with the same spacing. This principle quickly becomes apparent when *HotSpot-Calc* is used and the grid is changed from square to rectangular to triangular shapes, respectively.

When a hot spot sampling design is developed, the usual procedure involves first specifying the size of the hot spot that is desirable to detect in terms of L, which is one-half the length of the long axis. Unless a physical reason exists for expecting a hot spot of a particular shape, it is usually easiest to start with the model of a circular spot and then evaluate the effectiveness of the resulting plan in terms of the probability of detecting other potential shapes. For a circular hot spot, L is the radius of the circle (not the diameter).

After one decides on the hot spot size, the next step is to specify the acceptable risk of failing to detect a hot spot of the stated size. This probability is the false-negative risk, β.

The next step is to determine the grid spacing needed to achieve the desired false-negative risk for the specified hot spot size and shape. This determination is done with the help of values from a lookup table in the calculator that are based on the nomographs in Gilbert (9). These values present false-negative risk as a function of hot spot shape S and the ratio L/G.

Hypothesis Test. The hot spot sampling model can be described in statistical terms as a test of the null hypothesis (H_0) that the site is "clean" (i.e., that there are no hot spots of the specified size and shape) versus the alternative hypothesis (H_A) that one or more such hot spots exists.

Calculating Number of Samples. The number of samples required for hot spot sampling is simply the number of samples required to sample all grid-defined areas or all grid points on the site for the selected grid spacing. The number of samples required using a square grid can be approximated using eq 6:

$$n = A/G^2 \tag{6}$$

where n is the number of samples; A is the area to be sampled, in the square of the units for G (e.g., m^2); and G is the grid spacing, as defined previously. Thus, for a site that is 50 m by 100 m, with a square grid spacing of 2 m, the required sample size is $n = (50 \text{ m})(100 \text{ m})/(2 \text{ m})^2 = 1250$. For triangular grids, we must change the denominator of eq 6 to $0.886G^2$, and for rectangular grids (where the length is twice the width), we must change the denominator of eq 6 to $2G^2$.

Obviously, many samples are required to find small hot spots on a large site. However, because the objective of hot spot sampling is simply to determine if an analyte concentration in any individual sample exceeds a particular threshold concentration, it is sometimes possible to take advantage of compositing to reduce the overall analytical burden.

If a compositing scheme is used, care should be taken to use subsamples of equal mass and to thoroughly mix the composite. Soils consisting of large chunks of debris, rocks, tarlike material, or other materials that cannot be well mixed are poor candidates for compositing.

In addition, compositing will increase the required detector sensitivity by factors as high as the number of samples included in the composite. For example, what if the detection level for a contaminant were 10 μg/kg and five samples of soil were composited? If only one of the five samples contained this particular contaminant at 10 μg/kg, then the composite samples would be diluted to 2 μg/kg. Thus, detection sensitivity would have to be five times greater (i.e., lowered to 2 μg/kg) in order to detect this contaminant in the composited sample.

Equations and Approach. The equation used by *HotSpot-Calc* is $L/G = X$ for a given β probability of missing the hot spot (9). X was estimated from figures presented in Gilbert for each type of grid (i.e., square, rectangular, or triangular). These values are stored in a lookup table in the calculator and are used to calculate G (the required grid size) using the input of L (the length of the semimajor axis of the hot spot to be detected).

The following example demonstrates *HotSpot-Calc*. If one selects the shape of a circle (1.0), an acceptable probability of 20% for missing a hot spot, and a square grid for the sampling design, then $X = 0.51$ (from the lookup table in the calculator). By using this information and the formula in eq 7

$$L/G = X \tag{7}$$

one can solve for G for a desired L from the ratio value $L/G = 0.51$. If one wishes to detect a circular hot spot 10 ft in radius (i.e., $L = 10$ ft), then $10 \text{ ft}/G = 0.51$, and

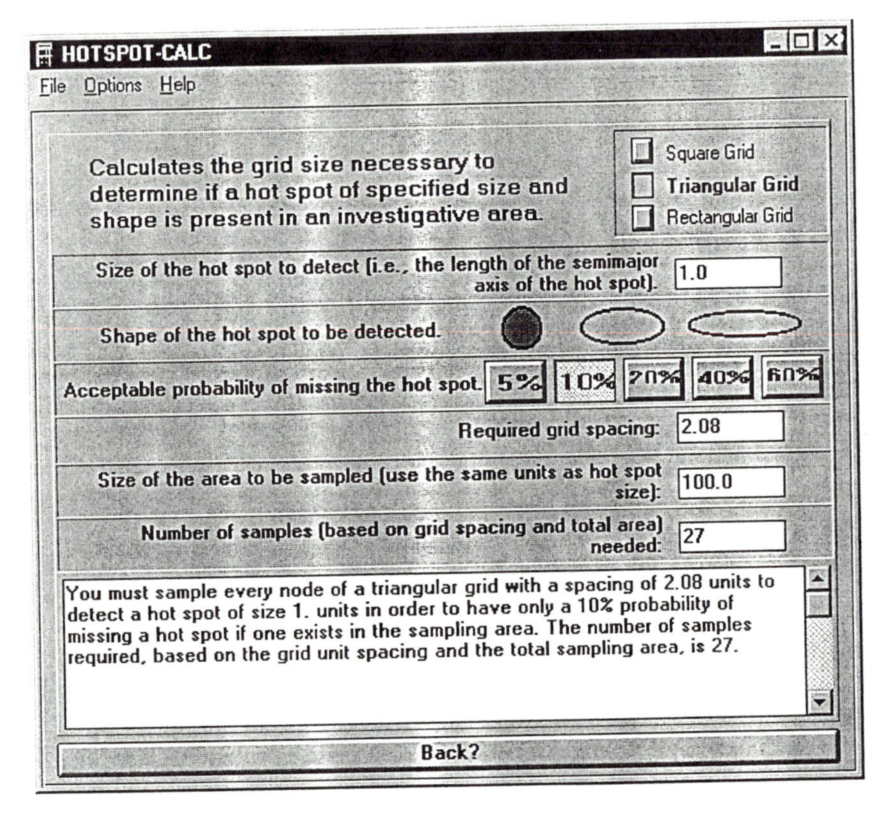

Figure 3. Example of HotSpot-Calc *using a triangular grid.*

so $G = 10$ ft/0.51 = 19.6 ft. In other words, one must use a square grid 19.6 ft between nodes to have a 20% chance of missing a circular hot spot with a radius of 10 ft (20-ft diameter).

Figure 3 illustrates the ease with which these calculations can be made using *HotSpot-Calc*. In this example the size of the hot spot to be detected has a radius of 1.0 (e.g., 1 ft, 1 m, 1 yard, or 1 mi), and it is assumed to be circular in shape. By using a 10% probability of missing it, a grid spacing of 2.08 is required (e.g., 2 ft, 2 m, 2 yards, or 2 mi, depending on the units used to describe the radius of the circle). If the boundaries of the area to be sampled are 25 units by 4 units, then the area is 100 square units. Because the number of samples needed for a triangular grid is (area)/[(0.88)(grid spacing)^2], then $n = 100/[(0.88)(4.3)]$, which is 26.5.

The effects of varying the shape of the grid, the probability of missing the hot spot, or the assumed shape of the hot spot can be assessed within seconds just by selecting any of the other buttons on the calculator. Table II illustrates some of the effects that are calculated by changing a series of these parameters. Two facts become apparent when Table II is assessed. First, the number of samples needed

Table II. Numbers of Samples Versus Hot Spot Shape and Probability of Missing It Using a Triangular Grid

Probability of Missing a Hot Spot (%)	*Assumed Shape of Hot Spot*	*Required Grid Spacing*	*Number of Samples for 100 Square Units*
10	Circle	2.08	27
20	Circle	2.13	25
40	Circle	2.44	19
60	Circle	3.13	12
10	Ellipse	1.64	42
20	Ellipse	1.72	38
40	Ellipse	2.08	27
60	Ellipse	2.44	19
10	Long ellipse	1.28	69
20	Long ellipse	1.43	56
40	Long ellipse	1.69	40
60	Long ellipse	2.04	28

NOTE: Required grid spacing data can be in any units.

Table III. Numbers of Samples Versus Various Grid Shapes Assuming a Circular Hot Spot

Probability of Missing a Hot Spot (%)	*Shape of the Sampling Grid*	*Required Grid Spacing*	*Number of Samples for 100 Square Units*
10	Triangle	2.08	27
10	Square	1.82	31
10	Rectangle	1.02	49

NOTE: Required grid spacing data can be in any units.

decreases as the acceptable probability of missing a hot spot increases. Second, as noted earlier, if the hot spot is circular, fewer numbers of samples are needed than when it is elliptical. The longer the horizontal axis is in the ellipse (i.e., the "skinnier" it is), the larger is the number of samples that will be needed for a given probability and grid shape.

In order to assess the effects of the shape of the grid on these calculations, we kept the probability of missing a circular hot spot constant and only the grid shape was varied; these results are listed in Table III. It is readily apparent that triangular grids are most efficient (resulting in the least number of samples) and that rectangular grids are the least efficient. This characteristic is an important fact that can be used in planning many sampling and analysis projects.

The following ideas are critical to successfully using the information from this calculator:

- The length L of the hot spot is the radius of a circle or one-half the length of an ellipse. People may forget this convention and think they are covered for a diameter of a circle or the whole length of an ellipse.

- The probability specified is the acceptable probability of *missing* a hot spot. It is not the probability of finding a hot spot.
- The major assumption, which is a big assumption, is that no misclassification errors occur. Thus, no false-positives or false-negatives are considered from the laboratory, the field, or *any* source.
- Every area or every node of the grid must be sampled. The nomographs used (9) are based on every node of a grid being sampled, but most people sample within a grid area defined by the nodes. This common practice should result in a somewhat less accurate estimate of numbers of samples needed because sampling areas defined by grid nodes will result in fewer samples being collected than if samples are collected at every grid node.

The equations are for a two-dimensional solution. If one is seeking a subsurface hot spot (which will be three dimensional), then this grid applies to a projection of the three-dimensional hot spot onto a two-dimensional horizontal plane. Thus, if the two-dimensional solution was 20 samples and one wanted to sample vertically at 1-m depths down to 3 m, then the three-dimensional sampling would require a total of 80 samples (20 from the surface, 20 at 1 m, 20 at 2 m, and 20 at 3 m).

Many aspects of hot spot sampling are not covered in *HotSpot-Calc*. Gilbert (9) provides details, for example, on how to calculate the probability that a hot spot was missed, or the potential size of a missed hot spot, given a particular grid size, as well as using a priori knowledge to refine the sampling design. These topics are not covered in the current version of the calculator.

The Third Principle

QC samples must be representative of the environmental samples being analyzed. Quality assurance (QA) is a system of activities that assures the producer or user of a product or a service that defined standards of quality with a stated level of confidence are met. *Quality control* (QC) is a system of activities that controls the quality of a product or service so that it meets the needs of users (*10*). Thus, QC consists of internal (technical) activities, such as the use of QC samples, to control and assess the quality of measurements whereas QA is the management system that ensures that an effective QC system is in place and working as intended (*11*).

In the discussion of the first principle of environmental sampling and analysis, the purpose of environmental sampling was established as the means to assess a small, but informative, portion of the environmental site population and then draw an inference about the rest of the environmental site population from the data gathered. QC samples are used in a similar way; they assess a small, but informative, portion of the measurement system population and then draw an inference about the environmental sample population. They provide an assessment of the kinds and amounts of bias and/or imprecision in the data that are obtained from the environmental samples. Thus, QC samples are used to assess the collection and measurement system in a manner similar to the way that environmental samples are used to assess the portion of the environment site from which they come. This very important concept is often overlooked, but it is critical to the understanding and application of the third and fourth principles of environmental sampling and analysis.

QC Sample Matrices

Relation to Environmental Sample. QC samples must represent, as closely as possible, the environmental samples they are being used to assess. Common sense dictates that when water samples are being analyzed, the QC samples should be water. When other matrices, such as soil, vegetation, or animal tissue, are analyzed, the QC samples should be of a similar matrix if at all possible. Unfortunately, suitable matrices are not always readily available. However, when QC samples in water are used as surrogates for soil, vegetation, or animal tissue QC samples, common sense tells us that the values obtained from the aqueous QC samples may often be very different from the values actually present in the other matrices. In order to help improve this situation, Radian International introduced the concept of QC Assessment Kits (*12, 13*). Initial QC Assessment Kits include blank and spiked soils packaged in convenient sets of 10 preweighed bottles. One can spike these blanks with custom-prepared standards or purchase certified prespiked standards. In time, QC Assessment Kits for other matrices will also be produced. The numbers of kits needed depends on the level of confidence desired in the QC data. These calculations are discussed under the fourth principle of environmental sampling and analysis.

How To Select from Many Different Kinds of QC Samples. Each of the many different kinds of QC samples is designed for a specific purpose. Some provide an assessment of bias, whereas others provide an assessment of imprecision. In addition, some are designed to assess laboratory-based variability and others are designed to assess overall variability (both sampling and analysis). QC samples designed for bias can be further divided into those that assess *low-level bias* (usually caused from contamination) and those that assess *normal-level bias* (usually caused from operation errors such as use of incorrect calibration standard concentrations, dirty detectors, or incorrect sample preservation) (*14*). Examples of the many kinds of QC samples that are used include method blanks, method spikes, matrix spikes, field spikes, solvent and reagent blanks, instrument blanks, trip blanks, laboratory control standards, replicate samples, and replicate analyses. Thus, it is not surprising that people become confused about which kinds of QC samples they need. The answer depends on what kinds of bias and imprecision they want to assess.

An expert system, *Practical Environmental QC Samples*, was written for an ACS short course on Practical Environmental Sampling and Analysis (*14*). An *expert system* is a computer program that emulates the logic of a human expert on a particular knowledge domain. When queried, the expert system gives the same answer as a human expert. Advantages of expert systems are that the answers provided are always consistent and can be readily accessed. Human experts are not always consistent, and they may not be available when needed. *Practical Environmental QC Samples* is one of several computer programs that are packaged in commercial software sold under the name *Practical QC.*[2]

Using Practical QC

Practical Environmental QC Samples may be used to plan the types of QC samples desired in a sampling and analysis project or to evaluate whether the QC data in a finished report meet specific needs (i.e., whether the report contains QC samples needed to provide desired information on bias and precision from sampling, laboratory, or overall sources of error). Figures 4–7 show how this expert system can be used to determine which kinds of QC samples are needed when interest is limited to assessing errors from normal-level bias arising from sampling activities.

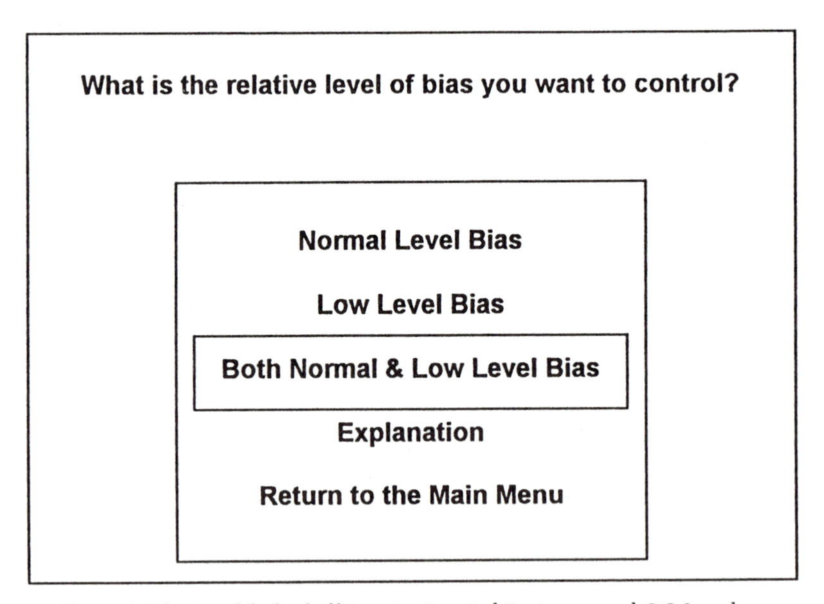

Figure 4. Selection of the level of bias using Practical Environmental QC Samples.

[2]For information about obtaining *Practical QC,* please contact Instant Reference Sources, Inc., 7605 Rockpoint Drive, Austin, TX 78731 (fax 512-345-2386). *Practical QC* is a DOS program, but a Windows icon is provided with its automatic installation program. It can be launched either from DOS or from Windows. Additional information may be viewed using the World Wide Web site URL at http://www.instantref.com/inst-ref.htm, and information can be requested using the message form included with the Web site.

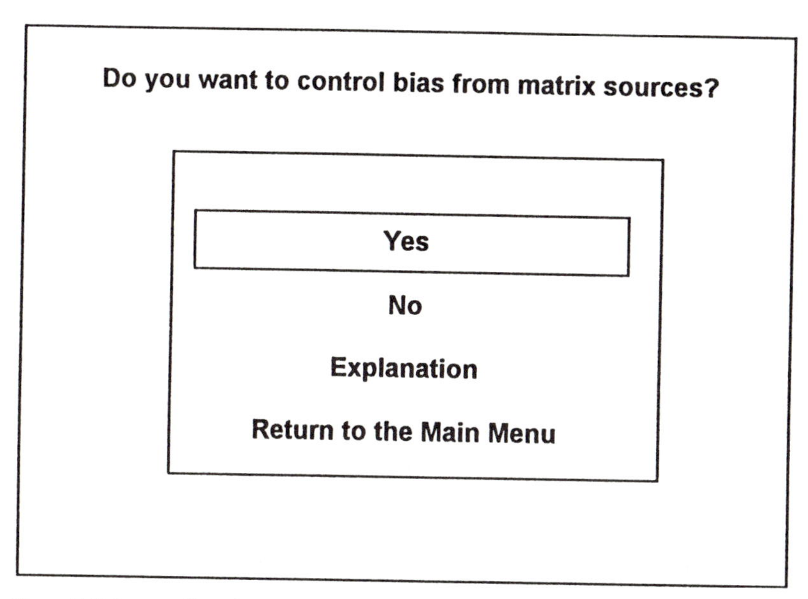

Figure 5. Selection of bias from matrix sources using Practical Environmental QC Samples.

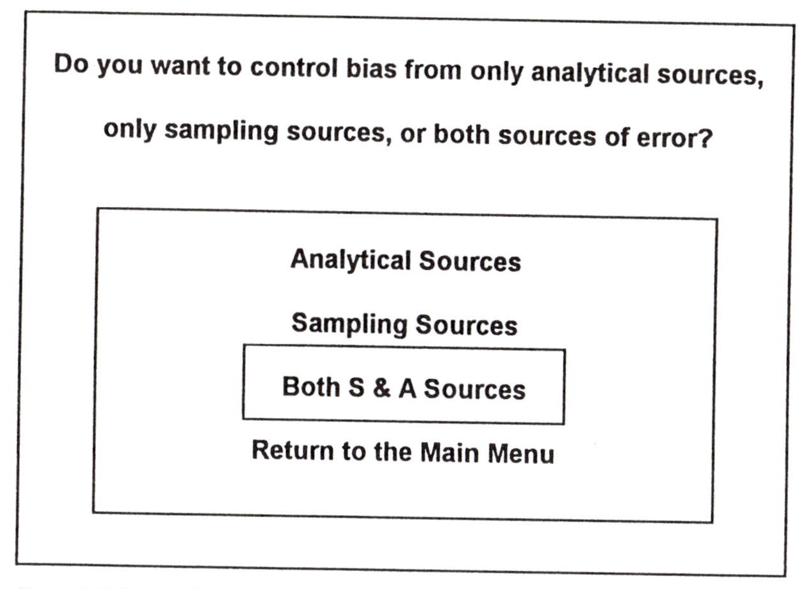

Figure 6. Selection of sources of bias to control using Practical Environmental QC Samples.

Recommendation: Your QC samples should include

Spiked (Fortified) Field Blanks

and

Spiked (Fortified) Laboratory Blanks

Both types of blanks must be analyzed, and the difference in variance between the spiked field blanks and the spiked laboratory blanks will be the estimated bias from sampling.

Because the spiked field blanks go through the entire process, from sampling through analysis, they are exposed to contributions from both sampling and analysis sources of bias. Subtracting variance (from spiked laboratory blanks) from variance (from spiked field blanks), bias from only the sampling process may be estimated by difference.

Figure 7. The Practical Environmental QC Samples *expert system recommendation.*

Figures 8–10 show how the expert system can be used to assess whether the QC samples in a report can be used to meet one's objectives. For example, if data were needed to assess both normal-level bias and low-level bias from laboratory and sampling sources, and the QC samples consisted of solvent blanks and matrix spikes, then only bias from laboratory sources would be covered. However, both normal- and low-level bias from the laboratory sources would be able to be assessed. The recommendations include what the QC samples should not be used for as well as what they may be used for.

The Fourth Principle

QC samples are used to provide an assessment of the kinds and amounts of bias and imprecision in data from analysis of the environmental samples. It was established earlier that QC samples are used to provide an assessment of the kinds and amounts of bias and/or imprecision in the data that are obtained from the environmental sample population. Thus, QC samples are used to assess the collection and measurement system in a manner similar to the way that environmental samples are used to assess the portion of the environment from which they come. Therefore, representative environmental samples are collected and analyzed to form conclusions about a par-

Select the Type of QC Sample You Have

Laboratory QC Samples	
Method Check	Quality Control Check Samples
MS or MSD	Matrix Spikes Prepared in the Laboratorry
Method Blank	Solvent and/or Reagent Blanks
Method Blank	Instrument Blanks
Method Blank	Glassware Blanks
Sampling QC Samples	
Contaminaton	Field Blanks
Contaminaton	Trip Blanks
Contamination	Material Blanks
Cross Contamination	Equipment Blanks
MS or MSD	Matrix Spikes Prepared in the Field
Background	Local Control Site Background Samples
Background	Area Control Site Background Samples
Both Lab and Sampling	
Replicate Analyses	Replicate Analyses of the Same Samples
Replicate Samples	Replicate Samples (Duplicates, etc.)
	Return to the Main Menu

Figure 8. The main menu selections from Practical Environmental QC Samples.

Solvent and/or Reagent Blanks

You can use data from solvent/reagent blanks to measure low level bias (usually contamination) from solvents/reagents used to prepare samples.

You usually can not use this data for normal level bias (procedural and operator errors) or for precision calculations from either laboratory or sampling sources

Figure 9. Recommended data usage from solvent blanks from Practical Environmental QC Samples.

Matrix Spiked (MS) Test Samples Spiked In The Laboratory

You can use data from laboratory matrix spikes to measure normal level bias from laboratory sources such as instruments and lab procedures.

You usually can not use this data for low level bias (contamination) or for normal level bias measurements from sampling sources or for precision calculations from either laboratory or sampling sources.

Figure 10. Recommended data usage from laboratory matrix spikes from Practical Environmental QC Samples.

ticular site, and representative QC samples are analyzed to form conclusions about the system that measures the environmental samples. This similarity between environmental sample usage and QC sample usage is often not appreciated or even recognized, but it is very important and absolutely critical when discussing the fourth principle of environmental sampling and analysis.

Basic QC Data Assessments Include False Positives and False Negatives

One of the most basic QC data assessments is to determine the presence of false-positive and false-negative conclusions in environmental analytical data. An analyte that is incorrectly concluded to be present in a sample is a false positive; these can cause regulatory and financial consequences for a laboratory's clients. One cause of false positives is misinterpretation of the identity of interfering analytes for the target analytes. When interferents are present in a sample, the method must be modified to eliminate them, but when they are present in the materials used to prepare or analyze samples (e.g., bottles, solvents, reagents, filters, columns, or detectors), their sources must be determined and the interferent removed if possible. Various kinds of QC samples [e.g., as determined from the *Practical Environmental QC Samples (14)* program] can be used to determine where, in the chain of events, the interferents are contributed, but the first step is to recognize their presence. Method blanks, which consist of a blank matrix similar to the samples but without the target analytes, are used to determine overall whether false positives are present in the materials and/or in the process used to prepare and analyze samples. Method blanks do not, however, identify the source of error.

A false negative occurs when an analyte is concluded to be absent in a sample, but in reality it is present at detectable levels. False negatives commonly occur from poor recovery of target analytes from a matrix, or from interferences that mask the target analytes. They are especially troublesome to government and regulatory personnel and also to scientists who work with risk assessments because they result in the conclusions that certain pollutants are absent when, in fact, they are present.

Statistical Evaluations from Batch Modes Are Useless

Most environmental analyses today are conducted in batch modes to facilitate cost-effective analyses. In doing so, one method blank (also called a laboratory blank) and one or two method spikes (or matrix spikes) are typically analyzed along with about 10–20 environmental samples. The resulting data for all of the environmental samples in that batch are accepted or rejected on the basis of those QC samples.

When used this way, the QC data of a batch does not provide a statistically sufficient amount of information for a user to draw an inference about the environmental samples. One or two QC samples, which is how these QC samples are grouped, do not provide enough information to predict the reliability of the measurement system characteristics with the other environmental samples that are grouped with them. The inference—that the environmental samples analyzed in conjunction with a method blank and one or two spiked method blanks (or matrix spikes) do not contain false positives or false negatives simply because the accompanying one or two QC samples did not contain them—is not necessarily correct. Thus, the present way of assessing QC data contains a basic flaw that is not usually recognized.

Alternatives to Batch Modes of QC Data Assessment. How can method blanks and method spikes (i.e., spiked method blanks) be used as representatives for the environmental sample population? The answer is to use a statistically valid number of QC samples. That number depends on the DQOs of a particular sampling and analysis project. As an example, the number of QC samples needed can vary from six (for an 80% probability that the associated environmental samples will not contain more than 25% false positives or false negatives) to 458 (for a 99% probability that associated environmental samples will not contain more than 1% false positives or false negatives).

The computer program *Success-Calc* was designed to determine the number of samples needed to detect a specified frequency of some characteristic occurring in the population (e.g., the percent defectives or percent contamination). In an environmental program it can be used for a number of different purposes. It can be used to design a QA program (i.e., the number of blanks and spikes needed to test for a percentage of problems in the sampling or analytical process), or it can be used to design an investigation program (i.e., the number of environmental samples needed to determine if some percentage of a site is contaminated). This calculator does not calculate the number of samples needed to estimate the frequency at which a characteristic occurs; rather, it calculates the number of samples required to decide when the true frequency of occurrence *exceeds* some predefined frequency using a specified decision rule.

Statistical Populations of QC Samples Are Often Attainable If Requested. Many of the QC or environmental samples needed for a statistical population are available (or can easily be made available), but they are not presently used in this way. Thus, increased costs associated with large numbers of samples may not be necessary and may in fact be minimal; costs may even be reduced with proper planning. For example, one should consider that a method blank is typically analyzed for each batch of samples. This procedure results in a large number of blank samples that may be usable for a statistical population of a method and matrix when gathered over the period of several weeks or months. In fact, method spikes or matrix spikes that do not contain all the analytes of interest may be counted as method blanks for those analytes of interest that were not included in the spikes. However, the key to obtaining a statistically usable population of sample data is that all significant parameters that can affect analytical method performance must remain constant. Significant parameters include the instrumentation and method, analytes, analyst, and the matrix.

Approach for Success-Calc

Sampling Objective. *Success-Calc* can be used to determine if the frequency at which some characteristic (e.g., false-positive measurements or contamination at a site) occurs is greater than a desired frequency. For example, this calculator will determine the number of samples needed to determine if the true rate of false-positive measurements due to laboratory contamination is greater than 5% with 95% confidence. Three pieces of information are needed:

1. the frequency of concern;
2. the confidence desired in concluding that the true rate exceeds the frequency of concern; and
3. the decision rule that will be used to conclude whether the true frequency exceeds the frequency of concern.

If the frequency of concern is $<10\%$, then the calculator uses an equation based on an exponential approximation to the binomial distribution that provides an approximate determination of the number of samples required (N). If the frequency of concern is $>10\%$, then an iterative approach is used that calculates the confidence achieved for some specified number of samples. In this case, the equation used takes the number (N) that a user enters and calculates the confidence (for a specified decision rule) with which one can correctly conclude that the samples could come from a population that has a higher frequency of occurrence than desired. This approach was used instead of an exact calculation in order to demonstrate the tradeoffs of modifying the numbers of samples, the decision rule, and the desired confidence. It thus allows one to evaluate whether the cost of these additional samples is worth the improved decision-making confidence.

Decision Rules. The approach used for *Success-Calc* also allows one to change decision rules while manipulating N and the frequency of concern. The decision rule for the objective is a statement of how many samples must exhibit the characteristic of concern (e.g., target analyte detections in blanks or environmental samples from a site) before one will conclude that the true frequency of this characteristic in the population exceeds the frequency of concern. The easiest, and least expensive, decision rule is the following: If none of the samples collected exhibit the characteristic, then the true frequency is less than the frequency of concern; if one or more samples collected exhibit the characteristic, then the true frequency is greater than the frequency of concern.

Some decision rules allow some samples to have the undesirable characteristic but still allow one to conclude that the true frequency is less than that of concern; such rules allow for "errors" due to a variety of sources, but they also require that more samples be collected. For example, if one uses the decision rule of zero false-positive detections in blanks to determine if the true frequency of false positives from blank contamination is less than 5%, with 95% confidence, then approximately 60 method blanks are required. Changing the decision rule to allow one false-positive detection in a blank, and still conclude that the true rate is less than 5%, requires approximately 90 samples. The equations used for this approach are presented in the next section.

Equations Used for Success-Calc

If the frequency of the characteristic that is desirable to detect is <10%, then the following equation is used:

$$n = \ln(\alpha)/\ln(1 - Y) \tag{8}$$

where ln is the natural log, α is (1 – the desired confidence level), and Y is the frequency to detect (Y must be <10%).

The 10% limit is based on comments by Cochran (*15*). Equation 8 is itself based on the exponential distribution and assumes that the characteristic to be detected occurs very infrequently, as opposed to the binomial, which can tolerate any frequency from 0 to 100%. Information on the exponential distribution model in the development of tolerance intervals is used by EPA in conjunction with guidance on evaluating gas pipelines for PCB contamination, but it is currently not published.

If the frequency of the characteristic that is desirable to detect is >10%, then the binomial eq 9 is used and iteratively solved for an appropriate n. In this case

$$P = n!/[r!(n - r)!\ (q^{(n-r)} \times p^r)] \tag{9}$$

where n is the number of samples in a sample collection, r is the number of samples with the characteristic to be detected, p is the true percentage of the population with the characteristic to be detected, q is the true percentage of the population without the characteristic to be detected, and $q = 1 - p$. In eq 9, P is the probability that a sample of size n can be collected from the population where there are truly p% items with the characteristic and only r samples have the characteristic (e.g., false positives or contamination). One first solves for P and then calculates $1 - P$, which is equivalent to the confidence in concluding that the true rate is less than p.

Decision Rules with *Success-Calc*. A decision rule is a summary statement that defines how a decision maker expects to use data to make the decision(s) identified in step 2 of the DQO process. In the same way that multiple decisions, for example, might pertain to multiple areas within a site, there also may be (and often are) multiple decision rules for different areas of the site or for different pollutants. Development of the decision rule involves the following three steps:

1. specification of the parameter that characterizes the population of interest;
2. specification of the action level for the study; and
3. development of an "if ... then" statement that describes the decision rule in terms of alternative actions.

The parameter characterizing the population of interest is a statistical parameter, such as the mean or 90th percentile or upper tolerance limit, for a particular analyte or measurement characteristic. For the calculators programmed in two parts of *DQO-PRO* (*Success-Calc* and *HotSpot-Calc*) the parameter of interest is the individual measurement results for each sample or grid point. For the other calculator (*Enviro-Calc*) the parameter of interest is the average concentration (e.g. the average concentration of a target analyte over the entire sampling site).

The best decision rule to use initially is usually that which requires the fewest samples, such as the "zero or one" decision rule described previously. However, if a different decision rule is used, then the probability P is calculated for each r allowed, and the resulting P values are summed. For example, if a decision rule of 1 or fewer "characteristic result" passes is selected, then the P for 1 must be calculated and added to the P for 0. If $r = 2$, then the P for 2 must be added to the P for 1 and to the P for 0 to obtain a total P. Then 1 minus this total P is taken to obtain the confidence.

The final confidence, n, r, and p define the sampling design that will meet the users' objectives (*16*). When a sampling design from this exercise is implemented, the decision based on the results of the sampling exercise will be that either the true frequency exceeds the frequency of concern or it does not. If it does (i.e., more samples reflected the characteristic than allowed), then it may be desirable to estimate the range of true frequencies possible, given the observed results. Or, if the number of samples with the observed characteristic was small, it may be desirable to determine what decreased confidence they have that the true rate is less than the frequency of concern. Both of these parameters are easily calculated using *Success-Calc*.

Success-Calc also determines the minimum and maximum percentage of the population with the chosen characteristic given that some number of samples collected indicated the presence of this characteristic. For example, it will calculate the minimum and maximum rates of false positives that could occur if some number of false positives were observed among the population of blanks analyzed. The number of samples analyzed is entered along with the number of samples having the chosen characteristic and the confidence level that is desired. The "Min–Max" button estimates the minimum and maximum frequency with which the characteristic could occur. This calculation is analogous to setting an upper and lower confidence level for a mean (*17*).

Calculating Upper and Lower Confidence Levels. The equations for calculating the lower confidence level (LCL) and upper confidence level (UCL) for the binomial distribution are

$$LCL = \{1 + [(n - r + 1) \times F(1 - \alpha/2;2n - 2r + 2,2r)/r]\}^{-1} \quad (10)$$

$$UCL = \{1 + [(n - r)/(r + 1) \times F(1 - \alpha/2;2r + 2,2n - 2r)]\}^{-1} \quad (11)$$

where the F statistic in each equation has the specified degrees of freedom, n is the number of samples collected, r is the number of samples with some characteristic, and α is 1 minus the desired confidence (e.g., $\alpha = 5\%$ for a desired 95% confidence).

Using *Success-Calc*. Figure 11 shows how *Success-Calc* can be used to estimate the number of QC samples (e.g., method blanks to assess false positives, or method or matrix spikes to assess false negatives in the data) that would be needed in order to ensure, with 95% confidence, that no more than 5% false positives or false negatives would be in the QC data from the laboratory, and thus indicate that the same assessment would be true for the associated environmental samples. The values for desired minimum rate of the characteristic of interest, the confidence level, and the decision rule are entered, and the bar labeled n at the bottom of the calculator is selected to obtain the calculation. Next, the bar labeled "Min–Max" is selected to obtain the maximum and minimum frequency rates that could be observed under the set of conditions entered. A maximum rate of 5% was desired, and 5% is the maximum rate that would occur; the minimum rate of occurrence could be as low as 0% false positives or false negatives. The decision rule of zero "defects" was used; this rule means that no false positives or false negatives can exist among the 59 method blanks or method spikes that were analyzed. If the same assessments were desired for QC data from sampling sources of error, then numbers of field blanks and spikes, or other appropriate QC samples (as determined, e.g., using *Practical QC*) would also be estimated using this program. (The estimated numbers of QC samples needed would be the same, but the type of QC samples required would be different.)

What if only one-half of the desired QC samples were available? By simply replacing n with 30 and selecting the button labeled "Probability", one would instantly discover that the confidence level for detecting a maximum rate of 5% false

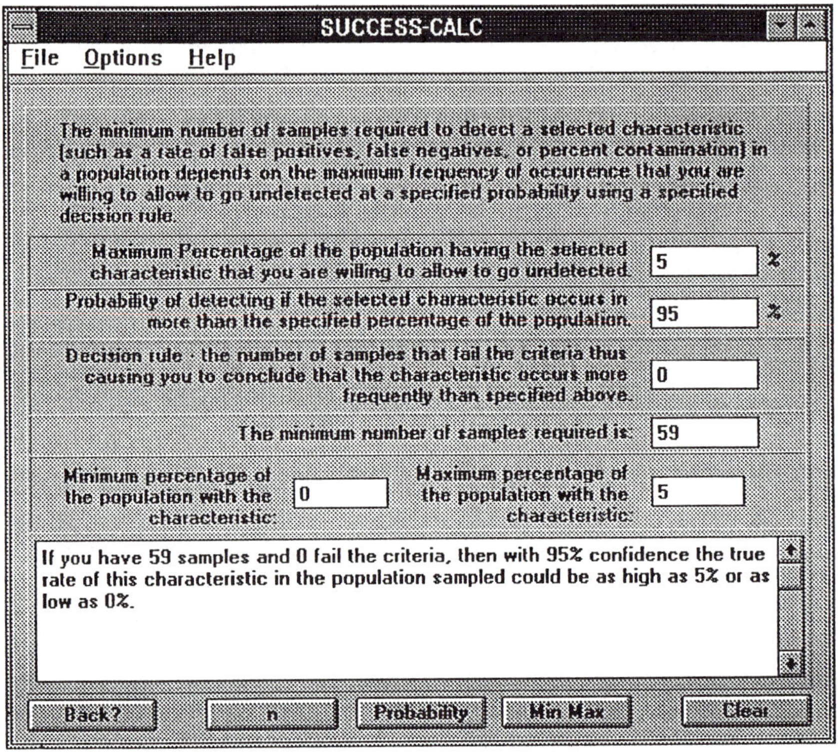

Figure 11. Calculation of numbers of QC samples needed using Success-Calc.

positives or false negatives is reduced from 95 to 79% and that the maximum rate of occurrence could be as high as 6% rather than 5%.

Of course, frequently the "zero" decision rule is not attained. If, after a QC sample set has been analyzed, false positives or false negatives were among the analyte identifications, and these were due to the measurement system rather than to an incorrect identification of the analyte, then one would want to know what the reduced confidence levels would be. Figure 12 illustrates how easy it is to obtain this revised assessment. By simply deleting the value for probability, changing the value for the decision rule from 0 to 3, and selecting the bar marked "Probability" at the bottom of the calculator, the new probability is calculated using the previously calculated value of *n*.

The preceding discussion indicated that in order for one to accumulate a statistically representative QC sample population over time, all significant parameters (such as the method and instrumentation, analytes, analyst, and matrix) must remain unchanged over the time that the QC sample population is accumulated. A desirable shortcut would be to accumulate QC samples for a method, analyte, and matrix on a laboratory-wide basis when multiple instruments and analysts are used.

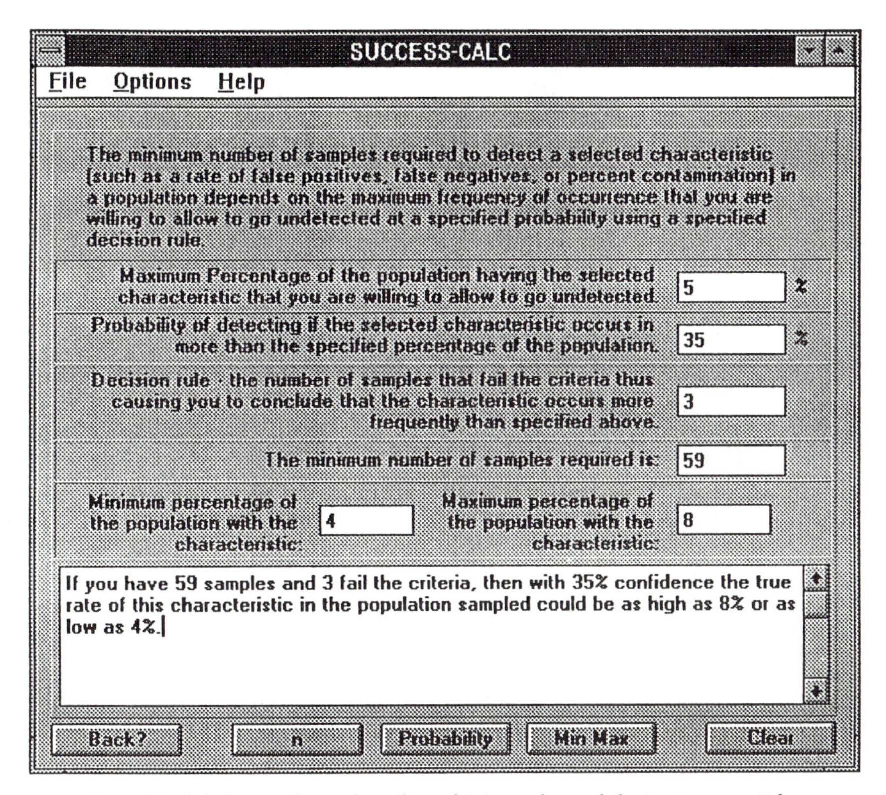

Figure 12. Calculation of revised numbers of QC samples needed using Success-Calc.

However, if different instruments and different analysts produce different populations of QC samples that are not identical (e.g., have different rates of false positives, false negatives, or other selected characteristics), then use of laboratory-wide QC samples will underestimate the numbers of QC samples needed for a desired rate of occurrence and level of confidence. The reason for this underestimation is that the true QC sample population would not be as large as the laboratory-wide number used in the calculation but, rather, some lesser number representing one of several consistent populations of QC samples.

Thus, if a laboratory does shortcut the process of QC sample accumulation by using data from multiple instruments or analysts, then to produce data with the confidence levels estimated by *Success-Calc,* it should document that all the QC data are equivalent as a single statistical population. This documentation may be done by comparing the rates of the selected characteristic (such as false positives or false negatives) from QC data generated from the different instruments or analysts within the laboratory (using the same method, analytes, and matrix) over the period of time used for QC sample population accumulation. This demonstration of equivalency is important because the probability is that the QC sample populations from different instruments or analysts will be different from each other and, therefore, not able to be combined for statistical evaluations unless the measured differences are shown to be small.

QC Assessment Kits

The numbers of QC samples needed in order to obtain reasonable confidence levels with low rates of false positives or false negatives are high in comparison with the modest QC budgets of many projects. When one enters the typically low number of method blanks and spikes that may be available from a typical project, especially if the project is small, then the confidence levels in the resulting data may be so low that one must reassess whether it is worth the cost of the sampling and analysis effort when the statistical probability that high rates (above 90%) of "correct" answers will be obtained is very low. This assessment is made even worse by the way QC samples are typically used; for example, a laboratory may analyze one method blank at the beginning of a day's run of samples and two method or matrix spikes. Under these typical scenarios, the probability of detecting more than 5% false negatives is only 10% and, even worse, the probability of detecting more than 5% false positives is only 6%.

Fortunately, as discussed already, a statistically useful number of QC samples may often be readily available just by grouping those for a particular method over some period of time (e.g., several weeks or months). However, in order to use such a grouping, nothing significant can change during that window of time. Significant changes that can occur over time, however, include the following:

- Changing or modifying instruments (such as cleaning detectors or changing chromatographic columns) can affect instrument detection levels and other measurement parameters.
- Analysts with varying degrees of experience and different analytical techniques can also affect results of a measurement system.
- Different matrices may have different artifacts and interferences and may also affect the recovery of target analytes differently.

Laboratories can usually readily document the consistent use of instrumentation and an analyst for a given period of time or a specific project. Environmental matrices, however, are more difficult and inconvenient to maintain consistency with over a period of time; this characteristic is especially true of soils. Thus, a consistent source of representative matrices is also important for an assessment of false-positive and false-negative conclusions from the analytical measurement system. Therefore, Radian International is developing a representative sandy loam soil in convenient QC Assessment Kits (*18*). Using these kits provides ongoing control of a major parameter (the matrix) needed to maintain consistency among a statistically relevant population of QC samples over time or for a project involving analysis of soils or aqueous sediments.

Because similar QC samples would be analyzed anyway, analyzing a group or batch of samples from a QC Assessment Kit will not significantly increase costs, but it will significantly improve the assumption of measurement process consistency because it removes the variability associated with unknown matrices and poorly homogenized samples. Time limitations of 3–6 mo are recommended as reasonable lengths of time over which to accumulate statistical populations of QC data from these kits or from any other grouped source of QC data. Method parameters should be consistent in laboratories that frequently use a given method for several weeks to several months, but this consistency must be documented in writing for the concept of using QC data grouped over time to be valid.

Conclusions

It is hoped that *DQO-PRO* and *Practical QC* will help people to more accurately assess the "goodness" of environmental analytical data as it pertains to their specific needs. Until now, this type of assessment has been relatively difficult to do accurately, but *DQO-PRO* should help remedy many of the problems associated with data of unknown quality. When combined with *Practical QC*, which helps to determine what kinds of QC samples to take, the four basic questions related to the four basic principles discussed in this chapter should be able to be answered so that environmental data of known quality can be produced or assessed. When this evaluation is done, the result may be data of good, bad, or mediocre quality, but at least its quality will be known and recognized as such. Recognition of the quality of data will be a big step forward for environmental science in general and environmental chemistry in particular.

References

1. Smith, J. S.; Eng, L.; Comeau, J.; Rose, C.; Schulte, R. M.; Barcelona, M.; Klopp, K.; Pilgrim, M. J.; Minnich, M.; Feenstra, S.; Urban, M. J.; Moore, M. B.; Maskarinec, M.; Siegrist, R.; Parr, J.; Claff, R. E. Chapter 34 in this book.
2. *Guidance for Planning for Data Collection in Support of Environmental Decision Making Using the Data Quality Objective Process;* interim final, Quality Assurance Management Staff. U.S. Environmental Protection Agency: Washington, DC, 1993; EPA QA/G–4.
3. Keith, L. H.; Crummett, W.; Deegan, J., Jr.; Libby, R. A.; Taylor, J. K.; Wentler, G. *Anal. Chem.* **1983,** *55,* 2210–2218.
4. Gilbert, R. O. *Statistical Methods for Environmental Pollution Monitoring;* Van Nostrand Reinhold: New York, 1987; pp 30–34.
5. Keith, L. H. *Instant EPA's Best Methods;* Instant Reference Sources: Austin, TX, in press.
6. Keith, L. H. *EPA's Sampling and Analysis Methods Database, 2nd ed.*; Lewis Publishers/CRC Press: Boca Raton, FL, 1996.
7. Keith, L. H. *Compendium of EPA's Sampling and Analysis Methods, 2nd ed.*; Lewis Publishers/CRC Press: Boca Raton, FL, in press.
8. Keith, L. H. *Environmental Sampling and Analysis: A Practical Guide.*; Lewis Publishers/CRC Press: Boca Raton, FL, 1991.

9. Gilbert, R. O. *Statistical Methods for Environmental Pollution Monitoring;* Van Nostrand Reinhold: New York, 1987; pp 119–131.
10. Taylor, J. K. *Quality Assurance of Chemical Measurements;* Lewis Publishers: Chelsea, MI, 1987.
11. Keith, L. H. *Environmental Sampling and Analysis: A Practical Guide;* Lewis Publishers/CRC Press: Boca Raton, FL, 1991; p 71.
12. Keith, L. H.; Patton, G. L.; Lewis, D. L.; Edwards, P. G.; Re, M. A. Presented at the 11th Quality Assurance and Waste Testing Symposium, Washington, DC, July 1995.
13. Keith, L. H.; Re, M. A.; Patton, G. L.; Edwards, P. G. Presented at Dioxin 95, Edmonton, AL, Canada, Aug. 1995.
14. Keith, L. H. *Practical QC;* Instant Reference Sources, Austin, TX, 1995.
15. Cochran, W. G. *Sampling Techniques, 3rd ed.;* John Wiley & Sons: New York, 1977.
16. Grant, E. L.; Leavenworth, R. S. *Statistical Quality Control, 6th ed.;* McGraw-Hill: New York, 1988; pp 201–208.
17. Hahn, G. J.; Meeker, W. Q. *Statistical Intervals: A Guide for Practitioners;* John Wiley & Sons: New York, 1991; pp 104–105.
18. *Analytical Reference Materials Catalog,* Radian Corporation: Austin, TX, 1996.

Chapter 2

Overview of the Sampling Process

Michael J. Barcelona

The involvement of chemists, and analytical chemists in particular, can substantially improve the result of sampling and analytical programs. The development of meaningful sampling protocols demands careful planning of the actual procedures used in sample collection, handling, and transfer. Critical aspects of sampling protocol development cannot be covered by traditional field and laboratory quality control measures. Criteria for sample representativeness must be developed with careful attention to the physical, chemical, and biological dynamics of the environment under investigation. Preliminary sampling and a well-conceived sampling experiment can provide the validation and experience necessary to design efficient sampling protocols that will meet program needs.

SAMPLING IN THE ENVIRONMENT for chemical analysis is a complex subject, and can be as varied and complicated as the objects that must be sampled to investigate the environmental effects and fates of chemical species on the planet. The subject does not lend itself to textbook treatments for students and has long been given diminished importance relative to the improvement and verification of analytical methods for environmental applications. Indeed, in reviewing environmental opportunities in chemistry, a recent National Research Council report (*1*) repeatedly cited the need for improved sensitivity and selectivity of analytical techniques and only in passing mentioned improved sampling for chemical analysis. However, that which cannot be reliably sampled is seldom worth the care and expense of analysis. Also, numerous analytical problems exist with chemicals in the environment at the parts-per-billion ($\mu g\ L^{-1}$) and parts-per-trillion ($ng\ L^{-1}$) concentration levels without delving further into the frontiers of ultratrace analyses of air, water, or land environments.

Many scientists, however, recognize the need for accurate and precise environmental sampling as well as analysis.

This chapter has two objectives for improved experimental design for field sampling problems. The main objective is to encourage the development of accurate and precise sampling protocols that go beyond the confusion of general methods, techniques, and procedures available at the present time. *Protocols* are thorough, written descriptions of the detailed steps and procedures involved in the collection of samples. This sampling validation objective is analogous to that described by Taylor (2) for analytical protocols to avoid the serious consequences of systematic error (i.e., bias or inaccuracy) on the results and conclusions of environmental studies. These studies may be undertaken to investigate contaminant distributions or the potential for exposure to contaminants for a variety of purposes.

Field handling and procedural blank determinations that permit meaningful evaluations of contaminant exposures have been the focus of several recent discussions. These determinations are particularly useful in interpreting the effects of systematic errors on interlaboratory analytical comparisons (*3-5*). This chapter supports the identification and control of sources of sampling error when such errors exceed those inherent in analytical determinations.

The more subtle objective of this chapter is to encourage a realistic appraisal of the practical limits that systematic sampling errors and resultant bias place on the purpose, results, and conclusions of studies in environmental chemistry. This objective has been included because although the actions of sampling may be deceptively simple, they must be carefully planned, refined, and documented if truly representative samples are to be provided for the purposes of the investigation. Furthermore, the results of environmental studies are frequently extended to purposes other than those of the original investigation. These objectives apply equally to specialized research as well as routine regulatory or compliance monitoring efforts because the results of research investigations are often generalized to a variety of environmental conditions.

Review of the Literature

Sampling for chemical analysis has been reviewed critically in the chemical literature. The excellent works of Kratochvil and co-workers (*6, 7*) provide a sound basis for chemists interested in recognizing and exploring chemical sampling problems. Their main emphasis is quite practical and the review article (*6*) includes valuable citations of past work for a variety of environmental matrices. These authors carefully point out the roles that statistics (*8*) and chemometrics (*9, 10*) can play in resolving sampling problems. They further underscore the need for the involvement of chemists, and more specifically analysts, in the planning, execution, and interpretation of the results of sampling and analytical efforts.

An environmental scientist's view of the sampling process is often quite different from that of a statistician. The scientist may be interested in representative samples of water from the hypolimnion of a particular type of lake. The statistician, on the other hand, may envision samples as a subset of the universe of all reducing surface water samples. In the environment, collecting samples (objects) from a largely uncharacterized universe of objects is often required. Only through some prior sampling experience can a sample or sample population be related to the universe or parent population that is the territory of statistical theory. This prior sampling experience can also permit some generalization of the results if the experiment is properly designed. The distinction between samples and parts of a parent population can be better illustrated by a reexamination of the types of objects and samples that are ultimately collected for analysis.

Kateman and Pijpers (*11*) categorized objects (e.g., a well-mixed fluid) from which samples are derived on the basis of the degree of homogeneity and the nature of the spatial change in a particular quality (e.g., dissolved lead content). The aim of their categorization was to lead into the corresponding types of samples and sampling strategies that are needed for analytical quality control. Their scheme has been expanded in Figure 1 to include temporal as well as the spatial change inherent in environmental sampling. Spatially heterogeneous objects or sample origins present a much greater challenge to accurate sampling than homogeneous objects that may only exist in the laboratory. This greater challenge comes from the need for both more specific criteria for representativeness in heterogeneous populations and for more detailed characterization of the conditions under which sampling takes place. For example, representative sampling of dissolved lead in a well-mixed solution in a beaker is rather simple compared to sampling dissolved lead in a variable mixture of reactive aqueous effluents entering a treatment plant operation. In the treatment plant, the effluent mixture is a much more complex object both in composition and extent as well as in variability in space or time. Useful criteria for representative effluent samples would necessarily include qualifiers of flow rate, process status, time, and perhaps other physical and biological variables. Kateman and Pijpers (*11*) provided a general review of the relationships between object types and theoretically optimal sample sizes, numbers, and frequencies for a variety of applications.

For each type of object or sample origin, corresponding types of samples or subsamples can be found that result from sampling, pooling, or compositing, as well as the reduction or preparation steps prior to analysis. Samples suitable for analysis must be representative parts of the object. Figure 2 contains an expanded overview of the nomenclature of sample types suggested by Kateman and Pijpers (*11*) for field sampling applications. Increments (or grabs) issue from the object or parent population. They are parts of the object, but because they are not representative parts of the object, they are not called samples. The *gross sample* (bulk sample) may be seen as a pool of two increments that is reduced or prepared as subsamples for analysis. Field control samples and laboratory

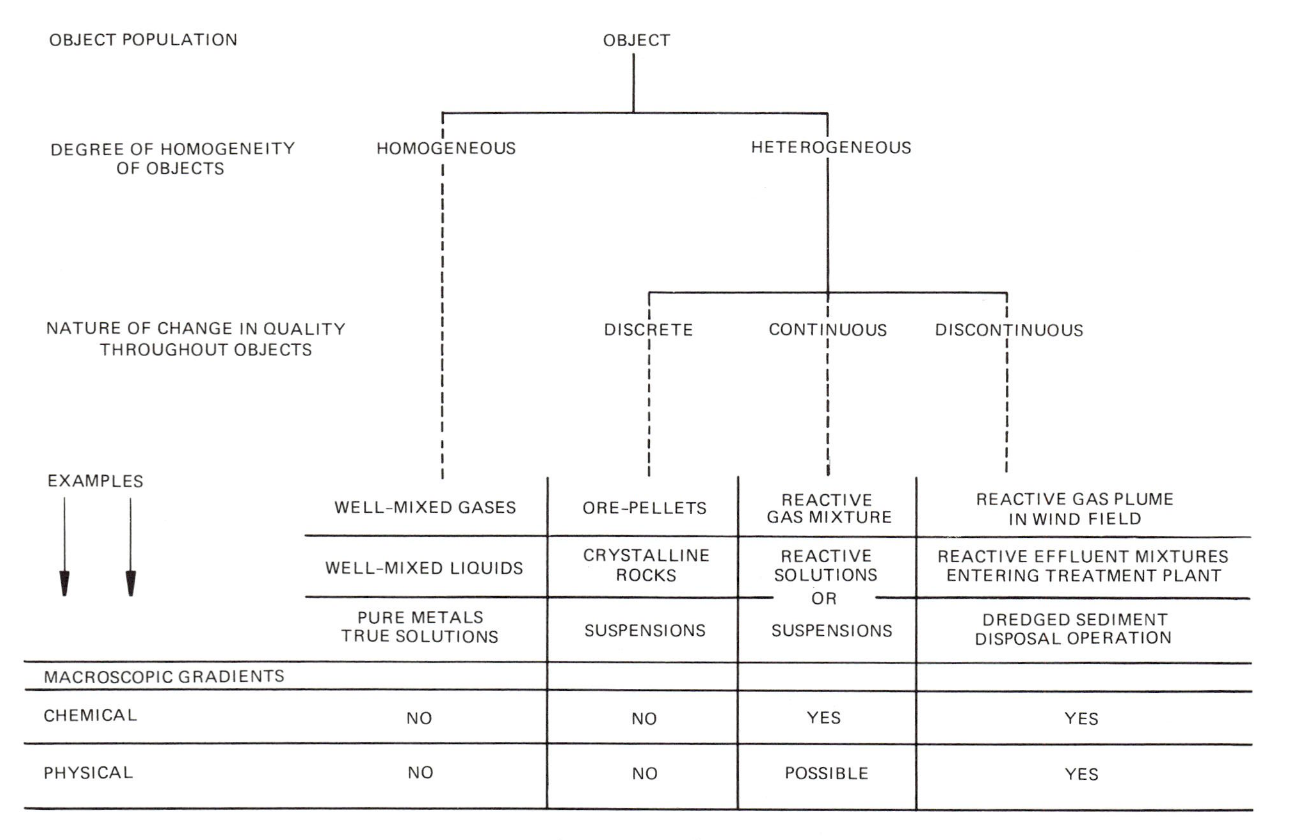

Figure 1. Types of macroscopic objects or sample origins.

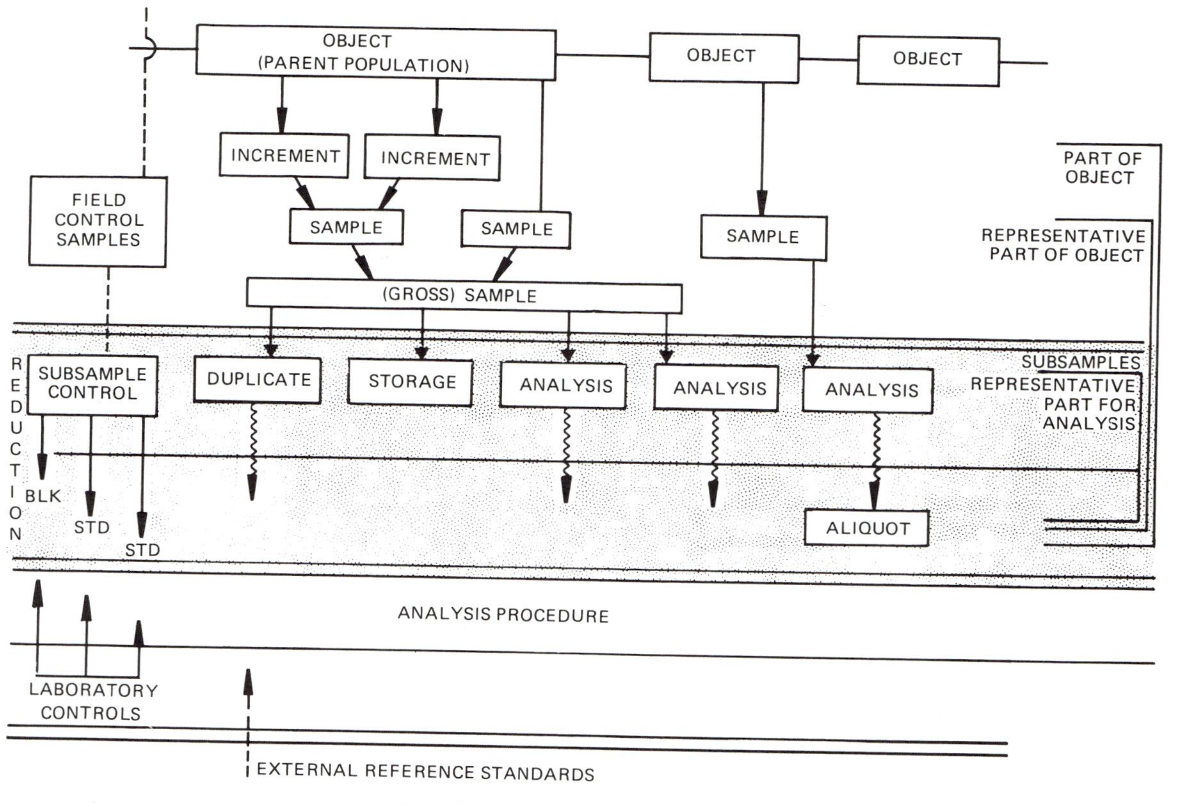

Figure 2. Sample nomenclature overview.

controls enter into the analysis stream with aliquots of the samples. Once these controls and the subsamples have been prepared, negligible sampling error is assumed to occur when aliquots are taken. For the purposes of the present work, the division between sampling and analytical errors is placed after the reduction step that is shaded in Figure 2.

Representativeness in sampling, therefore, presumes that the analysis of the sample or duplicate samples shows the same results as would the object itself. In more practical terms, the characteristic analyte or quality of the sample must be identical to, or minimally disturbed from, the object's quality. Sampling any object will be subject to random variations. In representative sampling efforts, however, we strive to identify and control systematic deviations, or determinate error, caused by sampling.

Planning for representative sampling should be made an integral part of the design of environmental studies. A number of references treat statistical sampling designs within study designs that are worth careful attention. These references include works that deal with biological (*12–14*), geochemical (*15*), and water quality monitoring studies. Among these, the methodology presented by Green (*12*) is particularly well-suited to environmental sampling for chemical analysis.

Green's methodology rests on a clear conception of the problem or question that needs to be answered. His suggestions included the following:

1. replicate samples within each combination of time, location, or other controlled variable;
2. an equal number of randomly allocated replicate samples;
3. samples collected in the presence and absence of a condition in order to test whether the condition has an effect;
4. preliminary sampling to provide a basis for evaluation of sampling design and statistical analysis options;
5. verification of the efficiency and adequacy of the sampling device or method over the range of conditions to be encountered;
6. proportional focusing on homogeneous subareas if the sampling areas have large-scale environmental patterns;
7. verification of sample unit size as appropriate for the size, densities, and spatial distributions of the objects that are being sampled; and
8. testing of the data to establish the nature of error variation to decide whether to transform the data, utilize distribution-free statistical analysis procedures, or test against simulated null-hypothesis data.

Green concluded his suggestions with seasoned advice: Once the best statistical method has been chosen and has provided the test of the hypothesis, accepting the result is preferable to rejecting it and searching for a "better" method.

More theoretical sampling design references may also be useful, and several have been reviewed in detail (*7*). The theoretical work of Gy (*19*) (dealing with solid sampling) is interesting, particularly one of his more recent publications (*20*). In the more recent work, he makes the distinction between sample handling and sampling as an error-generating operation. This statement is an acknowledgment that elements of a sampling operation may cause serious errors and cannot be treated strictly by statistics. Examples of such error-prone elements of the sampling operation are sampling locations and sampling mechanisms or materials. Unlike sample size, number, or sampling frequency, which are elements that can often be handled statistically, identifying and controlling the previously mentioned sources of systematic error are very difficult without prior sampling experience (*12, 20, 21*). Furthermore, in regard to location; sampling devices; or mechanisms, materials, and handling operations, the expertise of the chemist can be most fruitfully employed.

This area has been recognized by the American Chemical Society Committee on Environmental Improvement (*22*), which suggested the following minimum requirements for an acceptable sampling program:

1. a proper statistical design that takes into account the goals of the study and its certainties and uncertainties;
2. instructions for sample collection, labeling, preservation, and transport to the analytical facility; and
3. training of personnel in the sampling techniques and procedures specified.

To bolster these suggestions, the committee emphasized that all sampling procedures should be written into detailed protocols similar to the need for analytical protocols in a quality assurance program. Furthermore, the committee suggested documentation of decisions as to what methods, techniques, procedures, or materials are to be included in the protocols for sampling particular matrices for particular constituents.

These suggestions for sampling protocols are really expressions of professional accountability analogous to those expected from the peer-reviewed literature. Few chemical professionals would expect to publish a procedure for an organic synthesis or for the analysis of an exotic chemical species without carefully documenting the exact steps, preparations, and performance measures for such a procedure. Yet, the environmental literature contains numerous instances of striking phenomena "observed" in samples from one locale that were collected by largely undocumented procedures. Many research and monitoring efforts would materially benefit from improvement in the documentation of sampling and analytical work.

If an overall study program is viewed as a hypothesis to be tested by the scientific method of observation (i.e., sampling and analysis), which is followed by interpretation and reevaluation of the hypothesis, the value of detailed protocols can be readily appreciated. This parallel development of program purpose or hypothesis testing is depicted in Figure 3. The figure shows the progression from mere methods, techniques, or procedures to detailed protocols that in turn will often need to be refined to adequately test the hypothesis or achieve the purpose of a program. The value of experimentation will be to strengthen the results and conclusions of environmental sampling and analysis programs. This value is especially true for identification and control of systematic error. The literature of environmental chemistry also provides a number of examples where recognition of systematic sampling problems has led to significant advances in our understanding of environmental processes and chemical fates. The framework of a sampling protocol provides a basis for the application of growing sampling experience to a variety of present and future problems.

Elements of Environmental Sampling Protocols

There are far too many potential types and purposes of investigations in environmental chemistry to present a generally applicable strategy or formula for preparing sampling protocols. General guidance for combined sampling and analysis quality assurance (QA) program planning is available for a number of monitoring applications (*23-28*). However, guides that deal with protocols and QA procedures for air, water, wastewater, seawater, and hazardous wastes analyses are far more numerous than references that include the practical aspects of sampling. Thus, the need is for successively refined sampling experiments in specific sampling applications.

The historical strength of science as opposed to other fields of human endeavor (e.g., politics and religion) is the scientific or experimental method bolstered by peer review. Experimentation and experimental design skills are not widely taught in science or engineering curricula. For this reason, among others, environmental chemistry is often viewed by our basic chemical colleagues as a business of comparative, rather than absolute, measurements or observations. Much of the recent literature in environmental chemistry shows that this view is truly not the case, and examples of nearly absolute measurements of chemical contaminants in admittedly nonequilibrium systems are not hard to find. The fact that experimental design skills are necessary for reliable studies suggests that the subject should be taught as are other details of experimental or analytical work. A good source book for basic experimental design that is particularly useful for students is that of Wilson (*29*), among others (*30, 31*). A generalized sampling protocol for environmental applications presented here highlights the results of successful sampling experiments.

Table I contains an outline for a sampling protocol that would have general application in environmental research or monitoring. The protocol begins with the purpose of the overall program and the specific purpose or purposes of the

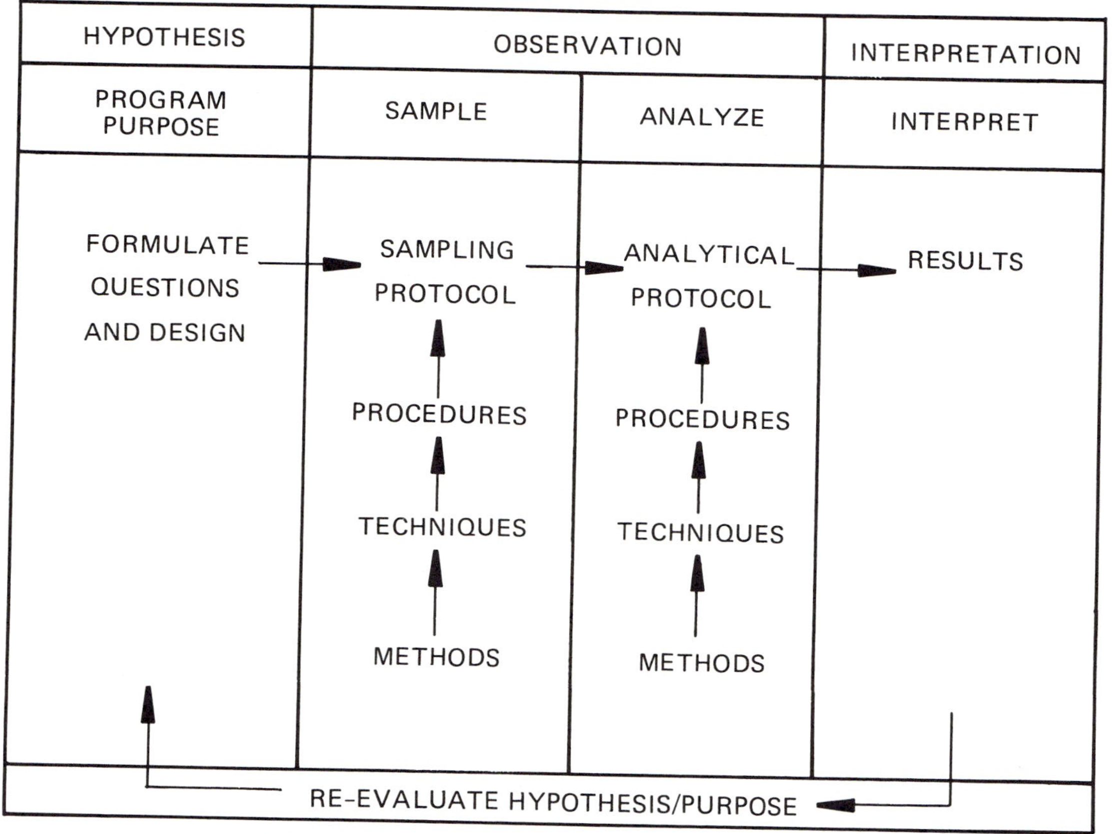

Figure 3. Relationship of program purpose and protocols to the scientific method.

Table I. Outline of a Generalized Sampling Protocol

Main Point (Program Purpose)	*Subelements*
Analytes of interest	Primary and secondary chemical constituents and criteria for representativeness
Locations	Site, depth, and frequency
Sampling points	Design, construction, and performance evaluation
Sample collection	Mechanism, materials, and methodology
Sample handling	Preservation, filtration, and field control samples
Field determinations	Unstable species and additional sampling variables
Sample storage and transport	Preservation of sample integrity

sampling effort. From this point, the specifics of what the samples are to be analyzed for and the questions "how many", "where", "when", and "how" are addressed in order.

A seemingly simple series of tasks in an initial sampling effort would be to first estimate the variability and mean value(s) of the analytes in the samples and apply statistics to estimate the number of samples and frequency of sampling necessary to achieve the acceptable confidence levels and to fulfill the program purposes. Then, the volumes or types of sample necessary for the specific determinations would be identified. Finally, graduate students or laboratory technicians would be sent out to the field armed with homemade or commercial sampling gear and having a firm resolve to "drown a few worms" and return with the needed samples.

This final step is often the weakest link in the sampling operations. Although a sufficiently large number of samples can normally account for random errors (32), serious systematic or determinate sources of error may be involved in the use of certain sampling devices. These problems mainly affect sampling accuracy or the relation between the analytical result (presuming that the analysis is perfectly accurate) and the actual composition of the environmental medium being sampled. As Lodge (33) pointed out, a clear distinction exists between sampling accuracy and representativeness. Sampling inaccuracy can often be minimized by sampling experiments and technological refinements. *Representativeness* is the correspondence between the analytical result and the actual environmental quality or the condition experienced by a contaminant receptor. Regardless of the purpose of the study or investigation, laboratory-oriented QA measures can only account for errors that occur after sample collection. The planning decisions, identification of sites, and procedures for sample collection must also be subject to QA and peer review.

Certainly, scientists must exercise care in the sample collection step of a sampling protocol and ensure that the sampling point and mechanism are not subject to serious systematic error. Accepting manufacturers' claims of the performance of "representative" sampling devices may be particularly imprudent without careful experimentation. Similar precautions may apply to the other elements in a sampling protocol.

Program Purpose

The goals or purposes of an environmental program or study are implicit in the task of sampling protocol preparation. The cost, time frame, and overall goals of a particular study may override efforts to carefully plan and conduct an adequate sampling operation. However, the long-term consequences of the quick answer should be considered. Environmental scientists can also argue that the cost of planning and conducting a limited sampling experiment can save considerable expense as well as "face" in studies that deal with trace chemical constituents of health concern. All individuals involved in the effort should understand the overall and immediate purposes of the study and recognize that the data must be well-documented as to quality.

Analytes of Interest

The selection of chemical constituent(s) of interest to a particular study may be categorized as primary and secondary. The *primary chemical constituents* may represent known species that have been identified in the sampling matrix or are required to be determined by regulation. Examples of these species include those required by drinking water regulations, maximum contaminant level rules, air quality regulations, or a variety of monitoring programs. *Secondary chemical constituents* may include transformation products of primary chemical species, environmental variables needed to characterize conditions or meet criteria for representativeness, and other chemical species that may be indicators of sample integrity. The secondary category is no less important than the primary grouping.

Consider a study that required a survey of priority pollutant compounds in groundwater. The concentrations of these trace compounds and species provide little insight into the bulk chemical composition and geochemical conditions in the environmental matrix. Many of the 129 priority pollutants undergo substantial chemical or biochemical transformations in air, water, and soils and result in the formation of other compounds and products (*34*). A more extreme example may result from restricting analytical determinations to the parent compound (e.g., aldicarb) in a pesticide formulation for an aquatic fate or transport study only to learn later that its transformation products are far more mobile, persistent, and toxic (e.g., the aldicarb sulfoxide and sulfone) (*35*). Allocating large amounts of funds and human resources for sampling and analysis of natural waters for selected or suspected toxic compounds is unrealistic without also considering pH, major cations and anions, total organic carbon, and alkalinity determinations. A more complete analysis in many cases could provide the means to check the consistency of selected samples by mass or charge balance methods and may aid in determining controlling reactions or conditions important for treatment or remediation efforts.

Careful consideration should also be given to the choice of analytical methods for specific determinations. This consideration is not only important for optimization of sample collection and handling procedures, but also to avoid matrix interferences for certain types of samples. Analytical methods developed and validated for drinking water or industrial wastewaters may not be applicable to samples that represent mixtures of these matrices. Also, the aims of research projects frequently require the use of more specific and sensitive types of analytical procedures than those in "standard" references.

The minimum sample volumes and types of sample preservation and handling procedures also depend on the detail and specificity of the proposed analytical program. Sample volumes necessary for specific analyses must be identified with careful consideration of representativeness. For example, both organic (*36*) and inorganic (*37*) chemical species in drinking water may vary significantly with the volume of the sample depending on whether sampling is done immediately on opening the tap or after allowing the water to flow for a time.

In addition to the specifics of analyte and analytical method selection, recognizing that physical, meteorological, or hydrologic variables may also need to be measured or determined is important. In many cases, the usefulness of chemical results is fully realized only when the evidence of environmental chemical processes can be linked to other controlling variables. This data is therefore essential to the interpretation of the chemical results and should be included in sampling protocol planning.

Sampling Location and Frequency

The sampling location in space and time can have a very real effect on the quality and usefulness of data in environmental chemistry. Site selection, of course, should be made primarily on the basis of the study goals and the nature of the environmental phenomenon or process under consideration. The optimum number, spacing, and frequency of samples at a coarse scale can best be estimated after a preliminary sampling experiment. Geostatistical and kriging approaches can be most helpful in working with this data, particularly where large numbers of heterogeneous samples may be needed to meet minimum confidence levels in sample results (*38-41*). This type of problem has been encountered frequently in geochemical studies of soils and subsurface solids. This problem is essentially the difficulty in obtaining a representative sample of a powdered solid material (*2, 6, 19, 20, 21, 42*) complicated by the problems of obtaining representative solid samples over a large geographic region. Geochemists have applied methods of hierarchical analysis of variance for preliminary geochemical studies to optimize the sampling designs of large-scale efforts (*43–45*). For example, optimum numbers, discrete depths, and locations have been chosen in lakes and within geographic zones of state or provincial regions in selected studies. The power of these approaches is that they allow the scientist and the program consumer (i.e., government agency or official) to qualitatively discuss the trade-offs involved in changes of such designs.

From a practical point of view, formal design plans have real potential in hazardous waste site assessments and in cleanup evaluations where the challenge of inhomogeneous sample matrices may be further complicated by spatial variability and the need to know suspect constituent concentrations at the parts-per-million (mg L^{-1}) level or below (*46, 47*).

Besides the numbers of sampling locations and samples to be collected, the frequency of sampling is the most significant cost multiplier in a sampling operation. Certain frequencies of sampling are set by regulation. More often in research and other investigations, relatively short-term studies are undertaken with little concern for the dynamics of the environment. If the purposes of the program include optimization of detection or sensitivity of the results of sampling operations to trends (i.e., long-term changes in environmental quality) or periods (i.e., short-term changes in environmental quality), this aspect of the sampling protocol needs very careful attention.

The distinction between long- and short-term changes is usually relative to the time period over which measurements are made. Periodic changes may be of the order of seconds at a sampling frequency of minutes. Trends are normally considered changes that may occur over several periods or sampling runs. If the environmental value or quality of interest varies with a certain frequency, the sampling frequency must be at least twice the frequency of that variation.

The problem of optimizing sampling frequency within time, program purpose, and cost constraints has been addressed for many environmental media (*18, 48–53*). The results of a "one-shot" or single-period sampling operation cannot be expected to represent average conditions in most instances.

Sampling Points. Samples may be collected from bridges, banks, stationary towers, or platforms as well as ships, airplanes, or passive personnel monitors. The important considerations here should be minimal disturbance of the samples. The design and construction of sample access points should disturb the local environment as little as possible or the collection of biased results may be inevitable. The literature for specific applications should be considered in the design of new operations.

Scale problems can also arise in the definition of the sampling point. Collection of a surface water sample from a lake or the ocean seemed to be the simplest operation one could perform with a bucket. In time, limnologists and oceanographers recognized that surface water samples collected with buckets were quite different in organic or trace metal content than samples collected by screens (*54*). The surface microlayer, enriched from fivefold to several hundredfold in lipids, Pb, Ni, Cu, and bacterial content, provides very different results from water samples taken just below the surface of water bodies (*55–57*). The selection of the sampling point must be done with an appreciation that certain microenvironments may either be created or traversed by sampling gear that may yield results quite different from the desired point in air, water, or soil environments. In the case of the sea-surface microlayer, the discovery of this microenvironment proved embarrassing to previous work; however, an exciting new area of research was opened with implications for air–sea gas exchange, bubble and aerosol formation, and chemical fractionation investigations (*58*). Too often, the slow recognition of systematic sources of error, like picking up the "slick" from the "surface" in water sample collection, may delay significant advances in environmental science. The value of a critically evaluated sampling experiment cannot be underestimated in such instances. Even though the chemical or biochemical constituents of such microenvironments may not be the object of specific studies or investigations, they may cause matrix effects or other analytical interferences for the chemical constituents of interest. Field control samples (i.e., blanks, spiked samples, and colocated samples) can be extremely helpful in evaluating these analytical effects on chemical results.

Sample Collection. *Sample collection* involves contact of the sample with the sampling device and its materials of construction. Often, the sample may contact the sampling technician or adjacent materials as well. The entire path of sample retrieval should be scrutinized to minimize systematic sources of error that cannot be accounted for by conventional laboratory-oriented QA and quality control (QC). The same care required for sample storage vessels should be extended to the selection of materials, methods, and devices for sample collection.

The history of trace metal sampling and analysis in the environment provides classic examples of the value of recognition and control of systematic sampling error. Basically, the modern environment has levels of Pb in the air, dust, and soil that may bias improperly collected samples by orders of magnitude. The pioneering work of Patterson and co-workers (*59, 60*) on environmental Pb sampling and analysis spurred an avalanche of research that resulted in the well-known "decrease" in levels of trace metals in the oceans over the past 20 years (*61–64*). These references show that previous attempts to unravel the geochemistry and environmental fates of trace metals were stymied because of biased sample collection and handling techniques. The contamination of samples of air, water, and biota by artifact trace metals had obscured the true environmental distributions and controlling processes. Far worse, however, was the delayed realization that not all environmental samples in geologic time had equivalent industrial metal levels.

Even though sample collection and handling techniques at the "clean-room" level may not be necessary for routine monitoring or investigative efforts, very careful consideration must be given to the choice of sampling mechanism and methodology for chemical contaminants in the parts-per-billion ($\mu g\ L^{-1}$) range. References 59–64 represent carefully documented attempts to control systematic error.

Here again, as in the case of recognizing the impact of the sea-surface microlayer on surface samples, the control of industrial Pb contamination has had a great effect on the results and interpretation of environmental chemistry research. It is sobering to contrast the modern practice for sampling seawater for Pb in its diligence and detail with the thousands of somewhat haphazardly collected samples of groundwater delivered under strict chain of custody to a variety of contract laboratories under various hazardous waste monitoring programs. We will undoubtedly discover subtle aspects of waste-constituent geochemistry in the future once sample collection details are as carefully documented as are laboratory QA and QC procedures (65).

Poor sample collection procedures can seriously bias chemical results. Documentation of sampling procedures in protocols can help to identify and control such errors. The core of the sampling operation is the mechanism or device used to collect the sample. Because we rarely know the true value for the concentration of a particular chemical constituent in the environment at any single time, we must either test sampling mechanisms under controlled laboratory conditions or intercompare them simultaneously in the field. In the laboratory, the concentration of the chemical constituent can be controlled, but the exact environmental conditions are difficult to simulate. In the field, depending upon whether loss (e.g., volatile compounds) or contamination (e.g., airborne Pb) is a likely source of systematic mechanism-related error, the highest or lowest result, respectively, is normally accepted as the most reliable for certain applications. Device-related errors cannot be accounted for by blanks, standards, or replicate control samples. The actual performance of sampling mechanisms must be verified in critical sampling applications. Numerous examples in the literature serve as a guide for future efforts. Many of these are provided in Table II. These examples demonstrate that sampling mechanisms and materials can contribute relatively large errors in comparison to analytical errors in environmental chemistry results, particularly for trace level concentrations (i.e., $< 1\ mg\ L^{-1}$) of chemical constituents.

Poor recoveries and positive bias caused by contamination are not the only problems that may arise from the use of certain types of samplers. The efficiency with which a person can control the operation of the sampler; maintain stable, reproducible operating conditions; and recognize a malfunction all play a significant part in the actual performance of environmental sampling equipment. Unfortunately, few manufacturers provide detailed operation or calibration instructions. More frequently, the principle of operation is referenced to a literature citation and specific precautions for safe use may be provided. Individual investigators, with the aid of the analytical staff, should plan to evaluate sampling performance in order to at least estimate the degree of systematic error as well as the routine precision that may be expected under

Table II. Selected Sampling Mechanism and Materials Evaluations

Environmental Matrix	*Sample*	*Reference*
Air	Aerosol metals	66
	Aerosol SO_4^{2-} and NO_3^-	95
	Precipitation	67–69
Water	Volatile organic compounds	70
Groundwater	Dissolved gases and volatile organic compounds	71
	Ferrous iron	72
	Miscellaneous	73
Seawater	Trace metals	74–76
	Surface microlayer	77
Soil	Pore water	78, 80
	Pore water organic compounds	79

field conditions. A general validation scheme for sampling devices is shown in the box on page 57. Although this scheme may be too involved for individual projects, large field programs would benefit from the use of such validation activities. If sampling errors are less than those that may be contributed by the analytical operation, reducing them further is probably not worth the effort in most cases.

Materials selection can be a potential pitfall for both sampling operations and achieving the purpose of an investigation. In situations where the sample remains in contact with a potentially reactive, sorptive, or leaching material, the opportunities for gross systematic errors exist. Sampling tubing (*70, 81, 82*), gaskets (*36*), plated surfaces (*37, 83*), and a variety of other such exposures have resulted in serious errors in sampling. The selection of a general sampling mechanism or a specific device made of appropriate materials for the application should be based on the most sensitive (i.e., labile, volatile, and reactive) chemical constituents under investigation. Once the sampling evaluation and experiment have been conducted, documenting the procedures and format field books to record deviations from the sample collection part of the sampling protocol is a relatively simple task (*84*).

Sample Handling, Field Determinations, Storage, and Transport

Most of the recent QA and QC guidance manuals available for the national environmental monitoring programs provide sound guidance for planning the procedures for sample preservation and handling. Because many sampling sites are exposed to rain, wind, sunlight, and temperature extremes, field transfers and

Sampling Device Validation Scheme Protocol

Laboratory: Operating range (ruggedness), recovery (accuracy), and precision

1. Generate known atmosphere, solution, etc.
2. Pump or collect multiple samples (and backup if possible) with device and collect controls by reference (direct) method.
3. Repeat step 2 holding concentration fixed and varying humidity, depth, lift, submergence, volume, or flow rate.
4. If possible, repeat step 3 over a range of concentrations for the expected range of variables.
5. Repeat steps 1–4 with interferences present, if possible.

Field: Estimation of recovery (accuracy) and precision

1. Use reference method, if possible, to establish the background, stable concentration.
2. Pump or collect multiple samples with device; if necessary, use backup samples to check breakthrough.
3. Repeat steps 1 and 2 over the range of field variables.

manipulations should be kept to an absolute minimum. Filtration procedures should be streamlined and protected so as to minimize sample transfers in the open air. Filter media selection and pretreatment steps should be planned and documented with the reduction of bias in mind (*85, 86*). A number of researchers performed filtration experiments in order to identify and control sample contamination or loss for trace metal determinations due to filtration apparatus or procedures (*87–90*). The detailed publications of Hunt and Gardner (*90–92*) are particularly useful toward minimizing systematic errors in filtration steps.

The sampling staff should be aware of routine laboratory procedures for both personnel and sample integrity protection. Gloves and eye protection provide valuable safeguards and can minimize artifact trace metals (*93*) or organic contamination of samples (*94*). Determinations of unstable chemical species or those that are difficult to preserve and store should be completed as soon as possible. A rapid field analytical method may be preferable to a laboratory method that requires extraordinary or complex sampling and handling procedures.

An effort should be made to handle and preserve field control samples (i.e., blanks, spikes, and colocated samples) in the same manner as the environmental samples. This precaution will allow more effective identification and control of postsample collection errors. Although not always possible, the experience of running the sampling device evaluation and sampling experiment should be useful in planning the sample handling and preservation procedures.

Once the samples have been delivered to the laboratory, the sampling operation extends to the point of sample reduction prior to analysis. The analyst who is consulted on sampling protocol development can materially improve the overall reliability of the results and conclusions of the program. A willingness to experiment (i.e., make and acknowledge errors) will provide the balance between statistical and experimental concerns needed to design and execute a cost-effective sampling effort.

Acknowledgments

I am thankful to a number of my past and present colleagues for their insight, help, and advice in the preparation of this chapter. The comments of the reviewers are also very much appreciated. Thanks also go to Pam Beavers who typed the manuscript and to Lynn Weiss who drafted the figures.

References

1. *National Research Council Opportunities in Chemistry*; National Academy Press: Washington, DC, 1985; p 18, pp 195–208.
2. Taylor, J. K. *Anal. Chem.* **1983**, *55(6)*, 600A–608A.
3. King, D. E. In National Bureau of Standards Special Publication 422; La Fleur, Philip D., Ed.; National Bureau of Standards: Gaithersburg, MD, 1976; pp 141–150.
4. Youden, W. J. *Statistical Techniques for Collaborative Tests*; Association of Official Analytical Chemists: Washington, DC, 1973.
5. Murphy, T. J. In National Bureau of Standards Special Publication 422; La Fleur, Philip D., Ed.; National Bureau of Standards: Gaithersburg, MD, 1976; pp 509–539.
6. Kratochvil, B.; Taylor, J. K. *Anal. Chem.* **1981**, *53(8)*, 924A–938A.
7. Kratochvil, B.; Wallace, D.; Taylor, J. K. *Anal. Chem.* **1984**, *56(5)*, 113R–129R.
8. Youden, W. J. *J. Assoc. Off. Anal. Chem.* **1967**, *50(5)*, 1007–1013.
9. Kowalski, B. R. *Anal. Chem.* **1980**, *52(5)*, 112R–122R.
10. Frank, T. E.; Kowalski, B. R. *Anal. Chem.* **1982**, *54(5)*, 232R–243R.
11. Kateman, G.; Pijpers, F. W. In *Chemical Analysis*; Elving, P. J.; Winefordner, J. D., Eds.; Wiley–Interscience: New York, 1981; Vol. 60, pp 15–69.
12. Green, R. H. *Sampling Design and Statistical Methods for Environmental Biologists*; Wiley: New York, 1979.
13. Bernstein, B. B. *J. Environ. Manage.* **1983**, *16*, 35–43.
14. Electric Power Research Institute. *Sampling Design for Aquatic Ecologic Monitoring*; Electric Power Research Institute: Palo Alto, CA, 1985; Vols. 1–5; EPRI EA–4302.
15. Garrett, R. G.; Goss, T. I. In *Geochemical Exploration*; Watterson, J. R.: Theobald, P. K., Eds.; Association of Exploration Geochemistry: Toronto, Ontario, Canada, 1978.
16. Liebetrau, A. M. *Water Resour. Res.* **1979**, *15(6)*, 1717–1725.
17. Ellis, J. C.; Lacey, R. F. In *River Pollution Control*; Stiff, M. J., Ed.; Ellis Horward: Chichester, England, Chapter 17, pp 247–274.
18. Sanders, T. G.; Ward, R. C.; Loftis, J. C.; Steele, T. D.; Adrian, D. D.; Yevjevich, V. *Design of Networks for Monitoring Water Quality*; Water Resources Publications: Littleton, CO, 1983; p 328.
19. Gy, P. M. *Sampling of Particulate Materials: Theory and Practice*; Elsevier: New York, 1979.

20. Gy, P. M. *Analusis* **1983**, *11(9)*, 413–440.
21. Brands, G. *Fresenius' Z. Anal. Chem.* **1983**, *314*, 646–651.
22. American Chemical Society Committee on Environmental Improvement. *Anal. Chem.* **1980**, *52(14)*, 2242–2249.
23. *Fed. Regist.* **1978**, *43(132)*, 29696–29741.
24. Riggan, R. M. *Technical Assistance Document for Sampling and Analysis of Toxic Organic Compounds in Ambient Air*; Battelle Laboratories, Columbus, OH; prepared for U.S. Environmental Protection Agency–Environmental Monitoring and Support Laboratory: Research Triangle Park, NC, 1983; EPA 600/4–83–027.
25. *Sampling and Analysis of Rain*; Campbell, S. A., Ed.; ASTM Special Technical Publication 823; American Society for Testing and Materials: Philadelphia, PA, 1983.
26. Peden, M. E. *Development of Standard Methods for the Collection and Analysis of Precipitation*; Illinois State Water Survey CR 381; U.S. Environmental Protection Agency–Environmental Monitoring and Support Laboratory: Cincinnati, OH, 1986.
27. Peden, M. E. In *Sampling and Analysis of Rain*; Campbell, S. A., Ed.; ASTM Special Technical Publication 823; American Society for Testing and Materials: Philadelphia, PA, 1983; pp 72–83.
28. Topol, L. E.; Levon, M.; Flanagan, J.; Schwall, R. J.; Jackson, A. E. *Quality Assurance Manual for Precipitation Measurement Systems*; U.S. Environmental Protection Agency–Environmental Monitoring and Support Laboratory: Research Triangle Park, NC, 1985.
29. Wilson, E. B. *An Introduction to Scientific Research*; McGraw–Hill: New York, 1952.
30. Box, G. E. P.; Hunter, W. G.; Hunter, J. S. *Statistics for Experimenters*; Wiley: New York, 1978.
31. Mace, A. E. *Sample-Size Determination*; Reinhold: New York, 1964.
32. Janardan, K. D.; Shaeffer, D. J. *Anal. Chem.* **1978**, *51(7)*, 1024–1026.
33. Lodge, J. P. In National Bureau of Standards Special Publication 422; La Fleur, Philip D., Ed.; National Bureau of Standards: Gaithersburg, MD, 1976; pp 311–320.
34. Kobayashi, H.; Rittmann, B. E. *Environ. Sci. Technol.* **1982**, *16*, 107A–183A.
35. Cohen, S. Z.; Creeger, S. M.; Carsel, R. F.; Enfield, C. G. In *Treatment and Disposal of Pesticide Wastes*; Krueger, R. F.; Seiber, J. N., Eds.; ACS Symposium Series 259; American Chemical Society: Washington, DC, 1984; pp 297–325.
36. LeBel, G. L.; Williams, D. T. *J. Assoc. Off. Anal. Chem.* **1983**, *66(1)*, 202–203.
37. Neff, C. H.; Schock, M. R.; Marden, J. I. *Relationship Between Water Quality and Corrosion of Plumbing Materials in Buildings*; State Water Survey; prepared for U.S. Environmental Protection Agency–Water Engineering Research Laboratory: Cincinnati, OH, 1987; EPA 600/S2–87–036.
38. Hughes, J. P.; Lettenmaier, D. P. *Water Resour. Res.* **1981**, *17(6)*, 1641–1650.
39. Flatman, G. T.; Yfantis, A. A. *Environ. Monit. Assess.* **1984**, *4*, 335–349.
40. Gilbert, R. A.; Simpson, J. C. *Environ. Monit. Assess.* **1985**, *5*, 113–135.
41. Brown, K. W.; Mullins, J. W.; Richitt, E. P.; Flatman, G. T.; Black, S. C.; Simon, S. J. *Environ. Monit. Assess.* **1985**, *5*, 137–154.
42. Grant, C. L.; Pelton, P. A. *Role of Homogeneity in Powder Sampling*; Remedy, W. R.; Woodruff, J. F., Eds.; ASTM Special Technical Publication 540; American Society for Testing and Materials: Philadelphia, PA, 1972; pp 16–29.
43. Tourtelot, H. A.; Miesch, A. T. In *Trace Element Geochemistry in Health and Diseases*; Freeman, J., Ed.; Geological Society of America Special Paper 155; Geological Society of America: Denver, CO, 1975; pp 107–118.
44. Connor, J. J.; Myers, A. T. *Role of Homogeneity in Powder Sampling*; Remedy, W. R.; Woodruff, J. F., Eds.; ASTM Special Technical Publication 540; American Society for Testing and Materials: Philadelphia, PA, 1972; pp 32–36.
45. Garrett, R. G.; Goss, G. E. In *Proceedings of the 7th International Geochemical Exploration Symposium*; Watterson, J. R.; Theobald, P. K., Eds.; Association of Exploration Geochemistry: Toronto, Ontario, Canada, 1978; pp 371–383.

46. Perket, C. L.; Barsotti, L. R. In *Multilaboratory Analysis of Soil for Lead*; Petros, J. K.; Lacy, W. J.; Conway, R. A., Eds.; ASTM Special Technical Publication 886; American Society for Testing and Materials: Philadelphia, PA, 1986; pp 121–138.
47. Loehr, R. C.; Martin, J. H., Jr.; Newhauser, E. F. Ibid., pp 285–297.
48. Madsen, B. C. *Atmos. Environ.* **1982**, *16(10)*, 2515–2519.
49. Shaw, R. W.; Smith, M. V.; Pour, R. J. *J. Air Pollut. Control Assoc.* **1984**, *34(8)*, 839–841.
50. Nelson, J. D.; Ward, R. C. *Ground Water* **1981**, *6*, 617–625.
51. Loftis, J. C.; Ward, R. C. *Water Resour. Bull.* **1980**, *16*, 501–507.
52. Sanders, T. G.; Adrian, D. D. *Water Resour. Res.* **1978**, *14*, 569–576.
53. Lettenmaier, D. P. *Water Resour. Res.* **1976**, *12(5)*, 1037–1046.
54. Garrett, W. D. In *Proceedings of the Symposium on Organic Matter in Natural Waters*; Hood, D. W., Ed.; Institute of Marine Science Publication No. 1; University of Alaska, AK, 1968.
55. Duce, R. A.; Quinn, J. G.; Olney, C. E.; Piotrowicz, S. J.; Ray, B. J.; Wake, T. L. *Science (Washington, DC)* **1972**, *176*, 161–163.
56. Zsolnay, A. *Mar. Chem.* **1977**, *5*, 465–475.
57. Morris, R. J.; Culkin F. *Nature (London)* **1974**, *250(5468)*, 640–642.
58. Macintyre, F. In *The Sea*; Goldberg, E. D., Ed.; Wiley-Interscience: New York, 1974; Vol. 5, *Marine Chemistry*; pp 245–299.
59. Patterson, C. C.; Settle, D. M. In National Bureau of Standards Special Publication 422; La Fleur, Philip D., Ed.; National Bureau of Standards: Gaithersburg, MD, 1976; pp 321–351.
60. Patterson, C. C.; Settle, D. M.; Schaule, B. K.; Burnett, M. In *Marine Pollutant Transfer*; Windom, H. L.; Duce, R. A., Eds.; Lexington Books: Lexington, KY, 1976.
61. Schaule, B. K.; Patterson, C. C. *Earth Planet. Sci. Lett.* **1980**, *47*, 176–198.
62. Burnett, M.; Patterson, C. C. In *Proceedings of an International Experts Discussion on Lead: Occurrence, Fate and Pollution in the Environment*; Branica, M., Ed.; Pergamon: Oxford, England, 1978.
63. Bruland, K. W.; Franks, R. P.; Knauer, G. A.; Martin, J. H. *Anal. Chim. Acta* **1979**, *105*, 233–245.
64. Windom, H. L.; Smith, R. G. *Mar. Chem.* **1979**, *7*, 157–163.
65. *Quality Control in Remedial Site Investigations*; Perket, C., Ed.; ASTM Special Technical Publication 925; American Society for Testing and Materials: Philadelphia, PA, 1986.
66. Milford, J. B.; Davidson, C. I. *J. Air Pollut. Control Assoc.* **1985**, *35(12)*, 1249–1260.
67. Slanina, J.; Van Raaphorst, J. G.; Zijp, W. L.; Vermuelen, A. J.; Rolt, C. A. *Int. J. Environ. Anal. Chem.* **1979**, *6*, 67–81.
68. Chan, W. H.; Lusis, M. A.; Stevens, R. D. S.; Vet, R. J. *Water, Air, Soil Pollut.* **1983**, *23*, 1–13.
69. Schroder, L. J.; Linthurst, R. A.; Ellson, J. E.; Vozzo, S. F. *Water, Air, Soil Pollut.* **1985**, *24*, 177–187.
70. Ho, J. S-Y. *J. Am. Water Works Assoc.* **1983**, *12*, 583–586.
71. Barcelona, M. J.; Helfrich, J. A.; Garske, E. E.; Gibb, J. P. *Ground Water Monit. Rev.* **1984**, *4(2)*, 32–41.
72. Stoltzenburg, T. R.; Nichols, D. G. *Preliminary Results on Chemical Changes in Ground Water Samples Due to Sampling Devices*; prepared for Electric Power Research Institute: Palo Alto, CA, 1985; EA-4118.
73. Gibb, J. P.; Schuller, R. M.; Griffin, R. A. *Procedures for the Collection of Representative Water Quality Data from Monitoring Wells*; Coop. Ground Water Report 7; Illinois State Water and Geological Surveys: Champaign, IL, 1981.
74. Segar, D. A.; Berberian, G. A. In *Analytical Methods in Oceanography*; Gibb, T. R. P., Jr., Ed.; Advances in Chemistry 147; American Chemical Society: Washington, DC, 1975; Chapter 2, pp 9–15.
75. Brewers, J. M. ; Windom, H. L. *Mar. Chem.* **1982**, *11*, 71–86.

76. Spencer, M. J.; Betzer, P. R.; Piotrowicz, S. *Mar. Chem.* **1982**, *11*, 403–410.
77. Van Vleet, E. S.; Williams, P. M. *Limnol. Oceanogr.* **1980**, *25*, 764–770.
78. Brown, K. W. In *Land Treatment, A Hazardous Waste Management Alternative*; Loehr, R. C.; Malina, J. F., Jr., Eds.; Center for Research in Water Resources, University of Texas: Austin, TX, pp 171–185.
79. Barbee, G. L.; Brown, K. W. *Water, Air, Soil Pollut.* **1986**, *29*, 321–331.
80. Evberett, L. G.; McMillion, L. G. *Ground Water Monit. Rev.* **1985**, *5(3)*, 51–60.
81. Barcelona, M. J.; Helfrich, J. A.; Garske, E. E. *Anal. Chem.* **1985**, *57(2)*, 460–464. (Errata, *Anal Chem.* **1985**, *5713*, 2752.)
82. Barcelona, M. J.; Gibb, J. P.; Miller, R. A. *A Guide to the Selection of Materials for Monitoring Well Construction and Ground Water Sampling*; State Water Survey Publication 327; U.S. Environmental Protection Agency: Cincinnati, OH, 1983; EPA 600/S2-84-024.
83. Andersen, K. E.; Nielsen, G. D.; Flyvholm, M.; Fregert, S.; Gruvberge, B. *Contact Dermatitis* **1983**, *9*, 140–143.
84. Barcelona, M. J.; Gibb, J. P.; Helfrich, J. A.; Garske, E. E. *Practical Guide for Ground-Water Sampling*; State Water Survey Publication 374; U.S. Environmental Protection Agency: Cincinnati, OH, 1985; EPA 600/S2-85-104.
85. Quinn, J. G.; Meyers, P. A. *Limnol. Oceanogr.* **1971**, *16(1)*, 129–131.
86. Maienthal, E. J. Presented at the 157th National Meeting of the American Chemical Society, Minneapolis, MN, April 1969.
87. Wagemann, R.; Brunskill, G. J. *Int. J. Environ. Anal. Chem.* **1975**, *4*, 75–84.
88. Truitt, R. E.; Weber, J. H. *Anal. Chem.* **1979**, *51(12)*, 2057–2059.
89. Laxen, D. P. H.; Chandler, I. M. *Anal. Chem.* **1982**, *54(8)*, 1350–1355.
90. Gardner, M. J.; Hunt, D. T. E. *Analyst (London)* **1981**, *106*, 471–474.
91. Hunt, D. T. E. *Filtration of Water Samples for Trace Metal Determinations*; Technical Report TR 104; Water Research Centre, Medmenham Laboratory: Marlow, Bucks, England, 1979.
92. Gardner, M. J. *Adsorption of Trace Metals from Solution During Filtration of Water Samples*; Technical Report TR 172; Water Research Centre, Medmenham Laboratory: Marlow, Bucks, England, 1982.
93. Berman, E. In National Bureau of Standards Special Publication 422; La Fleur, Philip D., Ed.; National Bureau of Standards: Gaithersburg, MD, 1976; pp 715–719.
94. Hamilton, P. B. *Anal. Chem.* **1975**, *47(9)*, 1718–1720.
95. Milford, J. B.; Davidson, C. I. *J. Air Pollut. Control Assoc.* *37(2)*, 125–134, in press.

Chapter 3

Legal Considerations in Sampling

Harry F. Klodowski, Jr.

Most sampling is done for a legal as well as scientific objective. If the legal goal is not recognized during planning, execution, and analysis of the sample, the results of a technically valid sampling scheme might not be admissible evidence in a courtroom and thus may not be useful. Sampling is used to estimate some value in a population. If the exercise is to be used as proof of a legal issue, the legal requirements are as important as the technical data quality objectives. Legal requirements depend on the relevant contexts: private, regulatory, or judicial. This chapter explains legal procedures relevant to use of scientific evidence, access for sampling, and confidentiality of data. Legal doctrines may influence where to sample, how to sample, sampling and analytical procedures and record keeping, and whether the data are confidential or a matter of public record.

SAMPLING CAN BE PERFORMED for regulatory, judicial, or private purposes. The legal objective can influence the sampling effort by specifying where to sample, defining the method of sampling, adding additional requirements to a valid technical sample design for evidentiary reasons, and determining whether the data are confidential.

Most sampling programs are driven by a legal requirement, and the legal objective of the effort should be defined early in the design of the sampling program. The legal objective may be regulatory compliance, in which case the regulatory program can define sampling methods as well as appropriate analytical methods. The legal purpose may be to develop evidence for civil or criminal litigation, in which case the litigation goal, or the issue in dispute, may influence data quality objectives (DQOs) and sample design, and the sampling protocol must also meet legal requirements for the introduction of evidence in court. Even when the legal objective is private data collection, without a regulatory requirement or threat of potential lawsuit, the purpose of the project influences sample protocol. Concerns about disclosure of data often influence the decision about whether to sample. The desire for confidentiality of data should be considered in planning the sampling effort.

When samples are collected and analyzed for a regulatory purpose or legal objective, it is necessary to have a legal framework for the project as well as the technical and scientific objective. This chapter will review implications of the legal objective for sampling, discuss common legal problems in sampling for industrial environmental management, review evidentiary requirements, and summarize current legal doctrine on confidentiality of environmental sampling data.

Regulatory Compliance

Comprehensive environmental legislation on air and water pollution, control of hazardous waste disposal, and the Superfund program for cleanup of waste sites have increased demand for sampling to monitor compliance with regulatory requirements. The regulatory program can specify the method of sampling as well as method of analyses. The water pollution control program is an example of a mature "command and control" regulatory program. A permit is required to legally discharge wastewaters. The permit sets discharge limits for chemical parameters and generally specifies the location, frequency, and type of sampling required to prove compliance with regulatory requirements. The law requires each discharger to submit monthly self-monitoring reports comparing permit limits to actual discharges. The monthly reports (called *discharge monitoring reports* or DMRs in the water program) are public documents. A similar permit and compliance-monitoring program is being established for air emissions.

The enforcement system rests on a system of accurate public disclosure of representative samples analyzed according to specified agency methods. Civil and criminal penalties can be imposed for violation of permit requirements. To make it easier for administrative agencies to introduce the data into evidence in proceedings to enforce the law, results reported in DMRs are made under oath and are given a special status under the rules of evidence, as admissions that can be used as evidence against the discharger without contradiction. The certification language on the DMR illustrates the legal consequences of the information:

> I certify under the penalty of law that I have personally examined and am familiar with the information submitted, and based on my inquiry of those individuals immediately responsible for obtaining the information, I believe the submitted information is true, accurate, and complete. I am aware that there are significant penalties for submitting false information, including the possibility of fine and imprisonment.

Because the entire regulatory process rests upon voluntary disclosure of effluent data, the agencies take allegations of tampering with samples very seriously. Tampering with *compliance-monitoring samples* results in a number of criminal enforcement actions every year (*1*).

The regulatory compliance-monitoring sample has elements of both judgmental and random sampling. The medium sampled is specified, but the sample is supposed to be random and representative, such as a requirement to sample lead in wastewater by 24-h composite samples consisting of hourly flow-weighed aliquots. The regulatory program can define a sampling method as well as an analytical method, and the regulations must be reviewed carefully to find any required procedures. The degree of detail of a regulatory specification on sampling is not consistent between environmental media.

In general, federal and state environmental regulations have provided more detail on analytical methods than sampling. For example, the appendices to the federal air pollution control regulations in the *Code of Federal Regulations,* Title 40, Part 50 (2), provide methods for analyses of *criteria air pollutants* that concentrate more on analysis than sampling. Similarly, in the water program, the appendices to the *Code of Federal Regulations,* Title 40, Part 136 (3), are very detailed on analytical methods and deal with sampling on a rather superficial basis. Figure 1, which is

- **Slight daily fluctuation in pollutant concentration and flow**
- **Recommendation: Grab samples (frequency depends on permit writer's judgment)**

- **Regular fluctuations in pollutant loading over the course of the day**
- **Very slight fluctuations in flow**
- **Recommendation: 24-h time-proportioned composites**

- **Irregular fluctuations in pollutant loadings over the course of the day**
- **Erratic fluctuations in flow**
- **Recommendation: 24-h flow-proportioned sample**

Figure 1. U.S. EPA guidance for water permit writers on type of sampling. (Reproduced from reference 17.)

from a recent U.S. Environmental Protection Agency (EPA) instructional manual, recognizes the problem of variability in the media sampled but provides little additional guidance on sampling. However, other provisions in the air regulations contain detailed requirements on sampling for compliance purposes and provide detailed specifications on the sample locations and quality assurance requirements for ambient air-monitoring stations (*4*). The hazardous waste program provides guidance, recommended procedures, or, in some cases, requirements for sampling waste streams according to specific procedures (*5*). The U.S. EPA has published (*6*) detailed guidance on statistical analysis of sample data to determine whether disposal sites meet cleanup criteria. The rules for the specific regulatory program must be reviewed to see if there are any binding requirements for the sampling effort.

When no rule specifies the place or method of sampling (or a required method of analysis), the sampler should review agency guidance, industry standards (*7*), and technical literature to develop and document a statistically defensible method of obtaining representative samples. Even when a method is specified, most programs allow the permittee to propose a reasonable alternative to a specified procedure if the alternative method has a sound technical justification.

Issues in Regulatory Sampling

Two trends in environmental regulation result in some common sampling problems. First, regulatory limits become more strict over time but never become less restrictive (*8*). The current practice in water pollution control is to set limits for so-called "toxic pollutants" at "no detectable" concentration of a parameter. The water program's antibacksliding and antidegradation rules do not permit an increase in a regulated pollutant in later permits (*9*). Second, pressure always exists to improve enforcement of an essentially voluntary system, and each new law contains measures to enhance enforcement. In addition to federal and state governmental enforcement actions for monetary penalties, injunctive relief for corrective measures, or criminal fines or imprisonment, most environmental laws also provide for civil administrative proceedings and permit citizen suit lawsuits for penalties and injunctive relief. For example, the 1990 Clean Air Act Amendments allow private citizens to sue for past violations of the air laws and also provide for a reward of up to $10,000 for information leading to a successful environmental prosecution (*10*).

These regulatory trends increase the importance of obtaining a truly representative sample. When sampling is done for a regulatory purpose, the regulatory program may specify the location and method of sampling. More often, no regulatory specifications govern the sampling. In many compliance-monitoring situations, the location of the sampling point is based on history and convenience rather than a considered judgment of whether samples from a particular location are representative of the discharge. Dischargers should determine if a sample location prevents a representative sample. If so, the sampling point should be moved to a more representative location or a more representative method should be used, after notice is given to the appropriate agency.

As emission limits decrease, common sources of contamination become more noticeable. Inadequate maintenance of sampling equipment and contamination through sampling and analytical practices may prevent the collection of a truly representative sample. In one example, sample contamination was responsible for apparent discharge violations for oil and grease and zinc exceeding water permit limits at a Midwestern site. The tubing that fed the continuous water sampler had not been replaced in years, and the sample was held in a plastic container that still contained the zinc oxide mold release agent used in manufacturing the container. In some compliance enforcement situations, a discharger is well-advised to use ultraclean sampling techniques and to audit the performance of routine sampling operations.

Judicial Proceedings

Litigation can result from agency enforcement of environmental regulations, but nongovernmental or private environmental litigation is exploding as a result of more regulation, citizen suit provisions, and the increasing costs of environmental compliance. Environmental lawsuits are filed by governments, private companies, and individual citizens seeking relief, including monetary damages, fines, corrective actions, and perhaps imprisonment of the offenders. Lawsuits can be filed in federal or state courts, or before agency hearing examiners or administrative law judges. In litigation, it is important to perform sampling in accord with the DQOs supporting the position of the client, but rules of evidence impose additional considerations for sampling.

The legal reasoning process differs from the scientific reasoning process. When the two different cultures meet in the courtroom, communication problems often arise. Scientists are trained to believe that numbers have precision, and lawyers assume language has precision. Each way of thinking is correct in its proper context. Scientists are accustomed to a formal, structured, and empirical method of finding truth. The legal system is flexible, unstructured, and practical, and a decision will be made based on the best proof available. Our legal system assumes that the truth will be found by allowing two adversaries to state their case before a judge or jury, who will decide factual issues on the basis of the evidence presented by the parties. Thus in most litigation a rule or a procedure does not usually determine the result. The outcome instead depends on the issue in dispute, the testimony and credibility of the witnesses, the other evidence admitted, and the skill of the advocates.

The legal and scientific systems search for truth in different ways. The basic legal framework—the right to an impartial judge, the right to a trial by a jury of peers, and the right to confront the accuser, cross-examine witnesses, and present rebuttal evidence—was developed in feudal England. Many elements of our legal system are at least 500 years old, and a lawyer practicing at the time of the American revolution would recognize the legal aspects of a modern trial. In our system, judges decide legal matters, and juries usually resolve factual disputes in civil cases and guilt or innocence in criminal matters. The case is prosecuted or defended by advocates, and the judge is the referee or gatekeeper. The judge is the gatekeeper because he or she decides, as a legal matter, what evidence is "admissible" or can be "introduced into evidence" or presented to the jury. The same rules of evidence technically apply in every litigation context: federal or state proceedings, agency hearings or trials before a judge, and civil or criminal lawsuits. In practice, rules tend to be relaxed in administrative proceedings and strictly enforced in criminal cases.

Rules of Evidence

The rules of evidence are a means of screening evidence considered reliable enough to present to the jury to decide disputed issues of fact. The rules of evidence came to the United States with common law of England. Some of the more arcane rules of evidence are best understood in the context of their history. Periodically, efforts have been made to modernize evidentiary doctrines, and the modern standard is the "Federal Rules of Evidence" (*11*) used in federal court proceedings. Many states do not have organized rules of evidence and rely on judicial opinions (*case law*) for rules of evidence.

There are two classifications of evidence: direct and circumstantial. Direct evidence is eyewitness testimony or observation without an inferential process; circumstantial evidence is indirect proof that relies on an inference to establish a material point. The different types of evidence are oral testimony on speech, actions, or conduct; physical evidence (also called "real" or "demonstrative" evidence); documents; and, in some cases, opinion testimony. Whatever its form, the evidence must meet three tests of admissibility: *relevance, materiality,* and *competence.* In general, relevant and material evidence will be admitted if competent. Exceptions to this general rule occur for reasons of policy, for example when the evidence is considered prejudicial, redundant, or unduly time-consuming (*12*).

The legal relevance of evidence measures its probative value, that is, whether the evidence makes a desired inference more probable than it would be without the evidence. The question "So what?" is a challenge to the relevance of evidence. The materiality of evidence is an evaluation of whether the inference to be proved by the evidence is an issue in the case. "Why is the evidence offered?" tests the materiality of evidence. Evidence can be admissible for one purpose but inadmissible for another. The competence of evidence is a judgment about the reliability or strength of evidence. A factual witness is competent to testify if he or she has personal knowledge (i.e., the capacity to observe, remember, and recount factual matters). Hearsay evidence is scrutinized because it may not meet the test of competence. Hearsay is a statement made by a person not present in the courtroom that is offered as proof of the fact asserted; the witness cannot be cross-examined and thus there is no opportunity to test the witnesses' perception and memory. Hearsay is not admissible in court proceedings, but many exceptions to the "hearsay rule" exist.

The question "Is the evidence what it is claimed to be?" tests the reliability or competence of evidence. Two of the procedural rules of competence designed to assure reliability of evidence are widely encountered in sampling: (1) laying a foundation and (2) authentication, for example, establishing the chain of custody of real evidence. Evidence must be authenticated by a witness before it is admitted in a process called laying a foundation. Testimony of a person with personal knowledge of what happened always satisfies this standard. The witness must testify that the demonstrative evidence, such as a map or picture, accurately reflects conditions on the site. For analytical data, testimony on the quality assurance and quality control measures including control samples should lay the foundation for the evidence. The court can take judicial notice (recognition of a fact without formal presentation of evidence) of a scientific test if the proper foundation is admitted into evidence. The proper foundation is usually testimony that the test is generally accepted in the scientific community, was in the proper working order, and was conducted or interpreted by a qualified person. Authentication is another element of the rule of competence designed to show evidence is reliable. When many people handle a specimen, how do we know that it has not been altered? The chain-of-custody record is required in sampling that might be used in litigation to prove that the evidence is authentic. The chain-of-custody form is designed to identify all persons who had possession of a specimen so that they can testify under oath if necessary to make the foundation. The chain-of-custody rules are not strictly applicable in compliance-monitoring sampling, but regulations on compliance monitoring usually require a written record of the time and place of the sampling and an identification of the sampler.

Opinion testimony was not admissible in feudal England, and opinion had a connotation of speculation or conjecture. The rule was that witnesses must testify to facts and not opinions or conclusions because the jury will draw its own conclusions from the facts. Some states require demonstration that an expert opinion is necessary before allowing expert testimony. All jurisdictions now allow opinion or expert testimony on an issue beyond the layman's common knowledge and experience. Modern codes, such as "Federal Rules of Evidence" Rule 702, allow expert testimony if it would assist the judge or jury. Federal practice reflects the trend of liberally allowing expert testimony (*13*). Experts can contribute the power to make inferences that a jury is not competent to draw. Experts can testify to what they have observed, what they know as experts (the state of the art or opinion in the specified field), and the inferences they have drawn. Experts can be cross-examined on qualifications (knowledge, skill, experience, training, or education), the basis of their opinion, and compensation for their testimony.

Access for Sampling

The issue of access for sampling is a common problem. Proper access to property for sampling is important because laws protecting private property allow property owners to refuse access to their property, file criminal charges against trespassers, or sue a trespasser for damages. The law also provides that a landowner can be sued if a person (including a sampler) is injured on the landowner's property. In criminal enforcement lawsuits, the constitutional protection against unreasonable searches creates special concerns for proper access to property for sampling.

When samples are to be collected from property owned by a third party, the sampler should request and obtain written permission to enter the property to collect samples as part of preparing for the sampling effort. In criminal investigations, a search warrant may be required. Failure to obtain the landowner's permission can result in delays in the work, additional costs of mobilization, bad feelings, and perhaps an outright refusal to grant permission to sample. In some regulatory programs, the agency has the right to enter any property to collect samples, and the agency can order an uncooperative landowner to allow access to perform sampling or can obtain a search warrant to provide access for sampling.

The sampling access agreement should define the type of sampling needed and the locations and time required to obtain the sample, and it should attempt to minimize any disruption of activities on the site. Property owners usually require assurance that the sampler (1) will repair all damages caused by the sampling, (2) will follow the appropriate safety precautions, and (3) has liability insurance for injuries to others or to his or her employees.

Private Sampling

In sampling for litigation, the place and method of sampling will be defined by the objective of the litigation and by technical concepts, including DQOs. For example, samples taken under the federal Superfund program reflect procedures believed to be required to assure that data are reliable enough to be used to determine responsibility for multimillion dollar waste site cleanups in a cost recovery lawsuit. The rules of evidence provide additional requirements for proper authentication of evidence and access for sampling. In private sampling, the place and method are not regulated, and the client decides what quality assurance and quality control are appropriate. The results of private sampling may be confidential.

Confidentiality of Environmental Sampling Results

The duty to disclose sampling results depends on the legal purpose of the sampling program, and it ranges from an absolute duty to disclose compliance-monitoring sampling data, to the absence of any obligation to reveal information privileged or protected from disclosure to others. The rules for disclosing compliance-monitoring and private sampling data are different from those that govern the disclosure of data in litigation.

No general duty exists to disclose all environmental data to the public or an administrative agency, but many environmental laws require certain data to be disclosed. In general, data required to be taken and reported to an agency for compliance-monitoring purposes are considered public information and cannot be confidential. For example, samples taken to fulfill discharge-monitoring requirements under the water permit program are public information. If the discharger takes more samples than are required, these data must also be reported. However, the data that are required to be disclosed are data taken at specified monitoring points and analyzed by agency-required methods (*14*). If the rules are taken literally, a sample analyzed by an unapproved method or at a different location would not have to be disclosed to the agency.

In most regulatory programs, raw material usage, emissions data, and process information used in permit applications to establish discharge or emission limits are public information by law. However, regulatory programs usually provide for protection of trade secrets or competitively sensitive business information by allowing submission of these materials to the agency as confidential information separate from the public record. Generally, a trade secret is information used in a business that gives the company a competitive advantage, such as a product formula or manufacturing process.

In agency and judicial lawsuits, the discovery process requires exchange of information before trial. Discovery is designed to expedite trials and encourage settlement by allowing the parties to evaluate the evidence that the other side will present at trial. The standard for disclosure of information in discovery is quite broad. The information does not have to be admissible evidence at the trial, but information that could lead to the discovery of admissible evidence is provided to the adverse party on request. Three exceptions to this duty to disclose are (1) confidential business information, (2) information protected by attorney–client or other privilege, and (3) the attorney work product doctrine. A party can request a court to protect confidential business information by limiting its disclosure to the public. The factors considered are the author and nature of the information, the extent to which the information is known outside the company, measures taken by the company to protect the information, the value of the information to the business, and the ability to independently recreate or reconstruct the information.

Information within the scope of the attorney–client privilege is protected from discovery in litigation. When legal advice is requested from a professional legal advisor, communications relating to that purpose made in confidence by the client are protected from disclosure unless the confidence is waived by telling people who did not need to know (*15*). Although the courts are reluctant to protect factual information from discovery, the results of an environmental audit directed by an attorney hired to provide a confidential legal analysis of the compliance status of a plant could possibly be protected from discovery.

The attorney work product doctrine protects material prepared by an attorney (or the attorney's agents) in anticipation of litigation from disclosure in discovery. The factual material assembled by the party's representative is given less protection than the conclusions, opinion, or legal theories of counsel. Legal opinion work product will not be disclosed, but factual information can be disclosed on a showing of substantial need and hardship (*16*). The legal purpose for the sampling will determine whether data from the effort are likely to be confidential, open to public disclosure by an agency, or subject to mandatory disclosure as part of discovery in a lawsuit.

Conclusions

The legal objective of the sampling effort should be considered early in the development of the sample plan. Legal provisions may require sampling at certain locations and may specify the sampling method. In samples used for litigation, samplers must be aware of obligations imposed by the laws of evidence, or the results of the sampling may not be admissible as evidence. The legal context of the sampling effort also determines whether the results are confidential or must be disclosed to adverse parties or to the public.

References

1. *Code of Federal Regulations,* Title 40, § 122.41(k)2, 1995 ed. provides that "Any person who knowingly makes any false statement, representation, or certification in any record or other document submitted or required to be maintained under this permit, including monitoring reports of compliance or noncompliance shall, upon conviction, be punished by a fine of not more than $10,000 per violation, or by imprisonment for not more than 6 months per violation, or by both."
2. *Code of Federal Regulations,* Title 40, Part 50, Appendices, 1995 ed.
3. *Code of Federal Regulations,* Title 40, Part 136, Appendices, 1995 ed.
4. *Code of Federal Regulations,* Title 40, Part 58, Appendices, 1995 ed.
5. U.S. Environmental Protection Agency. Office of Solid Waste and Emergency Response. *Test Methods for Evaluating Solid Waste; Volume II: Field Manual Physical/Chemical Methods,* 3rd ed.; Government Printing Office: Washington, DC, 1986; SW–846; PB88–239223, Part 4 of 4; p 9–1.
6. U.S. Environmental Protection Agency. Office of Policy, Planning and Evaluation. *Methods for Evaluating the Attainment of Cleanup Standards; Volume 1: Soils and Solid Media;* Government Printing Office: Washington, DC, 1989; EPA 230/02–89–042; p 2–4.
7. For example, *Specification and Guidelines for Quality Systems for Environmental Data Collection and Environmental Technology Programs;* American National Standard Institute. American Society for Quality Control: New York, 1994; ANSI/ASQC–E4–1994.
8. Friedman, F. B. *Practical Guide to Environmental Management,* 4th ed.; Environmental Law Institute: Washington, DC, 1992; p 29.
9. *Code of Federal Regulations,* Title 40, § 122.44 (l)2ii provides that "In no event may a permit with respect to which paragraph (1) (2) of this section applies be renewed, reissued, or modified to contain an effluent limitation which is less stringent than required by effluent guidelines in effect at the time the permit is renewed, reissued, or modified. In no event may such a permit to discharge into waters be renewed, issued, or modified to contain a less stringent effluent limitation if the implementation of such limitation would result in a violation of a water quality standard under section 303 applicable to such waters."
10. *U.S. Code,* Title 42, § 7413(f), 1994 ed.

11. "Federal Rules of Evidence" (PL 93–595) as amended 1995.
12. Most arguments over admissibility of evidence balance the probative value of the data against the possibility of prejudicial impact. For example, evidence of a witness's character is normally not admissible as evidence. Cleary, E. W. *McCormick on Evidence,* 2nd ed.; West Publishing: St. Paul, MN, 1976; pp 528–529.
13. The current standard for admission of expert opinion testimony is discussed in *Daubert v. Merrell Dow Pharmaceuticals,* 113 S. Ct. 1245, 1993.
14. *Code of Federal Regulations,* Title 40, § 122.41 (l)4ii, 1995 ed.
15. *U.S. v. Rockwell International,* 897 F.2d 1255, 1264, 3d Cir., 1990.
16. "Federal Rules of Civil Procedure" Fed. R. Civ. P. 26(b)3, 1995.
17. U.S. Environmental Protection Agency. *Training Manual for NPDES (National Pollutant Discharge Elimination System) Permit Writers;* Government Printing Office: Washington, DC, 1993; EPA 833–B–93–003.

Quality Assurance and Quality Control

Chapter 4

Defining the Accuracy, Precision, and Confidence Limits of Sample Data

John K. Taylor

The limits of uncertainty for data on samples include uncertainty due to their measurement and to the samples. Measurement uncertainty is controlled and evaluated by an appropriate quality assurance program. Sample and sampling uncertainty includes random and systematic components. The concept of quality assurance can be applied to the sampling operations to control and evaluate sampling uncertainty. The information necessary to assign statistically supported limits of confidence to sample data is discussed.

SAMPLE DATA CONTAIN A DEGREE OF UNCERTAINTY, and this uncertainty must be considered whenever the data are used. Ordinarily, measurements are made because information is needed to evaluate some property of a system so that decisions can be made concerning the system. The system or material of interest may be as small as a speck of dust or as large as the Earth. If the material of interest is small, then the sample to be analyzed—the entire object—is known with certainty. If the material of interest is very large, then only portions of the system may be examined. Clearly, some questions need to be answered. These questions include what and where to sample and how many samples will be required for a specific purpose. If, for example, the information to be found was the platinum content of the Earth, the results would be greatly influenced by the number and kinds of samples that were analyzed, and some degree of uncertainty would be associated with any conclusions drawn from the data.

Every measurement problem involving less than the entire universe of interest has some parallelisms to the example just mentioned. Researchers who

Published 1988 American Chemical Society

ponder such matters may almost despair in making any decisions involving measurement of samples. Unfortunately, many decisions are made in ignorance or contempt of the uncertainty of the sample data. In fact, "representative samples" are often used to make decisions even though no real evidence is presented to verify that the sample represents anything other than itself. These problems could be minimized or eliminated by proper design of measurement and sampling plans and by realistic evaluation and interpretation of the data obtained by using these plans. The following discussion is presented to provide guidance in these matters.

Data Requirements

Before measurements can be made, the concept of the problem to be solved and the model to be followed for its solution must be reasonably clear (*1*). What needs to be measured, the levels of concern, and the permissible tolerances for the uncertainty of the data must be clearly understood. On the basis of such information, intelligent decisions can be made on how many samples are required, how to measure them, and the number of measurements needed. With controversial issues, investigators must agree on what can and should be achieved in a measurement program prior to its initiation. Otherwise, not everyone will be satisfied with the outcome.

The goals and expectations of a sampling program must be realistic and can never exceed the measurement and sample limitations. Costs and benefits must be considered in the design of almost every measurement program.

As the detection limits of methodology are approached, the number of measurements can increase by an order of magnitude (actually requiring at least nine replicates) if quantitative results are required. This increase occurs because the limit of quantitation is about 3 times the limit of detection (LOD) (2). Because both figures of merit are calculated on the basis of single measurements, a concentration level equivalent to LOD can be the limit of quantitation for the mean of nine measurements. Accordingly, making decisions on data obtained at the limits of capability of the methodology is foolish.

Sources of Uncertainty

The total variance of measurement data (s^2_{total}) can be expressed in the simplest terms as

$$s^2_{\text{total}} = s^2_{\text{measurement}} + s^2_{\text{sample}} \tag{1}$$

where $s^2_{\text{measurement}}$ and s^2_{sample} are the sampling variances due to measurement and the sample, respectively. The measurement and sampling plans and operations must be designed and executed so that the individual components may be evaluated. The possible situations that can occur are presented in Table I (*3*).

Table I. Measurement Situations

	Situation	*Significance*
A.	Measurement variance	No
	Sample variance	No
B.	Measurement variance	Yes
	Sample variance	No
C.	Measurement variance	No
	Sample variance	Yes
D.	Measurement variance	Yes
	Sample variance	Yes

Situation A is confined almost exclusively to single-specimen analysis or where only semiquantitative data are required. Situation B largely pertains to the analysis of homogeneous materials. In most other cases, some degree of variability of samples is encountered. In situation C, single measurements of samples are sufficient. Unfortunately, situation D is prevalent, so both measurement and sample variability must be taken into account when evaluating sample data.

Measurement Uncertainty. Measurement uncertainty can be controlled and evaluated by an appropriate quality assurance program (*1*, *4*). When properly planned and executed, the measurement variance will be known so that the number of measurements can be minimized. Otherwise, sufficient replicate measurements must be made on a sufficient number of samples to evaluate both sources of variance. Ideally, measurement uncertainty should not exceed one-third of the total uncertainty tolerance. Otherwise, replicate measurements will be needed to reduce the uncertainty to acceptable limits.

Although the quantitative uncertainty of the measured values is emphasized, the need to verify the qualitative identification of the analytes that are reported should not be overlooked. The confidence of identification must approach certainty (*3*, *4*), which is achieved by a confirmation process as discussed in reference 2.

Sample Uncertainty. Sample uncertainty may contain systematic and random components arising from population and sampling considerations. Population considerations have been discussed in some detail by Provost (*5*) and will be mentioned only briefly here. Because every population has some degree of variability, a sound statistical sampling plan is necessary if the judgment of apparent differences is to be defensible. Because small differences of measured values with reference to baseline values, control areas, or norms are ordinarily of concern, the measurement of large numbers of randomly selected samples is inevitable.

Bias in samples from a population results from nonrandom sampling and from discriminatory sampling. Every population must be defined, and the act of definition biases the population, rightly or wrongly, by exclusion of certain individuals. For example, an analyte of interest may be looked for only in specific locations such as silt, and even in specific particle-size fractions. This sampling may be proper, but one must always consider how this might bias the data. Considerations such as in this example are usually based on the model employed, which should be periodically reconsidered for its adequacy.

Sampling Considerations

The sampling operations can provide both systematic and random components to the sampling uncertainty. Concerning the systematic components, the sampling equipment may not be able to experimentally realize the requirements of the model. For example, a respirable fraction of airborne particulates may be difficult to specify logically and even more so experimentally. Obviously, sampling equipment should be calibrated when factors such as flow rates, size discrimination, and temperature effects are important because errors in these factors could cause bias in sample data.

Some or all of the analyte of concern can be lost from samples because of absorption or reaction with container materials, sampling equipment, or sample-transfer lines. Deterioration arising from atmospheric contact, temperature instability, radiation effects, and interactions with other analytes can cause serious biases. The act of sample removal from its environment can disturb stable or metastable equilibria that could bias ensuing measurements. Stabilization of collected samples can be of major importance but difficult to realize. Subtle carry-over effects in sample containers and sampling lines that result from memory of previous samples can be of serious consequence.

Random components of sampling uncertainty can result from variability of all of the aforementioned sources of bias. Accordingly, considerable incentive exists to eliminate these sources of bias or, failing to do so, carefully controlling them. In addition, the sampling operations can have their own variability components due to carelessness or inability to keep all aspects in a state of statistical control.

Statistical Considerations

Systematic components of uncertainty from various sources are algebraically additive. This situation can be shown as follows:

$$B_{\text{total}} = B_1 + B_2 + \ldots B_n \tag{2}$$

where B_{total} is the total systematic uncertainty, and $B_1 \ldots B_n$ are the components of uncertainty from sources 1 ... n, respectively. Sources of bias can be conceptually identified, but quantifying their contributions may be difficult. This statement is valid for measurement bias and especially sample bias. In many cases, all that can be done is to develop a bias "budget" and estimate the bounds of each component. Of course, corrections should be made for bias whenever possible.

The random components of sampling uncertainty from several sources, expressed as variances, are additive. Thus,

$$s^2_{\text{sampling}} = s^2_1 + s^2_2 + \ldots s^2_n \tag{3}$$

where s^2_{sampling} denotes the total random components of sampling uncertainty, and $s^2_1 \ldots s^2_n$ denote components of sampling uncertainty from sources 1 ... n, respectively. Identifying the sources of some, if not all of these variances, may be possible, but evaluating the individual components may be difficult and certainly time-consuming. However, evaluating the overall value by suitable replicate experiments may be possible. This evaluation may be accomplished by taking replicate samples where sample population variability is expected to be negligible (e.g., in a narrowly defined sampling area). At least seven replicate sampling operations would be required, and the excess variance over that of measurement would represent sampling variance.

Sample population and sampling variances ($s^2_{\text{population}}$ and s^2_{sampling}, respectively) are additive and can be represented as

$$s^2_{\text{sample}} = s^2_{\text{sampling}} + s^2_{\text{population}} \tag{4}$$

This equation demonstrates that sampling variance must be kept relatively small or else it can seriously influence measurement.

The statistical treatment of population variance was discussed by Provost (5). The total uncertainty includes the sum of the contributions arising from random and systematic sources. The random component of uncertainty is evaluated by a statistical confidence interval, and the value obtained can be reduced by replication. The systematic component of uncertainty is independent of the number of replicates. Discussion of the procedures used for the computations is beyond the scope of this chapter.

Quality Assurance of Sampling

The two aspects of quality assurance—quality control and quality assessment—can be applied to sampling as well as to measurement. Quality control includes the application of good laboratory practices, good measurement practices, and standard operations procedures especially designed for sampling. The sampling operation should be based on protocols especially developed for the specific analytical problem. Strict adherence to these protocols and sampling protocols is imperative. Sample takers must be trained to follow the protocols faithfully. All required calibrations must be made on the basis of established schedules. Special care must be devoted to sample containers and to stabilization and protection of samples. A system for assuring positive identification of samples and documentation of all sample details must be operational. Chain-of-custody procedures, whether or not required externally, are necessary if sample integrity is to be defensible, which is almost always.

Quality assessment of the sampling process depends largely on monitoring for adherence to the respective protocols. Audits on a continuing basis are the best means to accomplish this purpose. A system that uses container, field, and laboratory blanks should be part of the sampling protocol. To be meaningful, the kinds of numbers of such protocols must be individually selected for each measurement program.

Because of their critical role in the quality assessment of the sampling operation, protocols must be carefully designed, and their adequacy should be under periodic review in any monitoring program.

Preplanning is especially important because any modification of the sampling operation during a measurement program could produce a different set of samples, which may not be compatible with previous sampling operations.

Conclusion

Three basic kinds of sampling plans are in common use (*3*). *Intuitive sampling plans* are based upon judgment, often by technical experts, and the interpretation of data is also based on judgment. Various experts can draw different conclusions from the same data set, and choosing between them may be difficult. *Statistically based plans* provide the basis for making probabilistic conclusions that are independent of personal judgment. However, the model upon which the data are based could be open to criticism. Often, statistical plans require more samples than are feasible for various reasons. In such cases, a *hybrid plan* may be used that includes intuitive simplifying assumptions. These assumptions must be clearly stated and considered when interpreting data, or else probabilistic judgments may be made that may not be fully justified. Of course, validated and evaluated data must be used no matter which approach to sampling is followed.

In addition to these sampling plans, protocol sampling, which follows legally or contractually mandated plans, is commonly used. These plans often require a representative sample, which may or may not be specified. Although such plans may be useful in practical situations, one must remember the empiricism upon which mandated plans are often based. Such data should not be used for other than the specified applications. The practice of specifying representative samples should be discouraged whenever possible because of the virtual impossibility of demonstrating representativeness.

Sampling and the associated measurements are often made to decide on compliance with some requirement. In such cases, answers must be obtained to the following questions:

- Is the mean value of the population within acceptable limits?
- Is a specified fraction of members of the population within acceptable limits?
- Are all members of the population within acceptable limits?

Actually, only the first two questions can be answered by sampling, or rather only by statistically sampling the population. The third question cannot be answered, except by examining every individual. No matter how many individuals have been examined and found to be in compliance, noncomplying individuals may exist. As an example, if as many as 3 million individuals were found to be within acceptable limits, then there is a 5% chance that one in 1 million members could be defective.

The more one looks critically at sampling, the more one should be convinced that sampling is not a trivial exercise. Accordingly, in all but the most simple situations, all aspects of sampling should be carefully planned, and sampling experts and statistical advisors should be used as necessary if meaningful and defensible conclusions are to be realized.

Abbreviations and Symbols

B_{total}	total systematic uncertainty
LOD	limit of detection
$s^2_{population}$	sample population variance
s^2_{sample}	variance due to the sample
$s^2_{sampling}$	total random components of sampling uncertainty
s^2_{total}	total variance of measurement data

References

1. Taylor, J. K. *Anal. Chem.* **1981**, 53, 1388A–1396A.
2. "Principles of Environmental Analysis"; American Chemical Society Committee on Environmental Improvement. *Anal. Chem.* **1983**, 55, 2210–2218.
3. Taylor, J. K. *Trends Anal. Chem.* **1986**, 5(5), 121–123.
4. Taylor, J. K. In *Environmental Sampling for Hazardous Wastes*; Schweitzer, Glenn E.; Santolucito, John A., Eds.; ACS Symposium Series 267; American Chemical Society: Washington, DC, 1984; pp 105–109.
5. Provost, L. P. Ibid., pp 79–97.

Chapter 5

Assessing and Controlling Sample Contamination

David L. Lewis

Most environmental sampling and analytical applications offer numerous opportunities for sample contamination. For this reason, contamination is a common source of error in environmental measurements. This chapter addresses the problem of assessing and controlling sample contamination and the resulting measurement error. Vulnerable points in the sample collection and analysis process as well as common sources of contamination are discussed. The different possible effects of contamination are also examined. Blanks are recommended as the most effective tools for assessing and controlling contamination. Different types of blanks and their respective uses and limitations are described. The applicability of control charts to blank measurements and their use in assessing and controlling contamination are also discussed.

CONTAMINATION IS A COMMON SOURCE OF ERROR in all types of environmental measurements. Most sampling and analytical schemes present numerous opportunities for sample contamination from a variety of sources. This chapter addresses the problem of assessing and controlling sample contamination and the resulting measurement error. The first part of the discussion examines the different points in the sample collection and analysis process at which contamination is likely to occur and identifies common sources of contamination for various measurement applications. The next portion deals with the different possible effects of contamination. The last part of the discussion examines the use of blanks to assess and control contamination. Different types of blanks and their respective uses are described. The applicability of control charts to blank measurements is also discussed.

3152–4/96/0085$16.50/0

Sources of Contamination

From an environmental sampling and analytical standpoint, *contamination* is generally understood to mean something that is inadvertently added to the sample during the sampling and analytical process. Although subsequent measurements may accurately reflect what was in the sample at the time the measurements were made, they do not give an accurate representation of the measured characteristic of the media from which the sample was taken.

Sample contamination may arise from myriad sources. To control contamination associated with a particular determination, potential sources of contamination must first be identified for the methods employed. Typically, an environmental sample may be contaminated at any of numerous points in the sample collection and analysis process. Contamination may be introduced in the field during sample collection, handling, storage, or transport to the analytical laboratory. After arrival at the laboratory, additional opportunities for contamination arise during storage, in the preparation and handling process, and in the analytical process itself. Common sources of sample contamination are summarized in Table I and discussed in greater detail in this section.

Equipment used for sample collection is a common route for introducing sample contamination in many types of environmental measurements. Sampling devices may be made of materials that contribute to sample contamination, or cross-contamination may occur as a result of improper cleaning of sampling equipment. Fetter (*1*), for example, described a variety of equipment-related sources of contamination encountered in groundwater sampling. He noted that

Table I. Potential Sources of Sample Contamination

Critical Steps in the Sampling and Analytical Process	*Contamination Sources*
Sample collection	Equipment and apparatus
	Handling (e.g., filtration, compositing, and aliquot taking)
	Preservatives
	Ambient contamination
	Sample containers
Sample transport and storage	Sample containers
	Cross-contamination from other samples or reagents
	Sample handling
Sample preparation	Glassware
	Reagents
	Ambient contamination
	Sample handling
Sample analysis	Syringes used for sample injections
	Carry-over and memory effects
	Glassware, equipment, and apparatus
	Reagents (e.g., carrier gases and eluents)

the first opportunity for contamination in subsurface investigations and in the installation of monitoring wells is during the process of drilling a borehole. He recommended steam cleaning of drilling rigs as a minimum measure to prevent introduction of contaminants such as gasoline, diesel fuel, hydraulic fluid, lubricating oils and greases, paint, and soil and scale from previous drilling operations. Organic-polymer-based drilling additives were cited as sources of chemical oxygen demand and total organic carbon contamination. Well-casing materials were also cited as potential sources of contaminants. These statements were supported by the work of Boettner et al. (2), which indicated that organic and organotin compounds may be leached from poly(vinyl chloride) and chlorinated poly(vinyl chloride) pipe and pipe cement.

Contamination and cross-contamination from sampling equipment is equally a problem in other types of environmental sampling. Ross (3) reported significant trace metal contamination of atmospheric precipitation samples collected in conventional polyethylene sample collectors washed only with deionized water. The highest instances of contamination were for manganese, cadmium, copper, and zinc. Rigorous acid washing of the sample collectors was shown to significantly reduce sample contamination. Collectors washed only in deionized water showed concentrations of individual elements as much as 50 times higher than corresponding acid-washed collectors. As another example, other studies (4-6) indicated that polytetrafluoroethylene (Teflon) and poly(vinyl fluoride) (Tedlar) bags can contribute significant amounts of hydrocarbon contamination to air samples collected in them.

Sample handling in the field is another potential source of sample contamination. For example, aqueous samples collected for soluble metals analysis are typically filtered and acidified in the field immediately after collection. Best et al. (7) reported significant nitrate contamination of surface water samples during field filtration. Filtration was performed with membrane filters and stainless steel filter holders that were rinsed with 5% nitric acid to minimize trace metal contaminants. Although the nitric acid rinse was followed by copious rinsing with deionized water, nitrate contamination as high as 3.5 mg/L was observed.

Acids and other chemical preservatives that may become contaminated after a period of use in the field offer another route of sample contamination during field handling (8). Sample exposure to the ambient environment should also be minimized to avoid airborne contaminants. Contamination from dry deposition of gaseous SO_2 and particulate sulfate has been cited as a source of bias in rainfall chemistry data generated during the period 1950–1980, which was prior to widespread use of collectors that open only during rain events (9). Lead and aluminum are also common airborne contaminants, especially in urban environments. Sample handling in industrial environments, often necessary during emissions testing and source characterization studies, presents even greater risks of contamination because of the considerably higher ambient levels of potential contaminants typically encountered.

Sample containers represent another major source of sample contamination. Plastic sample containers, for example, are widely recognized as a potential source of sample contamination in trace metal analyses. Moody and Lindstrom (*10*) examined 12 different plastic materials including conventional and linear polyethylene, polycarbonate, and several types of Teflon. They found significant levels of leachable trace elements in all of the materials examined. Linear polyethylene and the various Teflons were found to be the least contaminating bottles after cleaning. On the basis of their leaching studies, they recommended a cleaning procedure involving the use of both HCl and HNO_3, one after the other, for most trace element work.

Glass sample containers are typically used for collection of solids, sludges, and liquid wastes for organic analysis. Detergent or acid washing, followed by organic-free water rinsing and heating in an oven or muffle furnace, is recommended to minimize trace organic contamination from glass containers.

In selecting cleaning procedures for sampling equipment and sample containers, it is important to consider all of the parameters of interest. Although a given cleaning procedure may be effective for one parameter or type of analysis, it may be ineffective for another. When multiple determinations are performed on a single sample or on subsamples from a single container, a cleaning procedure may actually be a source of contamination for some analytes while minimizing contamination for others. This paradox may be caused by release of contaminants through cleaning procedures that are too rigorous or by contaminants introduced by the cleaning agent. For example, Bonoff et al. (*11*) reported contamination of water samples by zinc released from disposable filter cassettes following washing with 10% Ultrex nitric acid. Best et al. (*7*), in addition to reporting nitrate contamination introduced during field filtration, reported similar contamination resulting from the cleaning procedure used for the sample containers. During a pilot study preceding their investigation, high-density polyethylene sample bottles were cleaned by using a procedure that included a nitric acid rinse to minimize trace metal contaminants. Although the procedure used included repeated rinsings with deionized water following the acid rinse step, these rinsings were ineffective at reducing residual nitrate to acceptable levels. On the basis of the pilot study results, the cleaning procedure for field sample bottles was modified prior to subsequent field studies, and the nitric acid rinse was eliminated. Similar contamination problems have been observed in chromium analyses where chromic acid solutions were used to clean glassware and in phosphate analyses where phosphate-containing detergents were used.

Whereas sample containers represent one source of contamination during sample storage and transport, proximal storage of high- and low-level samples or samples and reagents has been identified as a potential additional source of contamination, particularly for volatile organic compounds (*12*). Levine et al. (*13*) confirmed that water samples could be contaminated by diffusion of volatile organic compounds through Teflon-lined silicone cap liners during sample shipment and storage. Although they found insignificant cross-contamination caused by storage of vials of organic-free water with saturated aqueous solutions of selected halocarbons, significant potential for contamination was observed for proximal storage of clean samples and neat reagents.

Numerous additional chances for sample contamination arise in the laboratory during sample preparation, handling, and analysis. Virtually every analytical technique presents its own special set of opportunities for sample contamination. Complex sample preparation techniques, such as those involving extraction and concentration of the analytes of interest, present proportionately more such opportunities than do simpler techniques involving fewer steps and less sample handling. Even simple, single-step preparatory techniques, however, often present opportunities for sample contamination. Bagchi and Haddad (*14*), for example, determined that precolumn cartridges and filter units commonly used for sample cleanup in ion chromatography can be a significant source of parts-per-billion contamination. They examined a popular, single-use cleanup cartridge and two types of filtration units designed to fit on a sample injection syringe. Their results indicated that both the cleanup cartridges and the filtration units were potential sources of sample contamination in inorganic ion analyses. Observed contaminants included lead, zinc, fluoride, chloride, nitrate, and sulfate ions.

Glassware and reagents are common sources of laboratory contamination in all types of analyses. Carry-over and memory effects from consecutive analyses of high- and low-level samples are also common to many types of instrumental methods, including gas chromatography, liquid chromatography, and many spectroscopic methods. Syringes and other devices used for sample injection are also common sources of contaminants. Sommerfeld et al. (*15*), for example, identified disposable plastic pipette tips as a source of iron and zinc contamination in analyses by graphite furnace atomic absorption spectroscopy. Even when the tips were acid rinsed to remove contamination, they found that the pipette tips could be recontaminated by contact with the aperture of the cool graphite tube.

Effects of Contamination

Chemical or physical properties of samples that cause errors in the measurement process are commonly known as *interferences*. Generally, two types of interferences are recognized: additive interferences and multiplicative interferences (*16*). *Additive interferences* are caused by sample constituents that generate a signal that adds to the analyte signal. Because they cause a change in the intercept but not the slope of the calibration curve, additive interferences have the most pronounced effect at low analyte concentrations. *Multiplicative interferences*, on the other hand, are caused by sample constituents that either increase or decrease the analyte signal by some factor without generating a signal of their own. Multiplicative interferences change the slope of the calibration curve but not the intercept.

Contamination is generally understood to mean something inadvertently added to a sample that leads to an erroneously *high* measured value. This definition, although generally implied, is not strictly correct unless we consider only contamination by the analytes of interest or by contaminants which the method employed cannot distinguish from the analytes of interest. This distinction is important because some types of contamination do not fit this general definition but instead lead to erroneously *low* measured values. These negative interferences may be multiplicative or they may be opposite in effect to additive interferences, and therefore change only the intercept of the response curve.

Multiplicative interferences are a common source of analytical error in many spectroscopic techniques, although matrix effects are a more common source of such error than contamination. Contaminants may, however, cause multiplicative interferences through adsorptive losses of the analyte of interest. These contaminants give erroneously low results. Adsorption acts as a multiplicative interference when a constant fraction of the analyte is adsorbed, regardless of analyte concentration (i.e., when relative bias is constant). When the amount of analyte is large compared to the available sites for adsorption to occur, the amount of analyte lost to adsorption tends to be constant, and relative bias decreases with increasing concentration. In such cases, adsorption causes a negative interference, opposite in effect to an additive interference. Bentonite drilling muds, for example, can absorb heavy metals in groundwater and lead to low measurements (*1*). Sample containers, although not usually considered contaminants in the normal sense, may also adsorb analytes from the sample and act as a negative interference.

Dilution is another example of an interference that causes the measured values to be erroneously low rather than high. This example is the special case in which the sample is contaminated by the solvent. Dilution occurs primarily in certain types of environmental sampling. Groundwater monitoring, in which monitoring wells are installed for the purpose of collecting groundwater samples, presents several opportunities for sample contamination by dilution. These opportunities may occur through addition of clean water to the well during well development, through improper well placement, through improper screening, or through cross-contamination between aquifers located at different depths (*1*). Many types of air sampling also present significant opportunities for sample dilution due to leaks in negative-pressure sampling systems allowing air leakage.

Regardless of the source or sources of sample contamination, the net effect is added inaccuracy in the measurement process. Like other types of measurement error, error due to contamination may be sporadic and represent special causes, or systematic and affect all measurements. Cross-contamination, such as that which often occurs during analysis when carry-over from high-level samples contaminates subsequent low-level samples, is a common source of sporadic contamination. Similarly, careless sample handling and dirty sampling equipment are often sources of sporadic contamination. Sporadic contamination most often affects the measurement process by introducing false positive results. A *false positive* is the error of concluding that an analyte is present in the media sampled when it is not. In the case of sporadic contamination where the contaminant acts as a negative interference, as in dilution or adsorption, false negatives may result. A *false negative* is the error of concluding that an analyte is not present when it is.

Contamination is a source of systematic error when the level of contamination is stable for all samples. Strictly speaking, however, stable, systematic error due to sample contamination is rare. Almost always, some element of sporadic error is associated with any source of contamination. In some cases though, the effect of this sporadic error component is small in comparison to the systematic error, or bias component. Thus, some types of contamination behave in a fashion that is primarily systematic. Systematic contamination increases the "background concentration" of the analyte of interest and thus affects the lower limit of the measurement process. Contaminated reagents are a common source of systematic contamination in many types of environmental measurements. Contaminated sample containers are another source of contamination that is often primarily systematic.

Use of Blanks To Assess and Control Contamination

The most commonly used analytical tools for assessing and controlling sample contamination are blanks. By conventional nomenclature, *blanks* are samples that do not intentionally contain the analyte of interest but in other respects have, as far as possible, the same composition as the actual samples (*17*). Additional descriptors, such as internal, reagent, field, solvent, and others, are used to indicate which of the various stages of the sampling and analytical process the blanks are considered to represent. Because blanks, by definition, do not intentionally contain the analyte of interest, their utility in assessing and controlling sample contamination is limited to contaminants causing additive interferences. In this regard, results for blanks are taken as a direct measure of the nonanalyte, or contaminant, signal for the corresponding samples.

Types of Blanks

Blanks play various roles in environmental measurements, depending on the analytical technique used and the goal of the blank measurements. Table II summarizes the types of blanks typically used in environmental measurements. The simplest blank, often called a *system blank* or *instrument blank*, is really not a blank at all in the sense of simulating a sample. Rather, a system blank is a

Table II. Summary of Blank Types

Common Name	*Other Names*	*Uses*	*Description*
		Laboratory blanks	
System blank	Instrument blank	To establish baseline response of an analytical system in the absence of a sample	Not a simulated sample but a measure of instrument or system background response
Solvent blank	Calibration blank	To detect and quantitate solvent impurities; the calibration standard corresponds to zero analyte concentration	Consists only of the solvent used to dilute the sample
Reagent blank	Method blank	To detect and quantitate contamination introduced during sample preparation and analysis	Contains all reagents used in sample preparation and analysis and is carried through the complete analytical procedure
		Field blanks	
Matched-matrix blank		To detect and quantitate contamination introduced during sample collection, handling, storage, transport, preparation, and analysis	Made to simulate the sample matrix and carried through the entire sample collection, handling, and analysis process
Sampling media blank	Trip blank	To detect contamination associated with sampling media such as filters, traps, and sample bottles	Consists of the sampling media used for sample collection
Equipment blank		To determine types of contaminants that may have been introduced through contact with sampling equipment; also to verify the effectiveness of cleaning procedures	Prepared by collecting water or solvents used to rinse sampling equipment

measure of the instrument background, or baseline, response in the absence of a sample. System blanks are often used in gas and liquid chromatographic methods to identify memory effects, or carry-over from high-concentration samples, or as a preliminary check for system contamination.

Solvent blanks are generally the next simplest type of blank and consist only of the solvent used to dilute the sample. Solvent blanks are used to identify or correct for signals produced by the solvent or by impurities in the solvent. Depending upon the analytical technique, the solvent blank may be used as a calibration blank. A calibration blank is used directly to set the instrument response to zero, or is used as one of a series of calibration standards, where the blank represents an analyte concentration of zero.

Another type of laboratory blank is the *reagent blank*. In addition to the solvent, the reagent blank contains any reagents used in the sample preparation and analysis procedure. These reagents may include color development reagents, reagents used in sample digestion steps, reagents used for pH adjustment, preservatives, or other reagents depending upon the analytical method. In methods where the solvent is the only reagent used in the sample preparation process, as in ion chromatography, for example, the composition of the reagent blank will be the same as that of the solvent blank. The distinguishing characteristic of the reagent blank, however, is that the reagent blank is carried through the complete analytical procedure in the same manner as an actual sample. This procedure should include all steps involved in sample preparation, such as cleanup, filtration, extraction, and concentration. The reagent blank thus provides a measure of contamination that may be introduced during sample preparation and analysis, whether from the reagents themselves, from glassware, or from other sources in the laboratory environment. Because it is carried through the complete analytical method, the reagent blank is also sometimes called a *method blank*.

Whereas laboratory blanks are reliable tools for assessing and controlling many types of laboratory contamination, they obviously address only a part of the overall measurement process. *Field blanks* must be used to provide information about contaminants that may be introduced during sample collection, storage, and transport. Like laboratory blanks, a number of different types of field blanks are available. The most common type of field blank, where the blank simulates the sample matrix, is sometimes called a *matched-matrix field blank*. This type of field blank is widely used in sampling involving aqueous matrices because deionized or distilled water is readily available and water blanks are easy to prepare.

Like reagent blanks, the distinguishing characteristic of matched-matrix field blanks is that they are carried through the entire sample collection and handling process so that the blank is exposed to the same potential sources of contamination as actual samples. Although exact duplication of the sample collection process for the blank sample may not always be feasible, the blank should be exposed to as many elements of the process as possible. In sampling groundwater monitoring wells, for example, preparing a blank that is exposed to the well casing and other potential contaminants of the well environment is generally not feasible. The blank sample can, however, be exposed to other elements of the collection process, such as bailers or peristaltic pumps used to obtain water samples from the well. Blanks should, of course, be treated in the same manner as other samples with regard to sample containers, field preservation, handling, and storage.

Although matched-matrix field blanks are most easily prepared for aqueous samples, they are usually prepared without too much difficulty in most air sampling applications as well. One method of preparing such blanks is by collecting samples of "clean" (e.g., breathing air or hydrocarbon-free air, as appropriate) compressed air with the normal sampling apparatus. Alternately, ambient air may be used by equipping the sampling system with an upstream purification device to remove the analytes of interest. An activated carbon canister, for example, may be used to remove trace organic compounds from ambient air to prepare air blanks for organic analyses.

Solid sampling applications typically present a greater challenge than air or water samples in preparing matched-matrix field blanks. If the media sampled is soil or soillike material, as might be encountered in characterization or remediation activities at a waste site, reagent-grade sand (i.e., silica) may be used to approximate the sample matrix. If the analytes of interest include organic compounds, the sand should be baked in an oven at 300 °C for several hours to remove any volatile species that may be present. As an alternate approach, a bulk sample may be collected from an adjacent area known to be free of the analytes of interest and used to prepare blank samples. Although this approach often provides a matrix more similar to that of the samples, the difficulty of ensuring that the blank samples are indeed free of target analytes is added. This difficulty is especially important in samples collected for metals analysis because most soils have significant naturally occurring background levels of numerous elements (*18*).

If the media sampled include solid or liquid waste materials, sludges, or slurries, a rigorous matched-matrix field blank may not be achievable. In such cases, a rough approximation is usually sufficient to identify most sources of contamination. Field blanks for these applications may be prepared by using uncontaminated soil or reagent sand, as described previously, or by preparing aqueous or solvent mixtures of soil or sand.

Another type of blank, similar in some cases to the matched-matrix blank, is the *sampling media blank*. This blank consists of the sampling media used for collection of field samples. This type of blank is used primarily in air sampling applications where filters or various types of solid adsorbents are often used in sample collection. Sampling media blanks typically involve no special preparation because the blanks are simply selected at random from the filters or traps available for sampling purposes. The blank sample media are handled and exposed to ambient conditions in the same manner as sample media used for sample collection, but no air is passed through them. After handling, sampling media blanks should be sealed in the normal manner and stored along with the other samples.

Equipment blanks are a special type of field blank used primarily as a qualitative check for contamination rather than as a quantitative measure. Equipment blanks are prepared by collecting water or solvents used to rinse sampling equipment prior to sampling. Analysis of this rinse solution then provides an indication of the types of contaminants that may have been introduced through contact with the sampling equipment. Equipment blanks may also be used to verify the effectiveness of equipment cleaning procedures and as a check for potential cross-contamination.

Nomenclature associated with blank samples is far from consistent in the literature, and distinguishing one type of blank from another is sometimes difficult except by context or by a more detailed description. Alternate names for the most common blanks are listed in Table II. Two terms frequently used ambiguously are *trip blank* and *field blank*. For example, sampling media blanks are also sometimes called trip blanks because they are typically prepared in the laboratory and act as a check for contaminants introduced during the trip from the laboratory to the field and back again. Other types of field blanks are also sometimes referred to as trip blanks. Matched-matrix blanks prepared in the laboratory and then taken to the field, for example, are sometimes called trip blanks. The term field blank is typically used in more general terms to describe any type of blank used to assess field contamination.

Regardless of the application or the type of blank used, researchers should recognize and minimize the potential for inadvertently introducing contamination during preparation of the blank or at any other point where the actual samples are not exposed to similar opportunities for contamination.

Use of Blank Results

When properly used, blanks can be extremely effective tools in assessing and controlling sample contamination and in adjusting measurement results to compensate for the effects of contamination. Used improperly, blank results can increase the variability of analytical data or be very misleading. An important part of using blanks effectively is understanding and recognizing their limitations. As mentioned previously, blanks are useful for detecting contaminants causing additive interferences but are ineffective in identifying interferences such as dilution or adsorption. Similarly, blanks cannot be used to spot noncontaminant error sources such as analyte losses due to volatilization or decomposition. Beyond these inherent limitations, the utility of blanks is determined largely by the manner in which they are used and the manner in which the results are interpreted.

Blanks serve both control and assessment functions in environmental measurements. In their control function, blanks are used to initiate corrective action when blank values above preestablished levels indicate the presence of contamination. Blanks are most often used in this control mode in laboratory operations where feedback is more nearly real-time. At the first sign of unusual contamination, analyses may be stopped until the source is identified and the contamination eliminated. If possible, affected samples may then be reanalyzed. When field blanks indicate possible contamination, resampling is usually more difficult and often impossible. Therefore, field blank data are generally used primarily for assessment rather than control. If field blank data are used for control, this control is generally accomplished only over relatively long periods of time. In their assessment role, both field and laboratory blank data may be used to define qualitative and quantitative limitations of the associated measurement data. Where appropriate, these blank data may also be used as a basis for adjusting data to compensate for background contamination. Any such adjustments, however, should be made with caution, and the average of multiple blank measurements should be used for a stable, "in control" measurement system.

Control Charts for Blanks

Whether blank data are used primarily for ongoing control or for retrospective assessment, Shewhart *control charts* (*19, 20*) provide the most effective mechanism for interpreting blank results. In the control mode, control charts can be used to detect changes in the average background contamination of a stable system. This detection is done by providing definitive limits, based on past performance, that signal when the level of contamination is greater than that which is attributable to chance causes. This signal allows corrective action to be initiated to identify and correct new or additional sources of contamination as they appear, before large numbers of samples are affected. In the assessment mode, control charts allow out-of-control periods to be easily identified so that corresponding sample data may be flagged or interpreted separately from the other data. By identifying out-of-control periods, control charts also allow more reliable estimates to be made of the average background contamination level under normal in-control periods.

Control charts have gained considerable acceptance for some types of quality control checks familiar to environmental chemists, such as recovery data for spiked samples and results for calibration check samples. Control charts have been less frequently applied, however, to results for blanks. One reason for this lack of use is that most references on development and use of control charts emphasize $\overline{X}$ and R charts, which are control charts for means and ranges, respectively. Because blanks are not commonly run in replicate, control charts for means and ranges are not usually applicable. Another problem in using control charts for blanks is dealing with "zero", or "not detected" results. This is primarily a problem in deriving the variability estimates necessary for developing control charts.

Dealing with values of zero or not detected results is easily overcome for many measurements simply by changing the reporting convention. In most spectroscopic methods, for example, even a zero analyte concentration produces a signal from which a concentration may be calculated by using the calibration function. In such cases, the calculated concentrations should be reported as calculated, even if the calculation yields a "negative concentration". In methods where a zero analyte concentration produces no measurable signal, such as in gas chromatography, where no peak is produced for integration, initial variability estimates may be developed by using results for low-level standards as discussed in the next section.

Control Charts for Individual Measurements

The problem of blanks not usually being run in replicate can be overcome by using a control chart for individual measurements. This special type of control chart is useful when no rational subgrouping scheme arises, when performance measures can only be obtained infrequently, or when the variation at any one time (within a subgroup) is insignificant relative to variation over time (between subgroups).

Although they share the same statistical basis, control charts for individuals (*X charts*) are different from control charts for means ($\overline{X}$ charts) and ranges (*R charts*) in the way the range is calculated and in the subgrouping scheme. For these reasons, individual control charts are interpreted somewhat differently than usual. In $\overline{X}$ charts, the chart reflects variability between subgroups (i.e., between means); in R charts, the chart is used to monitor variability within subgroups. In control charts for individuals, however, the range within a subgroup cannot be calculated because the subgroup size is one. Also, because individual measurements are plotted, a single chart combines all sources of variation.

The first step in preparation of an X chart for blanks is to tabulate historical data for blank measurements. This tabulation will consist of at least 20 individual results for the particular type of blank to be charted. After arranging the k results in chronological order, $k-1$ moving ranges are calculated, where the first moving range is the range between the first and second values, the second moving range is the range between the second and third values, etc. Next, the average of the moving ranges ($\overline{MR}$) is calculated, along with the average of the k measurements (X_{avg}). Before calculating the control limits for the individual values, the moving ranges are screened by first calculating the upper control limit for the moving ranges as $3.27\overline{MR}$. [The value 3.27 is the D_4 value for calculating control limits for ranges having $n = 2$, where n is the number of measurements in each subgroup (*20*).] Any moving ranges larger than the calculated control limit are removed, and then the average moving range is recalculated. Finally, the upper and lower control limits (UCL and LCL, respectively) for the individual values are calculated as

$$\text{UCL} = \bar{X} + 2.66\overline{\text{MR}} \tag{1}$$

$$\text{LCL} = \bar{X} - 2.66\overline{\text{MR}} \tag{2}$$

Although not tabulated in many tables of control chart factors, the value 2.66 used to calculate the control limits is the A_2 factor for calculating control limits for $\bar{X}$, where $n = 1$.

Unless the average blank value is substantially greater than zero, the LCL may be negative and thus will not be meaningful. Only the UCL can be used in these cases.

In traditional $\bar{X}$ charts, the underlying assumption is that variability within a subgroup is representative of the system variability. Control limits for $\bar{X}$ are thus derived by using the within-subgroup range to estimate the standard deviation from which the control limits are calculated. In individual control charts, the moving range between subgroups (i.e., between the individual points) is used to estimate the standard deviation. Because pairs of consecutive measurements are more likely to be affected by similar special causes than are results from different points in time, screening the moving ranges prior to calculating the control limits minimizes the contribution of these special causes. This screening prevents the control limits from being inflated by these special causes as would be the case if the standard deviation was calculated by using all the original data points.

In using this approach, the problem still arises of dealing with zero and not detected values in the blank data from which the control limits are to be calculated. In this case, the average moving range must be estimated by using alternate data. Results for low-concentration standard solutions provide the best substitute. Obviously, if the blanks of interest are, for example, matched-matrix field blanks, standards should be prepared in a similar manner by spiking the appropriate matrix with the analyte of interest. In either case, the concentration of the standard should be in the same range as the estimated detection limit (i.e., between 1 and 5 times the estimated detection limit). At this level, imprecision should be of approximately the same magnitude as that for blanks. The actual control limits are calculated by using the average moving range for the standards and the mean blank value for similar blanks.

One limitation that should be considered in using an X chart is the increased sensitivity of the limits to the distribution of the measurements. $\bar{X}$ charts are less sensitive than X charts to the distribution of individual measurements because mean values are used. Means tend to be normally distributed even when the individual values are from populations that are not normally distributed. Because contamination tends to induce positive errors in blank measurements, the resulting distribution is likely to be skewed toward positive values. If inspection of the results indicates positively skewed values, the measurement data should be transformed prior to developing the control chart, and the transformed data should be charted. A logarithmic transformation is generally most appropriate for environmental data.

Example of an X Chart for Blanks

As an example of the development and application of an X chart for blanks, consider the data in Table III. For the purpose of the example, let these data represent blank results for aqueous nitrate measurements. Assume that the results for blanks 1–20 represent historical data used to develop the example control chart, and the results for blanks 21–64 represent subsequent blank measurements. The completed control chart is shown in Figure 1.

The first step in developing the example control chart was to examine the distribution of the historical data in Table III. As shown in the frequency histogram in Figure 2, the raw results are significantly skewed as is often the case for blank data. Therefore, before proceeding further, the raw data were transformed by taking the natural logarithm of each value. This produced the transformed results listed in the third column of Table III. Figure 3 is a frequency histogram of the transformed data that shows significant improvement in the skewness of the distribution.

The next step in developing the example control chart was to calculate 19 moving ranges for the 20 chronologically ordered transformed results. These moving ranges are listed in the fourth column of Table III. The UCL for the moving range, 4.697, was obtained by multiplying the average moving range, 1.436, by 3.27. This moving range control limit was then used to screen the moving ranges prior to calculating the control limits for the blank measurements. Screening the moving ranges and removing any values exceeding the control limit prevent the control limits for the blank measurements from being inflated by values representing special causes. As indicated in Table III, the 10th moving range (blank number 11), 5.200, exceeds the moving range control limit. Therefore, this value was removed, and the average moving range was recalculated to yield a value of 1.300. Finally, upper and lower control limits for the blank measurements were calculated as the average of the transformed results (−4.273) plus and minus 2.66 times the average of the screened moving ranges (1.300), or −0.816 and −7.731, respectively.

The completed control chart for this example, shown in Figure 1, illustrates how control charts for blanks are effective tools both for ongoing control and for retrospective assessment of blank results. In the control mode, for example, out-of-control points like that for the 11th blank indicate unusual contamination from an assignable cause and should initiate corrective action to identify and eliminate the source of additional contamination. In the case of field blanks, analyzing the samples and plotting the results may not be possible until after all of the samples are collected. In such cases, control charts are still useful in

Table III. Nitrate Blank Results

Blank Number	*Nitrate Concentration*	*ln Nitrate Concentration*	*Moving Range*
Historical Nitrate Blank Results			
1	0.033	−3.411	
2	0.049	−3.016	0.395
3	0.002	−6.215	3.199
4	0.002	−6.215	0.000
5	0.008	−4.828	1.386
6	0.002	−6.215	1.386
7	0.014	−4.269	1.9468
8	0.016	−4.135	0.134
9	0.009	−4.711	0.575
10	0.009	−4.711	0.000
11	1.631	0.489	5.200[a]
12	0.063	−2.765	1.946
13	0.042	−3.170	0.405
14	0.022	−3.817	0.647
15	0.093	−2.375	1.442
16	0.022	−3.817	1.442
17	0.002	−6.215	2.398
18	0.002	−6.215	0.000
19	0.031	−3.474	2.741
20	0.004	−5.521	2.048
Average ln nitrate concentration		−4.273	
Average moving range			1.436
Moving range UCL			4.697
Average screened moving range			1.300
UCL			−0.816
LCL			−7.731
Subsequent Nitrate Blank Results			
21	0.006	−5.116	
22	0.028	−3.675	
23	0.021	−3.863	
24	0.020	−3.912	
25	0.031	−3.474	
26	0.002	−6.215	
27	0.002	−6.215	
28	0.026	−3.650	
29	0.040	−3.219	
30	0.726	−0.320	
31	1.111	0.105	
32	0.081	−2.513	
33	0.089	−2.419	
34	0.128	−2.056	
35	0.053	−2.937	
36	0.190	−1.661	
37	0.353	−1.041	
38	0.353	−1.041	
39	0.389	0.489	
40	0.389	−0.944	

Table III.—Continued

Blank Number	*Nitrate Concentration*	*ln Nitrate Concentration*	*Moving Range*
	Subsequent Nitrate Blank Results—Continued		
41	0.066	−2.718	
42	0.731	−0.313	
43	0.283	−1.262	
44	0.277	−1.284	
45	0.213	−1.546	
46	0.452	−0.794	
47	0.288	−1.245	
48	1.668	0.512	
49	0.051	−2.976	
50	0.056	−2.882	
51	0.253	−1.374	
52	0.054	−2.919	
53	0.097	−2.333	
54	0.672	−0.397	
55	0.221	−1.510	
56	0.206	−1.580	
57	0.293	−1.228	
58	0.128	−2.056	
59	0.431	−0.842	
60	0.180	−1.715	
61	0.108	−2.226	
62	0.052	−2.957	
63	3.586	1.277	
64	0.216	−1.532	

[a]This value exceeds the UCL for the moving range.

assessing the blank data by indicating both sporadic and systematic contamination problems and allowing the corresponding measurement data to be interpreted accordingly. The example control chart in Figure 1, for instance, shows a significant shift in background contamination during the course of the hypothetical sampling and analytical effort. Such a shift might be the result of a change in sampling or analytical procedures, a change in personnel, a new lot of sample bottles, or any one of a number of other possibilities. Identification of these types of changes in background contamination allows field sample data to be grouped and interpreted separately even if it is already too late to eliminate the new source of contamination.

Other Types of Control Charts for Blanks

X charts, like the example chart in Figure 1, are similar to $\bar{X}$ and R charts in that they are charts for individual quality characteristics. Applied to analyses of blank samples, the quality characteristic of interest is analyte concentration in the blank. Although such charts are powerful tools in detection and diagnosis

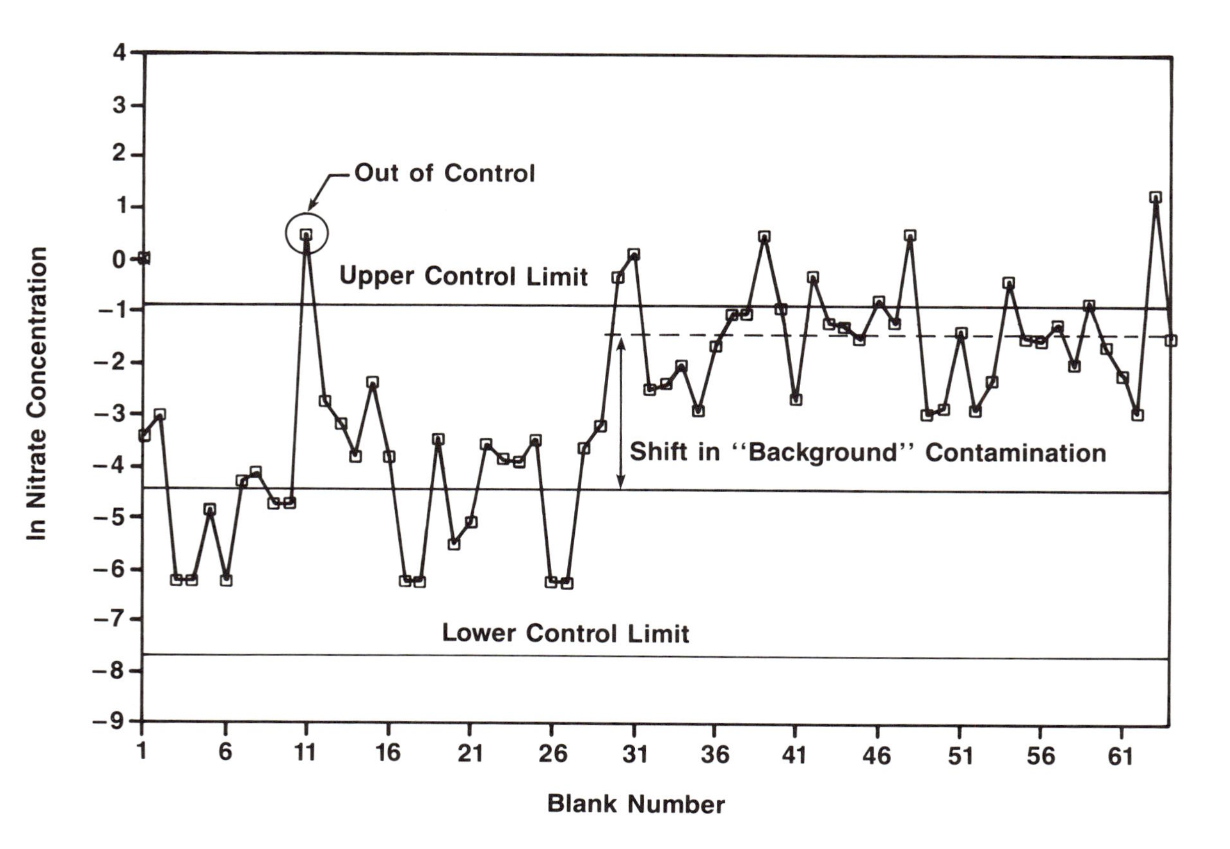

Figure 1. Example of an X chart for blank measurements.

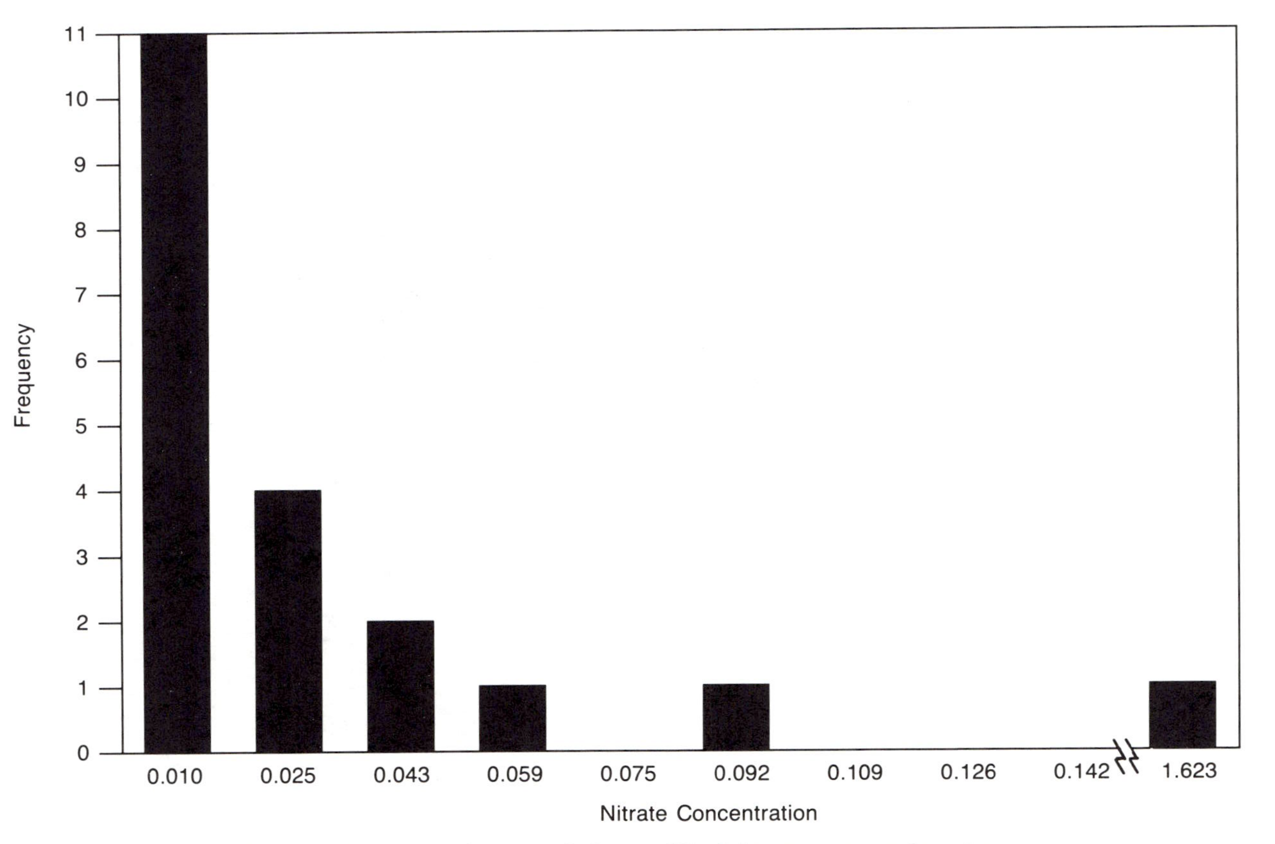

Figure 2. Frequency histogram for historical blank data prior to transformation.

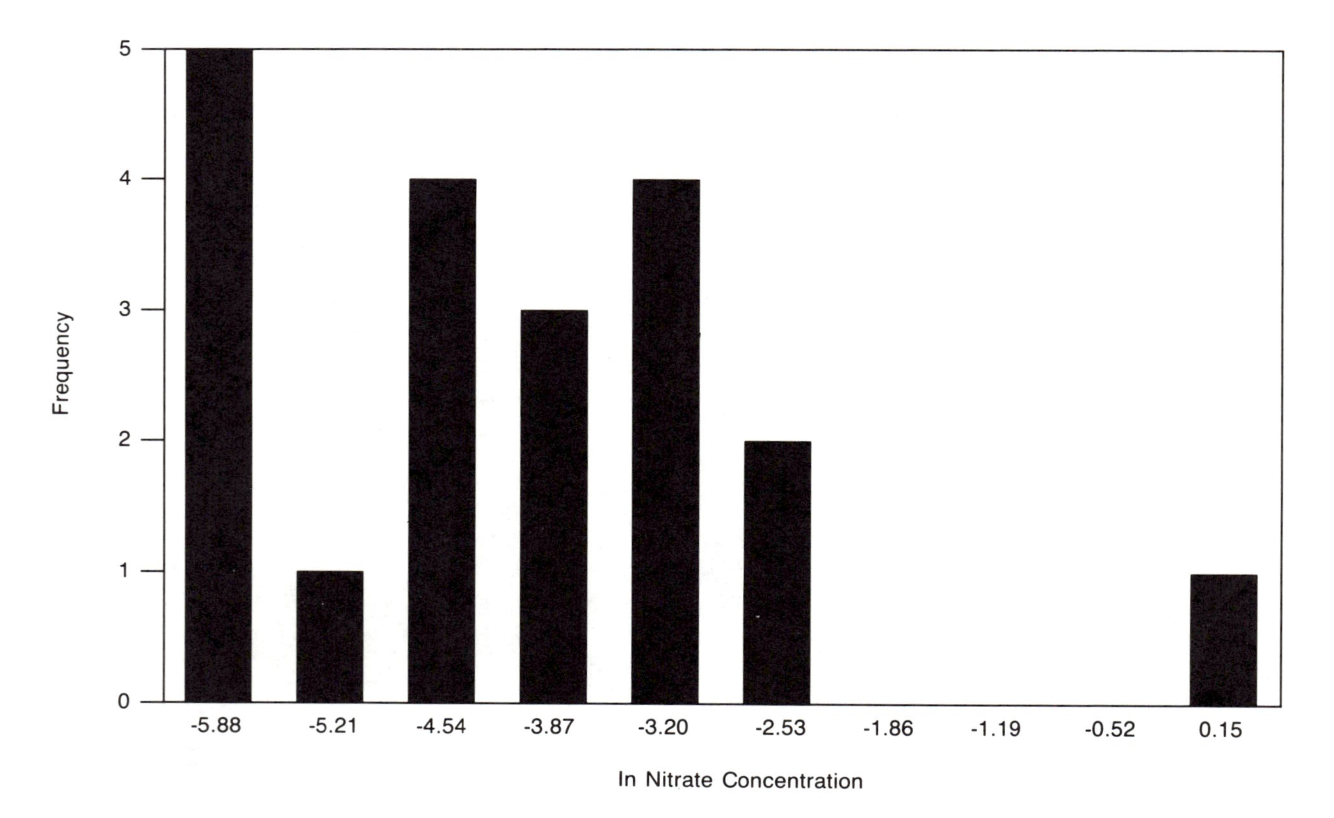

Figure 3. Frequency histogram for historical blank data after logarithmic transformation.

of quality problems in many applications, including measurement error caused by sample contamination, a separate chart is required for each quality characteristic monitored. This requirement presents a practical limitation to their applicability to measurement methods involving multiple analytes. Consider, for example, the U.S. Environmental Protection Agency's Method 624 for purgeable organic compounds. Maintaining separate control charts for each of the 31 target analytes that can be determined by this method would involve much work. In such cases, it may be worthwhile to focus primarily upon controlling contamination and giving up quantitative information about individual parameters in exchange for added ease of use.

If the user is willing to sacrifice information about average blank concentrations of individual analytes, one alternate approach that may be used involves redefining the quality characteristic of interest. For example, rather than charting the concentration of individual analytes, the total blank concentration may be used as a single measure of contamination. This single measurement may be calculated by summing individual concentrations or by summing the total signal and multiplying it by an average response factor. Although this approach permits the user to identify out-of-control periods, information is not provided for adjusting analytical data to compensate for background contamination.

Another approach, which may be appropriate when the primary emphasis is upon controlling rather than assessing contamination, is to use a control chart for nonconformities, or a c chart. Whereas X, $\bar{X}$, and R charts are control charts for variables, a *c chart* is a control chart for attributes. Rather than charting measured values of some specified quality characteristic, such as analyte concentration, c charts involve charting the number of nonconformities. Each instance of an article's deviation from specifications is a nonconformity. For blank measurements, each analyte detected in the blank or each analyte detected above a specified concentration would be considered a nonconformity. The total number of such nonconformities in each blank is the value plotted on the c chart. The center line is set at the average number of nonconformities per blank, $\bar{c}$, based on historical data. Three-sigma control limits are set at plus and minus 3 times the square root of $\bar{c}$ (i.e., $\bar{c} \pm 3(\bar{c})^{1/2}$).

Whereas the Gaussian distribution forms the basis for most commonly used control charts, c charts are based upon the *Poisson distribution*. The usefulness of c charts for a particular application thus depends upon the extent to which the Poisson distribution is an appropriate statistical model for that application. One of the required conditions for applicability of the Poisson distribution is that a large number of opportunities exist for nonconformities to occur. For this reason, c charts for blanks will generally be more applicable to methods involving large numbers of analytes than to methods involving only a few. Strict applicability of the Poisson distribution also requires that the nonconformities be independent of one another and have equal opportunities for occurrence. Because of these constraints, the use of c charts for blanks usually represents an approximation to strict theoretical applicability. Even so, the results obtained may be useful for practical purposes. Before adopting c charts for a particular application, a statistical quality control text (e.g., references 20–21) should be consulted for a detailed discussion of their use and limitations.

Assessing the Effectiveness of Control Charts for Blanks

The key role of control charts in controlling contamination is in detecting out-of-control points in an otherwise stable measurement process. Although results inside control limits do not indicate the absence of contamination, they are an indication that the effect of contamination is stable. In such a case, systematic adjustment of the measurement data using the average blank value to correct for the background contamination may be appropriate. Blank results that fall outside control limits provide a signal that some new source of contamination has entered the measurement system. Just as establishing the absence of any given analyte in a sample is analytically impossible, establishing the absence of contamination in a measurement process by analyses of blanks is also impossible. The best that can be achieved is to reduce the risk of not detecting contamination to an acceptable level.

In assessing this level of risk in using blanks to detect changes in contamination, the frequency of the blank measurements, the magnitude of change in the level of contamination that one desires to detect, and the amount of variability in the measurement system must be considered. Using three-sigma limits on the average of n measurements, the probability, P, of not detecting a bias (i.e., a change in level of contamination) of size b when the measurements are normally distributed and have a standard deviation of σ (22) is

$$P = \Phi\left\{3 - \left[\frac{(n)^{1/2}b}{\sigma}\right]\right\} - \Phi\left\{-3 - \left[\frac{(n)^{1/2}b}{\sigma}\right]\right\} \quad (3)$$

where Φ is the cumulative distribution function of the standard normal distribution. The probability of detecting a bias of size b in m independent tests (each based on the average of n measurements), or P_D, is

$$P_D = 1 - P^m \quad (4)$$

where m denotes any power of P.

Consider, for example, a case in which an X chart is used to monitor blank results for a particular analyte. In this case, historical data indicate that the average blank concentration of this analyte is 3 ppb, and the standard deviation is 4 ppb. What is the probability of detecting contamination greater than 10 ppb in a single blank? Because 10 ppb represents an increase of 7 ppb above background, $b = 7$. The subgroup size for X charts is one, so $n = 1$. In this case,

$m = 1$ also because a single measurement represents only a single point on the control chart. Therefore,

$$P = \Phi\left\{3 - \left[\frac{(1)^{1/2}(7)}{4}\right]\right\} - \Phi\left\{-3 - \left[\frac{(1)^{1/2}(7)}{4}\right]\right\}$$

$$= \Phi(1.25) - \Phi(-4.75)$$

$$= 0.894 - 0.000$$

$$= 0.894$$

and

$$P_D = 1 - (0.894)$$
$$= 0.106$$

Thus, the probability of detecting an additional 7 ppb of contamination is less than 11% for a single blank analysis. If the additional contamination is from a constant source, the probability of detection improves somewhat with repeated measurements. However, because the standard deviation is relatively large compared to the added contamination, 21 measurements are required to attain a greater than 90% probability of detection. On the other hand, a shift of 20 ppb for the same measurement system would have a greater than 97% probability of being detected in a single measurement. Figure 4 illustrates this relationship by showing probabilities of detecting unusual contamination in a single blank analysis for measurement systems having standard deviations of 2, 4, 6, and 8 ppb. Contamination levels as high as 30 ppb are illustrated.

Control charts are most effective for detecting contamination when measurement variability is small relative to the level of contamination to be detected. Many measurements may be required to detect small shifts in background contamination. Also, in order for such shifts to be reliably detected even through repeated measurements, the additional contamination must be persistent. An assumption in equations 3 and 4 is that the problem persists at the same level until corrected. Although reasonable for many sources of contamination, this model is not applicable in all cases. If contamination occurs sporadically at low levels, then detecting the changes in contamination levels is much more difficult. In such cases, the only reasonable approach is to work on identifying and eliminating the source or sources of contamination.

Conclusions

Environmental sampling and analytical efforts present numerous opportunities for sample contamination from a wide variety of different sources. Regardless of

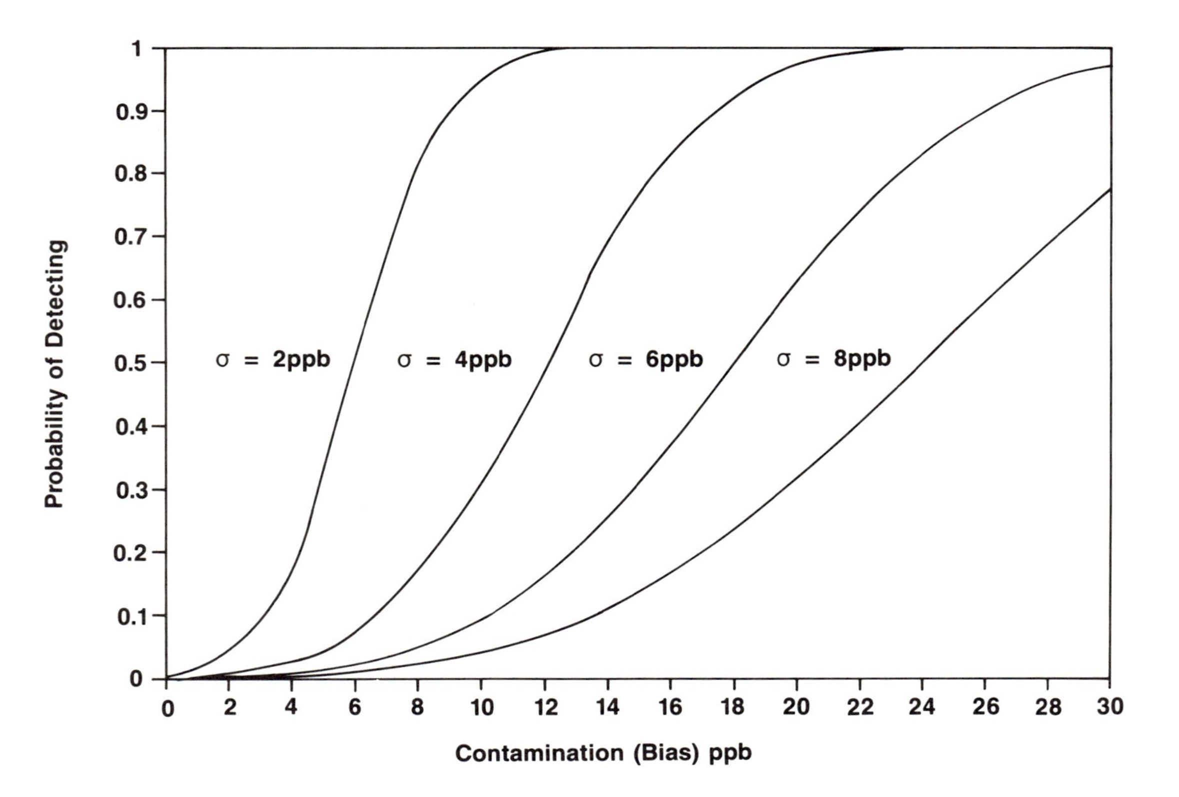

Figure 4. *Example of the relationship between measurement variability and probability of detecting unusual contamination (bias) in a single blank measurement.*

the source of contamination, the accuracy of the measurement process is affected. Because environmental measurements often address very low concentrations of analytes, contamination is an especially important source of potential error. To minimize error due to contamination, the potential sources of contamination must be identified and eliminated wherever possible.

Once a measurement system is established, appropriate types of blanks should be used to define background levels of contamination for the different parts of the sampling and analytical process. Blanks should also be used on an ongoing basis to assess and control contamination. In the assessment mode, the information provided by blanks may be qualitative or quantitative; blanks may be used as qualitative indicators of possible sample contamination or to derive quantitative estimates of background contamination levels. In the control mode, blanks are used to initiate corrective action when results above preestablished levels indicate unusual contamination.

Whether used primarily for assessment or for control, control charts should be used to maximize the effectiveness of blank measurements. Control charts for individual measurements, or X charts, are usually more appropriate for blanks than the more common $\overline{X}$ and R charts. Other types of control charts may be applicable or preferable under certain circumstances. Regardless of the type of control chart used, the risk of not detecting new sources of contamination in a measurement system depends upon the number of blank measurements, the magnitude of the effect of the new contamination, and the variability of the measurement system. This risk should be a primary consideration in developing the overall quality control strategy. By recognizing potential sources of contamination and using blanks to detect changes in background levels, reducing or correcting for contamination is generally possible, and the associated measurement biases can thus be reduced to acceptable levels.

Abbreviations and Symbols

A_2 factor for determining three-sigma control limits for $\overline{X}$ charts by using the average range, R
b size of bias
$\overline{c}$ average number of nonconformities per blank
Φ cumulative distribution function
D_4 factor for determining the upper three-sigma control limit for R charts by using the average range, R
k number of results
LCL lower control limit
m number of independent tests
$\overline{MR}$ average of the moving ranges
P probability of not detecting a bias
P_D probability of detecting a bias

s standard deviation
UCL upper control limit
X_{avg} average of k measurements

References

1. Fetter, C. W., Jr. *Ground Water Monit. Rev.* **1983**, *3*, 60–64.
2. Boettner, E. A.; Gall, G. L.; Hollingsworth, Z.; Aquino, R. *Organic and Organotin Compounds Leached from PVC and CPVC Pipe*; U.S. Environmental Protection Agency. U.S. Government Printing Office: Washington, DC, 1981; EPA 600/1-81-062.
3. Ross, H. B. *Atmos. Environ.* **1986**, *20*, 401–405.
4. Cox, R. D.; McDevitt, M. A.; Lee, K. W.; Tannahill, G. K. *Environ. Sci. Technol.* **1982**, *16*, 57–61.
5. Seila, R. L.; Lonneman, W. A.; Meeks, S. A. *J. Environ. Sci. Health* **1976**, *A11*, 121.
6. Kopczynski, S. L.; Lonneman, W. A.; Winfield, T.; Seila, R. *J. Air Pollut. Control Assoc.* **1975**, *25*, 251.
7. Best, M. D.; Drouse, S. K.; Creelman, L. W.; Chaloud, D. T. *National Surface Water Survey, Eastern Lake Survey (Phase I—Synoptic Chemistry) Quality Assurance Report*; U.S. Environmental Protection Agency, in press; EPA 600/4-86/011.
8. *Handbook for Analytical Quality Control in Water and Wastewater Laboratories*; Office of Research and Development. U.S. Environmental Protection Agency: Cincinnati, OH, 1979; EPA 600/4-79-019.
9. Fowler, D.; Cape, J. N. *Atmos. Environ.* **1984**, *18*, 183–189.
10. Moody, J. R.; Lindstrom, R. M. *Anal. Chem.* **1977**, *49*, 2264–2267.
11. Bonoff, M. B.; Filbin, G. J.; Gremillion, P. T.; Kinsman, J.; Mudre, J.; Schriner, C. "Field Operations Report"; Living Lakes, Inc. Aquatic Liming and Fish Restoration Demonstration Program; Living Lakes, Inc.: Washington, DC, 1987.
12. *Fed. Regist.* **1984**, *49*, 29–39; Appendix A to 40 CFR Part 136.
13. Levine, S. P.; Puskar, M. A.; Dymerski, P. P.; Warner, B. J.; Friedman, C. S. *Environ. Sci. Technol.* **1983**, *17*, 125–127.
14. Bagchi, R.; Haddad, P. R. *J. Chromatogr.* **1986**, *351*, 541–547.
15. Sommerfeld, M. R.; Love, T. D.; Olsen, R. D. *At. Absorpt. Newsl.* **1975**, *14*, 31–32.
16. *Trace Analysis: Spectroscopic Methods for Elements*; Winefordner, J. D., Ed.; Wiley: New York, 1976; Chapter 2.
17. Commission on Spectrochemical and Other Optical Procedures for Analysis. *Spectrochim. Acta, Part B* **1978**, *33B*, 248–269.
18. Shacklette, H. T.; Borngen, J. G. *Element Concentrations in Soils and Other Surficial Materials in the Conterminous United States;* U.S. Geological Survey Professional Paper 1270; U.S. Geological Survey: Reston, VA, 1984.
19. Shewhart, W. A. *The Economic Control of Quality of Manufactured Product*; Van Nostrand: New York, 1931.
20. Grant, E. L.; Leavenworth, R. S. *Statistical Quality Control*, 5th ed.; McGraw-Hill: New York, 1980.
21. Burr, I. W. *Statistical Quality Control Methods*; Marcel Dekker: New York, 1976.
22. Provost, L. P.; Elder, R. S. *Choosing Cost Effective QA/QC Programs for Chemical Analysis*; Office of Research and Development. U.S. Environmental Protection Agency: Cincinnati, OH, 1985; EPA 600/S4-85/056.

Chapter 6

Environmental Sampling Quality Assurance

Shrikant V. Kulkarni and Malcolm J. Bertoni

The field of environmental data quality assurance (QA) has evolved from a predominantly quality control (QC) protocol-compliance approach to a comprehensive quality systems approach. A quality system addresses the full range of planning, implementation, and assessment activities required to produce environmental data that satisfy the needs of data users, such as regulators or program managers. Emerging trends in environmental regulation and technology continue to pose new managerial and technical challenges in designing and implementing effective QA programs. The quality systems approach provides a robust framework for meeting these challenges, yet the effectiveness of QA programs can suffer from inattention to certain critical success factors. The most often overlooked success factor is the effective reporting of sampling QA information. Effective reporting requires that QA and QC information be consistently and clearly communicated not only to the analytical laboratory but also to the data user so that the data can be interpreted properly.

QUALITY ASSURANCE (QA) plays a critical role in the generation and use of environmental data. QA activities ensure that the environmental sampling and analysis process is verified and documented so that the uncertainties in the resulting data can be controlled and quantified. In this way, the information gained from QA activities allows a data user to determine whether the data are good enough to support their intended use.

In recent years the field of environmental data QA has evolved from a predominantly laboratory- and protocol-oriented quality control (QC) program into a more comprehensive systems approach to quality that focuses on clarifying and satisfying the data user's needs (*1*). QA practices of the past concentrated most efforts on standardized analytical procedures conducted in laboratories, where conditions could most easily be controlled. However, this focus addressed only one component of the overall sampling and measurement system. A more comprehensive approach to QA was needed to control the total error in environmental data arising from numerous data collection activities, from planning, through implementation, to assessment. The quality systems concept was developed to address the full range of management and technical activities needed to specify quality performance criteria and ensure that environmental data satisfy those criteria. This quality systems approach sharpens the focus on how error in data affects the decisions or estimates of interest to the end user of that data.

This chapter uses terminology that draws distinctions between different types of QA. *Environmental data QA* refers to the broadest view of QA: ensuring that environmental data satisfy the data user's needs. *Environmental sampling QA* refers to QA activities designed to ensure that technical procedures for collecting physical samples or specimens have been specified appropriately. Thus, sampling QA focuses on ensuring that a set of physical samples is adequately representative of the target population. *Environmental measurement QA* refers to QA activities designed to ensure that the process of measuring a physical sample is accurate. Thus, measurement QA focuses on ensuring that a measurement value is adequately representative of the physical sample being measured. These definitions of sampling and measurement QA have some unavoidable overlap, particularly with respect to procedures for collecting physical samples. Nonetheless, this overlap serves to highlight an important theme: QA that is focused on satisfying the data user's needs must pay adequate attention to both sampling and measurement issues, as well as the linkages between the two. The quality systems approach takes an integrated view of sampling and measurement so that the causes of errors can be identified and controlled effectively.

The quality systems approach provides a sound framework for developing effective QA programs in all kinds of organizations, yet many QA programs suffer from inattention to certain critical success factors. Some QA programs suffer from poor design: the organizational structure is not aligned with quality management principles, some key components of the quality system are missing, or QA policies and procedures are inadequate. Many QA programs, however, suffer from relatively simple problems with implementation. An important example is that sampling QA performance data and information often never go beyond the analytical laboratories and hence never get reported properly to the data user. QA reports provide crucial information that can help the data user interpret the sampling and analysis results properly and determine any limitations on their use. Greater awareness and understanding of key elements of the quality systems approach will lead to more successful QA practice.

This chapter presents an overview of the quality systems approach and identifies some critical factors for successful QA practice. A brief history of the evolution of QA is presented, leading to a discussion of some key emerging trends in environmental regulation and technology that will pose new managerial and technical challenges to the field of QA. Some fundamental principles underlying the quality systems approach are explained before an overview of the quality system developed by the U.S. Environmental Protection Agency (EPA) is presented. Finally, some factors important for successful QA practice are offered to motivate a critical evaluation of the reader's own data collection program.

The Evolution of Environmental Quality Assurance

The quality systems approach described later in this chapter evolved from a variety of different approaches to QA. This section describes the historical development of QA by looking at the QA program developed by the U.S. EPA, which is the lead federal agency for most environmental sampling activities in the United States. This brief history sets the stage for discussing emerging trends in environmental regulations and sampling and measurement technology that will affect the future development of QA programs.

Historical Development of Quality Assurance at the U.S. EPA

Understanding the quality systems approach requires an appreciation of how it evolved from previous approaches to QA. Over the past 25 years, the definition of environmental QA has changed significantly as the scope of QA activities has expanded. This expansion has occurred in three phases, corresponding roughly to the decades of the 1970s, 1980s, and 1990s.

In the 1970s, the EPA based environmental decisions on data generated without a formal institutionalized QA program. Data quality estimates were based solely on comparisons with standard or reference material measurements generated using an EPA-approved methodology. Consequently, QA activities were applied piecemeal and were restricted mainly to laboratory and field analytical measurements. QA activities ensured that the measurement system was well-defined, calibrated, and within specified control limits, based on analytical measurement QC that included the application of good laboratory practices and good measurement practices (2). QA protocols such as standardized test methods, calibrations, and performance audit procedures were essentially voluntary, and most of the data were reported with no indication of their quality (3).

The state of QA practice improved dramatically in the 1980s. From a theoretical perspective, Taylor (4) of the National Bureau of Standards [now the National Institute of Standards and Technology (NIST)] helped clarify the proper role of QA in measurement programs by defining QA in terms of operations and procedures undertaken to provide measurement data of a prescribed quality with a prescribed probability of being accurate. On the policy side, the EPA formally established its QA program in 1984 through Order 5360.1 (5), which spelled out the Agency's QA policy and mandatory program requirements. The primary goal of the EPA QA program at that time was to ensure that all environmental measurements supported by the EPA produced data of "known quality". The policy stated that the quality of data is considered to be "known" when all components associated with its derivation are thoroughly documented and such documentation is verifiable and defensible. This Agency-wide policy set in motion the development of environmental QA initiatives throughout the decentralized regional and program offices within the Agency.

Although the EPA's movement to a formal policy on QA represented a major step forward, systemic problems with the QA program remained through the 1980s. Many of these systemic problems arose from the fragmented, incremental enactment of various U.S. environmental laws and regulations since the 1970s. Most of the laws and regulations were written to require environmental media-specific pollutant control. This approach created problems for environmental QA because separate program offices grew out of the separate environmental laws. Consequently, the decentralized development and implementation of regulations gave rise to different approaches to QA. It became common to find that measurement methods required for the same analyte under the Clean Air Act (CAA), the Clean Water Act (CWA), the Resource Conservation and Recovery Act (RCRA), and the Federal Insecticide, Fungicide, and Rodenticide Act (FIFRA) were different or mutually incompatible.

Many differences among measurement methods were justified because of the differences in technical conditions encountered within the respective programs, yet it became clear by the end of the 1980s that the situation was becoming unwieldy. Research laboratories in the EPA Office of Research and Development (ORD) assisted in generating protocols for monitoring compliance with specific environmental regulations. Although the monitoring methods used the same basic technology for analysis independent of the pollutant matrix (e.g., atomic absorption for metals, or gas chromatography for volatile organic materials), the laboratories generated their own independent protocols and requirements for associated QA. The protocols differed not only in technical details to address program-specific sampling requirements and matrix interferences, but they also differed in their numbering systems and reporting requirements. Additionally, other federal agencies developed their own protocols when faced with the unique environmental problems associated with their activities, such as emissions from coal-fired electric utilities in U.S. Department of Energy (DOE) programs. The lack of universally accepted and validated protocols for detecting pollutants in any given matrix at the desirable concentration levels made the traditional QC-based QA activities unacceptable.

In the 1990s, this fractured QA landscape has given way to a more consistent and coordinated approach. The EPA's ORD has undergone a major reorganization to emphasize its risk assessment and exposure assessment approach to environmental protection. The ORD's Quality Assurance Division [QAD; formerly the Quality Assurance Management Staff (QAMS)] is attempting to institute a standardized quality systems approach within the EPA community. The QAD is attempting to harmonize its QA activities with other federal agencies through participation in interagency efforts to establish a commonly accepted quality system. These activities have been coordinated with similar efforts by professional organizations such as the American National Standards Institute (ANSI), the American Society for Quality

Control (ASQC), the American Society for Testing and Materials (ASTM), and the American Chemical Society (ACS).

A good example of this coordination effort can be seen in the area of water quality, in which partnerships have been established among a variety of governmental organizations, including the U.S. EPA, the U.S. Geological Survey, the U.S. Fish and Wildlife Service, the U.S. Army Corps of Engineers, the DOE, the Tennessee Valley Authority, the Potawatomi Indian Nation, the Delaware River Basin Commission, and agencies within several U.S. state governments (6). The technical challenges that are being addressed are not trivial: the water matrix alone has several subsets, such as deionized water samples, groundwater samples, surface water samples, brackish seawater samples, and wastewater samples, that have different technical requirements. Matrix effects can significantly confound chemical analysis at the low concentration levels often encountered in environmental sampling. Accordingly, members of the federal partnership are dealing with separate issues for the same analytes from different viewpoints. They are attempting to develop standard terminology and commonly accepted sampling and analysis technologies for the monitoring of pollutants, which can then be modified to suit each agency's sphere of activities. The diverse needs of these member organizations indicate how challenging it is to develop a nationwide unified strategy for water quality monitoring. Nonetheless, by starting from a common set of standards, consistency and comparability are enhanced.

The search for a common scientific platform grows from a need for defensible data to support environmental decision making. The rigid QC-based approach to QA that arose in earlier EPA monitoring programs is being replaced by an evolutionary, flexible, comprehensive quality systems approach. In essence, the QA community has recognized that planning for data of required quality is as important as implementing and assessing the data generated (7).

Emerging Trends in Environmental Regulation and Technology Affecting QA

Before a more detailed description of the quality systems approach is presented, it will be useful to consider some emerging trends in environmental regulation and technology that will shape the future evolution of quality systems. Just as the field of environmental sampling QA has begun to achieve more consistency in its methods for controlling and documenting the quality of environmental data, new developments in the way data are generated and used are necessitating changes in the practice of environmental QA. Perhaps the most significant emerging trend is in how environmental data are used to support regulatory development and compliance decisions. In the past, environmental regulations were organized around the control of pollution in specific media, such as air, water, or soil. Accordingly, many regulations focused on limiting the discharge of pollutants in a particular medium. Many popular pollution-control technologies, such as air-stripping in wastewater treatment facilities, basically transferred the pollutants from water to air. The water effluent from the facility met the regulatory discharge limits, but the airborne pollutants were still present in the ecosystem as a whole. Similarly, in the early 1980s, incineration technology was considered to be the most viable solution for solid waste treatment and reduction, particularly because of limitations on landfills. However, studies now indicate that some of the incinerator by-products, such as dioxins, furans, and other products of incomplete combustion, may present even greater hazards.

Currently, however, the focus of regulatory activity is shifting away from these traditional approaches and toward an integrated ecosystem management model. Under this new approach, pollutant discharges to all environmental media in a specific region are examined in relation to their impact on the local ecosystem, and a coordinated set of control measures is established to ensure ecological health. The current thinking in pollution prevention not only considers emissions in an integrated ecosystem but also weighs several other factors that impact on and reduce the pollution at the source. Under an environmental technology initiative (ETI) research project, an effort is under way to model an industrial park containing planned sites shared by firms committed to reducing energy use, reusing wastewater from one firm to another, or taking waste from a neighboring firm and using it as raw material. The EcoIndustrial Park (EIP) is an alternative to using the regulatory system to achieve environmental outcomes (8). Although matrix-specific pollutant monitoring will still be important, the concept of a matrix-specific regulatory system will need modification.

This new environmental management strategy presents several challenges for environmental sampling QA. First, a broader range of environmental data are required, including biological sampling and observational data. Moreover, the ecosystem management strategy requires integrated analysis of these multiple variables. This integrated analysis, in turn, requires a QA program that assures the comparability of data as well as the quantification of uncertainty in these diverse types of measurements and observations. Quantification and control of uncertainty become vital because of the multiplicity and complexity of interactions among variables and the potential need to detect subtle changes in ecological health indicators.

Another emerging trend that challenges old approaches to QA is the rapid development of new sampling technology. Many readers will be familiar with the QA challenges presented by increasingly sensitive analytical methods that allow reliable detection and quantification of pollutants at progressively lower concentrations. Perhaps the greater challenge to the discipline of environmental QA, however, is the trend toward real-time field measurement techniques and remote sensing. Problems in verifying the measurement system stem from difficulties in using a controlled but representative performance evaluation (PE) sample or reference sample for comparison under real-time field conditions. The Fourier transform infrared spectrometry (FTIR) technology, a new tool in field measurements and remote sensing, can be used and validated for emissions coming from a stack by comparing the data with that obtained using conventional methods. On the other hand, one might consider the problem of detecting and quantitating methane emissions from a lagoon. An ideal tool for this type of measurement is an "open path" FTIR, a technology that offers a snapshot concentration at a given instant in its path. However, validating these data or this technology using a reference or performance sample under the same conditions presents quite a challenge because external factors such as wind velocity, wind direction, and humidity, for example, can affect the measurement result. This type of complication does not render the FTIR data invalid or unusable. What it suggests is the necessity of developing some carefully controlled experiments to validate the data and to determine the components of measurement error that produce the greatest uncertainty in the measurement results.

In general, QA techniques and protocols must evolve along with the new data collection tools. As more measurement techniques move chemical analysis procedures out of controlled laboratory conditions and into the field, validation and documentation of sampling data and associated QA activities will be essential in establishing the defensibility of the data.

Advances in other technologies related to environmental sampling will also change the face of QA. Progress in information technology will continue to improve the capability to manage data efficiently and to share valuable information across organizations and programs on a real-time basis. In large field sampling and data collection programs such as Superfund site remediation activities, unforeseen problems and difficulties are more the norm than the exception. Timely review of information and quick turnaround response from the decision makers or the planners and designers of a study will lead to better corrective actions. For example, the statisticians and engineers who designed a sampling plan should have an opportunity to decide how to change a sampling location if the field crew encounters a physical obstacle (such as a large boulder) instead of soil when drilling a sample well for groundwater monitoring. Doing so ensures that the sampling plan integrity is not compromised and may also reduce the cost of revisiting the sampling site. Without mobile communications technology, such field adjustments would usually be impractical or very costly. Ultimately, the rapid information transfer of technology should make it easier to collect the data as planned the first time to support the decision to be made.

Environmental Data Quality Systems

The previous discussion of the past and potential future evolution of QA provides a useful context for this section, which presents an overview of the quality system developed by the U.S. EPA. First, some key principles underlying the quality system are discussed. This discussion is followed by an explanation of the main components of the EPA quality system. The quality mission provides the anchor for the organizational-level quality system components, which in turn support the project- or program-level components, which provide the day-to-day quality management tools for ensuring that the data satisfy their intended use.

Key Concepts Underlying the Quality Systems Approach

The quality systems approach must be understood in the context of a few key principles. The environmental data life cycle is presented as a model for understanding how QA programs control errors in data that unavoidably arise during planning, implementation, and assessment activities. To facilitate control of these errors, the QA discipline has developed a set of data quality indicators (DQIs) that characterize the quality of information contained in the data.

The Environmental Data Life Cycle. The process by which environmental sampling data are collected, analyzed, and used can be illustrated as an environmental data life cycle (Figure 1). The life cycle process is divided into four broad phases: planning, implementation, assessment, and reporting.

Planning. The planning phase is when the data user specifies the intended use of the data and plans the management and technical activities needed to generate the data. During this phase, the purpose and objectives of the data collection effort are clarified by answering some basic questions, such as:

- Why are the data being collected?
- What study questions need to be answered?
- What decisions will be supported using environmental data?

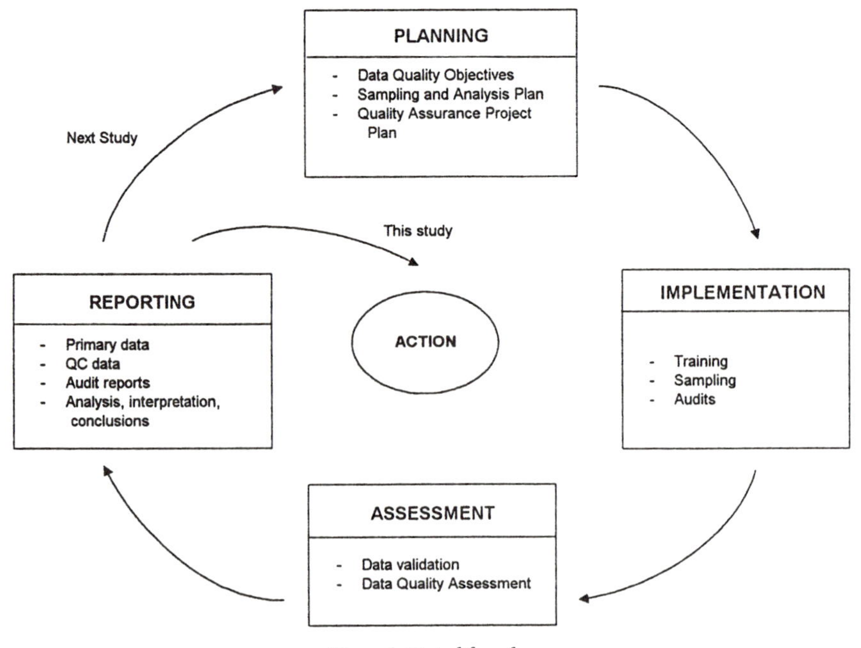

Figure 1. Data life cycle.

To ensure that the right type, quality, and quantity of data are collected to satisfy the data user's needs, the project objectives must be specified at the outset. The project objectives then provide the basis for specifying the technical details of how to perform the environmental sampling during later steps in planning. The end product of the planning phase is a specific and documented plan that describes the sampling and analysis protocols: why the data are being collected, where and when sampling will occur, how the samples will be collected and analyzed [such as standard operating procedures (SOPs)], and what QA protocols will be used to control errors and document the overall process. Many of the other chapters in this book address important details of what to include in the sampling and analysis plan under a variety of different circumstances.

Implementation. The implementation phase is when the plan is put into action by performing the sampling and analysis protocols in conjunction with associated QA protocols. During the implementation phase, the QA protocols become a focal point because they describe how the sampling and analysis methods are to be verified and documented, thereby ensuring that the resulting data are of known quality (*9*). These QA protocols may include, for example, field sampling documentation requirements, sample chain-of-custody procedures, technical systems audits, and laboratory QC procedures. The end products of the implementation phase are the environmental sampling results in the form of primary data and QC data, as well as documentation of how the sampling and analysis procedures actually were performed, including reports on any problems encountered.

Assessment. The assessment phase is when the results of the sampling and analysis are evaluated to determine whether the project objectives have been satisfied. This phase usually begins with a data validation step, whereby the documentation from the implementation phase is compared with the sampling and analysis plan to identify any discrepancies or anomalies. The crucial part of the assessment phase, however, involves analyzing and interpreting the data with respect to the data user's objectives for the overall data collection effort. Therefore, the assessment phase examines the data values and the uncertainties associated with those values to determine whether the sampling results are scientifically and statistically conclusive. The end products of the assessment phase are the validation reports, analyses, and conclusions of the data analyst.

Reporting. The reporting phase is when all of the relevant information about the data collection planning, implementation, and assessment are brought forward to the data user for consideration and action. Despite the critical nature of this phase, reporting is often treated as an afterthought, to the detriment of the entire data collection project. The presentation of sampling results should be given careful attention to avoid misunderstandings and misinterpretations. In particular, the report that is submitted to the data user must highlight relevant and appropriate QA information, such as QC sampling results and audit reports, so that the data user is made aware of the quality of information upon which decisions will be made or study conclusions will be drawn. If the QA information is buried in the back of an appendix, or not included at all, then the data user will lack important information about the uncertainties in the data and potential complications in the study that may affect the interpretation of the results. The end product of the reporting phase is a set of documents that summarize the results of the data collection project in a form that is accessible and meaningful to the data user. This documentation also provides an important resource for secondary data users who may be conducting related studies or planning future projects.

Errors in Environmental Sampling. Errors are unavoidable in environmental sampling and analysis, so QA protocols are designed to ensure that errors are controlled to levels that are tolerable to the data user. Error in data represents the difference between the measurement result and the true (but unknown) value of the thing being measured. When many measurements are combined to estimate a characteristic of the environment (a population parameter, in statistical terminology), then error refers to the difference between the estimated value (such as a mean pollutant level) and the true (mean) value. Many QA and QC protocols are designed to quantify the error in measurement data; many statistical procedures are designed to quantify the error in population parameter estimates calculated from measurement data.

It is useful to think of error as having different components. In one type of classification, error can be expressed in terms of random and systematic components, as described by Taylor (*10*). *Random errors* vary unpredictably in both magnitude and direction, and give rise to imprecision. *Systematic errors* vary in one direction, may or may not vary in magnitude, and produce bias.

In another type of classification, components of errors are expressed in terms of their source or cause. The broadest categories can be described as sampling error and measurement error. Sampling error in this context relates to errors in estimating population parameters due to variability in the population over space and time. Because any given sample will capture only those characteristics found at a given point in space and time, the estimate will contain sampling error to the extent that the sampling locations fail to adequately represent the population as a whole. Measurement error relates to the random and systematic errors that arise during the course of the measurement process, from physically obtaining a sample specimen through obtaining an analytical measurement result. Measurement error can be broken down further into components that represent each step of the measurement process, although from a practical standpoint this analysis is usually done only when a problem in the measurement system is detected.

In terms of the data life cycle, sampling error is most closely associated with the sampling design established in the planning phase; measurement error is most closely associated with problems in or inherent characteristics of the implementation phase. During the assessment phase, the data analyst evaluates the total error, which is the combination of sampling and measurement errors that determine how well the sampling and analysis results represent the true value of the population parameter under study.

Random errors can be expressed in terms of statistical variance and can be quantified using QC protocols. Smith et al. (*11*) have discussed quantitative indicators of error in terms of the design phase of sampling variance, the implementation phase of sampling variance (sample collection), and the analytical measurement phase of sampling variance. With this approach, QC samples can be designed that provide the information needed to calculate the components of error in the data due to different data collection steps. Table I describes the types of QC samples that allow the assessment of data quality.

The errors in sampling data collection (other than those due to an inadequate design) often can be ascribed to the following:

- misinterpretation of sampling plans;
- incorrect implementation of sampling plans with approved or validated protocols;
- disregard for or misunderstanding of the importance of adhering to sampling procedures;
- equipment calibration errors;
- contamination;
- sample storage and transportation errors; and
- sample documentation errors, including sample identification and chain of custody.

Many of these errors can be minimized by carefully implementing well-designed and detailed sampling and analysis plans and performing technical systems audits (TSAs) of the field sampling activities. In the authors' experience with many field technical audit activities, the field sampling team often had not seen or read the well-designed, reviewed, and approved sampling and quality assurance project plan (QAPP) that was prepared for their particular study. In interviewing the field sampling personnel during TSAs, auditors may find that the well-planned nuances and intricacies included in the project plan to answer specific questions raised by planners and data users are totally lost because of a lack of communication, or miscommunication.

Data Quality Indicators. Given the many ways in which error can arise in environmental sampling, the QA community has developed a set of data quality indicators (DQIs) that describe the various performance characteristics of a data set. DQIs are commonly described in terms of *precision, bias, representativeness, completeness,* and *comparability.* Commonly accepted definitions for DQIs, as supported by the EPA's ORD programs, are included in Table II (*11, 12*).

Bias is defined as a persistent or systematic distortion of a measurement process that caused errors in one direction. Bias can have a devastating effect on

Table I. QC Samples for Sampling Data Assessment

Sample Type	*Description*	*Assessment Parameter*
Collocated sample	One of two or more independent samples collected so that each is equally representative for a given variable at a common space and time.	Used to estimate overall precision.
Field duplicate	Two samples taken from and representative of the same population and carried through all steps of sampling and analysis in an identical manner.	Used to assess total variance of the method, including sampling and analysis.
Replicate sample	Two or more samples representing the same population, time, and place, which are independently carried through all steps of the sampling and measurement process in an identical manner.	Used to assess total variance.
Split sample	Two or more representative portions taken from a sample or subsample and analyzed by different analysts or laboratories.	Used to replicate the measurements of interest.
Spiked sample	A sample prepared by adding a known mass of target analyte to a specified amount of matrix sample for which an independent estimate of target analyte concentration is available.	Used to determine the effect of matrix on the method recovery efficiency.
Calibration check sample	A reference material used to check that the calibration is acceptable.	Used to ensure that the measurement system is under control.
Quality control sample	An uncontaminated sample matrix spiked with known amounts of analytes from a source independent from the calibration standard.	Used to establish intralaboratory or analyst precision.
Performance evaluation sample	A sample whose composition is unknown to the analyst.	Used to test whether the analyst can produce within given performance limitations.
Blind sample	A subsample submitted for analysis with a composition and identity known to the submitter but unknown to the analyst.	Used to test the analyst's or laboratory's proficiency.
Blank sample	A clean sample or a sample of matrix processed to measure artifacts in the measurement process.	Used to estimate contaminants.
Trip blank	A clean sample of matrix that is carried to the sampling site and transported to the laboratory for analysis without being exposed to sampling procedures.	Used to estimate contaminants.
Field blank	A clean sample carried to the sampling site and exposed to the sampling locations.	Used to check for analytical artifacts.
Sampling equipment blank	A clean sample collected in a sample container with the sample collection device and returned to the laboratory as a sample.	Used to check the cleanliness of the devices.

NOTE: QC is quality control.

Table II. Common Definitions of Data Quality Indicators

Data Quality Indicator	*Definition*	*Typical Expression*	*Typical Method of Assessment*
Precision	A measure of mutual agreement among individual measurements of the same property, usually under prescribed similar conditions.	Standard deviation, σ	Relative percent difference or relative standard deviation of measurement data
Bias	The systematic or persistent distortion of a measurement process that results in measurements consistently higher (or lower) than the "true" value.	Percent recovery; deviation from reference standard	Spiked sample recovery; performance evaluation sample result.
Representativeness	The degree to which data accurately reflect a characteristic of a population, the parameter variations at a sampling point, a process condition, or an environmental condition.		Scientific and statistical evaluation of sampling objectives, population of interest, sampling design, and associated sampling methods.
Completeness	A measure of the amount of valid data obtained from a measurement system compared to the amount that was expected to be obtained under normal conditions.	Percent	$\frac{(100 \times \text{number of valid observations})}{(\text{number of planned observations})}$
Comparability	The degree to which different methods (protocols), data sets, or decisions agree or can be represented as similar. It is a measure of the confidence with which one data set can be compared to another.		Scientific and statistical evaluation to determine whether sampling objectives, measurement methods, and data precision are sufficiently equivalent to allow valid comparison.

NOTE: Representativeness and comparability are typically expressed qualitatively.

decision making if it goes undetected or uncorrected. Bias can be estimated using data from performance evaluation samples or samples spiked with the analyte of interest.

Precision is the degree of variation among individual measurements of the same property. Precision is usually expressed as the statistical variance of a set of data. The precision of various components of the sampling and measurement process can be estimated using data from split samples and replicate measurements (*see* Table I). Obtaining estimates of the precision of a sampling and measurement system can be affected significantly by the sample collection methods, particularly when the system is dynamic. One can consider, for example, the collocation of sampling probes for measuring air contaminants. Collocated measurements in steady-state systems have been well-studied and used effectively in determining the precision of the measurement system, such as for stack emission measurements for regulatory development or compliance-monitoring purposes. However, a spray-paint booth system involves more dynamic interactions between the sampling method and the target population. During a particular spray-paint exposure study, the air flow patterns changed significantly when the painter entered the booth and began spraying. Even a slight change in the direction of the probe nozzle or movement by the operator changed the air flows and, consequently, the precision estimates, making it very difficult to achieve the objectives of the exposure assessment study. Fortunately, the design had called for taking a large number of paired observations to calculate the precision. Over the entire range of sampling configurations, the average precision turned out to be comparable to that obtained in a steady-state condition.

Representativeness is defined as the degree to which data accurately and precisely represent the frequency distribution of a characteristic in the population. A representative sample is defined in the EPA QAMS glossary (*12*) as a sample taken so as to accurately reflect the variable(s) of interest in the population. Because the true values of the population characteristics can never be known, representativeness cannot be quantified. Instead, representativeness is evaluated in qualitative terms based on scientific and statistical concepts of how well a population can be observed and characterized through sampling. Therefore, the determination of representativeness usually involves a judgment about whether the sampling strategy will adequately capture the population characteristics of interest, along with a statistical evaluation to determine whether the sampling strategy properly incorporates randomization and whether the number of samples taken will be adequate (*11*). The most important part of obtaining representative field samples is to plan all aspects of the field activity—from the statistical design of experiments; to the creation of suitable sampling protocols; to the collection, storage, and transportation of samples; to the suitable tracking of documentation. Samples collected from spatially homogeneous lake water will be more accurately representative of the target population than samples collected from spatially heterogeneous objects, such as debris in a field. For example, one might consider soil samples being collected from a field containing a large stockpile of used automotive tires and other debris. Representative sampling to estimate the extent of soil contamination and to model the possible contamination of groundwater in order to estimate the extent of cleanup will be very challenging. This challenge comes from the need for more specific criteria for representativeness in heterogeneous populations and for more detailed characterization of conditions under which sampling will take place. Barcelona (*13*) compares the difficulties in obtaining a homogeneous and a heterogeneous sample by considering both the representative sampling of dissolved lead in a well-mixed solution in a beaker in the laboratory and the representative sampling for dissolved lead in a variable mixture of reactive aqueous effluent entering a water treatment plant. In the latter case, useful criteria for a representative effluent sample will necessarily include qualifiers for flow rate, process status, time, and perhaps other physical and biological variables.

Comparability expresses the confidence with which one data set can be compared with another. In a large study such as the Chesapeake Bay Monitoring Program, which has extended over the entire Bay area over the period of the past 15 years, the comparability of the data from all the constituent studies will play a prominent role. Comparability extends beyond ensuring that measurement units are similar (e.g., in a metric system). Comparability is a qualitative indicator establishing the equivalency of study objectives, of primary measurement data among programs, and of QA and QC data among the programs. Sampling QA activities that help ensure the comparability of environmental data include careful preparation of the sampling protocols for field sampling activities based on the objectives of the study, and the use of comparable protocols in similar studies forms the basis for sampling QA activities.

Completeness is an expression of the number of sample measurements obtained in relation to the number of sample measurements planned. The significance of completeness is sometimes overemphasized or misunderstood. For example, in many routine applications completeness is calculated using a simplistic formula, such as:

$$\%\ \text{completeness} = \frac{100 \times \text{number of samples analyzed}}{\text{number of samples collected}}$$

A little reflection indicates that the proper approach is to express completeness in terms of the number of valid samples or valid data necessary to satisfy the data user's objectives. For example, the loss of critical samples or data, such as a duplicate sample for a critical measurement or a field QC sample to check the contamination, in a fairly large data set may still provide a satisfactory completeness indicator (e.g., 90% completeness) but may not allow data quality to be estimated. A judicious review of the data in terms of the data user's study objectives is important to determine whether the completeness objective has been attained.

Components of the U.S. EPA Quality System

The wide range of technical issues related to environmental sampling quality assurance explained in the previous discussion are addressed in a comprehensive and integrated manner by the quality systems approach, which ensures that environmental data are of known quality sufficient for their intended use. The EPA, ANSI, and ASQC have coordinated their efforts in establishing a standard framework for a comprehensive environmental data quality system (*14, 15*). This quality system provides the managerial blueprint for effectively specifying QA and QC procedures and performance criteria, implementing the data operations in accordance with the required performance, and assessing the data generation process and the quality of the resulting data. This section describes the EPA quality system, which is based on the ANSI–ASQC consensus standard for environmental quality systems (*14*). The EPA quality system comprises a mission and policy statement and components that are developed and applied at the organizational level and the project level, as shown in Figure 2.

Quality Mission. At the top level, a mission or policy statement establishes the fundamental purpose and authority of the quality system. For the EPA, this mission statement is expressed in EPA Order 5360.1 (*5*). More recently, the EPA has been updating its quality mission and policy statements to conform with its coordination efforts with ANSI and ASQC, as expressed in *EPA Quality Systems Requirements for Environmental Programs* (*15*). The EPA quality mission adopts the quality systems philosophy that each EPA program and regional office must be responsible for

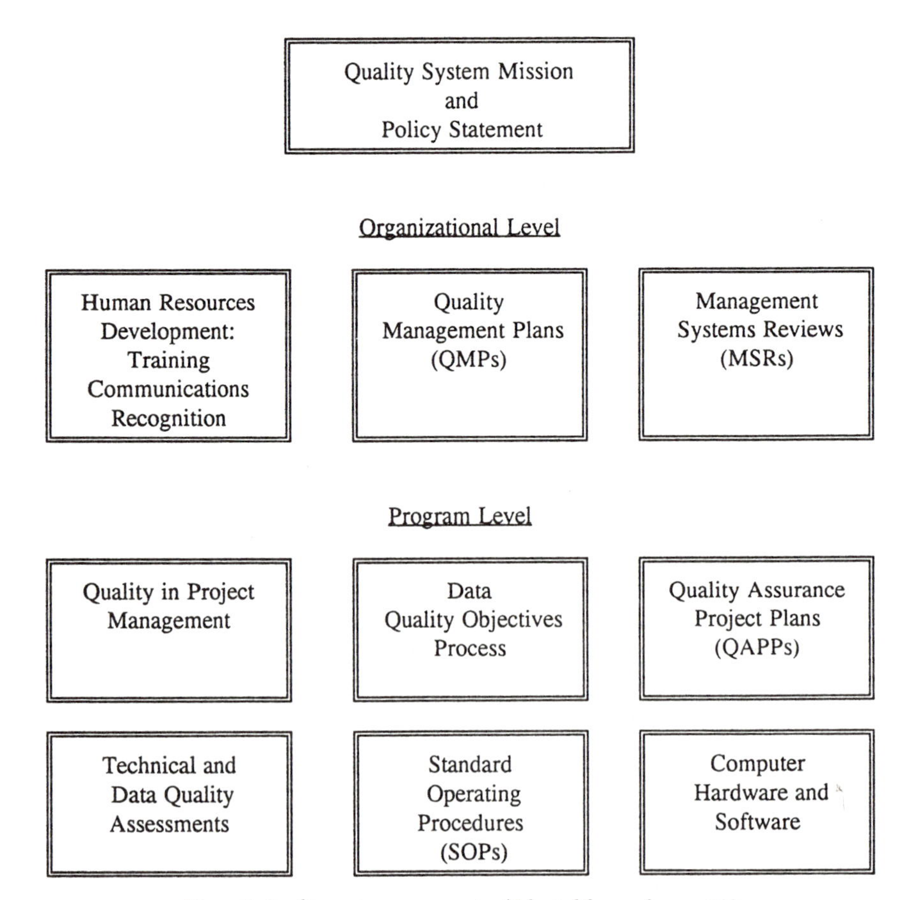

Figure 2. Quality system components. (Adapted from reference 15.)

adapting the quality system framework to their specific needs. The EPA plans to reissue its quality mission document in the form of a quality manual that will codify these principles and replace EPA Order 5360.1.

Organizational-Level Components. At the organizational level, the quality system includes management structures and procedures that sustain the long-term viability of the quality system and ensure that the required program-level components are maintaining their effectiveness. The three organizational-level components are as follows:

1. Human resources development: training programs, communications activities and systems, and recognition policies that are intended to maintain and improve the skills, motivation, and performance of personnel;
2. Quality management plan (QMP): a formal document that describes the organizational structure, functional responsibilities of management and staff, and required interactions for personnel involved in planning, implementing, and assessing data operations (*16*); and
3. Management systems reviews (MSRs): qualitative assessments of an organization's quality management structure, policies, practices, and procedures to determine their adequacy for assuring that the right type and quality of data are obtained (*17*).

Program-Level Components. At the program level, the quality system includes management and technical procedures and tools to ensure that specific data collection projects or monitoring programs work effectively and efficiently. The following six quality system components are employed at the program level:

1. Quality in project management: procedures for assuring quality in the planning and control of project scope, schedule, and budget;
2. Data quality objective (DQO) process: a systematic planning process that defines and documents the type, quality, and quantity of data required to satisfy a specified use;
3. Quality assurance project plan (QAPP): a formal document that describes in detail the specific measurement performance criteria and QA, QC, technical, and managerial oversight activities that must be implemented to ensure that the results of the data operations will satisfy the DQOs;
4. Technical and data quality assessments (DQAs): audits and other evaluations that examine and document the performance of technical systems and data with respect to performance criteria and DQOs specified in the QAPP;

5. Standard operating procedures (SOPs): approved documents that specify detailed methods, techniques, or protocols for performing routine or repetitive data operations or analyses; and
6. Computer hardware and software: performance specifications, procurement and configuration management, software development, and other information technology requirements needed to support the quality system.

This quality system framework is intended to be flexible enough so that each component can be adapted to fit an organization's particular circumstances and needs.

Two Key Quality Management Tools

Of the many quality system components, two quality management tools developed by the EPA are particularly helpful in performing planning and assessment activities at the project level: the DQO process and the DQA process. These processes should form an integral part of environmental sampling QA.

The DQO Process. The EPA has developed the DQO process to facilitate planning of environmental data collection activities (7). The DQO process is a seven-step planning procedure, based on the scientific method, that helps identify the type, quality, and quantity of data needed to satisfy the data user's needs (Figure 3). The outputs of the DQO process are qualitative and quantitative statements about why the data are needed, what the data should represent, and how much uncertainty is tolerable to the data user. The statements (DQOs) are used to develop a sampling design that meets the data user's performance requirements. In particular, the key results of DQO planning are expressed in the decision rule, which summarizes specifically what the data user wants to determine from the study, and the limits on decision errors, which express the data user's performance requirements for the data set as a whole. In the context of environmental decision making, these performance requirements are often expressed as tolerable limits on the probability of making an incorrect decision due to uncertainty in the environmental data.

The DQO process can be applied even when it is not clear what decision will be supported by environmental data. In the context of environmental studies, the objective often is to estimate the value of a parameter that describes some feature of a population. In this case, the data user's performance requirements may be expressed as a desire to encompass the true parameter value within a specified confidence (or probability) interval width with some stated level of confidence.

Translating the statement of an environmental problem into a meaningful set of DQOs can be very simple or very complex. A relatively simple problem might be to determine the effectiveness of a pollutant control technology, such as the use of a new sorbent for emissions control. For this example, the study objective is to estimate the percent reduction of emissions through use of a known amount of the sorbent and to specify the uncertainty in the estimate. An adequate statement of the

1. State the Problem to be Solved

Concisely describe the problem to be solved. Review prior studies and existing information to gain a sufficient understanding to define the problem

2. Identify the Decision to be Made

Identify what questions the study will attempt to resolve, and what actions may result.

3. Identify the Inputs to the Decision

Identify the information that needs to be obtained and the measurements that need to be taken to resolve the decision statement.

4. Define the Study Boundaries

Specify the time periods and spatial area to which decisions will apply. Determine generally when and where data should be collected.

5. Develop a Decision Rule

Define the statistical parameter of interest, specify the action level, and integrate the previous DQO outputs into a single statement that describes the logical basis for choosing among alternative actions.

6. Specify Tolerable Limits on Decision Errors

Define the decision maker's tolerable decision error rates based on a consideration of the consequences of making an incorrect decision.

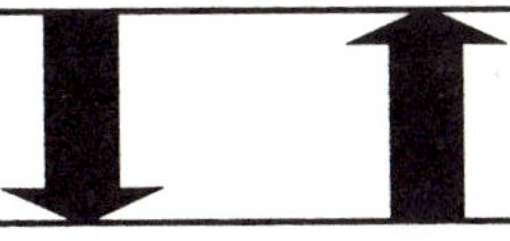

7. Optimize the Design for Obtaining Data

Evaluate information from the previous steps and generate alternative data collection designs. Choose the most resource-effective design that meets all DQOs.

Figure 3. The data quality objectives (DQO) process. (Adapted from reference 7.)

study objective would be to determine the percent removal of pollutant per unit volume within ±5% with 85% confidence under specified conditions. Additional clarifying statements would be included in the full set of DQOs.

A more complex situation would be to design a study like the Chesapeake Bay Monitoring Program, which has an objective of reducing certain types of pollution in the Chesapeake Bay by 40% by 1996. "Reducing Chesapeake Bay pollution levels by 40% by 1996" does not adequately and clearly state the sampling objectives. To characterize the current pollution levels with defined certainty is a difficult proposition because of the large number of variables involved. The Bay water extends from fresh river water to brackish water where it meets the sea, and it absorbs different pollutant-laden effluents, both natural and man-made, over a variety of geographic regions. Characterization of the water body and representativeness of the water samples have to be viewed in light of this information. Statistical theory can be applied to address the representativeness of a water sample in relation to the largely uncharacterized population of potential water samples, through the design of experiments. Because numerous environmental monitoring issues will be addressed in several studies, specific study boundaries can be defined according to the particular needs and conditions of each study. The DQO process helps identify how to break down the large problem into manageable parts. The results of the individual studies must then be combined into an overall composite picture to determine whether the expectation of reducing pollution by a given amount has been met.

The DQA Process. The EPA has developed the DQA process to facilitate the analysis of data to determine whether the data user's objectives (DQOs) have been satisfied (*18*). The DQA process is a five-step procedure that places commonly performed data analysis activities into a logical order. This systematic approach helps organize and focus the data analysis process in a manner consistent with the DQO process. The five steps of the DQA process are described in Figure 4.

The DQA process relies heavily on graphical and statistical methods for analyzing environmental sampling and analysis data. Graphical methods are emphasized so that the overall "picture" or "story" that the data are "telling" can be recognized by the data analyst and communicated to the data user. Statistical methods are emphasized because the presumption of the quality systems approach is that decision making is being strongly supported by science. Statistics provides the tools for making scientifically defensible inferences and estimates from uncertain data.

The DQA process helps ensure that the statistical methods used to analyze the data have a valid connection to the data user's objectives as well as the realities encountered in the sampling effort. In particular, the DQA process verifies the assumptions that must hold for the data user's performance requirements to be met. This verification helps ensure that the inferences and conclusions drawn from the data are truly consistent with the data user's DQOs. If assumptions are violated, perhaps because of some unexpected sampling and analysis results, the DQA process allows for iteration or "backtracking" to determine the best corrective action to take that will satisfy the data user's overall objectives. By keeping the focus on the

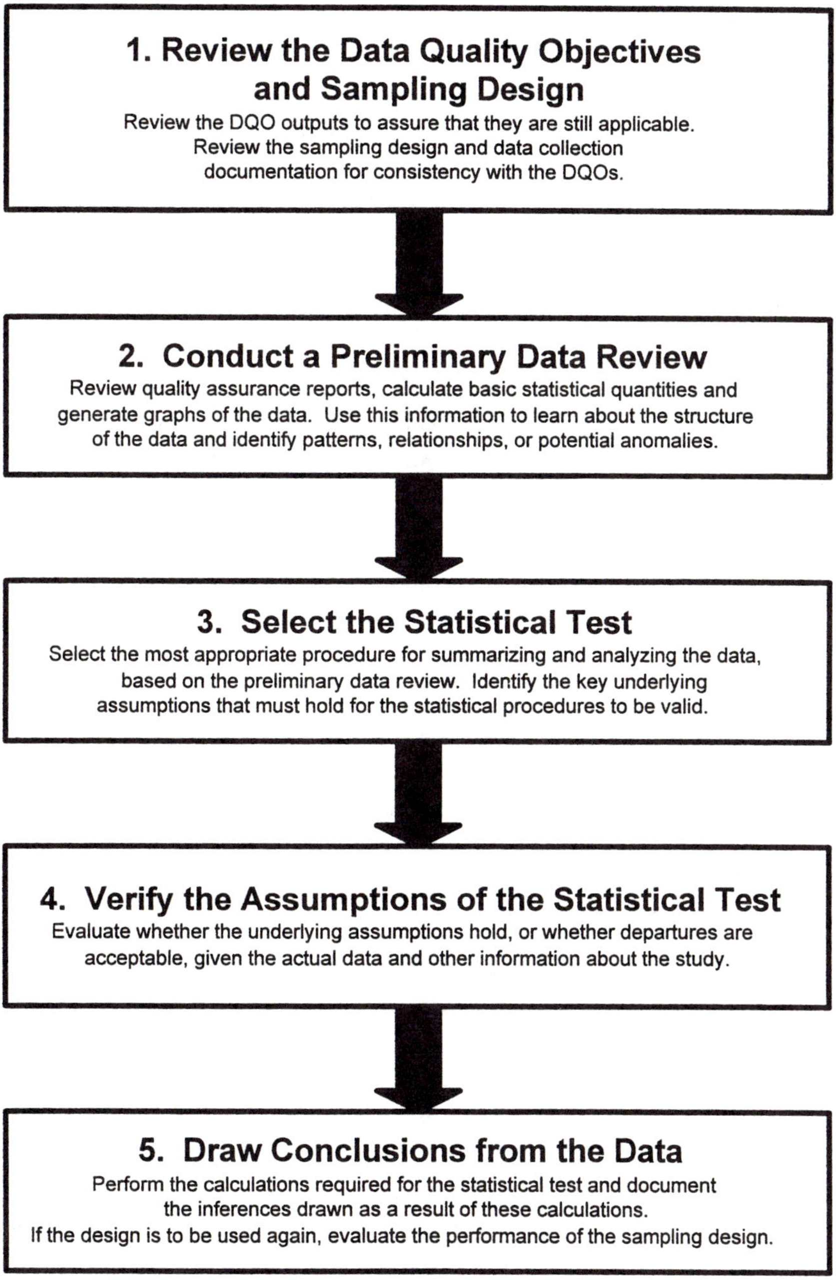

1. Review the Data Quality Objectives and Sampling Design

Review the DQO outputs to assure that they are still applicable.
Review the sampling design and data collection documentation for consistency with the DQOs.

2. Conduct a Preliminary Data Review

Review quality assurance reports, calculate basic statistical quantities and generate graphs of the data. Use this information to learn about the structure of the data and identify patterns, relationships, or potential anomalies.

3. Select the Statistical Test

Select the most appropriate procedure for summarizing and analyzing the data, based on the preliminary data review. Identify the key underlying assumptions that must hold for the statistical procedures to be valid.

4. Verify the Assumptions of the Statistical Test

Evaluate whether the underlying assumptions hold, or whether departures are acceptable, given the actual data and other information about the study.

5. Draw Conclusions from the Data

Perform the calculations required for the statistical test and document the inferences drawn as a result of these calculations.
If the design is to be used again, evaluate the performance of the sampling design.

Figure 4. The data quality assessment (DQA) process. (Adapted from reference 18.)

data user's objectives, the DQA process helps the data analyst prepare the analysis and documentation that will serve the data user's needs.

Key Factors for Successful Environmental QA Practice

Although the picture of an ideal quality system may be clear, the actual state of practice often falls short of the goal. This section addresses some critical factors that often determine the success of a QA program in practice.

Development and Documentation of Standard Protocols. Standard sampling protocols, often in the form of SOPs, provide a consistent and documented source of information. Sampling protocols for various environmental monitoring activities in various media are available (*19–24*) and are extensively discussed in this text and in the first edition of *Principles of Environmental Sampling* (*25*). The selection of proper sampling protocols is an important part of planning for sampling in environmental data collection. Planning tools and products, particularly the DQO process and QAPPs, are cornerstones of the quality system. They provide a sound basis for reviewing the sampling and test plans and QA plans with respect to their feasibility and their ability to yield the desired data. As the external QA management support contractor for a large multicontractor study, we found that major effort had to be dedicated to achieving uniformity in the planning stages so that at the end of the study it would be possible to obtain consistent and comparable information. It was educational for the sponsors and the external QA group to encounter such a diversity in planning in response to common objectives and requirements. Although the effort required to achieve some uniformity was significant, the QAPPs provided an essential vehicle for successfully coordinating the efforts.

Another important tool for ensuring the integrity and quality in sampling and field data collection is to ensure that validated procedures, usually SOPs used by the sampling team, are available and carefully followed, through the use of external and internal audits. Such audits ensure the ability to detect and correct any deviations from prescribed activities or SOPs for sample management, reagent–standard preparation, equipment calibration and maintenance, waste disposal, health, and safety (*26*).

The use of QC sample blanks and controls is important in the sampling program to ensure that the variance in the sampling and analysis methodology is inherent in the methodology and not due to external factors such as contamination (*27*). The definitions and the purpose of blanks and controls typically used in sampling QA are summarized in Table I.

Effective Planning. Sometimes QA fails in the planning phase of a project. Sufficient attention must be given to clarifying the purpose of the data collection effort so as to establish a sound basis for determining whether the quality of the data is sufficient for their intended use. Although the DQO process is an effective planning tool for documenting the rationale for the study and the criteria for evaluating data quality in relation to the decision or study objectives, too many projects currently proceed without first establishing DQOs. Consequently, the sampling design often does not properly satisfy all of the requirements of the data user. If DQOs are developed early in the planning phase, then establishing measurement performance criteria (DQIs) relevant to the data user's needs will be easier. This weakness in implementation does not stem from problems with the planning tools themselves but instead reflects inadequate attention to training and communication aspects of human resource development.

Proper Attention to QA for Field Sampling. Trends in environmental measurement technology are moving toward faster, real-time field measurements that are less expensive. These trends are largely in response to the need to search out "hot spots" of contamination efficiently and to the realization that total error often can be reduced most effectively by taking more samples, given the high natural variability usually found in environmental studies. Unfortunately, the pace of development of these new methods may be rushing ahead of the QA community's ability to establish appropriate QA protocols. Controlled studies of field method performance versus traditional laboratory methods are important steps in the development of field QA protocols, but they are not enough. Practical experience inevitably uncovers problems not anticipated during method development. QA-focused studies are needed that examine sources of variability and bias under real-world field conditions. Such studies, no matter how attractive, are part of the basic QA research and are not likely to be funded by result-oriented projects that use QA for support. Sampling errors from these studies are very likely to be so large that the precision attained in the analysis in the laboratories is not at all productive or cost-effective. Perhaps QA-focused studies are a potential area of research for interested government and standards organizations and the vested method developers.

It is critically important to perform field TSAs, performance evaluation audits (PEAs) of field data collection instrumentation, and audits of the data quality to ensure collection of data as planned with minimum deviations. Currently, this QA tool is underutilized, which exacerbates other weaknesses in field QA. The audit definitions and purposes they serve are summarized in Table III. Although the audits in themselves do not guarantee that all mistakes will be averted and errors will be minimized, they may offer an incentive for the sampling team to follow the protocols carefully. They may also help in data reduction and interpretation if certain abnormalities are observed. During the field audits for the large multicontractor study, it was found that different contractors or different teams from the same contractor offered amazingly different interpretations of the same protocols. Because of the commonality of reviews and audit activities, however, the data obtained were fairly uniform. The function of QA of the auditors was to observe that all the protocols were truthfully followed and to note any deviations. In the complex multicontractor studies that attempted to generate comparable data, the presence of common external auditors seemed to encourage all the teams to follow all the protocols thoroughly so that their results would not constitute an outlier set. Internal audits and reviews would serve a similar purpose.

Table III. Audits Relating to Sampling Data Collection

Audit Type	*Definition and Purpose*
Audit or review	A tool for assessing the quality of measurement activity and environmental data. It is a systematic evaluation to determine compliance with operational plans such as sampling plans, analysis plans, QA program and project plans, and plans for data reduction and reporting. A management review, however, includes the qualitative assessment of management of operational functions.
Technical systems audit	The most frequently performed audit; a qualitative on-site evaluation of measurement and data collection programs. A technical systems audit is an assessment of all facilities, equipment, personnel training programs, operation and maintenance procedures, calibration procedures, sampling and analysis procedures, QC procedures, recordkeeping procedures, and all other activities that may affect data quality. An internal or external technical systems audit conducted near the beginning of the program may indicate deficiencies and suggest corrective action that will conserve resources and enable the organization to meet its data quality goals.
Performance evaluation audit	A quantitative evaluation of the performance of a measurement system, conducted through analysis of independent reference materials of known value or composition. It is important that the value or composition of the material be certified, or at least verified, and be adequately documented. Usually the identity of the material is disguised so that the operator or analyst will treat it as a test program sample.
Audit of data quality	An assessment of the methods used to collect, reduce, interpret, characterize, and report a project's data and results. This data quality assessment is a detailed review of data recording and transfer, data calculation, documentation procedures, and attainment of data quality goals. Because of the cost and time limitations, the QA auditor may need to use a statistical approach to select the appropriate number of data sets to be evaluated.

NOTES: QA is quality assurance. Audits are often referred to as reviews.

Data Quality Assessment and Reporting. Data quality assessment and reporting are closely related and critically important steps in a successful QA program. DQA determines if the data quality satisfies the DQOs for a specific application. If the data assessment indicates that the data meet all of the quality objectives, then the decision maker may not be concerned with access to the QA data because the decision process can proceed as planned. If the assessment indicates that the data quality is totally unacceptable, then the decision maker may decide not to use the data and may either collect new data or find some other basis for decision making. Most data packages, however, fall somewhere in between these extremes. Data sets usually are not totally acceptable without certain restrictions. These restrictions, flags, or partial deficiencies may still allow decision makers to use the data with full awareness of the data quality shortcomings and the decision risk. Unfortunately, data users or decision makers rarely are trained in how to use partly deficient or flagged data.

Perhaps the most easily remedied weakness is data reporting. Ironically, improvement in this area would greatly improve the quality of decisions supported by environmental data. Key issues in data reporting include censoring, placement of QA information in project reports, and reporting of data uncertainty.

Censoring data is a significant problem. Usually the field sampling data will go to the analytical laboratories and become an integral part of the data package. Given the sequence of data collection events, the laboratories become de facto reviewers of sampling data and often inappropriately decide whether to use samples or not. In our experience, far too often when flags or problems are noted in the field sampling data (perhaps detected by use of blanks or QC samples described in Table I), the flags are not included in the analytical results or reports and rarely are discussed in the body of the report. Undocumented censoring, whether intentional or not, should not occur, especially in the laboratory.

When, where, and how to present the sampling data and the QA evaluation information should be a matter of concern. Usually, information on the QA evaluation or on the QA activities is presented as a separate report, as an appendix, or as a small isolated section in the report, as if to avoid attracting the reader's attention. An unfortunate side effect of this practice is that if the QA activities assisted in making the project successful, no record exists of its achievement, and an opportunity for the information to be used in future projects is lost. A candid reporting of the detection of failures through QA activities and subsequent corrective actions would form a valuable resource in planning the next phase of a program or similar project. Unfortunately, current reporting practices for QA–QC results do not insist on highlighting the strengths and weaknesses of data quality. For example, in a review of various final project reports published by NIST, the QA assessment information, particularly sampling QA information, rarely was included as part of the main body of the report.

One of the proposed solutions for presenting of QA data is to ensure that all data in a report or database contain estimates of uncertainty. These estimates could be in the form of probabilities of false-positive or false-negative decision errors in a hypothesis testing context, a confidence interval around the estimated population parameter in a parameter estimation context, or a probability interval for individual measurements when the intended use of the data has not been specifically established. The uncertainty estimates require that at least precision and bias estimates accompany the final results of research or monitoring studies. Thus, the uncertainty information needs to be documented where the project's final results appear, either in hard-copy form or in a computerized database.

Conclusions

The discipline of environmental data QA has matured, and a comprehensive standard framework for a quality system now exists. The quality system, however, needs to be viewed as a dynamic, evolving enterprise, consistent with the principle of continuous improvement. Emerging trends in environmental regulations and the technologies by which data are collected and analyzed will continue to spur change in QA protocols. Training and communication needs and institutional barriers should be evaluated and addressed to ensure that quality is actively managed, not relegated to an afterthought.

Perhaps the most glaring quality management deficiency today is the failure to ensure that QA information is properly reported and integrated into the decision-making process. Establishing standard formats for exchanging electronic data and integrating QA information into project reports are certainly positive steps toward overcoming these barriers.

Acknowledgments

The authors gratefully acknowledge the assistance of Elisabeth Oechsli, Kathryn A. Restivo, and Karin L. Johnson in providing editorial support.

References

1. Johnson, G.; Haeberer, A. F.; Warren, J. Presented at the U.S. Environmental Protection Agency's 13th Annual National Meeting on Managing Environmental Data Quality, San Francisco, CA, February 1993.
2. American Chemical Society Committee on Environmental Improvement. Subcommittee on Environmental Analytical Chemistry. *Anal. Chem.* **1980**, *52*, 2242–2249.
3. Stanley, T. W.; Verner, S. S. In *Quality Assurance for Environmental Measurements;* Stanley, T. W.; Taylor, J. K., Eds.; American Society for Testing Materials: Philadelphia, PA, 1983.
4. Taylor, J. K. In *Quality Assurance for Environmental Measurements;* Stanley, T. W.; Taylor, J. K., Eds.; American Society for Testing Materials: Philadelphia, PA, 1985; p 5.
5. *EPA Order 5360.1: Policy and Program Requirements to Implement the Mandatory Quality Assurance Program;* U.S. Environmental Protection Agency: Washington, DC, 1984.
6. Brossman, M. B. Presented at the 14th Quality Management Conference, Las Vegas, NV, April 1994.
7. *Guidance for the Data Quality Objective Process;* U.S. Environmental Protection Agency. Quality Assurance Management Staff: Washington, DC, 1994; EPA QA/G–4 (final).
8. Hileman, B. *Chem. Eng. News* **1995**, *May 29*, 34.
9. Kulkarni, S. V.; Haeberer, A. F. Presented at the 201st National Meeting of the American Chemical Society, Atlanta, GA, April 1991; paper CEI 3.
10. Taylor, J. K. In *Quality Assurance of Chemical Measurements;* Lewis Publishers: Chelsea, MI, 1987; p 9.
11. Smith, F.; Kulkarni, S.; Myers, L. E.; Messner, M. J. In *Principles of Environmental Sampling;* Keith, L. H., Ed.; ACS Professional Reference Book; American Chemical Society: Washington, DC, 1988; pp 157–168.
12. *Glossary of Quality Assurance Terms;* U.S. Environmental Protection Agency. Quality Assurance Management Staff: Washington, DC, 1993.
13. Barcelona, M. J. In *Principles of Environmental Sampling;* Keith, L. H., Ed.; ACS Professional Reference Book; American Chemical Society: Washington, DC, 1988; pp 3–23.

14. *Specifications and Guidelines for Quality Systems for Environmental Data Collection and Environmental Programs;* American Society for Quality Control: Milwaukee, WI, 1994; ANSI/ASQC E–4.
15. *EPA Quality Systems Requirements for Environmental Programs;* U.S. Environmental Protection Agency. Quality Assurance Management Staff: Washington, DC, 1993; EPA QA/R–1 (interim).
16. *EPA Requirements for Quality Management Plans;* U.S. Environmental Protection Agency. Quality Assurance Management Staff: Washington, DC, 1994; EPA QA/R–2 (interim).
17. *Guidance for Preparing, Conducting, and Reporting the Results of Management Systems Reviews;* U.S. Environmental Protection Agency. Quality Assurance Management Staff: Washington, DC, 1993; EPA QA/G–3 (interim).
18. *Guidance for Data Quality Assessment;* U.S. Environmental Protection Agency. Quality Assurance Management Staff: Washington, DC, 1994; EPA QA/G–9 (draft).
19. *Test Methods for Evaluating Solid Waste, 3rd ed.;* U.S. Environmental Protection Agency: Washington, DC, 1986; Chapter 9; SW–846.
20. *Characterization of Hazardous Waste Sites—A Methods Manual: Vol. II. Available Sampling Methods;* U.S. Environmental Protection Agency. Environmental Monitoring Systems Laboratory: Las Vegas, NV, 1985; EPA–600/4–84–075.
21. *A Compendium of Superfund Field Operations Methods;* U.S. Environmental Protection Agency: Washington, DC, 1987; EPA/540/P–87/001; NTIS PB88–181557.
22. *Radioactive Hazardous Mixed Waste Sampling and Analysis: Addendum to SW–846;* U.S. Department of Energy: Washington, DC, 1989.
23. *Compendium of Methods for the Determination of Toxic Organic Compounds in Ambient Air;*. U.S. Environmental Protection Agency. Environmental Monitoring Systems Laboratory: Research Triangle Park, NC, 1984; EPA/600/4–77–27a; Supplement: EPA–600–4–87–006, 1986.
24. *Handbook for Sampling and Sample Preservation of Water and Wastewater;* U.S. Environmental Protection Agency: Washington, DC, 1982; EPA–600/4–82–029.
25. *Principles of Environmental Sampling;* Keith, L. H., Ed.; ACS Professional Reference Book; American Chemical Society: Washington, DC, 1988.
26. Miller, M. S.; Tait, S. R.; Beiro, H.; Forsberg, D.; Carlberg, K. A. *Proceedings of the Fifth Annual Waste Testing and Quality Assurance Symposium;* U.S. Environmental Protection Agency: Washington, DC, 1989.
27. Black, S. C. In *Principles of Environmental Sampling;* Keith, L. H., Ed.; ACS Professional Reference Book; American Chemical Society: Washington, DC, 1988; pp 109–117.

Chapter 7

Defining Control Site and Blank Sample Needs

S. C. Black

Control sites and blank samples are essential components for any monitoring, surveillance, or analytical program. Correct use of blanks and quality control samples ensures that the quality of analytical results is well defined, and correct choice of control sites is necessary for setting background or normal levels of chemical constituents of concern. Definition of these items and guidelines for their selection are described. Some examples of their use in actual situations are presented.

THE ANALYSIS OF ENVIRONMENTAL SAMPLES for chemical constituents frequently involves detection of concentrations that are near the detection limit of the analytical methodology. To obtain quality data when levels are low requires a suite of quality control (QC) samples that check for contamination, allow background correction, and verify the calibration of the measurement instrumentation. These QC samples include the control and blank samples and control sites discussed in this chapter.

Controls and blanks are essential components for experimental studies as well as for sampling and analytical programs because firm conclusions cannot be drawn from such activities unless adequate controls or blanks have been included along with other essential QC measures that define the quality of the data. In some cases, poor data quality may be useful or may provide sufficient information for making a decision. In other cases, the data quality must be high. The process of deciding on data quality requires setting data quality objectives (DQOs) as described by the U.S. Environmental Protection Agency (EPA) (*1*). In one sense, *blanks* and *controls* are two names for a single process because they both have an important function in QC (i.e., defining background levels), and the name used depends on the purpose of the study. As with detection limits (2), for which the names and definitions have not been standardized, various organizations use a variety of descriptive names for the blanks. In another sense, *blank* has a specific meaning related to analytical procedures, and *control* relates to a specific type of sample, from the environment or from a population, against which the results of a procedure are judged. Therefore, the control could be described as a blank. In this chapter the common meanings of these terms are discussed, guidance for selection and use of blanks and controls is outlined, and appropriate examples are given. Blank sample selection will be discussed first, followed by controls, including analytical controls and control sites.

In many procedures, only a few blanks may be required; for example the solid waste analytical procedures recommended by EPA in the SW-846 manual (3) suggest three blanks, but other organizations suggest additional blanks for determining sources of contamination. For completeness, the listing that follows includes all of the types of blanks that have been described in the literature available to the author. After the listing, some considerations useful for the selection of appropriate blank samples will be discussed.

Blank Selection

Blanks may be defined as samples of analyte-free media (3) or as samples of media expected to have negligible or unmeasurable amounts of the substance(s) for which an analysis is being performed (4). They are necessary for measuring the uncertainty in a method that is due to random errors and, so, are used to set detection limits. They are also used to detect unwanted contamination in the various steps of a sampling and analysis program. In any process, the random errors are those errors that can be estimated by the use of standard statistical techniques and are expressed as the precision of a method. The other kind of error, which affects the total uncertainty of a process, is called systematic error. Systematic error cannot be estimated statistically and usually results in a consistent deviation (bias) in a final result. Discussions on the definition and use of the most common blanks are contained in references 4–9.

The random errors in a sampling and analytical procedure affect the precision of the methodology because precision is estimated by the standard deviations of those errors. Also, because the standard deviation is the square root of variance, and variances (s^2) can be summed, total precision can be estimated by summing the variance of each of those steps of the procedure when it can be determined. The calculation of the variance for each step can serve as a guide for selection of blank samples (e.g., choosing those with the largest effect on the final sum), and the square root of the sum yields the precision of the sampling and analytical method. The various errors that may occur in such a procedure are shown in eq 1

$$s_T^2 = s_a^2 + s_t^2 + s_s^2 + s_h^2 + s_p^2 + s_c^2 + s_m^2 \tag{1}$$

where the subscripts identify the variances as follows: T is total, a is spatial, t is temporal, s is sampling, h is transport and storage, p is preparation, c is chemical treatment, and m is measurement.

Except for the first two terms on the right side of eq 1, a blank can be devised for each variance to ensure that the variance is inherent in the methodology and is not due to contamination. These blanks have specific names to identify which part of a process is being characterized (6). A method blank is one of the most common types, and nearly every analytical procedure includes one, even though several other blanks can be used in analytical processes. Although the blanks used in sample collection and preparation have important functions, the method blank may be the most important because it can be used to estimate the lowest concentration of an analyte that can be measured, and because it also may affect the accuracy of a measurement in those instances when a blank correction is necessary (9).

The order of appearance and use of certain blanks in a sample collection and analysis program is described in this section. These blanks generally consist of distilled and deionized (DDI) water, type II reagent water, or organic-free water (obtained by passing tap water through a filter of one pound of activated carbon), although other blanks may be solvents or may be based on the matrix that contains the analyte of interest, such as soil, sludge, or sediment. Methylene chloride, for example, can be used in certain organic analysis procedures and can be either a matrix blank or a reagent blank. Hereafter, unless otherwise specified, the term "analyte-free media" will be used as a generic term to represent the various blank matrices.

To illustrate the selection and use of blanks, we will assume that a new sampling and analyzing procedure is being developed for determining the concentration of a newly recognized toxic organic compound in soil. The compound is known to be in the base/neutral group for extraction. The procedure is to draw a 2-m circle on the ground, take four cores (1 × 6 in.) equidistant on the circle and one core in the center, and combine the cores in a container that goes to a sample preparation area. At the sample preparation area the combined cores are placed in a ball mill that is operated for 1 h to thoroughly mix the samples. A suitable subsample is taken from the mixed cores and then replaced in the container for transfer to the laboratory. In the laboratory the sample is extracted, the extract is transferred to methylene chloride, and the solution is injected into a gas chromatograph–mass spectrometer (GC–MS) system for analysis.

In this procedure the sample can become contaminated from the coring tool, from the sampling environment, from the ball mill, from the spoon used to subsample, from the storage–transport container, from storage alongside other samples, from the reagents used for extraction, and from the injection device. The possible sources of contamination and the assignment of such sources to various blanks are illustrated in Table I. A prudent process when such a new procedure is first operated would be to collect a complete set of blanks to check all of these possible contaminant sources. After three to five repetitions, the blanks that do not contribute signif-

Table I. Information Derived from Blank Samples

	Source of Sample Contamination							
QC Blank Type	*Container*	*Sampling Equipment*	*Preservative*	*Transport and Storage*	*Preparation Equipment*	*Cross-contamination*	*Reagents*	*Analytical Instrument*
Rinsate	X	X		X		X		
Field	X		X	X		X		
Trip	X			X		X		
Preparation					X		X	X
Instrument								X
Reagent							X	X
Method					X		X	X

NOTE: QC is quality control.

icant errors to the analysis can be identified, and the number of blanks can then be reduced to, perhaps, a field blank and a method blank.

As a general rule, glass containers are used for samples in methods of organic analysis and plastic containers are used to hold samples for inorganic analysis.

Trip Blank

The *trip blank* (or container blank) is made up of analyte-free media used to fill a sample container; the container is then sealed, carried to the sampling site, returned to the laboratory unopened, and analyzed at the same time as the samples for which it serves as a blank (*3–5, 10*). This blank is primarily used for sampling of volatile organic materials to give an estimate of cross-contamination. For other types of analysis, it yields a measure of the amount of analyte contributed to a sample by the container. Generally, there should be one trip blank per day per medium sampled.

Equipment Rinsate Blank

After any sample-collecting equipment (e.g., spade, auger, siphon, or pitcher) has been used, it must be cleaned, generally by a standard operating procedure (SOP) developed for the sampling program. Following the cleaning, analyte-free media is poured into or over the collecting equipment and into a sample container and then returned to the laboratory for analysis (*3, 4, 10*). The purpose of this rinsate blank or equipment blank is to measure the effectiveness of the cleaning procedure in order to ensure that residue from the collection of a sample is not contaminating a subsequent sample. There should be one rinsate blank per set of samples.

Field Blank

The *field blank* is a sample container filled with analyte-free media that is covered, carried to the sampling site, and uncovered during sampling; after preservative is added, it is returned to the laboratory for analysis (*6*). Because the field blank not only determines the effect of the environment on the sample but also the contribution of the container and the preservative, if any, to sample contamination, it can be used for both purposes. A container blank should be tested separately only if contamination is found in the field blank. As a general rule, one field blank should be prepared per sampling team per day and per sample type.

Preparation Blank

The *preparation blank* (sometimes referred to as a sample preparation blank or sample bank blank) is required when such methods as stirring, mixing, blending, or subsampling are part of the procedure used to prepare the sample for analysis (*3, 8*). After a sample is processed in the apparatus (e.g., whiffle-splitter or blender), the apparatus is cleaned according to the SOP. After cleaning, the apparatus is rinsed with analyte-free media or the sample solvent, which then is analyzed along with samples. One of these blanks should be prepared for each shift or each day of operation.

Matrix Blank

The *matrix blank* is used to determine the presence of the analyte in the matrix when the sample matrix is not water but is instead, for example, an organic solvent, as may be encountered in a barrel of liquid wastes, the soil in samples that are being analyzed for Resource Conservation and Recovery Act (RCRA) constituents, or the particulate filter used in an air sampling system. The frequency of analysis should be equivalent to that for a method blank. In general, the matrix blank would only be analyzed if there were problems in the analysis not corrected by the various blank analyses. The matrix blank would then be used to determine whether the matrix was responsible for the problems in recovery or precision.

Calibration Blank

The *calibration blank* (or instrument blank) is used to estimate the background level of the analyte of interest in the analytical instrument. It consists of a volume of analyte-free media that is equivalent to the volume of sample that will be analyzed, and includes the same volume of acids as in a sample ready for analysis. This blank is introduced in the instrument and analyzed in the same manner as a sample. The result can be used to set the limit of detection for analysis because the limit of detection in many cases is determined by the "noise" of the instrument in the analytical region of interest; the result is generally used, however, for beginning the calibration curve for metals analyses (*3*). This blank is used each day for instrument calibration. Also, if analyzed after a batch of samples, it will check instrument contamination and any memory effects.

Method Blank

The *method blank* (or reagent blank) is one of the most important in any process. In general, the method blank consists of analyte-free media to which all reagents are added in the same volumes or proportions used in sample processing; it is carried through the complete sample preparation and analytical procedure (*3, 4, 8*). This blank detects contamination throughout the procedure, including sample preparation, reagents, and instrument. Several of these can be analyzed to determine the lower limit of detection for the analytical procedure, and then one for each 20 samples analyzed as a batch. The preferred outcome from analysis of method blanks is a less-than-detectable concentration of the analyte of interest. If one or more reagents are replaced with new stock, and the analysis is run, this sample becomes a reagent blank.

Case Studies of Blank Selection and Use

Many of the previously described blanks were used in various EPA studies in which the author was involved, namely, the Dallas Pb study (*11*), the Palmerton Zn study (*12*) and the Love Canal Monitoring Program (*13*). Included in the Dallas study was determination of the distribution of Pb in soil around a lead smelter and in a control area. The Palmerton Zn study was designed to determine the extent of Cd and Zn contamination around smelters in Palmerton, PA, which was on the National Priorities List for cleanup. The Love Canal study included monitoring of organic materials in air, water, soil, vegetation, and animals in the Love Canal area and in a control area.

In the Dallas Pb study, a field blank (used also as a rinsate blank), a sample bank blank (a combination of field, rinsate, and preparation blanks), an instrument blank, and a method blank were used. These blanks were prepared as indicated in preceding sections. For example, the field blank was DDI water used to rinse the soil collection equipment, the sample bank blank was a field blank used to rinse the ball mill and sieve that were used to homogenize the soil samples, the instrument blank was DDI water analyzed by atomic absorption spectroscopy (AAS), and the method blank was DDI water with added nitric acid from the stock that was used to extract Pb from the samples.

None of the 76 field blanks, 77 sample bank blanks, 148 instrument blanks, or 148 method blanks had a Pb concentration that exceeded the 0.25-μg/mL detection limit of the AAS procedure that was used. The results from these blank analyses suggested that the equipment cleaning procedures were adequate and that the reagent had no measurable Pb so that the soil samples were the only source of detectable Pb measured by the AAS.

In the Palmerton Zn study, the field blank was a sample container (polyethylene twirl-pack) filled with DDI water, carried to the sampling site, and sent, unopened, to the sample bank (properly this blank was a trip blank). Another blank was called the decontamination blank (rinsate blank), a sample container filled with DDI water used to rinse the soil-collecting device after cleaning. The sample bank blank was DDI water passed through the mixing equipment and the sieve after cleaning as specified by the SOP. The analytical procedure (AAS) included an instrument blank and a reagent blank or method blank (the HNO_3 and HCl used for extraction of metals from soil samples).

Recommendation

In many analytical methods, analysis of samples can be a very expensive procedure. A full suite of blanks adds to the expense, but they may be required for defining the quality of the sampling and analysis. However, collection and preparation of blanks is easy and comparatively inexpensive if done during the time of initial sampling and analysis. A prudent procedure to follow, then, would be to collect all necessary blanks prescribed by the SOP but analyze only the field blank and sample preparation blank (or the combination that is called the preparation blank herein) and the method blank. A calibration blank should be run just after calibration of the instrument. The other blanks would be stored and analyzed only if contamination of samples is detected. In volatile organic analysis, promptness may be necessary because of short permissible holding times. Method blanks would detect reagent and laboratory contamination, and when reagents are changed a few extra would be run. Periodic analysis of calibration blanks would track background and any buildup of contamination in the instrument. The other blanks would be available to aid in identification of contamination sources should a problem occur; otherwise they can be discarded at project end.

As a general rule, 5–10% of the analyses should be analysis of blanks. A good practice is to analyze one field and one method blank with each operating shift or with each 20 samples, whichever is more frequent. If target analytes appear in the blanks, then an increase in number and type of blanks may be necessary to determine the source and quantitate the value. Reference 7 suggests that 7–10 analytical QC samples may be required for runs of up to 20 samples, but this number includes duplicates, spikes, and calibration check samples. In any case, a lack of a fixed or prescribed number of QC samples per batch of analyzed samples should not invalidate the results without careful consideration of the circumstances under which the results were produced.

In those cases when an extraction device is used rather than collection of the sample itself [for instance, air filters, charcoal cartridges, ion exchangers, or a polymeric form of 2,6-diphenylene oxide (Tenax)], the field, storage, transport, and sample preparation blanks should be the extraction device. For particulate air sampling, as an example, an unused filter should be unsealed when the air sampling filter is collected and then treated as if it had been used for sampling. If the analyte were organic air pollutants, the sorbent would not be unsealed in the field. Also, when the analyte of interest is extracted from the filter by use of a solvent or by other extractive means, the unused filter should receive the same treatment. For many analytes, these extraction devices will give positive results on analysis so that a blank correction is necessary.

Discussion

The enactment of environmental regulations has forced analytical laboratories to pay increased attention to the quality assurance aspects of environmental sampling and analysis. Blank samples have received much consideration as a result. Although radioanalytical laboratories have used many of these QC samples routinely, a program analogous to the EPA's Contract Laboratory Program (CLP) has not been required. However, because of legal concerns, a program for analyses of radioactivity that is similar to EPA's CLP has been developed by EG&G/Rocky Flats (a Department of Energy contract corporation in Colorado). This program, the General Radiochemistry and Routine Analytical Services Protocol (GRRASP), specifies the work that must be done for each analysis before it will be accepted by EG&G.

GRRASP has 16 stated requirements. As with the CLP, the QC includes one laboratory control standard, one duplicate, and one blank for every 20 samples. The other GRRASP requirements are for the following: inspection criteria for incoming samples; chain of custody; sample turnaround time; minimum detectable activity; document control procedures; preparing and maintaining SOPs; procedures the SOPs are to cover; instrumentation specifications; instrument calibration; sample holding time; sample storage; QC procedures; chemical processing; chemical recovery; and data reporting.

Interestingly, EPA's CLP, used for volatile and semivolatile organic analyses, requires only a method blank; if any peaks show up in the blank analysis, the presence of the peaks is noted but blank subtraction is not done, contrary to GRRASP. It is unlikely of course in the organic analysis methods that, except for some plasticizers (e.g., phthalates), the blanks will contain measurable quantities of organic materials, so blank subtraction would be of little use.

Control Selection

Controls are basically of two types: those used in QC procedures to determine whether the analytical procedure is in control, and those used to determine whether a factor of interest is present in the population under study but not present in a control population. The population referenced here may be a group of environmental samples, a defined group of people, or any similar groups of things. The two types of controls are discussed in this section, and some suggestions for selection and use are provided.

In determining whether an analytical procedure is in control (i.e., that the procedure produces data of known quality), several QC techniques are needed. Among these are duplicate analyses (to determine precision), blank analyses (to determine contamination), and various types of standard samples (to determine bias). Blank analyses have already been discussed, and various standard samples will be discussed in the following section. Duplicate analyses are analyses of duplicate or split samples; they are not discussed in this chapter.

Calibration Check Standard or Sample

The *calibration check standard* or sample (CCS) is a type of control that is also called a QC calibration standard, calibration control sample (*8*), or continuing calibration verification (CCV) (*3*). In most laboratory analytical procedures, the CCS is a solution containing the analyte of interest at a low but measurable concentration that is introduced into the measuring instrument directly (i.e., it is not processed through the complete analytical procedure). The precise concentration of the CCS need not be known; the important point is that consistent (precise) results be produced by the analytical procedure. The CCS should be the first sample analyzed after the instrument is calibrated. Seven or more analyses should be done to establish a mean and standard deviation for construction of a control chart. A CCS should then be analyzed after each 20 samples or after each shift if fewer samples are analyzed per shift. The standard deviation of the CCSs is a measure of the instrumental precision unless the matrix containing the analyte is the same as that for the analyzed samples, in which case the standard deviation of the CCSs can be a check on the precision obtained from duplicate sample analyses.

Laboratory Control Standard

The *laboratory control standard* (LCS), as used herein, is a certified standard usually supplied by an outside agency. The LCS is used to determine whether an analytical procedure is producing results comparable to those of other analytical laboratories; it provides a control for any matrix effects. In contrast to the CCS, the LCS is processed through the complete procedure with the results being used to calculate the precision and bias of the method. A good source of LCSs is the National Institute of Standards and Technology (NIST; formerly the National Bureau of Standards), which has a variety of standard reference materials (*3, 14*) that contain certified concentrations of elements or compounds. The Environmental Monitoring and Support Laboratory of the U.S. EPA in Cincinnati, OH, is a useful source for organic or other standards in water or organic solvent matrices, and the Environmental Monitoring Systems Laboratory in Las Vegas, NV, can supply radioactive standards. If a suitable standard is not available, the laboratory should use highly pure reagents to prepare an in-house standard. Spectroscopic-grade reagents, for example, may be suitable for such a purpose. An LCS should be analyzed with every batch of samples until 7–10 results are available for calculation of the bias and precision of the method (*5*). If those results are within the control limits specified by the program protocol, then the frequency may be reduced to one per day. However, several LCSs should be analyzed any time the analytical instrument is recalibrated.

Matrix Control

The common name for *matrix control* is a field spike or matrix spike. For those sample matrices, such as sediments and sludges, in which a complex mixture of chemical and physical materials is liable to cause interferences in the analysis, a field spike may be required to obtain an estimate of the magnitude of those interferences. The losses from transport, storage, treatment, and analysis can be assessed by adding a known amount of the analyte of interest, about twice the level expected to be found in the sample, to the sample immediately after collection in the field or at the sample bank.

Discussion

An LCS with well-known characteristics, for example a sediment Standard Reference Material from NIST, may be too costly or the amount available may be too small to use for calibration checks on a frequent basis, in which case a standard prepared in the laboratory becomes useful. As mentioned, the concentration of the CCS need not be accurately known, unless it is to be used to determine method bias, because a consistent result as determined by control charting the results is all that is necessary for determining whether the measuring instrument is still in calibration. A matrix control (spike) is not needed for most measurements of radioactivity unless an extractive step is used or self-absorbance is expected to be a problem. For organic analyses, this control can be helpful because surrogate standards generally may not be as specific for the analytes of interest as would recovery of specific spikes.

Control Sites

Along with the controls used to measure the bias and precision of sampling and analysis, another very important control is the control site or control population. Just as the various blanks previously discussed serve as a baseline above which the analyte of interest must be detectable, the control site serves as a baseline against which a studied site or population is compared to determine whether a significant difference exists. In other words, if the results of a study of a given area or population are to be judged high, low, or insignificant, then they must be compared to similar results in some other area or population (i.e., the control site or population). For example, if the environmental impact of a given facility, such as a waste disposal site or a coal-fired power plant, is to be assessed, then the environmental levels of pollutants in the absence of the facility must either be known (e.g., because of environmental monitoring prior to construction and operation) or be estimated from analysis of a suitable control site. Also, if the contribution of pollutants from an urban area to the environment is to be assessed, then the contribution of the same pollutants from sources other than the urban area must be known.

In the laboratory, for example, a reagent blank is a control sample for the analyte of interest. But this blank has a very narrow interpretation because it sets the limit of sensitivity for an analytical methodology but provides no information on whether a sample contains more than "normal" amounts of an analyte. Samples from control sites are required to determine the normal analyte concentrations. The sites or populations that can be used as controls can be classified as local, area, national, or *background controls,* depending on the location selected.

Local Control Site. A *local control site* is a control near in space and time to the sample of interest. The reagent blank, for example, is a type of local control. In a groundwater monitoring program, for example, a control site is a well that is upstream of a suspect contaminating facility and provides control samples against which samples from a well downstream of the facility are compared. Suspect facilities (i.e., facilities suspected of contaminating the groundwater in their vicinity) could be either chemical plants using injection wells for waste disposal or commercial waste disposal sites. Factors to be considered in the selection of local control sites include the following:

- Local control sites should be upwind of the facility most of the time. A wind rose, which is a diagram summarizing statistical information about the wind, should be available so that air samplers can be placed upwind. A more accurate sampling arrangement would require taking control air samples only when the wind is blowing from the sampler toward the facility.
- Local control sites should be upgradient from the facility under study with respect to surface water and groundwater flow.
- The potable water source should not be affected by site effluents.
- Travel between the control site and the facility should be minimal because of possible transport of contaminants by vehicles.
- The control site should not be affected by a facility emitting the contaminants identical to those of concern at the facility being studied.

As an example of a local control, the groundwater monitoring strategy of the EPA requires monitoring of one upstream control well. Because groundwater is not normally a simple stream with narrow boundaries, EPA regulations require that three downstream wells be monitored, with the results being compared to the upstream well (*15*). For such a monitoring system to perform adequately, the assumption is made that at no time do the contaminants spread upstream. A much better local control would be intensive study of the background levels in the groundwater prior to construction of the facility under study. Unfortunately, either the background studies for the majority of suspect facilities do not exist, the previous analyses did not include the analytes of interest, or the data were of unknown quality. Another example of a local control is one that is frequently used in case–control studies, a type of epidemiological study. Such studies are used to determine the effects of exposure to certain substances. If an unusual frequency of illness were to occur among the employees of a company, then the cases would be all those affected employees having the same occupation in the facility, and the controls would be those employees in the facility not exposed to the hazardous material. The controls would also be matched on such variables as length of employment, age, sex, and time of beginning employment. A good description of this type of study is included in reference 16.

A less quantitative, but sometimes useful, local control is an implied one, such as used in the Palmerton Zn study. In this study, the distribution of Zn, its magnitude in surface soils around the smelter, and its change with distance in the predominant downwind direction were sufficient to implicate the smelter as the source. The distribution of the analyte–that is, the decreasing concentration with distance from the smelter–was the control, as no other source would produce the results found in the study.

Area Control Site. The *area control site* is in the same area (e.g., a city or county) as the pollution source but not adjacent to that source. The factors to be considered in site selection are similar to those for local control sites. All possible efforts should be made to make the sites identical except for the presence of the pollutant of concern. Among the factors to be considered are commercial activities, manufacturing, ethnicity, population size, income distribution, source of water, and traffic density.

An example of an area control site is the one selected in the Dallas Pb study (*11*). The purpose of that study was to determine whether the secondary Pb smelters in Dallas, TX, were a cause of elevated blood-Pb levels in children living near the smelters or whether the elevated blood Pb levels were due to deposited Pb from automobile exhausts. A control site was chosen in Dallas by using the following criteria: similarity in automotive traffic, similarity in population density, and similarity in ethnic background. Water supply was the same as for the study sites, and food habits were assumed to be similar because of the similarity in ethnic background. Because automotive emissions were a measurable source of Pb at the time of the 1982 study, the similarity in automotive traffic was a highly important criterion.

National Controls. *National controls* can be very useful if comparisons with the locally collected data are carefully performed. In general, national controls tend to be very broad, or less specific, than local or area controls, but they can identify anomalous results. These controls are U.S. average data such as those derived from census data, or from compilations in handbooks, or from various atlases. For example, if the study area data are not much different from the data in a local or area control, then national controls can be used to suggest whether the data from the more local control sites may be anomalous when compared to national data. Some of the factors to consider in selecting national controls are the following:

- Similar data should be selected. (National soil, water, and air monitoring programs as well as some bioassay data are available.)
- Areas should be similar. For example, farm area data should not be compared with data from industrial areas.
- Monitoring data should be upwind and upgradient from any possible sources such as waste disposal areas and smokestack industries.
- If possible, data from several control areas should be chosen.

The principal problem with national controls may be the lack of adequate data, either in the type of analyte reported, in the quality of the data, or in the estimates of uncertainty. Data of poor quality or of unknown quality should not be used other than as indicators for further data collection. If appropriate data are lacking, of course, then the use of national controls may be precluded as an option for a given study because of the time and expense required to conduct an adequate data collection program.

A good example of the use of national controls is in epidemiological studies on cancer incidence or causation. National age-adjusted cancer mortality data are frequently used as the basis on which the health effects of a given exposure to pollutants are estimated (*16*). This type of control should be used with caution, however, because national averages tend to obscure the differences that exist among countywide or even statewide data on cancer mortality.

Background Sites. A true *background site* is one that has not been affected by human activities. It represents a baseline against which studies of local or area contamination can be compared. Background sites may be difficult, if not impossible, to find because of transport processes that spread man-made contaminants throughout the biosphere. Even uninhabited locations such as remote islands, the polar regions, and continental badlands areas have been affected to some extent. For many man-made pollutants, however, carefully chosen areas can provide samples that may be considered to be background samples. For some naturally occurring substances, a background area may be more easily found. A good example is reported in reference 17, in which the background levels of uranium in soil were determined so that the spread of contamination from a uranium-processing facility could be measured.

Comparison Techniques

Usually, simple statistical techniques are sufficient when the results of a study area or group are compared with a suitable control. These techniques may include a Student's *t*-test, either one- or two-sided, or an analysis of variance (ANOVA). If the results of the test are ambiguous, the distribution should be tested for normality, or plotting the two data sets may be useful. Bar graphs are an effective plotting technique because any differences in magnitude or distribution between data sets become obvious. If both the statistical tests and the plotting techniques fail to show a difference, then additional sampling should be considered. Some elementary statistics are included in reference 8, but more complete statistical techniques will be found in the text by Brown and Hollander (*18*) and especially in the book by Gilbert (*19*).

Conclusion

The need for the various blanks and controls discussed herein is driven by the objectives of the study (DQOs), which specify the use of the data and the bias and precision required to meet those objectives. In most cases, these objectives can be derived by a careful study of the experimental design. Wherever in the procedure a possibility exists for introducing extraneous material into a sample collection, treatment, or analysis program, a blank can be devised to assess that possibility and measure the extent of contamination. Furthermore, if the results obtained from a sampling program need to be compared to a background, or to ambient levels, then a suite of control samples may be required to prove the presence of abnormal levels of the analyte. The basic principle for choosing a control site is the similarity of its parameters to those of the study site with the exception of the contaminant of concern.

Acknowledgments

The work described in this chapter was supported, in part, under contract number DE–AC08–94 NV11432 issued by the Nevada Operations Office, U.S. Department of Energy, to Reynolds Electrical and Engineering Company, Inc.

References

1. *Development of Data Quality Objectives, Description of Stages I and II*; U.S. Environmental Protection Agency. Quality Assurance Management Staff. Office of Research and Development. Government Printing Office: Washington, DC, 1986.
2. Klodowski, J. *J. Environ. Regul.* **1993,** 2, 295–306.
3. *Test Methods for Evaluating Solid Waste: Physical and Chemical Methods;* U.S. Environmental Protection Agency. Office of Solid Waste and Emergency Response. Government Printing Office: Washington, DC, 1990; Chapter 1, SW–846, Rev. 2.
4. *DOE Methods for Evaluating Environmental and Waste Management Samples;* U.S. Department of Energy. Government Printing Office: Washington, DC, 1993; DOE/EM–0089T, p 5–4.
5. Taylor, J. K. *Quality Assurance of Chemical Measurements;* Lewis Publishers: Chelsea, MI, 1987.
6. Shearer, S. D. In *Quality Assurance Practices for Health Laboratories;* Inhorn, S. L., Ed.; American Public Health Association: Washington, DC, 1978; pp 367–377.
7. *Upgrading Environmental Radiation Data;* U.S. Environmental Protection Agency. Office of Radiation Programs. Government Printing Office: Washington, DC, 1980; Chapter 5; EPA/520/1–80/012.
8. *Standard Guide for Good Laboratory Practice for Laboratories Engaged in Sampling and Analysis of Water;* American Society for Testing and Materials: Philadelphia, PA, 1991; Designation D3856–88.
9. American Chemical Society Committee on Environmental Improvement. *Anal. Chem.* **1983,** 55, 2210–2218.
10. *Analytical Laboratory Quality Assurance Guide: EM Environmental Sampling and Analysis Activities;* U.S. Department of Energy. Office of Environmental Management. Government Printing Office: Washington, DC, 1994; DOE/EM–0158T and DOE/EM–0159T.
11. Brown, K. W.; Beckert, W. F.; Black, S. C.; Flatman, G. T.; Mullins, J. W.; Richitt, E. P.; Simon, S. J. *The Dallas Lead Monitoring Study; EMSL-LV Contribution*; U.S. Environmental Protection Agency. Environmental Monitoring Systems Laboratory: Las Vegas, NV, 1983; EPA/600/X–83/007.
12. *Documentation of the EMSL-LV Contribution to the Palmerton, PA, Zinc Study;* U.S. Environmental Protection Agency. Environmental Monitoring Systems Laboratory: Las Vegas, NV, 1989; EPA/600/8–89/075.
13. *Environmental Monitoring at Love Canal;* U.S. Environmental Protection Agency. Office of Research and Development. Government Printing Office: Washington, DC, 1982; Vol. 1; EPA/600/4–82/030a.
14. *Standard Reference Materials: Handbook for SRM Users;* U.S. Department of Commerce. National Bureau of Standards. Government Printing Office: Gaithersburg, MD, 1985.
15. U.S. Environmental Protection Agency. *Code of Federal Regulations,* Title 40, Part 165, Subpart F, 1983; pp 506–510.
16. Rogers, E. M. *Pathway to a Healthier Tomorrow (Understanding Epidemiology);* pamphlet, Department of Health and Environmental Sciences. Dow Chemical: Midland, MI, 1980.

17. *1993 Site Environmental Report;* Fernald Environmental Restoration Management Corporation. Environmental Protection Department: Cincinnati, OH, 1994; FEMP–2342.
18. Brown, B. W., Jr.; Hollander, M. *Statistics: A Biomedical Introduction;* John Wiley & Sons: New York, 1977.
19. Gilbert, R. O. *Statistical Methods for Environmental Pollution Monitoring;* Van Nostrand Reinhold: New York, 1987.

Chapter 8

Effect of Equipment on Sample Representativeness

C. E. Koerner

Collected samples must be representative of the population being measured; this representativeness is imperative to ensure that the analytical data and the resulting conclusions are valid. To ensure the sample representativeness and preserve the integrity of the sample, one must challenge all of the assumptions made in the sampling process and examine the effect that the equipment and sampling protocol have on the sample. This chapter presents several factors that may affect sample representativeness and discusses the transfer mechanisms by which chemicals can migrate into or out of the sample matrix. The chapter then reviews the advantages and disadvantages of four primary materials used for environmental sampling.

Representative Samples

The goal of any environmental sampling activity is to collect a representative sample that closely resembles, or is a subset of, the population being measured. A *representative environmental sample* is one that is collected and handled in a manner that preserves its original physical form and chemical composition and that prevents changes in the concentration of the materials to be analyzed or the introduction of outside contamination. The importance of collecting a representative sample cannot be overemphasized, because if a sample is not representative of the larger population being studied, the resulting conclusions will not be valid no matter how well they are analyzed or interpreted.

The collection of a representative environmental sample is not a trivial effort, as can be shown by examination of the many types of variability (spatial, temporal, field, and analytical) that affect representativeness. This situation is particularly true for nonhomogeneous systems. The challenge is made more difficult because no easy methods exist for proving that a sample is representative. Traditional field and laboratory quality control practices are susceptible to reproducible biases and may not be able to detect situations that cause a sample to become nonrepresentative, such as the use of inappropriate field or laboratory equipment or the selective bias toward a particular portion of the population to be sampled (i.e., the sampler or analytical method collects or measures liquids better than solids). Because we know how to measure laboratory data quality, we tend to focus on those measurements and use them as an indicator of the reliability and accuracy of the data. This practice may greatly underestimate the error and allow a false sense of security, potentially leading to serious problems if incorrect conclusions are drawn.

To ensure sample representativeness and preserve the integrity of the sample, one must challenge all of the assumptions made in the sampling process. The person collecting the sample should always examine the effect that equipment and sampling protocol have on the representativeness of the sample. Thorough planning prior to any sampling activity is essential; there is no substitution for careful thinking and a detailed examination of all the equipment and sampling handling methods. Researching the nature of the chemical species of interest will also aid in determining factors that will influence the sampling results.

To collect a representative sample successfully, one must first clearly define the objectives of the sampling exercise. Defining objectives is best done using the data quality objective (DQO) process outlined in the U.S. EPA guidance manual for DQOs (*1*). This guidance manual describes the process for defining the sampling objectives, the quality of data to be collected, and the decisions to be made with the data. Without a clear definition of the sampling objectives, it is unlikely that a representative portion of the sample will be collected.

Prior to collecting a sample, it may be appropriate to determine whether a grab or composite sample is most suitable. The *grab sample,* or discrete sample, is one that is collected at a specific point in time and at a specific location. This type of sample is most analogous to a photograph, which takes a snapshot of the location where the camera was pointed at the moment the shutter release was pressed. Sample results may be different if the sample was collected from a different location or at a different time, just as the photograph will be different if the camera was pointed somewhere else or taken at a different time.

A *composite sample* is a mixture of samples collected from more than one sampling location or more than one sampling time. Composite samples give an "average" sample and may be used as an alternative to a larger number of individual grab samples. Composite samples are generally used in situations where different layers of materials are present (different phases or layers in a drum), under flow conditions where concentrations vary over time (surface water or air sampling), or when large bodies of waste appear to be homogeneous (large piles of similar waste material). It is not usually appropriate, however, to take composite samples for analysis of volatile organic compounds, because the mixing of the samples will cause the volatile compounds to be released. Before sampling, the sampler must also evaluate the effect of compositing samples to avoid diluting the samples below analytical detection limits or otherwise adversely affecting the ability to compare the results against regulatory action levels.

Examining the sensitivity of the media to be sampled is usually necessary prior to determining the type of sample and equipment to be used. Three different media (air or vapor; water; and soil or waste) are usually sampled in environmental investigations. The medium that has the lowest analytical detection limits (i.e., air) usually requires the most rigorous sampling protocol, the cleanest equipment, and the greater attention to detail. Conversely, the dirtier media samples (i.e., soil or waste) usually have a higher analytical detection limit and therefore require a relatively less rigorous sampling protocol.

Equipment Selection

To collect representative samples successfully, one must carefully consider and select the sampling equipment. The selection of equipment should be based on the equipment's ability to function effectively with minimal effect on the structure or concentration of the chemicals of interest. Obviously, the equipment should be constructed of materials that are inert with respect to these chemicals, neither leaching contaminants into the sample nor absorbing contaminants onto the walls of the equipment. Different operators should be able to use the same sampling equipment to reproducibly collect representative samples in order to minimize field variability within the samples. The design of sampling equipment should be kept as simple as possible, and the equipment should be easy to repair or maintain. Equipment should also be easy to decontaminate or be applicable for dedicated use to avoid residual sample adhering to the walls of the sampler (2).

The type of materials selected to construct sampling equipment can have a great effect on the integrity of the samples. Materials must be resistant to chemical reaction and degradation in the sampling environment (i.e., oxidation of metals and swelling or deformation of plastic). Because the physical strength of the material selected must also be suitable for the sampling activity, the sampler must be strong enough to obtain and hold the sample without breaking or deforming. Finally, the cost of the materials to be used in the construction of sampling equipment must be considered. The need for specialty materials may be so prohibitively expensive that it severely restricts the number or types of samples that can be collected. This restriction may be especially true if the sampler is designed so that it can only be used for the collection of a single sample before being disposed.

To understand the effect of equipment materials on sample representativeness, one must understand the mechanisms by which chemicals can be transferred to or from the equipment. The two most common transfer mechanisms are sorption and leaching. *Sorption* refers to either adsorption or absorption, or a combination of the two, and is often used when the specific mechanism is unknown. *Adsorption* is the adherence of contaminants to equipment material or to the micropores on the material surface. *Absorption* is the penetration of one substance into the inner structure of another (i.e., as in a sponge) (3). As the chemicals of concern move out of the sample matrix and adhere to the equipment wall, a false negative or a lower concentration in the sample is produced. If, at a later date, the conditions in the sample change, desorption or removal of the sorbed contaminants can occur, which causes a higher concentration in the sample, or a false positive.

Leaching is the process by which soluble contaminants are dissolved from the matrix of the equipment material by the sample liquid, or a previously sorbed compound is released. Leaching may occur, for example, with an acidic solution removing metals from glass, or organic solvents removing organic compounds from plastic equipment. Leaching may cause false positives or result in increased contaminant levels in the sample. Compounds that interfere with the analytical methods may also be leached, causing either false negative or positive results.

A third factor that affects the representativeness of samples is volatilization, or evaporation of a compound from the liquid, aqueous, or solid phase. The rate of volatilization can be significantly increased by poor sample design and by poor sampling protocol. Severe turbulence or mixing of the sample or introduction of air (as in an airlift pump for groundwater sampling) can cause many of the compounds of interest to volatilize from the sample (either liquid or solid) into the vapor, where they will not be analyzed and therefore not reported. Equipment whose operation involves generation of heat or the application of a vacuum may also significantly contribute to the loss of volatile organic compounds (or mercury). In all cases, the effects of the material used in the design of equipment should be considered to minimize the volatilization of the chemical species of interest.

Another mechanism that affects sample representativeness is the degradation of the compounds in the sample container, the sampling equipment, or storage container. Two examples of degradation are photodegradation and biodegradation. *Photodegradation* is decay by UV light and can cause rapid changes in the concentration of chemicals in sampling equipment, particularly in the air matrix. UV degradation is an effective remediation technique for destroying contaminants in air and aqueous solutions, so it should be equally effective in reducing the concentration of contaminants in the collected sample. Amber glass, 40-mL volatile organic analysis (VOA) vials are used for the collection of groundwater samples to minimize the penetration of UV radiation, and therefore the destruction of volatile organic materials in the sample. Clear glass sampling syringes, glass bombs, or poly(vinyl fluoride) (Tedlar; DuPont) bags may need to be shielded from the UV radiation in sunlight to avoid the reduction of contaminant levels.

Biodegradation, or the decomposition of a substance by microorganisms, can also cause a significant reduction in contaminant concentrations. Groundwater samples for the analysis of aromatic compounds can be affected by the presence of bacteria in the sample matrix. These bacteria use the aromatic compounds as a food source and can reduce the contaminant levels during the time between sample collection and analysis. In order to prevent biodegradation, these samples are preserved with hydrochloric acid to a pH less than 2.

Equipment Materials

Many types of material may be appropriate for sampling equipment. However, because of the limited scope of this chapter, the discussion will be focused on the four primary materials used for environmental sampling: Teflon fluorocarbon resins, poly(vinyl chloride) (PVC), stainless steels, and borosilicate glass.

Teflon Fluorocarbon Resin

Teflon is the registered trademark of the DuPont Company for its fluorocarbon resins. Teflon resins are chemically inert to almost all industrial chemical and solvents; they can be in continuous contact with another substance with little detectable chemical reaction taking place. The inertness of Teflon is due to (1) the very strong interatomic bonds between carbon and fluorine atoms; (2) the shielding of the carbon backbone of the polymer by fluorine atoms; and (3) the very high molecular weight (or long polymer chain length) compared to many other polymers. The two members of the family of Teflon resins that are most commonly used in the environmental industry are poly(tetrafluoroethylene) (PTFE) and fluorinated ethylene–propylene copolymer (FEP) (*4, 5*).

PTFE Teflon resin is a white-to-translucent (opaque) solid polymer made by polymerizing the tetrafluoroethylene (C_2F_4) monomer. It is a highly crystalline polymer with high thermostability. Its heat resistance, chemical inertness, electrical insulation properties, and low coefficient of friction in very wide temperature ranges make PTFE an outstanding plastic. When melted, PTFE does not flow like other thermoplastics, and it must be shaped initially by techniques similar to powder metallurgy. PTFE shows excellent resistance to corrosive agents and dissolution by solvents, with a maximum continuous service temperature of 260 °C (500 °F). However, the high cost of monomer preparation, purification, polymerization, and fabrication may cause the final PTFE product to be relatively expensive (*4, 5*).

FEP Teflon is a true thermoplastic that can be melt extruded and fabricated by conventional methods. FEP has a glossy surface and is transparent in thin sections, eventually becoming translucent as thickness increases. FEP has a maximum continuous service temperature of 205 °C (400 °F) (*5*).

Plastic materials are often used in preference to glass and metal because of their durability, flexibility, low weight, and relatively low price. Although plastic materials are often used in sampling equipment, plastics are not gastight (i.e., gases are able to diffuse through the walls of tubing and containers made of plastic). Permeation and diffusion occur in both directions, allowing oxygen into the sample as well as allowing chemicals of interest to escape. This limitation is particularly true for silicon, in which severe diffusion is observed for all gases. Therefore, when analysis for soluble gases in water is performed, a silicon-only seal or washer is not appropriate because the silicon will allow significant diffusion, even through the exposed seal area is only two percent of the surface area of a glass bottle (*6*). The low diffusion rate for Teflon makes it valuable as a liner for sample bottles or containers, often in combination with silicon, as in the septa of 40-mL VOA vials. However, no polymers, including Teflon, are gastight.

PVC

PVC is a linear chain compound produced by the polymerization of the vinyl chloride monomer. Rigid vinyl materials are primarily made up of high-molecular-weight vinyl chloride polymers and are unmodified by plasticizers or similar materials; the addition of plasticizers will increase the flexibility of the PVC product. Rigid PVC has sufficient structural strength, impact resistance, and hardness to replace metals in many forms. PVC has relatively good resistance to chemical attack but is subject to degradation by ketones, aldehydes, amines, and chlorinated alkanes and alkenes in the pure solvent form, although the effect aqueous solutions of these solvents have on the integrity of PVC is not exactly known (7).

Flexible PVC (Tygon; Norton Performance Plastics) is quite different from rigid PVC because of the addition of more than 25% of various phthalate esters, or plasticizers. These plasticizers give the PVC tubing its flexibility but are also the source of cross-contamination when the tubing is used for organic analysis. Phthalate esters can leach into the sample, where they are commonly detected by gas chromatography–mass spectroscopy analysis. They also cause the tubing to adsorb chemicals of interest, as discussed later in this chapter.

Stainless Steel

Stainless steel is one of any variety of steels alloyed with enough chromium to resist corrosion, oxidation, or rusting. Two types of stainless steel are generally used for environmental sampling activities: type 304 and type 316. Both are heavier and more costly than polymers.

Type 304 stainless steel is a chromium–nickel steel with general purpose corrosion resistance. It is nonmagnetic in the annealed condition, but slightly magnetic when cold worked. Type 304 stainless steel can be formed to most desired shapes with little difficulty (8).

Type 316 stainless steel is a chromium–nickel steel that cannot be hardened and that contains molybdenum; it has superior corrosion and heat-resisting qualities. Type 316 stainless steel has improved resistance to sulfur species and sulfuric acid and is commonly used for chemical handling equipment such as heat exchangers, condensers, evaporators, and piping. Type 316 stainless steel is also superior to other corrosion-resistant steels in applications involving seawater (8).

Most stainless steels, like nonferrous alloys and carbon steels, are attacked if exposed to a solution in which the oxygen content is not uniform, even though the solution might otherwise have no action on the metal. Usually this condition occurs when solid matter in the solution excludes oxygen from the surface of the metal in which the solid contacts. The attack is generally characterized by pits that occur under the deposit, while the rest of the metal surface may be bright. Type 316 is recommended for applications of this type because of its superior resistance to contact corrosion (*8*).

Both types of stainless steel are susceptible to corrosion following contact with stagnant natural fresh water for long periods of time. This corrosion is caused by growth of corrosive microbiological iron bacteria colonies; areas around welds are the most susceptible (*9*). Because stainless steel is most subject to oxidation near welds, the use of threaded joints is recommended. Corrosion of stainless steel may also occur at low pH levels or with high chlorine, sulfur, carbon dioxide, dissolved solids, or dissolved oxygen concentrations.

Borosilicate Glass

Borosilicate glass is a soda-lime glass containing about 5% boric oxide, which lowers the viscosity of the silica without increasing its thermal expansion. Such glasses have a very low expansion coefficient and high softening point (about 593 °C), with a continuous use temperature of 482 °C. The tensile strength is about 10,000 psi. Borosilicate glass transmits UV light in high wavelengths. Pyrex (Corning Glass Works) is a borosilicate glass that is commonly used for laboratory glassware and equipment (*3*).

Effect of Materials on Metals Sampling

The degree and rate of sorption of metals in aqueous solutions onto the surface of sampling equipment is influenced by the following factors: (1) the chemical form and concentration of the metal; (2) the characteristics of the solution, including pH, total dissolved solids, salinity, hardness, complexing agents, dissolved gases (especially oxygen, which can influence the oxidative state), suspended matter, and microorganisms; (3) properties of the container, such as chemical composition, surface roughness, surface cleanliness, surface area-to-volume ratios, and history of the container; and (4) external factors such as temperature, contact time, exposure to light, and agitation (*10*).

Sorption of Metals

Parker et al. (*11*) compared stainless steel types 304 and 316, PVC, and PTFE well casings with respect to the sorption of metals (arsenic, chromium, and lead) from aqueous solutions. These studies showed that PTFE clearly sorbed the least amount of metals, whereas the two stainless steel casings were the most sorptive. No loss of arsenic or chromium was noted in solutions exposed to either plastic casing after 72 h. This study also noted a large variability in the concentration of cadmium and chromium in samples exposed to both stainless steel casings and in the concentration of lead in samples exposed to type 316. This variation was attributed to surface oxidation of the stainless steel casings. However, all surfaces sorbed lead, with approximately 10% losses in PVC and 20% for the two stainless steel casings after 4 h.

Leaching of Metals

Hewitt (*12*) evaluated the leaching of nine metals (arsenic, barium, cadmium, chromium, copper, lead, mercury, silver, and selenium) from stainless steel types 304 and 316, PTFE, and PVC well casings after 1, 5, 20, and 40 days of exposure. PTFE generally leached the least amount of metals, with either type 304 or 316 stainless steel leaching the most. There was large variability in the leaching data with the greatest variance in those samples exposed to stainless steel. This variance was attributed to corrosion of the stainless steel casings. In some instances, the initial increases seen after 24 h were followed by a loss, indicating subsequent sorption of the leached analytes.

Effect of Materials on Organic Materials Sampling

Sorption of Organic Materials

In 1985, Barcelona et al. (*13*) performed research on the suitability of various types of plastic tubing for groundwater analysis. They concluded that the selection of the proper sampling tubing was critical to avoid materials-related errors. They demonstrated that serious bias of dissolved organic compound results occurs quite rapidly (within 5–10 min) because of sorption on exposed flexible tubing. Their data indicated an 80% loss of chloroform after only 1 h of contact with PVC tubing, with silicon rubber tubing a close second. They recommended PTFE tubing for most monitoring work, particularly for detailed analytical schemes, as it is least likely to introduce significant sampling bias or imprecision. It is also the easiest material to clean in order to prevent cross-contamination. Polypropylene was a second choice. Flexible PVC was not recommended for organic analysis because of the plasticizers and stabilizers (phthalates) that make up a sizable percentage of the material; documented interferences were likely with several priority pollutant classes. PTFE tubing was less sorptive than the other two tested; however, even PTFE quickly sorbed substantial amounts of several organic materials.

The Barcelona study (*13*) has often been referenced as justification for the preferential use of Teflon in any sampling activity over the use of any type of PVC. Although valid for flexible tubing, later studies have shown that rigid PVC is much less sorptive of organic materials than flexible PVC. Studies by Parker et al. (*11*) and Gillham and O'Hannesin (*14*), conducted under sterile conditions, have shown that the rate and extent of sorption of organic materials from aqueous solution are greater for PTFE than for rigid PVC. Parker et al. found that loss of trichloroethylene (TCE) was 10% after 8 h for PTFE and 60% after 6 weeks, whereas losses for PVC were only 6% after 1 week and 12% after 6 weeks. Parker et al. did not detect any sorptive losses of organic materials under sterile conditions for stainless steel.

Reynolds and Gillham (*15*) found that sorption to rigid PVC and PTFE was generally slow, with more than 5 weeks required to reduce most compounds to 50% of their initial value. Perchloroethylene (PCE), however, was sorbed at a much faster rate by PTFE than by PVC. Less than five minutes were required to reduce PCE levels to 90% of initial concentrations, compared to 1 day for PVC. These data indicate that groundwater samples with PCE may show a reduction in concentration after even short periods of contact with PTFE, indicating that rigid PVC is a better polymer choice than PTFE for sampling organic materials.

In these studies, uptake of the organic compounds into the polymer proceeded first by sorption–dissolution into the surface of the polymer followed by diffusion deeper into the polymer matrix (*16*). The reduced losses associated with rigid versus flexible PVC may be attributed to the higher density and greater crystallinity of rigid PVC. Berens (*17*) concluded that softening of PVC is only possible when exposed to nearly undiluted PVC solvents or swelling agents. Schmidt (*18*) found that gasoline would not appreciably swell PVC; he subjected sections of type 1 PVC well screen to several different grades of gasoline for 6 mo and found no detectable change in the screen when examined using a scanning electron microscope. Olsen et al. (*19*) also studied the permeation of 1,1,1-trichloroethane and toluene through PVC pipe using neat solvent and saturated aqueous solutions. They concluded that PVC pipe was an effective barrier to low levels of the swelling solvents but became permeable to swelling at high solvent levels (unless exposed to a nearly saturated solution of PVC or of a pure solvent) (*16*).

Bianchi-Mosquera and Machay (*20*) compared the sorption to, desorption from, and diffusion through PTFE and stainless steel miniwells (one-eighth in. O.D. tubing) for PCE and carbon tetrachloride. They found no significant difference between the concentration histories for PTFE or stainless steel for the low organic solute concentrations (39–51 μg/L), short exposure times, and adequate line flushing conditions of the experiment.

Leaching of Organic Materials

As previously discussed, the primary contribution to leaching of organic materials is from the phthalate esters that give plastic tubing its flexibility. Curran and Tomson (*21*) compared the leachability of organic compounds from several different rigid and flexible materials; they concluded that Teflon leached the least amount of contaminants and that rigid PVC was a very close second. The highest rates of leaching were from flexible PVC (Tygon) and silicon tubing. Phthalate ester plasticizers were found to leach from all flexible polymer tubing.

Reynolds et al. (*22*) demonstrated that sorption rates were dependent on flexibility of the polymer, water solubility of the compound, the ratio of solution volume to polymer surface area, and temperature. They exposed 10 materials (6 plastics, 3 metals, and borosilicate glass) to low concentrations of five halogenated hydrocarbons in water for periods up to 5 weeks. The borosilicate glass was the only material of the 10 tested that did not cause a reduction in solution concentration for at least one compound over the 34-day period. Stainless steel was the least reactive metal, causing the reduction of only two of the most halogenated compounds (bromoform and hexachloroethane). Aluminum caused reduction in four of the five compounds, and galvanized steel caused significant reductions for all compounds. Reynolds postulated that the polyhalogenated compounds underwent reductive hydrogenolysis in the presence of transition metals found in the metals. In this reaction, a hydrogen atom replaces a halogen substituent and oxidizes the transition metal. Parker et al. (*11*), however, did not observe any loss of organic materials with stainless steel when performing these tests under sterile conditions, indicating that biodegradation may have been a factor in the reduction.

The Reynolds data indicated that halocarbon concentrations declined rapidly after an initial delay period when in contact with metals, whereas concentrations declined more gradually when in contact with the polymers. The highest rate of halogen concentration decline for polymers occurred in latex followed by low-density polyethylene, polypropylene, nylon, PTFE, and rigid PVC, respectively. However, the rate of halocarbon decline for rigid PVC could not be distinguished from that of PTFE. The difference in sorption rates was assumed to be related to the different diffusivities of the compounds in the polymers.

Containers for Air Samples

The integrity and stability of air samples is very dependent on the chemical species being sampled. Because the scope of this chapter is only to deal with equipment materials, detailed sampling discussions are deferred to later chapters. As previously discussed, borosilicate glass is relatively inert to sorption and leaching of organic contaminants. However, glass is quite fragile and chemical compounds in the sample are subject to UV degradation if the sample container is not shielded from sunlight. Glass bombs typically have valved openings on either end of the bomb and are filled by pulling the air sample through the container, thereby avoiding cross-contamination from the diaphragms and walls of the sampling pump.

Tedlar bags made of poly(vinyl fluoride) are also frequently used to collect air samples. They are considered inert and are suitable for the analysis of sulfur compounds if the bag fittings are not metallic. These bags, however, are fragile and susceptible to leaks and loss of sample (due to crushing or expansion of the sample in unpressurized cargo compartments) during transport to the laboratory. Because they are clear, they are also subject to UV degradation and are usually transported in appropriately sized cardboard boxes to block sunlight. Unless the bags are filled using a rather cumbersome vacuum chamber, the potential exists for cross-contamination from the diaphragms of the pumps used to fill the bags.

Evacuated stainless steel canisters are also used for collecting air samples. The inside surfaces of these containers are treated or passivated, using a “summa” process to ensure an inert inner surface. The term “summa” was originally coined by the former Moletrics Corporation (Los Angeles, CA), but it apparently is not registered as a trademark. The process uses electropolishing combined with deactivation to produce a chemically inert inner surface. The presence of water vapor in the sample enhances the stability of the sample by forming a protective layer on the metal surfaces of the canister. The canisters are evacuated to a vacuum of approximately 28 inches of Hg. The vacuum draws the sample into the canister without cross-contamination from pumps. A flow control regulator can be used to obtain a composite, or integrated, sample over a longer period of time. The stainless steel canisters have an advantage of not being fragile or susceptible to UV degradation, although they are expensive and the cleaning procedures are more involved than for other containers. Some compounds, such as chloroform, have been shown to be stabile for periods as long as 18 mo (23). However, stainless steel canisters may not be appropriate for use if sulfur compounds are present because of the corrosivity of sulfur toward stainless steel.

Strength of Materials

The strength of materials used for the manufacture of sampling equipment can be of overriding concern, particularly when sampling for soils. Stainless steel is obviously the strongest material discussed in this chapter, although it may also be the most expensive and is heavier than most organic polymers. Glass is the least durable of materials considered and, because of this, may not be applicable under certain situations. Smaller diameter PTFE tubing is less flexible and is more susceptible to bending and kinking. Although PVC is not as strong as stainless steel, it is quite strong. The EPA *RCRA Groundwater Monitoring: Draft Technical Guidance* (24) indicates that PVC is suitable for well depths between 1200 and 2000 feet. The recommended depth for PTFE is from 225 to 375 feet. In groundwater well construction, there is also some question whether neat cement grout can properly bond to a Teflon well casing. Inadequate bonding could result in creation of a pathway for contaminants from the surface to migrate downward to the groundwater between the casing and the grout (25).

Contamination by Incidental Sampling Materials

A number of items commonly used during sampling investigations have the potential to add contamination to the sample. As discussed earlier in this chapter, every item that comes in contact with the sample must be examined to ensure that the integrity and representativeness of the sample are not compromised. Adhesive glues to join groundwater well casing materials have been shown (26) to contribute contaminants into the water. Therefore, glues should not be used in the construction of monitoring wells, and threaded joints are preferred.

Marking pens, particularly the Sharpie (Sanford) type, must be used with care during sampling activities. These pens contain volatile compounds, such as benzenes (i.e., ethyl benzene and chlorobenzene) and xylenes (26) that can evaporate and become entrained in air or groundwater samples if used too close to sampling activities. Black electrical tape commonly used for insulating and wiring in pumps and other equipment contains a high level of toluene (26), which can contaminate groundwater if submersible pumps with tape-wrapped electrical connections are used during well-purging activities. The adhesive on sample labels and the clear plastic tape frequently applied over sampling labels (to prevent the ink from running if the bottles become wet) also contains high levels of volatile organic compounds (26). The sampler should be cautious, therefore, when wrapping clear tape over the labels of 40-mL VOA vials, to avoid the cap and septa area. It may be more appropriate to evaluate the use of labels that are specially designed not to smear the writing when they become wet. Paints and greases on sampling equipment can also contribute metals and organic compounds if they are allowed into the sample.

Experiments performed by Canova and Muthig (27) have indicated that caprolactam and Santowhite (Monsanto) found in groundwater samples may be a result of nylon cord and latex gloves that were used during sampling of bailed wells. Caprolactam is used in the manufacturing of nylon cord, and Santowhite is an antioxidant used in latex gloves. These experiments show that alternate materials or sampling techniques should be considered to minimize the potential impact of nylon cord and latex gloves on the quality of groundwater samples.

Conclusions

The collection of samples that are representative of the population being measured is imperative to ensure that the analytical data and the resulting conclusions are valid. To ensure the sample representativeness and preserve the integrity of the sample, one must challenge all of the assumptions made in the sampling process and examine the effect that the equipment and sampling protocol have on the sample. As shown in the preceding discussion, no equipment materials are perfectly inert. Therefore, the sampler should work to minimize the contact with materials and select the equipment on the basis of its ability to function effectively with minimal effect on the structure or concentration of the chemicals of interest.

Rigid PVC has been shown to be best for collecting samples for the analysis of both metals and organic materials when lower concentrations of solvents are present. PTFE is best for metals and is approximately equivalent to rigid PVC for low concentrations of organic materials. PTFE is superior to rigid PVC when high concentrations of organic materials are present.

Stainless steel is not good at low pH levels or with high chlorine, sulfur, carbon dioxide, dissolved solids, or dissolved oxygen concentrations, because of corrosive effects under these conditions. Stainless steel is not as good as the organic polymers for metals because of the potential for oxidation to occur. Stainless steel has the advantage of being the strongest of the materials and is very easy to clean.

Borosilicate glass has been shown to be a superior equipment material because it does not leach or sorb contaminants of interest. However, borosilicate glass allows UV radiation to penetrate, potentially allowing contaminants of interest to photodecay. Glass is also less durable than the other materials and subject to breakage.

If flexible tubing is required, PTFE is superior to flexible PVC because flexible PVC may sorb organic materials or leach plasticizer compounds into the sample.

References

1. *Data Quality Objectives Process for Superfund: Interim Final Guidance;* U.S. Environmental Protection Agency. Government Printing Office: Washington, DC, 1993; EPA/540/G–93–071.
2. Miller, G. P. *HazMat World* **1993,** 6(11), 42–46.
3. *Hawley's Condensed Chemical Dictionary, 11th ed.;* revised by Sax, N. I.; Lewis, R. J. Van Nostrand Reinhold: New York, 1987.
4. *Kirk-Othmer Concise Encyclopedia of Chemical Technology,* 3rd ed.; Wiley-Interscience: New York, 1980; Vol. 11, pp 1–24.
5. *Teflon, A Performance Guide for the Chemical Processing Industry;* DuPont: Wilmington, DE.
6. Kjeldsen, P. *Water Resour. (Gr. Br.)* (Pergamon) **1993,** 27(1), 121–131.
7. "Report on Rigid Polyvinyl Chloride," NACE Technical Committee. Reprinted from *Corrosion,* **1956,** *12*(4).
8. "Type 304 Stainless Steel and Type 316 Stainless Steel"; Trent Weld Data Sheet. Trent Tube Division. Colt Industries: East Troy, WI.
9. Korbin, G. "Reflections on Microbiologically Induced Corrosion of Stainless Steels"; DuPont.
10. Masse, R.; Maessen, F. J. M. J.; DeGoeij, J. J. M. *Anal. Chim. Acta* **1981,** *127,* 181–193.
11. Parker, L. V.; Hewitt, A. D.; Jenkins, T. E. *Ground Water Monit. Rev.* **1990,** *10*(2), 146–156.
12. Hewitt, A. D. "Leaching of Metal Pollutants from Four Well Casings Used for Ground-Water Monitoring"; Special Report 89–32; U.S. Army Cold Regions Research and Engineering Laboratory: Hanover, NH, 1989.
13. Barcelona, M. J.; Helfrish, J. A.; Garske, E. E. *Anal. Chem.* **1985,** 57(2), 460–464.
14. Gillham, R. W.; O'Hannesin, S. F. "Sorption of Aromatic Hydrocarbons by Materials Used in Construction of Groundwater Sampling Wells"; *Ground Water and Vadose Zone Monitoring;* American Society of Testing and Materials: Philadelphia, PA, 1990; ASTM STP 1053, pp 108–122.
15. Reynolds, G. W.; Gillham, R. W. *Proceedings of the Second Canadian/American Conference on Hydrology, Banff, Alberta,* National Water Well Association: Dublin, OH, 1985; pp 125–132.
16. Parker, L. V. "Suggested Guidelines for the Use of PTFE, PVC, and Stainless Steel in Samplers and Well Casings"; *Current Practices in Groundwater and Vadose Zone Investigations;* Nielsen, D. M.; Sara, M. N., Eds.; American Society for Testing and Materials: Philadelphia, PA, 1992; pp 217–229; ASTM STP 1118.
17. Berens, A. R. *J. Am. Water Works Assoc.* **1985,** 57–65.
18. Schmidt, G. W. *Ground Water Monit. Rev.* **1987,** 7(2), 94–95.
19. Olsen, A. J.; Goodman, D.; Pfau, J. P. *J. Vinyl Technol.* **1987,** 9(3), 114–118.
20. Bianchi-Mosquera, G. C.; Machay, D. M. *Ground Water Monit. Rev.* **1992,** *XII*(4), 126–131.
21. Curran, C. M.; Tomson, M. B. *Ground Water Monit. Rev.* **1983,** 3, 68–71.

22. Reynolds, G. W.; Hoff, J. T.; Gillham, R. W. *Environ. Sci. Technol.* **1990,** *24*(1), 135–142.
23. Jayanty, R. K. M. et al. "Stability of Parts-Per-Billion Hazardous Organic Cylinder Gases and Performance Results of Source Test and Ambient Air Measurements Systems"; U.S. Environmental Protection Agency. Government Printing Office: Washington, DC, 1986; EPA/600/2–86/003.
24. *RCRA Ground-Water Monitoring: Draft Technical Guidance;* U.S. Environmental Protection Agency. Government Printing Office: Washington, DC, 1986; EPA/530–R–93–001.
25. Nielsen, D. M.; Schalla, R. In *Practical Handbook of Ground-Water Monitoring;* Nielsen, D. M., Ed.; Lewis Publishers: Chelsea, MI, 1991; pp 239–331.
26. Koerner, C. E. Radian Corporation, unpublished data, 1990.
27. Canova, J. L.; Muthig, M. G. *Ground Water Monit. Rev.* **1991,** *11*(3), 98–103.

Chapter 9

Extending the Concept of Data Quality Objectives To Account for Total Sample Variance

Thomas E. Barnard

The U.S. Environmental Protection Agency (EPA) has developed the concept of data quality objectives (DQOs) as an integral part of the overall environmental sampling process. Although firmly based in theory, DQOs have significant shortcomings when applied to field samples. Examination of the EPA's guidance reveals that of the five parameters used to measure data quality, only two can be quantified, and then only at the laboratory level. The underlying assumption that high-quality interpretation can only be derived from high-quality data is shown to be invalid. As a consequence, data may be rejected or qualified as having "poor quality" when in fact the magnitude of laboratory variance is small in comparison to total sampling variance. In this chapter, an attempt is made to extend the DQO concept to account for total sample variance and to shift the discussion from data quality to quality of conclusions. It is shown that, through the use of scientific inference, high-quality conclusions may be derived from "poor quality" data. Accordingly, there is a misguided focus on data quality at the expense of an appreciation of the overall quality of the study results and conclusions.

The measurement of environmental parameters and processes has been a challenge since the very beginnings of environmental science. It has been known, but only somewhat appreciated, that where we sample, when we sample, how we collect the sample, how we process the sample, and how we perform the analysis all contribute to measurement variability. The contribution of each of these factors to overall variability and the required level of consideration that each must be given in the design of a sampling program has been a matter of considerable debate.

Environmental sampling and analysis are conducted under a variety of legal regimes (*1*). Examples include permit monitoring, by which a permittee can be fined based upon his or her own data, or remedial investigations conducted under the Comprehensive Environmental Response Compensation and Liability Act (CERCLA). As a result of the need to collect legally defensible data, there is such an emphasis on sample handling procedures such as seals and custody forms that it is often forgotten that other factors may also contribute to the inability of the sample results to withstand legal and scientific rigor. The U.S. Environmental Protection Agency (EPA) has developed the concept of *data quality objectives* (DQOs) as an integral part of the planning processes of sampling and analysis (*2, 3*). The DQO concept is gaining acceptance, and its application is spreading to other areas in the environmental field.

As it is currently practiced, the DQO process focuses on only one aspect of measurement variance, analytical chemistry, with two possibly disastrous implications. Data can be considered to be of high quality even though field- or collection-level variance may be an order of magnitude greater than the analytical component variance. This assumption will in turn lead to a false sense of security concerning the conclusions. The other possibility is that data may be rejected or qualified because results of quality control samples are not within prescribed limits. If a sampling plan is properly designed in a statistical sense, then measurement error can be quantified and appropriately considered in the data analysis. The purpose of this chapter is to (1) briefly explain the concept of DQOs as they are currently used, (2) identify deficiencies in the application of DQOs to environmental data, and (3) develop a procedure for extending the focus from data quality to the quality of conclusions that are derived from the data.

Background

DQOs are defined by the EPA as qualitative and quantitative statements that specify the quality of the data required to support agency decisions during remedial response activities. The concept was developed as part of the remediation process under CERCLA when the EPA needed to make decisions regarding cleanup, remediation, and liability at contaminated sites. The purpose of DQOs is to ensure that all investigation activities are conducted and documented in a manner that ensures sufficient data of known quality are collected to support decisions concerning remedial action selection. Since they were first proposed, DQOs have been adopted as an essential tool in the planning of environmental investigations that are conducted in compliance with a variety of laws and regulations. DQOs are designed to serve as a guide throughout the sampling process. Two of the most common applications are in the selection of laboratory procedures and in the data validation process.

Data quality is defined by the EPA as the "degree to which data conform to a specified set of criteria: the degree of conformation allows for categorization into 'levels' of quality (valid, acceptable, rejected)" (*4*). Although there is no technique available to quantify data quality, EPA has provided five parameters that are used as indicators. They are precision, accuracy, representativeness, completeness, and comparability. The EPA's definition of these indicators are listed in Table I. The definitions of precision and accuracy are essentially the same as those used in the quality control field. The definition of representativeness is flawed. According to this definition if repeated measurements yield similar results then they are providing data that is representative of the population. This concept is wrong. Repeated measurements provide an estimate of the precision of a measurement and do not serve as indicators of accuracy or representativeness. *Representativeness* is a statistical concept that is a measure of how well a set of sample measurements yields information concerning the population. It cannot be determined for a single datum. Completeness is an accounting parameter that is a measure of the fraction of data that meet the requirements of the functional guidelines (*4, 5*). Because these guidelines are based on the instrument response to clean laboratory standards, their relevance to the response to real environmental samples is questionable. The comparability parameter concerns the question of whether two data sets yield similar estimates of the population parameter. Again this is really a question of the precision. In summary, the parameters presented in Table I as indicators of data quality do little to extend the measure of data quality beyond the standard precision and accuracy tools that are well established.

Despite the inability to quantify data quality, EPA has delineated levels of data quality and assigned corresponding types of analysis and data uses (Table II). There are at least three major problems with the information in Table II. First, it is based on the underlying assumption that laboratory measurements are of higher quality than field measurements and ignores the problems associated with chemical and biological changes that occur with the sample prior to analysis. Second, the underlying assumption that analytical methods employing gas chromatography followed by mass spectrometry (Contract Laboratory Program procedures specified in level

Table I. Parameters That Indicate Data Quality

Parameter	*EPA Definition*
Precision	Reproducibility of measurements under a given set of conditions.
Accuracy	Bias in a measurement system.
Representativeness	Can be assessed by collocated samples. By definition, collocated samples are collected so that they are equally representative of a given point in space in time. In this way, they provide both precision and accuracy information.
Completeness	Percentage of measurements made that are judged to be valid measurements
Comparability	Qualitative parameter expressing the confidence with which one data set can be compared with another

NOTE: EPA is U.S. Environmental Protection Agency.
SOURCE: Information is reproduced from reference 2.

Table II. Levels of Data Quality

Level	*Type of Analysis*	*Data Uses*
I	Field test kits	Site characterization
	Organic vapor analyzer	Personnel monitoring
II	Field portable gas chromatography	Monitoring during remediation
	X-ray fluorescence	Site characterization
III	RCRA characteristics	Risk assessment
	Non-Contract Laboratory Program	Evaluation of alternatives
	EPA methods	Engineering design
IV	Contract Laboratory Program procedures	Risk assessment
		Primary responsible party determination
		Engineering design
V	Special Analytical Services	Risk assessment
		Primary responsible party determination

NOTE: RCRA is Resource Conservation and Recovery Act.
SOURCE: Information is reproduced from reference 2.

IV) yield higher quality measurements than the ordinary gas chromatography methods (listed as level III) is unfounded. Mass spectrometry is used to confirm the analyte identification and does not impact either the accuracy or precision of the measurement. Third, in an unfounded jump in logic, the table specifies how different types of measurements can be interpreted. The last column of Table II overlooks scientific tools such as statistical analysis and environmental science, which must be utilized in inferring useful information from sampling data (*6*).

As a result of these shortcomings, Table II has been misused by both authors of sampling plans and regulators. It is frequently used as justification for the requirement to use more expensive level IV methods (generating extensive documentation) when level II or III measurements are adequate. It is used as justification for requirements that data users ignore "lower quality" data in the process of data evaluation. And finally, users of Table II tend to place an unfounded level of confidence in published reference methods at the expense of developing refined methods that are more suited for a particular site, sample matrix, or analyte.

Data validation is defined as "an independent systematic process for reviewing a body of data against a set of criteria to provide assurance that the level of quality of the data are known and documented. Data validation consists of data screening, checking, auditing, verification, and review" (*7*). In practice, data validation serves two important functions. First, it reviews the entire data collection, reduction, and management process and identifies any errors in the flow of data from the point of generation to the final laboratory report. Second, it compares analytical precision as measured by laboratory duplicates, spikes, and calibration standards against guidelines that are available from either the analytical method or documents such as EPA's functional guidelines (*4, 5*). These comparisons typically ignore the fact that studies have demonstrated the inability of commercial laboratories to consistently meet the guideline levels for accuracy, precision, and detection limit (*8–10*). In addition, application of more traditional measures of data quality such as ratios of cations to anions (calculated to measured total dissolved solids) and ratios of specific conductivity to total dissolved solids have shown far more errors in analytical measurements than were indicated by data validation (*11*).

In Table III the factors contributing to measurement error in environmental investigations are grouped according to their review during data validation (*12*). Although the scope of data validation will vary from project to project, Table III serves as a guide to the general practice. Experimental design considers the known properties of the attribute to be measured such as temporal and spatial variation, and it results in a sampling methodology that is consistent with the study objectives. This factor is almost never considered in data validation. Factors that are sometimes reviewed include sample collection procedures, subsampling of field samples at the laboratory, and pretreatment such as extraction and sample cleanup. Although this review consists of an assessment of the consistency of field procedures to those specified in the sampling plan, it does not provide an estimate of true sampling error. It is also important to note that spiking and splitting of samples for analysis usually occurs after these pretreatment steps. Data validation does include a thorough review of adherence to specified procedures for sample handling, instrument calibration, sample analysis, and data interpretation. In addition it reviews the paper trail for problems with transcription and association (linking the analytical results to field information). Data validation is limited in the scope of its review and the applicability of its conclusions. The review is focused on the analytical end of the investigation process and ignores many other components that contribute to measurement variability.

In summary, there are several problems with the DQO concept as it is currently used by environmental investigators. First, data quality is difficult to measure. Of the five parameters proposed by EPA as indicators of data quality, only accuracy and precision have any scientific basis. The remaining three—representativeness, completeness, and comparability—give no additional information regarding data quality. The data validation process does not consider all of the components of variance of environmental measurements. Indeed, the examination of a set of observations on a datum-by-datum basis ignores the scientific process of drawing conclusions from observations. Finally, there is a misplaced focus on data quality at the expense of consideration of the quality of scientific conclusion. In light of the these problems, a more fundamental approach to data quality is presented here.

Table III. Factors Contributing to Measurement Error as Reviewed During Data Validation

Not Reviewed	*May Be Reviewed*	*Always Reviewed*
Experimental design	Sample collection	Sample handling
	Subsampling	Instrument calibration
	Extraction	Measurement precision and accuracy
	Sample cleanup	Adherence to method
		Data transcriptions
		Data association

Extension of DQOs

A datum is an observation, fact, or quantity. The scientific process is that of drawing conclusions from data. Scientists use tools such as scientific reasoning, statistical analysis, and their knowledge of environmental processes and phenomena to develop valid conclusions from data. The formulation of valid scientific conclusions can only be the result of good science. Conclusions are published and then reviewed by other scientists. They are accepted, rejected, or refined. Environmental science is based on scientific conclusions that have been critically reviewed and are generally accepted. A goal of any investigation should be to obtain the highest possible quality conclusions. Regardless of the quality of the data, a sampling program that leads to conclusions that contradict environmental science must not be considered as valid.

How does the quality of a given data set relate to the quality of scientific conclusions that are derived from that set? The quality of conclusions is a difficult concept to evaluate. Statistics provides us with one measure of quality of conclusions, the confidence interval. Equation 1 gives the two-sided confidence interval for the population mean resulting from simple random sampling.

$$\mu = \overline{x} \pm t_{1-\alpha}\left(\frac{s}{\sqrt{n}}\right) \tag{1}$$

where μ is the estimate of the true population mean, $\overline{x}$ is the sample mean, t is the Student's t statistic, α is the probability of a type I error, s is the sample standard deviation, and n is the number of observations. Although s is a measure of data quality (sample precision), the width of the confidence interval can be controlled by either collecting more samples or allowing for a higher probability of a type I error. Equation 1 shows that it is possible to derive a high-quality conclusion (narrow confidence interval) from poor-quality data (high standard deviation) by taking more samples.

Another method of assessing the quality of conclusions is by determining the applicability of conclusions to other sites or systems. Low-quality conclusions are applicable only to a small area over a limited time and are of little use to environmental regulators and managers. High-quality conclusions can be applied to a variety of sites over a range of times. For example, a study conclusion may be that soluble lead (Pb^{2+}) from a specific waste discharge is being removed from a specific surface water through adsorption onto surfaces and sedimentation. If this conclusion can be extended to time periods other than that of the sampling and if they can be extended to similar discharges into surface waters of similar chemistry, then it must be considered a high-quality conclusion. The quality of the initial data then becomes irrelevant.

Once conclusions are formulated, hypotheses are developed that extend the application of the conclusions. In the long run, the quality of conclusions can be assessed by how well they stand up against further scientific investigations.

Table IV. Hypothetical Data for a Discharge Monitoring Study

Raw Data as Generated by Analytical Instrument	*Data Available to User after Data Validation*
–0.1119	<10
11.0406	11.0
1.7169	<10
1.2659	<10
6.0103	<10
6.4979	<10
12.4441	12.4
8.5209	<10
3.8570	<10
10.4404	10.4

NOTE: Values are in micrograms per liter.

Because the primary goal of any investigator is to develop high-quality conclusions, every decision regarding the design of the investigation should be evaluated against how it will impact the quality of the conclusions. The following examples illustrate that the relationship between data quality and quality of conclusions is not straightforward.

Data Censoring

Suppose that a surface water discharge permit requires that the mean lead concentration in the effluent be below 10 μg/L, which is also the method detection limit. Under the monitoring plan, 10 random samples are collected and analyzed every month. Under the terms of the permit, the data are validated by an independent party prior to being reported to the regulatory agency. The facility is considered to be in violation when the monthly mean exceeds 10 μg/L.

The results of one month's sample are presented in Table IV. The first column shows the results as they would appear in a direct reading instrument that is set to display four decimal places. The second column shows the results as they would be reported after data validation. Two changes have been made. First, the number of significant figures has been adjusted, and second, results below 10 μg/L have been censored. This example shows the common practice of not reporting analytical results that fall below the method detection limit and only indicating to the data users that the results were less than the detection limit.[1] This practice is based on the belief that a single observation has a very high variance associated with it and that there is a risk that the result will be misinterpreted. The counterargument is that a value of 5 ± 8 is a valid observation. As long as the value represents the analyst's unbiased estimate of the true value, it is more useful than <10. Additionally,

[1]There are a number of conventions for reporting measurements below the detection limit, such as treating censored values as 0, one-half the detection limit, or the detection limit. All of these methods introduce bias into the data and effectively result in degrading the quality of the data.

in environmental sampling, any single observation is of such little significance that the concern about misinterpretation is unwarranted. Although a variety of statistical procedures have been developed for estimated population parameters from censored data (*13–16*), all require assumptions concerning the population distribution and many biases can be introduced with their use (*17*). A better approach is to utilize all of the data.

Is the plant in compliance with the permit? This question would be difficult to answer if one had only the information in the right-hand column of Table IV. However, a simple one-sample statistical analysis is presented in Table V. The analysis shows that the 95% confidence interval of the mean (3.00–9.33) is completely below 10 μg/L, and the null hypothesis that the mean is above 10 can be rejected with a 1.1% probability of being wrong. This simple statistical tool allows for the attainment of high-quality conclusions from what the data validator would label as "poor-quality" data.

Matrix-Specific Detection Limit

The calculation of the analytical detection limit is an issue of controversy (*18–20*). Although a variety of formulae have been proposed for estimating the detection limit, the procedure in the EPA regulations (*21*) for water analysis uses eq 2.

$$\mathrm{MDL} = t\sigma_{\mathrm{A}} \tag{2}$$

where MDL is method detection limit and σ_A is the standard deviation of a replicate at or near the detection limit.

One of the major problems with the approach is that the standard deviation is determined by spiking a reagent blank or a relatively clean matrix. This procedure can be extended to include the effects of the sample matrix. However, this step is rarely done, and regulations frequently attempt to hold permittees to an MDL that

Table V. One-Sample Analysis of Data from Table IV

Statistical Measure	*Statistical Value*
Sample statistics	
Number of observations	10
Average value	6.17 μg/L
Variance	19.5 μg/L
SD	4.42 μg/L
95% CI for mean (9 df)	3.00–9.33 μg/L
Hypothesis test	
H_0	mean ≥ 10 μg/L (α = 0.05)
Computed *t* statistic	–2.74
Result	reject H_0 (α = 0.011)

NOTES: SD is standard deviation; CI is confidence interval; df is degrees of freedom; H_0 is null hypothesis; α is significance level. Analysis shows that the 95% CI of the mean is completely below 10 μg/L, and the H_0 that the mean is above 10 can be rejected with a 1.1% probability of being wrong.

does not consider matrix effects. In the analysis of real samples with more complex matrices, one would expect a larger standard deviation of replicates and therefore a larger actual detection limit.

A study was performed by Neserke and Taylor (22) that attempted to determine the actual detection limit of effluent from a wastewater treatment plant. A sample of about 18 gallons of effluent was taken from the Coors' wastewater treatment plant in Golden, CO. The sample was split into two subsamples that were spiked with low levels of cadmium, lead, mercury, and silver. The samples were sent to four commercial analytical laboratories with instructions to mix the samples with batches of other routine samples. This study differed from the normal technique for detection limit determination in several respects. First, the samples that were used for analysis had a complex matrix that provided for a more realistic estimate of the analytical standard deviation. Second, the samples were spiked at two levels rather than the one level that is typically used. This protocol again would provide for a more realistic estimate of the standard deviation of replicate measurements. Finally, four separate analytical laboratories were used. The result was that the detection limit calculated from the study would be best called an interlaboratory, matrix-specific detection limit (MSDL), probably a more useful parameter than the method detection limit that is presented in a published method.

The analysis of variance of the results for mercury are presented in Table VI. A model was selected that included all the components of variance that were significant at 5%. This resulted in a calculation slightly different from that of Neserke and Taylor, and interestingly the laboratory effects were significant. The matrix-specific detection limit was calculated by eq 2 with

$$\sigma_A = \sqrt{ms_{Residual}} \tag{3}$$

where $ms_{Residual}$ is the residual mean square error and

$$t_{0.99,51} = 2.41 \tag{4}$$

Therefore

$$\begin{aligned} MSDL &= 2.41 \times (0.00904)^{1/2} \\ &= 0.229\ \mu g/L \end{aligned} \tag{5}$$

Table VI. Analysis of Variance for Mercury in Spiked Wastewater Effluent

Source of Variation	*Sum of Squares*	*Degrees of Freedom*	*Mean Square*	*F Ratio*	*Significance*
Main effects					
Laboratory	0.167	3	0.0555	6.13	0.0012
Spiked Sample	0.179	1	0.1794	19.83	<0.0001
Residual	0.462	51	0.00904		
Total	0.807	55			

NOTE: All *F* ratios are based on the residual mean square error.

as opposed to a limit of 0.05 μg/L that the Colorado Department of Health was attempting to impose on the facility. The detection limit has serious legal implications as the regulators were attempting to impose an ambient water quality standard on the facility at the theoretical or method detection limit. In this example, data in the range of 0.05–0.23 μg/L would meet all acceptable criteria and would be considered high-quality data according to EPA criteria. Sample results in this range would be considered a permit violation, and Coors could be subject to fines. However, in consideration of all of the factors contributing to measurement error, these results probably would not pass a legal standard of reasonable doubt.

Data Quality vs. Environmental Science

The next example presents a situation in which the data that are considered to be of high quality suggest conclusions that violate basic environmental science. At the Rocky Flats Plant, located 20 miles northeast of Denver, CO, the upper 5 cm of the soil are sampled annually and analyzed for plutonium (23). The sample design consists of two rings with radii of one and two miles from the center of the plant. Each ring has 20 stations equally spaced around the circumference at increments of 18 degrees. Each year 10 subsamples are collected at each station and are physically composited. The plutonium concentration is measured for each composite by chemical extraction and precipitation on a planchet followed by α-spectrometry. The results are reported as picocuries per gram on a dry weight basis. In order to approximate a normal distribution a natural log transformation was performed on the data.

Some background information on the environmental behavior of plutonium is presented here. The major sources of plutonium contamination were fires and the excavation of a shallow drum burial ground. These all occurred in the 1950s and 1960s. The dominant transport mechanism of plutonium was atmospheric convection. The predominant winds in the area blow from the west and northwest. Plutonium is a heavy and highly insoluble molecule. Once it reaches the soil, it is not expected to move any appreciable distance. It probably precipitates or becomes bound to the soil as a charge species. The half-life of ^{239}Pu, the principal isotope at Rocky Flats Plant, is greater than 24,000 years.

The analysis of variance of the data from the years 1984 to 1992 is presented in Table VII. Two of the expected results are that the direction and the ring effects are significant components of the variance. The means plot in Figure 1 shows a clear increase in plutonium in the easterly stations for both rings. The significant ring effects are indicative of the limited distance over which the plutonium was carried before being deposited in the soil. The surprising result is that the year effects are significant. The means plot in Figure 2 shows that the yearly averages display a wavelike pattern. Figure 3 shows the yearly averages for the inner and outer rings. The pattern is apparent in both rings but perhaps stronger in the outer ring, for which the values are lower.

The significance of the year effects contradicts our knowledge of environmental science. The increases in plutonium cannot be explained with new sources of con-

Table VII. Analysis of Variance for Plutonium in Rocky Flats Soils

Source of Variation	*Sum of Squares*	*Degrees of Freedom*	*Mean Square*	F Ratio	*Significance*
Main effects					
Direction	582.2	19	30.6	72.6	<0.0001
Ring	61.3	1	61.3	145.0	<0.0001
Year	15.1	8	1.9	4.5	<0.0001
Interactions					
Direction × ring	102.2	19	5.4	12.7	<0.0001
Residual	131.8	312	0.42		
Total	892.5	359			

NOTE: All *F* ratios are based on the residual mean square error.

tamination. The decreases cannot be explained by any known environmental sinks. Several reasons for the significance of the year effect are postulated. Analyzing for low levels of plutonium in environmental samples is difficult. Natural background levels of radiation may interfere with the counting procedure. It is possible that the background radioactivity levels at the counting facility have undergone changes in radioactivity. The resultant correction of raw sample counts to account for background would introduce a bias in the data. Another possibility is that the sample collection or compositing procedures have undergone slight modifications and have affected the results. The important point is that data that have undergone a thorough review on a sample-by-sample or even a year-by-year basis and are considered to be of high quality lead to conclusions that violate our basic understanding of the site.

This contradiction leads one to question the quality of the data. These data were generated from a complex series of steps. Although each step is considered to be high quality, the sum of a series of high-quality steps does not necessarily lead to high-quality data.

Total Sample Variance

Any consideration of environmental data must consider all of the factors contributing to measurement error. Statistics provides us with the fundamental property that variance components are additive. This property allows us to develop expressions for *total sample variance,* such as in eq 6

$$\sigma_T^2 = \sigma_A^2 + \sigma_{Sub}^2 + \sigma_{Samp}^2 \quad (6)$$

where $\sigma^2{}_T$ is total sample variance, $\sigma^2{}_A$ is analytical or measurement variance, $\sigma^2{}_{Sub}$ is subsampling error, and $\sigma^2{}_{Samp}$ is sampling error, including natural variability of the attribute that is being measured.

The number of factors that contribute to total sample variance are dependent on the experimental design. As previously stated, the quality of the conclusions of

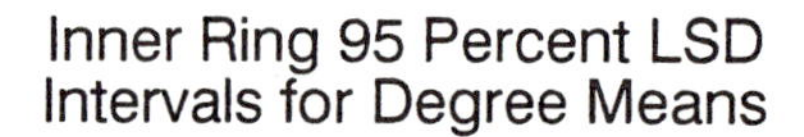

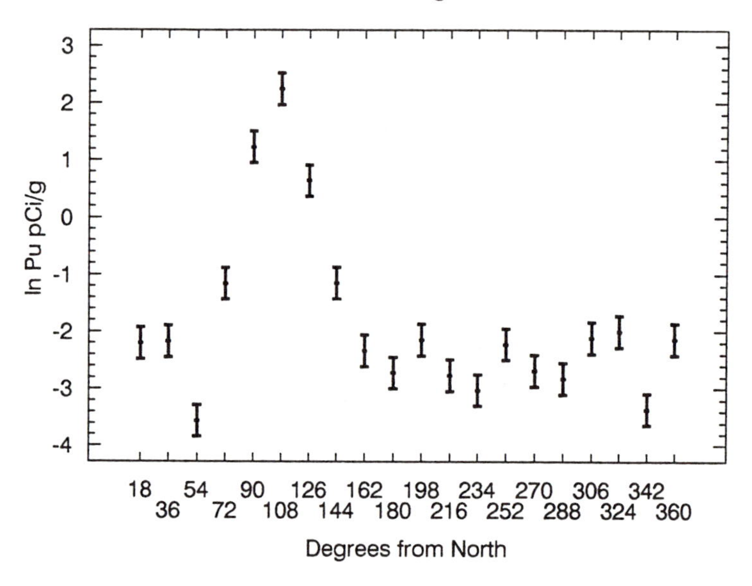

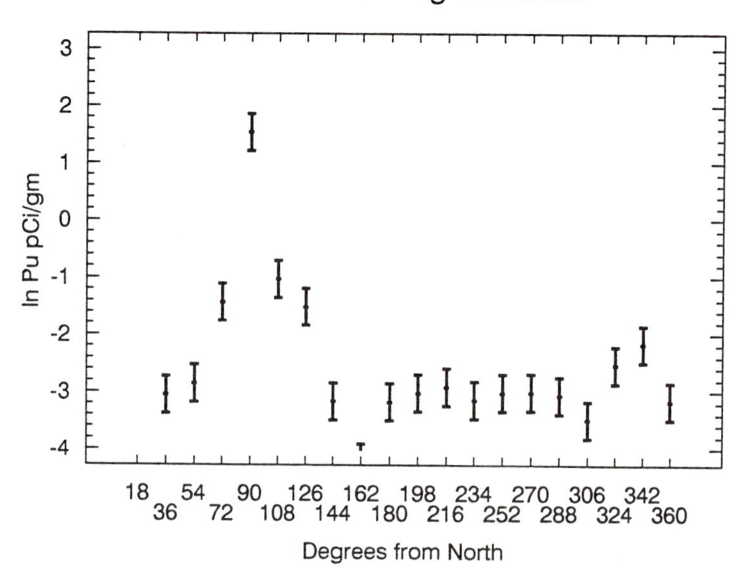

Figure 1. Means plot for the direction effect on soil plutonium at Rocky Flats Plant at Golden, CO. The inner ring is 1-mi radius from center of plan. The outer ring is 2-mi radius.

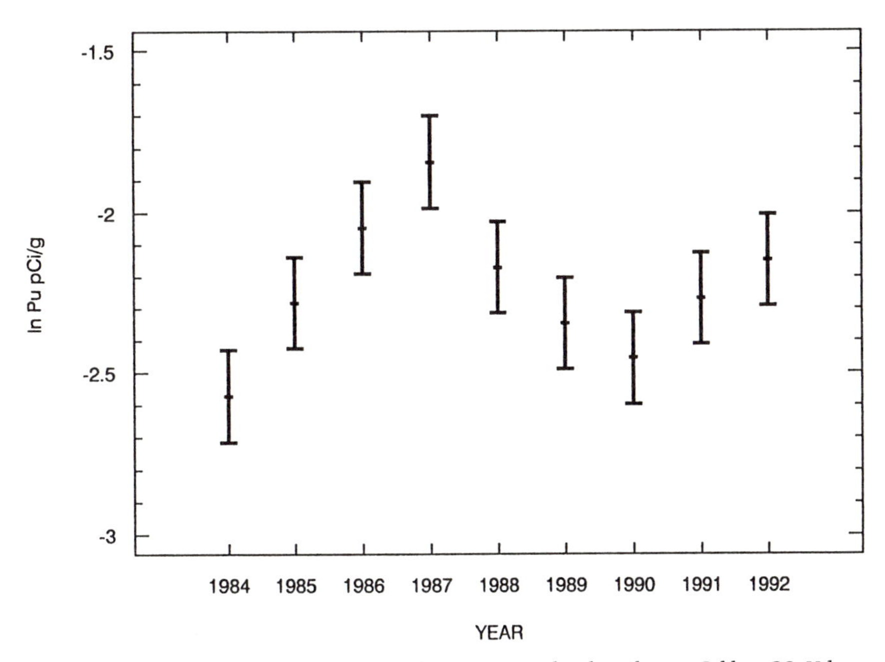

Figure 2. Means plot for the year effect on soil plutonium at Rocky Flats Plant at Golden, CO. Values are 95% LSD intervals for year means.

an environmental investigation is based not on the variance of a single sample but on the variance of the population estimates. For a three-level, completely nested design the variance of the estimate of the population mean is given in eq 7.

$$\sigma^2(\overline{x}) = \frac{\sigma^2_A}{rmn} + \frac{\sigma^2_{Sub}}{mn} + \frac{\sigma^2_{Samp}}{n} \qquad (7)$$

where $\sigma^2(\overline{x})$ is the estimate of the variance of the population mean, n is the number of samples collected, m is the number of subsamples per sample, and r is the number of replicate analyses per sample.

Although few environmental investigations are structured as a pure nested design, most sample protocols include some amount of subsampling and replicate analysis, and eq 7 provides insight into the relative contributions of the variance components. As the number of subsamples and replicate analyses increases, the relative importance of σ^2_A decreases. Another way of using eq 7 is to say that when the measurement error is high, the variance of the population parameters can be reduced by increasing the number of subsamples and laboratory replicates.

For reasons unknown to the author, few investigations are structured in a way that the components of variance can be easily quantified. Although a large amount of information exists about σ^2_A, there is a dearth of information on other compo-

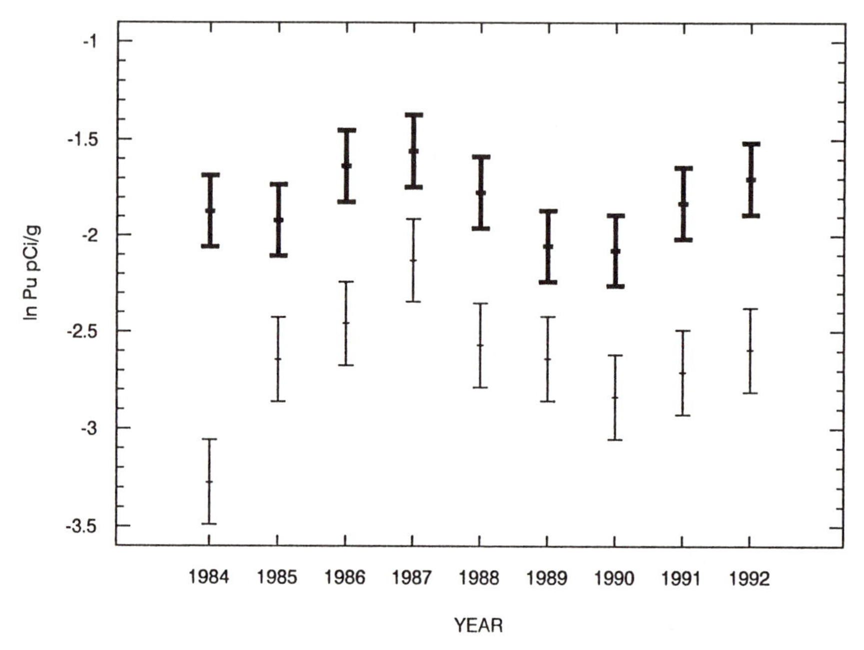

Figure 3. Means plot for the year effect on soil plutonium at Rocky Flats Plant at Golden, CO. Boldface means are the outer ring and the lighter lines are for the outer ring. Values are 95% LSD intervals for factor means.

nents. As a result, developers of sampling plans must make an extensive amount of assumptions to design efficient and cost-effective sampling programs. One approach to this dilemma is to review programs in which the same attribute has been measured over a long time with consistent and high-quality procedures. An example of this type of program is the Rocky Flats background geochemical characterization program. This program was designed to characterize the geochemistry of the soils and groundwater in the uncontaminated region west of Rocky Flats Plant. The program used what the EPA would call the highest quality analytical methods (similar to those of the Contract Laboratory Program, level IV on Table II). Consistent and high-quality field procedures were used, and the entire program had a high level of quality control and oversight.

The results of the calcium measurements in the groundwater over a period of several years are presented in Table VIII. The calculated variances are estimates of the total variance, which includes measurement, sample, and natural variability. When expressed as a percentage of the mean, the standard deviations range from a high of 66% for the weathered sandstone to a low of 25% for the weathered claystone. As a point of comparison, the EPA report for Contract Laboratory Program Routine Analytical Services (2) using inductively coupled plasma spectrometry states that the average relative percent difference between replicates was 6.0 for calcium. This value is based on performance evaluation sample results from 18 laboratories.

Table VIII. Background Calcium in Groundwater at Rocky Flats Plant

Parameter	*Colluvium*	*Rocky Flats Alluvium*	*Valley Fill Alluvium*	*Weathered Claystone*	*Weathered Sandstone*
No. of observations	20	67	43	19	37
Mean	99.5	38.7	90.4	53.7	36.4
SD	37.65	17.95	30.13	13.52	23.88
Variance	1417	322	908	183	570
Minimum	59.9	15.9	18.9	34.0	10.8
Maximum	186	84.6	120.0	73.4	91.9
Range	126.1	68.6	101.1	39.4	81.1
% RSD	38	46	33	25	66

NOTE: SD is standard deviation; RSD is relative SD.

The difference between the Rocky Flats and the EPA percent relative standard deviations (25–66% vs. 6%) is that the Rocky Flats data include sampling and subsampling errors (σ^2_{Samp} and σ^2_{Sub}). Although the figures presented here are not an exhaustive study, they do support the hypothesis that analytical error may be small in comparison to total sampling error. Thus data quality reviews that consider only the analytical errors are unrealistically focused.

Conclusions

Although the development and application of DQOs to the design of environmental sampling programs is logical, it lacks a fundamental basis. There is not a rigorous definition of, or an adequate way to measure, data quality beyond the common accuracy and precision techniques. Data validation provides a useful tool for reviewing the data generation process, identifying mistakes, reviewing adherence to methods, and comparing laboratory-level accuracy and precision to established guidelines. However, data validation does not provide a review of the entire data collection program or the quality of the conclusions that may be inferred from a data set. Misapplication of data quality guidelines can lead to the discarding of useful information. Furthermore, data validation can provide data users with an unwarranted level of confidence regarding a data set.

In a larger sense there is a misplaced focus on data quality at the expense of the quality of conclusions. There is no guarantee that high-quality data lead to high-quality conclusions. In applying tools of scientific analysis, environmental scientists may derive high-quality conclusions from poor-quality data. In the discipline of environmental sampling, in which our level of understanding of processes and the techniques available for investigations are constantly changing, the focus of a quality control–quality assurance program must be on the quality of conclusions. A well-conceived and cost-effective sampling program must incorporate our basic knowledge of environmental science and be based on an appropriate statistical design that considers all of the components of variance.

References

1. Klodowski, H. F. In *Principles of Environmental Sampling, 2nd ed.;* Keith, L. H., Ed.; American Chemical Society: Washington, DC, 1996; Chapter 3.
2. *Data Quality Objectives for Remedial Responses—Development Processes;* U.S. Environmental Protection Agency. Government Printing Office: Washington, DC, 1987, EPA–540 /G–87/003 OSWER Directive 9355.07 B.
3. *Data Quality Objectives for Remedial Response—Example Scenario;* U.S. Environmental Protection Agency Government Printing Office: Washington, DC, 1987, EPA–540/G–87/ 004 OSWER Directive 9355.07 B.
4. *Laboratory Data Validation Functional Guides for Evaluating Inorganic Analyses;* U.S. Environmental Protection Agency. Hazardous Site Evaluation Division. Government Printing Office: Washington, DC, 1988.
5. *Laboratory Data Validation Functional Guides for Evaluating Organic Analyses;* U.S. Environmental Protection Agency. Hazardous Site Evaluation Division. Government Printing Office: Washington, DC, 1988.
6. Barnard, T. E. In *The Handbook of Environmental Chemistry;* Einax, J., Ed.; Springer-Verlag: Heidelberg, Germany, 1995; Vol. 2/G, pp 1–47.
7. Wagner, K. Quantalex Inc., private communication, 1993.
8. Kimbrough, D. E.; Wakakuwa, J. R. *Environ. Sci. Technol.* **1992,** *26,* 2101–2104.
9. Kimbrough, D. E.; Wakakuwa, J. R. *Environ. Sci. Technol.* **1993,** *27*(13), 2692–2699.
10. Kimbrough, D. E.; Wakakuwa, J. R. *Environ. Sci. Technol.* **1992,** *26,* 2095–2100.
11. Wildeman, T. R.; Laudon, L. S.; Olsen, R. L.; Chappell, R. W. In *Chemical Modeling of Aqueous Systems II;* Melchior, D. C.; Bassett, R. L., Eds.; ACS Symposium Series 416; American Chemical Society: Washington, DC, 1990; pp 321–329.
12. Short, D. Diane Short & Associates, private communication, 1993.
13. Helsel, D. R.; Cohen, T. A. *Water Resour. Res.* **1988,** *24*(12), 1997–2004.
14. Newman, M. C.; Dixon, M. C.; Looney, P. M. B. B.; Pinker, J. E. III. *Water Resour. Bull.* **1989,** *25*(4), 905.
15. Porter, P. S.; Ward, R. C.; Bell, H. F. *Environ. Sci. Technol.* **1988,** *22*(8), 856–861.
16. Sanford, R. F.; Pierson, C. T.; Crovelli, R. A. *Math. Geol.* **1993,** *25*(1), 59–80.
17. Hinton, S. W. *Environ. Sci. Technol.* **1993,** 27(10), 2247–2249.
18. Currie, L. A. In *Detection in Analytical Chemistry: Importance, Theory, and Practice;* Currie, L. A., Ed.; ACS Symposium Series 361; American Chemical Society: Washington, DC, 1988; pp 1–62.
19. Kirchmer, C. J. In *Detection in Analytical Chemistry: Importance, Theory, and Practice;* Currie, L. A., Ed.; ACS Symposium Series 361; American Chemical Society: Washington, DC, 1988; pp 78–93.
20. Koorse, S. J. *J. Am. Water Works Assoc.* **1990,** *82,* 53–58.
21. U.S. Environmental Protection Agency. *Code of Federal Regulations,* Title 40, Part 136, Appendix B, 1985, Rev. 1.11. *Fed. Regist.* **1984,** *49*(209), 43430.
22. Neserke, G. E.; Taylor, H. *Water Environ. Res.* **1996,** *68*(1), 115–119.
23. "Rocky Flats Plant Site Environmental Report—1992"; EG&G Rocky Flats: Golden, CO, 1993; RFP–ENV–92.

Chapter 10

Pilot Studies for Improving Sampling Protocols

Walter S. Liggett and Kenneth G. W. Inn

Development of an environmental sampling and measurement protocol by means of experiments is feasible. The experiments require dissimilar plots, each of which is large enough for execution of a set of alternative protocols, but the experiments do not require that the property of interest be known for the plots. Thus, experimental development is possible even for parts of the protocol performed in the field. This chapter discusses the statistical framework for such experiments: what constitutes a proper set of alternative protocols, how the protocol responses can be used to estimate relative performance, and how the best protocol can be determined.

IN AN ENVIRONMENTAL STUDY, a sampling and measurement *protocol* is carried out on a unit of material to evaluate some *property* of the unit. Such a protocol permits study of changes in the property over many units. A unit of material is an area or volume small enough to be thought of as a point for purposes of the study. A unit may be called a *plot* or a *lot*. The property of interest in the study might be the amount of some substance in the soil within the plot, future leaching of a pollutant from the plot into a drinking water supply, the condition of the ecological community within the plot, or the habitat available in the plot for certain species. When applied to a plot, the protocol produces a *response* that is intended to be a measurement, perhaps on an unspecified scale, of the property of interest. Sometimes, a quantitative relationship between the property of interest and the response cannot be obtained because the property is only conceptual. This chapter discusses protocols with one-dimensional, continuous responses.

Properly chosen protocols provide responses with specificity and sensitivity adequate for the intended study. *Specificity* refers to the influence that unit proper-

Published 1996 American Chemical Society

ties other than the one of interest have on the protocol response. A protocol has adequate specificity if the response is sufficiently insensitive to spurious properties of the units. *Sensitivity* involves the difference observed between the responses from two different units and the variation observed when a response for a unit is obtained afresh. The size of one relative to the other is appropriate for comparing protocols with responses that may be on different scales. The sensitivity is adequate if the difference is sufficiently large relative to the variation.

Because the success of an environmental study depends on the choice of protocol, a pilot study for protocol development is often cost-effective. An experiment that leads to improvement of the sampling and measurement protocol is possible when available plots satisfy two conditions. First, each plot must be large enough to accommodate execution of several protocols without interference among protocols. Each of the responses obtained from a plot must be as if only one protocol were carried out on the plot. Second, the plots available must differ substantially in the property of interest. This condition does not mean, however, that the value of the property must be known for each plot. The purpose of this chapter is to present the statistical framework for such experiments: plans for application of protocols to plots, estimation algorithms, and experimental limitations.

Experiments of the type considered here have two major advantages. First, the sampling part of the protocol is included in its entirety. For example, a protocol for waste site characterization must include specification of where, within each plot, cores are to be extracted; the coring device; the section of the core to be used; the technique for grinding, mixing, and compositing sections of the core; the quantity of material to be sent to the laboratory; the technique to be used in preparing material for the analytical instrument; and the analytical instrument itself. All these specifications are included in the experiments considered. Second, finding a group of units that satisfy the two conditions described in the preceding paragraph is usually possible. Thus, experiments of the type considered here are flexible enough that they can even be used to adjust a protocol for new conditions. Examples of reasons for adjusting a protocol are changes in the type of soil, the biological activity, or the season.

The experiments considered here do not provide a scale for the protocol response. This limitation arises because the values of the property of interest are not known for the experimental units. Thus, although specificity and sensitivity can be improved, other experimental steps, perhaps included in the intended study, are needed for determination of the relation between the protocol response and measurement scales in general use.

Comparison in Environmental Studies

In contexts such as waste site characterization, the relation between the protocol response and established measurement scales is critical. For waste site characterization, the reason is that risk assessment is based on the actual amount of pollutant present. In other contexts such as the determination of the condition of ecological resources, the focus of the intended studies is generally comparison among locations and over time.

One might consider sampling a lot of particulate material for the purpose of determining the total quantity of some constituent (*1, 2*). Waste site characterization is one example. About the sampling of two-dimensional lots, Pitard (*2*) says, "The ideal solution would be to transform the two-dimensional lot into a one- or zero-dimensional lot. Most of the time this is too expensive to be considered. Practically, we must conclude that the sampling of two-dimensional lots made of particulate material is an unsolvable problem." By "unsolvable", Pitard (*2*) means that sampling *correctness* cannot be achieved, or in other words, that no protocol exists that gives all elements in the lot the same probability of being selected. As a consequence, measurement of the total quantity of a constituent may entail a large bias. Despite this difficulty, waste site characterization is usually thought of in terms of an established standard of cleanliness and measurement of remediation units to determine if the standard is met.

The experimental approach to protocol development presented here has potential as a way through this difficulty. A protocol development experiment that makes use of a group of remediation units from the waste site can be undertaken. Successful development will lead to a protocol with adequate specificity and sensitivity, if one exists. This protocol will provide good differentiation between cleaner units and more contaminated units although not the actual contaminant levels. A spatial statistics algorithm applied to the responses from this protocol would produce a contour map with the proper shape for choosing units for remediation even though the actual contaminant levels would be missing. After identification of the most contaminated units, a scale for the protocol responses could be determined by accurately measuring a few remediation units, perhaps starting with the most contaminated one. As prescribed by Gy (*1*) and Pitard (*2*), this measurement could be performed by removing all the soil in the remediation unit, transforming it into a one-dimensional lot, and sampling it correctly. Having obtained a scale for the protocol responses, one could then make the necessary remediation decisions on the basis of risk assessment. These steps, which consist of development of a protocol, application of the protocol, and calibration of the protocol, require correct sampling of only a few remediation units and thus reduce the cost of restricting measurement bias in waste site characterization.

Consideration of specificity and sensitivity before accuracy contrasts with the usual development of a sampling and measurement protocol. Measurement system performance is usually thought of in terms of the *bias* and *precision*. These terms underlie the process by which the EPA sets data quality objectives (*3*). These terms also guide the formulation of protocol development experiments in many fields. Barcelona (*4*) discusses such experiments for environmental studies; Bailar (*5*) suggests them in a broader measurement context; and Groves (*6*) considers them for surveys of human populations. Mostly, these experiments are aimed at reducing bias. Ideally, scientific models exist that provide the basis for protocol development by showing how the property of interest can be obtained bias-free from a practical protocol. Such scientific models underlie Gy's sampling theory (*1, 2*). However, sometimes this approach to protocol development is not feasible. In the case of sampling particulate materials, this limitation occurs when a correct sampling method cannot be adopted. In other cases, it occurs because the property itself is only conceptual. The alternative approach presented here does not require a scientific model but instead requires, as discussed already, experimental units that differ substantially in the property of interest.

Underlying Model

Selection of K protocols is a decision vital to the experiments considered here. Most important, the protocols included should be limited to those believed to respond selectively to the same property of study units. Another approach to protocol development exists. If the purpose of protocol development is regarded as providing a statistic for classifying unknown units into classes, and if groups of experimental units can be regarded as known representatives of these classes, then discriminant analysis can be used for protocol development (*7*). In this approach, protocols that respond to a variety of properties can be included, and the question of which property is best for discriminating among classes can be asked. This approach, which is the basis of some pilot studies performed as part of the U.S. EPA Environmental Monitoring and Assessment Program (*8, 9*), is fundamentally different from the one presented here.

In protocol selection, one should look first to whatever scientific theory helps connect the property of interest with a practical protocol. Even when considerable theory exists, choices arise for which there is no guidance. The experiments considered here fill in what is unknown about protocol choices. The potential benefits of such experiments have been discussed in the context of industrial quality improvement by Yano (*10*). What may be unknown about protocols includes both the influence of other unit properties on the response and the effect of various choices on sensitivity. The influence of other properties on the response, if strong enough, will be detected by the test for lack of specificity. Effects on sensitivity will be reflected in estimates of a performance measure.

The numerical assessment of protocol performance requires a mathematical model that connects the property of interest with the protocol responses. As discussed in the following section, conformance with this model is another aspect of protocol selection. A response denoted by y_{kji} results when protocol k is applied to unit j during replication i. Because the property of conceptual interest is not given for the units, its values can be expressed on a scale suitable for the model assumed for the y_{kji}. Let the value of the property for unit j be denoted by u_j. The model implies that the y_{kji} are related to the u_j by a linear calibration curve with intercept α_k and slope β_k. The calibration curve depends on the protocol, but the u_j values do not. The model is

$$y_{kji} = \alpha_k + \beta_k u_j + \sigma_k \varepsilon_{kji} \tag{1}$$

where σ_k is the standard deviation of the response and ε_{kji} is a random quantity with mean 0 and standard deviation 1. The repeatability is given by σ_k.

The amount of experimental effort depends on the number of protocols K, the number of units J, and the number of *replicates* specified for unit j, I_j. This relationship means that unit j must accommodate $K \times I_j$ sampling operations in such a way that the operations do not interfere with each other. As will be discussed subsequently, the protocols can be separated into blocks so that the number of sampling operations on a unit can be reduced.

The experiments considered here test the validity of eq 1. Thus, the rationale for the experiment requires that any knowledge of the connection between the protocols and the property of conceptual interest be used to limit the protocols to a group for which eq 1 is believed valid. If during the planning, eq 1 is thought to be valid and if during the statistical analysis, eq 1 is shown to be invalid, then the experiment will have been useful nonetheless. On the other hand, if eq 1 is known at the outset to be invalid, then the experiment may be uninformative. Consideration of this distinction in planning applies to both the mean response, which is given by the first two terms on the right side of eq 1, and the error, which is given by the last term.

Model for the Mean

Some studies involve, in addition to the property of real interest, unit properties expected to affect different protocols differently. One example is the evaluation of estuary condition based on sampling benthic organisms. In this example, the salinity of the water and silt-clay content of the bottom are properties that will generally influence different protocols in different ways. Of course, neither of these properties is of direct interest. In this case, if possible, protocols for the experiment should be selected only if they are insensitive to these peripheral properties. If this condition were to prove impossible, then one might consider confining the experiment and the intended environmental study to units with only a narrow range of salinity and bottom type so that the influence of these properties would not change very much from unit to unit. Performing an experiment of the type recommended here without regard for such peripheral properties might be a waste of effort because, as discussed later, sensitivity comparison depends on the validity of eq 1.

Because sampling of particulates as discussed by Gy (*1*) is important in environmental studies, we discuss this example in detail. Consider J lots composed of fragments. Let the fragments be divided into classes according to their mass; and let these classes be indexed by w. The mass of the fragments in class w is denoted by M_{jw}, and the average mass fraction of the analyte of interest for the fragments in class w is denoted by a_{jw}. Assume that under each protocol, the selection probability for fragments depends only on the fragment mass. Let the selection probability under protocol k for fragments in class w be P_{kw}. Gy (*1*) discusses mechanisms that might cause P_{kw} to depend on w. Under these circumstances, the mean of the sample mass fraction is

$$\frac{\sum_{w} a_{jw} M_{jw} P_{kw}}{\sum_{w} M_{jw} P_{kw}}$$

We ask, when does this expression have the same form as the mean response in eq 1? First, if P_{kw} does not depend on w, then the sampling fulfills Gy's definition of correctness and the mean mass fraction obeys eq 1. There are other situations, however. For example, consider a group of protocols that select only fragments in a narrow mass range. For such protocols, the selection probabilities P_{kw} are nonzero only for the same narrow mass range. If over this narrow mass range, M_{jw} and a_{jw} do not depend on w, then the mean mass fraction obeys eq 1. Adopting protocols sensitive only to fragments in a narrow mass range may or may not be reasonable in light of the property of real interest in the environmental study. This example shows that the protocols considered for the experiment do not have to be restricted to those that are correct but, on the other hand, do have to be restricted.

Model for the Error

The error in the response, which is given by the last term in eq 1, is modeled as being statistically independent from observation to observation and as having a standard deviation that depends only on the protocol. Of these two assumptions, the former is more important. In the experiment, the differences between replicate responses are an essential part of protocol comparison. Thus, the differences between replicates must reflect variation in all the error components incurred in the anticipated environmental study. This stipulation requires that each response be the result of carrying out the protocol from the beginning, that is, replicates cannot share sampling steps even though sharing would save effort. For example, in the case of measuring soils, if the protocol calls for compositing increments from a few locations randomly chosen within the unit boundaries, then to obtain a duplicate, the protocol must be executed again with new locations chosen independently. This type of duplicate is called a field duplicate by Van Ee et al. (3). Obtaining replicates through separate visits to the units is a reasonable precaution against actions by field personnel that cause the independence assumption to be violated.

Planning with the objective that each protocol have the same standard deviation for all units seems difficult. One can imagine protocol features that might equalize the standard deviation over the units. For example, although the same protocol must be used at every unit, a protocol can contain an objective criterion evaluated at the unit that determines such things as number of increments composited or depth of increment. The difficulty lies in seeing exactly how this criterion should be applied to achieve equal standard deviations. Fortunately, compensation for unequal standard deviations may be possible during the statistical analysis. First, small departures will not prevent large differences between protocols from being observed. Second, modeling the standard deviations as the product of separate protocol and unit effects is a generalization of the statistical analysis that may be possible. In the experiments considered here, the only violation of eq 1 that is treated as not crucial to the study conclusions is the possibility of standard deviations that vary with unit.

Protocol Optimization

When eq 1 is valid, the proper criterion for picking the best protocol is $|\beta_k/\sigma_k|$, which is larger for better protocols. This criterion has been part of the measurement literature at least since it was discussed by Mandel and Stiehler (*11*). An equivalent criterion, $\log(\beta_k^2/\sigma_k^2)$, is called a signal-to-noise ratio by Yano (*10*). The slope of the calibration curve, β_k, appears in this criterion along with the standard deviation of the response, σ_k, because the measurements must be put on a common scale before comparison. As shown by eq 1, the quantity $(y_{kji} - \alpha_k)/\beta_k$ is a transformation of y_{kji} to a common scale, the u_j scale. Because the standard deviation of this quantity is $|\sigma_k/\beta_k|$, the criterion $|\beta_k/\sigma_k|$ is appropriate.

Statistical Analysis

Estimation Algorithm

An estimate of $|\beta_k/\sigma_k|$ for each protocol follows from estimation of α_k, β_k, u_j, and σ_k from the responses y_{kji}. Because this process differs from ordinary regression in that the values of the u_j are also estimated, α_k, β_k, and u_j cannot be uniquely determined from the y_{kji} even when $\sigma_k = 0$. To obtain uniqueness, that is, identifiability, we impose the constraints $\Sigma_{j=1}^{J} I_j u_j = 0$ and $\Sigma_{j=1}^{J} I_j u_j^2 = N$, where $N = \Sigma_{j=1}^{J} I_j$. These constraints are not restrictions because, as discussed previously, the scale of the u_j is arbitrary.

The estimation employs an algorithm that might be called an iteratively reweighted singular value decomposition. Liggett (*12*) details the relation of this algorithm to maximum likelihood estimation. The algorithm consists of an initialization step followed by an iterative procedure that contains a step in which the β_k and u_j are recomputed, a step in which the σ_k are recomputed, and a step in which whether to continue (or to stop) is decided. We denote by $\tilde{\beta}_k$, $\tilde{u}_j$, and $\tilde{\sigma}_k$ the most recent values obtained as the iteration proceeds.

Initialization. Before the iteration, we estimate the α_k, compute initial values of the σ_k, and form averages over the replicates. We compute $y_{kj\cdot}$, the average of y_{kji} over i, and $y_{k\cdot\cdot}$, the average of y_{kji} over j and i. The intercept α_k is estimated by $y_{k\cdot\cdot}$. The values of the σ_k taken as most recent when the iteration begins are given by

$$\tilde{\sigma}_k^2 = \sum_{j=1}^{J} \sum_{i=1}^{I_j} (y_{kji} - y_{kj\cdot})^2 / (N - J) \tag{2}$$

β_k, u_j Step. In the first step of the iteration, we compute (or recompute) $\tilde{\beta}_k$, $\tilde{u}_j$ by applying the singular value decomposition to the matrix **A** with elements

$$A_{kj} = I_j (y_{kj\cdot} - y_{k\cdot\cdot}) / \tilde{\sigma}_k \tag{3}$$

where the $\tilde{\sigma}_k$ are the current values. Application of the singular value decomposition to this matrix reexpresses **A** in terms of three matrices (**U**, **D**, and **V**) as the product $\mathbf{UDV}^T$. The matrices **U** and **V** are orthogonal and the matrix **D** is diagonal. If $J \leq K$, then **U** is $K \times J$, and **D** and **V** are $J \times J$; otherwise, **U** and **D** are $K \times K$, and **V** is $J \times K$. Moreover, we have $D_{11} \geq D_{22} \geq \ldots \geq 0$. Computer software for the singular value decomposition is widely available. For example, Nash (*13*) provides pseudocode that is easily changed into a running program.

From the elements in the three matrices produced by the singular value decomposition, we obtain the values of β_k and u_j for this iteration. Our rationale is as follows. First, eq 1 gives for the elements of **A**

$$A_{kj} = (\beta_k / \tilde{\sigma}_k) I_j u_j + I_j \sigma_k (\varepsilon_{kj\cdot} - \varepsilon_{k\cdot\cdot}) / \tilde{\sigma}_k \tag{4}$$

Second, the matrix $\mathbf{UDV}^T$ has elements given by $\Sigma_m U_{km} D_{mm} V_{jm}$. Thus, because D_{11} is the largest element in **D**, we let $(\beta_k/\tilde{\sigma}_k) I_j u_j = U_{k1} D_{11} V_{j1}$.

Formulas for computation that take into account the constraints on the u_j are given by

$$\tilde{\beta}_k = U_{k1} D_{11} \tilde{\sigma}_k (\mathrm{N}^{-1} \sum_{j=1}^{J} V_{j1}^2 / I_j)^{1/2} \tag{5}$$

and

$$\tilde{u}_j = (V_{j1} / I_j) / (\mathrm{N}^{-1} \sum_{j=1}^{J} V_{j1}^2 / I_j)^{1/2} \tag{6}$$

σ_k Step. In the second step of the iteration, we recompute $\tilde{\sigma}_k$ using

$$\tilde{\sigma}_k^2 = \sum_{j=1}^{J} \sum_{i=1}^{I_j} (y_{kji} - y_{k\cdot\cdot} - \tilde{\beta}_k \tilde{u}_j)^2 / (N - 2) \tag{7}$$

where $\tilde{\beta}_k$ and $\tilde{u}_j$ are the current values.

Decision To Continue. The final step in the iteration is the decision whether to continue. The iteration is continued until $|\tilde{\beta}_k/\tilde{\sigma}_k|$ does not change appreciably between cycles. The values of $\tilde{\beta}_k$, $\tilde{u}_j$, and $\tilde{\sigma}_k$ when the iteration is stopped are the estimates, which we denote by $\hat{\beta}_k$, $\hat{u}_j$, and $\hat{\sigma}_k$.

Example

To illustrate this algorithm, we use measurements from an experiment on leaching of glass by high-purity water that was originally reported by Kingston et al. (*14*). For the purposes of this illustration, we adopt as the property of conceptual interest the tendency of glass to dissolve. In this experiment, 13 glass compositions were subjected in replicate to a glass leaching test. In our illustration, these 13 compositions correspond to the experimental units. These compositions differ in their tendency to dissolve. Of the elemental mass loss measurements made in the experiment, we consider three, those for the main glass constituents, Si, B, and Na. In addition, we consider the total mass loss, which was also measured. These measurements are given in grams per square meter. So that the elemental mass loss measurements might plausibly be considered to obey the model given by eq 1, we adjust them for the variation in concentration among the glass compositions. In this adjustment, the mass loss measurement for a particular element is divided by the concentration of that element in the particular composition and multiplied by the concentration in the base glass, which is the one we index by $j = 1$. We consider the three adjusted mass loss measurements and the total mass loss as four different protocols to be compared. The values are shown in Table I. For information on the uncertainty in these measurements, the reader should refer to Kingston et al. (*14*). Is it reasonable to consider these adjusted measurements as responses to the same property, the tendency to dissolve, as required in eq 1? The answer is yes under the simplest model of glass leaching, namely, that all the constituents of the glass leach into the leachant in the same proportion that they are present in the glass itself.

We estimate α_k, β_k, u_j, and σ_k from the responses in Table I. In the algorithm, $K = 4$, corresponding to the total mass loss ($k = 1$) and the adjusted elemental mass losses for Si, B, and Na ($k = 2, 3, 4$). The values of J and I_j can be obtained from Table I; thus $I_1 = 7$, but otherwise $I_j = 2$. As estimates of $(\alpha_k, \beta_k, \sigma_k)$, the algorithm gives (24.68, 11.20, 1.10) for total mass loss, (5.58, 1.70, 0.27) for Si, (1.54, 0.76, 0.02) for B, and (5.11, 2.52, 0.13) for Na. Consequently, the estimates of $|\beta_k/\sigma_k|$ are 10.2 for the total mass loss, 6.4 for Si, 42.4 for B, and 20.1 for Na. Provided that all four responses can be considered as measurements of a single property, the protocol based on the measurement of boron is the best because $|\hat{\beta}_k/\hat{\sigma}_k|$ is the largest when $k = 3$.

Using the estimates of u_j provided by the algorithm, we can plot the responses versus $\hat{u}_j$ for each protocol just as is routinely done in regression analysis. The 13 u_j estimates are –0.22, –1.00, –0.59, 0.80, 0.06, 3.53, 0.16, –0.43, –0.35, –0.26,

Table I. Adjusted Mass Loss Measurements Indexed by Protocol, Composition, and Replicate

j	i	*Total* k=1	*Si* k=2	*B* k=3	*Na* k=4
1	1	22.548	5.420	1.350	4.420
1	2	22.575	5.470	1.370	4.320
1	3	22.588	5.310	1.350	4.400
1	4	22.651	5.620	1.400	4.360
1	5	22.614	5.490	1.390	4.320
1	6	22.957	5.530	1.400	4.520
1	7	23.036	5.380	1.340	4.800
2	1	14.329	3.583	0.776	2.598
2	2	14.139	3.838	0.758	2.791
3	1	20.334	4.830	1.061	3.740
3	2	20.688	4.910	1.078	3.649
4	1	33.573	6.775	2.115	7.120
4	2	33.892	7.268	2.170	7.240
5	1	25.908	6.050	1.586	5.314
5	2	25.696	5.690	1.566	5.136
6	1	64.751	11.621	4.244	13.948
6	2	64.161	11.570	4.210	14.023
7	1	26.960	6.180	1.650	5.459
7	2	27.700	6.035	1.660	5.519
8	1	18.347	4.531	1.210	4.200
8	2	18.014	4.431	1.220	3.960
9	1	19.805	4.913	1.290	4.280
9	2	19.696	4.711	1.260	4.240
10	1	20.146	4.902	1.340	4.440
10	2	22.038	4.913	1.350	4.520
11	1	18.942	4.555	1.220	4.200
11	2	18.813	4.511	1.210	4.120
12	1	18.159	4.577	1.180	3.860
12	2	18.446	4.566	1.220	3.860
13	1	20.877	4.818	1.320	4.560
13	2	20.683	4.960	1.320	4.440

NOTE: Adjusted mass loss measurement values are in grams per square meter; *j* refers to composition, *i* refers to replicate, and *k* refers to protocol.

–0.42, –0.46, and –0.28. Instead of plotting y_{kji} versus $\hat{u}_j$, we plot $(y_{kji} - \hat{\alpha}_k)/\hat{\beta}_k$ versus $\hat{u}_j$. These plots are shown in Figure 1. One glass composition has a distinctly higher propensity toward leaching. Deviations from the model in eq 1 are shown in these plots as deviations from straight-line dependence. For total mass loss and Si, the plots show some indication of deviations from a straight line, and these deviations are somewhat similar. This similarity could be an indication of a second property that affects the responses. This possibility is discussed further in the next section. In Figure 1, the amount of scatter around a line with slope 1 is proportional to $|\sigma_k/\beta_k|$. In these terms, Figure 1 shows that the protocol based on boron is the best.

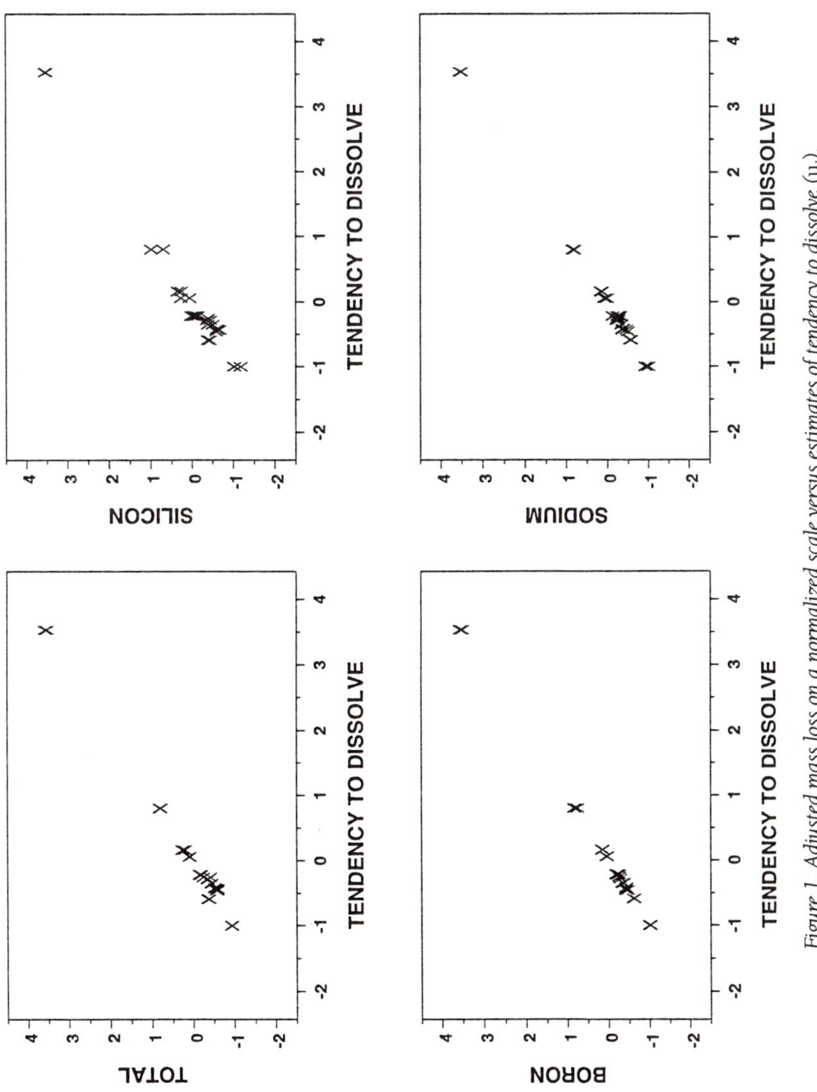

Figure 1. Adjusted mass loss on a normalized scale versus estimates of tendency to dissolve (u_j).

Response Surface Methodology

In contrast to this example, the choice among protocols in many pilot studies is more complicated than simply the choice of the best of K. In many cases, the choice among protocols can be thought of in terms of several factors, some of which can be assigned any value in a range and some of which can be assigned only values that correspond to discrete options. Examples of the former include the size to which the largest fragment is reduced by grinding, the time that the sample is mixed before a subsample is extracted, and the mass of the subsample. An example of the latter is the choice between two types of equipment. The ranges of the continuous and discrete factors together define an experimental region from which one would like to select the factor settings that give the best protocol. Besides practicality, the validity of the model given by eq 1 might, as discussed previously, lead to limitations on the experimental region. Experimental maximization of some criterion over an experimental region is the goal of response surface methodology (*15*).

For use in response surface methodology, the criterion

$$\eta = \log(\beta^2/\sigma^2) \tag{8}$$

which is equivalent to $|\beta_k/\sigma_k|$, is a good choice. Because η depends on the protocol, η can be thought of as a function of the factors that determine the protocol. If the factors are denoted by $x_1, x_2, \ldots, x_p$, then we can write $\eta(x_1, x_2, \ldots, x_p)$. The goal can be thought of as finding the factor settings that maximize this function.

In response surface methodology, the approach is to evaluate the function at various points in the experimental region, that is, at various settings of the factors, and then use these evaluations to determine a polynomial approximation to the function. The location of the maximum of the approximation can then be used as an approximation to the optimum factor settings. The points at which $\eta(x_1, x_2, \ldots, x_p)$ is to be evaluated correspond to the protocols that enter the experiment. Applying these protocols to experimental units as discussed, we obtain values for y_{kji} and thereby estimates of $\eta_k = \log(\beta_k{}^2/\sigma_k{}^2)$ using the estimation algorithm.

We fit a polynomial model to $\log(\hat{\beta}_k{}^2/\hat{\sigma}_k{}^2)$ instead of $|\hat{\beta}_k/\hat{\sigma}_k|$ for reasons involving the experimental error in the evaluation of η. In most cases, the experimental error in $\log(\hat{\beta}_k{}^2/\hat{\sigma}_k{}^2)$ will be largely due to the error in estimating $\sigma_k{}^2$, and therefore, the standard deviation of $\log(\hat{\beta}_k{}^2/\hat{\sigma}_k{}^2)$ can be approximated by the standard deviation of $\log(\hat{\sigma}_k{}^2)$. Because $\hat{\sigma}_k{}^2$ is a variance estimate with $N - 2$ degrees of freedom, the standard deviation of $\log(\hat{\sigma}_k{}^2)$ is approximately $[2/(N - 2)]^{1/2}$. Sometimes, adding 1 to $\hat{\beta}_k{}^2/\hat{\sigma}_k{}^2$ before taking the logarithm prevents very poor protocols from having too large an influence on the polynomial approximation. Box and Draper (*15*) provide more details on response surface methodology.

Because the number of protocols that can be evaluated in an environmental experiment is limited, the complexity of the polynomial to be used for modeling must be limited and the number of factors allowed to vary in the experiment must be limited. Although the polynomials used in response surface studies often contain quadratic terms so that a maximum within the experimental region can be portrayed, polynomials composed of just linear terms sometimes provide a useful model even though the maximum can only lie on the boundary of the experimental region. In fact, when the performance differences among protocols are very large, a polynomial with just linear terms may be sufficient to reveal important protocol improvements. Thus, although the response surface studies possible in sampling protocol development cannot be as elaborate as those in other areas, such studies can lead to large gains.

Blocking

In design of the experiment, an issue that might have to be resolved is a limitation on the number of independent responses that can be obtained from each experimental unit. Such a limitation might result from disturbance of the unit during sampling. Plot disturbance is familiar because it has been considered in other environmental sampling contexts. Reducing the number of independent responses is a problem because protocols carried out on different sets of units cannot be compared directly. Recall that in the estimation algorithm, we specified that $\Sigma_{j=1}^{J} I_j u_j = 0$ and $\Sigma_{j=1}^{J} I_j u_j^2 = N$ to make the estimates unique. Because of these specifications, the estimates of β_k from a set of units contain an unknown scale factor that changes from one set of units to another. This factor makes direct comparison of estimates of β_k obtained from different sets of units impossible.

A solution to this problem is blocking the experimental design. Box and Draper (*15*) describe blocking of experimental designs in response surface studies. Blocking is carried out in other contexts for exactly the reason it is suggested here, to resolve limitations on the number of observations that can be made while holding conditions nearly constant.

The following simple example illustrates blocking. Let the protocols that are to enter the experiment depend on two factors, x_1 and x_2; one of these factors might indicate how the soil sample is obtained, and the other might indicate how an analytical subsample is extracted. In the study, we consider four protocols that correspond to setting x_1 at –1 or +1 and to setting x_2 at –1 or +1. Let the criterion be approximated by

$$\log(\beta^2/\sigma^2) = a_0 + a_1 x_1 + a_2 x_2 \tag{9}$$

The coefficient a_0 depends on which set of units gives the estimate of β. We can carry out this experiment using two sets of units. At the first set, we implement the two protocols corresponding to $(x_1,x_2) = (-1,-1)$ and $(+1,+1)$. At the other set, we implement the other two protocols, the ones corresponding to $(x_1,x_2) = (-1,+1)$ and $(+1,-1)$. Because only a_0 depends on the set of units, we can estimate $a_1 + a_2$ from the first set and $a_1 - a_2$ from the second set. Solving for a_1 and a_2, we determine which setting of x_1 is best and which setting of x_2 is best. This allocation of four protocols, two to each set of units, is an example of blocking.

We can expand this example into the design of a small pilot study that involves four responses from each of eight units. The eight units are divided into two sets of four, and two protocols are implemented in duplicate on the units of each set. The algorithm for estimating $\eta = \log(\beta^2/\sigma^2)$ must be carried out separately for each set of units. Let the results for the two protocols implemented on the first set be η_{--}, η_{++} and the results for the second set be η_{-+}, η_{+-}. We obtain

$$\begin{aligned} a_1 &= (\eta_{++} - \eta_{--} - \eta_{-+} + \eta_{+-})/4 \\ a_2 &= (\eta_{++} - \eta_{--} + \eta_{-+} - \eta_{+-})/4 \end{aligned} \tag{10}$$

If a_1 is positive and large enough to be statistically significant, then the setting $x_1 = 1$ is best. Statistical significance can be based on the standard deviation of η given previously. Similarly, if a_1 is sufficiently negative, then the setting $x_1 = -1$ is best. In the same way, a_2 gives the best setting for x_2.

Lack of Model Validity

The previous section shows, on the basis of the model in eq 1, how to select the best protocol. This section shows how to check this model for agreement with the observed responses. This section also indicates, in general terms, how to proceed when there is lack of agreement. In most cases, however, lack of agreement must be resolved through scientific insight, for which there is no statistical method. In this way, this section is less prescriptive than the last one.

Consider first responses that show lack of validity of the model for the mean of the responses. As discussed in the section on the underlying model, such responses could result from a peripheral property of units that the protocols are not selective against. On the other hand, such responses could result from differences in the shapes of the calibration curves of different protocols, that is, from differences between protocols that cannot be modeled by just changes in α_k and β_k as in eq 1. Consider, for example, a case in which the amount of a volatile analyte is of interest and some of this analyte is lost during sampling. If the amount lost varies from one protocol to another and the amount lost is proportional to the amount present, then eq 1 will be valid. However, proportionality might not hold under various mechanisms such as saturation effects. In this case, the model in eq 1 fails because of nonlinear differences in the calibration curves rather than because of an interfering property. Another example is the measurement of ecological diversity. The various indexes of diversity are generally not linearly related to each other. Moreover, protocol changes, such as substitution of one device for catching organisms for another, might change the calibration curve in more complicated ways than just changing α_k and β_k.

Lack of validity of the model for the error is similarly ambiguous. When the standard deviation σ_k depends on the unit in violation of eq 1, the cause might be various unit properties that make sampling more or less difficult. On the other hand, measurement error is usually regarded as having constant relative standard deviation, in which case σ_k is proportional to the level of what is being measured and varies among units. The experimental data will generally be insufficient to distinguish such causes.

Model of the Mean

The model for the response mean can be tested in an approximate way with the lack-of-fit test used in regression analysis when replicates are present. Because the values of the u_j are not known, we use the estimates obtained previously. The test statistic is given by

$$\frac{\sum_{j=1}^{J} I_j (y_{kj\cdot} - y_{k\cdot\cdot} - \hat{\beta}_k \hat{u}_j)^2 / (J-2)}{\sum_{j=1}^{J} \sum_{i=1}^{I_j} (y_{kji} - y_{kj\cdot})^2 / (N - J)}$$

If we ignore the substitution of estimates for the u_j, then we can compare this statistic with F-ratio critical points for $J - 2$ and $N - J$ degrees of freedom. Computing this statistic for the glass leaching example, we obtain for total mass loss 18.99, for Si 8.36, for B 0.28, and for Na 1.09. These values are in general agreement with what is shown in Figure 1. Comparison of these values with the 5% critical point of the F distribution with 11 and 18 degrees of freedom, which is 2.37, shows that the first two values are so high that the null hypothesis is rejected. Thus, we conclude that the deviations from a straight line cannot be explained by the variation observed among the replicates. In addition, the third value, the one for B, is too low; in other words, the fit is better than one would expect on the basis of the variation exhibited by the replicates. As an explanation for the low value for B, we note the substitution of $\hat{u}_j$ for u_j. The problem with this substitution is discussed by Liggett (*12*). The values for total mass loss and Si are large enough to confirm the conclusion that some other property in addition to the tendency to dissolve affects the response. A scientific description of this other property is not available.

When the deviation from the mean response model is the result of a single interfering property of the units, the deviation may not be apparent in the results from any particular protocol but the deviation can be characterized by adding another term to eq 1 and fitting the more general model. As an alternative to eq 1, we fit

$$y_{kji} = \alpha_k + \beta_k u_j + \gamma_k v_j + \sigma_k \varepsilon_{kji} \tag{11}$$

The algorithm given previously can be modified to achieve this fitting. In the singular value decomposition step, we add $\tilde{\gamma}_k = U_{k2} D_{22} \tilde{\sigma}_k (N^{-1} \Sigma_{j=1}^{J} V_{j2}{}^2 / I_j)^{1/2}$ and $\tilde{v}_j = (V_{j2}/I_j)/(N^{-1} \Sigma_{j=1}^{J} V_{j2}{}^2/I_j)^{1/2}$, and in the variance estimation step, we compute

$$\tilde{\sigma}_k^2 = \sum_{j=1}^{J} \sum_{i=1}^{I_j} (y_{kji} - y_{k\cdot\cdot} - \tilde{\beta}_k \tilde{u}_j - \tilde{\gamma}_k \tilde{v}_j)^2 / (N - 3) \tag{12}$$

Although it is not exactly the case that the estimate of u_j is an estimate of the property of interest and the estimate of v_j is an estimate of the interference, the association may be strong enough to allow a scientific judgment on the cause of the interference. The estimate of γ_k is similarly useful in that it may indicate which protocols are most influenced by the interference.

Model for the Error

Consider now deviations from the model of the response random error. We proceed by first asking whether some units seem generally much more difficult to sample than other units. In other words, we ask whether for most of the protocols, the precision achieved at some units is much worse than at other units. To answer this question, we compute

$$(y_{kji} - y_{kj\cdot}) / \hat{\sigma}_k$$

for all protocols and units, and plot these versus the estimates of u_j. If the spread is much greater at some units than others, then we conclude that some units are much harder to sample or that something with a similar manifestation is occurring. Figure 2 shows such a plot for the glass leaching example. Caution must be exercised in interpreting this plot for two reasons. One reason is that for the so-called base glass ($j = 1$), many more replicates were performed. Thus, the spread for this glass can be expected to be larger even without any dependence of σ_k on j. Second, the quantity plotted, $(y_{kji} - y_{kj\cdot})/\hat{\sigma}_k$, is divided by the estimate of σ_k, which is inflated by any interfering second property. Thus, when, as in the glass leaching example, a second property is confirmed, plots like Figure 2 can be misleading. Further statistical analysis to remove these sources of confusion is possible. In any case, dependence of σ_k on j does not seem to be a problem in the glass leaching example.

If the difficulty in sampling varies considerably from unit to unit, then the possibility that the best protocol varies from unit to unit cannot be ignored. One way to explore this possibility is to arrange

$$\log[(I_j - 1)^{-1} \sum_{i=1}^{I_j} (y_{kji} - y_{kj\cdot})^2]$$

in a two-way table and perform median polish (*16*). Patterns in the residuals left after the row and column effects have been removed suggest an interaction between protocol precision and unit.

Scientific Investigation

The tests and plots suggested for checking the validity of eq 1 are, when lack of validity is demonstrated, only the beginning of a scientific investigation. The goal of

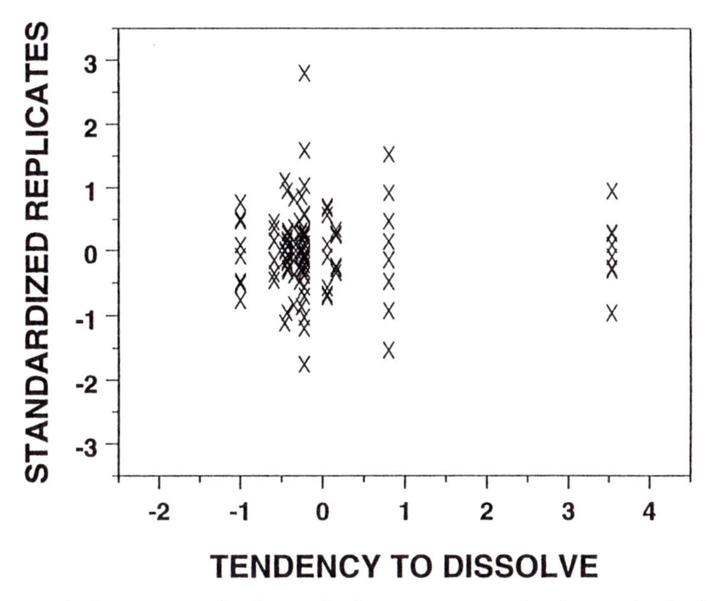

Figure 2. Replicates, centered and normalized, versus estimates of tendency to dissolve (u_j).

this investigation is a set of protocols that satisfy eq 1. As a step toward achieving this goal, these tests and plots might lead to an understanding of the problems with the protocols in the current study.

Conclusions

The benefits of the experiments described in this chapter entail special effort in carrying out some experimental steps. First, the selection of protocols for inclusion in the experiment requires considerable thought because this selection determines what can be learned. Second, as with any environmental field work, close adherence to the specified protocols requires careful planning. Third, valid conclusions require expertise in statistical analysis. Of course, the outcome of an experiment of the type considered here is not guaranteed. Nevertheless, the potential of such experiments is shown by successes in the development of measurement methods for product quality experiments (*10*).

Although one carefully planned experiment can lead to substantial improvement in a sampling and measurement protocol, additional experiments can be run to further optimize the protocol, as is often done in response surface experiments (*15*). Further experiments can refine the settings of the factors that define the protocol or help determine a protocol with better specificity. However, because of the effort involved, one should plan on one rather than a sequence of experiments and let the benefits of the first experiment justify another experiment.

Measurements specifically related to underlying properties of interest allow study results to be compared and integrated with other environmental information. Interestingly, the experiments discussed here can provide assurance of the validity of what is believed known about the relation between the chosen protocol and the property of interest. The reason is that when one has a set of protocols believed to be specific to a property and when the specificity of these protocols is not contradicted by the experimental data, one's understanding of the relation between the property and protocols is confirmed. This confirmation provides assurance.

References

1. Gy, P. M. *Sampling of Heterogeneous and Dynamic Material Systems: Theories of Heterogeneity, Sampling, and Homogenizing;* Elsevier: Amsterdam, Netherlands, 1992.
2. Pitard, F. F. *Pierre Gy's Sampling Theory and Sampling Practice, 2nd ed.;* CRC Press: Boca Raton, FL, 1993.
3. Van Ee, J. J.; Blume, L. J.; Starks, T. H. *A Rationale for the Assessment of Errors in the Sampling of Soils;* U.S. Environmental Protection Agency. Environmental Monitoring and Systems Laboratory: Las Vegas, NV, 1990; EPA/600/4-90/013.
4. Barcelona, M. J. In *Principles of Environmental Sampling;* Keith, L. H., Ed.; ACS Professional Reference Book; American Chemical Society: Washington, DC, 1988; pp 3–23.
5. Bailar, B. A. *Int. Stat. Rev.* (International Statistical Institute: Amsterdam, Netherlands) **1985**, *53,* 123–139.
6. Groves, R. M. In *Measurement Error in Surveys;* Biemer, P. P.; Groves, R. M.; Lyberg, L. E.; Mathiowetz, N. A.; Sudman, S., Eds.; Wiley: New York, 1991; pp 1–25.
7. Gnanadesikan, R. *Methods for Statistical Data Analysis of Multivariate Observations;* Wiley: New York, 1977.
8. *EMAP-Surface Waters 1991 Pilot Report;* Larsen, D. P.; Christie, S. J., Eds.; U.S. Environmental Protection Agency. Environmental Research Laboratory: Corvallis, OR, 1993; EPA/620/R-93-003.
9. Weisberg, S. B.; Frithsen, J. B.; Holland, A. F.; Paul, J. F.; Scott, K. J.; Summers, J. K.; Wilson, H. T.; Valente, R.; Heimbuch, D. G.; Gerritsen, J.; Schimmel, S. C.; Latimer, R. W. *EMAP-Estuaries Virginian Province 1990 Demonstration Project Report;* U.S. Environmental Protection Agency. Environmental Research Laboratory: Narragansett, RI, 1993; EPA/600/R-92/100.
10. Yano, H. *Metrological Control: Industrial Measurement Management;* Asian Productivity Organization: Tokyo, Japan, 1991.
11. Mandel, J.; Stiehler, R. D. *J. Res. Natl. Bur. Stand.* **1954**, *53,* 155–159.
12. Liggett, W. S. "Functional Errors-in-Variables Models in Measurement Optimization Experiments"; *1994 Proceedings of the Section on Physical and Engineering Sciences;* American Statistical Association: Alexandria, VA, 1995; pp 193–199.
13. Nash, J. C. *Compact Numerical Methods for Computers: Linear Algebra and Function Minimization;* Wiley: New York, 1979.
14. Kingston, H. M.; Cronin, D. J.; Epstein, M. S. *Nucl. Chem. Waste Manage.* **1984**, *5,* 3–15.
15. Box, G. E. P.; Draper, N. R. *Response Surfaces and Empirical Model Building;* Wiley: New York, 1987.
16. Tukey, J. W. *Exploratory Data Analysis;* Addison-Wesley: Reading, MA, 1977.

Chapter 11

The Sampling Theory of Pierre Gy

Comparisons, Implementation, and Applications for Environmental Sampling

Leon E. Borgman, John W. Kern, Richard Anderson-Sprecher, and George T. Flatman

The sampling theory developed and described by Pierre Gy is compared to design-based classical finite sampling methods for estimation of a ratio of random variables. For samples of materials that can be completely enumerated, the methods are asymptotically equivalent. Gy extends the finite sampling methods to situations in which complete enumeration of samples is not feasible. Gy's methods involve a set of sampling constants related to the heterogeneity of the material sampled; methods to estimate these constants from grouped data are given. Computer programs for the estimation of these constants are described, and environmental applications are discussed.

CLASSICAL FINITE SAMPLING THEORY, sometimes called design-based sampling, is most often associated with the design and analysis of sample surveys. The identical theory is also used, however, to guide the selection and analysis of samples in other applications. In particular, environmental studies are most often planned and interpreted from this standpoint (*see*, e.g., reference 1). A thorough description of classical sampling may be found in Cochran (2), and Thompson (3) describes applications of the classical theory to selected nonstandard problems.

The major advantage of classical random sampling is that it is fundamentally objective; assumptions made about the underlying population are, for practical purposes, nonexistent. An underlying characteristic of the classical paradigm is that it considers the real world (the population of interest) to be fixed and deterministic, and randomness is present only because of the sample selection process. When estimating a fixed population parameter, variation within the population is thus a hurdle to be surmounted, and probabilistic description of variation is neither necessary nor even meaningful. Whatever patterns of variability are present within the population are effectively removed or, more accurately, nullified (on average) by randomization of the sample.

The alternative to classical sampling is generally understood to be model-based sampling. Model-based sampling (like design-based sampling) actually consists of a variety of methods, the most practical and important variant for environmental samplers being geostatistical sampling (*see* reference 4 for a thorough general treatment and reference 5 for a discussion of geostatistical applications to environmental problems). The main distinction between model-based and design-based theories rests on the use or nonuse of a model to account for patterns of variability within the population. Geostatistical models describe the spatial covariance structure of variables of interest. Because these models are stochastic, they interject a degree of randomness into one's perception of the world itself. In other words, the world as observed through a window in space-time is no longer seen as a fixed fact but is rather viewed as a single realization of a random process.

Because the model-based approach views randomness as part of the population itself, random sampling is no longer necessary for a model-based sample design. In fact, it is generally not even desirable because regularly spaced observations usually provide the best information about the random process one assumes to be lurking behind the realized population. The price paid for allowing nonrandom sampling is that the patterns of variation in space and/or time (i.e., the character of the underlying process) must be adequately, although not perfectly, understood if estimates are to be reliable. The payoff is that model-based sampling is usually more efficient because it makes more complete use of information about the population.

Model-based sampling is now sufficiently well established among sampling theorists for opposing camps of design- vs. model-based statisticians to have formed, feuded, and, sometimes, made truces. More and more practitioners now recognize that both perspectives have value, depending on the actual problem at hand. Relevant articles of interest include Borgman and Quimby (*6*) and Brus and de Gruijter (*7*).

Are design-based and model-based sampling the only choices available? There exists, perhaps, a third way, as well. Pierre Gy, a French mining engineer, developed a sampling theory in relative isolation from the theoretical statisticians, and his theory is now being advanced as another alternative to the classical sampling theory (*8–10*). Like classical sampling, Gy's theory assumes a particular fixed state of nature, which the sampler wishes to describe with calculable confidence. Like model-based sampling (in particular, geostatistical sampling), Gy's theory attempts to address patterns in the variability of the population, ore body, or area to be sampled. Although Gy developed his theory in the context of a particular problem—the estimation of the grade (percentage mineral content) in a sample of ore—proponents view Gy's contributions as a general theory of sampling that offers improvements over standard methods in other contexts as well. In particular, whenever a sampled medium may be viewed as particulate (including fluids), Gy's theory may be applied. Environmental samplers who work with soils and waters are of course interested in any contributions that Gy's theory may offer.

The following pages describe certain important aspects of Gy's theory and compare it with classical sampling, thereby clarifying the theory itself and indicating what practitioners can hope to gain from this perspective. Lyman (*11*) compares Gy's variance estimator to that of Ingamells, including extensive data analysis. Much of the mathematics in Gy's theory is equivalent to that in classical design-based sampling theory, and the most important connections will be examined. Gy's work is most easily understood in the context in which it was developed, and the explanation that follows makes reference to a population of mineral particles containing varying amounts of ore. Examples are given for ways that this model can be used in situations that parallel problems of interest to environmental samplers.

Gy's Sampling Theory

One of the primary contributions of Gy's theory is its systematic identification of different aspects of population variability before sampling begins. Gy treats the world as deterministic, just as do classical samplers, but the choice of sample for Gy may depend on one's assessment of population variability. (This assessment may be made in part by studying the variogram of the variable being studied, so Gy's theory has connections with geostatistical sampling.) In particular, the method seeks to improve upon classical sampling by directly addressing the various sources of error rather than by simply relying on randomization to account for all potential variation. Pragmatically, this separation of error sources is often essential because some errors may represent actual biases, not just simple variation. After carrying out a presampling analysis, scientists may adopt whatever additional assumptions they believe are appropriate to organize the sampling process and to interpret results.

The essence of Gy's theory is to "divide and conquer". Once sources of error in sampling are identified, an attempt is made to minimize each type of error separately. Pitard (*10*) describes this part of Gy's theory in detail, and the following brief summary draws heavily from his exposition. Some error minimization is just good technician work. In other cases errors are intrinsic to the population being sampled. Many of Gy's procedures for accounting for variability are expressly motivated by properties of particulate sampling. Most notably, true simple random samples are not feasible in particulate populations. Also, units can occasionally be altered and rearranged by physical crushing and mixing. (Environmental populations can rarely be homogenized in this manner.) Finally, technicalities arise because the parameter of interest, the percent of a mineral in a body of ore, is a ratio instead of a simple additive measure.

The most basic errors identified by Gy result from the intrinsic heterogeneity of the world; that is, not all particles are the same, and unlike particles are unevenly distributed in space. To assess the heterogeneity in a population Gy asks two questions. First, how much do sampling units differ from each other (What is the *constitution heterogeneity*)?; and second, how are different types of units spread about or clustered within the population (What is the *distribution heterogeneity*)? Both types of heterogeneity affect the reliability of a sample.

Because the constitution heterogeneity is impossible to alter, the sampling error that is associated with it is termed the fundamental error (FE). This error can never be eliminated and is a major focus of much of Gy's theory. Estimation of the variance of the fundamental error is the portion of the theory that is outlined in most detail in the sections that follow.

A variety of errors are related to the interaction between the distribution heterogeneity and the sampling method used (always some form of cluster sampling). An important error of this type results from the interplay between uneven clumps of units in the population and sampling devices that grab clumps of units. This error is called the grouping and segregation error (GE). In brief, the grouping and segregation error is smallest when clustering is absent in both the population and the sampling procedure. Environmental samplers rarely have the luxury of being able to homogenize (mix) their populations, so this error will be present in most problems of interest. As is always true in cluster sampling problems, the desire to select small sampling clusters must be balanced against the practicality of sampling larger clusters.

When the population is itself a nearly linear flow of particles, then additional errors are associated with distribution heterogeneity. In this case the distribution heterogeneity may be addressed by using the variogram to describe variation in the population (not in a random process as in geostatistics). Stratified sampling can then be used as a remedial measure with strata selected according to the errors associated with trends (TrE) and cycles (CyE) that are identified in such a stream. Patterns in two or three dimensions are substantially more difficult to characterize, and Gy does not attempt to address such patterns.

A perfectly executed sampling plan would contain precisely the errors described previously. For a single lot, the ideal sampling error (ISE) would then be

$$\text{ISE} = \text{FE} + \text{GE} \tag{1}$$

and for a moving stream of particles the error would be

$$\text{ISE} = \text{FE} + \text{GE} + \text{TrE} + \text{CyE} \tag{2}$$

Gy also carefully delineates and probes errors that can enter a sample from sources other than those ideally described thus far. Survey statisticians have long recognized problems such as processing errors, nonresponse bias, and the effects of improper sampling. Analogous errors may arise in particulate sampling, and Gy has gone to great pains to describe, measure, and minimize errors of this sort. Among major error sources delineated by Gy are errors that arise from edge effects of sampling equipment and from similar mechanical problems (delimitation and extraction, or, jointly, mechanical errors); errors in preparing samples for laboratory analysis (preparation errors); and actual errors from the laboratory (analytic errors). Those familiar with quality control may note a similarity in spirit between this identification of error sources and similar exercises in the quality literature.

Three comments about these additional types of errors are relevant at this point. First, these errors are potentially important to the practitioner because they may bias observations, as mentioned previously. If they are recognized in time, most of these errors can be minimized, or even eliminated, by proper physical collection and handling. Second, because sampling is often done in stages, most of these errors can enter the problem many times, and the stage with the greatest error present will form a lower bound on the total sampling error. Third, the theory itself can only point to the existence of such variability; it cannot itself remedy errors at this level.

The error that cannot be removed by even the most careful technicians and the best instrumentation is the error intrinsic to the population variability, that is, the fundamental error. The following discussion focuses on fundamental error because it is always present and it is the only error that can be assessed independently of the sampling method. The fundamental error as defined by Gy is the relative error in estimating the grade (proportion of desired mineral), and it is thus a measure of the variation intrinsic in the population of available mineral particles. Its variance is the square of the coefficient of variation of the grade. If the grade is expressed as the ratio of two random variables on a set of sampling units, then the fundamental error may be estimated using methods given by Cochran (2). It can be shown that Gy's methods are equivalent to those given by Cochran, for a population for which complete enumeration is possible. However, Gy has developed methods for particulate sampling when complete enumeration of samples is not feasible or cost-effective.

Comparison of Gy's Theory and Classical Sampling

The description that follows uses the notation of classical statistics and the physical context of particulate ore sampling. Most environmental samplers will be accustomed to the notation used, but they will probably wish to translate physical variables into those used in their own areas of interest. For example, grade of ore may be analogous to the percentage of some chemical present in a particular medium.

Let L represent an ore body, where X is the total mass of the body and Y is the mass of the mineral of interest. The parameter to be estimated is $R = Y/X$, the grade of the ore body. In the language of classical finite sampling, this problem is the estimating of the ratio of two random variables. In finite sampling theory, it is assumed that a population is composed of N sampling units (U_i, $i = 1, 2, 3, \ldots N$) and that certain attributes of these units may be enumerated or measured. For the estimation of ore grade, the finite population consists of a set of fragments of ore. The mass of the mineral of interest contained in fragment i is denoted by y_i and the total mass of the fragment is given by x_i for $i = 1, 2, 3, \ldots N$. The notational conventions used for totals, averages and ratios are given in Table I.

An estimator of the ratio R and approximations of the first two moments of that estimate are given both by Cochran (2) and by Gy (8). Each uses a slightly different method of derivation to arrive at results, but the moments they obtain are equivalent up to the order of approximation. A summary of the derivation of these results is given.

Estimation of the Grade $\hat{R}$

In a finite population the statistical expectation operator is defined by averaging over all possible combinations (C_{Nn}) of samples, where

$$C_{Nn} = \frac{(N)!}{n!(N-n)!} \tag{3}$$

Cochran (3) shows that the expectation of $\bar{y}$ is given by

$$E(\bar{y}) = \frac{\sum \bar{y}}{C_{Nn}} = \bar{Y} \tag{4}$$

where the sum is over all possible samples. Using this result, it is clear that Ny/n is an unbiased estimator of Y. The natural estimator of R is based on the ratio of totals

$$\hat{R} = \frac{y}{x} \tag{5}$$

Table I. Notation for Two Random Variables Measured on a Sample of Size *n* from a Population of Size *N*

	Population of Size N	*Sample of Size n*
Total	$Y = \sum_{i=1}^{N} y_i$	$Y = \sum_{i=1}^{n} y_i$
Mean	$\bar{Y} = \frac{Y}{N},\ \bar{X} = \frac{X}{N}$	$\bar{y} = \frac{y}{n},\ \bar{x} = \frac{x}{n}$
Ratio	$R = \frac{Y}{X} = \frac{\bar{Y}}{\bar{X}}$	$\hat{R} = \frac{y}{x} = \frac{\bar{y}}{\bar{x}}$

Moments of the Estimated Grade $\hat{R}$

The moments of a ratio estimator are not obvious. Both Gy and Cochran find means and approximations of variances of $\hat{R}$. Brief derivations follow.

Define

$$\mu_x = E(x) = \frac{nX}{N} \qquad \mu_y = E(y) = \frac{nY}{N} \tag{6}$$

One may express $\hat{R}$ in terms of the relative variables u and v defined by

$$x = \mu_x(1+u) \qquad y = \mu_y(1+v) \tag{7}$$

Then, in terms of u and v, $\hat{R}$ is given by

$$\hat{R} = R\frac{(1+v)}{(1+u)} \tag{8}$$

Changing the denominator into a multiplicative factor and using Taylor's theorem gives

$$\hat{R} = R(1+v)(1-u+u^2-u^3+\ldots) \tag{9}$$

Because the expectations of both u and v are 0

$$E(\hat{R}) = R[1-\mu_{11}(u,v)+\mu_{20}(u,v)+\mu_{21}(u,v)+\ldots] \tag{10}$$

where

$$\mu_{ij}(u,v) = E[(u-\mu_u)^i(v-\mu_v)^j] \tag{11}$$

Equation 10 may be written in the form

$$E(\hat{R}) = R(1+S) \tag{12}$$

where S is given by

$$S = [-\mu_{11}(u,v)+\mu_{20}(u,v)+\mu_{21}(u,v)+\ldots] \tag{13}$$

S is equivalent to what Gy calls the fundamental bias

$$S = \text{Bias}(\hat{R}) \equiv \frac{E(\hat{R})-R}{R} \tag{14}$$

which is the relative bias in the estimate $\hat{R}$. Gy (8) and Cochran (2) both independently provided approximations of S using the first two terms in the series and writing the result in terms of the correlation between x and y. Matheron (*12*) in an examination of Gy's work used Laplace transforms to derive a general expression for the expectation of the ratio of two random variables raised to a power. This general result was used to derive Gy's formula. Further, Cochran (2) presents the exact results due to Hartley and Ross (*13*)

$$E(\hat{R}) = R\left(1 - \frac{\mu_{11}(\hat{R},x)}{\mu_x R}\right) \tag{15}$$

and

$$S = -\frac{\mu_{11}(\hat{R},x)}{E(x)R} \tag{16}$$

Only approximate formulas for the variance of $\hat{R}$ are available. Using eq 8 we compute the expectation

$$E(\hat{R}^2) = R^2(1+S') \tag{17}$$

where

$$S' = [\mu_{02}(u,v) - 4\mu_{11}(u,v) + 3\mu_{20}(u,v) + \ldots] \tag{18}$$

is obtained from the Taylor expansion of $1/(1 + u)^2$. Combining eqs 12 and 17 gives

$$\begin{aligned}\mathrm{Var}(\hat{R}) &= E(\hat{R}^2) - [E(\hat{R})]^2 \\ &= R^2(S' - 2S)\end{aligned} \tag{19}$$

Using the definition of $\mu_{ij}(u,v)$ given in eq 11, it can be shown (2) that, up to the second order moments of u and v

$$\mathrm{Var}(\hat{R}) \cong \frac{1}{\overline{X}^2}\left(\frac{N-n}{nN}\right)\sum_{i=1}^{N}\frac{(y_i - Rx_i)^2}{N-1} \tag{20}$$

The Fundamental Error

Gy (8) defines the fundamental error (FE) of estimation and the relative variance of the fundamental error, $\sigma^2(\mathrm{FE})$, as

$$\mathrm{FE} = \frac{\hat{R} - R}{R} \qquad \sigma^2(\mathrm{FE}) = \frac{\mathrm{Var}(\hat{R})}{[E(\hat{R})]^2} \tag{21}$$

respectively. This convention is slightly nonstandard in that the usual variance of the fundamental error is given by

$$\mathrm{Var(FE)} = \frac{\mathrm{Var}(\hat{R})}{R^2} = S' - 2S \tag{22}$$

In practice, this convention does not pose any difficulty because up to second order moments in u and v

$$\sigma^2(\mathrm{FE}) = \mathrm{Var(FE)} \tag{23}$$

From these forms, Gy derives the approximate formula

$$\frac{\mathrm{Var}(\hat{R})}{[E(\hat{R})]^2} \cong \left(\frac{N}{n}-1\right)\sum_{i=1}^{N}\left(\frac{x_i}{X}\right)^2\left(\frac{R_i - R}{R}\right)^2 \tag{24}$$

which is used in applications that follow.

In summary, Gy's fundamental error is exactly the relative bias in the estimate $\hat{R}$ of the ratio of random variables. This relative bias is equivalent to the bias given by Cochran. Further, the variance of the fundamental error as defined by Gy, σ^2(FE), is asymptotically equivalent to the usual variance of the fundamental error, Var(FE).

In applications where complete enumeration of the sample is possible, the methods given by Gy are equivalent to those of classical random sampling. Differences between Gy's methods and those of finite sampling lie in the methods developed for sampling of particulate materials after grouping into categories. These methods are used to reduce the cost of estimation when complete enumeration of the sample is not feasible. In these methods, estimators are developed that are similar to those applied to estimate the mean and variance of grouped data.

Application to Particulate Materials

To estimate R and the variance of $\hat{R}$ requires complete enumeration of the n units in the sample and measurement of x_i and y_i on each sampled fragment. In the case of particulate materials, this requirement is impractical. To overcome this problem, Gy derived a method of estimation of $\hat{R}$ and the variance that does not require fragment-by-fragment enumeration. Details follow.

Let L represent the population of particulate material with N fragments denoted by (U_i, i = 1, 2, 3, ... N). These fragments may be divided into classes $L_{\alpha\beta}$ with average volume V_α and average density Δ_β. If each fragment in the class $L_{\alpha\beta}$ is identified with an average fragment $F_{\alpha\beta}$, then an estimate of $\hat{R}$ and Var($\hat{R}$) based on the midpoints of the size and density classes may be used. This treatment is essen-

Table II. Definitions of Notation Used for Estimation of $\hat{R}$ and Var($\hat{R}$) Based on Size and Density Classes

Parameter	*Population of Size* N	*Sample of Size* n
Average volume	V_α	v_α
Average density	Δ_β	δ_β
Average mass	$\overline{X}_{\alpha\beta} = V_\alpha \Delta_\beta$	$\overline{x}_{\alpha\beta} = v_\alpha \delta_\beta$
Average ratio (grade)	$R_{\alpha\beta}$	$\hat{R}_{\alpha\beta}$

NOTE: Each fragment is represented by an average fragment ($F_{\alpha\beta}$).

tially the same as the computation of the mean and variance from a grouped frequency distribution. The necessary notation is listed in Table II.

One can consider eq 24 for the variance of the fundamental error. Summation on the index i is replaced with double sums on α and β as

$$\sum_{i=1}^{N}(R_i - R)^2 x_i^2 \cong \sum_{\alpha=1}^{r}\sum_{\beta=1}^{s} N_{\alpha\beta}(R_{\alpha\beta} - R)^2 \overline{X}_{\alpha\beta}^2 \tag{25}$$

Making this substitution in eq 24 and using the fact that $\overline{X}_{\alpha\beta} = V_\alpha \Delta_\beta$ gives

$$\begin{aligned}\frac{\mathrm{Var}(\hat{R})}{[E(\hat{R})]^2} &= \left(\frac{N}{n} - 1\right)\left(\frac{1}{X}\right)\sum_{\alpha=1}^{r}\sum_{\beta=1}^{s}\frac{N_{\alpha\beta}\overline{X}_{\alpha\beta}}{X}\left(\frac{R_{\alpha\beta} - R}{R}\right)^2 V_\alpha \Delta_\beta \\ &= \left(\frac{N}{n} - 1\right)\left(\frac{1}{X}\right) H\end{aligned} \tag{26}$$

H is defined to be a constant of constitution heterogeneity. The number of fragments $N_{\alpha\beta}$ in class (α,β) times the average particle mass gives the total mass in that size–density class.

Using this relationship

$$H \cong \sum_{\alpha=1}^{r}\sum_{\beta=1}^{s}\left(\frac{R_{\alpha\beta} - R}{R}\right)^2 \frac{X_{\alpha\beta}}{X} V_\alpha \Delta_\beta \tag{27}$$

Now an estimate of H based on the sample of n fragments is needed. One may estimate V_α and Δ_β with their sample equivalents v_α and δ_β, respectively. This procedure gives the estimate

$$x \cong \sum_{\alpha}\sum_{\beta} x_{\alpha\beta} \text{ where } \overline{x}_{\alpha\beta} \cong v_\alpha \delta_\beta \quad x_{\alpha\beta} \cong n_{\alpha\beta}\overline{x}_{\alpha\beta} \tag{28}$$

where $n_{\alpha\beta}$ may be known, or estimated based on average volume and mass. Depending on the degree of precision desired and the available resources, the number of grains in each volume density class may be counted or estimated based on the average size and density.

Defining δ_m as the density of the constituent of interest, δ_w as the density of the waste, and

$$\hat{R}_{\alpha\beta} \cong \frac{\left(\frac{1}{\delta_\alpha} - \frac{1}{\delta_w}\right)}{\left(\frac{1}{\delta_m} - \frac{1}{\delta_w}\right)} \tag{29}$$

an estimate of the ratio R is given by the weighted average

$$\hat{R}_2 = \frac{\sum_\alpha \sum_\beta x_{\alpha\beta} \hat{R}_{\alpha\beta}}{\sum_\alpha \sum_\beta x_{\alpha\beta}} \tag{30}$$

Substituting eq 28 into eq 27 and using eqs 29 and 30 gives the estimate

$$\hat{H} \cong \sum_\alpha \sum_\beta \left(\frac{\hat{R}_{\alpha\beta} - \hat{R}_2}{\hat{R}_2}\right)^2 \frac{x_{\alpha\beta}}{x} v_\alpha \delta_\beta \tag{31}$$

As with $n_{\alpha\beta}$, the average value of $\hat{R}_{\alpha\beta}$ could be estimated through assay if budget constraints allowed. In most environmental sampling scenarios, direct assay would be used. It should be noted that when estimated by eq 29, $R_{\alpha\beta}$ appears to depend only on α. However, the density of waste and mineral may vary with volume class depending on the degree of separation between ore and waste (percent liberation). As the fragment size decreases, the percent liberation of the constituent of interest is generally increased. This variation will be captured if $R_{\alpha\beta}$ is estimated by assay rather than by eq 29. Finally, because $X >> x$, an estimate of the variance of the fundamental error is

$$\begin{aligned} \mathrm{Var(FE)} &\cong \left(\frac{N}{n} - 1\right)\frac{1}{X}\hat{H} = \left[\left(\frac{X}{\overline{X}}\right)\left(\frac{\overline{x}}{x}\right) - 1\right]\frac{1}{X}\hat{H} \\ &\cong \left(\frac{1}{x} - \frac{1}{X}\right)\hat{H} \cong \frac{1}{x}\hat{H} \end{aligned} \tag{32}$$

Estimation of Physical Constants

Gy (9) developed a set of physical constants that can be used to estimate the variance of the fundamental error for mineralogical data. What follows is a method to estimate those physical constants from sample data. To facilitate the computation of these constants we introduce the usual dot notation for row and column sums typically associated with the analysis of variance.

$$x_{\alpha.} = \sum_{\beta=1}^{s} x_{\alpha\beta} \quad x_{.\beta} = \sum_{\alpha=1}^{r} x_{\alpha\beta} \tag{33}$$

$$x_{..} = \sum_{\alpha=1}^{r}\sum_{\beta=1}^{s} x_{\alpha\beta} = \sum_{\alpha=1}^{r} x_{\alpha.} = \sum_{\beta=1}^{s} x_{.\beta} \tag{34}$$

One can define the constitution heterogeneity for a given size class obtained by summing over the density classes

$$H_{\alpha} = \sum_{\beta} \frac{x_{\alpha\beta}}{x_{\alpha.}} \left(\frac{\hat{R}_{\alpha\beta} - \hat{R}_2}{\hat{R}_2} \right)^2 v_{\alpha} \delta_{\beta} \tag{35}$$

Two limiting cases can be identified for H_{α} given by complete homogeneity of the material sampled or complete heterogeneity. These limiting cases help to explain the method being used, and they also occur in certain applications.

1. Completely homogeneous ($\hat{R}_{\alpha\beta} \equiv \hat{R}_2$) for all α. In this case $H_{\alpha} = 0$ (no constitution heterogeneity).
2. Completely heterogeneous (completely liberated). In this case all of the material in class α can be separated into two density classes: $\beta = 1$, for pure mineral, with grade 1.0; $\beta = 2$, for pure waste, with grade 0.0.

In the completely liberated case, let $x_{\alpha 1}$ be the mass of the mineral in class α, and $x_{\alpha 2}$ be the mass of the waste in class α. Then, in this limiting case, let $c_{\alpha} = H_{\alpha}$

$$c_{\alpha} = R_{\alpha} \left(\frac{1 - \hat{R}_2}{\hat{R}_2} \right)^2 \delta_m + (1 - R_{\alpha}) \delta_w \tag{36}$$

where

$$R_{\alpha} = \frac{x_{\alpha 1}}{x_{\alpha 1} + x_{\alpha 2}} \tag{37}$$

is the ratio of the mass of the constituent of interest to the total mass in volume class α. In this case the liberation ratio (ℓ_α) is defined to be H_α/c_α, so that for the completely homogeneous case $\ell_\alpha = 0$ and for the completely liberated case $\ell_\alpha = 1.0$.

If we define

$$H^* = \frac{\sum_\alpha x_{\alpha.} v_\alpha c_\alpha \ell_\alpha}{\sum_\alpha x_{\alpha.} v_\alpha} \tag{38}$$

then the mineralogical factor c is given by

$$c = \hat{R}_2 \delta_m \left(\frac{1 - \hat{R}_2}{\hat{R}_2} \right)^2 + (1 - \hat{R}_2)\delta_w \tag{39}$$

and the liberation factor ℓ is the ratio of H^* and c

$$\ell = \frac{H^*}{c} \tag{40}$$

Finally the constitution heterogeneity H can be approximated by

$$\hat{H} \cong \sum_\alpha \frac{x_{\alpha.}}{x_{..}} v_\alpha c \ell \tag{41}$$

Letting v_{95} be the 95th percentile of the volumes, we can define the granulometric constant

$$g = \sum_\alpha \frac{x_{\alpha.} v_\alpha}{x_{..} v_{95}} \tag{42}$$

Then the variance of the fundamental error is estimated by

$$\sigma^2_{FE} = \frac{\mathrm{Var}(\hat{R})}{[E(\hat{R})]^2} \cong \left(\frac{N}{n} - 1 \right) \left(\frac{1}{X} \right) c \ell g v_{95}$$

$$\cong \left(\frac{1}{x} - \frac{1}{X} \right) c \ell g v_{95} \tag{43}$$

$$\cong \frac{c \ell g v_{95}}{x}$$

The final relation follows because $X >> 1$. In the context of environmental sampling, these constants H, ℓ, c, and g must be reinterpreted and estimated. François-

Bongarçon (*14*) noted that Gy's method, although potentially powerful, has failed to be widely applied even in mining applications because of the difficulty in adequately estimating the geological constants. Further research should be directed toward determination of appropriate physical constants in the environmental setting. Sinclair (*15*) emphasized the importance of characterization of heterogeneity (denoted as geologic and value continuity) for ore reserve estimation. Similar characterization is of equal importance in the environmental sampling context.

Summary of Gy's Basic Formula

Let $K = (c)(\ell)(g)$, where c is the composition (mineralogical) constant, ℓ is the liberation factor, and g is the granulometric constant. The constant c has units of mass per volume and the other constants are unitless. Then the basic formula advanced by Gy is

$$\sigma_{FE}^2 = \frac{\mathrm{Var}(\hat{R})}{[E(\hat{R})]^2} \cong \frac{Kv_{95}}{x} \quad (44)$$

Here, v_{95} is the 95th percentile of the fragment volumes and x is the sample weight. The symbol $\sigma^2{}_{FE}$ in eq 44 represents the square of the coefficient of variation of the fundamental error. This term is somewhat different from usual statistical notation, in which σ^2 is reserved for variances, but it is consistent with Gy's use of the term. Therefore, if K is known, one can estimate the square of the coefficient of variation of $\hat{R}$ as a function of the physical constants, x, and v_{95}. Alternatively, the weight of the sample, x, needed to achieve a specified coefficient of variation can be computed if v_{95} is known, or the size to which the material must be ground (i.e., the required v_{95}) can be calculated for a fixed weight of sample.

Other Applications

Gy's methods were developed specifically for application to particulate sampling. However, these methods may also be applied directly to other continuous materials, such as liquids. Although the physical constants developed empirically for minerals do not apply to fluids, the histogram methods can be applied directly. Environmental sampling for contaminants in liquid media is thus a natural area of application. In particular, Gy's methods suggest application to composite sampling. For example, monitoring a river for contaminant concentrations could be aided by Gy's methods, in that appropriate sizes of experimental units could be derived through a size analysis similar to that applied to particulates. Further research should include empirical experimentation to develop a set of physical constants for sampling of other than heavy metals.

Computer subroutines have been developed at the University of Wyoming to compute estimates of the ratio of random variables using the methods given by

Cochran (2) and Gy (8, 9). These subroutines also provide estimates of the constants, c, ℓ, and g. Some examples of the application of Gy's results follow.

Gy's Method Versus Finite Sampling

To compare Gy's approximate method to the classical finite sampling methods with complete enumeration, we simulated data representing 1000 soil fragments. The simulated population ratio, R, was assumed to have a lognormal distribution with expected value 0.05. The fragment masses, X, were assumed to be exponentially distributed giving many small fragments with a few larger fragments. Simulated bivariate data followed the model $Y = RX$. The simulated data were analyzed using the computer subroutines previously referenced. Results are included in the Appendix (*see* Tables AI–AIII).

By using the finite sampling methods where individual fragment-by-fragment enumeration was required, the estimate of the ratio was found to be $\hat{R} = 0.05206$ with an estimated relative variance, $\mathrm{Var}(\hat{R})/(\hat{R})^2$, of 0.003327. Using Gy's methods on the same data after data were cross-classified into size and density classes resulted in the estimated ratio $\hat{R} = 0.05249$ with an estimated variance of the fundamental error given by $\sigma^2(\mathrm{FE}) = 0.002601$. We consider these values to indicate relatively good agreement of the two methods, although results are conditioned on the particular realization of the simulated population. The physical constants derived by Gy were also calculated and are given in Appendix A, along with the other estimates and the cross-classified data.

A Mining Application

One application of the use of Gy's formulas is the determination of the sample size (mass) required to attain a desired relative precision in the estimation of the grade of a mineral of interest. This standard example, attributable to Ottley (*16*), gives the way in which Gy's formula is typically used in mining applications. Other more recent examples can be found in François-Bongarçon (*17*). It is anticipated that similar use can be made in environmental settings. Suppose it is anticipated that an ore of zinc contains 6.6% Zn as ZnS. If the ore can be crushed to a maximum size of 2 cm, what mass of sample is required to ensure that a 95% confidence interval gives an estimate of the grade with relative error ±10%?

An approximate 95% confidence interval for R is given by

$$\hat{R} \pm 2 \times \mathrm{SE}(\hat{R}) \tag{45}$$

The specified precision can be expressed by

$$\frac{2\mathrm{SE}(\hat{R})}{\hat{R}} = 0.10 \tag{46}$$

or equivalently as

$$\sigma_{FE}^2 = \left(\frac{0.10}{2}\right)^2 = \frac{(c\ell g)v_{95}}{x} \tag{47}$$

Gy substitutes for v_{95} using the 95th percentile of the diameters (d_{95}) and a shape factor f. Empirical studies have shown that in most mineralogical applications, $v_{95} = f(d_{95})^3$. In the present example, $f = 0.5$ and $d_{95} = 2$ cm, giving $v_{95} = 4$ cm^3. Gy recommends the granulometric constant $g = 0.25$ and the liberation factor $\ell = 0.05$. The mineralogical factor recommended by Gy is given by

$$c = \left(\frac{1-R}{R}\right)\left[(1-R)\delta_m + R\delta_w\right] \tag{48}$$

where some suitable constants are $\delta_m = 5.0$ for the density of mineral, $\delta_w = 2.6$ for the density of waste, and $R = 0.066 \times 1.5 = 0.099$. This gives $c = 43.34$ and $K = (c)(\ell)(g) = 0.54$. Substituting K and v_{95} into eq 47 gives a sample mass $x = 864$ g.

To improve the precision of estimates, the sample could be crushed further. To what diameter should the sample be crushed to give a relative error of 0.05% given the sample mass of 864 g? Again, solve eq 47, where $x = 864$ g is substituted. This gives $v_{95} = 1$ cm^3 or $d_{95} = 1.26$ cm, so the sample should be crushed to a diameter <1.26 cm.

The first example provides evidence of the similarity between estimates obtained through the use of classical sampling methods and Gy's methods. The second example gives an indication of the utility of Gy's specification of physical constants for sample size and handling determination. Gy has developed a method for converting the classical sample size determination problem into one of sample mass and sample handling procedures appropriate to achieve a specified precision. Pitard (*10*) provides many further details and examples in a modern context for these procedures.

An Application to Water Quality

A third application is found in the sampling of liquids. Suppose an estimate of the concentration of an organic contaminant such as polychlorinated biphenyl (PCB) flowing past a cross-section of river is desired. Water subsamples of volume v are to be taken at random locations in the cross-section and combined to some total volume v_t to estimate the concentration. What volume of subsample unit should be used and what total volume is required to give a specified coefficient of variation (σ_{FE}) for the estimate?

To answer this question, one may use Gy's methods, where each subsample unit (i.e., an increment of water and suspended particulate) is treated analogously to a fragment of solid material. Assume that the contaminant is found in solution and as a surfactant on suspended particulate material. If several sizes of subsampling units are used, then the set of subsample observations can be classified into a two-way table by volume and density. If there is little suspended particulate, then there will be just one density class. It is anticipated that the percentage of suspended particulate may vary with the volume and density of sample units. Then using Gy's basic equation

$$\sigma_{FE}^2 = \frac{(c\ell g)v}{v_t} \tag{49}$$

with a selected value of σ_{FE}, the volume of an individual subsample unit may be determined for a given total volume of sample, or, alternatively, a total volume of sample may be determined given a subsample unit volume. However, to apply eq 49, the constants c, g, and ℓ must be determined.

A basic field exercise may be used to determine these constants. Suppose a set of r samples of size n is taken where the volume of each subsample unit is intentionally varied so that v_1, v_2, ... v_r are the subsample volumes. Each sample unit is kept separate, and the volume and density are recorded. The set of $(n \times r)$ subsample units collected is then cross-classified based on volume and density. Sample units are combined within volume and density class and assayed for PCB content. If individual sample units are sufficiently large, and PCB concentrations are high, then individual sample units could be assayed. For volume–density class (α,β), the number of sample units in the class $n_{\alpha\beta}$, the average mass $x_{\alpha\beta}$, and the average PCB concentration is available. Based on this table, the estimates of c, g, and ℓ can be obtained from the formulas previously given in this chapter. These constants may then be used to determine the relationship between total sample volume (v_t), subsample unit volume (v), and σ^2_{FE}.

Conclusions

The methods of Gy and Cochran for ratio estimation have been shown to be asymptotically equivalent for samples that can be completely enumerated. Both are based on finite sample theory. Gy extends the procedure to treat data grouped into a two-way table of fragment volume and fragment density and provides a simple estimation procedure for estimating appropriate sample volume and fragment sizes to attain a specified relative error. A computer program, available from the authors, has been developed at the University of Wyoming to estimate the constants from a table of grouped data.

This chapter has shown certain equivalences between finite sampling theory and Gy's work for cases in which samples may be completely enumerated. In environmental settings and for the estimation of certain ores such as precious metals, further empirical study is required to improve the value of Gy's method for sampling materials that are not completely enumerable. The methods outlined in example 3 show how Gy's method can be implemented for the important problem of sampling liquid media. Future work is needed to determine the ultimate value of Gy's method in applications other than ore reserve estimation.

Appendix

Tables AI through AIII give the results of analysis of the simulated data using standard finite sampling theory and the method developed by Gy. The data were simulated by letting X, the sample mass, be distributed as an exponential random variable, and R, the grade, be lognormally distributed. The mass of mineral of interest Y was obtained as the product $Y = RX$. Table AI gives the results of standard finite

Table AI. Standard Variance Estimates Based on Finite Sampling Theory

Parameter	*Estimate*
Total of X	48021.5
Total of Y	2500.18
Estimated ratio	0.0520638
Estimated variance	0.901763×10^{-5}
Estimated squared CV	0.332676×10^{-2}

NOTES: Estimates are based on Cochran (2). X is fragment mass; Y is mass of mineral of interest; CV is coefficient of variation. Estimated squared CV in Table AI corresponds to estimated variance of the fundamental error in Table AIII [$(CV)^2 \cong \sigma^2(FE)$]. Estimated ratio in Table AI corresponds to approximate grade in Table AIII.

Table AII. Simulated Data Grouped by Size and Density Classes

Volume (cm³)	*Density Class (g/cm³)*				
	2.6378	*2.8382*	*3.0070*	*3.1647*	*3.3960*
8.9512	23.6117[a] 0.0286[b] 632[c]	25.4051[a] 0.1777[b] 50[c]	26.9161[a] 0.2800[b] 17[c]	28.3278[a] 0.3722[b] 12[c]	30.3988[a] 0.4883[b] 4[c]
31.9481	84.2732[a] 0.0308[b] 187[c]	90.6742[a] 0.1698[b] 18[c]	96.0669[a] 0.2917[b] 4[c]	101.1057[a] 0.3616[b] 1[c]	108.4972[a] 0.0000[b] 0[c]
56.7869	149.7934[a] 0.0361[b] 46[c]	161.1710[a] 0.1571[b] 4[c]	170.7565[a] 0.0000[b] 0[c]	179.7128[a] 0.0000[b] 0[c]	192.8510[a] 0.0000[b] 0[c]
77.9711	205.6735[a] 0.0283[b] 19[c]	221.2953[a] 0.1630[b] 2[c]	234.4567[a] 0.0000[b] 0[c]	246.7541[a] 0.0000[b] 0[c]	264.7936[a] 0.0000[b] 0[c]
105.5720	278.4796[a] 0.0317[b] 3[c]	299.6315[a] 0.0000[b] 0[c]	317.4518[a] 0.2655[b] 1[c]	334.1024[a] 0.0000[b] 0[c]	358.5276[a] 0.0000[b] 0[c]

NOTES: For these simulated data, the following units can be applied: volume in cubic centimeters, density in grams per cubic centimeter, mass in grams, and cell frequency in counts. Grade is unitless. Table AII represents intermediate calculations used to produce Table AIII.

[a]Average fragment mass (g).

[b]Average grade.

[c]Cell frequency (n).

Table AIII. Variance Estimates Using Volume and Density Classes of Pierre Gy

Parameter	*Estimate*
Approximate grade	0.0524973
Total mass	48027.0
H	124.920
c	87.9689
g	0.601487
ℓ	0.0422788
v_{95}	55.8410
σ^2(FE)	0.260103×10^{-2}

NOTES: Estimates are based on Gy (9). H is constitution heterogeneity constant; c is mineralogical factor; g is granulometric factor; ℓ is liberation factor; v_{95} is 95th volume percentile; σ^2(FE) is estimated variance of the fundamental error.

sampling methods, whereas Table AII gives the intermediate cross-classification of the data as is required using the Gy method. Table AIII gives the final results from application of the Gy method using estimation formulas developed by the authors. It should be noted that with as few as five volume and five density classes, reasonable agreement between the squared coefficient of variation and the variance of the fundamental error is obtained.

References

1. Gilbert R. O. *Statistical Methods for Environmental Pollution Monitoring;* Van Nostrand Reinhold: New York, 1987.
2. Cochran, W. G. *Sampling Techniques;* John Wiley & Sons: New York, 1977.
3. Thompson, S. K. *Sampling;* John Wiley & Sons: New York, 1992.
4. Cressie, N. A. C. *Statistics for Spatial Data;* John Wiley & Sons: New York, 1991.
5. Flatman, G. T.; Englund, E. J.; Yfantis, A. A. In *Principles of Environmental Sampling,* Keith, L. H., Ed.; American Chemical Society: Washington, DC, 1988; pp 73–84.
6. Borgman, L. E.; Quimby, W. F. In *Principles of Environmental Sampling;* Keith, L. H., Ed.; American Chemical Society: Washington, DC, 1988; pp 25–44.
7. Brus, D. J.; de Gruijter, J. J. *Environmetrics* **1993,** *4,* 123–152.
8. Gy, P. M. *Mem. BRGM* **1967,** 4(56), 42–51.
9. Gy, P. M. *Sampling of Particulate Materials, Theory and Practice;* Elsevier Scientific: New York, 1982.
10. Pitard, F. F. *Pierre Gy's Sampling Theory and Sampling Practice. Volume I, Heterogeneity and Sampling;* CRC Press: Boca Raton, FL, 1989.
11. Lyman, G. J. *Geochim. Cosmochim. Acta* **1993,** *57,* 3825–3833.
12. Matheron, G. *Rev. Ind. Miner.* **1966,** *46,* 609–621.
13. Hartley, H. O.; Ross, A. *Nature (London)* **1954,** *174,* 270–271.
14. François-Bongarçon, D. In *Proc. XVth World Mining Congr., Madrid, Spain;* World Mining Congress, 1992; p 21.
15. Sinclair, A. J. *Explor. Min. Geol.* **1994,** 2(3), 95–108.
16. Ottley, D. J. *World Min.* **1966,** 9(19), 40–44.
17. François-Bongarçon, D. *CIM Bull.* **1991,** (84), 75–81.

Sampling Waters

Chapter 12

Modern Sampling Equipment: Design and Application

Lorance H. Newburn

Automatic wastewater samplers are used extensively by water quality monitoring agencies and wastewater dischargers who must comply with permit requirements. The basic mode of operation of these samplers has changed very little since 1977. A sound understanding of the limitations of these designs and of the characteristics of the flow stream to be sampled is necessary in order to obtain a representative sample. Although studies have been conducted to evaluate samplers, few studies have been published on correct application. Some general guidelines for correct application are discussed.

BROAD-BASED NEED FOR AUTOMATIC WASTEWATER SAMPLERS was created within the framework of the Federal Water Pollution Control Act of 1972 (*1*). In order to achieve the Act's objective of restoring and maintaining the chemical, physical, and biological integrity of the nation's waters, wastewater effluents had to be accurately characterized both in quality and quantity. Wastewater sampling became an intensive activity by those discharging wastewater and by those agencies charged with compliance enforcement. The emphasis of this activity originally centered on conventional pollutants. The parameters of greatest interest were biological oxygen demand (BOD), chemical oxygen demand (COD), solids, conductivity, pH, oil and grease, fecal coliform, heavy metals, and a few organic compounds. In 1977, the Clean Water Act was amended (2). The scope of wastewater sampling was broadened to include toxic pollutants and the pretreatment of effluents by industries. The Consent Decree identified 129 toxic materials discharged by industries grouped into 34 categories, which became the target of a massive campaign to eliminate these substances from the nation's waters.

Water Sampling History

Traditionally, water was sampled by a simple dipping procedure that is very dependent on methodology. The sample collected may or may not be representative of the flow at the time the sample was taken. Also, water quality and flow rate in a given flow stream can vary considerably from one moment to the next. Collection of frequent aliquots is required to obtain an accurate representation of the flow over a given time period. As a result, manual sampling became a labor-intensive activity, and coupled with the increased concern for unbiased samples, those people interested in sampling found themselves increasingly dependent on automatic samplers.

Many manufacturers recognized the need for automated samplers, and a wide variety of designs was offered. With respect to sample representativeness, the basic difference in these designs was in the sample collection system. Samplers were offered that used low-, medium-, or high-speed peristaltic pumps, in which the pump operates from a timer, or pump rotations are counted to control volume. Also offered were samplers using vacuum pumps where a metering chamber is used to control volume; mechanical dippers filled by immersion; evacuated bottles, where each bottle has a tube connecting to the source; and pneumatic-ejection systems, where a submerged metering chamber is filled by the hydrostatic pressure of the liquid, and the resulting sample is forced to the sample container by gas pressure.

Automatic sampler use did not solve all the problems, however (3). The equipment varied considerably with respect to design, sample intake velocity, method of sample collection, versatility, and durability. Sampler users often lacked the skills necessary for proper application. It was not unusual for samples taken by one device to vary considerably from samples collected by another device in the same flow stream. Depending on the sampler selected, either overstating or understating the concentration of certain constituents was possible. An enforcement agency using one brand and model of sampler often obtained varied results from a discharger using the same brand and model. Apparently, samplers needed to be studied and methodologies needed to be standardized so accurate data could be obtained.

Sampler Evaluations

The first major sampler evaluation was published by Shelly and Kirkpatrick in 1973 (4). This study surveyed about 50 different samplers from 30 manufacturers for suitability in storm or combined sewer sampling applications. They concluded that no single unit could be universally applicable and suggested what improvements, if any, could be made that would make these units as ideal as possible. They also concluded that obtaining a representative sample requires more than the sampler itself. Their evaluation did not include side-by-side comparisons of commercial samplers but did suggest that such a comparison should be made. The U.S. Environmental Protection Agency (EPA) compared low-, medium-, and high-speed peristaltic pump units; a vacuum unit; an evacuated-bottle unit; a pneumatic-ejection unit; manual samples taken with a three-level Van Dorn-type sampler; and conventional dipping that followed U.S. Geological Survey field procedures. No statistical difference was found between vacuum samplers and medium- and high-speed peristaltic samplers.

A study comparing samplers was made by Harris and Keffer of the Region VII EPA office and published in June 1974 (5). They evaluated about 15 different samplers. The study compared vacuum-type samplers, which typically have a high initial intake velocity, and samplers that use peristaltic pumps having low to moderate intake velocities. The test data showed that vacuum samplers produced higher BOD and COD concentrations and disproportionately higher solids concentrations than samplers using peristaltic pumps. Vacuum samplers produced higher values than the manual flow-weighted grab samples, which were selected as the standard. They concluded that high-vacuum, high liquid intake velocity samplers were more effective in capturing solids than peristaltic pump samplers. They suggested that slower acting peristaltic pump samplers were either not capturing settleable materials or that particle settling velocities were higher than the liquid intake velocities after introduction to the intake line. The question was raised as to how vacuum samplers could produce higher results than their flow-weighted grab sample standards (Harris, D. and Keffer, W., personal communication). Some of the peristaltic pump samplers demonstrated 90%–95% sampling efficiency; the vacuum samplers demonstrated efficiency values in excess of 100%, and some samplers demonstrated efficiencies as high as 167%. One suggestion made to the authors was that if the strainer of a vacuum sampler was allowed to rest on the bottom of the flow stream where sediment had collected, the high intake velocity would scour those sediments from around the strainer; artificially enrich the sample; and produce higher than normal BOD, COD, and solids values. After some consideration, the authors agreed that this situation was probably true and that care should be exercised in positioning the strainer of a vacuum sampler or any other sampler having a high intake velocity so that no scouring effect would occur.

Another study was published in January 1975 by Barkley et al. (6). Whereas the Harris and Keffer report was based on comparisons in field use, this study was conducted in a laboratory under artificial conditions by using a tank with a mixer so that greater control could be exercised over the test media. Sixteen different samplers were put through an extensive series of tests and scored on 21 factors. Vacuum samplers were not studied in this series of tests, but some testing was done shortly afterward. The authors concluded that samplers with high intake velocities tend to scour sediments from around the strainer if care is not exercised in positioning the strainer above the sediments.

A separate study conducted by the U.S. Forest Service in Arcata, CA, dealt with contamination of successive samples in portable sampling systems (7). The results showed that a vacuum sampler using a metering chamber having greater wetted surface area produced more cross-contamination than a peristaltic pump sampler. This result contrasts a study conducted by the St. Anthony Falls Hydraulic Laboratory (*8*), which concluded that no evidence of cross-contamination was found in the same model of vacuum sampler used in the Forest Service study.

Another study (*9*) was conducted to determine the effect of certain variables on the recovery of volatile organic compounds. The authors concluded that high-speed suction-lift samplers cause volatile organic compounds to outgas. The authors recommended the use of moderate-speed peristaltic pumps to minimize this effect.

Although these studies may show the superiority of one sampler over another, the basic difference between these units is in liquid intake velocity and the wetted surface area of the liquid transport system. Vacuum sampler supporters argue that the higher intake velocity helps keep solids in suspension. Supporters of peristaltic pump samplers argue that intake velocities of 1.5–3 ft/s are more ideally suited for obtaining a representative sample because these samplers do not tend to scour sediments from around the strainer. As mentioned, data from the Harris and Keffer report (*5*) and research conducted by Barkley et al. (*6*) tend to support those favoring peristaltic pump samplers. Because of their greater wetted surface area, the metering chambers used in vacuum samplers may be a source of cross-contamination between samples (*7, 8*). The use of a metering chamber as a means of controlling sample volume is an advantage where the level of the stream being sampled changes considerably from one sample to the next. This situation would be a problem only when composite samples are collected because the sample could be biased by a variation in the volume of each aliquot placed in the sample bottle. Therefore, a number of factors need to be considered when selecting a sampler.

Ideal Sampler Features

Several attempts by those studying samplers have been made to define what would be considered the "ideal" sampler (*4–6*). As mentioned, some features are mutually exclusive, and a sampler incorporating all these features cannot be made. For those who must sample a number of locations, more than one type of sampler will probably be required. Features that the ideal sampler should have include the following:

1. alternating current (ac) or direct current (dc) operation with adequate dry battery energy for 120 h of operation at 1-h sampling intervals
2. suitability for suspension in a standard manhole and accessibility for inspection and sample removal

3. total weight including batteries under 18 kg (40 lb)
4. sample collection intervals from 10 min to 4 h
5. capability for flow-proportional and time-composite samples
6. capability for collecting a single 9.5-L (2.5-gal) sample or collecting 400-mL (0.11-gal) discrete samples in a minimum of 24 containers
7. capability for multiplexing repeated aliquots into discrete bottles
8. a single intake hose having a minimum ID of 0.64 cm (0.25 in)
9. intake-hose liquid velocity adjustable from 0.61 to 3 m/s (2.0–10 ft/s with dial setting)
10. minimum lift of 6.1 m (20 ft)
11. explosion-proof materials and electronics
12. watertight exterior case to protect components in the event of rain or submersion
13. exterior case capable of being locked and including lugs for attaching steel cable to prevent tampering and to provide security
14. no metal parts in contact with waste source or samples
15. an integral sample container compartment capable of maintaining samples from 4 to 6 °C for a period of 24 h in ambient temperatures ranging from −30 to 50 °C
16. with the exception of the intake hose, capability of operating in a temperature range from −30 to 50 °C
17. purge cycle before and after each collection interval with sensing mechanism to purge in the event of plugging during sample collection, followed by collection of complete sample
18. field repairability
19. interchangeability between glass and plastic bottles, particularly in discrete samplers
20. sampler exterior surface light in color to reflect sunlight

This list of features was developed before the amendment of the Clean Water Act, in which the emphasis in sampling shifted to toxic materials. One other feature, therefore, has to be included in the list: the ability to be readily adapted for toxic-pollutant sampling. For toxic pollutants, the sample should contact only tetrafluoroethylene (Teflon), glass, and medical-grade silicone rubber if a peristaltic pump is being used in the sample transport system.

Sampler Characteristics

According to the EPA's manual on complying with the sampling requirements published in 1979 (*10*), about 100 models of portable automatic sample collection devices existed. These devices varied widely in levels of sophistication, performance, mechanical reliability, and cost. No sampler on the market was considered ideal. Now, several samplers come close to being ideally suited. All of the major suppliers of portable samplers offer units that operate on ac or dc, can be suspended in a manhole, weigh less than 40 lb empty, offer sampling intervals from 1 min to at least 4 h, are capable of flow-proportional or timed-interval sampling, are convertible from sequential to composite sampling, are capable of multiplexing several samples into one discrete bottle, offer a 0.25-in ID suction line with a weighted strainer, have a minimum lift of 20 ft, are capable of being locked, have no metal parts in contact with the sample, operate over temperature ranges from −30 to 50 °C, and purge the sample line before and after sampling.

Samplers on the market differ, sometimes significantly, in sample intake velocity (none has an adjustable intake velocity), explosion-proof qualities, watertightness, insulation qualities, and ability to be repaired in the field. No manufacturer offers an ac- or dc-powered sampler that is certified to be explosion-proof. This accomplishment would require the addition of considerable weight to the sampler that would put it over the recommended 40-lb maximum. Pneumatically operated samplers currently being offered are explosion-proof. One sampler offers the advantage of exceptionally high lift (as high as 300 ft). Pneumatic-ejection units, however, may lead to cross-contamination of samples.

Watertightness is one definite requirement of a portable sampler. Manholes often surcharge, rivers rise, and samplers frequently flood. The waterproof qualities of commercially available samplers vary considerably, even though most manufacturers claim their samplers to be watertight. Moisture damage is the most frequent cause of sampler failure. If the purchase of a sampler is being considered, the purchaser should carefully examine the product beforehand and determine what kind of seal protects the electronic parts from moisture. One manufacturer rates their controller NEMA 6, which means the sampler can withstand submersion under 6 ft of water for at least 30 min.

Not all samplers are thermally insulated. Those that are insulated vary in insulation quality. The EPA research center in Cincinnati conducted insulation quality tests on several samplers and can provide data on their results.

Field repairability of samplers must necessarily be limited to replacement of expendable items or certain mechanical parts. As for the electronic parts, all samplers must be repaired by a skilled technician who has access to the proper test equipment. This requirement usually means shipping the unit back to the factory. Some manufacturers offer samplers that have removable control units; therefore, only the controller must be sent in for repair.

Two types of samplers are on the market: discrete and composite samplers. With a discrete sampler, an individual sample can be collected and retained in a separate container for future analysis. Most current manufacturers of discrete samplers offer units containing at least 24 bottles. With a true composite sampler, small aliquots can be taken at frequent intervals, usually over a 24-h period; collected in a single container; and held for future analysis. Of the two sampler types, the discrete sampler can determine what the constituents of the sample are and approximately when they were discharged. Discrete samplers are more widely used by water quality enforcement agencies for this reason. Discrete and composite samplers are available in various configurations depending on the particular application. Most samplers are dc-powered portable units, but some are ac-powered refrigerated units.

Most samplers are capable of gathering either timed-interval samples or samples collected proportional to flow. Timed-interval samplers have a fixed interval of time between each aliquot or sample. Flow-proportioned samples are based generally on equal increments of flow as measured by an associated flow meter. Most samplers require a simple contact closure signal from the flow meter, and some accept a 4–20-mA analog signal. A flow-proportioned composite sample, in which small aliquots are collected in a single container over small increments of flow, provides the most representative sample of the flow over a given time period. However, if the flow interval is too large, events such as the dumping of small tanks might pass by the sampler without being detected. For example, if the flow interval is 25,000 gal per aliquot and a 5000-gal tank is dumped, those 5000 gal could pass by the sampling point undetected.

Two other modes of operation of interest to individuals using discrete samplers include nonuniform time intervals and time override of flow-proportioned sample collection. Nonuniform time intervals give the user the option of programming a different time between each sample. This option is of particular interest to those studying hydrologic events, combined sewer overflows, and other events where extreme variation in flow or constituent concentration occurs. Each individual time interval can be programmed separately by the user. In the time-override mode, individual samples are composited into a single bottle at equal increments of flow volume as measured by the associated flow meter. However, at a programmed increment of time, the distributor is indexed to the next sample bottle, the next sample is placed in that bottle, and the compositing of samples is renewed. A time-override sample is used when the investigator desires to collect a sequential series of composite flow samples, each over a known, elapsed time interval while ensuring that at least one sample will be collected during the time interval, regardless of the flow volume.

Another important feature of a sampler is durability. A careful examination of the materials of construction is important to ensure that the sampler continues to perform as expected in highly humid, corrosive environments. All hardware should be stainless steel. Plated or painted hardware will not last very long in most sampling conditions. The case material should be of acrylonitrile butadiene styrene (ABS) plastic, glass fibers (Fiberglas), or some other material that is highly resistant to moisture and corrosive gases. As previously mentioned, the control unit must be able to withstand accidental submergence, and the exterior construction should be able to withstand the elements of nature without compromising the integrity of the sample. Sample bottles must be manufactured of a material that will not alter the composition of the sample. Glass bottles must be used in toxic-pollutant sampling applications. Polyethylene or polypropylene bottles are acceptable for most general-purpose applications.

Medical-grade silicone rubber tubing must be used in peristaltic pumps to avoid contamination of the sample by organic peroxides used in the manufacturing of conventional grades of silicone rubber. A test conducted by the EPA determined that short lengths of medical-grade silicone rubber do not alter samples in any way (*9*). When sampling for toxic pollutants, the suction line must be made of tetrafluoroethylene (Teflon). For general-purpose applications, the suction line can be poly(vinyl chloride), but should be food-grade to prevent the introduction of phenolic compounds.

User Knowledge

Probably the single greatest factor influencing the representativeness of the sample lies in the user and how the sampler is applied. To aid the user in complying with various federal and state regulations concerning water quality, a number of guides are available (*10, 11, 12*). But because of the many variations of conditions in which samplers can be used, coverage of them all in these guides is impossible. The collection of samples from a flowing waste stream requires a sound understanding of the problems inherent in sample collection if representative samples are to be obtained. Ideally, to get a true representative sample, the entire discharge should be collected, and then an aliquot of the thoroughly mixed discharge should be taken for analysis. This situation is impractical because millions of gallons of effluent are generally involved, and thorough mixing is impossible. Because of this incapability, the next best means of obtaining a representative sample is to collect a flow-proportioned composite sample. The key to representativeness of the sample is the method used to collect each aliquot.

Sample Intake Position

Before using an automatic wastewater sampler, the user should understand the characteristics of the flow stream and where in the stream the sample should be taken. The best discussion on this topic was contained in correspondence received in 1977 from Keffer, who authored one of the sampler studies (*5*) mentioned. Keffer pointed out that technical literature was lacking in standard guidance procedures for sampler installations or for sampling of any type. He stated that his opinion was not supported by carefully designed research activities but was a summary of his experience at annual sampling efforts at 400–500 facilities per year for over a 6-year period. He pointed out that the flow regime is a composite of a velocity distribution superimposed on a solids distribution, both of which vary with the physical condition of the site and vary in size, shape, and specific gravity of the solids particles. The shape and velocity of the flow stream determines the degree of turbulence. Those solids having a specific gravity much greater than water tend to settle in the bottom of the flow stream at low velocities or suspend near the bottom at higher velocities. Only where much turbulence occurs are these heavier solids uniformly distributed. Solids having densities much greater than water are generally not organic. Solids having specific gravities only slightly greater than water are usually organic and remain suspended in the flow but form layers or strata in smooth-flowing channels. Solids lighter than water and oils float and are almost always organic. Even where turbulence occurs in a flowing channel, oils tend to float. Thus, a flowing waste stream is generally highly stratified, nonhomogeneous, and presents a less than ideal medium from which to take a representative sample.

Keffer stated that from information currently available, a single-point intake is not likely to be satisfactory. Current assessment of state-of-the-art sampling methods suggested that a fixed intake located at 60% of the stream depth in an area of maximum turbulence and having an intake velocity equal to or greater than the average wastewater velocity at the sample point provides the most representative sample. This technique ignores contribution from bedload (i.e., sediment layers) or floatable solids. The selection of sampling at 60% of depth is based on the velocity and sedimentation concentration charts shown in Figure 1. Just below a hydraulic jump, the outlet of a flume or the nappe of a weir generally is the area of greatest turbulence in a flow stream. Attempting to sample at an intake velocity equal to or greater than the average wastewater velocity is difficult to achieve, and Keffer's routine practice was to maintain a transport velocity in excess of 2 ft/s. Placing the strainer of a high-velocity intake sampler on the bottom of a smooth-flowing stream where sediments have accumulated causes these sediments to be scoured from around the strainer. The sample is thus artificially enriched with solids.

Since 1980, two studies on positioning of sample intakes have been published. McGuire et al. (*13*) concluded that designers of water quality monitoring programs should realize the effects of sample intake positions on parameter concentration and should take into account the objectives of the study to determine the correct intake position. Reed (*14*), on the other hand, concluded that zones of natural mixing such as hydraulic jumps are unsuccessful in providing the necessary degree of mixture to obtain a truly representative sample. Therefore, some degree of calibration is necessary before accurate data can be obtained.

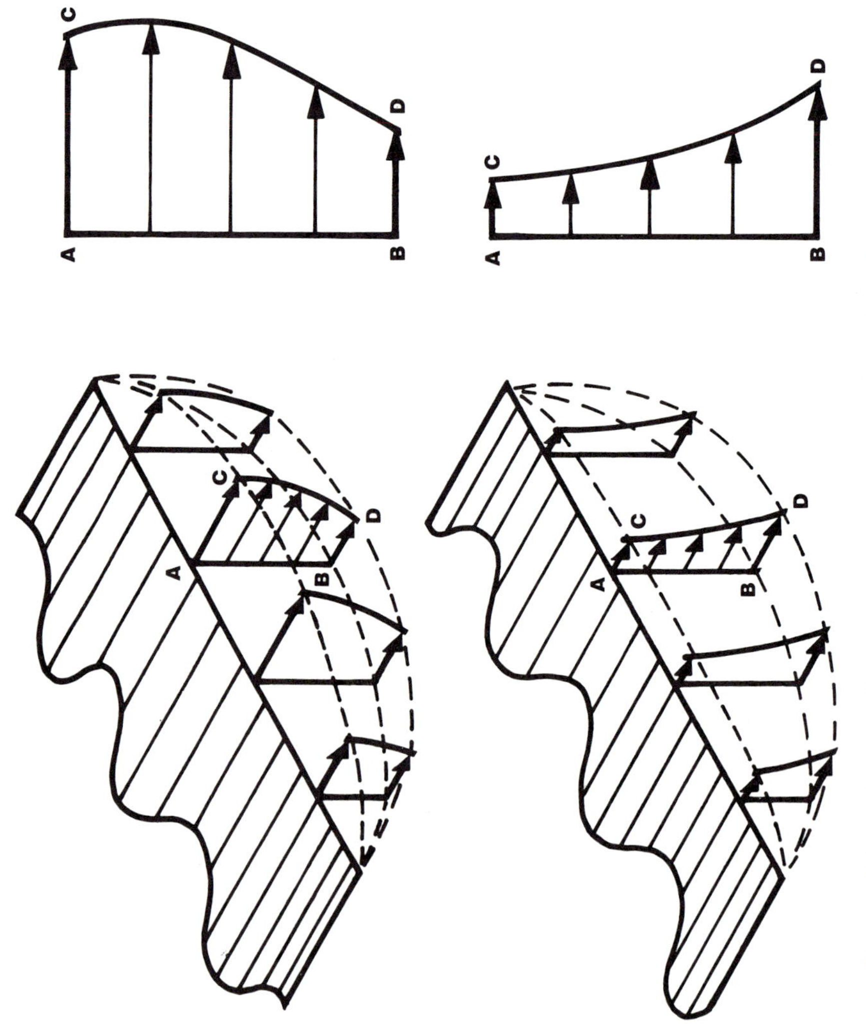

Figure 1. Velocity distribution (top) and sediment concentration (bottom) in a flowing stream or sewer.

References

1. "Federal Water Pollution Control Act Amendment of 1972"; Public Law 92–500, 1972.
2. "Clean Water Act of 1977"; Public Law 95–217, 1977.
3. Shelly, P. E. *Am. City & County* **1980**, *95*, 35–38.
4. Shelly, P. E.; Kirkpatrick, G. A. *An Assessment of Automatic Sewer Flow Samplers*; U.S. Environmental Protection Agency. U.S. Government Printing Office: Washington, DC, 1973; EPA 122-73-261. (Updated as *Sampling of Water and Wastewater*, 1976; EPA-600/4-77-039.)
5. Harris, D. J.; Keffer, W. J. *Wastewater Sampling Methodologies and Flow Measurement Techniques*; U.S. Environmental Protection Agency: Kansas City, MO, 1974; EPA 907-974-005.
6. Barkley, J. J.; Peil, K. M.; Highfill, J. W. *Water Pollution Sampler Evaluation*; Army Medical Bioengineering Research and Development Laboratory: Fort Detrich, MD, 1975; AD/A-009-079.
7. Thomas, R. B.; Eads, R. E. *Water Resour. Res.* **1983**, *19(2)*, 436–440.
8. Wood, A. "Test and Evaluation of a Portable Discrete Suspended Sediment Sampler for Manning Environmental Corporation"; External Memorandum No. 150; St. Anthony Falls Hydraulic Laboratory. University of Minnesota: Minneapolis, MN, 1977.
9. Ho, J. S. Y. *J. Am. Water Works Assoc.* **1983**, *75*, 583–586.
10. *NPDES Compliance Sampling Inspection Manual*; U.S. Environmental Protection Agency. U.S. Government Printing Office: Washington, DC, 1979; MCD-51.
11. *Handbook for Sampling and Sample Preservation of Water and Wastewater*; U.S. Environmental Protection Agency. U.S. Government Printing Office: Washington, DC, 1976; EPA-600/ 4-76-049. (Revised as EPA-600/4-82/029, 1982; addendum, 1983.)
12. Lauch, R. P. *Application and Procurement of Automatic Wastewater Samplers*; U.S. Environmental Protection Agency: Cincinnati, OH, 1975; EPA-670/4-75-003.
13. McGuire, P. E.; Daniel, T. C.; Stoffel, D.; Andraski, B. *Environ. Manage. (NY)* **1980**, 4, 73–77.
14. Reed, G. D. *J. Water Pollut. Control. Fed.* **1981**, 53, 1481-1491.

Chapter 13

Automatic Water and Wastewater Sampling

Elie M. Dick

This chapter is devoted to automatic water and wastewater samplers used in sampling discharges from point sources, such as industrial and municipal wastewater treatment plants, and non-point sources, such as storm water runoff. Sampling of groundwater or surface water, such as lakes, will not be discussed. Tips and guidelines will be provided to help the user in selecting and using automatic equipment. The ultimate objective is to ensure that accurate representative samples are collected. No previous knowledge of sampling and sampling equipment is assumed.

THE PRESENCE OF POLLUTION in our lakes, rivers, streams, and other water bodies is a major concern of all communities. In the United States and in many countries around the world, people have recognized the toll pollution has taken and continues to take on our water resources. This recognition has prompted governments, at the federal, state, and local level, to enact strict environmental regulations and to develop broad pollution prevention and remediation plans.

Many of these regulations and plans require monitoring programs designed to evaluate the physical, chemical, and biological characteristics of our water resources and wastewater discharges. Some of these characteristics are measured on site. However, to get a complete and more accurate picture, water and wastewater samples must be collected and sent to a laboratory for thorough analysis.

This chapter is devoted to automatic water and wastewater sampling for environmental analysis. Our focus will be on equipment used in sampling discharges from point sources, such as industry and municipal treatment plants, and non-point sources, such as storm water runoff. Groundwater sampling is a broad subject covered in other parts of this book, and therefore it will not be included in this chapter.

Because the ultimate objective is to always collect representative samples, tips and guidelines will be provided to help sampling professionals in achieving their goals.

Importance of Sampling

Environmental sampling has increased significantly since the enactment of the Federal Water Pollution Act of 1972 (*1*). Water and wastewater sampling has been very instrumental in characterizing the chemical, physical, and biological properties of our discharges. It has been used in setting environmental policies and regulations and in developing pollution prevention and remediation plans. Examples of sampling applications include the following:

- determining the current condition and trends of a water resource;
- identifying the source and impact of a specific pollutant;
- determining the effectiveness of a water pollution prevention program;
- providing a basis for issuing a permit to a discharger;
- ensuring compliance with permit requirements;
- collecting data for use in enforcement proceedings;
- measuring the performance of a municipal or industrial treatment plant; and
- levying fees on dischargers to wastewater treatment plants.

Sampling Plans

Before a sampling project is begun, the overall objective should be stated clearly, concisely, and preferably in writing. Often, the objective is dictated by regulations or government permits. In the United States, the U.S. Environmental Protection Agency (EPA) has established the National Pollution Elimination Discharge System (NPDES) and the National Pretreatment Program to regulate the discharge of polluted water and wastewater (*1–4*). To discharge, a source is required to have an NPDES permit that may prescribe specific sampling programs and other monitoring requirements. In this case, before setting the objective, the sampling investigator must have an understanding of the regulations and permit terms and conditions. This information can be obtained by referring to the permit documentation, by checking with the appropriate authorities, or by consulting with experts in the field.

Sampling projects can be costly and time-consuming. Even if the objective is clear, it is important to have a comprehensive written plan. The plan is an important document that offers four benefits. First, it will ensure that all important questions are asked and answered. Second, it helps in setting the best possible and workable strategies and tactics given the available information and limited resources. Third, it ensures that all individuals involved are following the plan. Finally, the plan helps the sampling project manager in evaluating field activities and in assessing the quality of the results. These points will be discussed throughout this chapter.

Types of Samples

The two basic types of water and wastewater samples are discrete samples and composite samples. Each type has its advantages and disadvantages. Figure 1 depicts the types of samples.

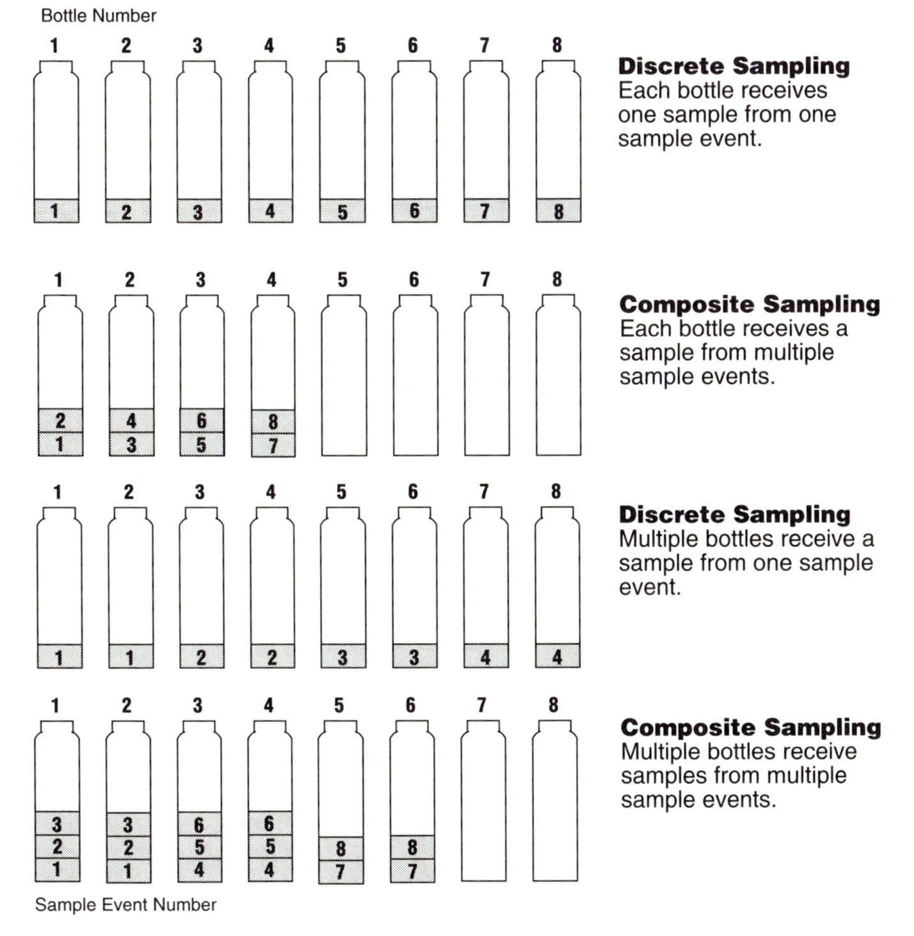

Figure 1. Types of water and wastewater samples.

Discrete Sample

A *discrete sample,* also known as a grab sample, is an individual sample collected within a short period of time and deposited in an individual container. A short period is generally defined to be less than 15 min. Manual discrete samples are obtained by dipping a container into the water or wastewater and filling it with the required volume. A discrete sample represents the conditions of the source at the time the sample was collected. It does not necessarily represent the source at any other time.

Discrete samples are recommended when the quality of the discharge is essentially constant. For example, a manufacturing facility that produces the same product for extended periods of time without change in its process is likely to have the same water quality discharge over time. If conditions are expected to change, then a series of discrete samples will be required. Obviously, this procedure will result in higher costs for sample collection and analysis. Discrete samples are also recommended when the quality of the discharge water fluctuates significantly. For example, a series of discrete samples is recommended when sampling batches of discharge wastewater with high and low pH. In this case, combining multiple samples in one bottle can conceal the peaks and valleys of the discharge. Finally, discrete samples are effective in auditing the results of composite sampling methods.

Composite Sample

A *composite sample* consists of a series of smaller samples collected over time and deposited into the same container. A composite sample can be obtained by combining two or more discrete samples. To some extent, composite samples represent the average characteristics of the source during the compositing period. They are particularly useful in calculating the average concentration and pollutant loading during the sampling period.

The advantage of composite sampling is lower cost for laboratory analysis compared with multiple discrete samples collected during the same period. Its disadvantage, however, is less information, which can lead to incorrect conclusions. For example, compositing samples from a stream with alternating high and low pH can produce a neutral sample.

Depending on the collection method, a variety of composite samples can be produced. A time composite sample is a series of samples of equal volume collected at equal time intervals and deposited in one container. This type of sample produces an average of the discharge, if the flow is relatively constant. If the flow varies, a flow-proportional composite sample is required. There are two ways to collect a *flow-proportional composite sample.* In the first, a series of samples of equal volume at equal flow volume intervals are collected and deposited in one container. In the second, samples are collected at equal time intervals, but the volume of each sample is made proportional to the flow volume during the corresponding sampling period. A *sequential composite sample* is a series of composite samples collected during time or flow subintervals. A wide variety of composite samples can be produced by changing the compositing method.

Manual Sampling

Discrete and composite samples can be collected either manually or automatically. Manual sampling consists of dipping a container in the stream and filling it with the required volume. The main advantage of manual sampling is its simplicity, but it has significant disadvantages:

- It is labor-intensive in applications in which a large number of samples must be taken over an extended period of time, such as sampling once an hour throughout the day at a manufacturing facility.
- It is not practical when sampling at odd hours or unpredictable times, such as sampling to comply with storm water runoff monitoring regulations (*4*).
- It is difficult, and sometimes impossible, to perform conditional or event sampling over an extended period of time. Examples include flow-proportional sampling, sampling when pH or other parameters fall outside certain limits, and sampling to catch illegal batch wastewater dischargers.
- It can produce inconsistent results due to human error or variations in sampling techniques.
- It can be dangerous, such as during storm events or during the sampling of streams containing hazardous chemicals.

Automatic Sampling Equipment

An *automatic sampler* is a device that can be programmed to collect samples for environmental analysis. A sampling program is simply a set of instructions that controls the tasks to be performed by the sampler. It specifies the types and volumes of samples to be taken, the conditions under which these samples are to be collected, and the containers into which they go. The conditions might be based on time, flow, liquid level, rainfall, pH, temperature, conductivity, dissolved oxygen, or any other parameter. The following are some examples of sampling programs.

- A program might specify that a series of 24 discrete 1-L samples be collected each hour throughout the day. This program provides 24 different snapshots of the source where the samples were collected.
- If conditions fluctuate throughout the day, then the preceding program may not produce an accurate picture at the sampling point. Better representative results can be obtained by collecting sequential composite samples. In this case, the sampling program can be modi-

fied to collect, every hour, a series of four, 250-mL samples 15 min apart that are combined into one bottle. This program thus produces a series of 24 different composite samples.

- Another example of a sampling program may call for collecting 1-L samples every time 5000 gallons of wastewater are discharged. This program can be very useful in estimating the total pollutant load in the effluent.

The enactment of the Federal Water Pollution Control Act of 1972, the Clean Water Act of 1977, the Water Quality Act of 1987, and the Storm Water Regulations of 1990 created a need for more advanced sampling programs (*1–4*). It is now well recognized that simply filling a bottle with water or wastewater and sending it to a laboratory for analysis is not acceptable. This approach can produce insufficient or erroneous data that can lead to costly decisions. These could include environmental damage, potential human health effects, as well as simple monetary losses.

In the 1980s, manual sampling lost ground to automatic samplers at a very rapid rate. Sampling professionals have confirmed that automatic samplers can improve the accuracy of their results and reduce labor costs associated with manual collection. It is not surprising that most of the sampling in the United States and the industrialized world is conducted with automatic samplers.

Types of Automatic Samplers

The two basic types of automatic samplers are portable samplers and permanent samplers.

Portable Samplers. Portable models are designed to be carried from site to site and generally are battery-powered. Typically, they are used in the field, in tight locations such as manholes, or at remote monitoring sites where power is not readily available. Figure 2 shows an example of a portable automatic sampler.

Permanent Samplers. These are designed for use in permanent or semipermanent applications. The majority of permanent samplers sold in the United States are of the refrigerated type. Refrigerated samplers are generally placed inside buildings, where AC line power is available. However, certain models are designed and built for outdoor monitoring applications. Figure 3 shows an example of a refrigerated automatic sampler.

Components of an Automatic Sampler

An automatic sampler generally consists of six main components: an electronic controller, a sample delivery system, a sample intake, a sample transport line, a sample storage system, and a power source. Each component performs a specific function

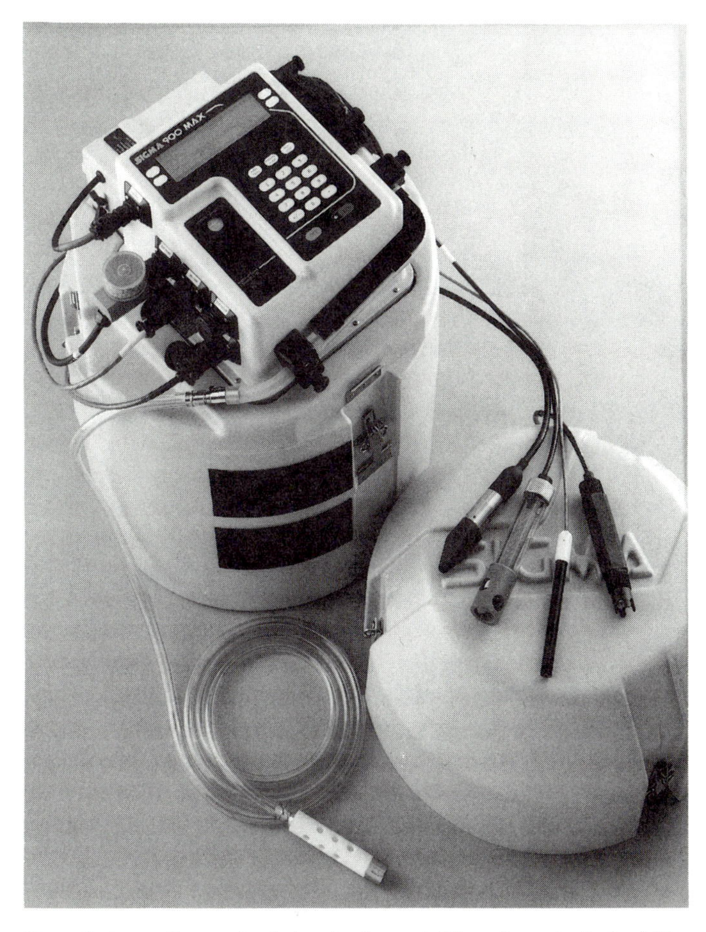

Figure 2. A portable sampler designed to be carried from site to site in the field.

and can affect the accuracy and precision of the sampling project. Figure 4 depicts the various components of an automatic sampler.

Electronic Controller. The *electronic controller* is the "brain" of the sampler. It is similar to a personal computer. It can be programmed to run any sampling routine. The controller activates the pumping system to collect samples and deposit them in the correct bottles, precisely as prescribed in the sampling program. In other words, it controls the date and time of sampling, the conditions and frequency of sampling, the volume of each sample, and the container in which each sample is deposited. In addition, some electronic controllers act as data loggers by storing the sampling program, the sampling data, and other parameters such as pH.

Figure 3. A refrigerated automatic sampler designed for permanent outdoor use.

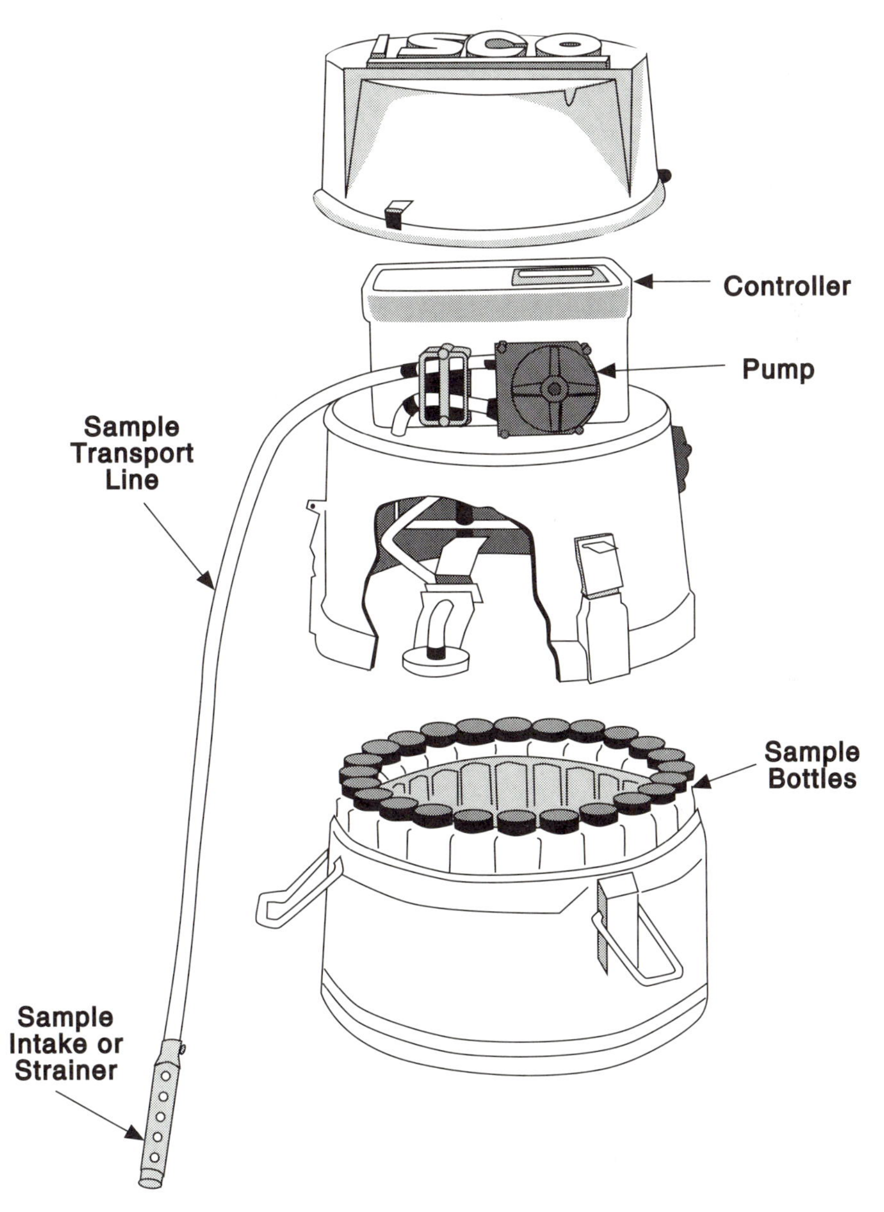

Figure 4. Components of a typical automatic sampler.

Many commercially available electronic controllers are advanced devices that can communicate with other instruments, such as open-channel flowmeters, liquid level meters, rain gauges, and a range of analyzers. Some are capable of interfacing with personal computers or printers for downloading or uploading sampling programs and sampling data. Stored computer data and hard copies are very useful. They can be used by enforcement agencies or by industrial organizations to provide proof of environmental compliance or violation of regulations. In addition, hard copies can be used to reduce labor costs, to improve accuracy in chain-of-custody documentation, and simplify data analysis and report generation.

With the increasing emphasis on controlling pollution from non-point source runoff, new samplers were introduced to help sampling professionals in their efforts to meet the new federal U.S. storm water regulations and to characterize runoff from storm water events (*4*). Controllers on these samplers have built-in software that can automatically take a discrete sample during the first 30 min of a storm event and flow-weighted composite samples afterwards. The first flush sample helps in identifying pollutants that run off immediately after the rainfall, while the composite samples are used in estimating pollutant loading into the receiving waters.

Three controller features must be considered before purchasing an automatic sampler. First, the controller must satisfy your present and possibly your future sampling needs. A controller with limited capabilities costs less but can be rendered useless if permit requirements are tightened or new regulations are enacted. Second, the controller must be proven to be environmentally sealed. This feature is particularly important, because samplers can get submerged in water, rained on, splashed with water, and exposed to dust and corrosive gases. Finally, the controller should offer ease of programmability. Complexity invites rule bending that can lead to erroneous results.

Sample Delivery System. The *sample delivery system* is simply a pump designed to transfer the sample from the liquid source to the bottles for storage. Commercially available automatic samplers offer three types of pumps: peristaltic pumps, vacuum pumps, and bladder pumps. The following overview of these pumps describes their advantages and disadvantages:

Peristaltic Pumps. Peristaltic pumps are the most commonly used pumps in sample delivery systems. Their principle of operation is simple. A motor rotates a two- or three-pinch roller that squeezes a tubing. This squeezing action draws the liquid from the source into the sample bottle for storage. Figure 5 shows the principle of operation of peristaltic pump samplers.

In the early days of automatic sampling, sample volume was determined on the basis of the duration of the roller rotation or the number of pump revolutions. This approach, however, produced inaccurate and nonrepeatable volumes. Today, the more advanced samplers use liquid presence detectors in conjunction with pump revolution counters to deliver high levels of accuracy and precision.

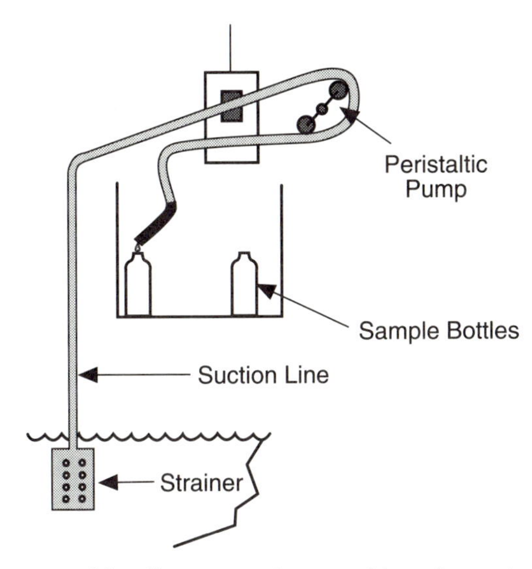

Figure 5. The rotation of the roller in a peristaltic pump delivers the sample to the container.

The advantages of peristaltic pumps in automatic sampling have been well-established. First, the sample touches only the sample tubing. The sampler can be programmed to automatically purge and rinse the tubing prior to sampling. Before the sampling activity is initiated, the tubing can be cleaned to meet the required standards. After sampling, it can be decontaminated or simply replaced. Second, peristaltic pumps are designed to produce the appropriate intake velocity. High intake velocities increase the potential for scouring high levels of solid particles, thereby overstating their presence in the source. On the other hand, low velocities allow solids to settle, thus understating them. Finally, the design of peristaltic pumps is simple. Their simple design reduces the likelihood of two factors that can have dire effects on the sampling project: field maintenance problems and pump failure.

Peristaltic pumps have two disadvantages. First, they have low lift capability of approximately 26 ft (8 m). Second, the vacuum created during operation pulls out any gases in the sample. This feature makes samplers with peristaltic pumps not suitable for collecting water or wastewater samples for volatile organic compound (VOC) analysis.

Vacuum Pumps. Another type of sample delivery system used on automatic samplers is vacuum pumping. This system uses a vacuum pump and a metering chamber. To collect a sample, a vacuum is applied to the metering chamber. This vacuum draws the water or wastewater through the sample intake, into the transport line, and then into the metering chamber. A conductive sensor continuously monitors the level of the liquid in the metering chamber. When the desired sample volume is reached, the applied vacuum is automatically interrupted, and further delivery of the liquid to the chamber is stopped. The final step consists of draining the liquid in the chamber into the sample container below.

In the early days of sampling, vacuum samplers were very popular in the United States and Canada. They were capable of delivering more accurate sample volumes than peristaltic samplers. This advantage, however, was lost following the development of liquid presence detectors and pump revolution counters. Vacuum samplers are still widely used in Europe, but peristaltic pumps are gaining acceptance.

There are four disadvantages for vacuum samplers. First, they have high intake velocities that tend to scour and collect high levels of solid particles at the sampling point. This feature overstates the presence of solids in samples. Second, the high vacuum required to draw the sample pulls out the dissolved gases and VOCs contained in the liquid. Third, the sample lines and the metering chamber are difficult to purge and clean. This feature increases the potential for cross-contamination between samples and can lead to inaccurate data. Finally, vacuum samplers have intricate mechanical designs making them unreliable in field applications. All these disadvantages have contributed to the low acceptance of vacuum samplers in the North American market.

Bladder Pumps. The third type of sample delivery system is the bladder pump. Its principle of operation is slightly more complex than either vacuum or peristaltic pumps. A bladder pump is essentially a pipe with a bladder on the inside and a source of compressed gas, generally air. Under the action of the compressed gas, the bladder expands and contracts, drawing the water or wastewater to the sample bottle.

Bladder pumps have been very popular in groundwater sampling for more than two decades. However, they were not incorporated into commercially available automatic water and wastewater samplers until 1993. Figure 6 depicts the principle of operation of bladder pumps.

Bladder pumps offer three significant advantages. First, bladder pumps do not allow air to come in contact with the liquid sample. This feature is extremely important in preserving the integrity of the sample, particularly VOC samples. Second, the expansion and contraction of the bladder are very gentle on the sample. This characteristic reduces the potential for turbulence, a factor that can alter the sample, and it thus makes bladder pumps ideal for sampling liquids containing VOCs. Finally, bladder pumps are capable of lifting samples up to 250 ft (76 m). This lift capability is substantially higher than peristaltic and vacuum pumps.

The main disadvantage of bladder pumps is their susceptibility to damage in waters containing high levels of solid particles. For this reason, bladder pumps are always equipped with screens in wastewater sampling applications. Obviously, screens will produce samples that are low in solids.

Sample Intake. The sample intake is the first point where the sample liquid enters the sampler. It is submerged in the water or wastewater and typically has a strainer to filter out large objects that can plug the transport line. The sample intake should be large enough to allow the water sample to enter without any obstruction,

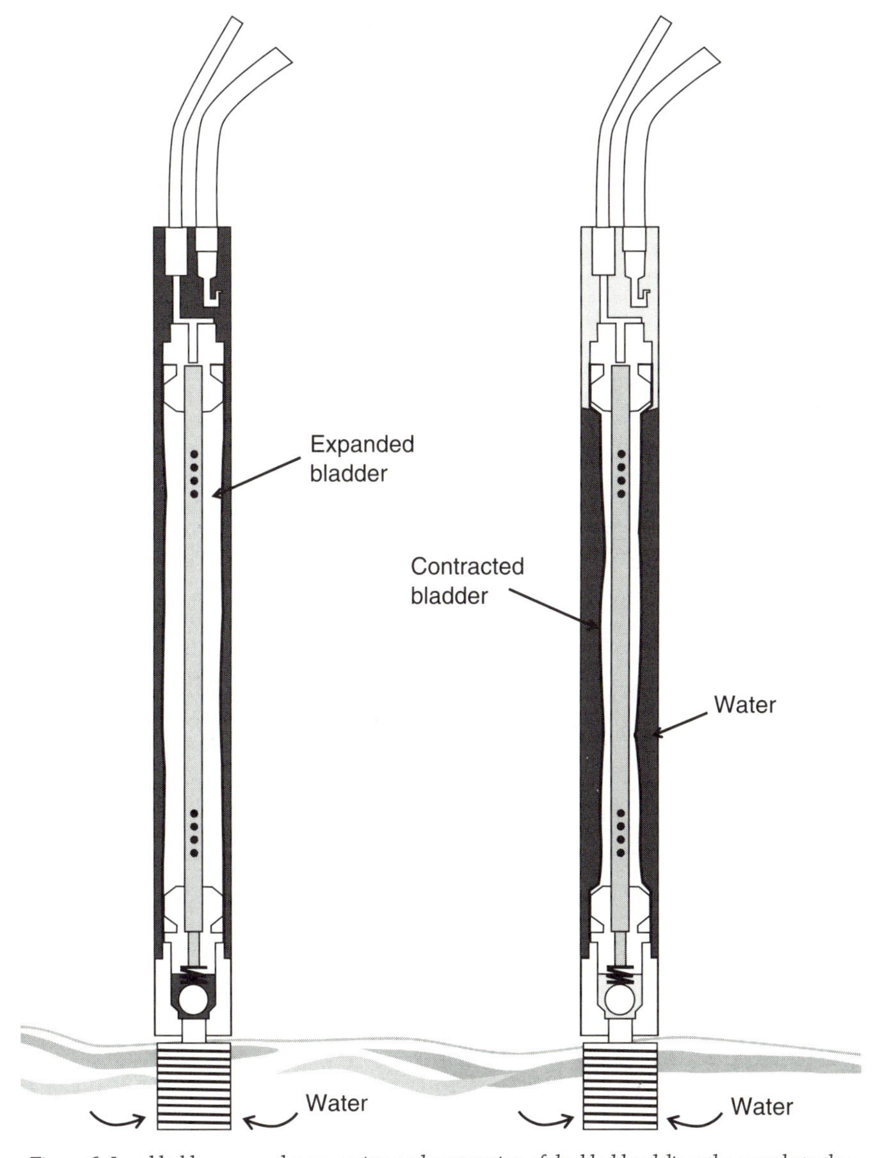

Figure 6. In a bladder pump, the expansion and contraction of the bladder deliver the sample to the container.

but small enough to maintain sufficient line velocity to prevent settling of solid particles. To prevent damage or deterioration, the sample intake should be compatible with the chemicals in the stream and be physically resistant to possible damage from flowing debris. The intake should be made of a material that will not alter the sample through leaching or sorption. Some sample intakes are constructed of stainless steel for use in monitoring priority pollutants, whereas others are made of polypropylene for general use. Some have small diameters for low flow applications, whereas others have large diameters for monitoring in large sewers.

Sample Transport Line. Generally, the *sample transport line* is a plastic tube connected to the sample intake. The water or wastewater sample travels through the transport line until it reaches the sample containers for storage. Like the sample intake, the inside diameter of the transport tubing should be large enough to prevent plugging, but small enough to maintain sufficient flow velocity to prevent settling of solid particles.

The sampling professional should consider the following precautions. First, the transport line should be free of kinks, twists, and sharp bends and should be as short as possible. All these can lead to plugging, low flow velocities, and high pressure drop. Worst of all they increase the possibility for incomplete purging of the transport line, thereby increasing the potential for cross-contamination between samples. Second, to prevent deterioration, the tubing should be compatible with the water or wastewater stream to be sampled. Chemicals and high temperature can cause failure of the transport line and interrupt the sampling process. Third, the tubing must be made of an inert material that will not alter the sample, chemically or physically, through leaching, absorption, or desorption. Finally, the transport line should be purged and rinsed just before collecting a new sample. Purging and rinsing reduce the potential for sample cross-contamination by removing any contaminants left in the sample line or that might have adhered to the tubing walls from previous samples. This step is especially important when monitoring for small fluctuations in pollutant levels.

Sample Storage. In general, *sample storage* consists of bottles that are placed in the base of the portable samplers or inside the refrigeration compartment of refrigerated models. Figure 7 shows bottles in a refrigerated sampler. Bags can also be used to store samples. However, they are generally more difficult to handle and ship to the laboratory for analysis.

Automatic sampler manufacturers offer a large number of bottle configurations. Three factors must be considered when a bottle configuration is selected. First, depending on the sampling requirements, bottle sizes vary from a few milliliters to several liters or even gallons. Second, the number of bottles varies from one to several dozens. Third, bottles can be made of polypropylene, polyethylene, Teflon, clear or amber glass, and with or without Teflon lids and linings. No single bottle configuration is suitable for all possible applications, and no one single sampler is designed to accept all possible bottle configurations. Before an automatic sampler

Figure 7. Refrigerated samplers with various bottle configurations.

with a particular bottle configuration is acquired, the sampling investigator should consult with permit requirements, with the appropriate regulations, and with the laboratory where the analyses are to be performed. Permits may dictate the type, frequency, and volume of samples to be collected and the pollutants that must be monitored. In addition, different laboratories require different sample volumes to perform the needed analyses.

The U.S. Environmental Protection Agency has published requirements for selecting the right type of containers and sample volumes (5). Table I shows a partial list of these requirements. It is important to check state and local requirements, as these may differ from those of the EPA.

Factors of extreme importance when considering sample containers are chemical compatibility, leaching, absorption, and desorption. These can compromise the integrity of the sample and can lead to inaccurate results. Finally, to preserve the samples, the bottles may have to be placed in a refrigerator or in a sampler base that can accept ice. Different samplers have different ice capacities and cooling capabilities. If required, the sample should be quickly cooled down to a temperature of 4 °C. This cooling reduces the potential for microbiological activity and chemical changes.

Power Source. The power source supplies the power needed to operate the sampler. Depending on the monitoring site, the power source can be AC line power or a

Table I. A Partial List of Parameters and Required Containers

Parameter	*Container Type*
Acidity	Plastic; glass
Alkalinity	Plastic; glass
Biochemical oxygen demand	Plastic; glass
Chemical oxygen demand	Plastic; glass
Chlorinated hydrocarbons	Glass; Teflon-lined cap
Chlorine, total residual	Plastic; glass
Coliform, fecal and total	Plastic; glass
Color	Plastic; glass
Cyanide	Plastic; glass
Fluoride	Plastic
Nitrate	Plastic; glass
Oil and grease	Glass
Oxygen, dissolved (probe)	Glass
Temperature	Plastic; glass

DC battery pack designed specifically for the sampler. Portable samplers are usually operated with a battery, whereas refrigerated units depend on AC power.

For portable applications, sampling professionals should use rechargeable batteries designed to withstand the harsh sampling environments. To prevent slow starts, project interruptions, and possible fines, one should always start with a new fresh battery or with a recently recharged one. This precaution is essential, because many types of batteries lose their capacities even during storage. A variety of inexpensive batteries are available at retail outlets. However, they tend to be bulky, can fail in some environments, or may not have the capacity needed to run the pump and the controller throughout the duration of the sampling program.

Recent Developments in Sampling

VOC Sampling

A recent development in sampling technology is the introduction of the first automatic sampler designed to collect water or wastewater samples containing VOCs. Until then, VOC sampling could only be conducted manually. The new portable sampler uses a bladder pump to deliver the samples. The pump is operated by a small battery-powered air compressor built into the sampler. The new VOC sampler uses an innovative design in its sample delivery system and sample transport line. For prevention of possible contamination, the sampler can be programmed to automatically purge and rinse both its sample transport line and each bottle several times before each sample is collected. For collection of a sample, the water or wastewater is injected into the bottle through a needle. The injection is conducted gently in a 360 (directional) degree stream. First, the stream removes all the air in the bottle, including any air bubbles on the bottle walls, and displaces them with the sample. Second, the sample water or wastewater is allowed to overflow to ensure that no air is entrapped inside the bottle. The third and last step consists of automatically sealing the sample by closing a valve made of Teflon and stainless steel. The date, time, and bottle number are stored in memory for future reference. Stored data can be retrieved using a personal computer to generate accurate reports and chain-of-custody documentation. Comparison tests, with manual grab sampling, peristaltic samplers, vacuum samplers, and with prepared standard solutions with known VOCs, were conducted to evaluate the performance of the new sampler. Results show that the new sampler delivers samples that are closest to the standard solutions. Field tests conducted in actual sampling applications helped fine-tune the new sampler and improve its operation in the harsh sampling environment.

Sample Volume Accuracy

Sample volume inaccuracy in peristaltic pump samplers was a problem. It was a direct result of changes in the liquid level of the stream to be sampled. Liquid presence detectors were introduced to solve this problem. However, early detectors were susceptible to excessive fouling and high maintenance. An innovative liquid presence detector was introduced in early 1990 that uses a noncontacting sensor (6). The detector uses a thin piezoelectric film installed in a housing wrapped around the sample transport line. Figure 8 depicts the new detector. Its principle of operation is sim-

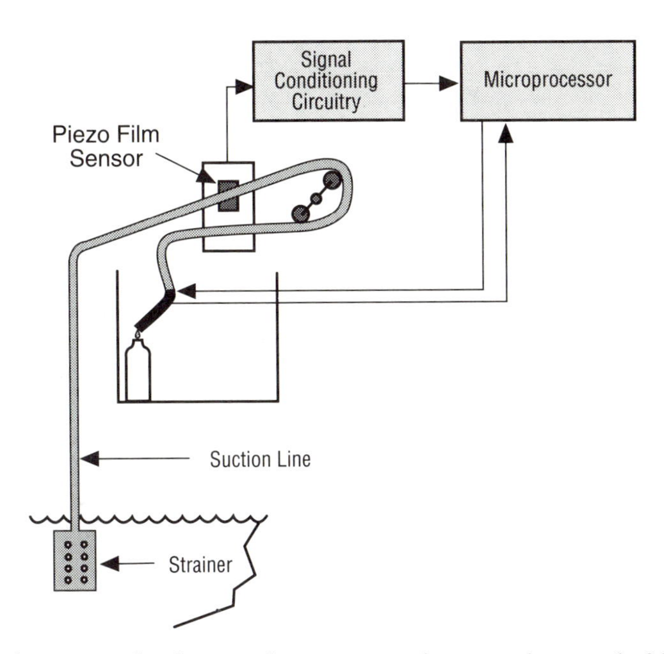

Figure 8. This innovative liquid presence detector continuously monitors the approach of the liquid inside the tubing. This information is used to deliver the exact sample volume.

ple. During the pumping operation, the detector monitors the approaching liquid inside the tubing and sends signals to the electronic controller. On the basis of this information and the pump revolution counter, the controller calculates the number of revolutions required to deliver the exact sample volume. The piezoelectric film in the new liquid presence detector does not contact the water or wastewater to be sampled and does not depend on the conductivity, pH, or other characteristics of the water for proper operation. This design eliminates fouling, maintenance problems, and periodic inspection, and improves representativeness of the sample. Figure 8 shows a detector used for monitoring the incoming liquid.

Refrigeration

Although most samplers are of the portable type, refrigerated samplers are gaining wider acceptance as a result of the significant improvements they have undergone. Several models are now available that are very rugged, can withstand the harsh outdoor environments, and are built to maintain the correct temperature for sample preservation in cold or hot outdoor applications. Figure 9 shows a refrigerated sampler designed for outdoor use.

Preserving Sample Integrity

Because the ultimate goal of a sampling project is to collect a sample that truly represents the source at the time of collection, every effort should be exerted to preserve the integrity of the sample during and after sampling. As sampling professionals can attest, this task is very difficult. It requires a carefully devised plan that starts even before acquiring the sampler. The following provides an overview of the steps needed to develop and implement a sample integrity preservation plan.

Sampling Equipment

The first step in the plan is to select the right automatic sampler. Often, samplers are purchased on the basis of price, with complete disregard to the sampling objective. As stated earlier, the method of collection can alter the sample. Both peristaltic and vacuum pump samplers can pull out the VOCs in the sample. Centrifugal pumps create enough turbulence and aeration to change the pH of the sample. Metering chambers in vacuum samplers are difficult to rinse and can cause cross-contamination between samples.

Some samplers have electronic controllers with limited programming and interfacing capabilities. As a result, samples are often collected under less than ideal conditions. Examples include flow-proportional sampling without open-channel flowmeters, parameter-based sampling without pH meters, and storm water runoff sampling without rain gauges. Certainly sampling can be performed under these conditions, but these cost-cutting efforts often lead to measurement variability and to samples that are not truly representative of the source stream.

Figure 9. A refrigerated sampler permanently installed outdoors.

Materials of Construction

The second step in the sample integrity preservation plan is to select the right materials of construction. Construction materials include sample bottles and sample tubing. Some bottles, lids, and linings can be chemically incompatible with the sample liquid and can degrade the sample. Additionally, some tubing can alter the sample by leaching, absorption, or desorption. Reuse of bottles and tubing helps reduce waste and protects the environment. However, used bottles and tubing must be professionally cleaned according to approved procedures and based on the analyte of interest (7).

Sample Handling

Ideally, once collected the sample must be shipped to the laboratory without further handling. Manual compositing of samples requires the transfer of samples from one bottle to another. The process of transferring samples tends to aerate samples, causing degassing of VOCs and change in the pH. It can cause oil and grease to adhere to the walls of the containers and can result in the loss of solid particles and even loss of the analyte. Compositing samples automatically eliminates the transfer process, thereby reducing the potential for inaccuracies.

Sample Preservation

Many samples start changing physically, chemically, or biologically almost instantaneously. Ideally, the samples should be completely analyzed immediately after collection (7). However, immediate analysis may not be practical or possible at all times. In this case, every effort must be made to preserve the samples. Sample preservation must be done according to permit requirements, regulations, or the recommendations of the laboratory at which the analyses are to be performed. Sample preservation consists of refrigeration, chemical fixation, or both. Refrigeration offers several benefits over chemical preservation. First, it eliminates the addition to the sample of any chemicals that can affect sample composition. Second, the added chemicals can interfere with the analytical methods. Third, addition of chemical preservatives can introduce impurities to the sample. Finally, rapid cooling down to 4 °C reduces microbiological activity and the potential for volatilization of dissolved gases and organic substances contained in the samples. Cooling can be accomplished by using portable samplers with large ice capacity or by acquiring specially designed refrigerated samplers. Small refrigerators sold at retail outlets are inexpensive. However, they are not designed for the harsh indoor or outdoor sampling environment and are not capable of maintaining sample temperature for proper preservation.

Sample preservation helps extend the holding time, but it cannot extend it indefinitely. Even if preserved, the samples must be shipped to the laboratory for immediate analysis to lower the risk of deterioration. The U.S. Environmental Protection Agency has published a comprehensive list of required preservatives and maximum holding times (5). Table II shows a partial list of these requirements.

Table II. A Partial List of Preservatives and Maximum Holding Time

Parameter	*Preservative*	*Maximum Holding Time*
Acidity and alkalinity	Cool to 4 °C	14 days
Biochemical oxygen demand	Cool to 4 °C	48 h
Chemical oxygen demand	Cool to 4 °C H_2SO_4 to pH < 2	28 days
Chlorinated hydrocarbons	Cool to 4 °C	7 days until extraction
Coliform, fecal and total	Cool to 4 °C	6 h
Oil and grease	Cool to 4 °C H_2SO_4 or HCl to pH < 2	28 days

Some parameters, such as temperature, pH, dissolved oxygen, and residual chlorine, cannot be preserved. They must be measured on-site during sampling. Parameter actuator loggers are instruments designed to continuously monitor various parameters in water and wastewater streams. Figure 10 provides an example of a parameter actuator logger mounted on an automatic sampler. Parameter actuator loggers automatically record the data and activate the automatic sampler to take a sample based on a preset program. These instruments offer several advantages. First, they have the capability to interface with personal computers or field printers to generate useful reports and chain-of-custody documentation. Second, they allow the user to record data at short intervals, to uncover problems, or spot trends. Third, they help in reducing labor cost and in eliminating human error.

Conclusion

Automatic water and wastewater sampling is a dynamic field. New samplers are continuously being introduced by equipment manufacturers. All efforts are being focused on enhancing the capabilities of the equipment, simplifying their use, and extending their survival rate in harsh environments. Some efforts are being directed at

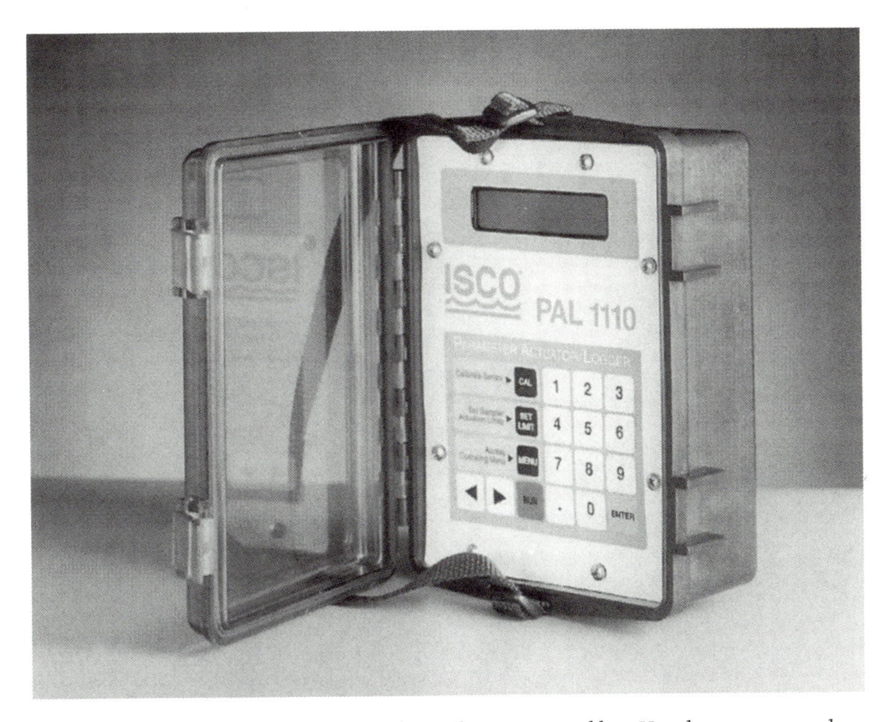

Figure 10. A parameter actuator logger designed to monitor and log pH and temperature and activate any automatic sampler based on a preset program.

improving user knowledge (*8–10*). Improving user knowledge is extremely important because one of the keys to a successful sampling project is a thorough understanding of the sampling equipment and the conditions of the stream to be sampled.

References

1. "Federal Water Pollution Control Act Amendment of 1972"; Public Law 92-500, 1972.
2. "Clean Water Act of 1977"; Public Law 95-217, 1977.
3. "Federal Water Quality Act of 1987"; Public Law 100-4, 1987.
4. U.S. Environmental Protection Agency. "NPDES Permit Application Regulations for Storm Water Discharges"; *Code of Federal Regulations,* Title 40, Parts 122, 123, 124, 1990.
5. "Required Containers, Preservation Techniques, and Holding Times"; *Code of Federal Regulations,* Title 40, Part 136, 1984; 49 FR 43260.
6. Stefanides, E. J. *Des. News* **1991,** *July 22.*
7. Macy, M. J.; Godfrey, B. *Environ. Test. Anal.* **1992,** *November/December.*
8. Thrush, C.; De Leon, D. B. "Automatic Stormwater Sampling Made Easy"; pamphlet from The Water Environment Federation: Alexandria, VA, 1993.
9. "NPDES Compliance Monitoring Inspector Training: Sampling"; U.S. Environmental Protection Agency. Office of Water Enforcement and Permits: Washington, DC, 1990.
10. "Storm Water Runoff Guide"; pamphlet from Isco: Lincoln, NE, 1992.

Chapter 14

Techniques for Sampling Surface and Industrial Waters

Special Considerations and Choices

James E. Norris

Special considerations and choices of sampling modes, sample locations, and sample treatment are described. The variations of these factors when applied to sampling industrial wastewaters, receiving streams, and their attendant biota and sediments are also addressed. Advantages and disadvantages of available options are discussed.

THE REGULATORY CLIMATE of the late 1980s led me to characterize analytical chemistry as the handmaiden of jurisprudence and political action in the first edition of this reference book. A more appropriate analogy in the mid-1990s might be that analytical chemistry is the reluctant bride, not merely the handmaiden. Unfortunately, those who use the results of the analytical art often do not fully understand its limitations, especially the limitations of sampling. This chapter addresses such a fundamental limitation: the validity of a chemical analysis limited by the validity of the sample chosen for characterization. No amount of statistical transformation of data, application of sophisticated analytical methods, or imposition of onerous quality assurance and quality control measures in the laboratory can transform a bad (nonrepresentative) sample into a good one. Sampling is an attempt to choose and extract a representative portion of a physical system from its surroundings. The subsequent analytical characterization of the sample defines certain properties of the sampled system. This characterization involves extraction of the sample from the physical system of interest, which is the focus of this chapter; preservation and shipment of the field-collected sample; and in-laboratory subsam-

3152–4/96/0259$15.00/0

pling of the field sample and subsequent analysis of the selected aliquot. The variables affecting the objectivity of this characterization are legion. This chapter addresses my experiences in the sampling of diverse water and wastewater streams over a period of 30 years.

This chapter is not meant to be all-inclusive, but rather addresses certain situations in which alternative sampling methods were evaluated and some conclusions about these alternatives were drawn.

Sediment Sampling

Industrial wastewaters and waters in receiving streams are never pure solutions. Substantial suspended particulates are generally the rule, and rarely are these suspensions stable. When stream velocity and agitation decrease, particulates settle to the bottom. Native bottom particulates and fine clays are also exposed to the components in wastewaters via processes such as absorption, adsorption, and occlusion, which bind these components to bottom materials. Thus, such a stream bottom is characterized by variously intermixed layers of native material and waste sediments. This composite system is of interest in characterizing the impact of the wastewater on the local environment.

Sampling of such sedimentary materials is inevitably an exercise in creative grab sampling. The techniques most used fall into two broad categories: *bottom grab sampling* (dredge sampling) and *core sampling*. Abundant variants and devices of each type are available. Bottom grab samplers have the advantage of obtaining a larger sample over a broader expanse of bottom. These samplers are generally easy to use. The greatest disadvantage of such samplers resides in the loss of finely divided particulates, which are carried away by outflowing water from the sampler. Indeed, for some streams, the bottom or near-bottom surface "fines" may be of most interest in stream quality characterization. In such circumstances, a core sampler represents a suitable alternative. When the closing mechanism or valve is engaged, waters in contact with the fines are entrapped and cannot carry the fines away. A major disadvantage of core samplers lies in the extremely small area of bottom encountered. As a general rule, more core samples of a bottom are required than bottom grab samples to provide sufficient and valid analytical data.

The aspect of maintaining the in situ vertical relationship of bottom sediment layers is generally academic. Bottom sediment layers are almost always so easily resuspended that sampling confounds and commingles such layers. This effect has been humorously compared to the Heisenberg uncertainty principle concerning the electronic level of matter. If, on the contrary, a compacted sedimentary bottom is sampled, or if an underlying clay layer is sampled, vertical integrity of layers can be maintained by using a core sampler having a split-spoon design. For the sorts of stream-bottom characterizations encountered in my experiences, core samplers have almost always been preferred; minimizing the loss of water-suspended fines has been important, and this technique meets that requirement.

Fish Sampling

No dissertation on techniques for sampling surface and industrial waters is complete without an excursion into fish collection and sampling. In the regulatory environment of the 1990s, toxicity-based effluent discharge limits and limits calculated from human health-based risk models are paramount, and the question of bioaccumulation arises logically. For example, the impact on indigenous fish species in receiving streams of both pesticide manufacturing discharges and pesticide agricultural runoff is often addressed in a fish collection and analysis campaign.

The collection of fish is the field of commercial fishermen. My experience has been that in all but the most limited of studies, the services of a commercial fisherman are an invaluable asset, especially when the study is limited to one or two varieties of fish in the indigenous population. However, when the sampling campaign is conducted by the investigator, some options can affect the success and ease of the collection effort. Among the various techniques for collection of fish, electric shockers and slat boxes present advantages over other alternatives such as hoop nets, gill nets, or trot lines. These alternative techniques are likely to obtain fewer samples per unit time (e.g., trot lines) or are likely to kill the specimen well before retrieval (e.g., gill nets). (As an alternative to repetitive stream collection of specimens, caged fish can be used for biouptake determinations when this use meets the objectives of the study.)

The preparation of the specimen is critical if reproducibility is required among multiple laboratories analyzing the fish tissue. If freezing the fish for preservation, shipping, and handling is necessary, then the whole fish should be frozen. A caveat should be noted here: If aluminum foil is used to wrap the fish for storage or transportation, the foil should first be washed with methylene chloride and thoroughly dried. The shiny side of the metal foil should not be in contact with the fish because this side is coated with a slip agent. If the fillet of the fish is chosen to ascertain bioaccumulation of chemical species, then the fillet should be manually removed, cubed, ground, quartered, mixed, and then split as samples for multilaboratory analysis.

Contrary to some popular protocols on the subject, use of a high-speed, high-shear blender to attempt homogenization prior to sample splitting is generally unwise. This technique almost always results in physical separation of fatty oils (lipids). If this step is undertaken, then splitting this nonhomogeneous multiphase sample into portions that truly represent the sampled mass will probably be impossible. Because so many organic compounds of regulatory concern are believed to accumulate preferentially in fatty tissue, lipid separation prior to sample splitting clearly should be avoided.

A protocol often found to be useful involves first freezing the fish sample, then grinding the frozen mass in the blender with dry ice. This protocol prevents melting of the tissue and mitigates the possibility of lipid separation.

Sampling of Industrial Wastewater Discharges

The sampling of wastewater discharges is the subject of numerous scholarly studies, papers, monographs, and books. Information on techniques and available sampling instrumentation can be found in the U.S. Environmental Protection Agency's *Handbook for Sampling and Sample Preservation of Water and Wastewater* (*1*). A more fundamental treatise has been published by the American Society for Testing and Materials (*2*). The three major sampling techniques available include *grab sampling, composite sampling,* and *continual sampling.*

In practice, grab sampling is almost always manual grab sampling. Composite sampling is usually accomplished by an automatic sampler taking periodic samples and compositing them in a jar or container. A continual sampler withdraws a sample constantly from a stream and accumulates the withdrawn volume for collection at a later time. In my experience, the most generally useful sampling technique has been composite sampling with an automatic sampler, which draws a constant sample volume at time intervals proportional to stream flow. In situations when organic compounds are those species of analytical interest, the automatic sampler commonly preferred is that which uses a peristaltic pump, tetrafluoroethylene (Teflon) tubing, and a cooled (4 °C) glass container for collection. Polyvinyl elastomer (Tygon) tubing should be avoided because of leachable phthalate plasticizers.

A continuous sampler, in concept, should give the most nearly representative sample of a water or wastewater flow. This statement is especially true for those automatic continuous samplers for which the pumping rate is proportional to stream flow. Such samplers, however, generally require large collection containers and present a more challenging routine maintenance problem than the somewhat simpler automatic composite samplers.

The automatic composite sampler alluded to is recommended because of its ruggedness, ample but modest size, and ease of maintenance. When sampling time intervals are proportional to stream flow, the compositing of constant-volume increments gives a sample nearly as representative as that obtained by the continuous sampler.

If one's objective is to characterize the stream over an extended period of time, then the grab sample is the least representative among the sampling choices available. If the grab sample is a manual grab sample, then it is also the simplest and easiest to obtain. In some situations, especially those dictated by regulatory protocols, the grab sample is the only choice available. For example, this situation occurs when sampling a wastewater for volatile organic compounds (VOCs) such as halocarbons and volatile aromatic compounds.

A single grab sample or a half-dozen grab samples over a 24-h period give only "snapshots" of the quality of the sampled stream. These samples can be representative at best of the stream condition during the several seconds of sampling. Where a stream is essentially constant-flow, spatially homogeneous in composition at any time and varying in composition only gradually over an extended period of time, a modest series of grab samples may give a fair approximation of stream characteristics. Real-world industrial wastewaters rarely meet the requirements of this idealized model.

Ironically, the officially sanctioned determination in wastewaters of chloroform, dichloroethanes, benzene, and dichlorobenzenes hinges significantly upon the least representative sampling technique chosen, the grab sample. In addition, effluent guideline limitations for such organic compounds consist of monthly averages and daily maxima. These limits imply that the measurements truly represent stream composition. The rationale for the regulatory requirement of grab samples is that VOCs are lost into the airspace of the 4 °C thermostated sample collection bottle if a continuous or automatic compositing sampler is used. Henry's law is usually considered, but one aspect rarely addressed is the dramatic decrease in partial vapor pressure of most VOCs in aqueous solution as the temperature drops from ambient levels to 4 °C.

The history of the sampled stream is of some importance in grasping the full significance of imputed VOC losses. Most wastewater discharges are warmer than their receiving streams. Indeed, many National Pollutant Discharge Elimination System (NPDES) compliance points measure a composite stream of warm, biologically treated effluent with warmed, once-through noncontact cooling water. For example, one can consider a composite stream at 35 °C (95 °F) exiting a large header and dropping 6 ft into a concrete-lined ditch basking under a summer sun at 100 °F on its way to a Parshall flume (i.e., an artificial channel or chute for a stream of water calibrated to indicate flow rate from the water level in the chute), and an NPDES outfall sampling point. Are VOC losses at 4 °C in a darkened sample collection bottle really significant compared with the VOC losses occurring in the sampled stream?

This situation can be considered from another perspective: When three or six grab samples in 24 h are used to determine VOC concentrations in a wastewater, does the researcher gain more objectivity of results by minimizing assumed Henry's law losses, or does the researcher lose more objectivity of results by taking less representative samples? My experience has been that more objectivity is gained by using automatically composited flow-proportional samples thermostated to 4 °C than by using several grab samples.

This choice does not exist for regulatory compliance monitoring because grab samples are specified. However, the regulation does not address the situation of sampling upstream from a compliance point for purposes of control and characterization. Therefore, this situation is when the automatic compositing sampler can be profitably used.

Thus, the choice of sample type depends upon the intent of the sampling campaign, the regulatory proscriptions, the nature of the sampled stream, and the importance of the results of analyses of the collected sample.

Sampling of Surface Waters: Receiving Streams

The Clean Water Act addresses the premier position of water quality standards in the regulatory scheme. More recently, discharge limitations have been calculated from human health-based risk models. A given industrial or municipal discharge may meet all defined limitations of the NPDES permit. If, however, water quality of the receiving stream is being threatened, and established water quality standards are compromised by such discharges, then the discharges must be further abated, notwithstanding the best available technology. The same can be said when risk models generate allowable concentrations that are being exceeded.

As far as the mechanics of sampling are concerned, most of the options discussed in the preceding section are, a priori, applicable to stream or river sampling. When this decision-making process is encountered, however, the peculiarities of stream sampling become manifest.

Particularly for a river that is a navigable stream, the employment of fixed samplers (e.g., continual or composite) is generally out of the question. My experience and that of many of my colleagues is that the most practical river sampling program is accomplished from a boat at known sampling locations. The sort of sample collected is almost always a manual grab sample or a series of manual grab samples composited prior to analysis.

As far as choice of sampling locations, the EPA *Handbook for Sampling and Sample Preservation of Water and Wastewater* (3) addresses the various techniques for choosing appropriate locations. One technique especially appropriate for sampling rivers for chemical constituents is the spatial gradient technique. By applying this technique, the distance between points on a transect or grid in a grid pattern can be determined. This approach is as good as any, but caveats do exist. The use of this technique must be tempered with a knowledge of many factors such as effluent plume dissipation, mixing zones, segregation of wastewater discharges in the stream, and tidal effects.

For a valid determination of the impact of an outfall on a receiving stream, an upstream control point must be selected. This control point must be sufficiently upstream to be isolated from the effects of the discharge.

For some streams and rivers, even those having considerable flow and which empty into bays, oceans, or large inland lakes, a measurable tidal effect occurs. I have observed a river at low annual flow that was influenced so much by an incoming tide 80 river miles downstream that a wastewater discharge plume spread upstream for one-half of a mile. Indeed, for such coastal river zones, laminar flow in the river often has more dense brackish water incoming under an outgoing freshwater flow. For these reasons, the choice of an upstream control point must be carefully made. A control point located nearby has the advantage of convenience of sampling, but one several miles upstream is better insulated from any influence of the discharge.

Aerial photography, including infrared techniques, provides insight into the phenomena of in-stream segregation of an effluent plume, variations of mixing zones as a function of stream flow, and impacts of tidal effects. Together with surface observations, these data provide guidance in the choice of downstream sampling locations. Such data are most valuable when a modest library of aerial photographs and surface observations extending over a period of one or two years is available. For example, little useful information is gathered from a midriver sampling point one mile downstream from an outfall if the discharge is hugging the near bank of the receiving stream. Such anomalies are easily discerned by aerial photography and careful surface observation.

Against such a backdrop of knowledge of local conditions, good choices of sampling points can be made, and the water quality of the river or stream can be more objectively assessed.

Conclusion

This work has attempted to give the worker in the field an overview of sampling options, their advantages, their disadvantages, and some elementary techniques to improve the validity of the sample collection and handling process.

In the final analysis, perhaps the most important single effort that can be undertaken to ensure appropriate sampling is a clear and open dialogue among client, sampler, and analyst. This dialog will help ensure that the fit between sampling and the final objective of the exercise is the best that can be obtained under the circumstances.

Acknowledgments

I acknowledge the support provided by BCM Engineers Inc. as well as the efforts and contributions made by K. Fitzgerald.

References

1. *Handbook for Sampling and Sample Preservation of Water and Wastewater;* U.S. Environmental Protection Agency. U.S. Government Printing Office: Washington, DC, 1982; EPA–600/4–82–029. (Addendum, 1983.)
2. *Annual Book of ASTM Standards;* American Society for Testing and Materials; Philadelphia, PA, 1986; Vol. 11.01, pp 130–139; Standard D3370–82.
3. *Handbook for Sampling and Sample Preservation of Water and Wastewater;* U.S. Environmental Protection Agency. U.S. Government Printing Office: Washington, DC, 1982; pp 195–200; EPA–600/4–82–029.

Chapter 15

Preservation Techniques for Organic and Inorganic Compounds in Water Samples

Jerry Parr, Mark Bollinger, Owen Callaway, and Kathy Carlberg

Preservation techniques are used to minimize changes between collection and analysis. Physical changes such as volatilization, adsorption, diffusion, and precipitation, and chemical changes such as photochemical and microbiological degradation are minimized by proper preservation. The preservation process encompasses both field and laboratory activities and includes techniques such as chemical addition, temperature control, and the choice of sampling containers. The lack of consistent, validated guidelines for preservation techniques and holding times for similar sample types has resulted in uncertainty among those who routinely collect and analyze samples. These numerous and often conflicting guidelines affect data quality, laboratory operations, and the cost of analytical determinations.

ANALYTICAL SAMPLES are in a chemically dynamic state at the time of collection. At the moment the sample is removed, the chemical processes that affect the sample may deviate from what occurs in situ for many reasons. Some of these reasons are obvious, others are more obscure. For example, a sample collected from a well source is exposed to conditions significantly different from the conditions underground. In the process of collection, the sample is often exposed to ambient light and its temperature most likely has changed. Consequently, photochemical reactions may take place, and the temperature-dependent kinetics of other types of reactions will be altered. Exposure to atmospheric conditions above ground will lead to changes in the dissolved gases in the sample. The presence of oxygen may initiate oxidation of some chemical species. As the sample is exposed to these changes in its new environment, the pH of the sample will likely change. Thus, the act of sampling may, for at least some finite period, alter the nature of the subset sample that is intended to be representative of the water source.

This dynamistic process does not end after the sample has been carefully transferred into some vessel for subsequent delivery to an analytical laboratory. In fact, the vessel may contribute to the process by adsorbing certain components of the sample. This adsorption permits additional loss of volatile compounds or introduces extraneous compounds into the sample. Complete stability of the sample through preservation cannot be totally achieved for every constituent in a sample, nor can all constituents be stabilized with the same degree of success. This situation is an increasing concern as the list of analytes determined by individual analytical methods grows longer and more complex, while detection limits reach lower levels.

The preservation process encompasses both field and laboratory activities and includes a variety of techniques. The practices and techniques recognized as resulting in the best stabilization of the sample are described in general terms in this chapter. Additional information concerning appropriate preservation techniques has been presented (*1–4*).

General Practices for Minimizing Changes

Preservation techniques are selected on the basis of their ability to minimize changes in order to best preserve the integrity of the sample after collection. Analytes present in a given sample can change for a variety of reasons. The most common changes that preservation techniques attempt to minimize are physical changes such as volatilization, adsorption, diffusion, and precipitation, and chemical changes including air oxidation, photochemical changes, and microbiological degradation. These changes are minimized with a variety of techniques, including sample container use, chemical additions, and temperature control.

Volatilization

Volatilization refers to the physical process in which volatile species can be lost to the atmosphere. The process is dependent on the vapor pressure of the analyte to be measured, the temperature of the sample, and the surface area.

For the purposes of this discussion, volatilization refers to the volatilization of organic molecules. Other species that undergo volatilization include gases such as hydrogen sulfide and hydrogen cyanide. However, volatilization of these species is controlled by preservation techniques described elsewhere.

Volatilization can be minimized by a simple technique. This technique involves containing the sample in a vessel with no headspace. Thus, any contact with air is prevented, and the development of an equilibrium of volatile compounds between the surface of the sample and the headspace based on vapor pressure is prohibited.

This technique is generally used for analyses of volatile organic compounds by gas chromatography or gas chromatography–mass spectrometry. However, this preservation process is sometimes overlooked in the measurement of other parameters (e.g., total organic carbon or total organic halides) that can include volatile components. Furthermore, as in most preservation techniques, the containerization process is of no use if the sample was improperly treated prior to preservation by agitation, vacuum filtration, or other procedures that volatilize the analytes of interest.

Adsorption and Absorption

Once a sample has been removed from its natural environment, the equilibrium between the sample and its environment is disturbed. Components in the sample can therefore undergo changes in response to the new environment. For example, components can adsorb, sometimes irreversibly, onto the walls of the sample containers. The two most common examples involve interactions of metal ions with glass surfaces and adsorption of oils onto container walls.

Metals can be irreversibly adsorbed onto glass surfaces. Therefore, for metals, the usual sampling approach is to collect the samples in plastic containers and thus eliminate the glass contact. Also, nitric acid is added to the sample to lower the pH to less than 2, which keeps the metal ions in solution. This situation is useful because the formation of metal hydroxides or hydrated oxides is insignificant under these conditions.

Oils present a more difficult problem. The adsorption of oils onto container walls cannot be easily prevented. Oils are likely to irreversibly adsorb onto the sides of plastic containers. Therefore, samples for organic parameters are collected in glass bottles. This *containerization* process allows the oils to be removed, typically with an organic solvent rinse. Another technique that is useful for samples that have low organic content is emulsification with a sonic probe. An aqueous solution of the oil in the sample is thus created.

Samples also have the propensity for absorbing gases from the atmosphere. Samples can absorb air components such as oxygen or carbon dioxide as well as vapor-phase species that might be present at the site (e.g., volatile organic compounds). Absorption of air components can have a significant impact on the sample by initiating air oxidation (e.g., sulfide to sulfate), chemical changes, or in the case of carbon dioxide, by changing the conductance and pH of the sample. (This process is a primary reason for field measurement of pH and conductance.) Absorption of other components can lead to false positive reporting of the absorbed component.

These changes can be minimized by expeditious preservation of the sample, and in the most extreme case, by eliminating exposure of the sample to the atmosphere. Field sample blanks can be used to determine if critical organic analytes are being absorbed into the samples from the air at the site.

Diffusion

Organic molecules such as phthalate esters and other plasticizers can diffuse through the walls of plastic sample containers or through bottle caps. This process is controlled by collecting samples in glass containers and by using bottle caps or liners that minimize this process. Tetrafluoroethylene (Teflon) liners are especially helpful in controlling contamination from diffusion processes.

Precipitation

Components in a sample may form salts that precipitate in the container. This process results from the interaction of components present in the sample due to a change in the sample environment (e.g., pH), or from the reaction of components in the sample with components in the air environment.

The most common occurrence is the precipitation of metal oxides and hydroxides resulting from reactions of metal ions in the sample with oxygen. Precipitation of these species is essentially eliminated by addition of nitric acid until a pH less than 2 results. The combination of a low pH and an excess of nitrate ion ensures that the metal ions stay in solution. Other acids (e.g., hydrochloric or sulfuric) tend to give anions that could enhance precipitation because of the low solubility of some chlorides and sulfates.

Chemical Changes

Components in samples can undergo a variety of chemical changes. Most of the highly specific preservation procedures were developed to control specific chemical changes. Some of these techniques are described in this chapter. Describing all of the techniques that could be required for every possible analyte is impossible. Most of these techniques are adequately described or referenced in the analytical method associated with that analyte.

Free chlorine in a sample can react with organic compounds to form chlorinated species (5). Obviously, this process is of most concern in samples containing free chlorine. Typically, samples of municipal drinking water and treated wastewaters are the most likely samples to contain free chlorine. For these sample types, the chlorine should be chemically removed by the addition of sodium thiosulfate.

Samples collected for determination of species such as cyanides or sulfides require preservation to ensure that the chemical equilibrium is strongly biased in one direction. Typically, these samples are preserved by pH control or by addition of an anion that will precipitate the component of interest. Thus, sodium hydroxide is added to samples collected for cyanide measurements.

This addition raises the pH to ensure that hydrogen cyanide gas cannot be evolved. Likewise, when sulfuric acid is added to samples collected for ammonia determinations to lower the pH, the stable ammonium ion is formed.

Photochemical Changes

Components in the samples can undergo changes associated with light-catalyzed reactions. The most common example is the photooxidation of polynuclear aromatic hydrocarbons. These changes are minimized by the collection of samples in amber glass containers.

Microbiological Degradation

Samples can contain organisms that may degrade organic components in the samples. These degradation changes are generally minimized by pH and temperature control or by chemical addition. Extreme pH conditions (low or high) and low temperatures are effective for minimizing degradation. Addition of toxic chemicals to the sample (e.g., mercuric chloride and pentachlorophenol) can kill the microorganisms and preserve the sample effectively, although these preservatives are not commonly used because of their inherent environmental hazard.

Preservation Process

Proper preservation of analytical samples requires planning of critically timed activities performed in both the field and in the laboratory. Both the field and laboratory personnel must work in concert to ensure that meaningful results can be obtained from the laboratory. These measures include not only field preservation, but additionally the proper documentation, packaging, shipping, and storage of the samples.

Planning

The first and most crucial step in ensuring that samples are properly preserved is the planning of the events that will occur between sampling and analysis. This process necessitates a discussion between the laboratory staff and the sample collection crew to address factors such as sample containers, on-site versus lab analyses, documentation, preservatives, contingency plans, cleaning of sample containers, and shipping logistics.

Sample Containers

In addition to preservation considerations, the decision as to the appropriate sample containers must be made by considering several factors:

- the cost of the containers and associated costs for shipping the samples to the laboratory,
- the cleanliness of the containers,
- the ability to factor in quality control activities (replicate samples),
- the ease of use for field applications.

For example, a 5-gal carboy might suffice on technical merits, but is obviously not practical for an analysis of fluoride. Generally, selecting bottles to meet the specific objectives of each project is more appropriate. Coding of these bottles by type, number assignment, color, or other means enhances the ability of the field crew to focus their efforts on sampling activities.

Communication between the laboratory and field crew should result in a clear understanding of who will supply the sample containers, how the sample containers will be cleaned and used, and what preservatives will be present in the bottles. For example, Table I illustrates the bottles required to collect a sufficient amount of properly preserved groundwater to measure the 260 chemicals listed in Appendix IX to the *Code of Federal Regulations*, Title 40, Part 264 (40 CFR 264).

Table I. Recommended Containers and Preservatives for Appendix IX Groundwater Monitoring

Sample Container	*Bottle Number*	*Preservation*	*Minimum Sample Size (mL)*	*Compound*	*Recommended Holding Time*
Glass (3 × 40 mL)	11	4 °C	40 each	volatile organic compounds	14 days
Glass (5 × 1 L)	12	4 °C	1000 each	semivolatile organic compounds, pesticides, dioxins	7 days until extraction, 40 days after extraction
Polyethylene	4	2 mL of 50% HNO_3 to pH <2	500	metals	6 months
Plastic	6	2 mL of 50% NaOH to pH <12 at 4 °C	500	cyanide	14 days
Plastic	7	1 mL in zinc acetate, 1 mL of 50% NaOH to pH <9 at 4 °C	250	sulfide	7 days
Plastic	1	4 °C	100	fluoride	28 days

On-Site Analyses and Related Activities

On-site analyses and other field procedures should be discussed with the laboratory staff. For example, if field filtration of groundwater samples is performed, the laboratory should supply bottles with nitric acid preservative for trace metal analyses. On the other hand, if the filtration is performed by the laboratory, the bulk samples without preservative should be returned to the lab.

Documentation

Sample labels and chain-of-custody records should clearly identify both the sample and the treatment that has been applied to the sample. The documentation activities include recording the date and time of collection, the sample name, and the preservatives added as well as any field treatment of the sample such as filtration.

Contingency Plans

The laboratory and the field crew should both recognize that well-planned activities can go astray. An understanding of the appropriate activities to perform if problems arise should be made clear. Contingencies for occurrences such as lost sample containers, low-volume samples, and interferences should be defined during the planning stage.

Sample Shipments

The timely shipment of samples to the laboratory is an activity that requires planning. The shipment schedule and method of shipment should be specified. Arrangements for answering questions that may arise (e.g., contact, phone number, and location) should be defined in advance of the sampling effort. Contingency plans should be discussed. Samples should be shipped to the laboratory as soon as possible to minimize the time between collection and analysis. This step is especially critical for the analysis of chemical constituents such as volatile organic compounds that have holding times of 1 week or less.

Coordination of Activities

Proper preservation requires a coordination of activities between the field crew and the laboratory staff. The field sampling crew has the responsibility for ensuring that the laboratory receives properly preserved samples. A number of factors must be addressed to ensure that this process is performed properly.

The field activities start with a check of all sample bottles, preservatives, and labels on the site to ensure that everything agrees with the sampling plan.

The field crew must understand the sample preservation requirements. They should know what on-site measurements and sample pretreatment steps are to be performed and should have the equipment to perform these tests. Tests for certain species, such as residual chlorine and pH, should be performed in the field. If dissolved metals are to be determined, the field crew should filter samples in the field before adding the nitric acid preservative. The field crew should have checks to ensure that the preservation process was performed properly. For example, if the process calls for addition of 2 mL of sulfuric acid to obtain a pH of less than 2, the field crew should check the pH after addition to make sure that the pH is less than 2.

All sample bottle labels, chain-of-custody records, and other pieces of documentation must be clearly completed. All chemical additions to the samples must be recorded. This documentation is as important to the overall success of the sampling and analysis effort as proper preservation of the samples. Laboratory decisions that affect the handling of the samples are dependent on knowledge of field procedures used in collection of the samples.

The responsibility of the field samplers does not stop when the sample bottles are filled. The field crew must ensure that the samples arrive at the laboratory in an expeditious manner. The shipment process involves packaging of the samples to prevent breakage and to conform to federal regulations. If samples are to be kept cold, proper shipping containers and packaging (e.g., ice or cold packs) must be used to maintain these conditions until samples are received in the laboratory.

The laboratory should be notified of incoming sample shipments. This notification should include method of delivery, expected arrival time, and airbill number if known. The laboratory and field crew should agree on how to handle unusual conditions such as international shipments, delivery after working hours, and holiday shipments.

The sample shipment should include all appropriate documentation. The laboratory must be able to schedule all analyses that require immediate attention because of sample holding-time constraints or rush reporting requirements based on documentation contained within the shipment.

The laboratory staff have the responsibility for ensuring that all laboratory aspects of the preservation process are achieved. This responsibility includes proper storage of the samples within the laboratory, performance of analyses within prescribed holding times, and clear documentation of all activities.

The laboratory process starts with a clear understanding of the sampling events and analytical requirements. The laboratory staff should be prepared to perform the various activities that will be necessary. These activities start with receiving the samples. All sample shipments should receive immediate attention. Laboratory receipt of samples involves unpacking the samples; validation of accompanying chain-of-custody documentation, if any; noting the condition upon receipt; and arranging for proper storage pending analyses. Any discrepancies or uncertainties noted at this stage should be clearly communicated to the appropriate laboratory and field staff for resolution.

Once the samples are scheduled for analyses, the laboratory has a continued responsibility for tracking all laboratory activities on each sample so that the analytical results can be correlated to the original sample collected in the field.

Holding-Time Considerations

A discussion of preservation techniques would be incomplete without addressing sample holding times because preservation techniques are used to maximize holding times. The American Society for Testing and Materials (ASTM) defines the holding time as "the period of time during which a water sample can be stored after collection and preservation without significantly affecting the accuracy of analysis" (*6*).

Holding-time constraints affect both field and laboratory operations. Samples must be shipped to the laboratory as soon as possible after collection to facilitate the laboratory in meeting holding times. The laboratory must have sufficient equipment and staff to guarantee appropriate scheduling of analytical tests to meet holding times. Obviously, the cost of field and analytical work can be greatly affected by very restrictive holding times.

Various regulatory agencies and other groups have attempted to determine the holding times that should be applied to water samples for specific analyses. The most frequently cited holding times are those promulgated under the Clean Water Act in 40 CFR 136 for the analyses of wastewater relating to National Pollutant Discharge Elimination System (NPDES) permits (*7*). Holding times for many of the parameters listed in 40 CFR 136 have also been established under other programs. For example, holding times are cited in many SW–846 methods, which are the analytical methods used to comply with the Resource Conservation and Recovery Act (RCRA) regulations, and rigid holding-time requirements are specified for laboratories performing work under contract to the U.S. Environmental Protection Agency in Superfund site investigations. In addition, methodologies specified by the ASTM, the U.S. Geological Survey, the American Public Health Association, the American Water Works Association, and the Water Pollution Control Federation also have holding-time requirements.

For the most part, the holding times established in these various programs are consistent. However, some inconsistencies do exist that can create confusion as to the selection of appropriate holding times. For example, holding times for volatile organic compounds range from 5 days from sample receipt for Superfund work, to 7 days from sampling for NPDES permits, to 14 days from sampling for RCRA groundwater analyses.

The basic premise that holding times are used to help maintain the integrity of a sample appears to have been lost in the establishment of general regulatory guidelines. The major concern over this trend is the lack of clear evidence to show that holding times have been selected based upon meeting the ASTM definition. In fact, evidence suggests that holding times can be extended for some parameters (*8*).

Because data reviewers can easily determine if holding times were achieved, the evaluation of the overall technical quality of laboratory data is increasingly based solely on this single criteria, which is of questionable scientific validity.

If holding times are used in judging the validity of analytical data, the holding times that are established must have some scientific basis. The maximum time that a sample can be held before compromising the integrity of the analysis is dictated by the matrix (e.g., surface water, groundwater, or wastewater), the properties and concentration of the substance being determined, and the preservation technique employed.

Holding times that are established without considering all of these factors can impose unnecessary constraints and costs on both field and laboratory activities and ultimately lead to the rejection of scientifically valid data.

References

1. *Handbook for Sampling and Sample Preservation of Water and Wastewater*; U.S. Environmental Protection Agency. U.S. Government Printing Office: Washington, DC, 1982; EPA-600/4-82-029.
2. *Test Methods for Evaluating Solid Waste*, 3rd ed.; U.S. Environmental Protection Agency. U.S. Government Printing Office: Washington, DC, 1986; SW-846.
3. *Standard Methods for the Examination of Water and Wastewater*, 16th ed.; U.S. Environmental Protection Agency: Washington, DC, 1985.
4. *Guidelines for Collection and Field Analysis of Groundwater Samples for Selected Unstable Constituents*; U.S. Geological Survey: Reston, VA, 1976.
5. Rook, J. J. *Water Treat. Exam.* **1974**, 23, 234.
6. *Standard Practice for Estimation of Holding Time for Water Samples Containing Organic Constituents*; American Society for Testing and Materials: Philadelphia, PA, 1987; ASTM D4515-85.
7. *Code of Federal Regulations*, Title 40, Protection of the Environment, 1986; Part 136, Table II.
8. Friedman, L. C.; Shroeder, L. J.; Brooks, M. G. *Environ. Sci. Technol.* **1986**, *20*, 826.

Chapter 16

Development of Stability Data for Organic Compounds in Environmental Samples

Michael P. Maskarinec and C. K. Bayne

Over the past several years, Oak Ridge National Laboratory has been involved in studies related to the stability of organic compounds in environmental samples, particularly with respect to preanalytical holding times. We have developed preservation techniques for volatile organic compounds in both water and soil. We have also studied the stability of selected semivolatile organic compounds in both matrices. With respect to volatile organic compounds in water, preservation with sodium bisulfate provides considerable stability (>28 days) and is considerably safer than hydrochloric acid. For volatile organic compounds in soil, methanol immersion is the optimum technique for levels above 1 μg/g, whereas sealed vials combined with closed-loop purge and trap are the most effective means of sampling at the nanogram per gram level.

MOST VOLATILE ORGANIC COMPOUNDS (VOCs) are not stable in environmental samples. Analytical chemistry laboratories analyze environmental samples as quickly as possible to ensure accurate measurements of the concentration at the time the analyte was sampled. With an increasing number of environmental samples, the time between collecting an environmental sample and its chemical analysis (holding time) may be too long; during this time the analyte may biodegrade (*1–3*) or may decompose. Regulatory agencies have specified holding times for classes of compounds (VOCs, semivolatile organic compounds, pesticides, and explosives) to standardize analytical laboratory procedures. For example, 40 CFR 136 (*Code of Federal Regulations; 4*) requires that VOC samples (both water and soil) stored at 4 °C must be analyzed within 7 days of collection. This requirement is very stringent for most analytical laboratories. Some authors have suggested (*5*) that holding times be extended to 14 days for acid-preserved samples. The U.S. Army Toxic and Hazardous Materials Agency (USATHAMA, *6*) recommends a 28-day holding time for VOCs preserved with sodium bisulfate, and a 56-day (until extraction) holding time for explosives stored at 4 °C. These recommendations are based on the Oak Ridge National Laboratory (ORNL) holding time studies (*7–11*). This chapter reviews ORNL holding time studies and presents a new definition for measuring holding times, called *practical reporting time* (*12*). Practical reporting times provide methods to estimate both the stability of an analyte as well as the consequences of measuring analyte concentrations beyond the regulatory holding time. In addition, this chapter suggests a simple protocol for establishing holding times for a particular matrix–analysis combination.

ORNL Holding Time Studies

Experimental Parameters

ORNL conducted holding time studies (*7, 8, 11*) for VOCs and explosives over a two-year period from August 1986 to September 1988. In these holding time studies, ORNL measured analyte concentrations on either 17 or 19 VOCs and four explosives in three water and three soil matrices. Four replicate concentrations were measured on days 0, 3, 7, 14, 28, 56, 112, and 365 for the VOCs and the four explosives. For VOCs in the three soil matrices, the four replicate measurements were made only on days 0, 3, 7, 14, 28, and 56 for the Tennessee and Mississippi soils, and on days 0, 3, 7, 14, 28, 56, and 111 for the sterilized USATHAMA soil. ORNL included two spiking levels and four storage temperatures. Tables I and II give the experimental parameters for the VOCs and explosives, respectively. The ORNL holding time studies represent 13,422 concentration measurements over long periods in a variety of matrices and under a variety of storage temperatures.

ORNL chose the three types of water matrices to assess the effect of varying water quality parameters on stability. The three water types used for this study are reagent-grade water (distilled), a groundwater (ground), and a surface water (surface). Reagent-grade water was obtained from Burdick and Jackson, Inc. (Muskegon, MI). The groundwater was drawn from well 1 at the ORNL Aquatic Ecology Facility (well depth, 205 ft; static water level below ground level, 30 ft). Surface water was taken from the headwaters of White Oak Creek on the Oak Ridge Department of Energy Reservation.

For soil matrices, ORNL chose a USATHAMA (now named the U.S. Army Environmental Center) soil (*13*), a Captina silt loam from Roane County, TN, and a McLaurin sandy loam from Stone County, MS. The USATHAMA soil is a THAMA reference soil (7% sand, 67% silt, and 26% clay) that contains no VOCs or semivolatile organic compounds. The Tennessee (8% sand, 62% silt, and 30% clay) and Mississippi (75% sand, 20% silt, and 5% clay) soils were furnished by the Environmental

Table I. Holding Time Experimental Parameters for VOCs

Matrix	*Experimental Factors*	*Factor Levels*
Water	17 analytes	benzene, bromoform, carbon tetrachloride, chlorobenzene, chloroform, 1,1-dichloroethane, 1,1-dichloroethene, 1,2-dichloropropane, ethylbenzene, methylene chloride, styrene, 1,1,2,2-tetrachloroethane, tetrachloroethene, 1,1,2-trichloroethane, trichloroethene, toluene, *o*-xylene
	Aqueous solution	Distilled water, groundwater, surface water
	Spiking target	50 μg/L, 500 μg/L
	Storage temperature	None at day 0, 4 °C, room temperature
	Storage time	0, 3 or 4, 7, 14, 28, 56, 112, 365 days
Soil	19 Analytes	benzene, bromoform, bromomethane, carbon tetrachloride, chlorobenzene, chloroethane, chloroform, 1,1-dichloroethane, 1,1-dichloroethene, 1,2-dichloropropane, ethylbenzene, methylene chloride, styrene, 1,1,2,2-tetrachloroethane, tetrachloroethene, 1,1,2-trichloroethane, trichloroethene, toluene, *o*-xylene
	Soil type	USATHAMA, Tennessee, Mississippi
	Spiking target	<100 μg/g
	Storage temperature	–70 °C, –20 °C, 4 °C for Tennessee and Mississippi soils
	Storage time	0, 3, 7, 14, 28, 56, 111 days for USATHAMA soil 0, 3, 7, 14, 28, 56 days for Tennessee and Mississippi soils

NOTE: The U.S. Army Toxic and Hazardous Materials Agency (USATHAMA) soil is a THAMA reference soil (7% sand, 67% silt, and 26% clay) that contains no volatile organic compounds (VOCs) or semivolatile organics. The Tennessee (8% sand, 62% silt, and 30% clay) and Mississippi (75% sand, 20% silt, and 5% clay) soils were furnished by the Environmental Sciences Division of Oak Ridge National Laboratory.

Table II. Holding Time Experimental Parameters for Explosives

Matrix	*Experimental Factors*	*Factor Levels*
Water	4 analytes	DNT, HMX, RDX, TNT
	Aqueous solution	Distilled, ground, surface
	Spiking target	50 μg/L, 1000 μg/L
	Storage temperature	4 °C, room temperature
	Storage time	0, 3, 7, 14, 28, 56, 112, 365 days
Soil	4 analytes	DNT, HMX, RDX, TNT
	Soil type	USATHAMA, Tennessee, Mississippi
	Spiking target	10 μg/g, 100 μg/g
	Storage temperature	–20 °C, 4 °C, room temperature
	Storage time	0, 3, 7, 14, 28 56, 112, 365 days

NOTE: DNT is 2,4-dinitrotoluene; HMX is octahydro-1,3,5,7-tetranitro-1,2,5,7-tetrazocine; RDX is hexahydro-1,3,5-trinitro-1,3,5-triazine; TNT is 2,4,6-trinitrotoluene.

Sciences Division of ORNL. Both soils were slightly acidic and low in organic carbons. The Tennessee soil had a higher cation-exchange capacity and microbial respiration rate than those of the Mississippi soil. The biodegradation and microbial activity have been examined (2–3) in the Tennessee and Mississippi soils for 19 organic compounds. The results showed that most chemicals depressed carbon dioxide efflux in the two soils when applied at 1000 μg/g soil, but this effect disappeared within a few days. These results cannot necessarily be extrapolated to microbial activity for the ORNL holding time studies, but they suggest that such activity is present.

Experimental Design

ORNL designed their holding time studies as a complete factorial experimental design for the factors in Tables I and II with four replicate measurements. Some minor variations on the complete factorial experiments are detailed in references 7, 8, and 11.

Sample Preparation

Water samples were dispensed into 1-L Tedlar gas sampling bags. The 1-L Tedlar air sampling bags with dual stainless steel fittings (hose/valve fitting and replaceable septum, catalog no. 231–01) were obtained from SKC, Inc. (Eighty Four, PA). The water was allowed to degas for 3 days, and the gas was removed from the bag. Appropriate volumes of each analyte were introduced through the septum port using gastight syringes. The contents of the Tedlar bag were mixed thoroughly by hand agitation for 3 min, after which the bags were allowed to sit for 30 min. After mixing, sample aliquots were collected in 40-mL (VOCs) or 7-mL (explosives) vials by gravity flow. These sample storage vials were borosilicate glass vials, with Teflon-faced silicone septa and screw caps with holes, purchased from Supelco (Bellefonte, PA). Teflon tubing (1/4 in. × 6 in.) was used to allow each vial to be filled from the bottom up to prevent mixing of the water with air. Each sample vial was completely filled with sample so that no headspace would remain after the sample vial was sealed. Each sample vial was sealed immediately with a Teflon-faced septum and screw cap with hole, and was stored at the appropriate temperature.

Soil samples were prepared by weighing 5-g aliquots for VOCs or 2-g aliquots for explosives into 40-mL borosilicate glass vials, with aluminum foil-coated, Teflon-faced silicone septa and screw caps with holes, purchased from Shamrock Glass Company (catalog no. 6–06 K). These vials were received fully assembled and pre-cleaned according to EPA 40 CFR 136 and EPA 40 CFR 141 regulations. Three days prior to spiking with the analyte stock solutions, the soil samples were wetted with 1.25 mL (VOCs) or 0.5 mL (explosives) of reagent-grade water (Burdick and Jackson, Inc.) and agitated with a vortex mixer for 30 s. The soil samples were then stored in the dark at room temperature. This preparation step allowed bacterial growth to come to a steady state.

For VOC analytes, water was introduced into a 1-L Tedlar gas sampling bag and allowed to degas for 3 days. This gas was then removed. The 1-L Tedlar air sampling bags with dual stainless steel fittings (hose/valve fitting and replaceable septum, catalog no. 231–01) were obtained from SKC, Inc. On the day that soil samples were spiked, excess air was removed from a water-filled Tedlar bag. Thirty soil samples were spiked with 1.25 mL of water and vortexed for about 30 s. These soil samples served as quality control blanks. Quality control blanks and soil samples were stored together in order to assess the possibility of cross-contamination. The Tedlar bag with the remaining water was weighed to the nearest gram on a double-pan balance using an empty Tedlar bag as a tare. Assuming 1.0 g/mL for water density, the remaining volume of water was estimated.

ORNL prepared stock solutions of VOCs for each soil type by dispensing the 19 VOCs standards into the Tedlar bag with the remaining water. Target compounds were received as methanolic solutions of 1800–2300-μg VOCs per milliliter methanol. Appropriate volumes of each VOC standard solution were introduced through the septum port using gastight syringes. The contents of the Tedlar bag were mixed thoroughly by hand agitation for 3 min, after which the bags were allowed to sit for 30 min. Aliquots of the VOC stock solution were then collected in 40-mL vials with soil samples by gravity flow at the rate of 0.20–0.25 mL aqueous VOCs per gram of soil. At this point, the soil samples were 80–100% saturated with water. Each soil sample vial was sealed immediately with a Teflon-backed, aluminum foil-faced septum and screw cap with hole, and agitated with a vortex mixer for 30 s. Soil samples were then analyzed or stored at the appropriate temperature.

For explosives analysis, the soil samples were spiked with 0.5 mL of each individual explosive stock solution on the day the holding time study was to begin. ORNL prepared daily explosive stock solutions as acetonitrile solutions. The explosive soil samples were then agitated with a vortex mixer for 30 s and stored at the appropriate storage condition.

To extract the explosives for chemical analysis, the soil samples were ultrasonically extracted with 10 mL of acetonitrile for 18 h in U.S. Environmental Protection Agency (EPA) volatile organic analysis (VOA) vials (40-mL borosilicate vials with septum closures). These vials were then centrifuged for 10 min. From each vial, 1 mL of extract was filtered through a 0.45-μm disposable Teflon filter into a 2-mL volumetric flask for the low-level concentration samples, or a 10-mL volumetric flask for the high-level concentration samples. Reagent-grade water (Burdick and Jackson, Inc.) was added to bring the volumetric flask to the proper volume. After mixing, aliquots were pipetted into autosampler vials. Blank sample aliquots were pipetted into the autosampler vials prior to the addition of stock explosive solutions. Blanks and samples were stored together in order to assess the possibility of cross-contamination.

Sample Analysis

ORNL conducted all volatile organic analyses by gas chromatography with mass spectrometric detection (GC–MS) according to standard EPA Contract Laboratory Program (CLP) methods (3), except for the use of daily external standards (instead of internal standards) to calculate results. Data were used without recovery or blank correction, as is customary with this method. Samples of higher concentration were analyzed by addition of a 2-mL aliquot to the instrument rather than the customary 5 mL. This protocol was followed to maintain instrument response within the linear range of the instrument.

Soil VOC samples were purged directly from the 40-mL vials into the trapping system. A double-threaded Teflon coupling machined in-house with O-ring seals was used. This protocol was followed to allow an assessment of the actual content of the vials without the complicating factor of sample weighing and transfer.

High-pressure liquid chromatography (HPLC) was the preferred analytical technique for explosive analytes because these analytes are thermally unstable (*14–20*). All water or soil explosive samples were eluted from an octadecylsilane (C_{18} or Zorbax-ODS; Mac-Mod, Inc., Chadds Ford, PA) reverse-phase HPLC column with a 50/25/25 mixture of water/acetonitrile/methanol (vol/vol/vol) flowing at 0.8 mL/min. The injection volume was 50 μL. An ultraviolet absorbance detector with a fixed filter (254 nm) was employed for quantifying the usual four analytes. The order of elution (increasing time) was octahydro-1,3,5,7-tetranitro-1,2,5,7-tetrazocine (HMX), hexahydro-1,3,5-trinitro-1,3,5-triazine (RDX), 2,4,6-trinitrotoluene (TNT), and 2,4-dinitrotoluene (2,4-DNT). Chromatograms were recorded on both a conventional strip chart recorder (backup document) and a recording integrator (primary document). Experimentally determined retention times, with windows of $\pm$0.3 min, were used for the initial identification of candidate explosive peaks. Peak areas obtained from the primary document were used for quantitation.

Identity confirmation for the test compounds was also provided by HPLC, but using a column (cyano groups chemically bounded to silica) that exhibits normal-phase behavior and therefore exhibits an almost inverted order of elution. In other words, the order of elution from the cyano column (increasing time) was 2,4-DNT, TNT, RDX, and HMX. A different eluant and flow rate (50/50 vol/vol of water/methanol; 1.5 mL/min) compared to the reverse-phase column were employed, but the monitoring wavelength remained the same. Data were collected using the Winchester disk drive of the data system, and chromatograms were printed off-line. Again, peak areas were used for quantitation.

ORNL Preservation Study

During ORNL's holding time study on VOCs, it became apparent that addition of hydrochloric acid to the samples, reducing the pH below 2, might inhibit both dehydrohalogenation and degradation of the aromatic compounds. Therefore, a set of preservation experiments was performed, using the same three water matrices stored under refrigerated conditions and analyzed at intervals of 0, 14, 28, and 56 days. No deterioration was noted in any of the compounds except styrene; moreover, the stability of styrene was greatly improved, with almost 80% remaining after 56 days. This short study indicated that the maximum holding time of VOCs in water can be increased to at least 56 days if samples are preserved with hydrochloric acid.

Preservation of water samples with HCl has its drawbacks. It is difficult to ensure that the pH of the sample is reduced to 2 without first measuring sample pH. It is also inconvenient to add a corrosive liquid during field sampling. Finally, HCl does have appreciable volatility and can be introduced into the instrumentation during purging. The possible detrimental effect on the analytical equipment cannot be tolerated. Therefore, there has been a general reluctance to require the use of HCl as a preservative for VOCs in water.

Because of these problems, and also because the database generated indicated that pH reduction is the primary factor involved in preservation, an attempt was made to identify other acids that might have the preservative effect of HCl without the associated drawbacks.

Two candidates were identified: sodium bisulfate and ascorbic acid. Both are noncorrosive (in the dry form), readily available, inexpensive, and nonvolatile. A study was carried out using these acids as preservatives, storing the samples at 4 °C; the data generated were compared to those obtained without preservation and with HCl preservation. Figure 1 shows the data for bisulfate preservation of ethylbenzene and styrene, two of the least stable aromatics. It is readily apparent that sodium bisulfate is as effective a preservative as HCl.

Figure 2 illustrates the data obtained for four EPA target ketones (those included on the target compound list) and carbon disulfide. These compounds were not included in the original VOC holding time study because of difficulty in obtaining standard compounds. Gradual reductions in the levels of carbon disulfide were evident during the 112-day study; the four ketones remained at or near their original concentrations.

Figure 3 depicts the groundwater data for ketones in line graph format. Boldface lines again indicate EPA CLP matrix spike recovery limits. One would expect

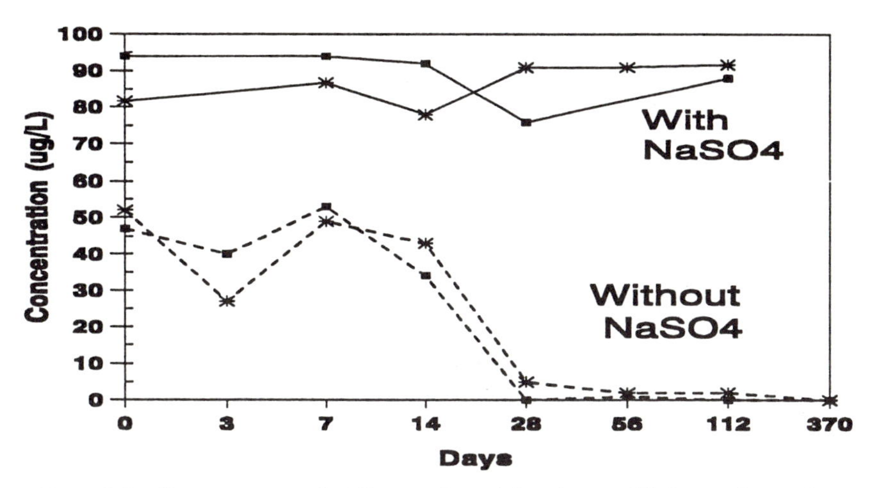

Figure 1. Bisulfate preservation of ethylbenzene (asterisks) and styrene (filled squares) in surface water stored at 4 °C.

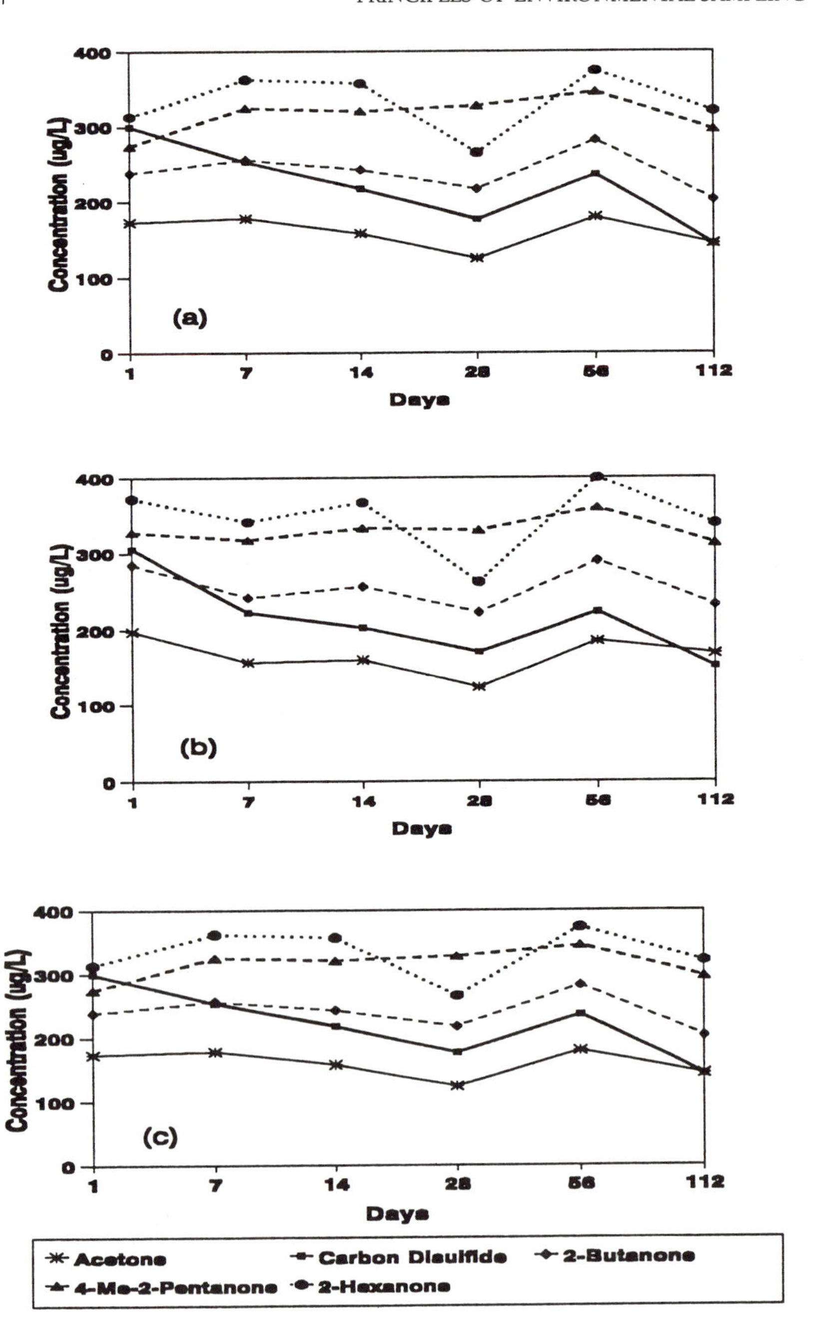

Figure 2. Bisulfate preservation of four ketones and carbon disulfide with an initial target concentration of 250 µg/L and stored at 4 °C. Matrix: (a) distilled water, (b) groundwater, and (c) surface water.

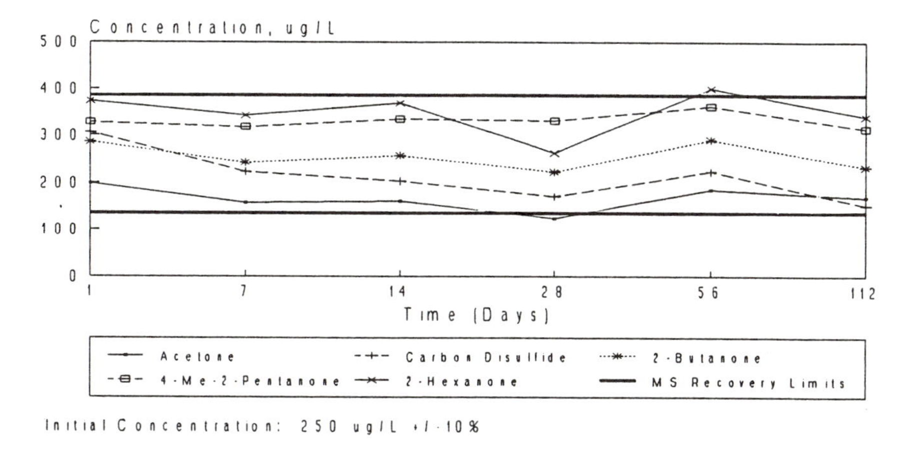

Figure 3. Four ketones and carbon disulfide in groundwater with bisulfate preservation.

greater variability from the more soluble ketones than from the more purgeable matrix spike compounds. However, virtually all the ketone data fall within matrix spike recovery limits.

Ascorbic acid was equally effective in preserving most VOCs studied. However, it was not possible to acidify the samples to a pH of 2 with this acid, and solubility problems were encountered before reaching a pH of 3. Additionally, the quantitation of bromoform proved difficult in the presence of ascorbic acid, with high standard deviations between replicate samples.

Throughout this preservation study, data obtained for the gases bromomethane and chloromethane were highly inconsistent because of instability of the standard compounds used for GC–MS quantitation. Aging of analytical standards is a problem that must be addressed before consistent data can be generated for these gases.

Studies on VOCs in soil revealed a very high tendency to lose these analytes quickly. Therefore, none of these studies was carried out for the full 365-day period. Several modifications were made in an attempt to improve stability, including (1) lining the cap with foil to reduce permeability, (2) closed-loop purge and trap analysis to reduce transfer losses, and (3) ultralow storage temperatures (–70 °C). None of these methods proved effective, however, and it can simply be stated that substantial losses of VOCs from soil samples are to be expected unless alternate preservation is used (e.g., methanol immersion).

Practical Reporting Time

Definition

ORNL analyzed the data from its holding time studies by calculating a statistically defined measure of holding time. The basic concept of holding times is to specify how long a sample can be held with reasonable assurance that the initial concentration has not changed significantly. The definitions of "reasonable assurance" and "changed significantly" are keys to holding time definitions. ORNL defined a holding time measure, called practical reporting time (PRT), that is based on statistical definitions of these terms.

A significant change in initial concentration is defined using statistical properties of the measurement system. A *critical concentration* (CC) is determined on the first day of the holding time study; the CC is the concentration below which there is only a 5% chance, due to measurement error, that a measured concentration will be observed. As the concentration decreases with time, the chance that the measured concentration for an individual sample will be below the CC increases. The PRT is defined as the day when there is a probability of 15% (an increase of 10%) that the measured analyte will be below the CC. Figure 4 illustrates an analyte concentration that is linearly decreasing with time, and measurement variation that follows a normal distribution.

PRTs are calculated using the following procedure: (1) an approximating model representing concentration versus time is fitted to the data by the method of least squares (*21*); (2) the one-sided 95% prediction limit for time zero is located for the approximating model (this limit is the CC); (3) a horizontal line is drawn from the CC until it intersects with the one-sided 85% prediction limit; and (4) a vertical line is drawn from this intersection (i.e., of the CC and the one-sided 85% limit) to the time-axis. This final intersection with the time axis is the PRT.

The PRT depends on the mathematical model used to approximate the degradation of an analyte concentration and the precision (standard deviation) of the ana-

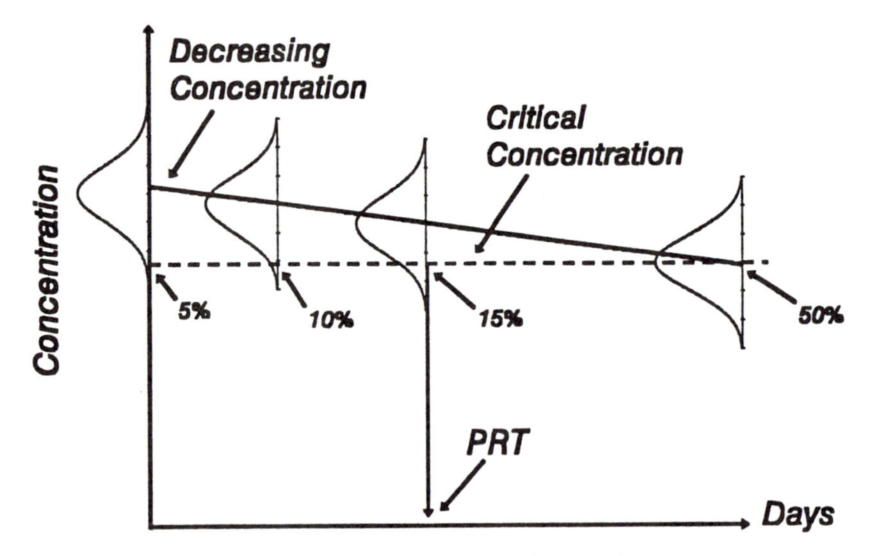

Figure 4. Practical reporting time (PRT) for an analyte with a linear decreasing concentration.

lytical measurement variation. For a given analyte, large measurement variations will give longer PRTs than those for smaller measurement variations. This result occurs because it is more difficult to detect changes in the initial concentration with larger measurement variation. The chance that a measured analyte is below CC will increase as an environmental sample is held longer past the PRT. The rate of this probability increase reflects the consequences of missing the PRT.

Approximating Models

PRTs depend on approximating models to represent the degradation of analyte concentrations with time. Approximating models can be either decreasing (most common), increasing, or a combination of a period of stability followed by a rapid decrease. Different approximating models may represent the same analyte under different environmental matrices and storage conditions. Five approximating models were used to represent analyte degradation (zero-order kinetic model, first-order kinetic model, log-term model, inverse-term model, and cubic-spline model). Zero-order and first-order kinetic models represent the degradation of concentration (C) or the natural logarithm (base e) of concentration ($\ln C$) by a line [$C = C_0 + (B)(D)$, or $\ln C = C_0 + (B)(D)$, respectively, where C_0 is initial concentration, B is the slope, and D is the number of days from the time an analyte was sampled]. These two models successfully approximated the data for 73% of the ORNL holding time experimental cases. To approximate more rapidly degrading concentrations, additional terms were added to the zero-order model. The log-term model adds a logarithmic term (e.g., $\ln D$), and the inverse-term model adds a reciprocal term (e.g., $1/D$). The log-term model and inverse-term model can approximate data with rapid concentration degradation for 18% of ORNL holding time cases. The coefficients for these four models can be estimated by the usual method of least squares (*21*). In addition, the least-squares analysis can estimate the precision (S, which is an estimate of the true measurement standard deviation, σ) for measuring a single analyte by the square root of the mean square error.

The four linear models were unable to approximate 9% of ORNL holding time experimental cases that had an initial constant-concentration plateau followed by degradation. An empirical model was applied for these cases, which had an initial constant-concentration for days less than day D_a, and a final concentration for days greater than day D_b, where D_a and D_b refer to any two sampling days that must be estimated for the cubic spline approximating model. The concentrations were modeled by a cubic polynomial between days D_a and D_b. The cubic spline starts with a value of the initial concentration at day D_a and ends with a value of the final concentration at day D_b. In addition, the cubic spline is continuous at day D_a and day D_b. Coefficients for the cubic spline are estimated by the method of nonlinear least squares (*21*); this model is used in most regulatory situations. Figure 5 illustrates the five approximating models.

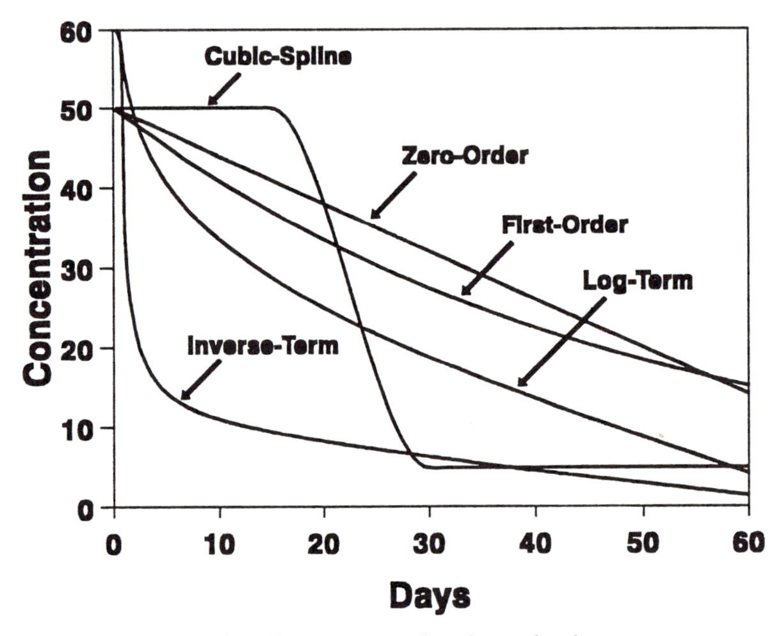

Figure 5. Five models used to approximate degradation of analyte concentration.

Results and Discussion

PRTs from the ORNL holding time studies were used to develop (1) a quadratic equation for estimating PRTs, and (2) a graphical method to assess the risk for samples held past the PRT.

Sigma-to-Slope Ratios

Holding time studies at other laboratories may have different spiking levels, environmental matrices, storage conditions, and measurement errors. For zero-order and first-order kinetic models, these holding time results can be related to the new PRT definition through the ratio of the single measurement precision (S, which is an estimate of the true measurement standard deviation, σ) to the absolute value of the slope ($|B|$) of the line that approximates the concentration change (i.e., $S/|B|$, sigma-to-slope ratios). Figure 6 plots the observed PRT values versus sigma-to-slope ratios ($S/|B|$) for ORNL holding time cases approximated by zero-order and first-order models. Sigma-to-slope ratios >385 indicate that precision values are divided by small slope values. These small slope values would not be significantly different from zero at the 5% significance level. Their corresponding PRT values would be set to the maximum experimental time (56, 111, or 365 days). Tables III–VI list the sigma-to-slope ratios with the positive (increasing slope) and negative (decreasing slope) signs for the ORNL holding time experiment.

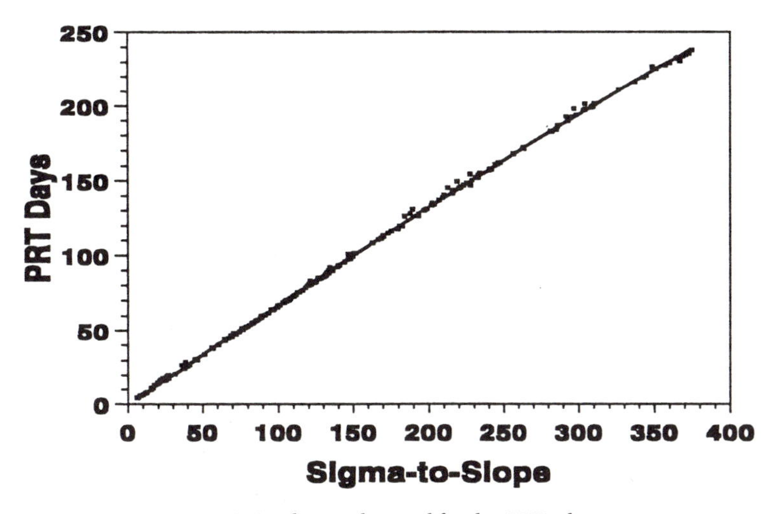

Figure 6. Quadratic polynomial fitted to PRT values.

A quadratic polynomial fitted by the method of least squares to PRT versus the sigma-to-slope ratio represents 99.95% of the variation in the PRT values. Predicted PRT values are also shown in Figure 6 for the quadratic approximation

$$\mathrm{PRT} = -0.3051 + 0.6894\left(\frac{S}{|B|}\right) - 0.0134\times 10^{-2}\left(\frac{S}{|B|}\right)^2$$

For example, Table III shows that $S/B = -151$ for benzene at 50 µg/L concentration in groundwater stored at 4 °C. This number is first converted to a sigma-to-slope ratio by taking the absolute value, $S/|B| = +151$. The corresponding PRT value can then be approximated either graphically by Figure 6 or by the quadratic polynomial. Approximate PRT values are rounded down to the next whole number of days:

$$\mathrm{PRT} = -0.3051 + 0.6894(151) - 0.0134\times 10^{-2}(151)^2$$

$$\mathrm{PRT} = 100.7 \approx 100 \text{ days}$$

The actual PRT value is 101 days calculated from the mathematical formula. Smaller sigma-to-slope values have shorter holding times. For example, Table III shows $S/B = -94$ for benzene at 50 µg/L concentration in surface water stored at 4 °C. The quadratic polynomial approximate value is PRT = 63 days and the exact calculation is PRT = 62 days. The margin of error is small between the quadratic polynomial and the exact calculations in Figure 6 with a maximum difference of 5.6 days at PRT = 198 days.

Table III. Sigma-to-Slope Ratios for VOCs in Environmental Water Samples Used in the ORNL Holding Time Study

		Distilled Water		*Surface Water*		*Groundwater*	
VOC	*Concn (μg/L)*	*4 °C*	*Room*	*4 °C*	*Room*	*4 °C*	*Room*
Benzene	50	NS	NS	–94	–131	–151	–227
	500	233	–256	369	NS	–246	–286
Bromoform	50	NS	–225	–76	–40	126	–121
	500	351	–150	NS	–242	NS	NS
Carbon tetrachloride	50	NS	NS	–144	—[b]	NS	–189
	500	NS	–207	–308	–72	NS	NS
Chlorobenzene	50	NS	–337	–97	–180	NS	–188
	500	NS	–79	NS	–221	NS	–202
Chloroform	50	–304	–229	–87	–282	NS	219
	500	217	–240	NS	NS	–286	–375
1,1-Dichloroethane	50	NS	NS	–82	–130	NS	NS
	500	304	NS	NS	NS	–309	–366
1,1-Dichloroethene	50	NS	152	–120	–107	349	–213
	500	246	–103	NS	–83	–175	–118
1,2-Dichloropropane	50	NS	NS	–128	NS	133	184
	500	170	220	223	NS	NS	NS
Ethylbenzene	50	NS	–372	—[b]	–79	—[b]	–45
	500	NS	–89	—[b]	–198	–296	–105
Methylene chloride	50	NS	–326	–108	–148	292	NS
	500	244	–146	–345	–199	–357	–364
Styrene	50	NS	–263	—[b]	—[a]	—[a]	–134
	500	NS	–81	–140	–92	NS	–120
1,1,2,2–Tetrachloroethane	50	—[b]	—[a]	–282	–12	132	–16
	500	—[a]	—[a]	NS	–61	NS	–12
Tetrachloroethene	50	NS	–203	–83	–132	NS	–297
	500	–209	–38	–146	–57	–128	–65
1,1,2-Trichloroethane	50	NS	–15	–131	–263	65	147
	500	303	—[b]	NS	NS	NS	–140
Trichloroethene	50	210	–298	–80	285	NS	NS
	500	207	–86	NS	–193	–343	–163
Toluene	50	NS	–373	–85	–46	–100	–71
	500	NS	–122	–127	–204	–310	–234
o-Xylene	50	NS	–371	–202	–51	–119	–27
	500	NS	–173	NS	NS	NS	–112

NOTE: ORNL is Oak Ridge National Laboratory; VOC is volatile organic compound; NS means slope is not significant. Room and 4 °C refer to water sample storage temperatures.

[a]Practical reporting time (PRT) values and days past the PRT (15% < γ% ≤ 50%) are <7 days, where γ% refers to risk probability.

[b]*See* Table VII.

Table IV. Sigma-to-Slope Ratios for VOCs in Environmental Soil Samples Used in the ORNL Holding Time Study

	USATHAMA Soil		*TN Soil*			*MS Soil*		
VOCs	*–70 °C*	*–20 °C*	*–70 °C*	*–20 °C*	*4 °C*	*–70 °C*	*–20 °C*	*4 °C*
Benzene	—[a]	–112	—[a]	–27	—[a]	—[a]	—[a]	—[a]
Bromoform	NS	95	22	38	–6	NS	NS	—[a]
Bromomethane	—[a]	NS	—[a]	—[a]	—[a]	—[a]	—[a]	—[a]
Carbon tetrachloride	—[a]	NS	—[a]	–17	–8	—[a]	—[a]	—[a]
Chlorobenzene	—[a]	NS	NS	NS	—[a]	—[a]	—[a]	—[a]
Chloroethane	—[a]	—[b]	—[a]	–22	–8	—[a]	—[a]	—[a]
Chloroform	—[a]	NS	—[a]	—[b]	—[a]	—[a]	—[a]	—[a]
1,1-Dichloroethane	—[a]	NS	—[a]	–19	–7	—[a]	—[a]	—[a]
1,1-Dichloroethene	—[a]	NS	—[a]	–17	—[a]	—[a]	—[a]	—[a]
1,2-Dichloropropane	—[a]	NS	—[a]	—[b]	–6	—[a]	—[a]	—[a]
Ethylbenzene	—[a]	—[b]	—[a]	–36	—[a]	—[a]	—[a]	—[a]
Methylene chloride	—[b]	NS	NS	NS	–15	NS	—[a]	—[a]
Styrene	NS	NS	NS	NS	—[a]	NS	–23	—[a]
1,1,2,2-Tetrachloroethane	NS	NS	24	26	–6	NS	NS	—[a]
Tetrachloroethene	—[a]	—[a]	—[a]	–21	—[a]	—[a]	—[a]	—[a]
1,1,2-Trichloroethane	–56	NS	NS	NS	–7	—[b]	—[a]	—[a]
Trichloroethene	—[a]	–89	—[a]	—[b]	—[a]	—[a]	—[a]	—[a]
Toluene	—[a]	–104	—[a]	—[b]	—[a]	—[b]	—[a]	—[a]
o-Xylene	—[a]	–64	NS	NS	—[a]	—[b]	—[a]	—[a]

NOTE: Concentrations of VOCs are all <100µg/g. NS means slope is not significant; –70 °C, –20 °C, and 4 °C refer to soil sample storage temperatures.

[a]PRT values and days past the PRT (15% < $\gamma\% \leq 50\%$) are <7 days.
[b]*See* Table VII.

Table V. Sigma-to-Slope Ratios for Explosive Compounds in Environmental Water Samples Used in the ORNL Holding Time Study

		Distilled Water		*Surface Water*		*Groundwater*	
Explosive	*Concn (µg/L)*	*4 °C*	*Room*	*4 °C*	*Room*	*4 °C*	*Room*
DNT	50	—[a]	—[b]	–24	—[a]	—[b]	—[a]
	1000	167	197	–74	–110	NS	–124
HMX	100	–100	–91	–25	–40	–105	–86
	1000–2000	—[b]	—[b]	–141	–170	NS	NS
RDX	50	NS	–134	–38	–31	NS	—[b]
	1000	NS	–76	130	99	NS	NS
TNT	50	–108	–10	—[b]	—[a]	–27	—[a]
	1000	360	NS	—[b]	—[a]	–122	—[b]

NOTE: NS means slope is not significant. Room and 4 °C refer to water sample storage temperatures.

[a]PRT values and days past the PRT (15% < $\gamma\% \leq 50\%$) are <7 days.
[b]*See* Table VII.

Table VI. Sigma-to-Slope Ratios for Explosive Compounds in Environmental Soil Samples Used in the ORNL Holding Time Study

		USATHAMA Soil			*TN Soil*			*MS Soil*		
Explosive	*Concn (μg/g)*	*−20 °C*	*4 °C*	*Room*	*−20 °C*	*4 °C*	*Room*	*−20 °C*	*4 °C*	*Room*
DNT	10	NS	−367	—[a]	116	−182	—[a]	NS	NS	−110
	100	169	169	231	233	NS	NS	120	181	240
HMX	10	NS	NS	NS	228	NS	−41	125	136	NS
	100	NS	NS	NS	104	97	91	70	89	88
RDX	10	NS	NS	—[b]	144	−193	—[b]	NS	NS	−216
	100	NS	−293	NS	104	132	106	86	108	114
TNT	10	−144	−84	—[a]	NS	−67	—[a]	NS	—[a]	—[a]
	100	−309	−24	—[a]	NS	−83	−24	NS	−247	−78

NOTE: NS means slope is not significant. Room, 4 °C, and −20 °C refer to soil sample storage temperatures.
[a]PRT values and days past the PRT (15% < γ% ≤ 50%) are <7 days.
[b]*See* Table VII.

Estimating PRT values based on the sigma-to-slope ratios is very useful for short-term holding time studies. If a chemist wishes to estimate the PRT value for a compound but the experimental time is limited, then a short-term experiment could be conducted to estimate the slope of the model and the standard deviation of a single analytical measurement if the chemist is willing to assume a zero-order or first-order kinetics model. The best procedure is to run replicate concentration measurements at the beginning and end of the experimental time. The chemist may wish to run additional measurements on the third and seventh days to detect rapid degradation, and additional measurements in the middle of the experimental time to detect lack of fit. From this experiment, the slope and the measurement standard deviation can be estimated. By using the sigma-to-slope ratio, a PRT value can be estimated for the compound using either Figure 6 or the quadratic polynomial. This procedure can be used despite analyte, concentration level, sample matrix, or storage condition. The chemist can calculate PRT values for any analyte and conditions for which zero-order and first-order models are used to approximate the analyte concentration degradation. Longer holding time experiments will give additional confidence that the correct assumptions have been made.

Days Past PRT

A discussion of the effect on the risk (γ) probability if samples are held past the PRT is in order here. The risk probability is the probability that an analyte concentration measurement is less than the critical concentration. This risk probability will increase as samples are held past the PRT. The decision maker must decide if the increased risk probability is unacceptable. The rate of increase of the risk probability will depend on the sigma-to-slope ratio. Figures 7 and 8 are nomographs for increasing risk probabilities for days past the PRT value. This nomograph is based on the sigma-to-slope ratios estimated for the ORNL holding time study. The holding time

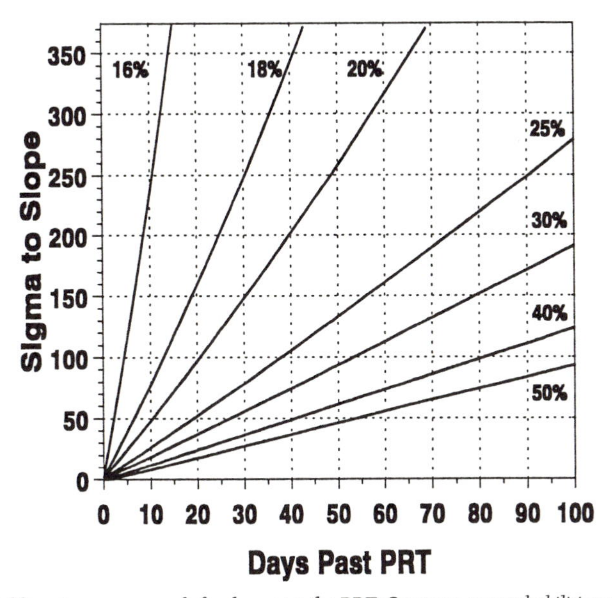

Figure 7. Holding time nomograph for days past the PRT. Contours are probabilities of an analyte measurement being less than the critical concentration.

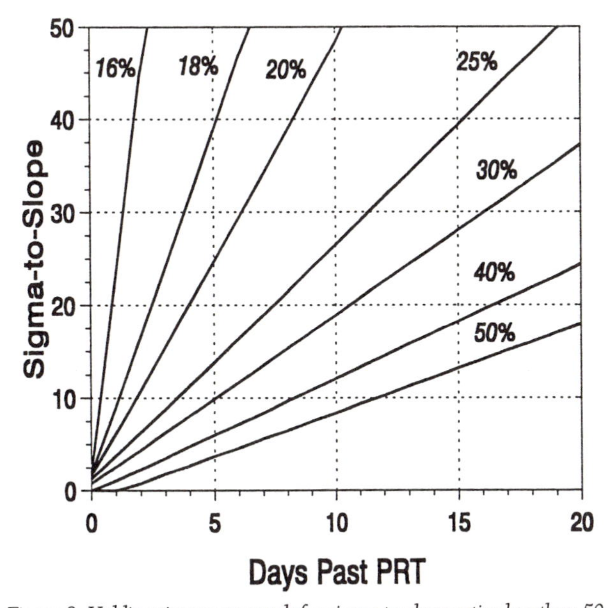

Figure 8. Holding time nomograph for sigma-to-slope ratios less than 50.

nomograph is used in conjunction with the sigma-to-slope ratios in Tables III–VI. For example, the PRT value is 101 days for benzene at 50 μg/L concentration in groundwater stored at 4 °C with a sigma-to-slope value of $S/|B| = +151$. Figure 7 shows that at 10 days past the PRT value the risk probability is a little less than 0.17. This increase may be considered acceptable. But, 30 days past the PRT value, the risk probability increases to 0.20, and this increase may be considered unacceptable. This risk probability means the analyte concentration has degraded so that the chance is 1 in 5 that an analysis of an environmental sample will give a concentration below the critical concentration. Figure 8 is an enlargement of a section of the nomograph for cases with sigma-to-slope ratios less than 50.

The nomograph illustrates the risk probabilities of concentration changes (below the CC level) rather than the actual concentration change for a single measurement. In addition to measurement precision and slope, the actual concentration change would require the value for the initial concentration, which may vary for different holding time studies. Different holding time experimental designs would affect the risk probability calculations through the number of degrees of freedom, and the variances and covariances of the intercept and slope. PRT calculations for the ORNL holding time study show this methodology can be extrapolated to results from a large class of experimental parameters for analytes, spiking levels, matrices, and storage temperatures.

PRT Values for Special Cases

Tables III–VI list the sigma-to-slope ratios for analytes modeled by the zero-order and first-order kinetic models. Entries designated NS are for cases that have a nonsignificant slope (i.e., no concentration degradation) at the 5% significance level, and the PRT values are the maximum experimental time. Entries designated by a superscript *a* are modeled by the log-term, inverse-term, or cubic spline models and have PRT values < 7 days. In addition, the number of days past PRT is < 7 days for a probability value of $\gamma\% = 50\%$. These cases represent rapid degradation of analyte concentration. Entries designated by superscript *b* indicate the number of days past PRT is > 7 days for a probability value of $\gamma\% = 50\%$. All 27 cases designated by superscript *b* are modeled by the cubic-spline model except 50 μg/L of DNT in groundwater stored at 4 °C, which was modeled by a log-term model.

PRT values and days past the PRT for these special cases are listed in Table VII. Two cases in Table VII show unusual jumps in the number of days past PRT. Ethylbenzene spiked at 50 μg/g in USATHAMA soil stored at –20 °C shows a jump from 6 days to 90 days, and toluene spiked at 50 μg/g in Mississippi soil stored at –70 °C shows a jump from 2 days to 54 days as the $\gamma\%$ probability changes from 30 to 35%. These unusual increases are due to the cubic spline leveling off to a constant concentration. As a result of these cubic spline behaviors, the $(1 - \gamma)\%$ prediction limits that are $<70\%$ become parallel to the critical concentration limit. These parallel lines never intersect with the critical concentration limit for $\gamma\% > 30\%$.

Table VII. Days Past PRT Corresponding to γ% Probability

Analyte	Spike	Matrix	Storage Temp	PRT Day	γ% Probability						
					20%	25%	30%	35%	40%	45%	50%
Environmental Water Samples											
Carbon tetrachloride	50 μg/L	Surface	Room	17	3	4	6	7	8	10	11
Ethylbenzene	50 μg/L	Surface	4 °C	13	0	0	1	1	1	1	2
	50 μg/L	Ground	4 °C	19	2	4	5	6	7	8	9
	500 μg/L	Surface	4 °C	19	0	1	2	2	3	3	4
Styrene	50 μg/L	Surface	4 °C	10	0	1	1	1	1	2	2
1,1,2,2-Tetrachloroethane	50 μg/L	Distilled	4 °C	9	0	0	1	1	1	1	2
1,1,2-Trichloroethane	500 μg/L	Distilled	Room	8	0	1	1	2	2	2	3
DNT	50 μg/L	Distilled	Room	29	3	5	6	8	9	11	12
	50 μg/L	Ground	4 °C	2	1	2	3	4	5	7	10
HMX	1000 μg/L	Distilled	4 °C	20	2	3	4	5	5	6	6
	1000 μg/L	Distilled	Room	29	1	2	3	3	4	5	5
RDX	50 μg/L	Ground	Room	134	7	13	18	23	28	32	37
TNT	50 μg/L	Surface	4 °C	18	0	1	1	1	1	1	1
	1000 μg/L	Surface	4 °C	32	2	4	5	7	8	9	10
	1000 μg/L	Ground	Room	0	4	6	8	9	10	11	12
Environmental Soil Samples											
Chloroethane	70 μg/g	USATHAMA	−20 °C	19	2	3	4	4	5	6	7
Chloroform	60 μg/g	Tennessee	−20 °C	29	1	2	4	7	12	16	19
1,2-Dichloropropane	60 μg/g	Tennessee	−20 °C	47	2	5	7	9	9	9	9
Ethylbenzene	50 μg/g	USATHAMA	−20 °C	21	2	4	6	90	90	90	90
Methylene chloride	60 μg/g	USATHAMA	−70 °C	20	1	2	3	3	4	4	5
1,1,2-Trichloroethane	50 μg/g	Mississippi	−70 °C	0	4	7	9	10	12	13	15
Trichloroethene	50 μg/g	Tennessee	−20 °C	40	2	3	4	5	5	6	7
Toluene	90 μg/g	Tennessee	−20 °C	42	0	1	1	2	2	2	3
	50 μg/g	Mississippi	−70 °C	2	1	1	2	54	54	54	54
o-Xylene	40 μg/g	Mississippi	−70 °C	14	1	2	2	3	4	4	5
RDX	10 μg/g	USATHAMA	Room	18	1	2	3	3	4	4	5
	10 μg/g	Tennessee	Room	7	2	4	5	5	6	7	7

NOTE: The risk probability (γ%) is the probability that an analyte concentration measurement is less than the critical concentration.

Summary and Conclusions

ORNL has conducted major holding time studies that included 17 or 19 VOCs and 4 explosives in both water and soil matrices using different storage conditions. The ORNL holding time studies represent 476 experimental cases with 13,422 concentration measurements over long time periods. These studies indicated that sodium bisulfate is an effective preservative for extending the holding time period for VOCs in water. In addition, a concept called practical reporting time (PRT) was introduced, which provides a technically valid approach for assessing the data usability of samples analyzed after the regulatory holding time has expired. Development of the

PRT definition also gives a statistically quantitative method to measure the holding time period for different analytes and matrices. In addition, PRTs can be estimated with short-term experiments and can be used to determine the time period during which sample stability will not affect the analytical results.

References

1. Moore, A. T.; Vira, A.; Fogel, S. *Environ. Sci. Technol.* **1989**, *23*, 403–406.
2. Walton, B. T.; Anderson, T. A. *Chemosphere* **1988**, *17*, 1501–1507.
3. Walton, B. T.; Anderson, T. A.; Hendricks, M. S.; Talmage, S. S. *Environ. Toxicol. Chem.* **1989**, *8*, 53–63.
4. *Fed. Regist.* **1979**, 40 CFR Part 136, Proposed rules 44, No. 233:69534, Dec. 3.
5. Bottrell, D. W.; Fisk, J. F.; Hiatt, M. *Environ. Lab.* **1990**, *2*, 29–31.
6. USATHAMA. *U.S. Army Toxic and Hazardous Materials Agency Quality Assurance Program;* U.S. Army Toxic and Hazardous Materials Agency: Aberdeen Proving Ground, MD, 1990.
7. Maskarinec, M. P.; Bayne, C. K.; Johnson, L. H.; Holladay, S. K.; Jenkins, R. A. *Stability of Volatile Organics in Environmental Water Samples: Storage and Preservation;* Oak Ridge National Laboratory: Oak Ridge, TN, 1989; ORNL/TM–11300.
8. Maskarinec, M. P.; Bayne, C. K.; Johnson, L. H.; Holladay, S. K.; Jenkins, R. A.; Tomkins, B. A. *Stability of Explosives in Environmental Water and Soil Samples;* Oak Ridge National Laboratory: Oak Ridge, TN, 1991; ORNL/TM–11770.
9. Maskarinec, M. P.; Johnson, L. H.; Holladay, S. K.; Moody, R. L.; Bayne, C. K.; Jenkins, R. A. *Environ. Sci. Technol.* **1990**, *24*, 1665–1670.
10. Maskarinec, M. P.; Johnson, L. H.; Bayne, C. K. *J. Assoc. Off. Anal. Chem.* **1989**, *72*, 823–827.
11. Maskarinec, M. P.; Bayne, C. K.; Johnson, L. H.; Holladay, S. K.; Jenkins, R. A.; Tomkins, B. A. *Stability of Volatile Organics in Environmental Soil Samples;* Oak Ridge National Laboratory: Oak Ridge, TN, 1992; ORNL/TM–12128.
12. Bayne, C. K.; Schmoyer, D. D.; Jenkins R. A. *Practical Reporting Times for Environmental Samples;* Oak Ridge National Laboratory: Oak Ridge, TN, 1993; ORNL/TM–12316.
13. Lang, K., U.S. Army Toxic and Hazardous Materials Agency, personal communication, 1990.
14. Jenkins, T. F.; Bauer, C. F.; Leffett, D.C.; Grant, C. L. *Reverse Phase HPLC Method of Analysis of TNT, RDX, HMX, and 2, 4–DNT in Munitions Wastewater;* U.S. Army Cold Regions Research and Engineering Laboratory: Hanover, NH, 1984; CRREL Report 84–29.
15. Jenkins, T. F.; Leffett, D.C.; Grant, C. L.; Bauer, C. F. *Anal. Chem.* **1986,** *58*, 170–175.
16. Bauer, C. F.; Grant, C. L.; Jenkins, T. F. *Anal. Chem.* **1986,** *58*, 176–182.
17. Jenkins, T. F.; Walsh, M. E. *Development of an Analytical Method for Explosive Residues in Soil;* U.S. Army Cold Regions Research and Engineering Laboratory: Hanover, NH, 1987; CRREL Report No.87–7.
18. Jenkins, T. F.; Grant, C. L. *Anal. Chem.* **1987,** *59*, 1326–1331.
19. Maskarinec, M. P.; Manning, D. L.; Harvey, R. W.; Griest, W. H.; Tomkins, B. A. *J. Chromatogr.* **1984**, *302*, 51–63.
20. Maskarinec, M. P.; Manning, D. L.; Harvey, R. W. *Application of Solid Sorbent Collection Techniques and High-Performance Liquid Chromatography with Electrochemical Detection to the Analysis of Explosives in Water Samples;* Oak Ridge National Laboratory: Oak Ridge, TN, 1986; ORNL/TM–10190.
21. Draper, N.; Smith, H. *Applied Regression Analysis,* 2nd ed.; John Wiley & Sons: New York, 1981.

Chapter 17

Solid-Phase Extraction Basics for Water Analysis

Craig Markell and Donald F. Hagen

Solid-phase extraction is increasingly replacing liquid–liquid extractions for the preparation of environmental water samples. The reasons for this replacement include reductions in labor, solvent usage, and turnaround time, all of which lead to increased efficiency. This chapter discusses the principles, selection, and use of reverse-phase and ion-exchange mechanisms for extracting trace analytes from water samples, gives practical troubleshooting and method development tips for the use of solid-phase extraction, and reviews the status of solid-phase extraction in accepted U.S. Environmental Protection Agency methods. Overviews of three typical methods are given.

SOLID-PHASE EXTRACTION (SPE) has been used for environmental analysis of water samples since the pioneering publication of Junk et al. (*1*) in 1974. In that work, a group at Iowa State University found that a column packed with XAD resin not only extracted a variety of organic compounds from a water sample, but also released those compounds in a more concentrated form when the resin was eluted with an organic solvent. Since that time, numerous publications have appeared (2–6) that are variations on the same basic theme. This principle has formed the basis for a revolution in the extraction of organic compounds from water or aqueous matrices, which were more commonly extracted using liquid–liquid extraction (LLE). The purpose of this chapter is to review the strengths and limitations of the use of SPE for water analysis, give a snapshot of the currently accepted and practiced applications, explain some of the principles and fundamentals, give some practical tips on the topics of troubleshooting and method development, detail some example methods to illustrate the actual use of SPE, and predict the advances that will certainly take place over the next few years.

3152–4/96/0297$15.00/0

Several forces are driving the replacement of LLE with SPE. Foremost is the elimination of the large volumes of organic solvents used to extract water samples. Depending on the analyst and method, from 200 mL up to 1 L of toxic solvents may be used for a single LLE. The solvent is usually methylene chloride because of its density, volatility, and solvent strength, but other equally hazardous solvents may be used in alternate methods. These large volumes certainly accomplish the job of extracting organic compounds from water, but this method entails the twin risks of worker exposure and emission of the spent solvent. Most people agree that the unrecovered discharge of solvent up the fume hood is a bad practice, but few people think about the fact that the 1-L water sample is saturated with the organic extractant after the analysis. Using methylene chloride as an example, this saturation factor means that the extracted sample contains 1.6% by weight of the methylene chloride, or 16 g, which is often flushed down the sink. Laboratories that control their solvent discharges have the added expense and trouble of recycling or disposal. The contradiction of generating significant amounts of hazardous waste during an analysis intended to guarantee a clean environment has methods development labs scrambling to find better methods. In addition to using less solvent, SPE methods are often faster and less labor-intensive, both of which add up to a more efficient and economical analysis, and delays due to emulsions are never a factor in SPE. With the intense competition between environmental labs, any way to streamline sample preparation is of great interest.

Accepted Methods

As of the summer of 1994, a number of private, industrial, and government laboratories had installed SPE methods as standard in the analysis of water samples. Generally, the most respected single source of analytical methods for environmental analysis is the body of methods developed by the U.S. Environmental Protection Agency (EPA). Within the EPA, the research group having contributed the most to the development of SPE methods is the Environmental Monitoring Systems Laboratory (EMSL) in Cincinnati, OH, where most of the drinking water SPE methods were developed. In order to get a snapshot of where SPE methods are with regard to acceptance and application areas, Table I was developed. The list in Table I is constantly expanding, which indicates the importance the Agency is putting on converting LLE methods to SPE. Between the late 1980s and 1994, the trend was to convert most of the drinking water methods (500 series) to SPE. In 1993, acceptance of SPE in certain wastewater and hazardous waste analyses began, and that trend appears to be gaining momentum in the program offices. To clarify the terminology applied to EPA methods, "accepted" will be defined as any method published by the Agency that allows the use of SPE in the extraction step, regardless of *Federal Register* status. A subset of methods accepting SPE technology are the methods that are officially recommended for the analysis of certain regulated analytes by virtue of *Federal Register* announcements.

Table I. EPA Methods Allowing the Use of SPE

Method	*Analyte*
	Drinking and Source Water Analysis
Method 506	Phthalates; adipates
Method 507	Organonitrogen; organophosphorus pesticides
Method 508	Organochlorine pesticides; PCBs
Method 515.2	Acid herbicides
Method 525.2	Semivolatile organic compounds
Method 548.1	Endothal
Method 549.1	Paraquat; diquat
Method 550.1	Polynuclear aromatic hydrocarbons
Method 552.1	Haloacetic acids; dalapon
Method 553	Benzidines; nitrogen-containing pesticides
Method 554	Carbonyl compounds
Method 555	Acid herbicides
	Wastewater Analysis
Method 1664	Oil and grease; TPH
	SW–846 Analysis
8061	Phthalates; adipates
3535	Organochlorine pesticides
	Superfund Analysis
QTMs	Phenols; pesticides; PCBs

NOTE: SPE is solid-phase extraction. QTMs are quick turnaround methods. PCBs are polychlorinated biphenyls. TPH is total petroleum hydrocarbons.

SOURCE: Data for drinking and source water are from references 7 and 8; for wastewater are from reference 9; for SW–846 are from reference 10; and for Superfund are from reference 11.

Critical Issues

Although the conversion of LLE methods to SPE is often quite successful, several critical issues must be considered that explain why SPE acceptance is not more general for all water types and all analytes. The first critical issue is whether the sorbent is able to extract the analyte of interest from the required volume of sample without breakthrough, which results in low recoveries. Some analytes, particularly the more water-soluble ones, simply are not sufficiently retained on the sorbent.

Another critical issue is the nature of the water sample with respect to dissolved or suspended solids. Dissolved solids can interfere with the sorption of the desired analyte by competing for active sites on the sorbent. An example of this type of interference would be a water sample that contains a relatively high percentage of solvent, such that the solvent actually displaces the analyte from the sorbent. Suspended solids are more of a concern because of their tendency to plug the pores of the sorbent bed, thus reducing the flow rate and limiting the volume of sample that can be passed through the sorbent in a reasonable amount of time. Figure 1 is an illustration of the effect suspended solids have on flow rate when SPE with a vac-

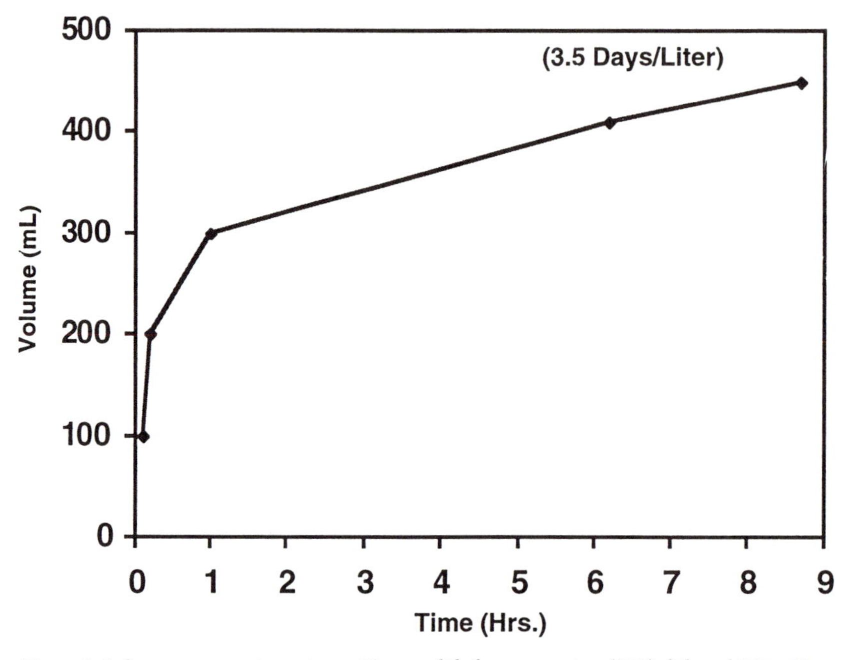

Figure 1. Lake water extraction using a 47-mm solid-phase extraction (SPE) disk and 26 in. Hg. The volume shown is that extracted through the disk, or the filtrate.

uum device is used. Although the water sample used for Figure 1 was not visibly turbid, it had a greenish color, indicating the presence of biological materials, such as algae. These suspended solids can also act as a competing "sorbent" for analytes in the sample. For example, polychlorinated biphenyls (PCBs) can exist in equilibrium between the water and suspended solids in the water. In that case, the fate of the solid-sorbed PCBs depends on sample treatment and elution technique.

A final critical issue is the technique with which SPE is done. As with all methods, good recoveries and good relative standard deviations (RSDs) depend on the analyst's technique. Because a new technology is being adopted, new techniques need to be learned. Even a procedure as simple as sodium sulfate drying may have to be modified to be successful with a new method. This factor really comes under the heading of method optimization and troubleshooting. These issues will be discussed in more detail, because the intention of this chapter is to provide a practical discussion of how SPE can be successfully adopted and what the limitations are.

Mechanisms of SPE

For the purpose of this discussion, we'll define three basic types of solid-phase interactions: reverse phase, normal phase, and ion exchange. *Normal phase* refers to polar interactions between sorbent and analyte, which are maximized in a nonpolar liquid (matrix), such as hexane. In the environmental laboratory, normal-phase interactions are of great utility in cleanup steps, such as the well-known Florisil column cleanups that remove polar species from sample extracts. Unfortunately, water is strongly sorbed to normal-phase sorbents, aggressively competing for active sites and making this interaction of little use for extracting analytes from aqueous matrices.

Reverse Phase

Reverse phase refers to nonpolar interactions between sorbent and analyte, which are maximized in polar matrices, particularly water. Typical sorbents used for reverse phase include bonded silicas, particularly heavily loaded C_{18} bonded silicas, and hydrophobic resins such as poly(styrene–divinylbenzene) (SDB). The reverse-phase mechanism is certainly the most useful for replacing LLE, because it actually depends on the same principle: migration and partitioning of the analyte from the water into an organic phase. In addition, reverse-phase extractions tend to be (1) subject to fewer interferences than ion exchange, (2) more forgiving to minor variations in "recipe", and (3) easier to perform with fewer steps. The most critical factor to the success of a reverse-phase extraction is the partition ratio of the analyte between the sorbent and the water, which is determined by the affinity of the analyte for each phase.

Generally, the analyte's water solubility has more to do with the reverse-phase retention of the analyte than any other factor, although the strength of interaction between the analyte and sorbent should not be overlooked. Other researchers have found (*12*) a strong correlation between molecular volume and recovery. Experience has shown (*13*) that good recoveries can be expected if the analyte has a water solubility of less than about 0.5%, assuming 0.5 g of C_{18} bonded silica and a 1-L sample. In a practical sense, these data suggest that most analytes of environmental interest should work very well using a reverse-phase system. Analytes that may pose problems are those that have high water solubility, such as many metabolites, ionic analytes (salts), and some of the modern pesticides that contain a high percentage of polar functional groups. Table II summarizes results expected from several classes of compounds. Although there might be a few exceptions, "good recoveries" means 70% or greater recovery is obtained by most analysts. "Borderline" indicates that some of the analytes in this class are not well-retained, or that recoveries may be a function of sorbent, matrix conditions (pH or ionic strength), or sample volume.

As indicated in Table II, one need not always accept poor recoveries. By using methods to be discussed later, marginal or even poor recoveries can sometimes be dramatically increased by modifying the aqueous matrix, reducing the sample volume, or changing the sorbent. For example, in work reported at the 1993 Pittsburgh Conference, Markell et al. (*13*) showed how the recovery for octanoic acid could be increased from less than 10% to over 70% by simply changing the pH, ionic strength, and sorbent used in the reverse-phase extraction.

Table II. SPE Performance of Several Common Analyte Classes Using Reverse-Phase Extraction

Analyte Class	*SPE Performance*
Dioxins	Good recoveries
Organochlorine pesticides	Good recoveries
PCBs	Good recoveries
Organophosphorus pesticides	Good recoveries
PAHs	Good recoveries
Phenolics	Good recoveries[a]
Acid herbicides	Good recoveries[a]
Triazines	Good recoveries
Phthalates and adipates	Good recoveries
Surfactants	Borderline
Carbamates	Borderline
Explosives	Good recoveries[a]
Small ($\leq C_6$) acids and amines	Poor recoveries

NOTE: Extractions assume a sorbent mass of ~0.5 g, a 1-L sample, and no modification of the matrix. PAHs are polyaromatic hydrocarbons. In general, good recoveries mean ≥70% obtained by most analysts. Borderline indicates that some of the analytes in this class are not well retained, or that recoveries may be a function of sorbent, matrix conditions (pH or ionic strength), or sample volume.
[a]Good recoveries can be obtained for most analytes in this class using optimized matrix conditions and sorbent.

A typical reverse-phase extraction comprises the following steps, which are self-explanatory:

1. Wash the sorbent with the eluting solvent to remove possible interferences.
2. Condition the sorbent to wet the pores and solvate the functional groups.
3. Pass the sample through the sorbent.
4. Elute the sorbent with organic solvent to remove adsorbed analytes.

Ion Exchange

Ion exchange refers to electrostatic interactions between ionic groups of opposite charge, one bonded to the sorbent and the other being a part of the analyte. Typical sorbents are ion-exchange functional groups bonded to either silica or SDB. These separations are usually carried out in water, although a mixture of water plus organic solvent is often used, especially in the elution step when the analyte is insoluble in pure water. Ion exchange can be a very useful technique when a high degree of selectivity is desired, as with a matrix high in other organic compounds that are neutral. The sorbent containing the adsorbed analyte and other organic compounds can be washed with organic solvent, which removes the neutral organic compounds but does not disrupt the ionic interaction. Ion exchange is sometimes the only way to accomplish the extraction of ionic analytes that are simply too water-soluble for reverse-phase extractions, such as the haloacetic acids.

The weakness of ion exchange is the possibility that other ionic species in the matrix will compete with the analyte for active sites and thus cause low recoveries. This possibility is even more likely when we consider the nature of residue analysis when the analyte is present in very low concentrations and other ionic species may be present at levels several orders of magnitude higher than the analyte. Fortunately, what makes the analysis feasible is that the affinity with which many environmental analytes adsorb onto ion-exchange sites is significantly larger than the affinity with which many ions commonly found in water adsorb. *Capacity,* or the number of ion-exchange groups per gram of sorbent, also plays a role in recoveries when competing species are present. Another potential weakness in ion exchange is the complexity of the extraction, as compared to reverse-phase extractions. The following is a typical recipe for an ion-exchange extraction:

1. Wet the sorbent with methanol. (This step is optional, depending on the hydrophobicity of the sorbent bed.)
2. Condition the sorbent with the proper counterion. (This conditioning may require more than one step.)
3. Wash the sorbent with water.
4. Pass the sample through the sorbent.
5. Wash the sorbent with organic solvent to remove organic interferences (optional).
6. Elute with the appropriate solvent.

A preliminary washing step could be incorporated, although some care must be taken to ensure that counterions with a high affinity for the sorbent are not introduced. For example, the addition of sulfate to a strong anion exchanger might reduce the capacity for the subsequent extraction. In addition, the sample often needs to be brought to the proper pH for the analyte to be ionic for the extraction. Although this complexity should not be considered a serious problem, with ion exchange more variables need to be considered during method development or troubleshooting than with reverse phase.

Sorbents

The amount and type of sorbent are what really govern the recovery obtained from SPE. Simply stated, the analyte must be totally retained by the sorbent, then quantitatively eluted from the sorbent, for quantitative recoveries. Lack of retention can be caused by two phenomena, which are related. If the analyte undergoes a partitioning process between the sorbent and the sample liquid, as in the chromatographic process, then it will eventually reach the bottom of the sorbent bed and exit with the liquid. This phenomenon is called *breakthrough*, and the volume at which the first analyte molecules begin to emerge is called the *breakthrough volume*. If the sample volume is less than the breakthrough volume, good recoveries are expected. Another aspect that can cause low recoveries is saturation of the sorbent with respect to the analyte or other species in the sample, which leads to *overloading*. One way to differentiate the concepts of breakthrough and overloading is that one molecule of an analyte could partition its way through the sorbent and result in breakthrough, even though the sorbent is not nearly saturated (i.e., not overloaded) with respect to that analyte. In practice, most low recoveries seen in environmental SPE are caused by breakthrough, as opposed to overloading.

Reverse-Phase Sorbents

Both breakthrough and overloading are related to the capacity of the sorbent for the analyte and related species, in the presence of the sample liquid. Capacity is very important in reverse-phase SPE and is related to both the amount of sorbent and to the strength of the interaction between analyte and sorbent. For bonded silica sorbents used in a reverse-phase mode, capacity is usually determined by the carbon loading, although surface area and functionality certainly play a role. The C_{18} bonded silicas used in environmental analysis are typically 15–20% octadecyl groups by weight, with the remaining weight being the high-surface-area (350 m^2/g) silica substrate. One reason that resin-based sorbents have higher capacity than silica-based sorbents is that they are 100% organic and thus have more organic functionality to do the extracting. An exception to this general rule occurs with secondary interactions, by which some functionality of the analyte interacts with the exposed silica sites, usually in an ion-exchange mechanism. A common example is a positively charged amine interacting with a negatively charged site on the silica. These secondary interactions are extensively used in the bioanalytical area but are usually not important in environmental reverse-phase extractions unless the analytes are nitrogen-containing pesticides or metabolites.

Ion-Exchange Sorbents

Ion-exchange sorbents can be either silica- or resin-based and can have a variety of exchange sites. The fundamental interaction is between the bonded exchange group and the ionic analyte, although reverse-phase interactions between hydrophobic areas of the analyte and hydrophobic areas of the sorbent can be important, especially if the sorbent is resin-based. If the bonded groups are anionic, the sorbent is a cation exchanger, whereas anion exchangers are cationic sorbents. The terms "strong" and "weak" exchangers refer to the nature of the exchange group. If the bonded group is a strong acid or base, such as sulfonate or quaternary ammonium, it is a strong exchanger. Weak acids or bases, such as carboxylate or amine, are weak exchangers. The usual way to choose an exchanger is to use weak exchangers for strong analytes and strong exchangers for weak analytes to facilitate the elution of the analyte under mild conditions. Strong analytes bonded to strong exchangers often require harsh eluants or high volumes of eluant for quantitative elution. Resin-based exchangers are often reported to have three to five times the capacity of bonded silica exchangers and would be preferred for applications needing high capacity.

Selection of sorbent can be governed by a variety of factors. Table III compares the features of resin- and silica-based sorbents. For environmental analysis using reverse-phase mechanisms, the usual choice is to opt for the highest capacity sorbent in order to capture analytes of a wider polarity range. This desire for higher capacity sorbents tends to point one to resin sorbents, such as SDB, for methods that are intended for the extraction of a wide range of pollutants. C_{18} silica, on the other hand, works very well when the analyte list includes only nonpolar analytes, such as organochlorine pesticides or dioxins. Silica-based sorbents also tend to be cleaner with respect to coextracted interferences than resin-based sorbents, although cleaner forms of resins are becoming available.

Formats

Originally, and through the late 1980s, SPE sorbents were only available in the form of packed tubes, commonly called either cartridges or columns. Still the predominant format for SPE, tubes contain from 0.1 to 10 g of sorbent, with usual sizes being 0.1 to 1 g. Materials of construction are usually polypropylene tubes with polyethylene frits to hold the sorbent in place. Also available are glass or polytetrafluoroethylene (PTFE) tubes with inert frits for users who are concerned with contamination from the polyolefin materials. The tube format is standard in bioanalytical labs but has had a more difficult time being accepted in environmental applications. Two reasons seem to account for this difficulty: (1) slow flow rates necessitated by slow kinetics and (2) susceptibility to plugging by suspended solids. Both of these problems are exaggerated by the large volumes that need to be analyzed in environmental applications. A typical environmental sample is 1 L, which can take much more than 1 h to extract because typical flow rates through tubes are often determined to be 10 mL/min or less. Plugging is a situation that is unavoid-

Table III. Comparison of Resin- and Silica-Based Sorbents

Sorbent Feature	*C_{18} Silica*	*SDB Resin*
pH Stability	1–8	0–14
Carbon loading (%)	15	93
Functionality	Aliphatic	Aliphatic + aromatic
Secondary interactions	Silica	None
Capacity/weight	Lower	Higher (3–5×)

Note: SDB is poly(styrene–divinylbenzene).

able with any SPE format. It can be successfully dealt with and will be discussed in detail later. On the other hand, many laboratories have developed very effective methods using tubes, which have the advantage of being economical.

An alternative shape for SPE sorbents was introduced in 1989 with a new technology that enmeshes the sorbent in a PTFE web of fibrils to form a disk (*5*). Typical dimensions for disks are circles of 47 or 90 mm diameter that are about 0.5 mm thick. A typical sorbent loading is 90% by weight. The disks are considered by many to be a more appropriate format for environmental analysis because the flow rates can be much faster than tubes while maintaining quantitative recoveries. In 1993, a slightly different disk was introduced that uses a glass fiber filter impregnated with functionalized silica (*14*). Devices to use tubes and disks range from homemade apparatuses to vacuum boxes to highly automated sample preparation stations. Because of the increased use of SPE, a constant flow of improvements can be expected over the next few years.

Troubleshooting

When an analyst begins using a totally new extraction method, the analytical results, in terms of recovery, will sometimes inevitably fail to meet the expectations of the analyst. The biggest single index of success in a SPE is the percent recovery. Although opinions vary on what an acceptable recovery is, most analysts are content with a method that can consistently produce at least a 70% recovery on spiked samples. For difficult analytes, 30–50% recoveries are sometimes considered good enough. Single-digit RSDs are desirable, but lower spiking levels or difficult matrices may yield RSDs of 25% or more. One caveat to keep in mind is that SPE results usually cannot be expected to be better than the standard method being replaced. More than one analyst has complained of low recoveries, only to find that the standard technique gives comparable results. Although a wide variety of reasons exist for low recoveries, many of which are unrelated to the SPE, examination of a data set can often give clues to the cause of the problem. This section will discuss some commonly seen trends and the possible causes. The main focus will be on reverse-phase extractions, because most of the environmental extractions use that mechanism.

Low Recoveries for Well-Retained Analytes

Low results for all analytes in a data set often indicate a procedural problem. Incorrect conditioning of the solid phase, excessively high flow rates, insufficient elution volume, and measurement or math errors may all be manifested in this manner. Conversely, a data set with generally good recoveries and a few low recoveries probably indicates good technique with a chemically related problem such as insufficient retention of the problem analytes.

If early eluting gas chromatograph (GC) analytes seem to have lower recoveries than later eluting compounds of the same data set, the cause is often volatility. The procedure should be examined for steps in which volatile compounds could escape from the spiked sample, the solid-phase device, or the eluant. Any solvent reduction steps after elution are especially suspect; they can be checked by spiking the eluant resulting from a blank extraction, then finishing the sample preparation.

Good recoveries in deionized (DI) water, but problems in real samples, may be indicative of matrix effects. Factors like pH, ionic strength, or organic content of the sample versus the DI water should be considered. In an ion-exchange extraction, the presence of strongly retained, competing ions of the same charge that saturate the ion-exchange sites should be considered. Matrix modification can often overcome problems of this type.

Low Recoveries for Poorly Retained Analytes

When the low recovery of an analyte is caused by insufficient retention, which causes breakthrough, matrix modification steps should be considered. Matrix modification refers to the chemical adjustment of pH or ionic strength in the sample solution, both of which are fairly common in LLE. By understanding and adjusting the chemistry of the system, recoveries of 10% or greater can sometimes be raised to acceptable levels through one or more of the following steps:

- pH adjustment to suppress ionization of target analytes;
- increasing the ionic strength of the matrix, or "salting out"; or
- increasing the ratio of sorbent weight to sample volume.

pH Adjustment. Adjustment of pH operates on the principle that ionic analytes tend to remain with the aqueous phase, whereas neutral analytes tend to partition into the organic phase. A simple example is the addition of a mineral acid to a water sample containing carboxylic acids, bringing the pH down to 1 or 2. This step neutralizes or protonates the organic acids in the solution and can dramatically increase the recoveries. Many phenoxyacid herbicides can be made to partition into either water or organic solvent, depending on whether they are ionic or neutral, respectively. The opposite of pH reduction is the addition of base to raise the pH to the point at which amines are neutral. Normally, lowering of pH can be done on either bonded silicas or resin sorbents, but raising the pH beyond 8–10 for extended periods of time should be done only on resin phases to avoid dissolution of the base silica.

Ionic Strength. *Salting out* is another common technique in LLE to drive the analyte into the organic phase, and it can be very effective in SPE. The principle is to increase the relative polarity of the matrix as compared to the polarity of the analyte, thus changing the partition coefficient and increasing the driving force for the analyte to partition into a less polar organic phase. The most commonly used salts are sodium chloride and sodium sulfate, although others have been used very successfully. The best results are expected when the matrix is saturated with the salt being used, although this condition can increase the viscosity of the sample and generally make a mess. A more reasonable approach is to make several salt concentrations and plot the analyte recovery versus the salt concentration. In this way, the analyst can decide on a reasonable recipe to use and still get acceptable results. For example, 10% sodium chloride might be sufficient to increase recoveries of a set of phenolic analytes to acceptable levels, and it is much easier to work with than saturated (~35% by weight) sodium chloride. The optimum conditions for salting out occur when the analyte of interest is a neutral or neutralized compound. If salting out is attempted with an ionic analyte, lower results may be observed because of the increased interaction of the ionic analyte with the ionic modifier (*13*).

Sorbent. Another method of increasing the recovery for difficult compounds that exhibit low breakthrough volumes is to choose a sorbent that maximizes the strength of the sorbent–analyte interaction, thus exhibiting a higher capacity for the analyte. If the best available sorbent still results in low recoveries, a brute force solution to breakthrough is to maximize the ratio of sorbent weight to sample volume. This maximization can be done with less sample, more sorbent, or both. When more sorbent is used, more elution solvent will be necessary and the analyst should be conscious of the economy of the situation versus the more traditional LLE.

By using one or more of these strategies, methods can be developed for compounds that initially give low recoveries and appear to be undoable by reverse-phase SPE. As a word of caution, some analytes are simply inappropriate for reverse-phase extractions. A well-designed experiment that varies a few conditions will quickly show if reverse-phase extraction is inappropriate. Troubleshooting or method development can often benefit from the technique of plotting the recovery versus the parameter being investigated. For example, plotting flow rate versus recovery might yield a plot similar to Figure 2a, plotting pH versus recovery might yield a plot similar to Figure 2b for a carboxylic acid analyte, or plotting sample volume versus recovery might yield a plot similar to Figure 2c. This technique can give a clear indication of the bounds within which an extraction should succeed.

Suspended Solids in Samples

Many of the water samples received in the laboratory contain suspended solids that can cause problems with plugging of the solid-phase device. The result of plugging is a gradual decrease in flow rate, to the point at which SPE is no longer a practical means of extracting further sample. Severity of plugging is a function of the concentration and size of suspended solids, the porosity of the sorbent bed or frit, the sample volume, and the surface area of the top of the sorbent bed. Problems can be encountered with almost all types of waters: drinking water, groundwater, surface water, and wastewater. At the microscale level, the pores of the SPE bed can become plugged by the suspended solids in the sample, effectively reducing the pore size

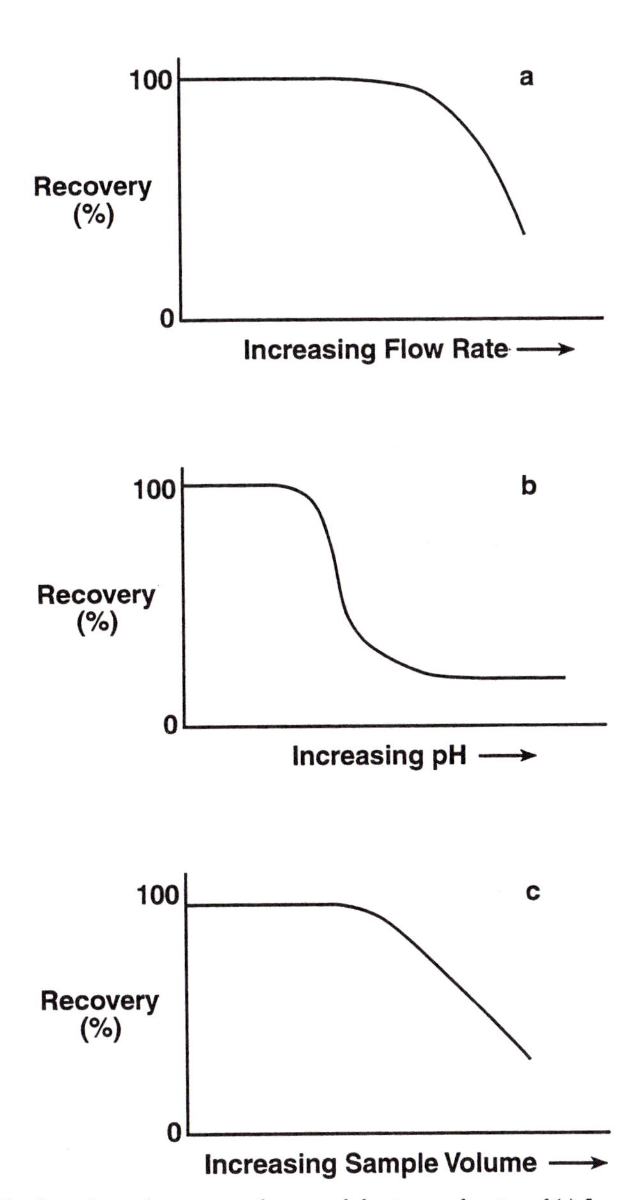

Figure 2. The three plots indicate expected recovery behavior as a function of (a) flow rate, (b) pH, and (c) sample volume variations.

and restricting the flow through those pores. As the plugging progresses, the flow decreases not in a linear manner, but in an exponential manner, as shown in Figure 1.

Of course, if the sample contains a significant amount of suspended solids, problems can be expected in any type of extraction, including LLE. This discussion will be limited to those samples that range from clear to turbid and may contain a small layer of sediment in the bottom of the bottle. As a general rule, gels of biological origin (e.g., wastewater from meat packers or rendering plants) are the worst, along with any sample that is very turbid and will not settle with standing. Fine clays can be difficult, as can stagnant surface waters with high algal activity. At the other end of the spectrum, samples with sand tend to be relatively easy to filter because of the large particle size. This discussion will address many of the solutions for plugging that have been found effective in laboratories using SPE technology.

Prefiltration

The first solution to plugging is very simple and effective but may ultimately cause problems with the results. Prefiltering the sample through a fine-pore filter (0.5–1 μm) prior to the SPE will probably eliminate any problems with plugging. The filtered water sample is then extracted using SPE, and the analysis concludes as usual. This approach is very viable if the analytes of interest do not totally or partially adsorb on the suspended solids in the sample, or if the analytical results desired are only for the fraction of the analyte dissolved in the water. Hydrophobic analytes such as dioxins and PCBs are well known to equilibrate between the water and the particulate material in the water. In some cases, the fraction on the particulate may represent the majority of the analyte. Most environmental analysts want the total analyte concentration in the sample and not just that in the dissolved portion. This requirement means that prefiltration actually creates a second sample for analysis, namely, the solids that were filtered out. Although separation of the suspended solids from the liquid may be necessary for samples high in suspended solids, dealing with the whole sample using SPE is actually a reasonable approach and can be much more efficient in terms of effort once one deals with the plugging. Another similar strategy is to prefilter the sample and save the prefilter with the sediment, extract the filtered water sample by SPE, and recombine the sediment with the SPE sorbent during the final elution step, which could be done by placing both (1) the prefilter with the sediment and (2) the SPE sorbent into a vial with a minimum amount of solvent, or into a Soxhlet extractor.

In-Place Prefilters

A compromise between prefiltering the sample and running the sample without prefiltration is placing a prefilter in place above the sorbent bed, running the sample through the combination, and eluting the combination. Although this strategy is possible with both tubes and disks, this discussion will be confined to disks, where most of the work has been done. In essence, the disk is placed on the filtration apparatus, a glass fiber or other prefilter is placed on top of the disk, and the reservoir is clamped in place. As a slight modification of this method, a filter-aid material can be placed on top of the disk, or even placed on top of the combined filtration and SPE disks. The benefits of this approach are that it adds little cost, is easy to do, and does a reasonable job of collecting and eluting the whole sample. Typical improvements in flow rate are from two to ten times faster, depending on the sample and the prefilter chosen.

Larger Surface Area Sorbent

The brute-force approach to dealing with plugging is simply to use a larger surface area sorbent device. Usually, a 47-mm disk will give much faster flows and less plugging than an SPE tube, and a 90-mm disk will be much faster than a 47-mm disk. Even though a 90-mm disk has about four times the area of a 47-mm disk, the decrease in flow time from a 47-mm to a 90-mm disk will often be from one to two orders of magnitude because the plugging is exponential rather than linear. Another potential benefit of increasing the surface area is that such an increase will usually be accompanied by a similar increase in the amount of sorbent, the capacity of which could be important for wastewater with relatively high amounts of dissolved organic compounds. Because this approach usually collects the suspended solids on the top of the SPE device as a filter cake for subsequent elution, the larger diameter device will spread the filter cake into a thinner layer, allowing more effective contact between the eluting solvent and the sediment particles. The disadvantage of larger diameter SPE devices is that the cost will certainly be higher than for corresponding smaller devices. As with any new method, the analyst simply has to calculate the cost-to-benefit ratio of the old versus the new method, especially in view of escalating solvent and solvent disposal costs.

Decanting

Finally, one of the most overlooked means of overcoming problems associated with many types of suspended solids is to allow the solids to settle in the bottom of the sample bottle (as opposed to vigorously shaking the sample) and then gently to decant the majority of the supernatant sample into the sample reservoir. When most of the sample has passed through the sorbent, but before the reservoir goes dry, the remaining material in the bottle can be added to finish the extraction. This technique is surprisingly effective and adds nothing in terms of time or cost to the analysis.

Example Procedures

This section details three typical SPE procedures being used for environmental analysis. These methods have been selected to illustrate the use of SPE for a variety of analytes and matrices and include a reverse-phase drinking water procedure, an ion-exchange drinking water procedure, and a reverse-phase procedure for dirty water matrices.

Method 525.2 (7, 8)

Method 525.1 was developed in the late 1980s as a way to monitor a large number of regulated analytes in drinking water, without the drawbacks of LLE. The original analyte list, which has been expanded in Method 525.2, includes polyaromatic hydrocarbons (PAHs), PCBs, phthalates and adipates, organochlorine pesticides, triazines, and organonitrogen pesticides, all of which lend themselves to a straightforward reverse-phase procedure.

In this method, the sorbent is a 47-mm C_{18} bonded silica disk or a 1-g tube, and the sample is 1 L of finished drinking water or source water. The general procedure is as follows:

- Wash the sorbent and system with final eluting solvents.
- Condition the sorbent with methanol.
- Pass the water sample through the sorbent; the flow rate depends on the format of the SPE device.
- Elute the sorbent consecutively with 5 mL ethyl acetate, 5 mL 1:1 ethyl acetate:methylene chloride, and 5 mL methylene chloride; rinse the sample bottle.
- Dry the eluant with anhydrous sodium sulfate.
- Concentrate the eluant to 1 mL, and analyze by GC–MS.

This method is capable of reaching detection limits below the parts-per-billion level and is much more efficient than the LLE method it replaces. The elution steps in Method 525.2 include ethyl acetate, which is a more effective eluting solvent for a water-saturated sorbent particle. Method 525.1 only used methylene chloride.

Method 552.1 (7, 8)

This method is one of the first ion-exchange procedures for an analyte set that cannot be done by a reverse-phase procedure because of excessive water solubility of the compounds. The analytes are the haloacetic acids (monochloro-, monobromo-, dichloro-, dibromo-, trichloro-, and bromochloroacetic acid) and dalapon, all of which are small carboxylic acids with high water solubility. Essentially, the pH of the sample is adjusted to a point at which the acids are all ionic. The anionic acids are passed through a bed of resin-based, quaternary ammonium anion exchange particles. Once the sample has been extracted, the analytes are eluted with 10% sulfuric acid in methanol. This eluant at once provides a combination of (1) low pH to protonate the acids, (2) an abundance of a strongly competing counterion (sulfate), and

(3) methanol to disrupt any remaining reverse-phase interaction between the analyte and the sorbent.

In this method, the sorbent is a 47-mm strong anion exchange disk or tube, and the sample is 100 mL of drinking or source water, adjusted to pH 4.5–5.5. The procedure is as follows:

- Condition the sorbent with methanol, HCl, and NaOH to affix the proper counterion (hydroxide) to the ion-exchange sites.
- Pass the water sample through the sorbent.
- Rinse the sorbent with methanol; this rinse will help dry the resin and will remove any adsorbed neutral organic compounds. The methanol will not disrupt any ion-exchange interactions.
- Elute the acids with 10% sulfuric acid–methanol.
- Continue by derivatizing the acids and analyzing by GC with electron capture detection.

Results are generally quantitative, although certain samples that contain high levels of sulfate have been found to result in low recoveries. The sulfate ions have a high selectivity for the quaternary ammonium groups and aggressively compete for the ion-exchange sites, thereby "poisoning" the sorbent.

Method 3535 (9)

This final example is a method intended for wastewaters instead of drinking water samples. Method 3535 was accepted by EPA's Office of Solid Waste in July 1994 as a way to prepare samples for Method 8081, the analysis of organochlorine pesticides in water samples or Toxicity Characteristic Leaching Procedure (TCLP) leachates. Although Method 3535 probably will not receive *Federal Register* approval until mid-1996, the Office of Solid Waste allows the use of the new SW–846 method for most samples, except samples for mandatory testing. The general principles of Method 3535 are very similar to those of Method 525.2 described previously, but because the water samples could contain very high concentrations of analytes and some suspended solids, a 90-mm C_{18} bonded silica disk is used because it has four times the capacity and surface area of a 47-mm disk.

In Method 3535, the sorbent is a 90-mm C_{18} bonded silica disk (packed tubes have not been validated yet). The sample is 1 L of groundwater, wastewater, or TCLP leachate. The procedure is as follows:

- Wash the disk and system with acetone and methylene chloride.
- Condition the disk with methanol.
- Pass the sample through the disk.

- Elute the disk with a small (~5-mL) aliquot of acetone, followed by two, 15-mL aliquots of methylene chloride, rinsing the bottle also.
- Dry and concentrate the eluant, and analyze it by GC–ECD.

Generally, the results obtained with this method are 80–100% recoveries with single-digit RSDs.

Future Developments

Clearly, SPE is already playing a key role in the analysis of a variety of organic compounds in drinking and source water, and the trend toward adapting the technique to "dirty" water samples, such as wastewaters, is beginning to accelerate. The technique is also beginning to be investigated as a way to extract TCLP leachates, obtained after tumbling a solid material with a low-pH buffer and filtering to determine if the solid contains materials that make it a hazardous waste (*15*). Other clear trends are the growing influence of SPE disks and the use of resin-based sorbents to allow effective extraction of polar analytes. As in Method 552.1, more ion-exchange mechanisms are being exploited as an alternative to reverse-phase techniques. Extending the range of analytes to include metals will be very possible with the availability of ion-exchange sorbents.

Senseman et al. (*16*) reported that site sampling and storage on the SPE sorbent is an attractive possibility and even may provide enhanced storage stability of several pesticides. For ultratrace work, volumes beyond 1 L are being used for hydrophobic analytes (*6*). An automated system was reported by Liska et al. (*17*) for use in on-line extractions. Beyond water samples, several publications have demonstrated that SPE can be used for solid samples, such as soil or food (*18–20*). The technique generally used is to extract the solids with a water-miscible solvent, filter that solvent, and then use water to dilute the solvent to a point at which the solvent content is less than 10%. The solution, whose volume could be as much as 1 L, is then passed through a sorbent bed to extract the analytes of interest.

As a result of the current high level of activity by researchers, the next few years will not only reveal a number of improvements in formats, sorbents, and methods, but will also further define the areas in which SPE will be an attractive technique for sample preparation of environmental water samples.

References

1. Junk, G. A.; Richard, J. J.; Grieser, M. D.; Witiak, D.; Witiak, J. L.; Arguello, M. D.; Vick, R.; Svec, H. J.; Fritz, J. S.; Calder, G. V. *J. Chromatogr. A* **1974**, *99*, 745–762.
2. Richard, J. J.; Junk, G. A. *Mikrochim. Acta* **1986**, *I*, 387–394.
3. Bellar, T. A.; Budde, W. L. *Anal. Chem.* **1988**, *60*, 2076–2083.
4. Liska, I.; Krupcik, J.; Leclercq, P. A. *J. High Resolut. Chromatogr.* **1989**, *12*, 577–590.

5. Hagen, D. F.; Markell, C. G.; Schmitt, G. A.; Blevins, D. D. *Anal. Chim. Acta* **1990,** *236,* 157–164.
6. McDonnell, T.; Rosenfeld, J.; Rais-Firouz, A. *J. Chromatogr. A* **1993,** *629,* 41–53.
7. U.S. Environmental Protection Agency. *Methods for the Determination of Organic Compounds in Drinking Water;* Supplement 3, 1996; Environmental Monitoring and Systems Laboratory: Cincinnati, OH, 1988.
8. U.S. Environmental Protection Agency. *Methods for the Determination of Organic Compounds in Drinking Water;* Supplement 2, 1992; Environmental Monitoring and Systems Laboratory: Cincinnati, OH, 1988.
9. U.S. Environmental Protection Agency. n-*Hexane Extractable Material (HEM) and Silica Gel Treated* n-*Hexane Extractable Material (SGT-HEM) by Extraction and Gravimetry;* Draft Method 1664; Government Printing Office: Washington, DC, 1994.
10. U.S. Environmental Protection Agency. Office of Solid Waste and Emergency Response. *Test Methods for Evaluating Solid Waste,* Third Edition; Government Printing Office: Washington, DC, 1986.
11. U.S. Environmental Protection Agency. Contract Laboratory Program. *Draft Statement of Work for Quick Turnaround Analysis;* Government Printing Office: Washington, DC, 1992.
12. Larrivee, M. L.; Poole, C. F. *Anal. Chem.* **1994,** *66,* 139–146.
13. Markell, C. G.; Wisted, E. E.; Hagen, D. F. Presented at the Pittsburgh Conference, Atlanta, GA, 1993; paper 1037.
14. Blevins, D. D.; Schultheis, S. K. *LC–GC* **1994,** *12,* 12–16.
15. Markell, C. G.; Song, L.; Pieper, R. M. "Solid Phase Extraction of TCLP Leachates"; Presented at the 16th Annual EPA Conference on Analysis of Pollutants in the Environment, Washington, DC, 1994.
16. Senseman, S. A.; Lavy, T. L.; Mattice, J. D.; Myers, B. M.; Skulman, B. W. *Environ. Sci. Technol.* **1993,** *27,* 516–519.
17. Liska, I.; Brouwer, E. R.; Ostheimer, A. G. L.; Lingeman, H.; Brinkman, U. A. Th.; Geerdink, R. B.; Mulder, W. H. *Int. J. Environ. Anal. Chem.* **1992,** *47,* 267–291.
18. Klaffenbach, P.; Holland, P. T. *J. Agric. Food Chem.* **1993,** *41,* 396–401.
19. Redondo, M. J.; Ruiz, M. J.; Boluda, R.; Font, G. *Chromatographia* **1993,** *36,* 187–190.
20. Price, S.; Warwick, J.; Fellegy, M. "Investigation of Solid Phase Extraction Disks in Food Sample Preparation for Pesticides Analysis"; Presented at the California Pesticide Residue Workshop, Sacramento, CA, 1994.

Chapter 18

Sampling Waters

The Impact of Sample Variability on Planning and Confidence Levels

U. M. Cowgill

This chapter addresses generic difficulties associated with water sampling: problems of sampling related to the physical state of the sample (e.g., ice, snow, and dew), how these disconcerting situations are resolved or circumvented, and how these considerations affect proposed sampling protocols and desired confidence limits of subsequently obtained chemical results. Types of contamination expected from sampling tools composed of various materials, and sorption and leaching of compounds from such materials are considered. In addition, examples of variation in chemical composition are presented in relation to replication, sampling frequency, and sampling location in various water bodies. Lakes, rivers, groundwater, extensive ice development, stratigraphy encountered in dew depositions, and variations found during a 3-h rainstorm are among the examples shown.

SAMPLING IN ENVIRONMENTAL ANALYSIS is akin to dishwashing in analytical chemistry. If the dishes are dirty, all analytical activity thereafter is for nought. Similarly, nothing is gained from the chemical endeavor if the samples are not representative of the environment from which they are taken. The first portion of this chapter is devoted to details demanding attention if proper samples are to be gathered. The next portion concerns specific types of samples and associated problems to be considered in sample collection. The remainder of the chapter addresses how these various topics affect proposed sampling protocols and desired confidence limits of subsequently obtained chemical results. The majority of the examples are drawn from inorganic chemistry because few studies have been devoted to the variability of trace organic compounds in relation to sampling. Further discussion of organic compound sampling may be found in the works of Keith (*1–3*). A good general review of sampling for chemical analysis may be found in the publications of Kratochvil and co-workers (*4*, *5*).

Problems Associated with Sampling

Much of the material presented in this chapter was gathered in the process of trying to discover the best way to sample particular bodies of water or water in a physical state other than the liquid. Such data are not usually published. But this kind of problem does reflect the kind of cogitation that is a necessary preamble to any extensive and successful research effort. The Linsley Pond Study (North Branford, CT) (*Arch. Hydrobiol.*, 1970–1978) involved a very lengthy preliminary planning stage wherein proper sampling could be identified only by careful chemical analysis. Sampling had to be adequate for precise measurement of major as well as minor constituents. Such requirements had to apply not only to water but to ice, snow, rain, mud, plants, and microscopic biota. Most extensive geochemical studies involve detailed comprehensive preliminary work. Many of the examples used to illustrate sampling pitfalls originated from those various preliminary studies carried out on a variety of water bodies. Much of the unpublished work referred to in this chapter was generated in various places and over several decades; therefore, these data will be referred to by date rather than by place.

The purpose of sampling dictates the nature of the sample to be gathered, the equipment used to acquire the sample, the size of the sample, and the frequency of sampling. Generally, the purpose of water sampling is to acquire a representative sample of the body of water being studied. Once such a representative sample has been obtained, the investigator may be interested only in studying its general chemical composition and may therefore have no interest in seasonal changes. Thus, the necessity of frequent sampling is limited.

Contamination from Sampling Equipment

Table I shows the type of contaminants contributed to water samples by materials used in sampling tools and monitoring-well construction. Poly(vinyl chloride)-threaded (PVC-threaded) or poly(vinyl chloride)-cemented (PVC-cemented) joints may have contaminants that leach into contiguous water. Such contaminants may be removed below the limits of detection in distilled water by steam cleaning. The source of the steam should be distilled water. Five separate steam washes will remove undesirable substances (Cowgill, U. M., *Symposium on Field Methods for Ground Water Contamination Studies and Their*

Table I. Contaminants Contributed to Water Samples by Materials Used in Sampling Devices or Well Casings

Material	*Contaminants Contributed Prior to Steam Cleaning*
PVC-threaded joints	Chloroform
PVC-cemented joints	Methyl ethyl ketone, toluene, acetone, methylene chloride, benzene, ethyl acetate, tetrahydrofuran, cyclohexanone, three organic Sn compounds, and vinyl chloride
Polytetrafluoroethylene (Teflon)	Nothing detectable
Polypropylene or polyethylene	Plasticizers and phthalates
FRE	Nothing detectable
Stainless steel	Cr, Fe, Ni, and Mo
Glass	B and Si

SOURCE: Adapted from Cowgill, U. M., *Symposium on Field Methods for Ground Water Contamination Studies and Their Standardization*, American Society for Testing and Materials, in press, and Cowgill, U. M., unpublished data, Dow Chemical Company, 1985.

Standardization, American Society for Testing and Materials, in press). Similarly, steam cleaning of fiberglass-reinforced epoxy material (FRE), polypropylene, polyethylene, and glass will substantially reduce the quantities of substances that leach out of these materials into the contiguous water (Cowgill, U. M., Dow Chemical Company, unpublished data, 1985). As a result of this experience, I recommend that no sampling devices or well casings should be used without prior, thorough, steam cleaning.

Two incidents involving the use of PVC casing with solvent-welded joints are noteworthy. Many contracting firms have installed wells employing PVC-cemented joints around industrial plant sites. Substances originating from that cement have been detected in groundwater a decade after installation (Cowgill, U. M., Dow Chemical Company, unpublished data, 1985). Another incident occurred when the Midland, MI, water treatment plant changed their method of water treatment. When this finished water was initially transported to the Dow Chemical Company's Aquatic Toxicology Laboratory, a quarter mile of cast-iron pipe was destroyed and replaced with PVC pipe containing cemented joints. After 84 days, 1.3 mg/L of methylene chloride, 3 μg/L of chloroform, and 6 μg/L of tetrahydrofuran were detected in the water despite the fact that 7,257,600 gal of water (pH 8) had passed through this quarter mile of pipe.

Additionally, the chemical contribution to water transported through soldered pipes cannot be ignored. Tin and lead are the most common contaminants encountered under such circumstances. Water containing high quantities of calcium tends to extract the lead preferentially. Tin, however, continues to be removed in small amounts for decades (6).

Sorption and Leaching of Contaminants by Sampling Tool Materials

Barcelona et al. (*7, 8*) published a study comparing the maximum sorption of dilute (400-μg/L) halogenated hydrocarbon mixtures in water by various plastics. PVC exhibited a maximum sorption of 622 $\mu g/m^2$, and polytetrafluoroethylene (PTFE) (Teflon) sorbed only 237 $\mu g/m^2$. Hunkin (Hunkin, G. G., Sr., Ground Water Sampling Inc., personal communication, 1985–1986) employed similar but not identical procedures outlined by Barcelona et al. (*7*) and found that the maximum sorption by FRE was 203 $\mu g/m^2$. This figure was obtained by exposing 6 g of FRE particles for 72 h to 400 μg/L of individually halogenated hydrocarbons similar to those used elsewhere (*7*). Barcelona et al. (*7, 8*) did not use exposure times in excess of 60 min. The data revealing the actual loss of halocarbons from solution are quite interesting. In 72 h, the sorption on the FRE sample was 20.25% (Hunkin, G. G., Sr., Ground Water Sampling Inc., personal communication, 1985), and sorption values of PTFE and PVC in 1 h were 38% and 98%, respectively. These data were calculated on the basis of the halocarbon concentration in the fluid expressed in terms of sample surface area. When the data are expressed in terms of organic concentration loss, the percent sorption is slightly lower. The presence of organic carbon has an effect on sorption that apparently is material-dependent (*8*). The depletion of halocarbons from the solution is more dependent on the type of material than on the tubing diameter (*8*). However, when a constant flow rate is used, losses are more likely to increase rather than decrease as the tubing diameter becomes larger (*7, 8*). Clearly, sorption is a function of the mass as well as the surface.

This discussion shows that the use of PTFE or PVC compromises the value of total organic carbon as an indicator of pollution. This result has not been found for FRE (Hunkin, G. G., Sr., Ground Water Sampling Inc., personal communication, 1986). Thus, growing evidence suggests that thermoplastic materials sorb many priority pollutants efficiently. Furthermore, this sorption may be so effective that detection of undesirable compounds in water may be delayed when synthetic sampling tools are suspended in wells and have long contact with well water, or when sampling devices used for continuing sampling of surface water studies are made of thermoplastic materials. Barcelona et al. (*7, 8*) showed that PTFE tubing brought about an adsorptive loss of chlorinated hydrocarbons of 21% in less than 1 h. Similar losses have been observed for FRE in 72 h (Hunkin, G. G., Sr., Ground Water Sampling Inc., personal communication, 1986).

Miller (*9*) noted that the extent of adsorption of metals at low concentrations on container walls is determined by metal concentration, pH of the sample, length of contact with the container, sample and container chemical composition, and the presence of dissolved organic carbon as well as complexing agents (*10*). Robertson (*11*), in his elemental study of seawater, reported that adsorption of In, Sc, and Ag could be significantly reduced when the sample was stored in glass or polyethylene. This reduction occurs by adjusting the pH of the sample to 1.5 with HCl. Robertson (*12*) also pointed out that PVC contained Zn, Fe, Sb, and Cu; that polyethylene contained Sb; and that these metals could leach from such materials into the contiguous water. Boettner et al. (*13*) reported the leaching of alkyltin and organic pipe cement from PVC and chlorinated PVC into contiguous water at pH 5 and 37 °C. Dibutyltin dichloride and dimethyltin dichloride were also detected in this study. Miller (*9*) noted that adsorption and subsequent leaching of Pb were more rapid when Cr(VI) was present in the solution along with six volatile organic compounds than when Pb and Cr(VI) were alone in solution. The Cr(VI) did not adsorb or leach from PVC, polyethylene, or polypropylene. The intent of this discussion is not to engage in a lengthy review but only to point out that contaminant leaching from sample tools is quite complex and requires serious attention by those involved in sampling water for trace quantities.

Replication

The number of replications required to characterize a water body is determined by the purpose of sampling. The purpose falls into two general categories: (1) sampling for the purpose of description (i.e., the chemical composition of the water body), and (2) sampling for the purpose of monitoring (i.e., the substances to be monitored have already been identified). Replication in the second category must be described in terms of acceptable arithmetic means, standard deviations, and confidence limits. In an area to be monitored, descriptive sampling is necessary not only to discover what is present but to obtain chemical background information as well.

Before proceeding further, replication should be defined. If three aliquots are analyzed from a 1-L sample, and if three 1-L samples are collected and an aliquot is removed from each liter and analyzed, then the results of these two approaches will differ. The first approach is a good way to check the instrument and ascertain that it is functioning properly. However, these data are biased and nonrandom in a statistical sense from the standpoint of the water body, but not from the standpoint of evaluating the entire analytical technique. The second approach is true *replication*. Any variation among the three aliquots gathered from three separate samples is the experimental sampling error. This type of replication provides the measure of sampling precision. Furthermore, the built-in randomization ensures the validity of this measure of precision. In addition, replication of this sort will help to illuminate gross errors in analytical measurements. The first approach could not be used in this manner.

The first aproach (three aliquots per liter) to sampling will usually result in a smaller percent coefficient of variation than the second approach (*14*), and may occasionally exhibit a smaller standard deviation. However, the second approach will provide a result more representative of the body of water being studied at the time of sampling. Because the second approach results in the ultimate objective of all water sampling, it is the most likely approach to achieve needed results over time. Table II illustrates this point with some results from Linsley Pond in North Branford, CT (Cowgill, U. M., unpublished data, 1965–1980).

An examination of the data shown in Table II reveals that the percent coefficient of variation is always lower when three aliquots per liter are analyzed than when one subsample is analyzed from each of three 1-L samples. This study also shows that the concentrations are neither consistently higher nor lower but that as the concentration of the substance of interest declines, the variation between the two sampling regimens increases. Significant differences are encountered between the two systems. The second approach, however, does not indicate the true value or even how close to the true value either analysis may be. This point is discussed further by McBean and Rovers (*14*).

Another approach to the problem is to gather larger samples. Table III shows the results of this endeavor. All the samples were gathered from each side of a boat with a Van Dorn bottle outfitted with a graduated-steel tape (Cowgill, U. M., unpublished data, 1965–1980). [Further discussion on available types of Van Dorn bottles may be found in reference 15.] No significant difference arose between 20- and 30-L samples taken from either side of the boat. In addition, the percent coefficient of variation becomes progressively smaller. The data shown from either side of the boat involving 20- and 30-L samples are, statistically speaking, no different from the results obtained by subsampling each of three 1-L samples replicated 10 times. Results obtained from either side of the boat contain too much discrepancy in the smaller volume samples. To illustrate the problems of replication, the worst possible case has been selected, that of the euphotic zone, or the zone of a productive lake in which much of the living activity takes place. In addition, the depth varies by 2.5 m. The conclusion is that replicated 20-L samples provide reasonable results regardless of which side of the boat the samples are collected from or how small the concentrations of the substance of interest may be. Another point is that each body of water has a "personality of its own", and this kind of descriptive sampling needs to be

Table II. Chemical Results Obtained from Various Sampling Techniques on Linsley Pond

Number of Subsamples Replicated 10 Times	*Ca (mg/L)*	*Mg (mg/L)*	*Sr (μg/L)*	*P (μg/L)*	*Pb (μg/L)*
Three aliquots per liter	53 (4.5)	30 (8.3)	130 (20)	20.4 (5.1)	2.4 (0.5)
One aliquot from each of three 1-L samples	29.7 (5.0)	11.8 (3.8)	64 (20)	72 (30)	13 (4)
χ^2, $P<$	0.02	0.005	0.001	0.001	0.01

NOTE: The depth sampled was from the surface to 2.5 m below the surface. The numbers in parentheses denote the standard deviation.

Table III. Chemical Results Obtained by Collecting Samples from the North and South Side of the Boat

Sample Size	*Ca* (*mg/L*)	*Mg* (*mg/L*)	*Sr* (*μg/L*)	*P* (*μg/L*)	*Pb* (*μg/L*)
		South Side			
Three 2-L samples	55	56.2	155	100	5
	(10.0)	(10.0)	(25)	(25)	(3.0)
Three 5-L samples	40	22	120	110	10
	(11.0)	(10.0)	(21)	(25)	(3.1)
Three 20-L samples	30	12	75	75	15
	(3.0)	(2.1)	(10)	(10)	(2.4)
Three 30-L samples	31	14	70	80	18
	(2.8)	(2.1)	(10)	(8)	(2.1)
		North Side			
Three 2-L samples	76	30	95	135	22
	(16.0)	(10.0)	(15)	(17)	(4.5)
Three 5-L samples	34	18	104	85	6
	(7.0)	(5.0)	(15)	(8)	(5.5)
Three 20-L samples	29.8	11	68	72	13
	(4.1)	(2.0)	(9)	(9)	(3.0)
Three 30-L samples	31.4	15	72	78	15
	(3.8)	(2.4)	(11)	(9.3)	(2.8)

NOTE: The numbers in parentheses denote the standard deviation. The sampling stratum is the same as that in Table II.

carried out to characterize the chemical composition of the body of water properly.

In the case of sampling for monitoring purposes, the substances of interest have been identified, and their background values (e.g., mean, range, and variance) have been ascertained. The data are assumed to have been examined for normality (i.e., the data exhibit a normal distribution over time). In this case, the investigator may wish to specify the conditions of variability that will be tolerated. Thus, the estimated standard deviation may be set within a certain percentage of its true value at a particular confidence level. Or, the investigator may specify an accuracy of a sample mean by specifying the percent coefficient of variation that will be tolerated at a particular confidence limit. Further discussion on methods used to estimate the optimum amount of replication to characterize such water may be found in Berg's book (*16*). The volume of the sample in monitoring regimes is partly determined by the substance being monitored as well as its expected concentration. The volume should always be of sufficient size to avoid the north–south side of the boat dilemma.

Frequency of Sampling

When engaging in descriptive sampling, the frequency of sampling will be determined by the purpose of sampling. In the case where seasonal variation is of interest, weekly sampling will be necessary. To illustrate the extent of variation, Table IV depicts variation in some elements in Linsley Pond over a 1-year period of study (*17*). Explanations for these oscillations have been described in detail elsewhere (*17, 18*); however, these variations represent the period of human disturbance, the chemical composition of the ice, the chemical composition of the density current, the thawing of the ice, the periods of vegetative decay, the periods of vegetative growth, and the algal bloom.

When the purpose is monitoring for some group of substances, the frequency of sampling is often specified by permits or other types of regulatory action. In the absence of regulatory criteria, sampling frequency may be based on historical data illustrating the variation of some variable of interest through some specified period of time. The rule of thumb is that the length of the record should be at least 10 times as long as the longest period of interest. Thus, if the longest period of interest is 1 year, then 10 years of data would be required. Berg (*16*) provided a detailed discussion of this approach.

Types of Blanks

The types of blanks required for proper quality assurance may be described as equipment blanks, field blanks, and sampling blanks. *Equipment blanks* are obtained as follows: Prior to departure from the laboratory, all equipment should be soaked in the best grade of double-distilled (glass) water available. This soaking water should then be stored in appropriate glass vessels for analysis. *Field blanks* are obtained as follows: Prior to the collection of a sample, all equipment used to collect that sample and that will have contact with that sample must be soaked in the best grade of double-distilled (glass) water available. This water should then be stored in appropriate glass vessels for analysis. *Sampling blanks* are obtained as follows: During the period that a sample is being gathered, bottles containing the best grade of distilled water available should be exposed to the air. This exposure should extend from the beginning to the end of sampling. At the termination of sampling, the bottles containing blanks should be sealed, labeled, and placed in 4 °C chests for eventual analysis.

These three types of blanks may become of paramount importance in the event that erratic results are obtained from sample analysis. These blanks, on analysis, may identify unsuspected contaminants associated with distilled water purity, improper dishwashing procedures, contamination problems associated with travel, and latent air contaminants that may have been sorbed by the samples during collection. This last problem has been encountered most frequently when groundwater samples were involved. The sampling blank was found to contain substances that in the past had been absent from groundwater samples.

Table IV. Striking Weekly Changes in Some Elemental Concentrations in Various Strata in Linsley Pond Observed during a 53-Week Study

Date	*Strata* (*m*)	*Ti* (*kg*)	*Fe* (*kg*)	*Mg* (*kg*)	*Sr* (*kg*)	*Pb* (*kg*)	*Br* (*kg*)	*Mn* (*kg*)	*Hg* (*g*)
Sept. 14, 1965	0–14.5	—	—	5400 (100)	29 (2)	5.8 (1)	—	—	—
Sept. 20, 1965	0–14.5	—	—	8300 (100)	49 (5)	10.5 (2)	—	—	—
Nov. 1, 1966	8–11	—	—	—	—	—	—	580 (11)	—
Nov. 8, 1966	8–11	—	—	—	—	—	—	40 (4)	—
Feb. 6, 1966	0–14.5	—	—	—	—	—	—	—	200 (15)
Feb. 13, 1966	0–14.5	3 (0.2)	180 (10)	—	—	—	300 (5)	—	900 (90)
Feb. 20, 1966	0–14.5	26 (2)	1100 (50)	—	—	—	750 (10)	—	—
June 26, 1966	0–2.5	—	—	—	—	—	—	—	198 (10)
July 3, 1966	0–2.5	—	—	—	—	—	—	—	42 (5)

NOTE: The quantities in each stratum represent the total amount found in that layer. The numbers in parentheses denote the standard deviation.

Nature of Samples and Problems Associated with Sample Collection

This section will be devoted to problems associated either with the type of water body being studied (e.g., lakes, rivers, streams, oceans, or aquifers) or the physical state of the sample (e.g., snow, ice, rain, fog, or dew).

In many water bodies, the stratum from which the sample is gathered is of utmost importance from the standpoint of chemical composition. Data presented in Table V illustrate this point.

In lakes shallower than 5 m, sufficient wind action exists so that mixing can be assumed to occur. Therefore, neither chemical nor thermal stratification is likely to be encountered. In lakes deeper than 5 m, chemical stratification as well as thermal stratification may be found. The examples used in Table V for Linsley Pond represent the height of the density current (February 20, 1966), the spring homothermal period (March 13 and 20, 1966) and the period of eutrophication (anoxic mud surface) during the summer months (July and August 1966). Cedar Lake, CT, is not known to stratify. Thus, when sampling such water bodies, something should be known about the morphometry and morphology of lakes, reservoirs, and estuaries prior to developing any kind of sampling regimen.

Rapidly flowing rivers that are shallow fail to exhibit any sort of consistent, recognizable chemical stratification. The example presented for this type of river is the Jordan River, which was sampled in the summer of 1963 in the northern Jordan Valley of Israel (*6*). A rapidly flowing river of considerable depth may undergo chemical stratification unaccompanied by any pronounced thermal stratification. The Ohio River at Pittsburgh, PA, was examined over an 8-year period (Cowgill, U. M., University of Pittsburgh, unpublished data, 1970–1978).

The data presented in Table V show that the type of monitoring to be carried out in such water bodies should determine the stratum from which a sample is to be collected, and that the position should be strictly adhered to for the entire period of sampling.

Similar types of problems have been encountered in groundwater sampling, especially when much of the well casing has been screened. Data presented in Table VI were gathered sporadically over a period of 14 years from a 33-m-deep well that had 5 m of the casing screened. The original purpose of sampling this abandoned but not closed drinking water well was to see if measuring any chemical interchange between the groundwater and nearby lakes was possible. Thus, prior to all sampling, a minimum of 10 bore volumes was removed. Each bore volume was sampled for temperature, pH, and conductivity. When these three variables approached stability after freshwater had entered the well casing as noted by a sudden drop in temperature, sampling at various depths within the screened area was carried out. The closed bailer (*19*) used in this study was 2 m long.

Table V. Variation in Chemical Composition in Relation to the Sampling Position in Various Water Bodies

Date	*Strata* (m)	*Mn* (μg/L)	*Fe* (μg/L)	*P* (μg/L)	*Ca* (mg/L)	*Ti* (μg/L)	*K* (mg/L)	*Na* (mg/L)
			Linsley Pond, CT (stratified)					
Feb. 20, 1966	0–2.5	300	400	38	27	—	—	—
		(35)	(38)	(4)	(3)	—	—	—
	8–11.0	1000	5000	75	30	—	—	—
		(83)	(450)	(8)	(3)	—	—	—
Mean for March 13, 1966	0–2.5	250	480	59	28	—	—	—
and March 20, 1966		(24)	(50)	(6)	(3)	—	—	—
	8–11	2000	900	64	29	—	—	—
		(100)	(83)	(6)	(3)	—	—	—
Mean for July	0–2.5	400	300	55	31.3	—	—	—
and August		(43)	(35)	(6)	(4)	—	—	—
	11–14	6000	2700	264	34.0	—	—	—
		(550)	(100)	(20)	(2)	—	—	—
			Cedar Lake, CT (unstratified)					
April 17, 1966	0–2.5	—	—	32	—	4.1	—	—
		—	—	(3)	—	(0.1)	—	—
	2.5–5.0	—	—	34	—	4.2	—	—
		—	—	(3)	—	(0.8)	—	—
July 17, 1966	0–2.5	—	—	15	—	2.2	—	—
		—	—	(2)	—	(0.4)	—	—
	2.5–5.0	—	—	16	—	2.1	—	—
		—	—	(2)	—	(0.5)	—	—
Aug. 7, 1966	0–2.5	—	—	16	—	2.0	—	—
		—	—	(2)	—	(0.3)	—	—
	2.5–5.0	—	—	15	—	1.9	—	—
		—	—	(2)	—	(0.2)	—	—

Continued on next page

Table V.—Continued

Date	Strata (m)	Mn (μg/L)	Fe (μg/L)	P (μg/L)	Ca (mg/L)	Ti (μg/L)	K (mg/L)	Na (mg/L)
			Jordan River, Israel					
June–August, 1963	0–0.5	—	934	—	51.3	—	0.82	3.5
(weekly)		—	(88)	—	(5)	—	(0.1)	(0.5)
	2–2.5	—	950	—	52.0	—	0.88	3.8
		—	(85)	—	(5)	—	(0.1)	(0.5)
			Ohio River, Pittsburgh, PA					
1970–1978	0–0.5	—	1.2[a]	—	55	—	2.1	6.4
(biweekly)		—	(0.2)[a]	—	(6)	—	(0.1)	(1.0)
	8–10	—	4.3[a]	—	80	—	4.2	10.5
		—	(0.4)[a]	—	(7)	—	(0.3)	(1.5)

NOTE: The numbers in parentheses denote the standard deviation.
[a]These values for Fe are expressed in milligrams per liter.
SOURCE: Adapted from references 6, 17, and 18, and U. M. Cowgill, unpublished data, 1970–1978.

Table VI. Variation in Chemical Composition in Relation to Sampling Strata in a 33-m Drinking Water Well Containing a 15-m Screen

Strata (*m*)	*Mo* ($\times 10^{-2}$ $\mu g/L$)	*V* ($\times 10^{-4}$ $\mu g/L$)	*Bi* ($\times 10^{-2}$ $\mu g/L$)	*Hg* ($\mu g/L$)	*Be* ($\times 10^{-2}$ $\mu g/L$)
19–21	15 (2.5)	0.8 (0.2)	8.5 (1.7)	1.6 (0.4)	0.7 (0.2)
24–26	10 (2.0)	30 (4.0)	4.5 (1.0)	18.3 (1.7)	18 (3.0)
29–31	28 (3.5)	1115 (97.3)	18.2 (2.0)	3.0 (0.5)	11 (1.0)

NOTE: The numbers in parentheses denote the standard deviation.

Most of the variability is reasonably consistent within the depth sampled (i.e., no percent coefficient of variation is greater than 28.5), yet the variation between depths is statistically significant in most cases. The cause of this variation is natural. The substrata vary chemically. Careful sampling, which avoided any excessive disturbance of the recharged water column, showed consistent variation over time and was later confirmed to originate from the substrata by sampling the various substrata and leaching them with natural water.

Position of sample collection is also important in the collection of ice and snow. During the 53 weeks of the Linsley Pond Study, the pond was ice-covered from January 16 through March 6. During that time, the thickness of the ice varied from 5 to 25 cm. In the open water, the variation in chemical composition was statistically significant in the vertical direction and varied little within an ice stratum in the horizontal direction. To obtain enough material to make chemical analysis precise enough, ice was gathered weekly so that on melting, the ice provided a volume of 250–300 L of water. Table VII shows the variation in chemical composition of ice in relation to depth of ice. It should be noted that there is a poverty of data on the chemical composition of ice from lakes and ponds. Maksimovich and Yashchenko (*20*) noted that the mineral content of ice in lakes and ponds varied between 22 and 90 mg/L. The sources of the chemical composition of ice are the chemical composition of the surface water from which ice forms; entrapped dust that contributes to the concentration of Fe, Ti, and Mo (*17, 18*); entrapped plankton; and the rate at which formation occurs.

The elements Ca, Si, Al, P, Ba, Sr, and Mn are concentrated in the ice in reference to the surface centimeter of water, but this concentration–depth relationship is not the case for Fe and Ti. How ice is to be sampled is largely dependent upon the purpose of sampling. If the concern is the geochemical effect of melting on the receiving water, then adequate results can be obtained by bulk sampling in a number of localities on the lake. If the interest is how the composition of the surface water is related to that of the ice in contact with it,

Table VII. Variation in Chemical Composition of Ice in Relation to the Thickness of the Ice, and Chemical Composition of Water in Relation to Depth of Water

Thickness or Depth (cm)	*Ca*	*Si*	*Al*	*Fe*	*P*	*Ba*	*Sr*	*Mn*	*Ti*
				Ice Thickness					
0–5	29,000 (1500)	30,000 (3000)	800 (80)	600 (55)	38 (4)	67 (6)	62 (6)	200 (20)	18 (2)
5–10	27,500 (2000)	33,000 (3000)	600 (65)	500 (60)	42 (4)	46 (6)	59 (6)	150 (15)	22 (2)
10–15	28,000 (2000)	32,000 (3000)	800 (80)	550 (60)	50 (5)	67 (6)	60 (6)	180 (20)	18 (2)
15–20	28,000 (1000)	25,000 (1500)	500 (55)	430 (40)	40 (4)	65 (6)	60 (6)	100 (10)	12 (1)
				Water Depth					
0–1	1800 (150)	4800 (300)	180 (15)	600 (60)	9 (1)	8 (1)	3 (0.5)	32 (3)	18 (2)

NOTE: All values are in micrograms per liter. Numbers in parentheses denote the standard deviation.

then the ice in a series of strata should be sampled and the ice samples' composition compared with that of the surface water.

The increase of Ca, Si, Al, Fe, Ba, Sr, Mn, and Ti in the surface ice is in part related to entrapped atmospheric dust. The chemical composition of ice varies with the depth of the ice more than with the horizontal component, may vary significantly from day to day, and varies significantly on a weekly basis (*17, 18*) during the period of ice cover.

The chemical composition of snow is highly variable. In severe winters, snow can contribute substantially to the elemental composition of a water body. Ideally, snow should be sampled at the time of snowfall because its composition changes considerably on standing. However, if snow is sampled immediately upon deposition, conditions tend to reflect only various chemical changes in the atmosphere. The chemical composition during a snowfall changes in much the same way as during rainfall. This situation will be discussed later in this section. If snow hasn't occurred for some time, the initial snowfall tends to be, from a chemical viewpoint, quite concentrated. As the snowfall progresses, the composition is diluted until, by the end of a long snowfall, the composition becomes a rather poor grade of distilled water containing only a few elements (Cl, Br, I, S, B, Na, K, Ca, Mg, and N) that can be detected with ease. Thus, if sampling is done shortly after a 0.5-m snowfall, a distinct chemical variation will be encountered between the snow that fell initially and the snow that recently fell. This discussion refers to snowfalls within 100 km of industrialized areas. High-altitude snowfalls are chemically much more dilute. This summary on the chemical (elemental) composition of snow is the result of sporadic investigations that have occurred in lake basins at various altitudes in diverse places (Cowgill, U. M., unpublished data, 1963–1981).

The chemical composition of rain is highly variable and the comments just made about snow also apply. Table VIII shows the chemical results of a 3-h rainfall collected in March 1978 on the roof of a 12-story building in Pittsburgh, PA. The rain was collected in polytetrafluoroethylene (Teflon) rain collectors that were covered with very fine nylon netting to prevent the entrance of soot in the samples. Collections were made at 10-min intervals. The most pertinent data are shown in the table.

These data show that the rain has been reasonably successful in cleaning the atmosphere of its contaminants with the exception of sulfate and total nitrogen. Pittsburgh is a limestone region, and certainly some of the Ca and Mg in rain originates from limestone dust, but the remainder of the contribution arises from the local cement industry. The major point is that reports of the chemical composition of a particular rainfall are meaningless if the aliquot of the rainfall analyzed is not timed and dated. As may be observed from examination of the data in Table VIII, the composition varies with time. Reporting the average composition of a rainfall is meaningless because 70 min of rain having a pH of 2.4 will damage vegetation and buildings. The fact that the final pH approached that of normal rain would not be useful if the purpose

Table VIII. Variation in Chemical Composition of a 3-h Rainstorm in Pittsburgh, PA, in March 1978, and Variation in Chemical Composition of Uncontaminated Rain

Time (min)	*pH*	*Cl*	SO_4	*B*	*Na*	*K*	*Mg*	*Ca*	*N*
				Pittsburgh Rainstorm					
0–10	1.2	10.1	29.3	0.18	5.8	0.61	2.75	3.05	4.7
		(2)	(4)	(0.03)	(1)	(0.1)	(0.5)	(0.6)	(1)
10–20	1.5	8.8	25.8	0.15	5.2	0.54	2.40	2.68	4.2
		(2)	(4)	(0.03)	(1)	(0.1)	(0.5)	(0.5)	(1)
50–70	2.4	6.5	18.8	0.12	3.7	0.39	1.77	1.90	3.0
		(1)	(4)	(0.02)	(1)	(0.1)	(0.5)	(0.5)	(0.5)
120–150	3.9	4.6	12.0	0.08	3.1	0.28	1.00	1.40	1.8
		(1)	(3)	(0.01)	(0.05)	(0.1)	(0.2)	(0.5)	(0.5)
170–180	5.2	4.2	7.6	0.08	2.4	0.25	0.38	0.25	0.86
		(1)	(2)	(0.01)	(0.05)	(0.1)	(0.1)	(0.1)	(0.1)
				Uncontaminated Rain					
	—	5.1	1.7	—	3.1	0.2	0.3	0.2	<0.009

NOTE: All values are in milligrams per liter. The numbers in parentheses denote the standard deviation.
SOURCE: Adapted from reference 21 for uncontaminated rain data only.

of collection is to monitor and avoid damage to living things and structures of civilization. Further discussion on adequate precipitation sampling can be found elsewhere (22–24).

Great interest has been expressed recently on the chemical composition of fog and dew. With the possible exception of the first monolayer of seawater, no physical state of water is more difficult to sample than fog and dew. Dew was collected with a kind of glass vacuum pump from three arbitrarily selected strata of a grass field. A total of 10 acres was collected on an early summer morning (Cowgill, U. M., Yale University, unpublished data, 1968). These 10 acres provided 1 L of dew from each of the three strata. Three replicates of each strata were analyzed. Table IX illustrates some of the problems encountered with dew collection and analyses. The samples had to be concentrated over P_2O_5 to detect K, Cl, S, and N with ease. The biggest single problem is that reproducibility is rather poor in the sense that the percent coefficient of variation is much greater

Table IX. Chemical Composition of Three Strata of Dew Extracted from a Grass Field

Date	*Strata*	*K*	*Cl*	*SO_4*	*N*
July 15, 1968	top	40	100	1200	30
		(20)	(50)	(500)	(16)
	middle	37	575	985	39
		(20)	(300)	(510)	(17)
	bottom	79	1000	1962	56
		(35)	(500)	(990)	(30)
July 25, 1968	top	65	975	1500	75
		(30)	(500)	(700)	(40)
	middle	69	1000	1783	95
		(31)	(500)	(800)	(48)
	bottom	54	1400	2000	98
		(25)	(700)	(990)	(50)
Aug. 15, 1968	top	74	1000	1700	85
		(35)	(500)	(810)	(42)
	middle	55	1000	1400	85
		(25)	(500)	(750)	(42)
	bottom	100	930	975	78
		(42)	(410)	(500)	(40)
Aug. 29, 1968	top	82	1500	1565	85
		(40)	(750)	(700)	(42)
	middle	80	1600	1511	90
		(40)	(800)	(711)	(42)
	bottom	55	755	800	44
		(22)	(400)	(410)	(20)

NOTE: The grass field consisted of *Phleum pratense L.* (Timothy) and *Bromus inermus L.* (Bromegrass). All values are in micrograms per liter. The numbers in parentheses denote the standard deviation.

than would normally be acceptable. The percent coefficient of variation did not improve noticeably with time.

Some variation among the strata can be seen, although the bottom strata are probably influenced by some upward movement from the soil to the dew. These samples were collected at the same time in the morning (5:00 a.m.). However, a great deal more data need to be collected before any conclusions can be reached concerning the chemical composition of dew. The Cl, S, and N probably originated from the atmosphere, and the K originated from the grass. Since these data were gathered, various synthetic materials that absorb many times their weight have become available. The use of such material to collect data from a predetermined number of stations might provide a successful alternative to the vacuum cleaner approach to the sampling of dew.

A similar approach has been used to collect fog. Unfortunately, samples collected simultaneously 200 m apart provided a 200% spread on elements such as S, N, and Ca. Until duplicate samples can be collected simultaneously and upon analysis provide some reasonable degree of concordance, the chemical composition of fog will remain elusive.

Finally, the oceans present sampling problems not unlike those of stratified lakes. The precautions that have been noted under that heading should be adhered to in marine sampling. The chemical composition of near-shore waters is chemically far more variable than any composition thus far encountered in freshwater bodies. Some work (25, 26) is in progress to estimate atmospheric contributions to ocean waters. Further discussion may be found in Broecker and Peng (27).

Planning and Desired Confidence Level of Chemical Results

Before engaging in a research project in which the accuracy will be dependent upon proper sampling, the purpose of sampling must be clearly identified. Once this purpose has been identified, the items to be studied must be selected, and the concentration range expected for these items must be described. At this point, sampling tools that do not contain and will not sorb the items of interest must be procured. The degree of replication must be discussed not only in the light of the desired precision but also with the view of the confidence limits within which the desired data should fall. Frequency of sampling will largely be determined by the purpose of sampling. Finally, background data must be gathered from the water body to be studied. Once such data have been collected, more information will be available about the items of interest; their concentration range; their seasonality; and their mobility within the system to be studied, and an appropriate research effort can be described, with proper sampling becoming the cornerstone of the effort.

References

1. Keith, L. A. *Environ. Sci. Technol.* **1981**, *15*, 156–162.
2. Keith, L. A. *Advances in the Identification and Analysis of Organic Pollutants in Water*; Ann Arbor Press: Ann Arbor, MI, 1981; Vol. 1, pp 3–479.
3. Keith, L. A. *Advances in the Identification and Analysis of Organic Pollutants in Water*; Ann Arbor Press: Ann Arbor, MI, 1981; Vol. 2, pp 481–1170.
4. Kratochvil, B. G.; Taylor, J. K. "A Survey of the Recent Literature on Sampling for Chemical Analysis (Final Report)"; NBS Technical Note 1153; National Bureau of Standards: Washington, DC, 1982.
5. Kratochvil, B. G.; Wallace, D.; Taylor, J. K. *Anal. Chem.* **1984**, *56*, 113R–119R.
6. Cowgill, U. M. *Int. Rev. Gesamten Hydrobiol.* **1980**, *65*, 379–409.
7. Barcelona, M. J.; Helfrich, J. A.; Garske, E. E. *Anal. Chem.* **1985**, *57*, 460–464.
8. Barcelona, M. J.; Helfrich, J. A.; Garske, E. E. *Anal. Chem.* **1985**, *57*, 2752.
9. Miller, G. D. *Proc. Nat. Symp. Aquifer Restor. Ground Water Monit. 2nd* **1982**, 236.
10. Massee, R.; Maessen, F. J. M. J.; DeGoeij, J. J. M. *Anal. Chim. Acta* **1980**, *127*, 181-193.
11. Robertson, D. E. *Anal. Chim. Acta* **1968**, *42*, 533–536.
12. Robertson, D. E. *Anal. Chim. Acta* **1968**, *40*, 1067–1072.
13. Boettner, E. A.; Ball, G. L.; Hollingsworth, Z.; Aquino, R. *Organic and Organotin Compounds Leached from PVC and CPVC Pipe*; U.S. Environmental Protection Agency. U.S. Government Printing Office: Washington, DC, 1981; EPA-600/1-81-062.
14. McBean, E. A.; Rovers, F. A. *Ground Water Monit. Rev.* **1985**, *5*, 61–64.
15. Lind, O. T. *Handbook of Common Methods in Limnology*; C. V. Mosby: St. Louis, MO, 1974; pp 26–31.
16. Berg, E. L. *Handbook for Sampling and Sample Preservation of Water and Wastewater*; U.S. Environmental Protection Agency. U.S. Government Printing Office: Washington, DC, 1982; EPA-600/4-82-029.
17. Cowgill, U. M. *Arch. Hydrobiol.* **1970**, *68*, 1–95.
18. Cowgill, U. M. *Arch. Hydrobiol.* **1976**, *78*, 279–309.
19. Gibb, J. P.; Schuller, R. M.; Griffin, R. A. *Proceedings of the Sixth Annual Symposium on the Disposal of Hazardous Wastes, Chicago, Illinois, March 17-20, 1980*; U.S. Environmental Protection Agency: Cincinnati, OH, 1980; pp 31–38; EPA-600/9-80-010.
20. Maksimovich, G. A.; Yashchenko, R. V. *Zh. Khim. Kii* **1963**, Abstract 21 E 51.
21. Gorham, E. *Geochim. Cosmochim. Acta* **1955**, 7, 231–239.
22. Asman, W. A. H.; Jonker, P. J. In *Deposition of Atmospheric Pollutants*; Georgic, H. W.; Pankrath, J., Eds.; D. Reidel: Amsterdam, 1982; pp 115–123.
23. Campbell, S.; Scott, H. In *Quality Assurance for Environmental Measurements*; ASTM Special Technical Publication 867; Taylor, J. K.; Stanley, T. W., Eds.; American Society for Testing and Materials: Philadelphia, PA, 1985; pp 272–283.
24. Raynor, G. S.; Hayes, J. V. In *Sampling and Analysis of Rain*; ASTM Special Technical Publication 823; Campbell, S. A., Ed.; American Society for Testing and Materials: Philadelphia, PA, 1983; pp 50–60.
25. Green, D. R. *National Research Council of Canada Publication No. 16565*; Marine Analytical Chemistry Standards Program; National Research Council: Halifax, Canada, 1978.
26. Duce, R. A. *Special Environmental Report 12*; WMO No. 504; World Meteorological Organization: Geneva, Switzerland, 1979.
27. Broecker, W. S.; Peng, T-H. *Tracers in the Sea*; Lamont–Doherty Geological Observatory. Columbia University: New York, 1982. (For Atlases containing chemical data [GEOSECS], see pp 670–672.)

Chapter 19

Fundamentals and Considerations for Field Filtration of Environmental Water Samples

M. P. Floyd

Environmental water samples are often filtered in the field to remove suspended mineral and biotic matter from the sample. The purpose of filtration is to allow water to be chemically analyzed in the absence of suspended matter or to allow the filtered matter itself to be analyzed. Water sample filtration tends to be the least understood part of field sampling efforts and often receives the least amount of attention for quality assurance–quality control practices. Filtration media used for environmental water sampling can be categorized as depth filters, screen filters, and membrane filters. There is a wide variety of filters available in each category as well as a relatively large selection of filtration equipment that can be used with the various filters. Filters and related equipment must be selected with care to ensure that sample representativeness will be maintained. Filter properties that should be carefully considered include particle retention capacity, filter rating (precision), hydraulic efficiency, microscopic properties, and chemical composition. Field protocols for water sample filtration should also be carefully considered to ensure sample representativeness. Consideration should be given to filter media and equipment preconditioning and cleaning requirements, proper handling and operation of filters and filtration equipment, and proper use of pressure or vacuum to cause the water sample to pass through the filter. Filtration equipment, filter media, and field protocols should be critically evaluated by testing to ensure that sample representativeness will be maintained.

SAMPLES OF SURFACE WATER, GROUNDWATER, AND WASTEWATER often require special preparation in the field prior to submittal to the laboratory. Sample preparation frequently consists of the addition of preservative or extractive chemicals to the sample and cooling of the sample to ensure its stability. Sample preparation

3152–4/96/0337$15.75/0

may also include filtration. Sample filtration is usually performed in the field as soon as possible after sample collection to allow immediate preservation of filtered water samples or filtered material.

Solid matter is sometimes filtered from water samples to allow analysis of the "dissolved" fraction of the water sample. Solid matter is also filtered from a sample where it would otherwise foul instrumentation or hinder laboratory analysis of the water sample.

Solid matter is sometimes removed from a water sample by filtration to allow analysis of the solid matter itself by

- on-filter analysis of captured matter by X-ray diffraction, X-ray fluorescence, gravimetry, microscopy, or other means;
- on-filter culturing of captured organisms followed by microscopic examination; and
- off-filter analysis of captured matter after destruction or digestion of the filter, or after removal of matter from the filter by elutriation or other means.

Water sample filtration is one task of many undertaken in the field. However, it tends to be the least understood part of the sampling effort. Filtration often receives the least amount of attention for quality assurance–quality control practices.

This chapter provides a basic understanding of the general types of filters and filtration systems available and considerations necessary for their use. Detailed information and specifications for individual filtration applications are not provided because

- the array of filtration products available is very large;
- filtration applications can vary between projects for the same analyte based on data quality specifications; and
- regulatory requirements for water sample filtration vary with time and between jurisdictions.

The goal of this chapter is to promote more understanding of water sample filtration and to prompt the filter user to critically evaluate each filtration application.

Filtration Media

Various texts provide a significant amount of information on the nature and function of sample filtration systems. Brock (*1*) provides an extensive discussion of the types of filters and filtration equipment available and their use for a wide variety of applications. Information presented in this section is a brief summary of that available in the literature.

Filters used for environmental water sampling can be considered to fall into three basic categories. A filter can be described as a depth filter, screen filter, or membrane filter.

Depth Filters

Depth filters usually consist of a mat of fibrous matter, although some are constructed of arrays of fused spheres. Depth filters used for water sample filtration are usually mats of fibers up to about 0.1 in. thick. Most depth filters are designed to capture particles throughout their entire thickness.

Materials that make up depth filters used for water sample filtration vary widely and commonly include fiberglass, cellulose, and plastics. Less common materials include ceramics, hemp, silk, fur, jute, and asbestos. Fibers in depth filters can be compressed or felted, woven, glued, or fused together. A photomicrograph of a fiberglass depth filter is shown in Figure 1.

Particles are captured by a depth filter when they become entangled in the filter's labyrinth of fibers or elements, as shown in Figure 2. Particles are separated from the water sample when they encounter a passage that is smaller in size than the particle, or when the shape of the passage is not compatible with the orientation and geometry of the particle. Particle capture can occur because of inertial effects as individual flow paths of the water sample passing through the filter turn past fibers. Particle capture can also occur because of electrostatic or chemical interaction of particles with the filter.

Depth filters are often employed for filtration of invertebrate aquatic organisms and phytoplankton for examination or analysis. Depth filters are also used for the removal of suspended mineral and plant matter from water samples that are to undergo chemical analysis. Some depth filters are capable of removing submicrometer-size particles from water.

Screen Filters

Screen filters used for environmental water samples consist of a mesh of fibers or strands woven or cast in a regular pattern to create uniform size openings. Screen filters are sometimes referred to as net filters and may be constructed of metal or plastic. Screen filters capture most particles on their surface and are usually used for straining out invertebrate aquatic organisms larger than about 50 μm.

Membrane Filters

Membrane filters usually consist of a relatively thin sheet or film that is porous or is perforated to allow fluids to pass and retain particles above a specific size on their upper surface, as shown in Figure 3. Membrane filters are usually thin in comparison to depth filters and are often less than 0.010 in. thick.

Figure 1. Photomicrograph of the surface of a fiberglass depth filter at a magnification of 3000×. (Courtesy of Millipore Corporation.)

Membrane filters are typically designed to capture almost all particles above a specific size on the filter's surface. Depth filters contrastingly are typically designed to capture relatively large quantities of particles through their entire thickness. Membrane filters are sometimes referred to as "surface filters" because of their ability to capture most particles on their upper surfaces. Particle capture by membrane filters can occur by electrostatic or chemical interaction with the filter material, as with depth filters.

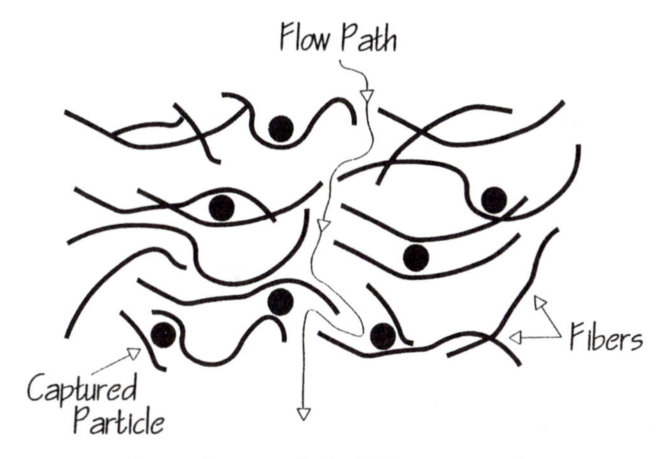

Figure 2. Diagram of a depth filter in cross section.

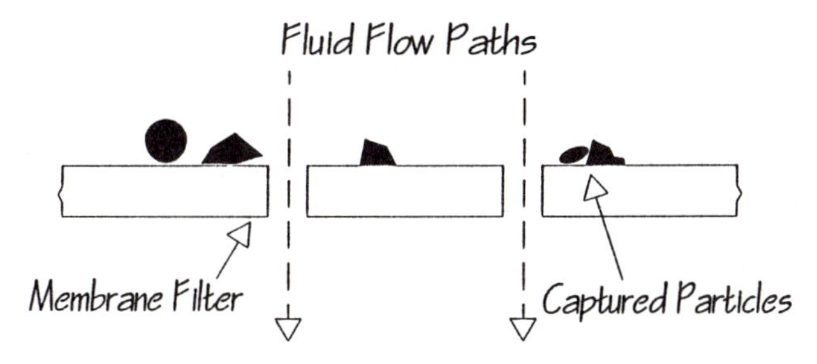

Figure 3. Diagram of a membrane filter in cross section.

The design, manufacture, and use of membrane filters is a complex subject with entire texts devoted to the discussion. Membrane filters used for environmental water sampling are usually employed for filtration of suspended mineral and plant matter, and microorganisms down to submicrometer sizes.

Membrane filters are commonly made from polycarbonate, polyester, cellulose acetate, cellulose nitrate, and polytetrafluoroethylene (PTFE). Other materials include poly(vinyl chloride), metal, regenerated cellulose, nylon, polypropylene, and acrylonitrile. Some membrane filters are a composite of materials to provide optimal filtration and strength properties.

Pores or perforations in membrane filters are formed by various processes including chemical casting, stretching, sintering, and etching. Cylindrical holes are created in some membrane filters by nuclear particle bombardment followed by etching. Membrane filters with perforations created by nuclear bombardment and etching are sometimes referred to as track-etched filters. A photomicrograph of the surface of a mixed cellulose–ester membrane filter is shown in Figure 4. A photomi-

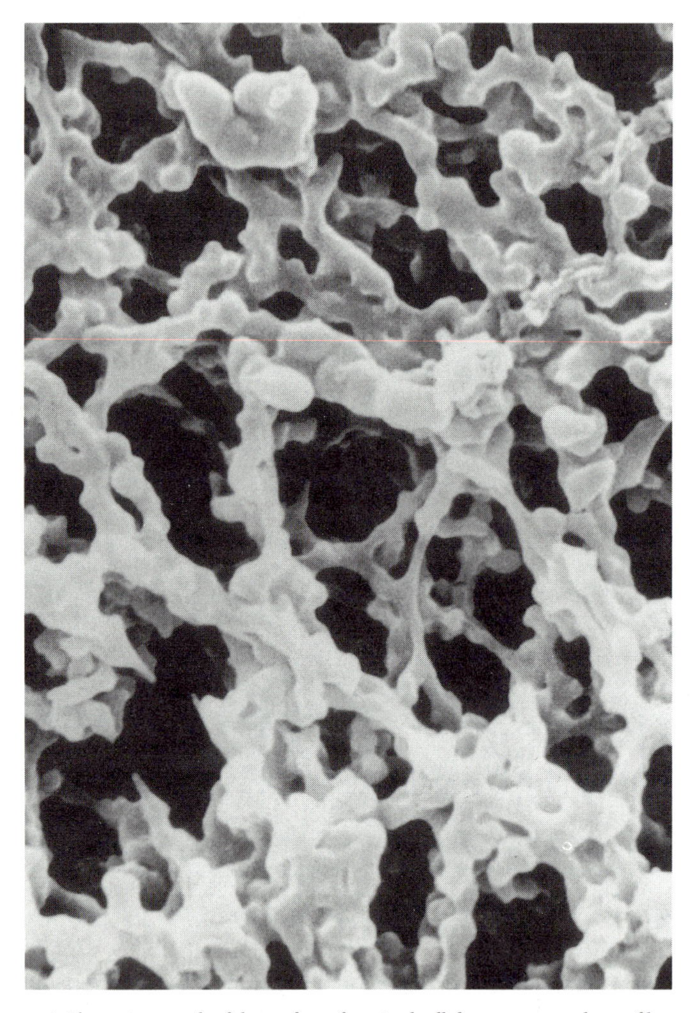

Figure 4. Photomicrograph of the surface of a mixed cellulose–ester membrane filter at a magnification of 10,000×. (Courtesy of Millipore Corporation.)

crograph of the surface of a track-etched polycarbonate membrane filter is shown in Figure 5.

Comparison of Filtration Media

No ideal filter exists for all water sample filtration applications. Each type of filter can have distinct advantages and disadvantages for individual applications.

Screen filters are generally used for filtration of relatively large particles and organisms in comparison to depth and membrane filters. Screen filters are some-

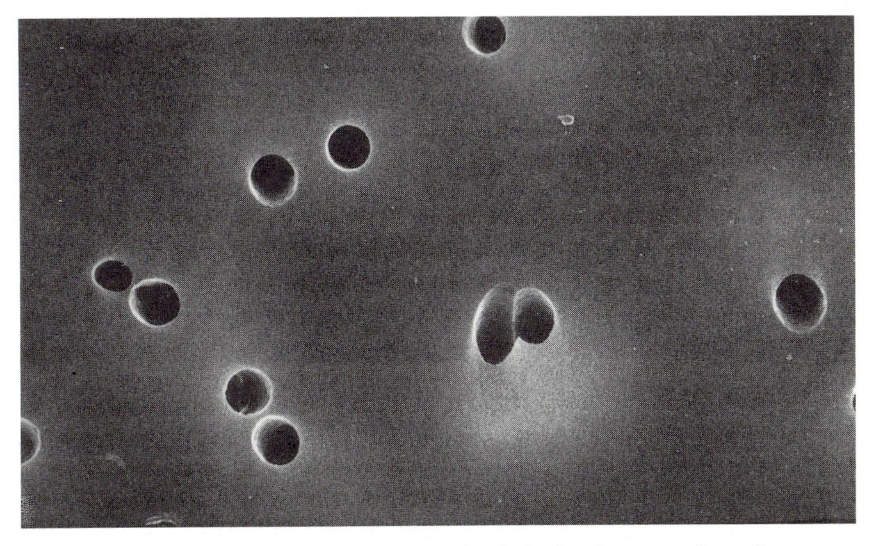

Figure 5. Photomicrograph of the surface of a track-etched polycarbonate membrane filter at a magnification of 1880×. (Courtesy of Poretics Company.)

times used for removing large particles from a water sample before the sample is passed through a depth or membrane filter. This preliminary level of filtration is sometimes referred to as *prefiltration*.

Depth and membrane filters are used widely and sometimes interchangeably for environmental water sampling. Differences in the performance of depth and membrane filters are summarized in this section.

Particle Retention Capacity Prior to Clogging

Depth filters are generally regarded as having a higher particle retention capacity prior to clogging than membrane filters with an equivalent minimum retention size. A depth filter can capture particles throughout its entire thickness. A membrane filter is usually designed to retain most particles on its surface and thus can often clog more quickly.

The time that a filter can adequately function before becoming clogged is also partially dependent on the density of perforations or pores open to flow. The cross-sectional area of some membrane filters that is open to flow can be small compared to depth filters with an equivalent minimum retention size. Some track-etched membrane filters have open areas of only a few percent of the total cross-sectional area. However, the open area of other membrane filters can exceed 50% and is comparable to depth filters.

The ability of some depth filters to handle greater particle loads prior to clogging can be an important attribute in the field when turbid or large-volume water samples are filtered. Filter clogging can require filter replacement during the filtration of an individual sample. Filter replacement can increase labor costs and increase the consumption of filters. Filter replacement can also increase the chance of sample contamination and particulate bypass, depending on the type of filter and filtration equipment used. Depth filters are sometimes used for prefiltration of water samples to be passed through a membrane filter to reduce the potential for clogging of the membrane filter.

Filter Rating

Filter rating (or filter precision) is defined for this discussion as the ability of a filter to retain particles larger than a specified size, and allow pass-through of smaller particles. The ability of a filter to retain particles above a specified size is usually one of the most carefully considered properties of a filter. Analyses such as bacteriologic assays can require complete retention of microorganisms above a specific size to ensure that representative counts of the target organism can be achieved. The retention of particles smaller than the specified minimum retention size is sometimes ignored.

Some depth filters may not completely retain all particles above a specified minimum retention size, depending on the individual filter. Some particles larger than the minimum retention size may pass through a depth filter depending on the position and orientation of individual fibers or fused spheres in the particle-flow path. Some depth filter manufacturers report that certain depth filters are capable of the complete retention of particles exceeding specified minimum retention sizes.

Some depth filters can potentially retain a significant number of particles smaller than the specified minimum retention size. Retention of smaller particles can occur where fibers or fused spheres that comprise the depth filter intersect one another and create small pockets or flow path openings that can trap smaller particles. Retention of small particles can also occur because of the tortuosity of the flow path through the filter, and related inertial effects.

The capture of particles smaller than the specified minimum retention size by a filter does not present a problem for many applications. However, the retention of smaller particles can be of concern when water samples undergo filtration for dissolved fraction analyses.

Some discussion is given in the literature to the potential importance of colloidal transport of trace elements, macromolecules, hydrophobic organic compounds, radionuclides, and other matter in aqueous systems, especially groundwater (2–4). Deciding on the minimum retention size for filtering groundwater samples for dissolved fraction analysis remains controversial. There is even controversy as to whether groundwater samples should be filtered at all because of the possible capture of colloidal material (5). Low-flow groundwater sampling techniques may provide an alternative to sample filtration (6). Analysis of both filtered and unfiltered fractions of a water sample are performed in some cases because of uncertainties about filtration.

Some membrane filters are capable of providing a relatively "precise" level of filtration for environmental water sampling. Track-etched membrane filters are often considered to be capable of the highest level of filtering precision, both in terms of retaining particles larger in size than the specified minimum retention size, and allowing pass-through of particles smaller than the minimum retention size.

Membrane filters such as cellulose acetate, cellulose nitrate, silver, and PTFE may, in some circumstances, provide almost complete retention of particles larger than a specified minimum retention size. However, there sometimes can be a potential for retention of a significant amount of particles smaller than the minimum retention size, depending on the individual membrane filter used. Smaller particles may be retained because of the "reticulated" structure of some membrane filters and the presence of a significant number of openings that are smaller than the specified minimum retention size.

Pass-through of 100% of particles in a water sample smaller than the specified minimum filtration size may not always be possible, regardless of the type of filter used, because of possible chemical and electrostatic interaction between the filter material and particles, hydraulic and inertial effects on the surface and within the filter, and secondary filtration. Secondary filtration occurs when particles larger than the minimum retention size become trapped by the filter and create flow restrictions that can retain smaller particles. The potential for secondary filtration becomes greater when large-volume, turbid water samples are filtered and accumulations of particles occur in or on the filter. The potential impact of secondary filtration can be lessened by reducing the amount of water passed through an individual filter or by prefiltration.

Hydraulic Efficiency

The hydraulic efficiency of a filter can be considered in terms of the rate of flow that results when a specific pressure (or vacuum) differential is applied across the filter. Filters that have high hydraulic efficiency allow a relatively high rate of flow through the filter under a specific pressure differential. Filters with low hydraulic efficiency allow only a relatively low rate of flow under the same pressure differential.

Depth filters are sometimes more hydraulically efficient than membrane filters with an equivalent minimum retention size. The reason for this higher efficiency is that depth filters sometimes have a higher percentage of area open to flow. Higher efficiencies can mean that less vacuum or pressure is required to force a water sample through a filter. Use of low pressure or vacuum in some filtration systems can be an advantage when the following are concerns:

- pressure- or vacuum-related chemical stability of the sample,
- potential chemical interaction of the sample with the atmosphere or drive gas, and
- potential release of dissolved gases from the water sample.

Higher hydraulic efficiency can also help to reduce the time required for filtration. Some filters, especially depth filters, may consolidate or compact when relatively high differential pressures are applied to force a water sample through the filter. Filter compaction may adversely affect a filter's hydraulic efficiency as well as its particle retention properties.

Microscopic Examination

Depth filters may not always be suitable for filtration of water samples when retained particles are to undergo microscopic examination. Significant numbers of particles to be examined, such as asbestos, viruses, or bacteria, may pass into the body of some depth filters and become difficult or impossible to view under a microscope.

Some membrane filters, such as track-etch filters, provide a relatively smooth and uniform background for examination and particle counting under the microscope. Membrane filters with a reticulated structure and fibrous depth filters may provide a more complex background for microscopic examination and may make trapped particles more difficult to see.

Fiber Release

Some depth filters have the potential for release of filter fibers into the filtered sample. The release of fibers may provide unwanted interference or bias in the analysis of a filtered sample. The physical release of filter material into a water sample is generally not a concern with the use of membrane filters.

Filtration Equipment

Filtration equipment is defined for the purpose of this chapter as a device or combination of devices to

- hold the filter,
- supply pressure and/or vacuum to drive a water sample through the filter when the use of gravity alone is insufficient,
- contain the sample before and after filtration, and
- transfer pressure or convey the sample from one part of the filtration apparatus to the other.

Commercially available filtration equipment varies widely in design, configuration, and use. Some examples of field filtration equipment available for environmental water sampling are discussed in this section.

Vacuum Filtration

Vacuum filtration equipment typically consists of a manually or electrically powered vacuum pump; a borosilicate glass, metal, or plastic disk filter holder; and vessels or chambers to hold the water sample before and after filtration. Figure 6 is an example of a vacuum filtration apparatus.

The use of vacuum or subatmospheric pressure to cause water to move through a filter limits the pressure differential that can be imparted across the filter. Limiting the pressure differential can increase the time required for filtration, especially when filters with low hydraulic efficiency are used or large-volume turbid water samples are filtered. Pressure differential limitations may also result in more frequent filter clogging.

Vacuum filtration may not be appropriate when the application of vacuum can cause the release of dissolved gases from a water sample. The release of dissolved gases from a water sample can, in some cases, adversely affect the sample's chemical stability and representativeness.

Some types of vacuum filtration equipment have prefiltration chambers that are open to the atmosphere. Open filtration chambers may exacerbate potential problems with exposure of the sample to the atmosphere, including the adsorption and desorption of gases. Open prefiltration chambers can also potentially allow entry of contaminants into the sample, especially during windy field conditions.

Vacuum filtration of water samples can sometimes be accomplished within a self-contained disposable filtration system, such as that depicted in Figure 7. This

Figure 6. Example of a vacuum filtration apparatus: (a) prefiltration chamber; (b) disk filter holder; (c) postfiltration chamber; (d) electrical power vacuum pump.

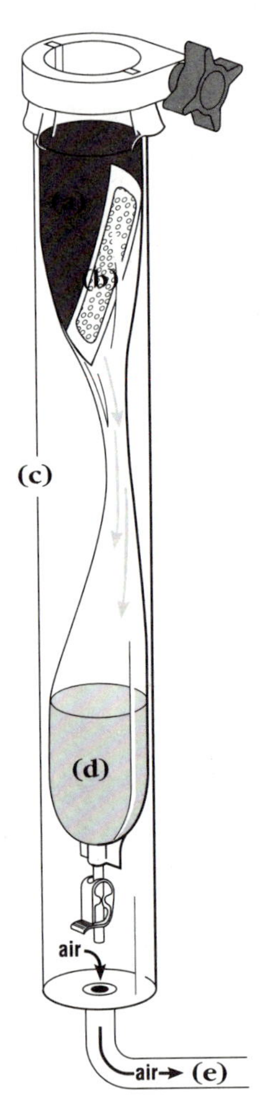

Figure 7. Diagram of a self-contained disposable filtration system: (a) prefiltration chamber; (b) filter; (c) support column; (d) postfiltration chamber; (e) applied vacuum. (Reproduced with permission from Gelman Sciences.)

system consists of a clear, flexible plastic bag partitioned by an integral filter into pre- and postfiltration chambers. The disposable filter bag is suspended in a clear plastic column for support. The filter operates when vacuum is applied to the lower portion of the column outside the bag.

A self-contained and disposable filtration system, such as the one-vacuum system described, can offer advantages for filtration of environmental water samples when chemical determinations for the filtered water sample are to be performed. One advantage is that all wetted components of this type of filtration system are normally disposable, thus eliminating the need for cleaning the filtration apparatus between samples. Another advantage is that the opportunity for contamination of the sample by dust or handling is somewhat reduced because filtration is performed within an enclosed system.

Pressure Filtration

Several types of pressure filtration systems are commercially available. Many pressure systems can also be operated using vacuum; however, operation by pressure is usually preferred by most environmental engineers and scientists.

Pressure filtration of water samples can be accomplished with the use of a disk filter holder and pump such as that shown in Figure 8. Pressure filtration can also be accomplished by using a cartridge filter in place of a disk filter. Cartridge filters are discussed in a following section.

Disk filters and their holders used for pressure filtration systems vary in composition and size. Stainless steel disk filter holders are typically used when sterilization of the filter holder is required for microbiologic sampling. Plastic filter holders are often used when filtered water samples are to be submitted for trace element or "metals" analyses. Clear plastic filter holders can allow visual examination of the filter during use to monitor the accumulation of solids, ensure proper position of the filter within the holder, and detect filter bypass or rupture.

Pressure necessary to drive a water sample through a filter in a pressure filtration system can be applied by an electrically powered pump. Electrically powered peristaltic pumps are often used with disk and cartridge filters. Peristaltic pumps require the use of flexible tubing within the pump head that may, in some cases, be chemically incompatible with the sample. The chemical compatibility of various materials used in environmental water sampling is discussed in Chapter 8 of this book.

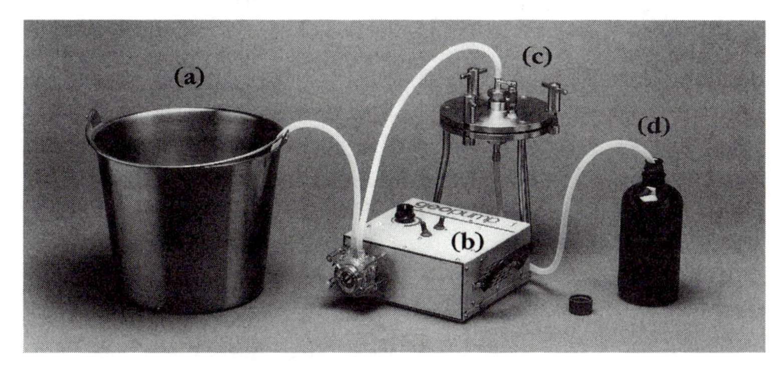

Figure 8. Example of a disk filter system with a peristaltic pump: (a) prefiltration container; (b) peristaltic pump; (c) disk filter holder; (d) postfiltration sample container.

Manual pressure pumps, syringes, or bottled air or specialty gases can be used to drive water samples through disk or cartridge filters. Figure 9 is a disk filtration system with a manually powered air pump.

Exposure of water samples to the atmosphere or some specialty gases during pressure filtration processes can chemically bias some samples. Exposure of samples to atmospheric oxygen during handling and filtration can adversely affect the representativeness of water samples collected from anoxic or oxygen-limited environments (7).

A disk filter holder or cartridge filter can sometimes be connected to the discharge line of a sampling pump for direct, in-line pressure filtration of a water sample. In-line filtration can minimize exposure of the sample to the atmosphere during filtration and reduce the potential for chemical changes and contamination.

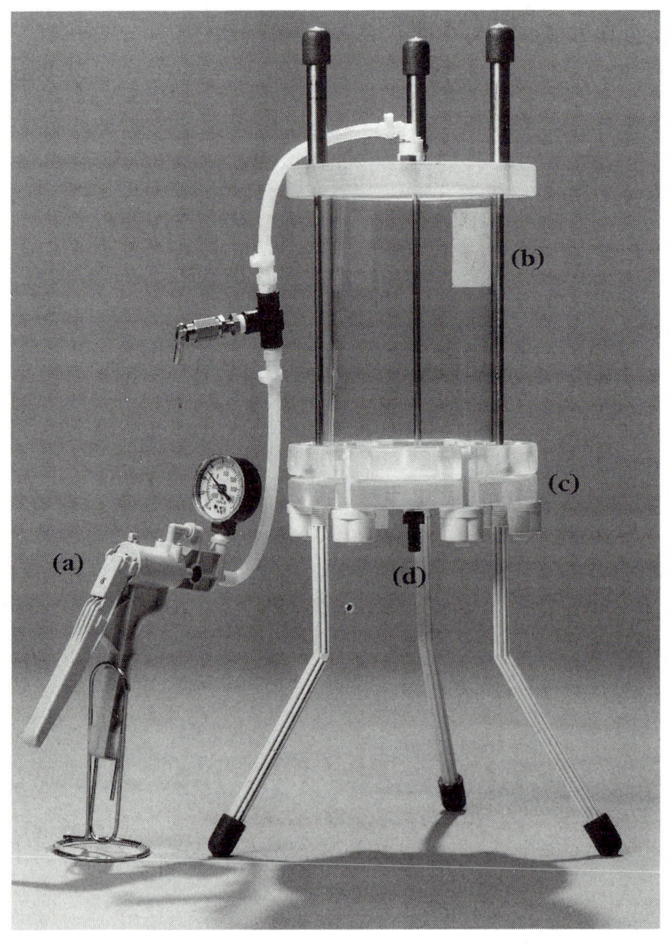

Figure 9. Example of a disk filtration system with a manually powered air pump: (a) manual air pump with gauge; (b) prefiltration chamber; (c) disk filter holder; (d) discharge point (postfiltration).

Cartridge Systems

Some membrane and depth filters are available in the form of disposable or reversible cartridges. A diagram of a disposable cartridge filter is shown in Figure 10.

Some cartridge filters can be used with either pressure or vacuum filtration equipment, although cartridge filters are most commonly used in the field for pressure filtration. A manual air pressure filter apparatus with a cartridge filter is shown in Figure 11.

Filters contained within cartridges are protected from potential damage and contamination. Some cartridge filters are factory-sterilized and packaged in sterile wrapping for microbiologic sampling.

Some cartridge filters contain pleated filter elements. Pleating greatly increases the area of the filter that can be contained within a cartridge. Increased filter area is an important consideration when turbid and large-volume water samples are filtered.

Cartridge filters can be used for direct, in-line pressure filtration of water discharged from sampling pumps or bailers. Use of cartridge filters in this manner can eliminate or reduce the need for filtration equipment that requires cleaning and minimize exposure of the sample to the atmosphere and air-borne contaminants.

A possible disadvantage with some cartridge filters is that the filter element contained in the cartridge may not be clearly visible to the user. A filter that is not visible cannot be examined for visual signs of flaws or contamination before use and cannot be readily monitored during use.

Cartridge filters can be more difficult to precondition prior to use than disk filters. Filter preconditioning usually involves washing or soaking filters prior to use to

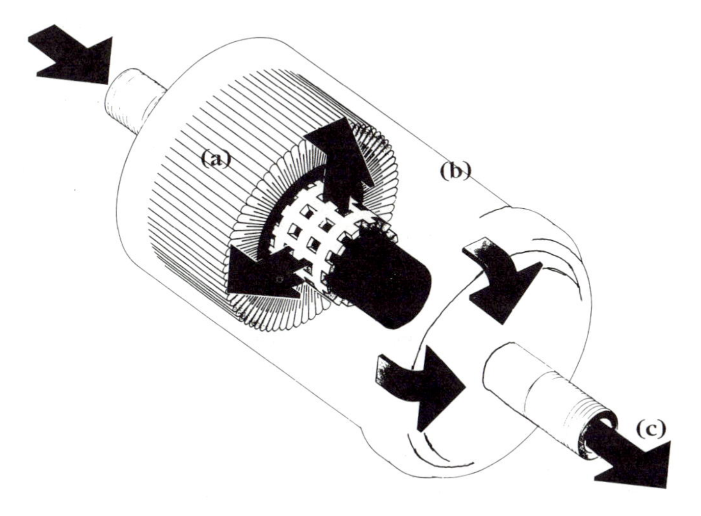

Figure 10. Diagram of a disposable cartridge filter: (a) filter element; (b) housing; (c) direction of flow. (Reproduced with permission from QED Ground Water Specialists.)

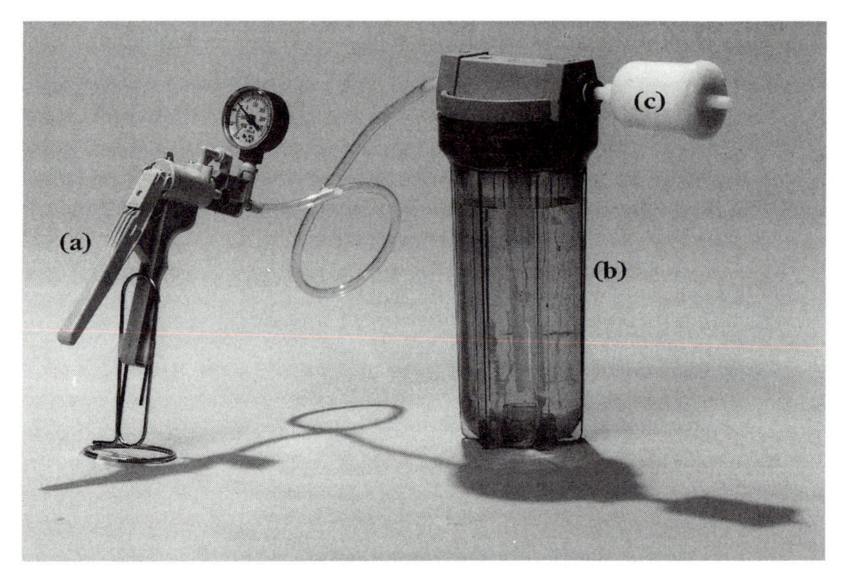

Figure 11. Example of a manually powered air pressure apparatus with a cartridge filter: (a) manual air pump with gauge; (b) prefiltration chamber; (c) disposable cartridge filter.

remove possible contaminants from the manufacturing process or leachable substances. Some filters must be preconditioned by soaking in water prior to use if they are subject to swelling during hydration.

Criteria for Selection of Filters and Filtering Equipment

There are many practical and technical considerations for selecting water sample filters and filtration equipment for use in the field. The box on pp 353–354 summarizes some of the more critical considerations. Of the various considerations listed, the minimum retention size of a filter is usually given the most attention. Many published analytical methodologies and guidelines specify the minimum retention size for various analyses (*8–13*).

Minimum retention sizes for biologic analyses are generally well founded by research and long practice and are based on the size of the organism to be filtered. Many standard methodologies for inorganic analysis of water for major cations, trace elements, and radionuclides specify a minimum retention size of 0.4–0.45 μm when analysis of the dissolved fraction of the water sample is to be performed, as discussed by Hem (*14*). The practice of filtration of water samples to a minimum retention size of 0.4–0.45 μm for many inorganic analyses is not well founded. The 0.4–0.45-μm minimum retention size does not necessarily represent the size threshold between dissolved matter and finely divided plant, animal, and mineral matter in aquatic matrices. Additional research by regulatory and standards organizations is needed.

Considerations for Selection of a Field Filtration System

- What analyses are to be performed? Will analyses be performed on the filtered water sample, on material captured by the filter, or both?
- Does an alternative to filtration exist, such as "low-flow" purging and sampling of groundwater?
- If captured particulates are to be analyzed, what method(s) will be used?
- What is the nature of the water to be filtered? What is its turbidity level, temperature, and expected chemical composition?
- Should a net, depth, or membrane filter be used? What type of net, depth, or membrane filter should be used? Should a cartridge filter, disk filter, or other form of filter be used?
  - What is the required minimum retention size?
  - Is "precise" filtration required? Is the retention of matter smaller than the specified minimum retention size a concern?
  - Is sloughing or growthrough of filtered material a concern?
  - Is the release of filter fibers into the sample a concern?
  - Is the filter material chemically compatible with the material to be sampled?
    - Will the filter degrade or swell when exposed to the sample?
    - Will the filter contaminate or chemically bias the sample? Are filter materials compatible with the sample, including any "additives" such as inks, dyes, wetting, binding, and antistatic agents?
  - Will the filter be affected by sample temperature?
  - Is the filter bacteriostatic? Is it toxic to organisms to be filtered? Is it sterilized?
  - Is electrostatic reaction of particles with the filter a concern?
  - Should the filter be hydrophilic or hydrophobic?
  - Is the hydraulic efficiency of the filter a concern?
  - Does the filter require preconditioning?
- What field conditions will be encountered?
  - Will it be dusty, rainy, or freezing?
  - Will portage of filtration equipment be required?
  - Will chemical safety be a consideration?
  - Will electrical power be available?

Continued on next page

Considerations for Selection of a Field Filtration System—*Continued*

- What brand of filter should be used?
  - Is the manufacture of the filter adequately controlled?
  - Does the filter manufacturer use adequate quality assurance and quality control practices to reasonably verify that filter specifications are met on a continuous basis?
  - Is the filter provided free of contaminants from the manufacturing process?
  - Is the filter provided in packaging that prevents contamination during handling and shipment?
  - Will spacer materials provided with disk filters contaminate the filter?
- What method should be used to drive the water sample through the filter?
  - Should air, compressed specialty gas, or hydrostatic pressure be used to drive the sample through the filter?
  - Is the possible interaction of the sample with the atmosphere a concern?
  - What is the acceptable range of pressure/vacuum that can be used?
  - How will pressure/vacuum be controlled?
  - Can a hand pump or an electrically powered pump be used?
- What is the cost of filtration?
  - What is the unit cost of filters?
  - How many filters will be consumed?
  - How much does the filtration apparatus cost?
  - How much time is required to precondition filters?
  - How much time is required to prepare or precondition the filter apparatus?
  - How much time is required to filter each sample?
  - How much time is required to clean filtration equipment between each sample?
- Is the method of filtration prescribed by regulatory requirements or standardized methodologies?

Filtration Procedures

Specifications for equipment and field procedures for filtration of environmental water samples should be included in a field sampling plan or manual. Filtration equipment and procedures should be project- or study-specific and should be understood by field personnel who will perform the filtration.

The box on pp 356–358 lists general good operating procedures for water sample filtration that should be considered for the development of field protocols.

Evaluating Filter Performance and Filtration Procedures

Filtration equipment and procedures should receive some degree of evaluation prior to initial use to ensure sample integrity and representativeness. Periodic evaluations should also be performed to ensure and document continued proper performance.

Filtration equipment and procedures can be evaluated in numerous ways. Some of the many possible methodologies are listed in Table I.

Conclusion

Failure to understand the advantages and limitations of various filters, filtration equipment, and filtration methods can result in inappropriate applications and

Table I. Selected Methodologies for Assessing Filter Performance and Filtration Procedures

Method	*Purpose*
Filtration and analysis of split samples	Evaluate variability caused by filtration under the same or varying operating conditions.
Filtration and analysis of blank samples	Determine possible "positive" bias imparted by filtration.
Digestion and analysis of filter media	Determine the potential for positive chemical bias.
Analysis of filtration equipment wash water or cleaning solutions	Evaluate equipment cleaning procedures.
Filtration and analysis of spiked samples	Evaluate possible bias imparted by filtration equipment and procedures.
Microscopic examination of filter surface	Evaluate particle retention properties of the filter. Evaluate possible filter damage or degradation.
Double filtration and microscopic examination of surface of second filter	Evaluate particle retention properties of first filter.

List of Good Operating Procedures for Filtration of Environmental Water Samples

- Manufacturer's directions for use of filters and filtration equipment should be followed when applicable and proper.
- Regulatory requirements and standard methodologies should be followed when applicable and proper.
- Filtration equipment should be preconditioned prior to initial use and at regular intervals, as necessary. Filtration equipment preconditioning for trace element analysis of water is discussed in reference 8.
- Filter media should be preconditioned before use, as necessary. Some filter manufacturers provide specific recommendations for filter preconditioning.[1]
- Procedures for proper cleaning or sterilizing of filtration equipment between samples should be developed and followed. Solutions used for cleaning should be effective, should be compatible with filtration apparatus materials, and should not be a potential source of analytical interference.[1]
- Water to be sampled should be used for final rinsing of filtration equipment immediately before sample collection, when possible. Depending on the method used to clean filtration equipment, droplets of residual wash and/or rinse water may be left on equipment surfaces after equipment cleaning is completed. Residual rinse water can sometimes act to dilute or contaminate a water sample introduced into a filtration system after it is cleaned, if sample water is not used for final rinsing.[1]
- Sample water should not be used for final rinsing of portions of filtration equipment downstream from the filter, if filtered water is to undergo analysis. Exposure of this portion of a filtration device to unfiltered sample water may cause particulate contamination of the filtered sample. The effects of residual wash or rinse water in portions of the filtration system downstream of the filter can be reduced or eliminated by wasting the first portion of the sample passed through the filter system in order to flush the system.[1]
- Filtration equipment and filters should be transported in the field in sealed containers to prevent contamination and damage. Containerization of filtration equipment during transport may also be necessary for safety reasons where hazardous materials are sampled. Some field personnel choose to transport filtration equipment and filters in clear plastic bags that are changed between samples.

Continued on next page

List of Good Operating Procedures for Filtration of Environmental Water Samples—*Continued*

- Filter directionality should be observed. Filter performance is in some cases dependent on the direction of flow through the filter. Filter rupture may occur when filter directionality is not observed. Some disk filters differ in color or texture between sides. This difference can be an important consideration for microscopy. Most cartridge filters have a flow direction indicator cast or printed on the body of the cartridge.
- Disk filters should usually not be handled directly so as to avoid contamination and possible physical damage. Some manufacturers recommend the use of flat forceps specifically designed for handling disk filters.
- Disk filter spacer materials must not be confused with filter media. Some disk filters are packaged with spacers between individual filters. There are reports that spacers have mistakenly been used in place of filters, even though the spacers differed in appearance.
- Filters should generally be used for only one sample and not reused for other samples because of the possible carryover of interferences.
- Filters and filtering equipment should be visually inspected prior to each use. Filtration equipment and filters should be inspected for visual signs of contamination and physical damage. Damaged filtration equipment may present a safety hazard to field personnel where pressure and vacuum filtration is performed.
- Filters and filtration equipment should be carefully monitored during use to ensure proper operation, to detect the accumulation of particles and possible clogging of the filter, and to detect possible filter bypass or rupture.
- Filtration should always be performed within pressure/vacuum limitations specified by the filter and filtering equipment manufacturer(s). Failure to operate within pressure/vacuum limitations can affect the particle retention properties of the filter and/or cause rupture of the filter. Excessive pressure can result in filtrate being pulled through the body of the filter. Failure to observe equipment pressure/vacuum limitations can result in injury to personnel.
- Pressure/vacuum limitations for the filtration of aquatic organisms should be carefully considered and followed. Use of excessive pressure/vacuum during filtration may destroy some aquatic organisms.

Continued on next page

List of Good Operating Procedures for Filtration of Environmental Water Samples—*Continued*

- Disk filters should be properly supported within filtration equipment in accordance with manufacturer's recommendations for expected pressure differentials. Improperly supported filters can distort or rupture during filtration.
- Particle retention properties of a filter can be a function of controllable factors such as the velocity of water passing through the filter, the pressure differential applied to the filter, and water temperature. It may be advantageous to minimize variation of these factors between samples to reduce variability and possible related bias.
- It may be advantageous to use filters produced within the same manufacturer's quality control/production lot or batch number for each study to reduce possible filter performance variability.
- Filters may require changing during the filtration of turbid samples. Care should be exercised when changing filters, especially disk filters. Improper or careless changeout of disk filters can introduce contaminants into the filtration system or allow particulate bypass. In some cases, the entire filtration apparatus may require cleaning between disk filter changes to prevent contamination of the sample. Filtration systems are sometimes set up in parallel to avoid filter changeout. Prefiltration may also be used to prevent filter clogging and the need for filter changeout.

[1]Discussions on the need for, and practices of, filter preconditioning and filtration equipment cleaning and flushing for various analyses are contained in Shelton (*13*) and referenced technical memoranda by the U.S. Geological Survey that are referenced therein.

misrepresentation of water quality data. Standardized analytical methods developed by many government agencies and standards organizations are frequently not specific or detailed enough to relieve the filter user from critically evaluating each application.

Filters and filtration equipment should be carefully selected to take into account project data requirements and practical considerations such as ease of use and cost. Proper procedures for filtration should be developed, documented, and understood by field personnel for each application. The performance of filters, filtration equipment, and field procedures should be evaluated prior to initial use and periodically thereafter to ensure that project data requirements are met.

Disclaimer

The content of this chapter should not be interpreted as an endorsement or criticism of any manufacturer's filter or filtration equipment for any application. Discussions and illustrations in this chapter are general in scope and do not cover the entire spectrum of water sample filtration because of space limitations. The inclusion or absence of any filter, filtration equipment, or filtration technology is not an endorsement or criticism of any manufacturer's product. Filter and filtration equipment manufacturers should be consulted concerning the proper selection and use of any filter and related equipment.

References

1. Brock, T. D. *Membrane Filtration: A User's Guide and Reference Manual;* Science Tech: Madison, WI, 1983.
2. Kearl, P. M.; Korte, N. E.; Cronk, T. A. *Ground Water Monit. Rev.* **1992,** *Spring,* 155–161.
3. Puls, R. W.; Clark, D. A.; Bledsoe, B.; Powell, R. M.; Paul, C. J. *Hazard. Waste Hazard. Mater.* **1992,** 9(2), 149–162.
4. Fetter, C. W. *Contaminant Hydrogeology;* McConnin, R. A., Ed.; Macmillan Publishing: New York, 1993; pp 149–157, 164.
5. Braids, O. C.; Burger, R. M.; Treca, J. J. *Ground Water Monit. Rev.* **1987,** *Summer.*
6. Backhus, D. A.; Ryan, J. N.; Groher, D. M.; MacFarlane, J. K.; Gschwend, P. M.; *Ground Water* **1993,** *31*(3), 466–479.
7. Herzog, B.; Pennino, J.; Nielsen, G. In *Practical Handbook of Ground-Water Monitoring;* Nielsen, D. M., Ed.; Lewis Publishers: Chelsea, MI, 1991; pp 475–499.
8. *Standard Methods for the Examination of Water and Wastewater,* 19th ed.; Greenberg, A. E.; Clesceri, L. S.; Eaton, A. D.; Franson, M. H., Eds.; American Public Health Association; American Water Works Association; Water Environment Federation: Washington, DC, 1995.
9. U.S. Environmental Protection Agency. *Methods for Chemical Analysis of Water and Wastes;* Government Printing Office: Cincinnati, OH, 1983; EPA–600/4–79–020.
10. *1996 Annual Book of ASTM Standards, Water and Environmental Technology;* American Society for Testing and Materials: Philadelphia, PA, 1996; Vol. 11.01, Water (I and II); PCN:01–110194–16.
11. U.S. Environmental Protection Agency. *Test Methods for Evaluating Solid Waste;* Government Printing Office: Cincinnati, OH, 1986; SW–846, PB88–239223.
12. *ASTM Standards on Ground Water and Vadose Zone Investigations,* 2nd ed.; Furcola, N. C. et al., Eds.; American Society for Testing and Materials: Philadelphia, PA, 1994.
13. Shelton, L. R. *Field Guide for Collecting and Processing Stream-Water Samples for the National Water Quality Assessment Program;* U.S. Geological Survey: Sacramento, CA, 1995; USGS Open File Report 94–455.
14. Hem, J. D. In *Study and Interpretation of the Chemical Characteristics of Natural Water,* 3rd ed.; U.S. Government Printing Office: Alexandria, VA, 1985; pp 59–61; USGS Water-Supply Paper 2254.

Chapter 20

Planning and Design for Environmental Sampling

L. J. Holcombe

The planning and design for environmental sampling includes considerations for the frequency of sampling and location of sample points. An example groundwater study is described to illustrate the major steps involved in planning and design. Two different technical objectives were addressed, resulting in two different sampling designs. The first approach represents an environmental assessment, for which sampling design should include sufficient data to characterize the site, including background characterizations and historical data. The second approach represents a data collection effort for the calibration and testing of groundwater models, for which sampling design should include information on sampling and spatial variability. The contrasting designs presented in this chapter can be used for other groundwater and environmental sampling designs.

EVERY SUCCESSFUL SAMPLING PROGRAM requires a planning and design step prior to collection and analysis of the samples. The design must be based on the technical objectives of the program and the specific conditions of the site. In regulatory parlance, technical objectives are often referred to as data quality objectives or DQOs. Technical objectives dictate, in the least, the precision and accuracy needed and the type of data interpretation or data display needed. The specific characteristics of the site define the stratification of the target population as well as the variability within a given stratum. These two attributes, technical objectives and specific characteristics of the site, form the basis of any sampling design.

The subsurface sampling program presented here illustrates two designs for the same site. The study site is a coal ash impoundment at an electric generating facility. The results from the two different site characterizations have been reported earlier (*1, 2*). In the first design, the objective was to perform an environmental assessment of the coal ash pond. The second design had as its objective to collect data for calibration and testing of groundwater models. Comparing these two designs and their results will provide guidance for planning subsurface and groundwater site investigations.

Site Description

The site described in this chapter is located at a 400-MW, coal-fired steam generating station. The site has an ash pond system consisting of two unlined settling basins separated by a dike (Figure 1). The ponds were constructed in 1973 and have a total surface area of about 60 acres.

Fly ash and bottom ash represent the largest volume of materials going to the pond. Over the 16-year operating life, about 500,000 cubic yards of ash have been routed to the ponds.

The site is located in a province of northeasterly trending belts of moderate- to high-grade metamorphic rocks. In general, the stratigraphy consists of metamorphic bedrock overlain by saprolite, a mantle of weathered bedrock that has developed in place. Groundwater flow in the bedrock aquifer of the region is predominantly controlled by the presence of fractures and weathered veins. The permeability of the unfractured metamorphic rocks is extremely low. Groundwater flow in the saprolite is affected by the structure and texture of the parent rocks. Groundwater tends to accumulate, and springs typically emerge, where the saprolite contacts the lower permeability unfractured bedrock.

A total of 44 monitoring wells and two lysimeters were installed at the site as part of the two investigations (Figure 2). The number and placement of the monitoring devices was based on the need to include the important flow paths into and out of the pond and was determined from a conceptual hydrogeologic model of the site, as developed from a preliminary placement of 11 wells. The majority of wells were installed downgradient of the primary pond to characterize the flow paths between the pond and the river. The primary pond appears to be the source of both recharge and discharge for the shallow groundwater system. Conceptually, pond water is being forced downward into the shallow bedrock beneath the pond by the higher hydraulic head in the pond (about 723 ft). Based on wells placed at multiple depths within the pond, the direction of groundwater flow reverses locally in the pond as a result of the hydraulics and geometry of the pond.

The major element concentrations in the coal ash are fairly typical of wet-sluiced eastern coal ash. Total and leachable concentrations of trace elements in the ash are also typical of reported values for coal ash (*1*). The coarser fractions of the ash, including bottom ash and pyrite rejects, settle out nearer the discharge line outfall. This settling tends to affect groundwater chemistry in the immediate vicinity, particularly where the pyrite rejects are concentrated.

Most of the soils in the area of the ash ponds can be characterized as residual granite or gneissic bedrock. Ambient groundwater compositions reflect the granitic

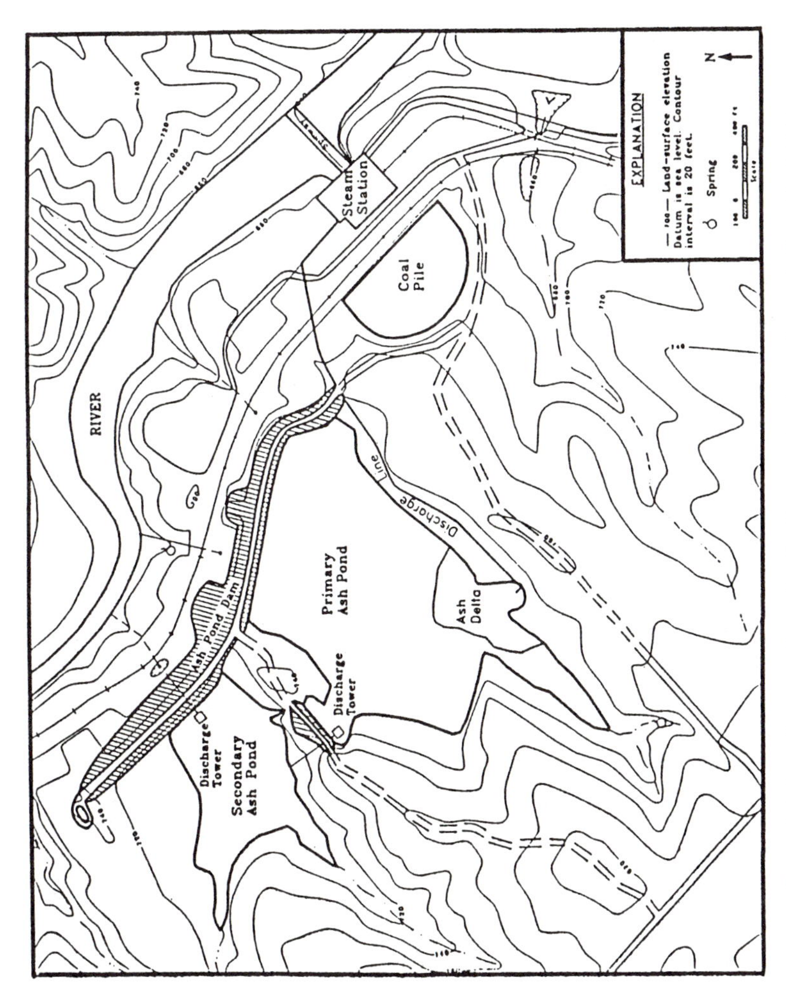

Figure 1. Ash pond system. (Reproduced with permission from reference 1. Copyright 1991.)

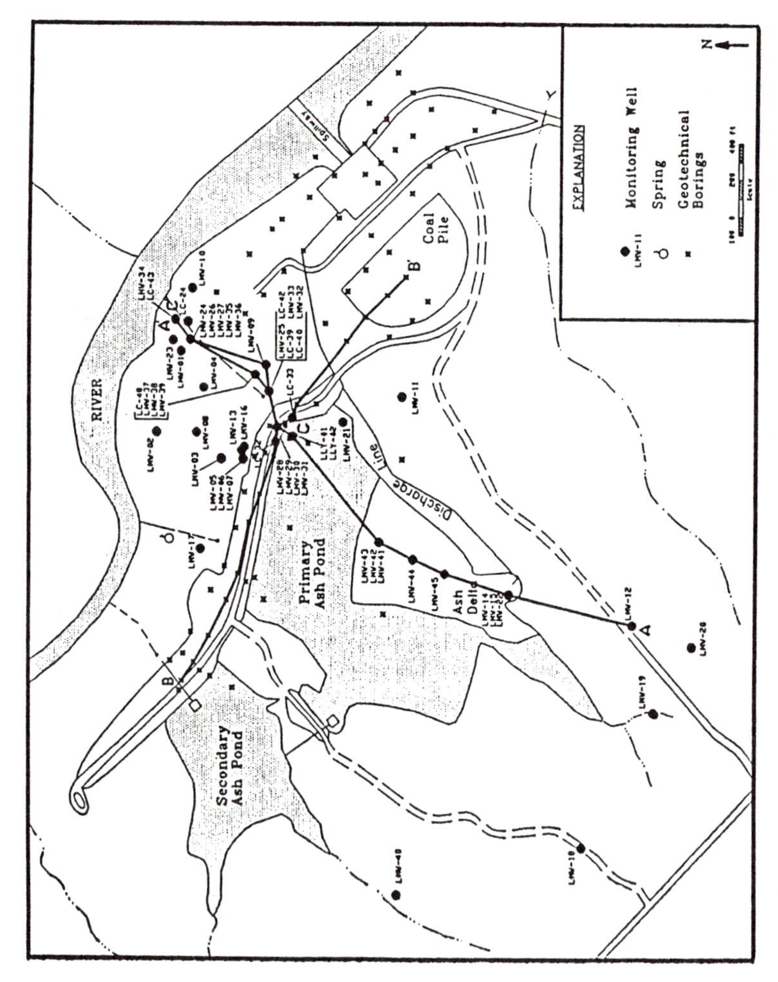

Figure 2. Geotechnical borings and monitoring well locations. (Reproduced with permission from reference 2. Copyright 1993.)

geologic environment [i.e., generally lower pH (about 6) and lower concentrations of calcium, magnesium, and carbonate than would be found in sedimentary rock aquifers (3)].

Technical Objectives

Assessing Environmental Impact

The technical objective of the first study at the site was to assess environmental impacts of the ash pond. The data from this characterization will be used to assist federal and state agencies in making prudent regulations for the management of these wastes and will assist utility managers in selecting appropriate practices for handling these materials. Currently, utility ash ponds are regulated as solid or special waste impoundments under state guidance.

One key indicator of environmental performance is the comparison of groundwater compositions between ambient or background groundwater and downgradient groundwater. In order to make these comparisons, we needed measurements that were precise enough to differentiate between sampling variability and actual differences in the two groundwater domains. A second key indicator of environmental performance is comparison of water quality to regulatory limits. Therefore, the measurements had to be accurate so comparisons could be made to regulatory limits.

Design. The key design considerations for this technical objective included the following:

- characterization of shallow groundwater flow and water quality;
- determination of sampling and analytical variability; and
- measurement of elemental concentrations.

The hydrogeologic characterization of the shallow groundwater included:

- reviewing available hydrogeologic data for the site;
- collecting geologic cores to define the lithology and stratigraphy of deposits surrounding the pond and to measure physical and geochemical properties of the unconsolidated deposits;
- installing monitoring wells and piezometers to define hydraulic gradients, to measure in situ aquifer properties, and to characterize the groundwater chemistry; and
- measuring hydraulic gradients in the ash pond to characterize the flux of water into and out of the pond.

Figure 3 presents the sampling locations for the shallow groundwater characterization.

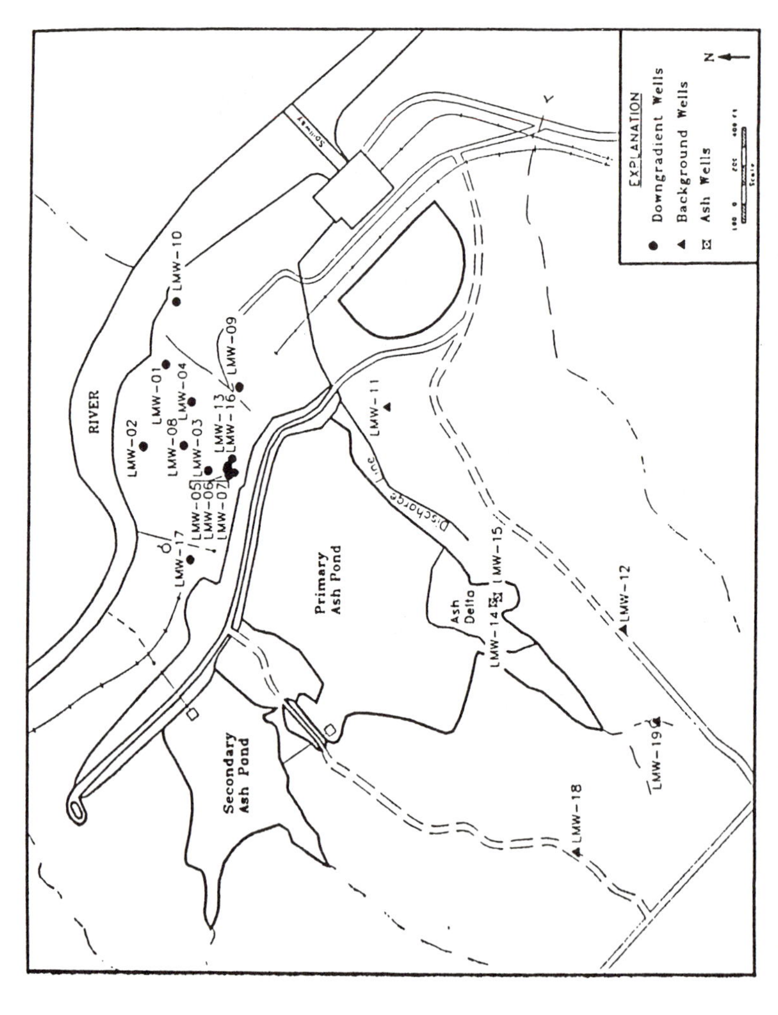

Figure 3. Sampling locations for shallow groundwater characterization. (Reproduced with permission from reference 1. Copyright 1991.)

An estimate of sampling and analytical variability was made using a prescribed set of quality control (QC) measures. Sampling QC included analysis of field equipment rinse blanks, and collection and analysis of duplicate samples for each matrix. Analytical QC included analyzing QC check samples or laboratory control standards, running analytical blanks, analyzing duplicates (splits), and analyzing matrix spikes and matrix spike duplicates. Additionally, the laboratory quality assurance (QA) consisted of analyzing the following: (1) audit samples, (2) some samples by two analytical techniques, and (3) some samples in a different laboratory. The components of total variability were determined from the variability between locations, duplicate samples, and analytical splits.

Water samples were analyzed for a range of elements, anions, and water quality variables. Field measurements were made of electrical conductivity, total dissolved solids, redox potential (Eh), temperature, pH, and alkalinity. Anions analyzed included (1) those major ions associated with coal ash, such as chloride, sulfate, and sulfite; and (2) trace ions associated with low-volume waste streams, such as bromide, phosphate, and thiourea. Elemental analyses included the elements listed under the primary and secondary drinking water standards, elements typically associated with coal ash (calcium, magnesium, boron, and strontium), elements associated with low-volume wastes (copper, chromium, and iron), and other elements picked up by the emission spectrophotometric technique.

Results. Once the analytical data were validated and tabulated (electronically), a statistical test, Duncan's multiple-range test, was used to group the concentrations according to well locations. The assumptions underlying the Duncan multiple-range test are those of the analysis of variance (statistical test for grouping samples with similar means). Because these assumptions include normal distribution and equivalent variances, a nonparametric Duncan multiple-range test was applied. The method ranks and categorizes the data and performs statistical tests on ranks and categories.

Monitoring wells were grouped into three categories—upgradient, beneath the ash pond, and downgradient—and differences were determined in analyte concentrations between each group for all the analytes using the statistical tests just described. Of the constituents found in elevated levels in ash wells relative to background wells, only a handful were significantly elevated in wells downgradient of the ash pond. Calcium, total dissolved solids, magnesium, strontium, and sulfate concentrations were significantly higher in the downgradient wells than in background groundwater samples. Figures 4 and 5 present isoconcentration contour plots for calcium and sulfate, respectively, in the shallow downgradient groundwater. These were calculated using a semivariogram analysis performed by a contouring program (CPS).

Data Set Collection for Model Validation

The technical objective for this second study at the same site was to provide a high-quality data set for validating geohydrochemical transport models. Model validation

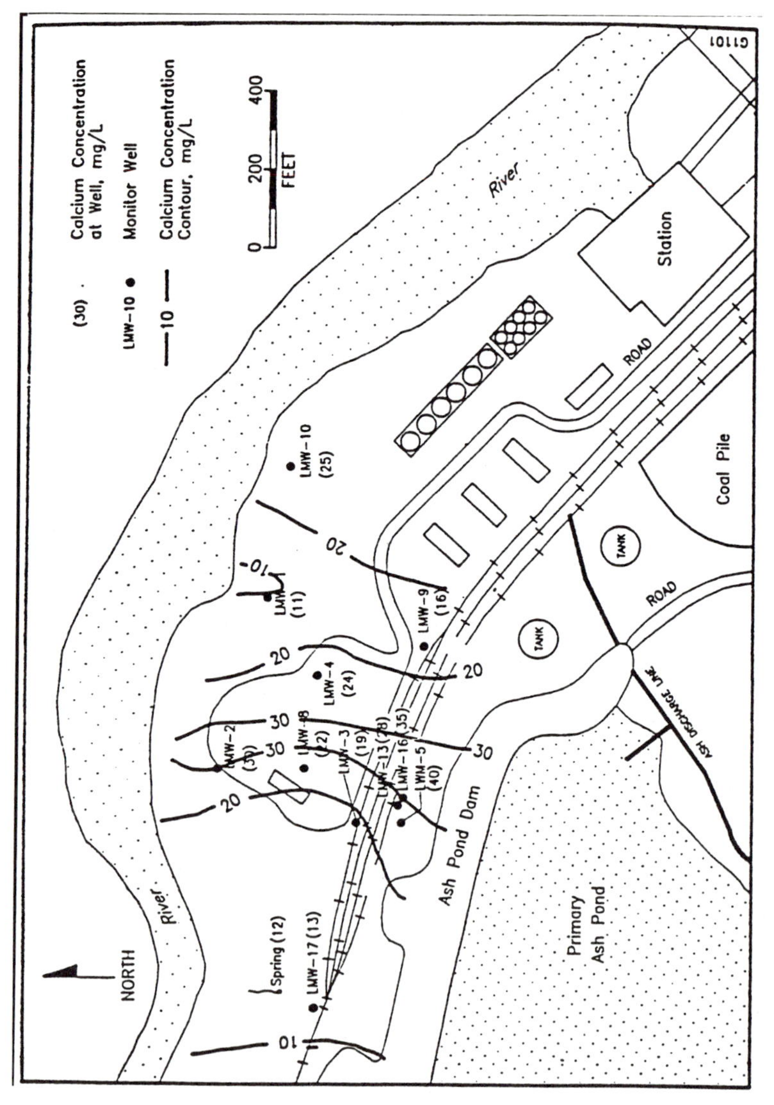

Figure 4. Isoconcentration contours for calcium. (Reproduced with permission from reference 1. Copyright 1991.)

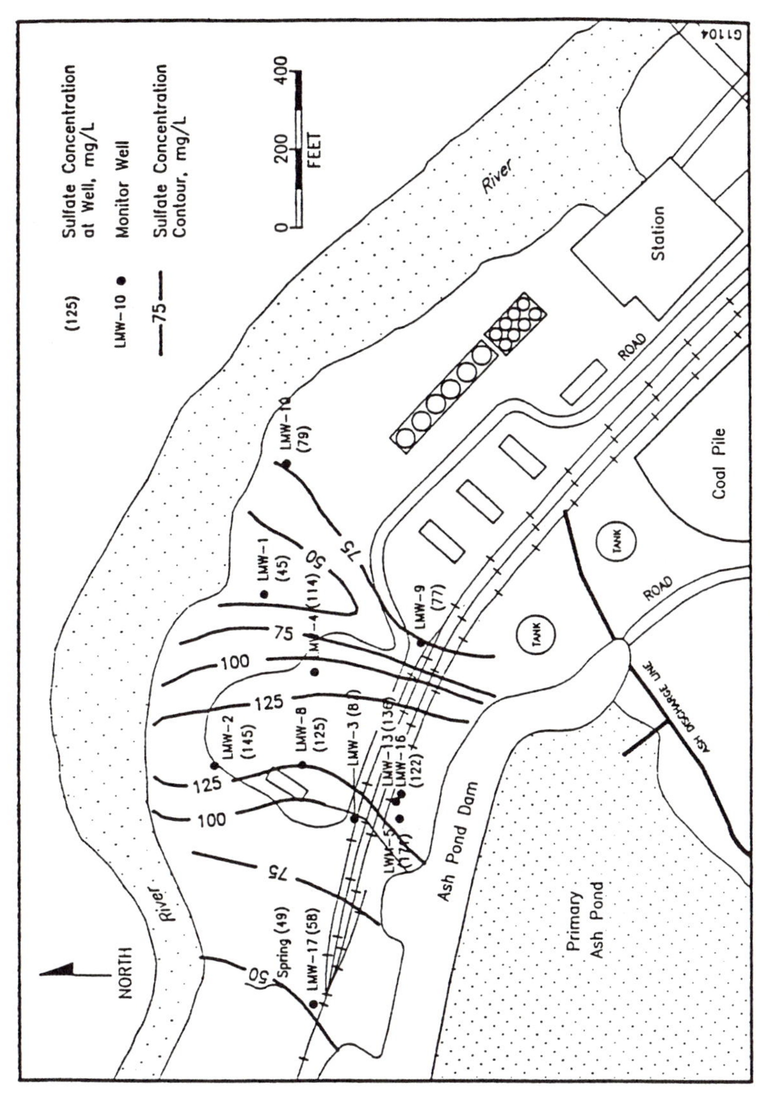

Figure 5. Isoconcentration contours for sulfate. (Reproduced with permission from reference 1. Copyright 1991.)

is defined here as an integrated test of the accuracy with which geohydrochemical models simulate the hydrologic and geochemical processes occurring in the subsurface environment. The process consists of confirming that the conceptual and numerical models accurately represent the processes occurring in the real world.

The actual process of model validation was not a part of the objectives of this study, only the generation of a data set to be used in validation. The users of this information will be model developers and those wishing to evaluate or compare the accuracy of models.

Several subsurface physical and chemical attributes could be used as a basis of comparison for model validation. In this study the key indicators selected were the downgradient groundwater concentrations of constituents associated with the coal ash, because these concentrations are of prime importance to environmental assessments. The hydrogeologic characterization of the site (both upgradient and downgradient) and the background geochemistry and source (ash pond) geochemistry were also provided as input to the models so that all models could generate the downgradient groundwater concentrations for validation.

Design. The key considerations for design in this study were

- creation of a three-dimensional concentration profile;
- estimation of small-scale spatial variability; and
- measurement of chemical species (not just total elemental concentrations) in the groundwater.

Three-dimensional characterizations were required in order to fully define important mechanisms of transport occurring in the subsurface. Much of the dissolution, attenuation, and transport phenomena occur vertically beneath and below the pond; these phenomena would not be accurately characterized in two dimensions.

A detailed hydrogeologic and geochemical characterization of the bedrock valley downgradient of the ash pond began in late 1991 and lasted through the summer of 1992. The investigation obtained a more complete vertical and lateral characterization of aquifer properties, groundwater chemistry, and soil geochemical properties of unconsolidated deposits within this bedrock valley. This characterization also obtained a more complete geochemical picture of source waters within the ash deposits.

Figure 6 presents the monitoring well locations installed along the projected axis of the bedrock valley.

A nest of multiple wells all in close proximity, termed a *key station,* was installed downgradient of the pond along the bedrock valley transect to obtain an estimate of the natural hydrogeologic and geochemical heterogeneity in the system. Three monitoring wells were completed at the same depth in alluvial deposits; the wells were located within approximately 5 ft of each other. The well drilling and completion methods were performed as similarly as possible for each of the wells.

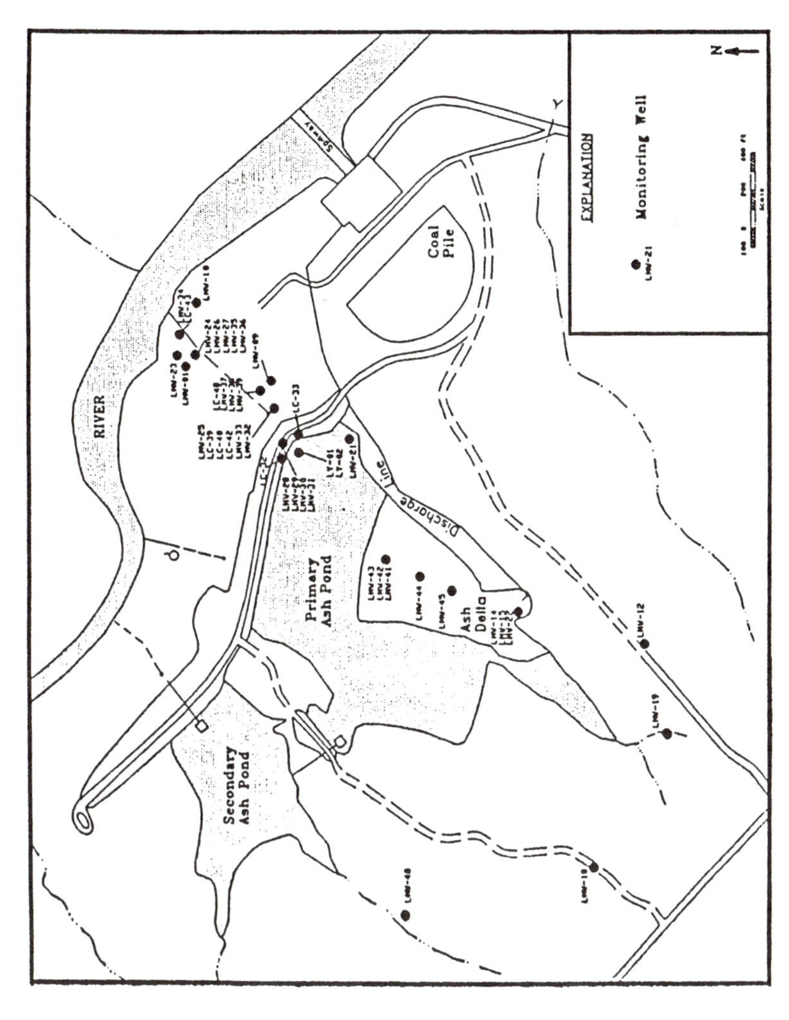

Figure 6. Transect well locations. LC is bore hole. (Reproduced with permission from reference 2. Copyright 1993.)

The validation exercise focused on two sets of chemical constituents: (1) those constituents found in the downgradient groundwater in measurable levels in the previous study, and (2) water quality parameters and ions that are required for aqueous equilibrium calculations (models).

The constituents chosen to validate against were calcium, sulfate, iron, and magnesium. The other water-quality parameters included those from the previous study. Because some key analytes may be present in more than one valence state (e.g., iron), it was necessary to precisely measure both the specific redox couples [e.g., Fe(II) and Fe(III)] and the Eh and pH of the water samples.

Results. The general spatial trend of dissolved constituents in the groundwater beneath the site can be described as higher dissolved constituents in the ash nearest the discharge sluice, with decreasing concentrations downgradient of the sluice. In vertical section (Figures 7 and 8), the concentrations of sulfate and calcium, in particular, show the influence of the pond water on the shallow downgradient groundwater. The vertical concentration profile was developed from monitoring wells placed at different depths along the sampling transect. In general, deeper downgradient wells contain slightly higher concentrations of the ash constituents, calcium and sulfate, than the shallow downgradient wells.

The key station data were used to construct a confidence interval that defines the range of small-scale aquifer variability. The differences in concentration between well locations may then be checked against this confidence interval to determine whether a real difference exists between the measurements. Real differences, as opposed to differences due to random small-scale variability, should be discernible by transport models.

Most of the concentrations of calcium and magnesium in the downgradient wells were outside the range of small-scale variability; iron concentrations were all within this variability. Sulfate concentrations were within the range of small-scale variability in approximately one-half of the locations. These results indicate that some well locations and analytes show real spatial differences in the groundwater chemistry, which might be resolved by a model, whereas differences for other analytes or locations were within the noise of small-scale variability.

Summary

The key considerations for designing environmental sampling programs include

- intended use of the data, and intended data users;
- key indicators to be measured, and methods of interpreting the data; and
- accuracy and precision requirement, including a QA–QC program sufficient to define accuracy and precision statistics.

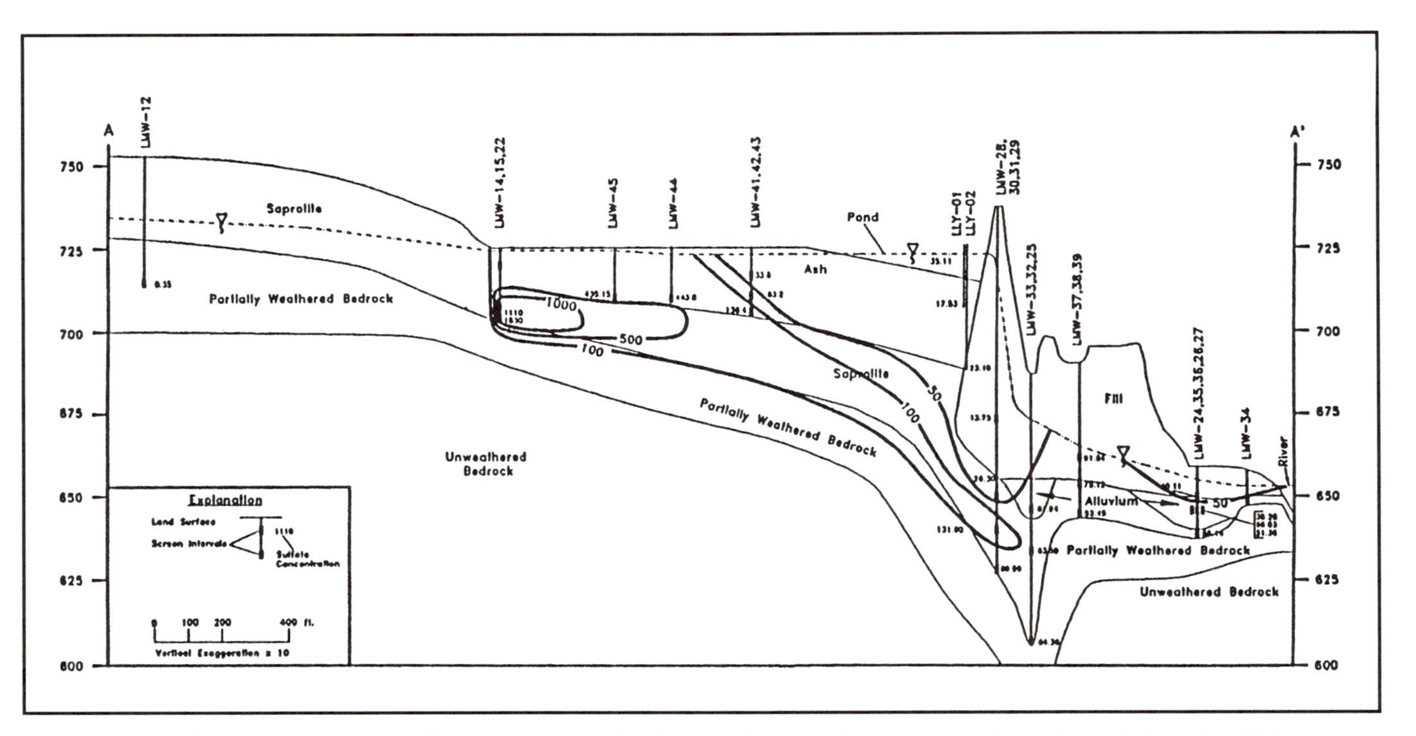

Figure 7. Isoconcentration plot of sulfate along transect. LLY is lysimeter. (Reproduced with permission from reference 2. Copyright 1993.)

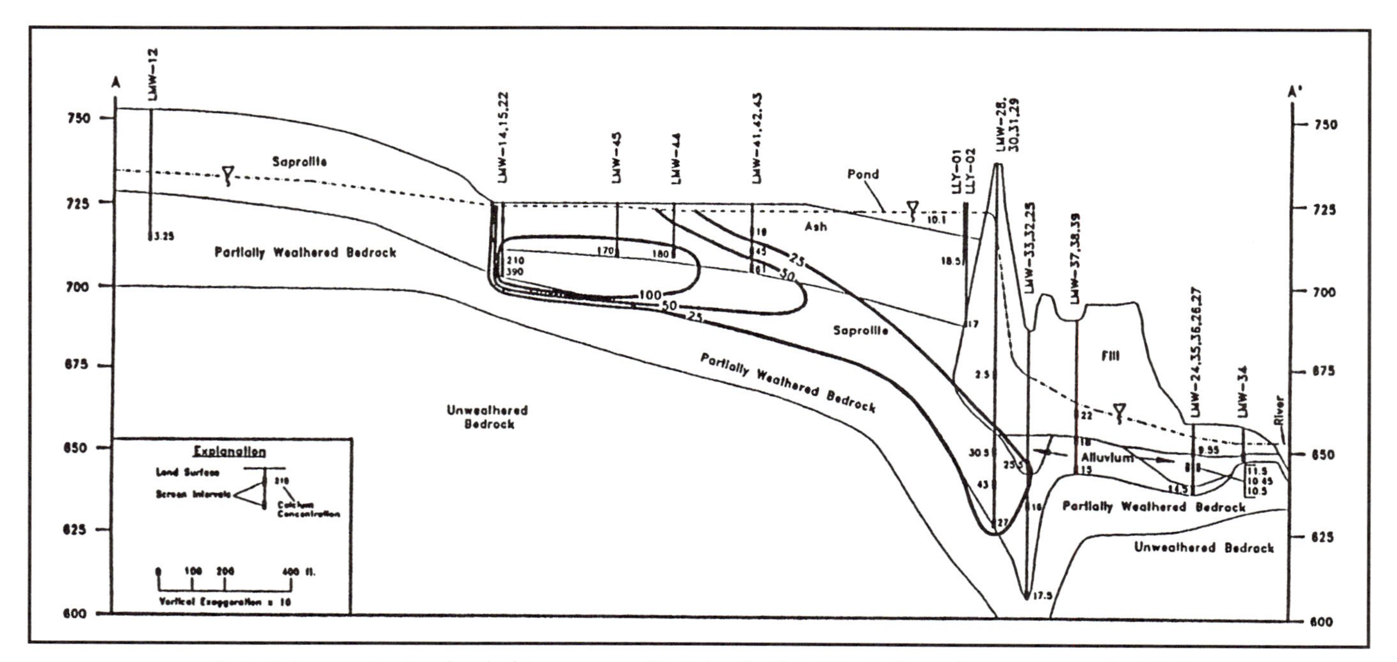

Figure 8. Isoconcentration plot of calcium transect. (Reproduced with permission from reference 2. Copyright 1993.)

In the case study presented here, the data users were different between the two technical objectives. Defining what data the users required was essential to sampling design. These technical objectives in turn determined the choice of key indicators, in this case the key measurement parameters. The technical objectives also dictated the precision requirements, or more importantly, the types of measurements needed to adequately define the resultant precision.

Acknowledgments

This work was the result of research on coal ash management funded by the Electric Power Research Institute (EPRI). I thank John Goodrich-Mahoney and Ishwar Murarka of EPRI for their assistance on this research. Also, co-researchers on the project and co-authors of earlier reports were James Erickson of GeoTrans, Inc.; John Fruchter and John Thomas of Battelle Pacific Northwest Laboratory; and Andrew Weinberg and Rebecca Coel of Radian.

References

1. Holcombe, L. J.; Thompson, C. M.; Rahage, J. A.; Erickson, J. R.; Fruchter, J. S. *Comanagement of Coal Combustion By-Products and Low-Volume Wastes: A Southeastern Site;* Electric Power Research Institute: Palo Alto, CA, 1991; EN–7545.
2. Holcombe, L. J.; Weinberg, A.; Coel, R. J.; Erickson, J. R.; Fruchter, J. S.; Thomas, J. *Model Validation Data Set for Utility By-Product Management Sites, Volume I and II;* Electric Power Research Institute, Reports Distribution Center: Palo Alto, CA, 1993; file report.
3. Stumm, W.; Morgan, J. J. *Aquatic Chemistry;* Wiley-Interscience: New York, 1970.

Chapter 21

Sampling Groundwater Monitoring Wells

Special Quality Assurance and Quality Control Considerations

Robert T. Kent and Katherine E. Payne

Quality assurance and quality control for sampling groundwater monitoring wells begins by defining the hydrologic and geochemical characteristics of the aquifer. Variations in lithology, permeability, and geochemistry of naturally occurring waters within the aquifer should be identified and taken into account when designing and sampling the groundwater monitoring wells, and when drawing conclusions based on the analytical data of the collected sample. Any conclusions should also take into account the limitations of the accuracy of the analytical test method used to quantify the monitored species. Other factors to take into account are sources of sample alteration that include the degree of well purging during sampling, contamination or alteration of the sample by the sampling device, and absorption of air contaminants by the sample during collection.

QUALITY ASSURANCE (QA) AND QUALITY CONTROL (QC) PLANS for groundwater monitoring should provide for documentation of potential sources of sample alteration during collection. This documentation includes the extent of well purging prior to sampling, careful selection of the device used to purge or sample the well, and implementation of the appropriate preservation and handling techniques of the samples collected. An effective QA and QC plan begins with well design and is maintained through review of the analytical data. These and other considerations are discussed in order to present new data and fresh opinions based on the in-house and field testing of well sampling techniques, interpretation of analytical data obtained from monitoring systems, and a review of available pertinent literature.

Hydrogeologic Controls on Groundwater Monitoring

The initial and perhaps most important QA and QC consideration in sampling groundwater monitoring wells is developing a thorough knowledge of the physical and chemical characteristics of the aquifer system. This knowledge includes identifying variations in lithology of the aquifer, the directions and velocity of groundwater flow, and the spatial and temporal variations in groundwater quality. The following are examples of ways in which a knowledge of the aquifer system may influence the design of the groundwater monitoring system, the achievable representativeness of a groundwater sample, and the interpretation of the sample analytical results relative to the hazardous waste management unit.

Under regulations set forth by the Resource Conservation and Recovery Act (RCRA), owners or operators of hazardous land disposal facilities are required to monitor the uppermost permeable limit underlying the facility. The lithology of uppermost sediments along the southern Gulf Coast of the United States consists of interbedded, discontinuous strata of interdeltaic silts, clays, and sands. In the course of monitoring a hazardous waste facility, one well within the facility monitoring system may be screened in a coarse sand and another well in the same monitoring system screened in a clayey sand, where both wells monitor the uppermost permeable unit underlying the regulated facility. Water samples collected from such wells have been shown to exhibit a significant variation in concentration of inorganic monitoring parameters as a result of the naturally occurring variations in lithology of the permeable zone. Where the variations in sediments within a monitored zone include variations in carbon content, significant variations may occur in the amount of adhesion of any organic waste constituents present in the monitored zone to the aquifer sediment. Proper documentation of lithologic variations within a monitored zone is crucial to the correct interpretation of sample analytical data.

Variations in the permeability of the aquifer have been demonstrated to affect the quality of the groundwater sample collected from the monitoring well. Wells of similar construction and design screened in variable lithologies may exhibit variations in recovery rates subsequent to purging due to the innate variations in permeability of the monitored unit. As a result, the concentrations of monitored species, both organic and inorganic, may vary between wells because of variations in aeration or chemical reduction of the sample in the wellbore during the recovery period. For example, a monitor well completed in a well-sorted coarse sand may be sufficiently purged within 15 min; thus, prompt collection of a water sample is allowed. On the other hand, a monitor well that is completed in a sandy silt may take several hours to recover sufficient water to collect a sample. Where the recovery rate of a well is believed to affect water quality results, the suspected influence should be verified by sampling the slowly recovering well. The degree of influence may then be more or less quantified and considered in the overall interpretation of the water quality data.

The occurrence of vertical gradients of flow between permeable strata within an aquifer system, if not properly accounted for, may result in the monitoring of water quality in multiple zones within one well. Figure 1 shows a single-screen completion across a multilayered aquifer. The upper aquifer is contaminated, but the lower aquifer is not. Under static conditions, the lower aquifer has a higher hydraulic head than the upper aquifer, and vertical cross flow in the well occurs from the lower aquifer to the upper aquifer. Assuming that the transmissivities of the aquifers are 2000 gal/day/ft and the storage coefficient is 0.0003, over 40 days of pumping this well at a rate of 15 gal/min would be required before a sample of the contaminated water in the upper aquifer could be obtained. To avoid cross flow between permeable strata, the screen section of any one monitoring well should be set to monitor one discrete permeable stratum. Identification of vertical components of flow in the aquifer system can be made with clustered or nested wells, which consist of a group of wells installed in the same immediate vicinity and screened at variable depths.

In addition to defining the hydraulic characteristics, the chemistry of the monitored zone(s) should be thoroughly defined prior to interpreting sample analytical data obtained as part of a RCRA monitoring program. A naturally occurring spatial variation in the salinity of connate waters is commonly found along the Texas Gulf Coast. During the course of monitoring a regulated waste facility in the Gulf Coast area, total dissolved solids were measured at 30,000 mg/L in one well. Water quality in wells located at the other end of the facility showed total dissolved solids of approximately 500–800 mg/L. On the basis of a statistical comparison of the concentrations of chromium between wells, the well yielding saline-quality water was interpreted to have been affected by the waste management unit. Monitoring wells were later installed at points between the freshwater and saline wells. A measurement of specific cations and anions, total dissolved solids, and specific conductivity revealed a naturally occurring spatial variation in fresh and saline-quality water in the region, in contrast to any plume of contamination originating from the waste management unit.

The degree of alteration in the chemistry of the groundwater sample during sample collection (e.g., oxidation, precipitation, and adsorption) has been shown to be influenced by the initial chemistry of the groundwater (e.g., initial Eh, which is a measure of the available electrons in solution; pH; redox buffering capacity; and pH buffering capacity). Through laboratory simulation of sampling from a very shallow water table (less than 18 ft below ground), the amount of iron precipitation due to aeration of the sample during collection has been demonstrated to be significantly reduced in waters with a lower initial pH (*1*). Mixing of well water with the atmosphere causes aeration and oxidation of

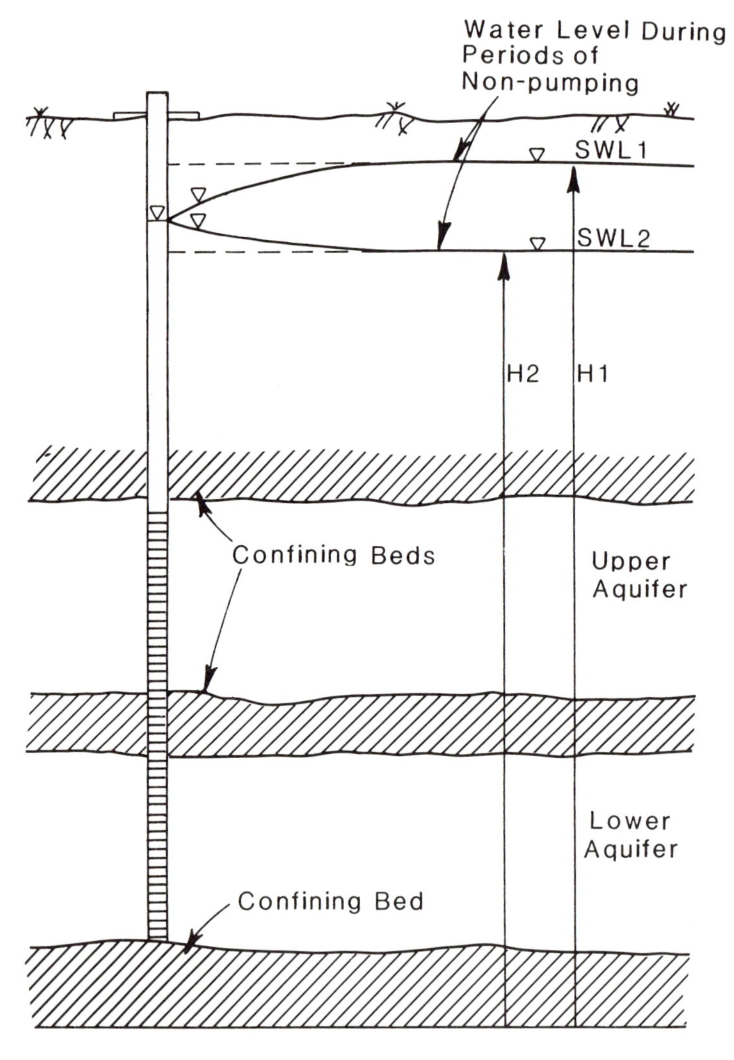

Figure 1. Two-layer aquifer system.

ferrous iron to ferric iron. Ferric iron rapidly precipitates as iron hydroxide and can adsorb other monitoring constituents including arsenic, cadmium, lead, and vanadium. Therefore, aside from the amount of aeration that may occur as a result of the sample collection procedure, the amount of iron precipitation resulting from aeration is dependent upon the innate quality of the groundwater.

In addition to considering the hydrologic and geochemical controls of the aquifer upon the groundwater sample, the flow behavior of the monitored species should be taken into consideration when designing the monitoring well. Figure 2 shows a cross section of a surface impoundment containing soluble

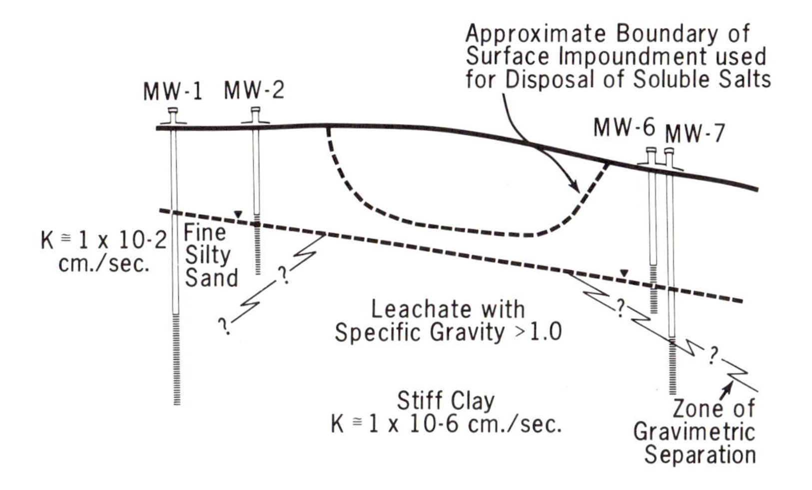

Total Dissolved Solids content of samples from monitor wells

WELL	DEPTH	TDS
MW-1	45	830
MW-2	25	450
MW-7	48	95,000
MW-8	30	4,800
MW-9	30	3,900
MW-10	48	78,000

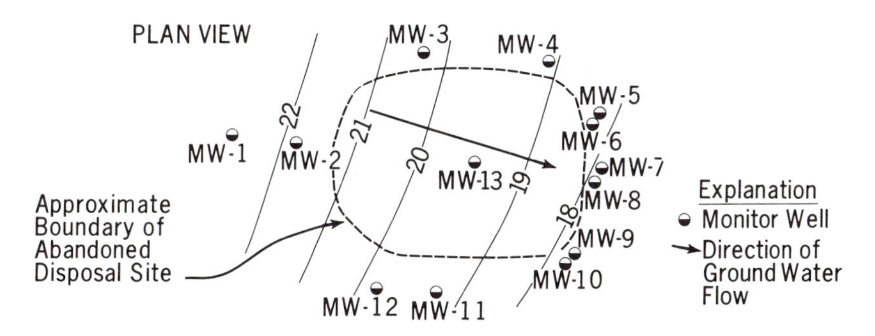

Figure 2. Flow behavior of leachate from a surface impoundment.

salts. As indicated by chemical analysis of groundwater from shallow and deep cluster wells, the flow of leachate originating from the surface impoundment was concentrated along the bottom of the aquifer because of the greater density of the leachate relative to the density of the unaffected groundwaters. Without consideration of the flow behavior of the leachate and without investigation of the vertical extent of the aquifer, the contaminants may have gone undetected.

In conclusion, a thorough investigation of the hydrologic and geochemical conditions of the aquifer will result in a more representative monitoring system and will result in a more correct interpretation of groundwater quality data.

Sampling Strategy

The goal of a QA and QC program for sampling wells is to minimize alteration of the collected sample by specifying methods of sample collection and preservation and by documenting sampling procedures. These specified methods allow an interpretation of the data that takes into account variations or alterations occurring as a result of the sampling process. Areas where variations have been observed include purge time, sample holding time, sample collection methods, and sample preservation methods. In this section, specific examples are presented of potential sources of sample alteration that may occur during well sampling.

Well Purging

The purpose of well purging is to eliminate stagnant water in the wellbore and adjacent sand pack that may have undergone chemical alteration. This elimination allows the collection of a sample that is representative of the in situ quality of groundwater near a particular well.

Various methods for determining the necessary extent of well purging have been recommended. The U.S. Geological Survey (2) recommended pumping the well until temperature, pH, and specific conductance are constant. Schuller et al. (3) recommended calculation of the percent aquifer water pumped versus time based upon drawdown in the well. The U.S. Environmental Protection Agency (EPA) recommends removal of three well-casing volumes prior to sampling.

The extent of well purging will vary with the hydraulic properties of the water-bearing unit being monitored. Without proper consideration of the flow characteristics of the monitored unit, the integrity of the sample collected after purging could be compromised. Giddings (4) illustrated that emptying the wellbore of a well screened in a low-yield unconsolidated aquifer may result in a steep hydraulic gradient in the sand pack. This steep hydraulic gradient, in turn, can lead to the addition of clays and silts to the produced water; turbulent flow into the well; and a possible loss of volatile organic compounds in the produced water. Similarly, wells screened in very low-yield bedrock with fracture flow may also be bailed dry. If the water-bearing fractures or higher permeable zone is located near the static water table, the water will refill the well by cascading into the wellbore, and volatile compounds will be lost from the water. Purging of a high-yield aquifer that has major water-producing fractures or a highly permeable unit at the bottom of a screened section in an unconsolidated aquifer may result in limited purging of water higher up in the wellbore and subsequent sampling of water that has been stagnant in the well.

Removal of stagnant water from the wellbore before sampling is necessary to ensure that a representative sample is obtained. However, equally important are the hydraulic processes resulting from well purging. To minimize turbulent flow and sample alteration, each monitor well should be tested to determine the necessary extent and the appropriate rate of well purging prior to preparing a sampling program. Determination of the necessary extent of well purging can be based upon equilibration of groundwater indicator parameters during well evacuation. Figure 3 indicates changes in nine analytical parameters with pumping in a well that had been idle for 6 months prior to pumping. When pumping began, the partially reduced water surrounding the pump was discharged first. As pumping continued, formation waters were drawn into the

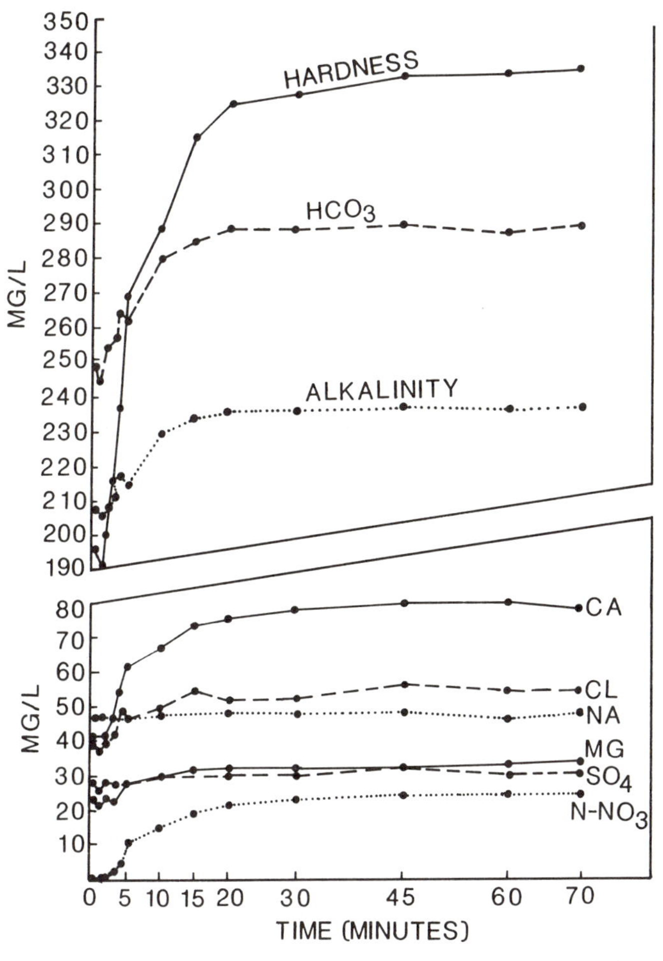

Figure 3. Concentrations of chemical parameters versus pumping time. (Reproduced with permission from reference 3. Copyright 1981 R. I. Chapin.)

wellbore, and the rate of change in the concentration of the chemical parameters decreased. Because of the low discharge rate of this well, the mixing of wellbore water and formation water continued for 45 min (5). To obtain a representative sample, the QA and QC sampling protocol for this well would specify a minimum period for well evacuation of 45 min prior to collection of the sample when the existing pump is run at the same flow rate used in the test.

Sampling Device

Commonly used sampling devices include electrical submersible pumps, positive-displacement bladder pumps, bailers, and suction-lift pumps. Choosing a sampling device is dependent upon site-specific criteria including compatibility of the rate of well purging with well yield, well diameter, limitations in the lift capability of the device, and the sensitivity of selected chemical species to the mechanism of sampling delivery.

Aeration or degassing of a sample can occur during withdrawal of a sample from a monitoring well. The introduction or loss of volatile organic compounds or gases (e.g., O_2, N_2, CO_2, and CH_4) in the groundwater sample can affect the solution chemistry of the sample and result in the change in speciation of both volatile organic compounds and other analytes of interest. The degree of aeration or degassing of the groundwater has been shown to vary with the type of sampling device employed.

A field evaluation of sampling devices was conducted in association with ongoing remedial action at the Savannah River plant in Aiken, SC (6). The electric submersible pump was chosen over various modified bailers, the bladder pump, and others because of its accuracy, precision, reliability, its ability to evacuate a well, and its moderate cost. However, levels of organic compounds at the Savannah River plant can be as high as 200,000 ppb. In this case, detection of organic compounds near analytical detection limits (i.e., 1–10 ppb) was not a criterion for choosing a sampling device. Where detection of organic compounds at low levels is desired, evaluation of sampling devices to determine the least potential alteration should be conducted.

Bailers are commonly used for both purging and sampling water from small diameter, shallow wells because of their relative low cost, portability, and ease of maintenance. Disadvantages of the bailer as a sampling device are mixing and the potential aeration or degassing of the sample during sample collection. The aeration is the result of repeated submergence and removal of the bailer during sampling, which may result in turbulent flow of water in the wellbore. Further aeration can occur as a result of pouring the collected sample from the top of the bailer into the sample bottles. Aeration of a sample when using a bailer can be minimized by gently lowering the bailer into the water when collecting the sample. Aeration or degassing can be further reduced with a bailer modified to include a bottom-draw valve. The device allows emptying of the bailer at a slow controlled rate; thus, aeration of the sample, which occurs during decanting, is avoided. Slight improvements in sample representativeness between conventional bailers and bottom-draw-type bailers have been documented by Barcelona et al. (*7*).

Field and laboratory testing of suction-lift and gas-displacement pumps indicate these pumps are consistently below average in terms of the accuracy of the sample delivered compared to other devices (*8–10*). The suction-lift pump employs application of a strong negative pressure that can cause degassing of the water sample. Gas-displacement pumps, typically air- or nitrogen-lift, can cause gas stripping of carbon dioxide and result in a change in initial pH of the carbon dioxide or cause gas stripping of volatile compounds.

Absorption of Air Emissions

In addition to the loss of volatile compounds due to degassing during sample collection, recent data indicate that organic compounds may be absorbed from the atmosphere into the water sample when decanting from the bailer to the sample bottle. In one case, 11 monitoring wells were sampled to determine the lateral and vertical extent of nitrotoluene and dinitrotoluene (DNT) isomers relative to surface impoundments containing DNT and DNT process byproducts. During collection of the samples, corresponding field blanks were collected at each monitor well to monitor potential absorption of organic compounds from the air by the collected water sample. The field blanks consisted of distilled water; the water was passed between two glass sample bottles approximately six times at the well site prior to collection of the well water sample. Water quality results for both the well sample and the field blank are included in Table I. The average percent variation in concentration of 2,4-DNT and 2,6-DNT in the groundwater, excluding outside values, as measured by the field blank, was 6% and 7%, respectively. The percent variation represents the potential DNT available for absorption, as indicated by concentrations measured in each corresponding field blank.

Sampling Device Construction Materials

Most of the wells presently used in RCRA monitoring programs are constructed of rigid poly(vinyl chloride) (PVC). Several studies have been completed to investigate the absorption and release of organic compounds by rigid PVC. Preliminary studies have led the EPA to recommend the use of well construction materials made exclusively of polytetrafluoroethylene (PTFE) or stainless steel as opposed to PVC (*11*). In fact, the quantities of the organic compounds absorbed by the PVC are low (<1 ng/cm^2) (*12*). A recent study by Reynolds and Gillham (*13*) documents absorption of organic compounds by PTFE tubing. In particular, the uptake of the compound tetrachloroethylene was noted within 5 min of exposure to the solution; quantification of the amount of absorption by the tubing was not reported (*13*). On the basis of these and other studies (*14*),

Table I. Organic Analyses of Groundwater and Corresponding Field Blanks for 2,4-Dinitrotoluene and 2,6-Dinitrotoluene

Well Number	Parameter	Concentration (mg/L) Well Sample	Field Blank	Percent Variation
1	2,4-DNT	ND	ND	0
	2,6-DNT	ND	ND	0
2	2,4-DNT	1.88	0.014	0.7
	2,6-DNT	3.77	0.008	0.2
3	2,4-DNT	0.017	0.019	112.0
	2,6-DNT	0.024	0.013	54.0
4	2,4-DNT	0.098	0.002	2.0
	2,6-DNT	0.356	0.001	0.3
5	2,4-DNT	0.001	ND	0
	2,6-DNT	0.003	ND	0
6	2,4-DNT	0.306	0.001	0.3
	2,6-DNT	0.188	ND	0
7	2,4-DNT	0.011	ND	0
	2,6-DNT	0.083	ND	0
8	2,4-DNT	0.050	0.001	2.0
	2,6-DNT	0.356	0.001	0.3
9	2,4-DNT	0.004	0.002	50.0
	2,6-DNT	0.050	0.001	2.0
10	2,4-DNT	ND	ND	0
	2,6-DNT	ND	ND	0
11	2,4-DNT	ND	0.006	0.6
	2,6-DNT	ND	ND	0

NOTE: The abbreviation ND denotes not detected.

no real justification has been presented for the replacement of rigid PVC sampling devices with sampling devices utilizing PTFE tubing.

Distinguishing between the use of sampling materials constructed of rigid PVC versus materials constructed of flexible PVC is important. Plasticizers are added to PVC resin to yield a more flexible product. These plasticizers include phthalate esters, which have been shown to leach into water. Rigid PVC pipe with a National Sanitation Foundation (NSF) listing would not be expected to contain any more than 0.01 wt % of plasticizers (*15*). Flexible PVC can contain from 30 to 50 wt % of plasticizers. When monitoring for low levels of organic compounds, materials constructed of flexible PVC (e.g., tubing and sample bottles) can be appropriately substituted with PTFE, which does not require the addition of plasticizers for flexibility.

Sample Preservation

Water samples may undergo change with regard to their physical, chemical, and biological state during transport and storage. To preserve the integrity of a sample after collection, the samples are generally refrigerated or preserved by the addition of acid or alkaline solutions.

Despite these practices of stabilizing samples, a potential exists for alteration of a sample during transport and storage. Particular practices and areas of disparity that may contribute to the variance of apparent water quality during the sample holding period are as follows:

1. delaying filtering and preserving of samples until samples reach the laboratory,
2. aeration of the sample during filtration,
3. failure to filter samples prior to the addition of acid for preservation, and
4. the lack of necessary temperature reduction for successful stabilization of the sample during transport.

A field experiment has shown that the delay of preservation of samples can lead to variation in water quality analysis (*16*). In the experiment, multiple samples were collected from one monitoring well installed at an anaerobic lagoon and one monitoring well installed at an inactive sanitary landfill. Once collected, the samples were divided into four sets; the first set was preserved immediately, and the remaining sets were preserved 7, 24, and 48 h after collection by the addition of acid. Each of the collected samples was analyzed for calcium, iron, potassium, magnesium, manganese, sodium, and zinc within the EPA prescribed holding times specified for each of the parameters. Iron showed the most dramatic change in concentration. Seven hours after collection, the measured concentration of iron in the sample collected from the well located at the lagoon was 0.33 mg/L; the concentration of iron in the sample collected from the same well and preserved immediately was 11.6 mg/L. The change in iron concentration from the sample collected at the landfill showed a change from 5.74 to $<$0.08 mg/L between 0 and 7 h after collection and before preservation. Significant changes were also noted for magnesium, manganese, and zinc.

One possible explanation for the sample alteration is the aeration of the sample during transfer from the sampling device to the sample bottle or from the sampling device to a holding vessel prior to filtration, and prior to fixation of the metals by addition of acid. When groundwater is in a reduced state, the addition of oxygen via aeration can cause oxidation of ferrous iron to ferric iron and subsequent precipitation as ferric hydroxide. Once allowed to form, much of the ferric hydroxide will be removed by filtering prior to analysis.

A recent laboratory experiment measured the precipitation of iron from a collected sample with different filtration methods and different sampling devices. The filtration methods tested included on-line filtration, vacuum filtration following transfer from a holding vessel, and the same vacuum filtration procedure after a 10-min holding time. Sampling mechanisms used included a bailer, peristaltic pump, bladder pump, air- and nitrogen-lift pumps, and a submersible electrical pump. With each sample mechanism used, the samples handled by on-line filtration exhibited higher dissolved iron concentrations than samples transferred to holding containers prior to filtration. The 10-min holding period appeared to have no consistent effect on the concentration of measured iron as compared to immediate filtration from the holding vessel (*17*). The study indicates that the turbulence and associated aeration of the sample during filtering can significantly alter sample quality. In fact, the study indicates that aeration of the sample during filtration has at least as much impact on sample quality as the sampling device itself.

Many monitor wells are completed in low-yield, clay-rich sediments. Completing these wells in such a fashion that water samples can be collected free of sediment is impractical and in some cases impossible. EPA recommends field acidification of samples collected for metals analysis to a pH less than 2 (*18*). Acidification of unfiltered samples can lead to dissolution of minerals from clays in the suspended solids. Table II indicates that the measured concentrations of calcium and magnesium in samples acidified prior to filtration are directly related to the concentration of suspended solids. On the other hand, concentrations of calcium and magnesium in unacidified samples show no correlation to dissolved solids (*19*). This lack of correlation does not imply that samples should not be acidified, but rather that samples should be filtered prior to acidification. Otherwise, constituents of interest that may occur naturally in the formation matrix may be dissolved when acidified and result in a sample that is not representative of the water contained in the aquifer.

EPA states that preservation of samples by refrigeration requires that the temperature of collected samples be adjusted to a temperature of 4 °C immediately after collection and during shipment. The cooling rates of water samples chilled by ice and the temperature maintenance ability of frozen blue ice were recorded in order to observe the effectiveness of different types of ice in cooling samples and to determine the effort required to maintain sample bottles at 4 °C. In this study, 10 250-mL bottles and 12 500-mL bottles were filled with tap water. The initial temperatures of the samples were recorded. Thermistors (electronic thermometers) were inserted through small holes drilled

Table II. Addition of Acidic Preservative Prior to Filtering

Sample Number	*Suspended Solids (mg/L)*	*Acidified*		*Unacidified*	
		Ca (mg/L)	*Mg (mg/L)*	*Ca (mg/L)*	*Mg (mg/L)*
1	22,000	2442	55	44	18
2	18,500	1980	54	73	16
3	9700	1452	34	95	13
4	8600	1452	36	78	15
5	5200	915	33	134	20
6	3400	827	47	284	36
7	3100	704	27	101	19
8	2200	453	33	134	27
9	1900	286	18	78	13

in the center of each sample bottle lid. The bottles were placed in a 48-qt cooler and covered with two bags of ice, and their temperatures were monitored. Readings were taken every 10 min until the monitored bottles reached their desired temperature of 4 °C. These bottles were transferred to a precooled ice chest filled with blue ice. As indicated in Figure 4, the temperature of the samples dropped to 4 °C within 3 h. As shown in Figure 5, the blue ice was successful in maintaining the bottles less than 4 °C for 24 h. In contrast, when ambient temperature samples were placed in a 48-qt cooler and covered only with blue ice, the samples did not reach 4 °C (Figure 6).

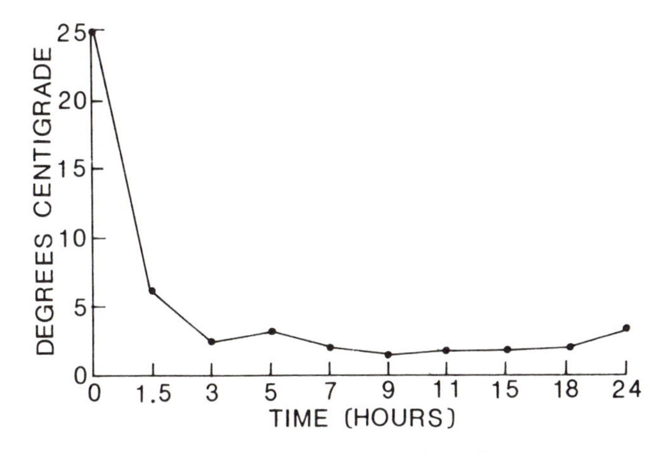

Figure 4. Field refrigeration of samples with water ice.

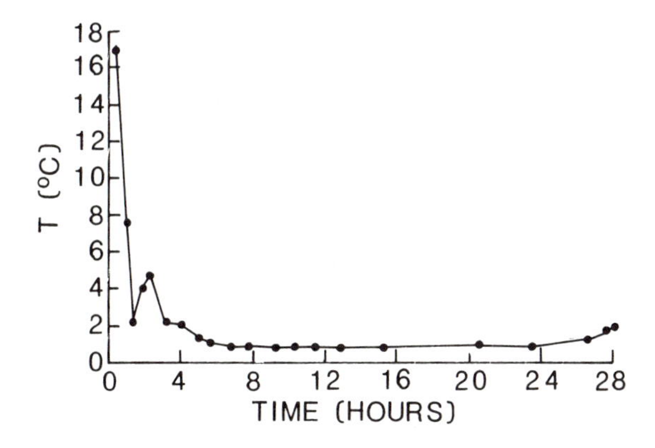

Figure 5. Bottles placed in crushed ice chilled to 4 °C and transferred to an ice chest prechilled with blue ice.

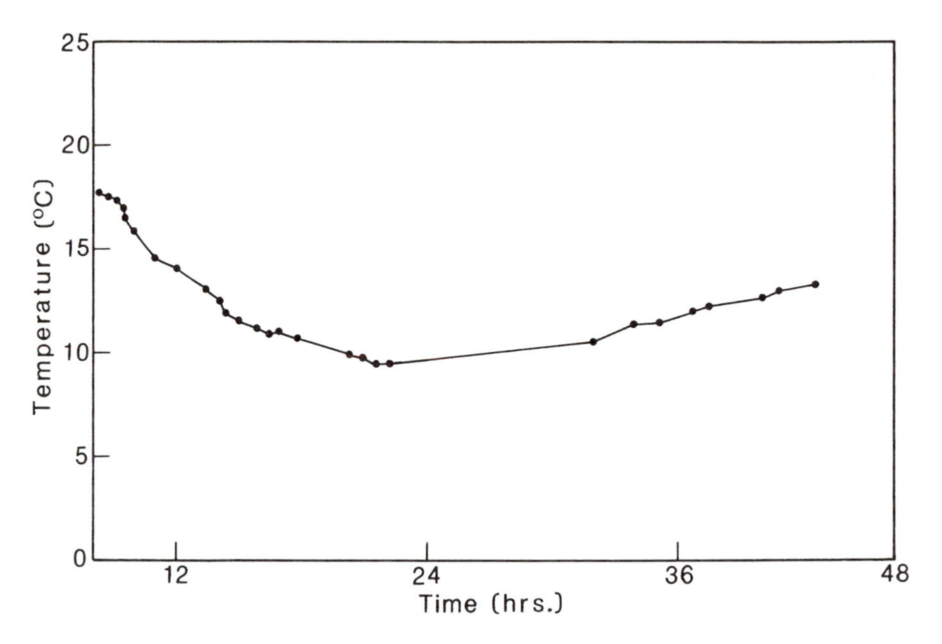

Figure 6. Field refrigeration of samples with blue ice.

This experiment suggests that when using blue ice for refrigeration, the samples must be initially chilled with wet ice. Samples can then be transported to the laboratory in either blue ice or wet ice. However, when using wet ice for an extended period, additional ice may need to be added to the ice chest to maintain the recommended temperature of 4 °C.

In conclusion, multiple avenues for sample alteration exist during collection, including the method of well purging, the method and device used to sample the wells, and the method of sample preservation. The alteration of a sample that occurs during sampling may be more or less quantified by collecting replicate samples a few days apart (*20*) (e.g., collecting samples from the monitoring system on Monday, collecting another set of samples on Thursday, and comparing the variability in analytical data between the two sets). When the variation is significant, the groundwater sampling protocol should be tested in the field (e.g., comparing variations in analyses resulting from different methods of filtration) to determine the source or sources of sample alteration.

Laboratory Test Methods

The limitations in accuracy of the test method used to quantify the analyte should be considered in any statistical or intuitive interpretation of the analytical data. Of particular concern is the accuracy of the gas chromatography–mass spectrometric (GC–MS) procedure. The method is commonly used to quantify specific organic waste constituents in groundwater samples collected from RCRA monitoring wells. The accuracy of the data obtained by the GC–MS method is dependent upon the experience of the analyst in identifying and quantifying the detected compounds. Furthermore, limitations occur in accuracy associated with the method of analysis, including efficiency of the solvent used in the extraction and the efficiency of the column itself. The degree of attainable accuracy is reflected in the surrogate-spike recovery limits established by the EPA for specific compounds. The surrogate spike consists of an aliquot of each sample that is spiked with a particular known compound of known concentration. The spike is analyzed to determine if the GC–MS equipment is operating at an acceptable level of accuracy. The EPA advisory limits of recovery for particular compounds range from 35% to as much as 140% of the known concentration of the aliquot. The variability resulting from the GC–MS analytical method may far outweigh the variability resulting from the sample collection method.

Summary

The development and implementation of a sampling and analysis plan for monitoring hazardous waste sites is a requirement of federal regulations. Although implementation of a plan does not guarantee that the analytical results accurately represent the in situ quality of the groundwater, such a plan may result in recognition of the potential changes that have occurred between collection and analysis. This information, when properly considered in reviewing the analytical data, may lead to a more accurate interpretation of the data once received.

The initial QA and QC consideration when sampling groundwater monitoring wells consists of developing a thorough knowledge of the hydrogeologic system. Documenting factors such as how the recovery rate of the well influences the sample or identifying lithologic variations in the monitored strata that may affect water quality will result in a more accurate interpretation of the sample analytical data. The inherent limitations of the accuracy of the analytical test method used to quantify the monitored species should also be considered when evaluating the sample analytical data.

Abbreviations

DNT dinitrotoluene
GC–MS gas chromatography–mass spectrometry
PTFE polytetrafluoroethylene
PVC poly(vinyl chloride)
QA quality assurance
QC quality control
RCRA Resource Conservation and Recovery Act

References

1. Stolzenburg, T. R.; Nichols, D. G. *Proceedings, 6th National Symposium and Exposition on Aquifer Restoration and Ground Water Monitoring*; National Water Well Association: Dubin, OH, 1986; p 231.
2. *Guidelines for Collection and Field Analysis of Ground Water Samples for Selected Unstable Constituents*; U.S. Geological Survey: Washington, DC, 1976; Book 1, Chapter D-2.
3. Schuller, R. M.; Gibb, J. P.; Griffin, R. A. *Ground Water Monit. Rev.* **1981**, *1(1)*, 46.
4. Giddings, T. *Ground Water Monit. Rev.* **1984**, 253–256.
5. Chapin, R. I. M.S. Thesis, University of Texas at Austin, 1981.
6. Muska, C. F.; Colven, W. P.; Jones, V. D.; Scogin, J. T.; Looney, B. B.; Price, V., Jr. *Proceedings, 6th National Symposium and Exposition on Aquifer Restoration and Ground Water Monitoring*; National Water Well Association: Dubin, OH, 1986; pp 235–246.
7. Barcelona, M. J.; Helfrich, J. A.; Gibb, J. P. *Ground Water Monit. Rev.* **1984**, *4(2)*, 36.
8. Nielsen, D. M.; Yeates, G. L. *Ground Water Monit. Rev.* **1985**, *5(2)*, 83–99.
9. Schuller, R. M.; Gibb, J. P.; Griffin, R. A. *Ground Water Monit. Rev.* **1981**, *1(1)*, 44.
10. Barcelona, M. J.; Helfrich, J. A.; Gibb, J. P. *Ground Water Monit. Rev.* **1984**, *4(2)*, 38.
11. *RCRA Ground-Water Monitoring Technical Enforcement Guidance Document*; U.S. Environmental Protection Agency: Washington, DC, 1986.
12. Miller, G. D. *Proceedings, 2nd National Symposium on Aquifer Restoration and Ground Water Monitoring*; National Water Well Association: Worthington, OH, 1982; pp 236–245.
13. Reynolds, G. W.; Gillham, R. W. *2nd Annual Canadian/American Conference on Hydrogeology*; University of Waterloo, Waterloo, Ontario, Canada, 1985.
14. Current, G. M.; Thomson, M. D. *Ground Water Monit. Rev.* **1983**, *3(3)*, 68–71.
15. Barcelona, M. J.; Gibb, J. P.; Miller, R. A. *A Guide to the Selection of Materials for Monitoring Well Construction and Ground Water Sampling*; Illinois State Water Survey: Champaign, IL, 1983; p 40.
16. Schuller, R. M.; Gibb, J. P.; Griffin, R. A. *Ground Water Monit. Rev.* **1981**, *1(1)*, 46.
17. Stolzenburg, T. R.; Nichols, D. G. *Proceedings, 6th National Symposium and Exposition on Aquifer Restoration and Ground Water Monitoring*; National Water Well Association: Dubin, OH, 1986; p 216.
18. *Test Methods for Evaluating Solid Waste*, 2nd ed.; U.S. Environmental Protection Agency: Washington, DC, 1982.
19. Kent, R.; McMurtry, D.; Bentley, M. *Water Resources Symposium No. 12*; Center for Research in Water Resources: Austin, TX, 1985; 461–487.
20. Splitstone, D. E. *Proceedings of the National Conference on Hazardous Wastes and Hazardous Materials*; Hazardous Materials Control Research Institute: 1986; pp 8–12.

Chapter 22

Groundwater Sampling

James S. Smith, David P. Steele, Michael J. Malley, and Mark A. Bryant

Because more than 50% of this nation's water supply comes from groundwater, groundwater quality has become a priority. Accurate and precise analytical measurements cannot truly represent groundwater quality if the sampling alters the sample. The sampling of groundwater has numerous variables that affect the analytical results of the sample. These variables include the purging volume; the time interval between purging and sampling; the sample treatment, such as filtration; the sample storage (e.g., oxidation); the purging methodology; and placement of the purging device relative to placement of the sampling device in the well. Accurate groundwater measurements that can be interpreted over a time interval can be obtained if the sample regimen is consistent and developed according to the chemical information desired and the physical characteristics of the groundwater system.

GROUNDWATER IS ONE OF OUR most important natural resources. As scientists, we are beginning to learn the characteristics of groundwater and how to protect this precious resource. This chapter is concerned with the reevaluations of groundwater sampling techniques based on many years of experience in sampling and interpretation of the analytical results.

Groundwater is sampled to determine its quality and how the quality changes with time. Researchers want to find out what chemical species are present in the groundwater and how much of each species is present. Later, the researchers will want to determine if these qualitative and quantitative aspects of groundwater have changed. If groundwater sampling and analysis are successful, then the results can be used to develop explanations or hypotheses concerning the effect of time on groundwater quality.

This situation sounds easy enough to accomplish. First, a well is drilled to the groundwater under investigation. Appropriate well construction techniques are used to avoid and prevent contamination of the aquifer to be studied. Then the well is developed to fulfill the U.S. Environmental Protection Agency's (EPA's) criteria for water turbidity (*1*). Three or more well water volumes are purged from the well. The groundwater is sampled, and the samples are sent to a laboratory for analysis. Finally, the scientists can relax and wait for the analytical results.

Nothing is wrong with this scenario as long as the analytical variability and the variability of sampling do not affect the results and therefore the scientists' ability to determine the actual changes in groundwater quality over a period of time. Experience has taught scientists to recognize the analytical variability, especially when the results are near the limit of quantitation. Despite following accepted practices, scientists have found that the variability in the sampling of groundwater is the major source of error in the measurement of groundwater quality. This broad generalization about groundwater sampling and analysis means that unless special care is taken to minimize the variability of the sampling procedure, the interpretation of groundwater quality measurements taken over a period of time is probably wrong. If this situation is true, then serious problems are associated with groundwater protection as it is currently practiced. This chapter adds additional information to the paper by Bryden et al. (2).

Groundwater Properties

Groundwater is a very complex matrix. Although each aquifer represents a different water quality, our experience has shown that all groundwater has two unique properties for natural water systems:

- The movement of groundwater through the aquifer precludes the transport of chemicals by particulates. Only substances soluble in the groundwater matrix are mobile within the aquifer.
- Groundwater is nearly oxygen-free. In other words, the dissolved ions tend to be in their most reduced state. High iron content of groundwater indicates that the water in the aquifer is carrying ferrous iron at neutral pH values. Ferrous iron groundwater cannot be aerobic groundwater.

In other words, a groundwater sample cannot contain any particulate matter, and must be protected from air at all times if the sample is to be truly representative. If groundwater samples that contain particulates and have been exposed to the air are analyzed, then the results obtained do not represent the water within the aquifer.

How critical are these properties to the analysis of groundwater quality? On a macroscopic scale, we have observed unit pH changes as ferrous iron is oxidized by the oxygen in the air to the insoluble ferric iron. Corresponding changes in specific conductance occur also. But more important is how particulates in the sample and our oxidation of the sample affect "trace" analysis of organic compounds and metals. What does trace analysis mean?

This question appears silly to the environmental scientist accustomed to parts-per-billion data. We have become anesthetized with parts-per-billion thinking. A trihalomethane concentration of 100 ppb is considered a large amount. For toxicity evaluations, this statement is true. But the measured amount is very small (i.e., 100 ng/mL). This amount (100 ng) can be translated to one ten-millionth of the amount of sugar that can be placed on a dime. The point is that the measurement of low turbidity [5 nephelometric turbidity units (NTU)] is not an appropriate indicator of particulate-free water for the detection of parts-per-billion levels for pollutants. Therefore, the turbidity criteria are appropriate for well development only and cannot be used as an indicator parameter for particulate-adsorbed pollutants that are suspended in the groundwater sample. This situation is like using a truck weighing station to indicate the presence or absence of a dime. We have observed too many results in which the quantitative amount exceeds the solubility of the substance. The analytical work is correct, but the conclusion that the substance is in the groundwater is dubious.

No matter how much care is taken in well construction to avoid contamination from the surface or near-surface area, trace amounts will get into the aquifer region of the well. Maximum care should be planned and executed for the removal of surface and subsurface contamination before well construction.

Techniques for Proper Groundwater Sampling

The best possible analysis of groundwater constituents results from the removal of particulates from all samples and in keeping all unpreserved samples oxygen-free until analysis. Even then, special instructions to the laboratory may be required to obtain accurate results. For instance, groundwater having a high ferrous iron content should be analyzed for the acid–neutral semivolatile fraction instead of the prescribed methodology of EPA's Method 625 (3) or contract laboratory program's method (4). This situation is due to the acidic compounds being entrapped into the iron precipitate when sodium hydroxide is added to the sample before the base–neutral extraction. Following the base–neutral extraction, the acidification of the sample does not release the acid compounds, and poor recoveries are observed.

The toughest question to answer is "How should groundwater be sampled to obtain a sample that accurately represents the aquifer?" Our recommendations include use of the following:

- purge-volume test,
- consistent sampling protocol,
- anaerobic sampling and sample handling conditions, and
- filtration of all samples.

Purge-Volume Test

We used EPA's recommended three well water volumes for purging a well before sampling. In connection with this protocol, we monitored pH and specific conductance as indicator parameters in the field. In almost every case, we obtained a constant pH and specific conductance reading before the three well water volumes were removed. Therefore, we have always thought that we have fulfilled EPA's and our own quality criterion that the stagnant well water be replaced by fresh groundwater from the aquifer. We have assumed that the chemical parameters or the quality of that fresh groundwater is a constant. However, working with several groundwater systems that are contaminated with volatile organic solvents demonstrated that this assumption is grossly inaccurate.

By performing a purge-volume test to analyze for the parameter of most concern, we found some very interesting phenomena involving volatile organic solvents. For example, near the Delaware River, where the saturated zone is connected to the river and the tides associated with the river, more than six well volumes must be purged before the concentration of trichloroethylene becomes constant (*see* Table I, sample numbers 1–5).

In a more complex mixture of volatile organic solvents, the purge-volume test of a well showed three trends. Two solvents started at relatively high concentrations and decreased with purging to a "not detected" value. Two other solvents that were "not detected" at the three well water volumes purged were detected at the four well volumes purged and increased until constant values were obtained near the 10 well water volumes purged. The fifth volatile organic solvent detected in this well remained at a constant concentration value

Table I. Purge-Volume Test Results for Trichloroethylene

Sample Number	*Volume Purged (Well Water Volumes)*	*Concentration (ppb)*	*Time after Purging*	*Water Column Location*
1	0	not detected		
2	3	75		
3	6	150		
4	12	170		
5	24	170		
6		170		bottom
7		100	4 min	top
8		100	4 h	
9		10	24 h	

throughout the test. The information gained from the purge-volume test was very important in the site groundwater evaluation. Any inconsistency in well purging would not only influence the quantitative result, but also would change the qualitative suite of chemicals observed. If the well purge-volume test was omitted, then the interpretation of this well's data would probably draw the familiar conclusion of laboratory error(s).

The purge-volume test is valuable because it allows choice of the best purging conditions to obtain the most consistent analytical results. This test is especially important in the sampling and analysis of groundwater for volatile organic solvents.

Consistent Sampling Protocol

The word that keeps making its way into our recommendations on groundwater sampling is *consistency*. Consistency is the key to obtaining data that can be compared over a time interval. Consistency includes purge volume, purge rate, time between purge and sampling, level of water sampled, and sample handling.

The purge volume is selected after the purge-volume test. This purging of the well will be the same every time the well is sampled. The purge rate will also be the same for each sampling. The maximum purge rate is advantageous. Ideally, the well should be emptied first, and then after the groundwater achieves equilibrium, the second well volume should be purged.

One of the least consistent elements of groundwater sampling is the time between the purging of the well and the sample removal from the well (*see* Table I, sample numbers 7–9). Within 24 h after purging, we observed no trichloroethylene in samples from a well containing groundwater that contained 200 μg/L of trichloroethylene immediately after completion of purging. This observation strongly suggests that sampling should occur as soon after purging as possible. In any sampling protocol, the time between the completion of purging and the sampling of a well must be constant each time the well is sampled.

One of the most perplexing problems that we have encountered concerns the depth within the well that water is purged and sampled. This problem occurs in wells where the groundwater cannot be removed completely by the purging pump. The water column in the well decreases, but the well is not pumped to dryness. With the purging pump in the bottom of the well, we observed a case in which the concentration of trichloroethylene was 50% greater at the bottom of the well as compared to the concentration near the top of the water column in the well (*see* Table I, sample numbers 6 and 7). Even though several possible explanations can be offered for this observation, we feel that the lack of mixing within the well water volume contributed the major effect. The result of this experience is that the position of well water sampling should be as close as possible to the point of well purging. From sampling to sampling, the sampling point within the well water column must be the same.

Anaerobic Sampling, Sample Handling Conditions, and Sample Filtration

We believe sample handling must be drastically changed in order to obtain the most representative sample of the groundwater in the aquifer. The sample must be free of particulates and must not be exposed to air. Any sample that is not preserved should be placed into a sample container without any headspace similar to the present volatile organic analysis (VOA) sample handling protocol. Every sample except the VOA sample can be pressure-filtered under a nitrogen atmosphere through a 0.45-μm polyvinylidene fluoride or polytetrafluoroethylene filter medium. Organic samples are included in those parameters to be filtered.

The best situation would be the use of the purging pump as a sampling pump, too. To sample, insert a 0.45-μm filter into the polytetrafluoroethylene (Teflon) tubing outlet line. The filter would not aerate the sample if the outlet tube is always full of groundwater sample. This methodology delivers a filtered sample that is exposed to a minimum amount of air.

Groundwater monitoring is at the stage where the variations in the chemical data due to sampling can be reduced. The benefit of this work is a more standardized sampling methodology for groundwater. This standardization will produce a consistency in sampling that will allow time variations of groundwater parameters to be seen above sampling variations. This procedure will produce more accurate analytical data and a truer representation of the water within an aquifer. Application of this procedure leads to a better understanding of groundwater and a more cost-effective program to protect this precious resource. Groundwater quality data can thus be interpreted accurately; pollutant migration, seasonal fluctuations of parameters, and groundwater cleanup technologies can then be understood. Research on groundwater sampling should be the top priority in groundwater environmental protection.

References

1. "RCRA Ground-Water Monitoring Technical Enforcement Guidance Document"; U.S. Environmental Protection Agency. U.S. Government Printing Office: Washington, DC, 1986.
2. Bryden, G.; Mobey, W.; Robine, K. *Ground Water Monit. Rev.* **1986**, *6(2)*, 67–72.
3. *Fed. Regist.* **1984**, *49(209)*, 153–174.
4. "U.S. EPA Contract Laboratory Program Statement of Work for Organics Analysis Multi-Media–Multi-Concentrations"; Exhibit D, October 1986; U.S. Environmental Protection Agency: Washington, DC, 1986.

Sampling Air

Chapter 23

Sampling for Organic Chemicals in Air

Robert G. Lewis and Sydney M. Gordon

Airborne organic chemicals may be classified on the basis of vapor pressure. Volatile organic chemicals (VOCs) are present in ambient air in the gas phase, and nonvolatile organic chemicals (NVOCs) exist predominantly in the aerosol phase. Semivolatile organic chemicals (SVOCs), with vapor pressures from 10^{-2} to 10^{-8} kPa, may be substantially distributed between the gaseous and aerosol (particle-associated) phases. Vapor pressure classification will largely dictate the choice of sampling methodology. This chapter attempts to address the problems associated with sampling for organic chemicals in air. The emphasis is on ambient (outdoor) air, with limited discussion of the special requirements of non-workplace indoor air sampling. Sampling methods for stationary and mobile sources are not included. Overall aspects of sampling and individual sampling methods for VOCs, SVOCs, and NVOCs are presented. The principal focus is on collection of organic compounds from air, although continuous monitoring methods, which include analytical detection, are covered. Whole-air collection, cryogenic trapping, sorbent-based sampling, and real-time monitoring methods are described for VOCs. For SVOCs and NVOCs, sampling methods employing particle filters and sorbent traps are also presented. Sample integrity and artifact problems related to air sampling are discussed along with potential breakthrough during sampling. Specific requirements for determining polar VOCs are covered. Particular difficulties associated with separating and maintaining the integrity of phase-distributed SVOCs are also discussed. The authors have surveyed the current literature and have included an extensive list of references that provides comprehensive coverage of the state-of-the-art of sampling methodology.

ORGANIC CHEMICALS by far account for the majority of pollutants found in air. The 1985 edition of the *Toxic Substances Control Act Chemical Substance Inventory* (*1*) listed 63,000 compounds, more than 90% of which were organic chemicals. About 10,000 more have been added since 1985. Thousands of these may exist in air. The first serious attempt in the United States to control organic pollutants in the air came with the passage of the Clean Air Act Amendments (CAAA) of 1990. Title III, Section 112 of the CAAA listed 189 hazardous air pollutants (HAPs) to be controlled. This list of HAPs was incorporated into the Clean Air Act (CAA) under Title I, Part A, Section 112 (2). Some 166, or 88%, of the 189 HAPs are organic chemicals or mixtures of organic chemicals. The listed HAPs represent a broad spectrum of chemical classes that cover a wide range of volatility, polarity, chemical reactivity, and other chemical and physical properties. Their measurement in air presents a formidable task to those concerned with monitoring and regulating air pollution. The U.S. Environmental Protection Agency (EPA) is currently assessing the availability of methodologies to monitor the CAA HAPs (3, 4).

Volatility

Vapor pressure is the most important property governing the approach to sampling organic chemicals in air. Organic chemicals can be conveniently categorized by volatility into three groups: *volatile, semivolatile,* and *nonvolatile.* We define the ranges of saturation vapor pressures (p_o, at 25 °C) associated with these categories as $>10^{-2}$, 10^{-2} to 10^{-8}, and $<10^{-8}$ kPa, respectively (5, 6). Volatile organic chemicals (VOCs) may be further subdivided into very volatile organic chemicals (VVOCs), those VOCs having vapor pressures greater than ca. 15 kPa.

These classifications are based on empirical observations of air-sampling behavior. Compounds with $p_o < 10^{-2}$ kPa are difficult to recover by thermal desorption of Tenax (2,6-diphenyl-*p*-phenylene oxide polymer) (as well as other *sorbents*) or from canisters. Those with $p_o < 10^{-8}$ kPa tend to be retained with particulate matter on particle filters. The nonvolatile classification also agrees with the theoretical prediction by Junge (7) that compounds with p_o below 10^{-8} kPa will be mostly in the particle-associated phase in ambient air. Other vapor pressure classifications have been offered. For example, Bloemen and Burn (8) suggest 0.13 kPa (1 Torr) as the upper limit for VOCs, and Appel (9) classifies semivolatile organic chemicals (SVOCs) as those with vapor pressure in the range of 10^{-4} to 10^{-8} Torr (ca. 10^{-5} to 10^{-9} kPa).

Boiling points have also been used to classify airborne organic chemicals. VOCs are variously defined as having boiling points below 60, 100, or 150 °C, whereas nonvolatile organic chemicals (NVOCs) are those that boil above ca. 300 °C (*10, 11*). Boiling points, however, are not reliable predictors of volatility. Many solids have a propensity to sublime and, therefore, possess solid-state vapor pressures at ambient temperatures that are larger than their boiling points would predict. The boiling points of many compounds are difficult or impossible to determine because they decompose before they boil at atmospheric pressure. Several important SVOC air pollutants have boiling points above 400 °C and vapor pressures between 10^{-4} and 10^{-7} kPa. A more realistic upper boiling point for VOCs may be 180 °C (*12*). Even with this criterion, however, nine CAA HAPs with p_o above 10^{-2}

kPa have higher boiling points. The most notable discrepancy is for 1,3-propane sultone, which has a p_0 of 0.26 kPa and an estimated boiling point of 320 °C at atmospheric pressure.

Although vapor pressure is the main criterion used to define airborne organic chemicals, the boundaries between the volatility groups have been set somewhat arbitrarily. Moreover, the boundaries are not sharp, but may be regarded as "gray zones" that delineate a gradual transition from one volatility category to the next. Compounds with vapor pressures that fall into these zones may frequently be collected using sampling methodology associated with either of the categories that bracket the zone.

VOCs are generally present in the atmosphere in the gaseous state only, and many are found in urban outdoor air at concentrations in the 0.1–10-$\mu g/m^3$ range. Indoor air in residences, offices, public access buildings, and transportation vehicles often contains VOCs at levels an order of magnitude higher than those outdoors. SVOCs are usually in the ambient outdoor air at picogram to nanogram per cubic meter concentrations, but they may be present at high nanogram to microgram per cubic meter levels in indoor environments. SVOCs may exist in air distributed between the vapor and particle-associated phases, making air sampling more complicated. NVOCs are generally found at still lower concentrations in air.

Vapor pressure data have been published for many organic compounds (*3, 13–17*). They are variously presented in units of kilopascal (kPa), pascal (Pa), or millipascal (mPa), or in millimeters of mercury (mmHg, Torr), which is equivalent to 0.133 kPa. Bidleman (*18*) has developed a method to estimate the vapor pressures of nonpolar organic chemicals by capillary column gas chromatography (GC) using *n*-alkanes as reference compounds.

The individual organic compounds listed under Title I, Section 112 of the CAA (Title III of the 1990 CAAA) may be classified as 99 VOCs (including 31 VVOCs), 51 SVOCs, and 5 NVOCs. In addition, there are 11 complex mixtures: cresols (VOCs); xylenes (VOCs); hexachlorocyclohexanes (SVOCs); 2,4-D (2,4-dichlorophenoxyacetic acid) free acid, esters, and salts (SVOCs and NVOCs); 4,6-dinitro-*o*-cresol and salts (SVOCs and NVOCs); polychlorinated biphenyls (SVOCs and NVOCs); dibenzofurans (SVOCs and NVOCs); glycol ethers (VOCs and SVOCs); toxaphene (SVOCs); coke oven emissions (VOCs, SVOCs, and NVOCs); and polycyclic organic matter (SVOCs and NVOCs). Fifty-two of these are polar and 29 are moderately or highly reactive. The organic HAPs from the CAA are listed in Tables I–V.

Other Important Properties

Polarity, water solubility, and reactivity are other important properties that impact on sampling methodology. Hydrocarbons (including halogenated hydrocarbons) are generally regarded as nonpolar compounds and can be readily characterized at ambient air levels using currently available methods. *Polar organic compounds* typically contain oxygen, nitrogen, sulfur, or other heteroatoms and may be categorized as either ionizable or polarizable. Ionizable organic compounds include alcohols, phenols, amines, and carboxylic acids; polarizable organic compounds include ketones, ethers, nitro compounds, nitriles, and isocyanates. Polar compounds are generally more difficult to recover from sampling devices and present special analyt-

Table I. Properties of Clean Air Act Very Volatile Organic HAPs

Compound	*CAS No.*	*Vapor Pressure (kPa at 25 °C)*	*Boiling Point (°C)*	*Water Solubility (g/L at °C)*	*Customary Classification*	*Reactivity in Air*
Acetaldehyde	75–07–0	127	21	33.0 / 25	Polar	
Acrolein	107–02–8	29	53	>100 / 21	Polar	Reactive
Allyl chloride	107–05–1	45	45	19.5 / 20	Nonpolar	
1,3-Butadiene	106–99–0	267	–5	Insoluble	Nonpolar	Reactive (?)
Carbon disulfide	75–15–0	35	47	<1 / 20	Nonpolar	
Carbonyl sulfide	463–58–1	493	–50	>100 / 20	Polar	
Chloroform	67–66–3	21	61	0.85 / 20–24	Nonpolar	
Chloromethyl methyl ether	107–30–2	30	59	Reacts	Polar	Reactive
Chloroprene	126–99–8	30	59	Slightly soluble	Nonpolar	
Diazomethane	334–88–3	373	–23	Reacts	Polar	Highly reactive
1,1-Dimethylhydrazine	57–14–7	21	63	Reacts	Nonpolar	Reactive (?)
1,2-Epoxybutane	106–88–7	22	63	>100 / 17	Polar	Reactive
Ethyl chloride	75–00–3	133	13	>100 / 20	Nonpolar	
Ethyleneimine	151–56–4	21	56	Miscible	Polar	Reactive (?)
Ethylene oxide	75–21–8	147	11	Miscible	Polar	Reactive
Ethylidene dichloride	75–34–3	31	57	<1 / 20	Nonpolar	
Formaldehyde	50–00–0	360	–20	>100 / 20.5	Polar	
Hexane	110–54–3	16	69	<1 / 16.5	Nonpolar	
Methyl bromide	74–83–9	240	4	Slightly soluble	Nonpolar	Pesticide
Methyl chloride	74–87–3	507	–24	Slightly soluble	Nonpolar	
Methyl iodide	74–88–4	53	42	10–50 / 18	Nonpolar	
Methyl isocyanate	624–83–9	46	60	Reacts	Polar	Highly reactive
Methyl *tert*-butyl ether	1634–04–4	33	55	Soluble	Polar	
Methylene chloride	75–09–2	47	40	10–50 / 21	Nonpolar	
Phosgene	75–44–5	160	8	Slightly soluble	Polar	Reactive (?)
Propionaldehyde	123–38–6	31	49	50–100 / 18	Polar	
Propylene oxide	75–56–9	59	34	400 / 20	Polar	Reactive
1,2-Propyleneimine	75–55–8	15	66	>100 / 19	Polar	Highly reactive (?)
Vinyl bromide	593–60–2	147	16	Insoluble	Nonpolar	
Vinyl chloride	75–01–4	427	–14	Slightly soluble	Nonpolar	
Vinylidene chloride	75–35–4	67	32	5–10 / 21	Nonpolar	

NOTE: Clean Air Act very volatile organic hazardous air pollutants (HAPs) have vapor pressures >15 kPa.
SOURCE: Data are from reference 3.

Table II. Properties of Clean Air Act Volatile Organic HAPs

Compound	*CAS No.*	*Vapor Pressure (kPa at 25 °C)*	*Boiling Point (°C)*	*Water Solubility (g/L at °C)*	*Customary Classification*	*Reactivity in Air*
Acetonitrile	75–05–8	9.86	82	>100 / 22	Polar	
Acetophenone	98–86–2	0.13	202	6.3 / 25	Polar	
Acrylamide	79–06–1	0.07	125/25 mm	>100 / 22	Polar	Reactive
Acrylic acid	79–10–7	0.43	141	>100 / 17	Polar	
Acrylonitrile	107–13–1	13.33	77	716.0 / 25	Polar	
Aniline	62–53–3	0.09	184	1.0 / 25	Polar	
o-Anisidine	90–04–0	0.01	224	<0.1 / 19	Polar	Reactive
Benzene	71–43–2	10.13	80	1–5 / 18	Nonpolar	
Benzyl chloride	100–44–7	0.13	179	Reacts	Nonpolar	Reactive (?)
Bis(chloromethyl) ether	542–88–1	4.00	104	Reacts	Polar	Reactive
Bromoform	75–25–2	0.75	149	<0.1 / 22.5	Nonpolar	
Carbon tetrachloride	56–23–5	12.00	77	<1 / 21	Nonpolar	
Catechol	120–80–9	0.03	240	>100 / 21.5	Polar	
Chloroacetic acid	79–11–8	0.09	189	>100 / 20	Polar	
Chlorobenzene	108–90–7	1.17	132	<1 / 20	Nonpolar	
o-Cresol	95–48–7	0.03	191	25.9 / 25	Polar	
Cumene	98–82–8	0.43	153	Insoluble	Nonpolar	
1,2-Dibromo-3-chloropropane	96–12–8	0.11	196	<0.1 / 18	Nonpolar	
1,4-Dichlorobenzene	106–46–7	0.08	173	<1 / 23	Nonpolar	
Dichloroethyl ether	111–44–4	0.09	178	Reacts	Polar	Reactive (?)
1,3-Dichloropropene	542–75–6	3.71	112	<0.1 / 16.5	Nonpolar	
Diethyl sulfate	64–67–5	0.04	208	Reacts	Polar	Reactive (?)
N,N-Dimethylaniline	121–69–7	0.07	192	<1 / 21	Polar	
Dimethylcarbamyl chloride	79–44–7	0.65	166	Reacts	Polar	Highly reactive
N,N-Dimethyl-formamide	68–12–2	0.36	153	>100 / 22	Polar	
Dimethyl sulfate	77–78–1	0.13	188	>100 / 20	Polar	Reactive (?)
1,4-Dioxane	123–91–1	4.93	101	>100 / 20	Polar	
Epichlorohydrin	106–89–8	1.60	117	50–100 / 22	Polar	Highly reactive
Ethyl acrylate	140–88–5	3.91	100	4.2 / 20	Polar	
Ethylbenzene	100–41–4	0.93	136	<1 / 23	Nonpolar	
Ethyl carbamate	51–79–6	0.07	183	>100 / 22	Polar	
Ethylene dibromide	106–93–4	1.47	132	<1 / 21	Nonpolar	Pesticide
Ethylene dichloride	107–06–2	8.20	84	5–10 / 19	Nonpolar	Pesticide
Hexachlorobutadiene	87–68–3	0.05	215	<0.1 / 22	Nonpolar	
Hexachloroethane	67–72–1	0.05	Sublimes at 186	<1 / 21	Nonpolar	

Continued on next page

Table II. Properties of Clean Air Act Volatile Organic HAPs—*Continued*

Compound	*CAS No.*	*Vapor Pressure (kPa at 25 °C)*	*Boiling Point (°C)*	*Water Solubility (g/L at °C)*	*Customary Classification*	*Reactivity in Air*
Hexamethyl-phosphoramide	680–31–9	0.01	233	>100 / 18	Polar	
Isophorone	78–59–1	0.05	215	0.1–1 / 18	Polar	
Methanol	67–56–1	12.26	65	>100 / 21	Polar	
Methyl chloroform	71–55–6	13.33	74	<1 / 20	Nonpolar	
Methyl ethyl ketone	78–93–3	10.33	80	>100 / 19	Polar	
Methylhydrazine	60–34–4	6.61	88	<1 / 24	Nonpolar	Highly reactive
Methyl isobutyl ketone	108–10–1	0.80	117	1–5 / 21	Polar	
Methyl methacrylate	80–62–6	3.73	101	15.9 / 20	Polar	
Nitrobenzene	98–95–3	0.02	211	1.9 / 25	Polar	
2-Nitropropane	79–46–9	1.33	120	1.7 / 20	Polar	
N-Nitroso-*N*-methylurea	684–93–5	1.33	124	<1 / 18	Polar	Reactive
N-Nitrosodimethyl-amine	62–75–9	0.49	152	>100 / 19	Polar	Reactive
N-Nitrosomorpholine	59–89–2	0.04	225	>100 / 19	Polar	
Phenol	108–95–2	0.03	182	50–100 / 19	Polar	
1,3-Propane sultone	1120–71–4	0.27	180/30 mm	0.1	Polar	Reactive (?)
β-Propiolactone	57–57–8	0.45	Decomposes at 162	37.0 / 20	Polar	
Propylene dichloride	78–87–5	5.60	97	<0.1 / 21.5	Nonpolar	Pesticide
Quinoline	91–25–5	0.01	238	<0.1 / 22.5	Polar	
Styrene	100–42–5	0.88	145	<1 / 19	Nonpolar	
Styrene oxide	96–09–3	0.04	194	<1 / 19.5	Polar	Highly reactive
1,1,2,2-Tetrachloro-ethane	79–34–5	0.67	146	<0.1 / 22	Nonpolar	
Tetrachloroethylene	127–18–4	1.87	121	<0.1 / 17	Nonpolar	
Toluene	108–88–3	2.93	111	<1 / 18	Nonpolar	
o-Toluidine	95–53–4	0.01	200	5–10 / 15	Polar	
1,2,4-Trichlorobenzene	120–82–1	0.02	213	<1 / 21	Nonpolar	
1,1,2-Trichloroethane	79–00–5	2.53	114	1–5 / 20	Nonpolar	
Trichloroethylene	79–01–6	2.67	87	<1 / 21	Nonpolar	
Triethylamine	121–44–8	7.20	90	Soluble	Polar	Reactive (?); strong base
2,2,4-Trimethyl pentane	540–84–1	5.41	99	Insoluble	Nonpolar	
Vinyl acetate	108–05–4	11.06	72	Insoluble	Polar	
o-Xylene	95–47–6	0.67	144	Insoluble	Nonpolar	
m-Xylene	108–38–3	0.80	139	Insoluble	Nonpolar	
p-Xylene	106–42–3	0.87	138	Insoluble	Nonpolar	

NOTE: Clean Air Act volatile organic HAPs have vapor pressures between 10^{-2} and 15 kPa.

SOURCE: Data are from reference 3.

Table III. Properties of Clean Air Act Semivolatile Organic HAPs

Compound	*CAS No.*	*Vapor Pressure (kPa at 25 °C)*	*Boiling Point (°C)*	*Water Solubility (g/L at °C)*	*Reactivity in Air*
Acetamide	60–35–5	9.6×10^{-3}	222	>100 / 22	
4-Aminobiphenyl	92–67–1	8.0×10^{-6}	302	<0.1 / 19	
Benzidine	92–87–5	1.3×10^{-6}	402	< 1 / 22	
Benzotrichloride	98–07–7	1.0×10^{-2}	213	Reacts	Reactive
Biphenyl	92–52–4	5.2×10^{-5}	254	Insoluble	
Bis(2-ethylhexyl)phthalate (DEHP)	117–81–7	1.3×10^{-8} [a]	384	<0.1 / 22	
Caprolactam	105–60–2	1.3×10^{-4}	139/12 mm	>100 / 20.5	
Captan	133–06–2	1.3×10^{-7}	479	<1 / 20	
Carbaryl	63–25–2	1.9×10^{-7}	331	40 /	
Chloramben	133–90–4	6.3×10^{-7}	350	<0.1 / 22	
Chlordane	57–74–9	1.3×10^{-6}	175/2 mm	<1 / 23	
2-Chloroacetophenone	532–27–4	1.6×10^{-3}	245	<1 / 19	
Chlorobenzilate	510–15–6	2.9×10^{-7}	415	<0.1 / 22	
m-Cresol	108–39–4	5.3×10^{-3}	202	10–50 / 20	
p-Cresol	106–44–5	5.3×10^{-3}	202	<1 / 21	
Dibutylphthalate	84–74–2	5.6×10^{-6}	340	<1 / 20	
3,3′-Dichlorobenzidine	91–94–1	3.5×10^{-7}	402	Insoluble	
1,1-Dichloro-2,2-bis(*p*-chlorophenyl)ethylene (DDE)	72–55–9	4.3×10^{-7}	350	<0.1 / 22	
Dichlorvos	62–73–7	7.1×10^{-3}	140/20 mm	0.01 /	
Diethanolamine	111–42–2	1.3×10^{-3}	269	>100 / 14	Reactive (?); strong base
3,3′-Dimethylbenzidine	119–93–7	3.9×10^{-8}	300	<1 / 19	
Dimethyl phthalate	131–11–3	1.2×10^{-3}	282	<1 / 20	
2,4-Dinitrophenol	51–28–5	1.3×10^{-6}	Sublimes on heating	<1 / 19.5	
2,4-Dinitrotoluene	121–14–2	4.9×10^{-4}	300	<0.1 / 17	
1,2-Diphenylhydrazine	122–66–7	1.1×10^{-2}	220	Insoluble	Reactive (?)
Ethylene glycol	107–21–1	6.7×10^{-3}	198	>100 / 17.5	
Ethylene thiourea	96–45–7	2.0×10^{-7}	450	1–5 / 18	
Heptachlor	76–44–8	3.1×10^{-5}	145/1.5 mm	<0.1 / 18	
Hexachlorobenzene	118–74–1	2.9×10^{-7}	324	<1 / 20	
Hexachlorocyclopentadiene	77–47–4	5.3×10^{-3}	234	<0.1 / 21.5	Reactive (?)
Hexamethylene-1,6-diisocyanate	822–06–0	2.5×10^{-4}	255		Reactive (?)
Hydroquinone	123–31–9	9.6×10^{-7}	218	10–50 / 20	
Maleic anhydride	108–31–6	6.7×10^{-6}	202	Soluble	Reactive
Methoxychlor	72–43–5	1.9×10^{-7}	447	<1 / 23	
4,4′-Methylenediphenyl diisocyanate (MDI)	101–68–8	1.3×10^{-4}	538	Insoluble	Reactive (?)

Continued on next page

Table III. Properties of Clean Air Act Semivolatile Organic HAPs—*Continued*

Compound	*CAS No.*	*Vapor Pressure (kPa at 25 °C)*	*Boiling Point (°C)*	*Water Solubility (g/L at °C)*	*Reactivity in Air*
Naphthalene	91–20–3	6.5×10^{-3}	218	<1 / 22	
4-Nitrobiphenyl	92–93–3	5.3×10^{-8}	340	Insoluble	
4-Nitrophenol	100–02–7	1.7×10^{-7}	279	<0.1 / 21	
Parathion	56–38–2	1.3×10^{-6}	375	<1 / 23	
Pentachloronitrobenzene	82–68–8	3.2×10^{-4}	328	<1 / 22	
Pentachlorophenol	87–86–5	1.5×10^{-5}	310	<1 / 20	
p-Phenylenediamine	106–50–3	3.5×10^{-5}	267	Soluble	
Phthalic anhydride	85–44–9	2.9×10^{-5}	285	Reacts	Reactive
Propoxur (Baygon)	114–26–1	1.0×10^{-7}	400	Slightly soluble	
Quinone	106–51–4	1.3×10^{-3}	201	Slightly soluble	
2,3,7,8-Tetrachloro-dibenzo-*p*-dioxin	1746–01–6	6.0×10^{-8} [b]	495	<1 / 25	
2,4-Toluenediamine	95–80–7	4.3×10^{-6}	292	1–5 / 21	
2,4-Toluene diisocyanate	584–84–9	1.3×10^{-3}	251	Reacts	Reactive
2,4,5-Trichlorophenol	95–95–4	2.9×10^{-3}	252	<0.1 / 18	
2,4,6-Trichlorophenol	88–06–2	1.5×10^{-4}	245	<1 / 21	
Trifluralin	1582–09–8	1.3×10^{-5}	140/4.2 mm	<0.1 / 22.5	

NOTE: Clean Air Act semivolatile organic HAPs have vapor pressures between 10^{-2} and 10^{-8} kPa.
[a]Vapor pressure value from reference 306.
[b]Vapor pressure value from reference 307.

SOURCE: Remaining data are from reference 3.

Table IV. Properties of Clean Air Act Nonvolatile Organic HAPs

Compound	*CAS No.*	*Vapor Pressure (kPa at 25 °C)*	*Boiling Point (°C)*	*Water Solubility (g/L at °C)*	*Reactivity in Air*
2-Acetylaminofluorene	53–96–3	1.5×10^{-13}	444	<0.1 / 20.5	
3,3'-Dimethoxybenzidine	119–90–4	4.3×10^{-14}	458	<0.1 / 20	
4-Dimethyl aminoazobenzene	60–11–7	9.1×10^{-11}	407	<1 / 22	
4,4'-Methylenebis(2-chloroaniline)	101–14–4	5.2×10^{-17}	517	<1 / 25	
4,4'-Methylenedianiline	101–77–9	2.3×10^{-11}	393	<1 / 19	

NOTE: Clean Air Act nonvolatile organic HAPs have vapor pressures $<10^{-8}$ kPa.

SOURCE: Data are from reference 3.

Table V. Properties of Clean Air Act Complex Mixture Organic HAPs

Compound	*CAS No.*	*Vapor Pressure (kPa at 25 °C)*	*Boiling Point (°C)*	*Water Solubility (g/L at °C)*	*Reactivity in Air*
Cresol/cresylic acid (isomer mixture)	1319–77–3	0.04	202		
Xylenes (isomer mixture)	1330–20–7	0.89	142	<1 / 22	
Hexachlorocyclohexane (all isomers)	58–89–9	7.5×10^{-6}	323	<1 / 24	
2,4-Dichlorophenoxy acetic acid (2,4-D) including salts and esters		1.3×10^{-5} to 1.3×10^{-11}	135 / 1 mm		
4,6-Dinitro-*o*-cresols and salts	534–52–1	$\leq 10^{-5}$	312	Slightly soluble	
Polychlorinated biphenyls		$\leq 2.5 \times 10^{-6}$	>315	0.08 to 0.001 ppm	
Dibenzofurans		2.7×10^{-6} to 1.1×10^{-10}	>305 (mp)	2.0×10^{-7} to 4.0×10^{-10}	
Glycol ethers		0.003 to 1.45	120–249	10–100 / 22	
Toxaphene (chlorinated camphene)	8001–35–2	1.5×10^{-6}	155 / 0.4 mm	<1 / 19	
Coke oven emissions		1.0×10^{1} to 2.0×10^{-13}	80–525	Insoluble	
Polycyclic organic matter (POM)		1.0×10^{-2} to 2.0×10^{-13}	218–525	Insoluble	

SOURCE: Data are from reference 3.

ical problems because of their chemical reactivities, affinities for metal and other surfaces, and water solubilities. These problems are more severe with ionizable compounds. Polar derivatives of organic compounds are often more toxic than the parent compounds, especially in polynuclear aromatic hydrocarbons (PAHs). It is believed that oxy- and nitro-PAHs may account for a major part of the mutagenicity of airborne particulate matter.

A recent study has shown (3) that organic chemicals, commonly classified as either polar or nonpolar, do not exhibit the expected high degree of correlation with their electronic polarizabilities. Instead, nonpolar organic compounds are characterized by relatively low water solubilities, whereas compounds normally considered to be polar are characterized by relatively high water solubilities. Classifying organic compounds on the basis of their solubility in water, therefore, provides a more realistic distinction between polar and nonpolar compounds than does classification based on polarizability.

Reactive compounds (i.e., compounds that decompose readily or react with water, oxidants, or sampling media) are even more difficult to determine in air. Monitoring methods for these compounds generally take advantage of their reactivities by allowing them to react with special collection media and measuring the product or response that results. Several of the HAPs listed in the CAA are probably too reactive to exist in ambient air (e.g., diazomethane and methylhydrazine).

Sampler Design and Operation

Organic compounds may be sampled in air by several means, as shown in Table VI:

- entrapment on a filter, sorbent, or combination of the two;
- reaction with a treated substrate;
- condensation in a cold trap (freeze-trapping);
- whole air collection; or
- real-time detection.

Monitoring strategies are further classified on the basis of area (stationary, fixed) or personal (on or near the individual) sampling methods, *active* (air sampling using a pump or vacuum-assisted critical orifice) or *passive* (sampling by diffusion, gravity, or other unassisted means) sampling modes, and direct (instantaneous) or indirect (batch, laboratory-based) analysis methods. The most commonly used methods are *sorbent sampling* and *whole air collection,* with whole air collection being applicable to VOCs only. The sampling interval may be *integrated* over time (e.g., 8 or 24 h), or it may be *continuous, sequential,* or *instantaneous* (grab sampling). Active (pumped) sampling rates may be low flow ($<$10 L/min), medium flow (10–100 L/min), or high flow ($>$100 L/min). For diffusion-controlled passive samplers, the effective sampling rate is defined by Fick's first law of diffusion as the product of the diffusion coefficient (D) of the compound of interest and the cross-sectional area (A) of the sampling face of the device divided by the length (L) of the diffusion path, or $(D \times A)/L$. These devices typically have low flow rates (e.g., between 0.1 and 50 mL/min) and require sufficient air movement across the sampling face in order to maintain constant (and known) sampling rates. The higher the sampling rate of a diffusional passive sampler, the larger the minimum face velocity required.

Table VI. Classification of Air Sampling Procedures

	Indirect		
Sampling Mode	*Grab (Area)*	*Integrative (Area or Personal)*	*Direct Continuous (Area)*
Active (area or personal)	Canisters Plastic bags Gas syringes Cryogenic condensation	Filters Solid sorbents Canisters Plastic bags	Photoionization detector Automated GC with whole-air injection Direct air sampling MS–MS
Passive (area or personal)		Diffusional badges Dust-fall sampling	

NOTE: Monitoring strategies are classified on the basis of area (stationary, fixed) or personal (on or near the individual) sampling methods; active (air sampling using a pump) or passive (sampling by diffusion, gravity, or other unassisted means) sampling modes; and direct (instantaneous) or indirect (batch, laboratory-based) analysis methods. GC is gas chromatography; MS–MS is tandem mass spectrometry.

Because ambient air concentrations of SVOCs and NVOCs are often at nanogram per cubic meter levels and below, high-volume sampling is frequently required. VOCs in air are typically found in the microgram per cubic meter range so that low-volume sampling is usually adequate. Many SVOCs and NVOCs are present at sufficiently high concentrations in indoor air and in outdoor air near sources to permit the use of low- or medium-volume sampling. For most indoor applications, sampling rates cannot exceed 25 L/min without affecting the normal ventilation rates of the room or building. For occupied indoor air spaces, samplers must also be quiet and unobtrusive. Indoor air samplers are also more subject to tampering, especially by small children, and thus require better security. Because indoor air is generally less well mixed than outdoor air, sampler placement is more critical. Sampler placement is particularly important for diffusional passive monitors, which may not sample properly under the quiescent conditions found in many enclosed air spaces.

Integrative sampling systems depend on physical or mechanical entrapment of vapors or particles, or both, from the air over an interval of time. The collected compounds are then extracted or removed unchanged for analytical measurement. Sampling periods of 24 h are most commonly used in environmental monitoring. This practice provides an average daily concentration at a single-sample analytical cost. Sometimes 12-h intervals are needed to determine diurnal changes in concentrations. Shorter sampling intervals are generally not useful for environmental monitoring purposes unless repeated sequentially throughout the day to approximate continuous monitoring. The analytical costs associated with repeated, short-term monitoring are usually prohibitive. Long-term monitoring periods (i.e., 1 week) may produce data of lower quality due to sample degradation, breakthrough losses, or other sampling artifacts. Sampling intervals of 8 h are common in occupational monitoring.

Reactive samplers differ from the former only in that the collection medium contains a reagent or reagents that will react with the compound of interest to produce products that can be more easily determined or that may be more efficiently retained by the sampler. Such collectors are usually designed for specific individual compounds or classes or compounds.

Grab samples are simply those obtained by allowing an evacuated vessel or syringe to be filled rapidly with the air to be monitored. Measurements obtained in this manner give only an indication of what was present at the sampling site at the time of sampling. However, they can be useful for screening purposes and provide preliminary data needed for planning subsequent monitoring strategies.

Continuous air samplers are designed to give real-time or near real-time assessment of the concentration of a compound in air. The detector is an integral part of the sampling device and provides a continuous reading of the contaminant concentration. A variation of this type of sampler is the sequential or batch sampler, which sequentially accumulates and analyzes the sample. Unless the collection cycle is lengthy, air concentrations determined by a sequential sampler more nearly represent real-time than time-averaged values.

Air samplers are commercially available that sample at rates ranging from less than 1 mL/min to more than 1700 L/min. Because vapor pressures of airborne organic compounds may span a range over many decades, sampling efficiencies can vary widely with the rate of sampling. More important than the absolute sampling rate, however, is the face velocity or the ratio of flow rate to the size of the trap. For example, a sampler drawing air through a 500-cm^2 particle filter at 1 m^3/min has the same linear air velocity (33 cm/s) as one that pulls air through a 4-cm^2 filter at 8 L/min. Both, incidentally, would operate at face velocities too high to adequately retain most organic compounds, no matter what the physical state.

An important factor that must be considered in the design of a sampler for airborne organic compounds is the possible existence in air of both gaseous and suspended particulate forms of the compounds. Traditional air sampling systems that are designed to separately trap both states will not maintain the integrities of the two physical states due to volatilization and sorption artifact occurrence during sampling. This matter will be discussed in more detail in the section on SVOCs.

The selection of a sampling system for airborne organic chemicals will depend not only on the chemical and physical properties of the compounds sought, but also on the intended purpose of the monitoring effort. Outdoor monitoring to study long-range transport of air pollutants may require higher sampling volumes and longer sampling intervals than monitoring to characterize emissions. If the pollutant sources are variable, temporal or directional monitoring may be needed. Monitoring in support of human exposure assessment may require stationary indoor and outdoor sampling, as well as breathing-zone measurements. The level of time resolution in human exposure monitoring studies will depend on whether the pollutants of concern elicit chronic or acute health effects. Sample size will also be dictated by the sensitivity of the analytical finish. Sampling rates, portability, ruggedness, power requirements, and field burden (both operator and subject, in the case of exposure monitoring) all must be factored into the selection of the appropriate monitoring devices. Finally, acquisition and operation costs often become deciding factors, although analytical cost is usually the principal element limiting the scope of monitoring programs directed at organic pollutants.

Sorbent Sampling

Collection of organic chemicals from air onto solid sorbents is the most widely used air sampling methodology. Many types of sorbents trap chemicals by adsorption, absorption, or chemisorption. The efficiency of a sorbent for air sampling depends primarily on its capacity (usually defined by the breakthrough volumes of the analytes sought) and its ease of desorption. Other properties that affect overall performance are resistance to air flow (pressure drop), ability to trap particles, tendency toward artifact formation, and degree of water collection.

The capacity of a sorbent depends on the number of active or sorptive sites available to the air sampled. Consequently, the surface-to-volume ratio of the sorbent is an important criterion that must be considered. The greater this ratio, the more contact area containing active sorptive sites is available to entrap compounds from the influent air. For granular solids, larger mesh sizes should generally result in higher sampling efficiencies. However, smaller particles offer more resistance to air flow (greater pressure drops across the trap), which may not permit a sufficient volume of air to be sampled in the allotted time to meet analytical needs. Porous and macroreticular solids, of course, offer far more surface area for a given particle size. Liquid sorbents coated on solid supports offer an essentially infinite contact area. Retaining the liquid on the support is sometimes difficult at high flow rates, however, and such methods are rarely employed. Polyurethane foam does not offer the concentration of sorptive sites that an equal volume of granular sorbent does, but its very low resistance to air flow makes it a popular sorbent for SVOCs.

Inorganic sorbents (e.g., silica gel, alumina, or Florisil) are somewhat hydrophilic, as is activated charcoal. These sorbents tend to collect water from the air to the extent that they may eventually become deactivated (i.e., all of the sorptive sites become occupied with water) and cease to sample efficiently. Organic polymeric sorbents (e.g., Tenax-GC, Amberlite XAD-2, or Chromosorb 102) are hydrophobic and thus less likely to exhibit problems associated with atmospheric moisture. Hydrophilic sorbents generally do a better job at collecting polar organic compounds, but difficulties in separating co-collected water from the analytes of interest and adverse effects of water on gas chromatographic separation and analytical detection limit their use. In addition to its hydrophilicity, activated charcoal is further limited by the fact that it is difficult to desorb organic compounds from it after they have been collected. This strong sorbate bonding, referred to as irreversible sorption, requires the use of highly polar solvents, such as carbon disulfide, for desorption. Compounds with a high degree of aromaticity or chemical functionality may be impossible to recover from activated charcoal. Graphitized carbon and carbon molecular sieve sorbents, which have been introduced in more recent years, are less hydrophilic and easier to desorb than activated charcoal and have larger capacities than organic polymer resins.

The effects of the geometry of the collection bed on sampling efficiency are obscure. Although the volume of a sorbent bed 3 mm thick (in the direction of air flow) and 15 cm in width is the same as that of a bed 30 cm long and 1.5 cm in diameter, the linear air velocity would be much higher through the 30-cm-long bed at a given flow rate. Although contact times would be identical for the two, the capacity or breakthrough volume may be lower for the 3-mm-thick bed because of the shorter distance required for diffusion through the bed and the larger exposed surface area of the exit face.

Sorbents of various types and their application to sampling organic compounds of varying chemical and physical properties will be discussed in more detail in the succeeding sections.

Volatile Organic Chemicals

Air sampling for VOCs is normally conducted using grab, continuous, or integrative techniques (*19*). Probably the two most widely used procedures for collecting VOCs from the atmosphere are those based on sampling with solid sorbents or evacuated containers. Because they are small and easy to transport, sorbent cartridges are used for area or personal exposure monitoring in either an active or passive sampling mode. The active mode uses a mechanical pump to draw the air through the sorbent cartridge; in the passive mode, diffusion into the sorbent results in entrapment of the compounds of interest. Evacuated containers are usually limited to fixed-location sampling, and air samples are collected using either a pump to pressurize the container or by allowing the preevacuated container to fill to near-atmospheric pressure by using a critical orifice or other flow controller to maintain constant air flow.

Most analytical methods favored in the characterization of VOCs in air are complex and require the use of relatively large laboratory-based instruments to measure concentration profiles (*10*). Several techniques have, however, been developed recently that allow direct real-time (or near real-time) analysis of VOCs in air. Although these techniques have not been widely applied, they show considerable promise for the characterization of VOCs in air, and have the distinct advantage that they eliminate the need for the sample preconcentration step that is required with sorbent or canister samples.

Sorbent Sampling

Sampling with solid sorbents continues to be widely used to obtain VOCs from the atmosphere (6, 20, 21). Figure 1 is a schematic of a typical sorbent sample collection system for air. In practice, a suitably large volume of air (2–100 L) is pumped through a sorbent bed containing 0.1–2 g of sorbent. The VOCs present in the air are adsorbed on the sorbent surface, thus undergoing separation from the air matrix. Solid sorbents used for sampling VOCs generally fall into three major categories: (1) macroreticular and porous resins (e.g., Tenax, Porapak, and XAD), (2) inorganic sorbents, and (3) activated charcoal and various carbon-based sorbents such as graphitized carbon blacks (Carbotrap) and carbon molecular sieves (e.g., Ambersorb or Carbosieve).

The inorganic sorbents (e.g., silica gel, alumina, Florisil, and molecular sieves) have relatively high polarity, thus permitting the efficient collection of polar VOCs. However, they also have a strong affinity for water, and this affinity causes them to undergo rapid deactivation and limits their usefulness.

Activated charcoal is a nonpolar sorbent and has been used to sample low-molecular-weight organic compounds. Because it exhibits irreversible sorption for these compounds, carbon disulfide or other polar solvents are used to recover the analytes. Solvent extraction suffers from the disadvantage that the collected compounds are diluted in a relatively large volume of liquid. As a result, when detection limits in the low-nanogram per gram or subnanogram per gram range are required,

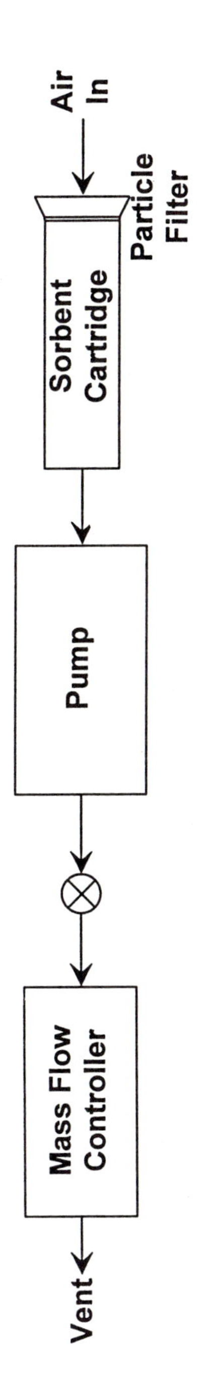

Figure 1. Schematic of a typical sorbent sample collection system for air.

solvent desorption cannot generally be used. By contrast, with thermal desorption the sorbent is simply heated in an inert gas stream, and the entire sample is transferred to the analytical system, thus limiting the sample volume needed to obtain a full analysis. Although it also collects water from the air, activated charcoal has nevertheless found increasing acceptance as a sorbent for passive sampling. Several studies have reported (22–24) the use of activated charcoal badges for diffusive air sampling. De Bortoli et al. (22) showed that results obtained with the passive charcoal badges compare favorably with those found for nonpolar compounds using active samplers packed with Tenax or Porapak and recovered by thermal desorption.

Single-Sorbent Sampling: Tenax. Collection of VOCs on the porous polymeric resin Tenax, followed by thermal desorption for their recovery, has enjoyed wide use (*25–29*). It has the advantage of high thermal stability (to about 300 °C) along with a low affinity for water, so that sample collection is not compromised by water collected from the atmosphere, and large volumes of air can be sampled. In addition, the cartridge containing the Tenax (or any other solid sorbent) is the first element in the sampling train (Figure 1), in contrast to canister-based sampling systems, in which the canister is the last element in the sampling train. Tenax does, however, have significant drawbacks. These include low capacity for VVOCs and polar VOCs (i.e., premature breakthrough), high backgrounds due to contamination, poor batch-to-batch reproducibility, and artifact formation due to chemical reactions that occur during sampling or thermal desorption (*30–33*). These reactions can give rise to erroneous measurements of atmospheric concentrations or even lead to the incorrect determination of compounds not present in the air sampled (*33*). Nevertheless, with careful attention to detail and the use of the proper procedures, all these problems can be reduced or even eliminated in certain cases.

The breakthrough volumes for many VVOCs and polar VOCs on Tenax are known (*25, 26, 34*). To sample for one of these compounds using Tenax, it is first necessary to determine both the amount of sorbent needed and the maximum sample volume that can be tolerated. The result of this calculation will show whether the required detection limit can be reached. For some chemicals, even this approach will not work, but this limitation is not unique to Tenax.

For most chemicals, background levels and poor batch-to-batch reproducibility with Tenax can be reduced by precleaning (e.g., by solvent extraction and thermal purging) and testing samples of the sorbent batch until the desired level of cleanliness is obtained. Artifact formation is most commonly associated with the analysis of three compounds—benzaldehyde, acetophenone, and phenol—that are produced by the reaction of the Tenax itself with strong oxidizing agents (Cl_2, O_3, NO_x, and SO_x) (*31, 33*). These compounds should therefore not be analyzed from Tenax unless special precautions are taken to remove the oxidizing agents from the air stream during sampling (*31*).

Walling (*35*) proposed an empirical approach to identify inconsistent results when using Tenax for sampling. The technique, called distributed air volume (DAV) sampling, can reveal the presence of complications associated with sampling but cannot indicate their origin. In the DAV approach, a set of four Tenax cartridges simultaneously sample the same atmosphere at different flow rates over the same period. Except for the lighter compounds, which are expected to break through the bed, the concentration values obtained from the set must agree within established analytical tolerances in order to be considered consistent.

Early indications were that application of this criterion to data obtained in the field results in as much as 50% of the data in a sample set being rejected (*36*). A later investigation that also used this approach was the six-building air quality study sponsored by EPA (*37*). About 220 indoor and outdoor Tenax air samples were collected in triplicate at widely differing flow rates to provide 10-, 15-, and 20-L air samples over a 12-h collection period. Thermal desorption of the Tenax tubes was followed by gas chromatography–mass spectrometry (GC–MS) analysis. No significant difference was observed for 25 of 29 target VOCs. The exceptions were those compounds that had breakthrough volumes less than 20 L, and even then the differences followed expectations. This result confirms and extends the results of an earlier study (*38*) of indoor air under controlled conditions, which concluded that no differences could be observed among 10 VOCs measured simultaneously at three concentration levels on Tenax and in evacuated canisters. Several subsequent studies (*29, 39–41*) that compared Tenax with other sorbents for preconcentration of VOCs in air also found little or no difference.

Multisorbent Sampling. Tenax collection efficiencies for VOCs of molecular weight lower than C_6 are often unacceptable for quantitative purposes (*42*). Efforts to increase the range of compounds that can be monitored using sorbents have provided some of the impetus for the development of carbon-based polymeric sorbents, such as Ambersorb and Carbosieve. These materials have high retention for volatile and polar compounds, but they also collect water and retain the less volatile VOCs too strongly for quantitative thermal desorption (*21*). This property has led to further attempts to minimize the various problems associated with sorbent sampling of VOCs by combining several sorbents in one sampling cartridge. For example, Hodgson and Girman (*34*) evaluated a multisorbent cartridge containing Tenax-TA, Ambersorb XE-340, and activated charcoal and found that the combination effectively trapped VOCs over a wide boiling point range. Other multisorbent combinations that have been evaluated include a Tenax and Carbotrap sampler (*42*), a glass bead, Tenax-GC, and Ambersorb XE-340 sampler (*43*), a glass bead, Tenax-GC, Ambersorb XE-340, and charcoal sampler (*44*), a HayeSep DB (porous polymer), Carboxen 1000, and Carbosieve S-II sampler (45), and various carbon-based multisorbent samplers (*41, 46–54*).

The commercial availability of carbon-based multisorbent samplers (*55, 56*) containing the graphitized carbon black sorbents (e.g., Carbotrap C and Carbotrap B) and the carbon molecular sieves (e.g., Carboxens and Carbosieve S-III) has spurred work on the use of these sorbents for sampling VOCs over a wide range of volatility and polarity (*41, 46–54*). In practice, such samplers consist of a glass or stainless steel cartridge that is packed with the sorbents. As shown schematically in

Figure 2, the cartridge contains, in order of increasing retentivity, the graphitized carbon black sorbents Carbotrap C (weakest sorbent, on the cartridge inlet side) and Carbotrap B (stronger sorbent) for higher molecular weight VOCs ($\geq C_6$); and the carbon molecular sieve Carbosieve S-III (strongest sorbent) for low-molecular-weight VOCs (ca. C_2). Although their use has provided a means of extending the VOC range, they do not avoid the adverse effects on GC performance due to co-collected water vapor that is always present in air samples at concentrations that are generally 6 to 7 orders of magnitude higher than those of the target analytes (*36*). This problem has been effectively overcome, as described in the following section, by including a water management step in the sample transfer procedure that makes use of a dry nitrogen purge.

Another problem with multisorbent samplers arises because most of the trace-level work conducted with them to date has utilized sorbent combinations that were made up somewhat empirically. Ideally, the procedure for optimizing conditions for sample collection with carbon-based sorbent combinations should be based on a rational approach that can provide a theoretical prediction of the prospects of successfully collecting target analytes (*57*).

Several studies have demonstrated that VVOCs can be effectively retained and thermally released using tubes packed with carbon-based multisorbent combinations. O'Doherty et al. (*50, 51*) used tubes packed with the carbon molecular-sieve Carboxens (Carboxen 569, 1000, 1001, 1002, and a mixture of Carboxen 1000 and 1003) to trap and preconcentrate the very volatile replacement chlorofluorocarbons (i.e., hydrofluorocarbons and hydrochlorofluorocarbons) from sample volumes of several liters, without the need for cryotrapping using liquid nitrogen. The range of replacement chlorofluorocarbons examined varied in boiling point from –48.4 to –9.8 °C. Using tubes packed with Carbotrap C, Carbotrap B, and Carbosieve S-III,

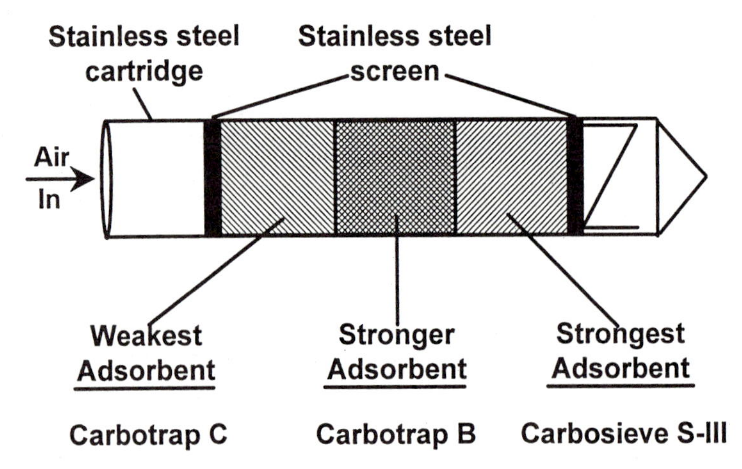

Figure 2. Composite multisorbent cartridge containing carbon-based sorbents.

Bishop and Valis (46) were able to collect a diverse mixture of VOCs in laboratory tests. Compounds included the nonpolar cyclohexane, trichloroethylene, toluene, and chlorobenzene, as well as the polar VOCs ethanol, methyl acetate, 1-nitropropane, and tetrahydrofuran. In a practical application of carbon-based multisorbent sampling, Ciccioli et al. (49) analyzed 1-L air samples collected in the center of Rome, Italy, and from a forested area. More than 140 different VOCs were identified and quantitated, covering wide ranges of volatility (C_3 to C_{14}) and polarity (alkanes, alkenes, and aromatics to alcohols, aldehydes, and ketones). Similar applications have been reported by Kruschel et al. (54) and Helmig and Greenberg (58).

As part of an evaluation of a portable sequential tube sampler, Pollack et al. (52) conducted an experiment that was designed to determine the distribution of VVOCs and VOCs among the sorbents in a multisorbent bed. Three tubes were packed with the individual sorbents used in the multisorbent tube, namely, Carbotrap C, Carbotrap B, and Carbosieve S-III. The tubes packed with the individual sorbents contained the same amount of that sorbent that is normally used in the multisorbent configuration. Then, the tubes were connected in series using Swagelok unions to form a three-stage tube "train". A multicomponent standard gas mixture, containing compounds that ranged from C_2 to C_9, was loaded onto the tube train by drawing the standard through the tubes using a pump with a mass flow controller. A 1.2-L sample volume was used. After sample loading was completed, the tube train was taken apart, and the tubes were analyzed individually to determine the distribution of each component among the three sorbents.

Some of the results of this experiment are summarized in Table VII, which shows the percent recovery of the compounds from each of the tubes in the train, in their order of elution from the GC column. The values were determined by taking the total peak area for a compound from the tubes and relating that total to the fraction collected in any one tube. The total areas measured for each compound in the tubes were generally within about 20% of the areas obtained with a sample of the same volume loaded onto a single multisorbent tube packed with the three sorbents. The data indicate that none of the sorbents was able to trap the C_2 hydrocarbons (ethane, ethylene, and acetylene) at ambient temperatures. However, the Carbosieve S-III did succeed in capturing some of the C_3 and C_4 components. The C_5 through C_7 compounds were collected on the Carbotrap B, and Carbotrap C took care of the higher molecular weight compounds. It is clear that, in many instances, a compound was not collected quantitatively by a single sorbent but was generally distributed between either Carbosieve S-III and Carbotrap B or Carbotrap B and Carbotrap C. This distribution pattern suggests that tubes containing the three carbon-based sorbents considered in this work are probably more likely to quantitatively capture a wider range of VVOCs and VOCs than tubes packed with only two of the sorbents.

Water Management Techniques and Measurement of Polar VOCs. All air samples taken in "real-world" atmospheres contain water vapor, and the relative humidity can be as high as 90–100%. These samples require special water manage-

Table VII. Distribution of VVOCs and VOCs Among Carbon-Based Sorbents in Three-Tube Train

	% Recovery by Individual Tube [a]		
Compound	*Carbosieve S-III*	*Carbotrap B*	*Carbotrap C*
Ethylene	0	0	0
Acetylene	0	0	0
Ethane	0	0	0
Propylene	<u>100</u>	0	0
Propane	<u>100</u>	0	0
Isobutane	<u>100</u>	0	0
1-Butene	<u>100</u>	0	0
n-Butane	<u>100</u>	0	0
cis- and *trans*-2-Butene	<u>100</u>	0	0
3-Methyl-1-butene	22	<u>78</u>	0
Isopentane	0	<u>100</u>	0
1-Pentene	0	<u>87</u>	13
n-Pentane	0	<u>100</u>	0
Isoprene	0	<u>96</u>	4
cis-2-Pentene	0	<u>100</u>	0
n-Hexane	0	<u>100</u>	0
cis- and *trans*-2-Hexene	0	<u>100</u>	0
1,1,1-Trichloroethane	0	<u>100</u>	0
Benzene	0	<u>85</u>	15
Carbon tetrachloride	0	<u>100</u>	0
Cyclohexane	0	<u>100</u>	0
2- and 3-Methylhexane	0	<u>93</u>	7
Trichloroethylene	0	<u>100</u>	0
n-Heptane	0	<u>84</u>	16
Toluene	0	<u>85</u>	15
2-Methylheptane	0	25	<u>75</u>
n-Octane	0	6	<u>94</u>
Tetrachloroethylene	0	<u>100</u>	0
Ethylbenzene	0	41	<u>59</u>
m- and *p*-Xylene	0	14	<u>86</u>
Styrene	0	15	<u>85</u>
o-Xylene	0	8	<u>92</u>
n-Nonane	0	9	<u>91</u>
Isopropylbenzene	0	0	<u>100</u>
n-Propylbenzene	0	14	<u>86</u>
1,3,5-Trimethylbenzene	0	7	<u>93</u>

NOTE: VVOCs are very volatile organic compounds; VOCs are volatile organic compounds.
[a]Underlined values indicate tube containing major fraction of component.

SOURCE: Data are from reference 52.

ment procedures in order to maintain sample integrity. Potential problems include ice accumulation and possible plugging in the thermal desorption focusing device and adverse effects on the chromatographic separation process.

These water vapor problems have, for the most part, dictated the need to incorporate methods to remove moisture prior to analysis. Standard canister-based methods for monitoring VOCs in air, notably EPA's Compendium Method TO-14 (*59*), include a Nafion dryer in the air sample flow path (Figure 3) to reduce moisture. As the air sample is passed through the dryer, the water vapor content of the sample is significantly reduced, the amount being removed from the sample depending on the conditions established in the dryer (*36*). Pleil et al. (*60*) have shown that heating a Nafion tube dryer and flushing with dry nitrogen results in the removal initially of up to 99% of the water vapor for a period of time before the removal rate approaches equilibrium. This period is normally sufficient to process a typical air sample. Janicki et al. (*61*) have shown that a Nafion tube inserted in a container filled with molecular sieve 5A effectively removes water vapor from humidified gas streams even under continuous operation for more than 10 days.

Many nonpolar VOCs, such as those listed in Method TO-14, pass through this type of dryer unaffected (*60, 62*). However, polar components, such as alcohols, ketones, or aldehydes, are removed with the water vapor, thus preventing satisfactory measurement of these species in air samples. This limitation has prompted numerous recent efforts to develop alternative methods for removing moisture from air samples prior to GC analysis (*48, 49, 52, 63–67*).

The use of a dryer to remove moisture prior to analysis, as in EPA Method TO-14, permits the inclusion of a desorption step after cryogenic preconcentration. In

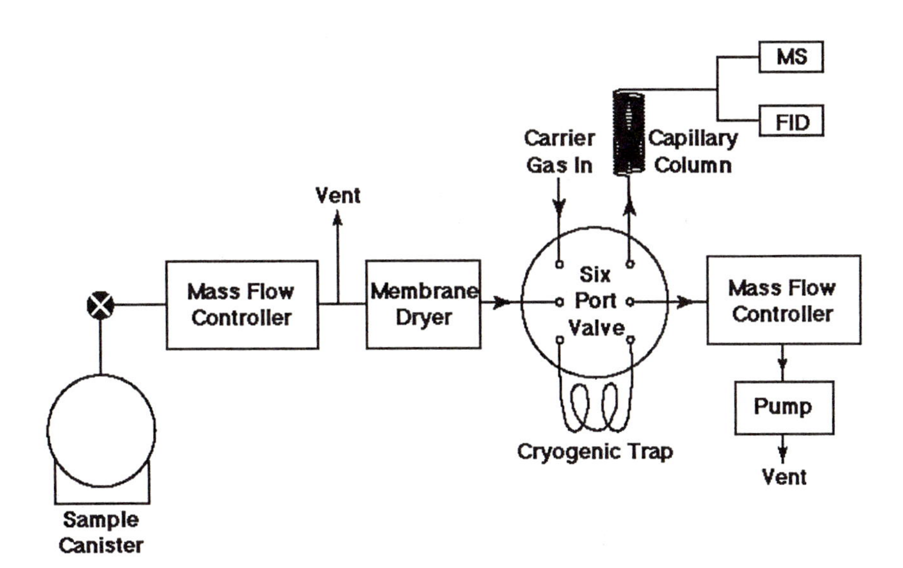

Figure 3. Schematic of a typical canister sample recovery system.

this step, the cryotrap is ballistically heated (to 150 °C) to rapidly vaporize the condensed sample and refocus it for injection into the GC. Because this desorption is nonselective, the sample retains water that is transferred to the GC column. A recently developed alternative technique makes use of a controlled desorption trap (*66, 67*). In this procedure, a slow temperature ramp and flow-regulated desorption of the cryotrap take advantage of differences in the vapor pressures of VOCs and water in the sample at low temperatures. VOCs, collected on a primary cryotrap, are allowed to vaporize at a temperature that limits the partial pressure of water vapor. A secondary cryotrapping step is required to refocus the slowly desorbed sample prior to injection into the GC. The result is a significant reduction in the mass of water in the sample injection volume, compared with rapid heating without water removal, and a water management technique that permits the quantitative analysis of polar and nonpolar VOCs in a humidified air sample.

An increasingly popular method is based on the use of a multisorbent bed in a combined preconcentration and water management system. A commercially available two-stage carbon-based trap, consisting of Carbotrap B and Carbosieve S-III, is placed in the air sample flow path (*48, 52, 63*). The sorbent bed traps and concentrates the VOCs in the sample. Then, the trap is subjected to a dry purge in which dry nitrogen is passed through it at room temperature. Water vapor is preferentially lost during the dry purge, after which the trap is heated and back-flushed with helium to thermally desorb the volatile compounds onto the GC column for analysis. More than 90% of the water present is removed by this procedure.

Tests of the effectiveness of the two-stage trap for transferring polar VOCs to the GC have been conducted using humidified air samples containing the compounds of interest at levels below 10 parts per billion by volume (ppbv) (*63*). These tests indicated that, for the most part, the transfer efficiency of the target compounds from the humidified sample to the GC via the trap was 70–100%. A notable exception was vinyl acetate, for which the recovery efficiency was only 10%.

Instead of eliminating the water vapor prior to analysis as a means of dealing with the moisture problem, an alternative approach (*64*) has explicitly incorporated the water vapor into the analytical scheme. This approach focused on the capability of the quadrupole ion trap mass spectrometer for operating in the chemical ionization mode using the water vapor in the sample along with water from a separate vial as the reagent gas. Although the potential of this technique has been clearly demonstrated, the water vapor present in the air sample passing through the GC degrades chromatographic performance to such an extent that special techniques, such as those described already, must still be incorporated into the analysis to remove the moisture.

Continuous Sorbent-Based Sampling Procedures. Requirements established under the Clean Air Act Amendments of 1990 directed the EPA to collect ambient air measurements for ozone precursors in 22 ozone nonattainment areas in the United States. Automated, continuous GC techniques that provide continuous 24-h profiling of the changing concentrations of VOCs in air over time are currently the only cost-effective way to directly meet the stringent sampling and speciation requirements for these Photochemical Assessment Monitoring Stations in the monitoring networks (*68, 69*).

Although conventional capillary GC instrumentation does not lend itself directly to continuous measurement, on-line collection and analysis systems have been developed that integrate a large fraction of the monitoring period (e.g., about 45 min out of every hour) (*47, 58, 68, 70–75*). A typical sequence for on-line continuous GC sampling and analysis, due to Woolfenden et al. (*71*), is shown in Figure 4. Air is drawn into the on-line sorbent trap, which is packed with Carbotrap B and Carbosieve S-III, for a fixed period (ca. 45 min) at a constant flow rate. Thereafter, the trap is heated rapidly and flushed with an inert gas stream to desorb the analytes onto the head of the GC column as a narrow band. Sample collection for the next run continues while analysis of the previous collection is being performed.

The practical and logistical difficulties associated with the field use of liquid nitrogen or other cryogens to trap VVOCs and VOCs constitute a major problem (*70, 71*). Conventional approaches to this issue require cold trapping and subsequent subambient operation of the GC oven. The use of cryogens is expensive and can be difficult to supply reliably for long-term unattended operation at remote sites (*70*). One system that overcomes this limitation includes a small Peltier-cooled sorbent trap. This device was originally designed for refocusing analytes desorbed during off-line analysis of sorbent tubes with the Perkin-Elmer ATD-400 Automatic Thermal Desorption capillary–GC system; it was later modified for use in whole-air sampling (*70, 71, 76*). The unit, which uses 100 mg or less of sorbent, is first cooled to –30 °C before ambient air is drawn through the trap. After a preset sampling time, the trap is automatically connected to the GC via a six-port valve and is rapidly heated to about 325 °C to desorb the analytes onto the GC column. The Peltier-cooled trap has been extensively tested for on-line air sampling and analysis of hydrocarbons in the volatility range C_2 to C_{10} (*48, 52, 72, 77*).

Portable Sorbent-Based Sampling Devices. Long-term integrative sample collection is the preferred approach for measuring relatively constant VOC emissions and for assessing average exposure, as in studies of chronic health effects (*19*). However, to obtain detailed information on temporal or spatial changes, as in the case of intermittent emissions or acute exposures, short and frequent sampling periods in the various locations in which exposure may occur are often more appropriate. To this end, two recently developed portable air sampling devices provide the means to collect VOCs on a schedule best suited to identify and characterize exposure occurrences.

The Perkin-Elmer STS 25 sequential tube sampler measures temporal variability of VOCs by automatically collecting air onto a series of multisorbent tubes (*52, 71, 78*). The device is portable and self-contained, and can be used to sample air sequentially onto a series of up to 24 sorbent tubes, one at a time, for periods of 0.1 s to 10 h per sample. Air is drawn through each tube in turn using a personal sampling pump. When not being sampled, diffusion-limiting caps ensure that each tube

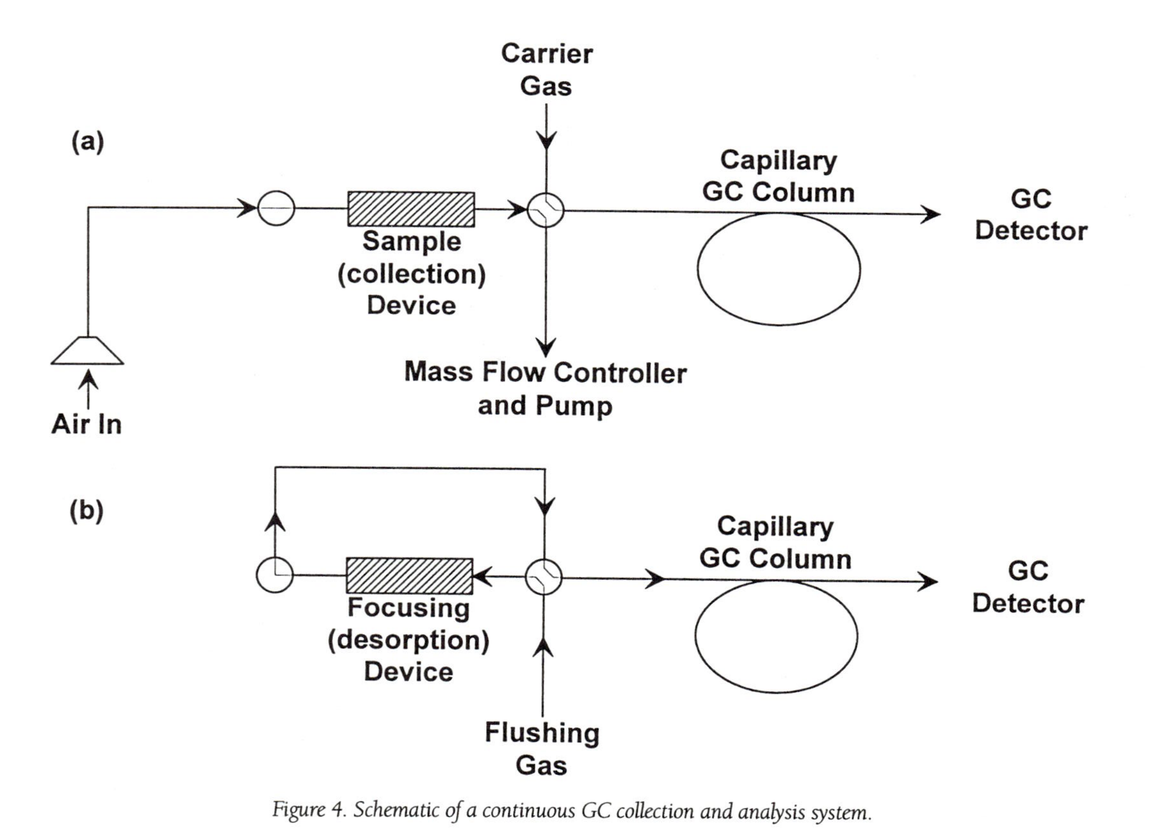

Figure 4. Schematic of a continuous GC collection and analysis system.

is effectively sealed. The system has been carefully evaluated for the collection of nonpolar VOCs (from Method TO-14) ranging in volatility from Freon-12 to 1,2,4-trichlorobenzene (*52, 78*). Detection limits are in the low- to sub-nanogram per gram range, and sampling volumes are limited by breakthrough for the most volatile compounds of interest. Principal applications include all situations in which the variation of pollutant concentration with time is of interest.

The Total, Isolated by Microenvironment, Exposure (TIME) monitor is a small, portable air sampler that is designed to independently measure VOCs in four microenvironments—indoor residential, indoor workplace, outdoors, and in transit—and determine their relative contribution to the total exposure (*52, 79, 80*). It makes use of an electronic shadow sensor, which provides information on the microenvironments by measuring vertical distances. Based on the microenvironment identified by the shadow sensor, software incorporated into the device controls a valve system that switches the sampled air stream to the multisorbent sampler tube associated with that microenvironment. The mobility pattern component of the monitor identifies the subject's location once every 30 s, thus providing a detailed record of the time spent in various microenvironments by the participant during sampling. Thus, at the end of a fixed sampling period, the sensor provides the subject's exposure by microenvironment and, by addition, the subject's total exposure. The device can operate for at least 8 h at a time using a rechargeable battery pack. This method has considerable practical significance because the microenvironmental exposure measurements can be used to focus risk reduction strategies on those microenvironments in which the risk is greatest.

Active and Passive Sorbent Sampling. Although sampling for VOCs with solid sorbents poses certain problems, personal exposure monitoring requires sampling devices that are so small that samplers based on other techniques are largely precluded (*21*). Many total human exposure studies have used sorbent tubes with small, battery-operated, low-flow pumps. Sampling flow rates are typically in the range of 10–50 mL/min and sampling volumes are in the range of 2–15 L for most sorbents. Several manufacturers offer compact, self-contained pumps with rechargeable battery packs for power, a precisely adjustable flow controller, and a pouch for the cartridge. These systems are suitable for fixed-site sampling as well as for personal exposure monitoring.

In an effort to further reduce the size, weight, and cost of the monitor, increasing attention has been paid to the development and use of passive (diffusion-controlled) sampling devices (PSDs). A device that has been widely used, especially in some major indoor air studies, is the 3-M organic vapor monitor (OVM 3500) passive diffusive badge (*22, 23, 81, 82*), which has been used by Otson et al. (*82*) for short-term exposures (days) and by Shields and Weschler (*81*) for long-term exposures (weeks). As described by Shields and Weschler (*81*), the sampler consists of a diffusion screen and a charcoal sorbent pad assembled in a disk-shaped plastic holder. The diffusion screen creates a concentration gradient from its surface to the carbon sorbent pad. Sampling is begun by removing the monitor from its aluminum container and continues throughout the sampling period. VOCs in the air diffuse into the monitor at known constant uptake rates. At the end of the sampling period, the monitor is capped to terminate sampling and is then stored in a refrigerator until analyzed. In field tests, the OVM 3500 passive monitor has been shown (*81, 82*) to provide reliable measurements of selected airborne VOCs at concentrations ranging from about 2 to 6,000 μg/m^3. For analysis, the badges are usually extracted with carbon disulfide, and the organic compounds present in the extract are separated and identified using standard GC–flame ionization detector (FID) procedures. The OVM 3500 badge may not be suitable for outdoor applications because deactivation of the charcoal sorbent would likely occur at relative humidities above 80%.

Cao and Hewitt (*83*) have determined the buildup with time of artifacts on four adsorbents commonly used for passive sampling of VOCs in air (Tenax-TA, Tenax-GR, Carbotrap, and Chromosorb 106). Artifact buildup on Chromosorb 106 was very high, and they concluded that this material cannot be used for passive sampling in air. The buildup on Carbotrap was also high, especially in the GC range corresponding to hydrocarbons smaller than C_5, but was very low on Tenax-TA and Tenax-GR. As a general rule, storage and exposure times should be as short as possible in order to keep the background buildup on solid sorbents as low as possible.

Lewis et al. (*84*) have developed and evaluated a thermally desorbable, high-efficiency PSD for both short-term and long-term collection of VOCs. The PSD is a small stainless steel cylinder (3.8-cm o.d. × 1.2 cm) and is packed with a small amount of solid sorbent (e.g., Tenax or other sorbent). Two configurations of the PSD have been developed. The high-rate PSD has an effective sampling rate of 75–100 mL/min, and detection limits are ca. 1 ppbv with 1- to 3-h exposure times. The low-rate PSD is designed to permit 24-h exposures and has an effective sampling rate that is lower than that of the high-rate unit by a factor of ca. 20 (*85*).

A new sample preparation technique, solid-phase microextraction (SPME), that combines sampling and preconcentration of VOCs in air in one step and allows direct transfer of the analytes into a GC has been described recently by Chai and Pawliszyn (*86*). SPME uses a 100-μm fiber coated with poly(dimethylsiloxane) in a modified syringe for easy handling, protection of the fiber, and direct injection into the GC. To sample with the SPME device, the fiber is lowered into the gas atmosphere by depressing the syringe plunger. VOCs partition into the polymeric coating of the fiber until equilibrium is established (in a matter of minutes). The plunger is then withdrawn and the fiber is retracted into the needle. The syringe needle is used to pierce the septum of the GC injector, and the analytes are characterized in the usual way. The method allows the detection of nonpolar VOCs at parts per trillion (ppt) to parts per billion (ppb) concentration levels with an ion-trap MS, and it has a precision of 1.5–6% relative standard deviation. Storage stability of compounds in the coating is, however, limited to no more than 1 h, even when stored in a refrigerator.

Reactive Sampling. Both active (*59, 87, 88*) and passive (*89*) collection modes have been used to conduct *reactive sampling* for certain VOCs such as formaldehyde and other carbonyl compounds, ethylene oxide, amines, diazomethane, isocyanates, and phenols.

Several methods are available for the determination of airborne aldehydes (*90*). One example is the use of the high-efficiency passive sampling device, developed by Lewis et al. (*84*), with glass-fiber filters coated with 2,4-dinitrophenylhydrazine (DNPH) (*91*). This PSD has a sampling rate of 103 mL/min for formaldehyde, about 100 times greater than that of the standard Palmes tube.

Another passive sampler that has been widely used to monitor formaldehyde is the PF-1 diffusion tube (Air Quality Research) (*92*). This sampler, which consists of a sodium bisulfite impregnated disk, has been used in indoor air surveys (*93, 94*). Sampling begins when the seal is removed from the glass tube and continues throughout the sampling period (5–7 days). Formaldehyde in the air diffuses into the tube at a known constant rate and is collected on the filter at the bottom of the tube. At the end of the sampling session, the tubes are capped and stored in a refrigerator until analyzed. Analysis is by the standard chromotropic acid method, in which absorbances are measured on a spectrophotometer by National Institute of Occupational Safety and Health (NIOSH) Method 3500 (*88*).

The collection of formaldehyde emissions by passive diffusion into a liquid forms the basis of a standard test method [American Society for Testing and Materials (ASTM) Method D 5014) (*95*)]. Formaldehyde is absorbed into an aqueous solution of 3-methyl-2-benzothiazolinone hydrazone hydrochloride (MBTH) contained in a sampler consisting of a glass vial with a septum cap that retains a Knudsen disk. Formaldehyde passes from the air into the MBTH solution via the Knudsen disk at a constant rate. After collection, the resulting azine is oxidized to form a blue cationic dye, the concentration of which is measured colorimetrically. This method is suitable for personal or area monitoring of formaldehyde in indoor air over the range from 0.025 to 14 parts per million (ppm).

Aldehyde levels in indoor and outdoor air have also been measured using active sampling methods, including DNPH impingers used according to EPA Method TO-5 (*59*) and DNPH-coated silica gel cartridges used according to ASTM Method D 5197 (*87, 95*). Aldehydes and ketones readily form a stable derivative with the DNPH reagent. The derivative is removed from the cartridge with acetonitrile and is analyzed for aldehydes and ketones using high-performance liquid chromatography (HPLC) with a UV absorption detector. The method is suitable for the detection of formaldehyde concentrations ranging from high ppb to low ppm levels. High concentrations of ozone have been shown (*96*) to interfere negatively in this method because of its reaction with both the DNPH and its hydrazone derivatives in the cartridge. The extent of this interference is a function of the temporal variations of both the ozone and the carbonyl compounds during sampling. Artifacts have also been observed (*97*) as a result of exposure of the DNPH-coated cartridges to direct sunlight, which should therefore be avoided.

Two continuous measurement methods have been developed for the determination of formaldehyde in air. The first, ASTM Method D 5221, makes use of a commercial analyzer that has been modified such that formaldehyde vapor is absorbed in acidified pararosaniline (triaminotriphenylcarbinol hydrochloride) with subsequent addition of dilute sodium sulfite (*95, 98*). The resulting colored product has a strong visible absorption band that is measured colorimetrically. This method is applicable to formaldehyde concentrations that range from 10 to 500 $\mu g/m^3$. The second method, which has a detection limit of 0.2 ppbv (0.25 $\mu g/m^3$), relies on the Hantzsch reaction; that is, the cyclization of a β-diketone, an amine, and formaldehyde, to produce a fluorescent derivative of formaldehyde (*99*). Both of these continuous monitoring techniques are discussed further under ***Real-Time Measurements***.

The active collection mode has also been used as the basis for the development of reactive sampling techniques for several other VOCs. Examples include the use of solid sorbent tubes containing XAD-2 polymeric resins coated with octanoic acid for the VVOC diazomethane in workplace atmospheres by NIOSH Method 2515 (*88*), Sep-Pak C_{18} cartridges impregnated with phosphoric acid for the trace level determination by GC of aliphatic amines (*100*), carbon molecular sieve coated with hydrobromic acid for ethylene and propylene oxides by NIOSH Method 1614 (*88*), impingers containing 1-(2-methoxyphenyl)-piperazine (2-MPP) in toluene for isocyanates by NIOSH Method 5521 (*88*), and impingers containing sodium hydroxide for phenol and methylphenols according to EPA Method TO-8 (*59*). Recently, Nishioka and Burkholder (*101*) developed a method for phenolic compounds, based on the use of a 200/400 mesh granular Amberlite AG-1 anion exchange resin. Analysis is by GC–MS or GC–FID. The method is also applicable to chlorinated phenols, which are analyzed, after methylation, by GC–ECD.

Canister Sampling

Design and Use. The use of passivated stainless steel canisters to collect integrative whole air samples for VOCs is an attractive alternative to sorbent sampling, and canisters have become one of the most widely used sampling methods for these compounds. The canisters are passivated by an electropolishing technique known as SUMMA-polishing (*36*). Besides its present-day widespread use for the collection of VOCs, early applications of canister-based technology included the measurement of atmospheric halocarbons (*102*) and biogenic compounds (*103*). Procedures for cleaning and certifying canisters for whole-air sampling have been summarized in EPA Method TO-14 (*59*). Extensive field use of canisters has confirmed that this method produces data of acceptable quality as measured by completeness (90% overall), precision (within ±25%), and accuracy (within ±25%) for most target VOCs (*104*).

Canister sampling is carried out by allowing the air to enter a preevacuated container either via a critical orifice or mass flow controller, or by using a pump to fill the canister to a pressure of a few atmospheres (*36, 59*). For analysis, an aliquot (100–500 mL) of air is withdrawn from the canister and cryofocused into a GC attached to a mass-selective detector (MSD), ion-trap detector (ITD), or FID (*36, 59*). Detection limits are generally below 1 $\mu g/m^3$ for most target VOCs. Figure 3 is a block diagram of a typical canister sample recovery system for analysis of VOCs in whole air.

Most canister air sampling is carried out using 6-L units. In the passive sampling mode, a mass flow controller is set at 3–4 mL/min so that ca. 5.5 L of air can be collected over a 24-h period (*36*). Canister sampling can be extended to periods up to 1 week using a timer and solenoid-controlled valving system (*105*). With pump-driven samplers, ca. 14 L of air can be sampled into a 6-L canister at 11 mL/min over a 24-h period.

SUMMA canisters have advantages over plastic (Tedlar) bags for whole-air sampling. They display relatively good storage stabilities for VOCs (including some polar compounds) with vapor pressures above 10^{-2} kPa, greatly reduced problems due to contamination and artifact formation, and the absence of breakthrough effects, and they allow for multiple analyses (*21, 36, 63, 106*). Major disadvantages associated with the use of canisters include their cost and bulkiness, the limited air volumes that can be sampled, difficulties experienced in recovering less volatile and more polar VOCs, and the co-collection of water. Water is present in relatively large amounts in air samples and can cause serious problems, such as ice formation that can clog the cryogenic trap. In the analysis of nonpolar compounds, the sample aliquot is normally passed through a dryer membrane (Nafion tubing) to remove the water. As noted earlier for multisorbent samplers, this process also effectively removes many of the polar VOCs (*36*). Kelly et al. (*63*) have developed a method for removing moisture from canister samples in which a two-stage carbon-based sorbent trap, consisting of Carbotrap B and Carbosieve S-III, is used to separate the water from the sample prior to GC analysis.

Pleil et al. (*107*) have developed a 12-syringe sequential sampler as a variation of the standard canister sampler. The system, which is portable and battery-operated, contains SUMMA-polished syringes (150 mL each) that can be consecutively filled at rates ranging from 2 to 90 min per syringe. Analysis is by a fully automated GC–MSD technique that is designed to sequentially analyze the contents of each syringe. The sampler is analogous to the sequential sorbent-tube sampler described earlier, and is valuable in situations where volatile organic levels undergo large changes in short periods of time, and time-averaged samples would not reliably represent peak exposures.

A spatially resolved air monitoring strategy for deducing the origin of toxic VOC emissions has been developed (*108*). To determine the correlation between wind direction (transport) and the identity and concentration of specific airborne pollutants, the technique relies on a wind direction sensor to route whole air, sampled at a constant rate, into either one of two canisters. When the wind comes from the suspected emissions area, the sample is directed into the "In" sector canister; otherwise, the sample is collected in the "Out" sector canister. Differences observed between the "In" sector sample and the "Out" sector sample at a specific monitoring site are attributed to candidate VOCs emanating from the "In" sector source. Field tests of the technique show good correlation with expected VOC emissions, which were measured independently using canister grab samples taken from the target areas.

Raymer et al. (*109*) have used 1.8-L canisters in a portable spirometer to collect primarily human alveolar breath for subsequent GC–MS determination of VOCs. Measurements of CO_2 indicate that >97% of the breath collected is alveolar in origin. Clean air for inhalation is provided by two organic vapor respirator cartridges, and sample collection is effected in about 2 min. With the 1.8-L canisters as the breath collection devices, the data suggest that the system can be used reliably for VOCs with volatilities greater than that of *p*-dichlorobenzene.

Whitaker et al. (*110*) have developed a new personal whole-air sampler (PWAS) that can be worn to monitor personal exposures to VOCs. The PWAS consists of a 1-L canister, a mass flow controller, two 1.3 A-h batteries, and an electronics module with a digital display for the sampling set point, actual flow rate, and battery voltage. The sampler, which weighs 3.35 kg, fits into a laptop computer carrying case and is able to collect a 900-mL sample at a linear flow rate for 12–16 h. Laboratory tests of the PWAS have shown that the sample flow rate is unaffected by temperature or the activity of the subject wearing the sampler. Recoveries of 89% or better have been obtained for several nonpolar VOCs, and precision tests have yielded a coefficient of variation less than 10% for all test analytes. The unit can be used as an extremely compact microenvironmental sampler or as a personal whole-air sampling system.

Canister Storage Stability. Experimental studies have shown that the nonpolar VOCs that the TO-14 list comprises (*59*) are stable in humidified passivated canisters for at least 7 days, and some of these compounds are stable for up to 30 days (*36, 111, 112*). More recently, the stabilities of some polar VOCs were evaluated in unpolished and SUMMA-polished canisters at ppb levels under dry and humid conditions (*63, 106*). Kelly et al. (*63*) found that the polar VOCs of interest in their study exhibited satisfactory storage stability over a period of at least 4 days in polished canisters under humidified conditions. However, Pate et al. (*106*) obtained stable results for polar compounds for up to 31 days when using humidified passivated canisters. The difference between the two data sets may be due to the fact that Pate et al. supplied a much larger amount of water (that apparently exceeded the saturation limit) to their canisters than was present in the canisters used by Kelly et al.

The recognition that not all VOCs are equally stable in canisters under all possible sampling temperatures and relative humidity conditions is based on purely empirical evidence (*36, 63, 106, 111, 112*). Coutant (*113, 114*) has pointed out that it is impractical to experimentally investigate the stability of all possible combinations of important nonpolar and polar VOCs at all concentration levels of interest and under all realistic sampling conditions. A more reasonable approach is to develop guidelines for future evaluation and application of canister sampling technology based on inferences drawn from the fundamental processes that govern the stability of whole air samples.

Coutant (*113, 114*) has developed a model to assess the potential for physical adsorption as a mechanism for loss of nonpolar and polar VOCs from the vapor phase in canister samples using the principles embodied in the Dubinin–Radushkevich isotherm and ideal solution behavior in the adsorbed phase. The model attempts to explain canister performance by taking into account the chemical and physical nature of the canister surface, the water content of the sample, the temperature of the sample, the pressure in the canister, the properties of the target compounds, and the competitive effects occurring with multicomponent adsorption. The general form of the isotherm is given by:

$$\ln(w/w_o) = -B/\beta^2\left[RT\ln(P/P_0)\right]^2 \quad (1)$$

where w is the amount of gas adsorbed on the canister wall, w_o is the amount of gas that may be adsorbed at saturation, B is a constant that is characteristic of the surface, β is the electronic polarizability of the sorbate, R is the gas constant, T is the absolute temperature, P is the vapor pressure concentration, and P_0 is the equilibrium vapor pressure for the pure sorbate at temperature T. In terms of eq 1, wall adsorption largely depends on compound polarizability, volatility (as represented by the equilibrium vapor pressure of the compound), temperature, and the vapor phase concentration of individual compounds. As a result, it may be used to estimate losses of sample components due to their adsorption on canister walls, where every gaseous compound in the canister competes for adsorptive sites. Coutant (*113, 114*) has developed a computer program for predicting adsorption behavior and vapor phase losses for approximately 80 compounds, including water. A major prediction of the model, which is generally consistent with experimental observation, is that water vapor will displace the more volatile nonpolar and polar compounds from a canister surface at relative humidities in the range of 1–20%. The model further indicates that the sample pressure should be set as high as possible without causing precipitation of liquid water within the canister.

Alternative Canister Deactivation Process. Despite the widespread use of SUMMA-polished canisters for monitoring VOCs in air, certain polar compounds have been found to exhibit reduced recoveries after storage. Restek Corporation has developed a surface deactivation process, referred to as the Silcosteel process, in which a high temperature reaction is used to bond micron-thick layers of pure, flexible fused silica directly onto the inner surfaces of stainless steel tubing for GC columns (*115*). Recently, Holdren and co-workers (*116, 117*) examined the effectiveness of using this process to provide better surface passivation of canisters for specific classes of VOCs. Canisters with the standard electropolished inner surfaces were tested and compared with those treated with the Silcosteel process and with various deactivation reagents that were added to further deactivate the surfaces. They found that the Silcosteel process can be successfully applied to the inner surfaces of canisters (*116*). These treated canisters exhibited storage characteristics for a test set of nonpolar and polar VOCs similar to those characteristics for new, untreated SUMMA-polished canisters. However, changes observed in storage stability were found to be statistically significant, and whereas some of the test compounds (including acrylonitrile and methyl *tert*-butyl ether) exhibited low variability, several others (including toluene, acetaldehyde, and 2-butanone) showed large changes. The Silcosteel process is a promising technique for deactivating stainless steel surfaces. Recent improvements have resulted (*118*) in the development of fused silica-lined canisters that yield data for a wide range of VOCs that are very comparable with the data obtainable with SUMMA canisters.

Real-Time Measurements

With continuous samplers that are designed to give a real-time measure of the concentration of a compound in air, the detector is an integral part of the sampling device and provides a continuous reading of the compound concentration. Samplers that sequentially trap and analyze the sample provide a means for obtaining near real-time analysis. Some real-time monitoring approaches make use of portable instruments, whereas others are based on larger and more sophisticated devices that generally provide more specific information on a wider range of compounds.

GC-Based Devices. Total VOC concentrations can be measured with handheld portable photoionization detectors (PID) such as the HNu DL-101 or the Sentex Scentogun (*119*). The PID uses UV light to ionize the sample. Positive ions so formed migrate toward a negatively charged collector, generating a current proportional to the total concentration of the ions. The extent of ionization or types of species ionized can be changed by using UV lamps of different energies. The higher the energy of the lamp, the larger the number and variety of compounds that can be ionized.

Portable gas chromatographs (PGCs) are capable of near real-time ambient air monitoring and of providing more information than total VOC monitors. These instruments are commercially available (e.g., Photovac or HNu) equipped with a 10.6-eV photoionization detector for enhanced sensitivity. Berkley et al. (*120*) have evaluated the Photovac 10 S70 photoionization PGC and obtained a benzene detection limit equivalent to 0.03 ppbv, with comparable sensitivities for other aromatic compounds and chloroalkenes and a linear response over a wide concentration range (0.5–130 ppbv).

Berkley et al. (*121*) compared PGC data with EPA Method TO-14 canister data obtained in two field studies and conducted a side-by-side laboratory evaluation of five commercially available PGCs before testing them further at a field bioremediation site (*122*). The PGC–canister comparison showed acceptable agreement between the two data sets. PGC analysis can be applied to VOCs that ionize below 10.6 eV and are sufficiently volatile to elute isothermally from the column at 50 °C or below. The GCs included in the multi-instrument evaluation were the Sentex Scentograph PC, the Photovac 10 S+, the HNu 311, the SRI 8610, and the MSI 301. VOC concentrations at the field site were slightly above ambient background levels. Several concurrent collocated canister samples were also collected in the course of this study and analyzed by EPA Method TO-14 to determine correct concentrations and further assess the reliability of the portable GC data. All of the GCs performed essentially as expected and were able to detect compounds at the levels encountered with reasonable accuracy.

The advantages of the Microsensor Technology M200 microchip GC for field analysis have been evaluated and discussed (*123, 124*). The M200 provides very fast analyses and highly reproducible retention data. This portable instrument contains two independent GC modules, each with its own miniature whole-air sample injector, narrow-bore column, and microdetector. The major limitation of the unit for field analysis of VOCs has been the relatively high detection limit (≥100 ppbv) of the built-in thermal conductivity detector (*123, 124*). Recently, Carney et al. (*125*) modified the device by adding an automated sorbent tube concentrator that uses a three-phase carbon-based multisorbent trap at ambient temperatures. Concentration enhancements of 750-fold have been achieved (*126*), thus lowering detection limits of the microchip GC to the 10-ppbv range.

An EPA van-based GC capable of on-site sequential sampling and analysis of VOCs in ambient air was used to draw air for 14 min each hour through a Nafion dryer tube (to remove water) and into a cryogenic trap at –170 °C (*127*). The trap was rapidly heated to 150 °C, and the VOCs were flushed onto the GC column for analysis. The trap was then rapidly recooled to resume sampling while the analysis was carried out. The entire process was automated and performed repetitively, and total sampling and analysis time was 64 min per cycle. The sequential gas chromatograms so generated have been used in a new monitoring strategy, known as temporal profile analysis (TPA), to sort VOCs into source-related groups and thus identify the specific VOC emission sources affecting a particular area.

MS-Based Devices. Several MS-based methods have shown promise in ambient air and indoor air studies of VOCs. These instruments range in complexity from the highly compact ion mobility spectrometer (IMS) to tandem mass spectrometers, such as the triple-stage quadrupole and ion-trap MS systems with various direct air sampling inlets.

The real-time monitoring capabilities, portability, and ppb-level sensitivity of IMS for components in the vapor phase have prompted the development of portable field monitors by Environmental Technologies Group and Graseby Dynamics (*128*). The technique has been used to measure aliphatic and aromatic amines (*129*) as well as to detect explosives (*130*) and illicit drugs (*131*). Although IMS instruments are capable of providing quantitative data, they provide little information on the identity of the ions in a peak. This information may be obtained by using a combined IMS–MS system.

Recently, Limero et al. (*132*) described the development and evaluation of a total non-methane hydrocarbon analyzer, based on IMS technology, to provide early detection of air-quality degradation aboard Space Station Freedom. IMS was favored for this application because photoionization detection and other conventional total hydrocarbon analyzer techniques are not very sensitive to important potential spacecraft contaminants such as Freons. A breadboard instrument has been constructed and preliminary testing completed. It was designed to study instrumental parameters as they relate to spacecraft operations. It has also been used to determine response times and instrument stability, and to evaluate the effectiveness of various algorithms in quantifying binary and tertiary mixtures.

The Sciex TAGA 6000 E triple-stage quadrupole (tandem) MS, equipped with a direct sampling atmospheric pressure ionization inlet (API–MS/MS), has been used for the detection of a wide variety of compounds at trace levels in real time. Examples of these studies include the determination of amines in indoor air from steam humidification (*133*), the development of a method for the continuous measurement of dimethyl sulfide in air (from oceanic emissions) at low parts-per-trillion levels (*134*), and the analysis of human breath to determine the biological half-lives of expired VOCs (*135*).

The radio-frequency quadrupole ion trap MS offers an attractive alternative to the triple-stage quadrupole, embodying most of the desirable characteristics of a field-deployable MS system for real-time characterization of VOCs in air, namely, high sensitivity, high specificity from its capability for multiple stages of tandem-in-time MS (e.g., MS/MS, MS/MS/MS, etc.), user friendliness, and reduced size, weight, and cost (*136*). Recent developments in ion-trap technology (*137–141*) and in direct air sampling techniques (*142–149*) now open up the possibility of developing small field-deployable tandem MS systems with exceptionally high sensitivity and specificity for direct air monitoring.

The 3DQ Discovery ion-trap MS system from Teledyne represents a breakthrough in MS technology (*137, 140, 141*). The 3DQ is a small, lightweight, benchtop, ion-trap system that uses newly patented filtered noise field (FNF) waveform technology to perform selective ion storage. This ability to eject all unwanted ions and only accumulate ions of interest results in more efficient utilization of the ion capacity of the trap, leading to greatly improved sensitivities for very low level components in complex mixtures. The technique also provides a means for performing multistage MS (i.e., MS^n) with exceptionally high efficiency, making compound identification unambiguous. The FNF ion-trap technology has also been incorporated into an enhancement unit for use with the large Finnigan MAT research-grade, ion-trap MS. The combination system has been evaluated with an atmospheric sampling glow discharge ionization (ASGDI) source (*142–144, 147, 149*), a countercurrent-flow tubular membrane (*147, 149*), and a pulsed direct air sampling module (*150*) to detect vapors of various nonpolar and polar compounds in air.

These direct air sampling, ion-trap techniques hold great promise for the measurement of trace gaseous constituents of air. They are especially applicable to compounds such as polar VOCs for which adequate monitoring methods are not yet available, and to those circumstances in which VOCs need to be measured in real time; for example, in microenvironments in which the concentrations of hazardous pollutants change rapidly.

Other Techniques. A continuous monitor for gaseous formaldehyde, which has a detection limit of 0.2 ppbv, has been developed by Kelly and Fortune (*99*). The monitor relies on the Hantzsch reaction (i.e., the cyclization of a β-diketone, an amine, and formaldehyde) to produce a fluorescent derivative of formaldehyde. This reaction provides high selectivity, high reagent stability, and low reagent cost, while the glass coil scrubber that serves as the collection device gives high efficiency and simplifies the design. Signal response rise and fall times are about 80 s each, with a lag time of 2 min. Repetitive calibrations during extended field operations indicate a reproducibility of about ±7%. The instrument was field-tested during a 10-day study that was part of the Atlanta Ozone Precursor Study (*151*).

A commercially available continuous monitor for formaldehyde in air uses an automated wet-chemical colorimetric analyzer with a continuous signal output (*95, 98*). As described in ASTM Method D 5221, the analyzer consists of a portable unit that contains a small pump, an analytical module, and a colorimeter. An acidified pararosaniline scrubber solution is pumped through the reference cell in the colorimeter, then diluted with water. The diluted solution then passes into an air scrubber coil, where formaldehyde is quantitatively absorbed from the air sample stream. After the liquid and air are separated, the sample solution is mixed with dilute aqueous sodium sulfite reagent. The alkylsulfonic acid chromophore is allowed to develop, and the resulting absorbance of the solution is continuously measured as it passes through the sample cell. The detection limit of the method is about 10 $\mu g/m^3$.

A highly selective and portable chemical sensor has been developed to measure chlorinated compounds in the gas phase without interference from hydrocarbons, oxygen-containing organic chemicals, or inorganic vapors (*152*). Called the RCL sensor, the device consists of a bead of heated material (a proprietary rare earth based ceramic) whose conductivity changes on contact with chlorinated organic chemicals or hydrogen chloride. The response time of the sensor is about 2 min, and it exhibits a linear response over the range of 0.2–10 ppm. Thus, it is highly suited to the task of characterizing chlorinated hydrocarbon emissions in the workplace or at hazardous waste sites. Work is continuing to improve the intrinsic sensitivity of the device.

Semivolatile and Nonvolatile Organic Chemicals

General Description of Methodology

The term "semivolatile organic chemical" is used broadly to describe an organic compound that is too volatile to be collected on a particle filter but not sufficiently volatile to be efficiently recovered from sorbents by thermal desorption or from air sampling canisters. They can be generally classified as those compounds having vapor pressures between 10^{-2} and 10^{-8} kPa at 20–25 °C. Compounds with lower vapor pressures are considered nonvolatile.

In numbers of different compounds, far more SVOCs and NVOCs than VOCs may exist in the atmosphere, although air concentrations are generally much lower. Polynuclear aromatic hydrocarbons (PAHs), pesticides, polychlorinated biphenyls

(PCBs), polychlorinated dibenzo-*p*-dioxins (PCDDs), polychlorinated dibenzofurans (PCDFs), and phthalate esters are SVOCs and NVOCs that have been considered among the most important air pollutants.

Filter Sampling

In general, collection of SVOCs and many NVOCs onto air filters is not an efficient process. Except for four-ring and larger PAHs, SVOCs sorbed onto or occluded inside airborne particulate matter probably do not constitute an appreciable fraction of the total SVOC content of the air except in areas near strong sources. Regardless of the distribution of SVOCs between the vapor and particulate states in air, it is difficult to assess the relative concentrations by filtration. Air flowing at only a moderate rate through a filter may strip much of the SVOCs from the particulate matter after it is collected. On the other hand, collected particulate matter may act as a sorbent for gaseous SVOCs passing through the filter. Desorption during sampling will, in most cases, largely outweigh sorption so that the total quantity of SVOCs collected on the filter will likely be less than that initially contained on the particles. For example, as long ago as 1974 Beyermann and Eckrich (*153*) found that 90–100% of *p,p*′-DDT (1,1,1-trichloro-2,2-bis(*p*-chlorophenyl)ethane), *p,p*′-DDD (1,1-dichloro-2,2-bis(*p*-chlorophenyl)ethane), dieldrin, and lindane passed through a glass fiber filter (GFF). In 1976, Lewis (*154*) added 19 organochlorine and organophosphate pesticides with p_o values ranging from 10^{-7} to 10^{-8} kPa to 10-cm GFFs coated with airborne particulate matter and pulled unfiltered ambient air at 280 L/min through the filters for 24 h. The mean spiking level corresponded to 100 ng/m^3. The face velocity across the filter was 60 cm/s, about the same as that for the current PS-1 air sampler. The result of this experiment was the loss of 95–100% of the pesticides from the filters after corrections for background. Similar experiments using the PS-1 sampler with chlorinated dibenzo-*p*-dioxins showed that 1,2,3,4-tetrachlorodibenzo-*p*-dioxin (1,2,3,4-TCDD) (p_o = 4.7×10^{-7} kPa) was nearly quantitatively (ca. 95%) vaporized under essentially the same conditions (*155*). Van Vaeck et al. (*156*) passed 250 m^3 of nitrogen at 250 L/min through 20-cm × 25-cm GFFs that had already been used to sample 500 m^3 of ambient air. Losses of 21–70% of octacosane through pentacosane (p_o = 3×10^{-4} to 4×10^{-7} kPa) were observed. Significant fractions of the compounds may have already been stripped from the filters during sampling. Recently, Storey et al. (*157*) concluded from studies with 11 semivolatile PAHs and *n*-alkanes that adsorption onto mineral oxide surfaces such as glass- or quartz-fiber filters is not important in urban air. However, they note that it may be in rural or remote atmospheres.

The rate and extent of loss of particulate-associated organic compounds from an air filter during sampling will be directly proportional to the face velocity. For example, Appel et al. (*158*) found that filter samples collected with an air sampler operating at 11 cm/s contained an average of 30% more total carbon (in μg/m^3) than those obtained with a standard Hi-Vol sampler operated at 50 cm/s.

Cotham and Bidleman (*159*) backed up GFFs loaded with urban air particulate matter with clean GFFs and passed clean air (SVOC- and particle-free) through them to study the stripping of particle-associated semivolatile organochlorine compounds. They found that substantial portions of the SVOCs were transferred to the backup filters (e.g., 20% for *p,p'*-DDT).

Although PAHs are generally believed to be introduced into the atmosphere as particles, and are often occluded deep within particles, published literature indicates that substantial losses from volatilization occur during sampling. Davis et al. (*160*) and Guerin et al. (*161*) reported 40–90% evaporative losses of pyrene (p_o = 3.1×10^{-6} kPa) and fluoranthene (p_o = 6.5×10^{-7} kPa), and small losses of less volatile PAHs from standard Hi-Vol filters at ca. 500 L/min (face velocity ca. 17 cm/s). Likewise, PAHs with $p_o > 10^{-9}$ kPa exhibited heavy losses from vapor-spiked urban particulate matter in dynamic studies performed with GFFs (*162*).

Rounds et al. (*163*) conducted evaporative desorption experiments with *n*-alkanes and PAHs on unspiked particulate-loaded Teflon membrane filters (TMFs) exposed to slow nitrogen purges for up to 28 days. The nitrogen was humidified and passed at flow rates ranging from 1 to 4 L/min through 71-mm TMFs that were loaded with automobile-exhaust particulate matter to correspond to 28 μg/m^3 air concentrations. The mean face velocity was only 1.5 cm/s, much less than that for most air samplers used for SVOC monitoring. Desorbed gases were trapped downstream on Tenax-TA and analyzed by thermal desorption. Observed losses of nonadecane ($p_o \approx 10^{-5}$ kPa) were ca. 20% and those for fluoranthene (p_o = 6.5×10^{-7} kPa) were ca. 40% after 131 m^3 of nitrogen over 28 days.

The empirically determined high rates of evaporative losses of SVOCs from spiked filters and particulate matter are in contrast to theoretical predictions made by Zhang and McMurry (*164*). Assuming equilibrium conditions between gas- and particle-phase concentrations, constant temperature, constant air concentrations, and that the sampled gas-phase concentration in contact with the particulate matter on the filter was equal to that equilibrium concentration corresponding to collected particulate concentration, they predicted that evaporative losses from filters during sampling would be small. These conditions are, of course, entirely unrealistic. Because sampling periods typically range from 8 to 24 h, significant changes in temperature (except for indoor monitoring) and air concentrations during sampling are most likely to occur, which is likely to have a large effect on the rate of volatilization. Even at constant temperature and face velocity, the rate of volatilization will be strongly dependent on the pollutant concentration in the air being sampled, increasing with decreasing concentration (*165*).

Evaporative losses from filters after sampling can also be significant, even for compounds of relatively low volatility (*165*). Exposed filters should be extracted as quickly as possible after sampling and the extracts stored at −10 °C until analysis.

The use of filters for airborne NVOCs is less prone to sampling error. However, significant portions of NVOCs with vapor pressures as low as 10^{-11} kPa have been found to pass through or be lost from particle filters during hot weather (*166*). Therefore, it is always advisable to use a backup vapor trap. GFFs or quartz-fiber filters (QFFs) have seen the greatest use for collection of particulate matter for analysis of SVOCs and NVOCs. QFFs have been shown to be more artifact-free. These filters are normally precleaned by baking out at 500–600 °C for several hours prior to use. TMFs and Teflon-impregnated glass fiber (TIGF) filters are also popular and may be less artifact-prone than GFFs or QFFs. Watts et al. (*167*) found that TIGF filters performed better than GFFs or QFFs for collection of urban air samples for use in mutagenicity testing based on the significantly better sonication extraction efficiency of TIGF filters. The extracts of TIGF filters used to sample ambient air in The Netherlands were reported by de Raat et al. (*168*) to be slightly less mutagenic than those from GFFs due to lower retention of more volatile PAHs by the TIGF filters. This behavior was believed to be the result of higher pressure drops through the TIGF filters. Nucleopore (*169*) and silver membrane filters (*170*) have also been used for low-volume sampling. Filter material is most often subjected to solvent extraction. However, thermal desorption has been employed to a limited extent (*171*).

Gases may also be sorbed by particulate matter collected on a filter during the process of air sampling (*172–174*). Because VOCs are likely to be in the sampled air at concentrations that are orders-of-magnitude higher than those of the SVOCs, their contribution to any resultant positive artifact is expected to be greater despite their lower retention efficiencies. For example, Hart and Pankow (*175*) found that less than 1% of the organic chemicals sorbed from air onto the second of two QFFs or TMFs in series could be accounted for as PAHs or semivolatile *n*-alkanes (C_{16}–C_{31}). Appel et al. (*176*) found that GFFs exhibited a much greater tendency to adsorb polar compounds, particularly dicarboxylic acids, than nonpolar air pollutants.

Filter-sorbed VOCs normally would not be apparent when typical extraction and analyses procedures for SVOCs are followed. However, they would constitute the major portion of volatilizable carbon when the filter is heated. The type of filter material used will affect the magnitude of the adsorption artifact. McDow and Huntzicker (*173*) found 24–75% more volatilizable organic carbon (removable at 275 °C) on GFFs than on QFFs. An average of 21% more was found on QFFs than TMFs.

Sorbents for SVOCs and NVOCs

SVOCs and NVOCs are normally collected from air onto solid sorbents, most often in combination with a particle filter. Popular sorbents for SVOCs include polyurethane foam (PUF), Amberlite XADs, Chromosorbs, Porapaks, and Tenax. There has also been limited use of inorganic sorbents such as Florisil and silica gel. The polyether type of PUF (upholstery type) has enjoyed wide usage for pesticides (*177–179*), PCBs (*177–182*), PAHs (*162, 183–185*), PCDDs/PCDFs (*155, 186–189*), and other SVOCs. The popularity of PUF is based on its low cost, ease of handling, and low air flow resistance (Figure 5). The latter property makes PUF ideally suited for high-volume air sampling. High-volume samplers utilizing particle filters backed up by PUF form the basis of several EPA methods for SVOCs. Method TO-4 (*59*) for organochlorine pesticides and PCBs is taken from Lewis et al. (*177, 178*) and employs a PS-1 sampler that pulls air at 225 L/min through a 10-cm-diameter QFF

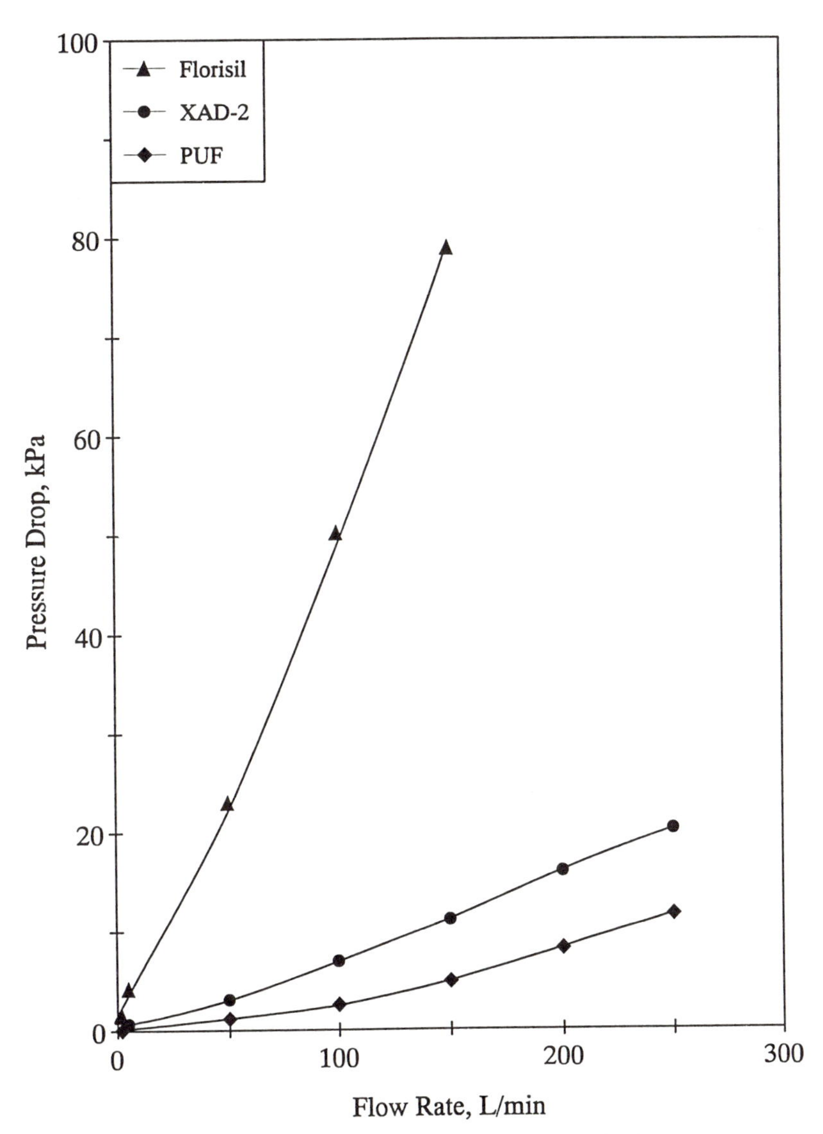

Figure 5. Pressure drop versus flow rate for three types of sorbents.

followed by a glass Soxhlet extractor cartridge containing a 65-mm-o.d. × 76-mm-long PUF plug. The method is applicable to a wide variety of pesticides (e.g., organophosphorus, carbamate, triazine, and pyrethroid). Both Method TO-9 for PCDDs and TO-13 for PAH call for the same sampler and flow rates (*59*). Method TO-13 provides the option for use of 16- to 20-mesh XAD-2, which is required for efficient trapping of some lighter PAHs. Method TO-9 has also been demonstrated to be efficient for PCDFs (*187–189*).

The Hi-Vol sampler, which is designed to sample air through a 20-cm × 25-cm GFF at flow rates of 1100–1700 L/min, has also been modified to collect SVOCs from ambient air by adding two or three 76- to 90-mm-o.d. × 76-mm-long PUF plugs or two 20-cm × 25-cm × 1.3-cm thick PUF pads in series (*185, 190, 191*). These modified Hi-Vols have been used for sampling PCBs (*190*), PAHs (*160, 185, 191*), and PCDDs/PCDFs (*192*) at flow rates of 500–600 L/min up to a total sample volume of 1000 m^3 of air. Bidleman and Olney (*180*), Keller and Bidleman (*184*), Trane and Mikalsen (*183*), and Marty et al. (*193*) employed similar high-volume samplers operated at 400–750 L/min for collection of PCBs and PAHs. Hawthorne et al. (*194*) used a PM-10 high-volume sampler with two 18-cm × 23-cm × 5-cm thick PUF sheets behind a standard Hi-Vol filter to collect semi- and nonvolatile phenols, alkanes, and PAHs from ambient air at 1100 L/min. Polychlorinated naphthalenes (PCNs) have been collected from ambient air onto PUF by high-volume sampling (*177*) and by medium-volume sampling (*195*). Phthalate esters have also been successfully trapped on PUF by high-volume air sampling (*196*).

PUF is also used in low-volume samplers for pesticides and PCBs. A portable sampler utilizing a small, battery-powered pump to draw air at 4 L/min through a 22-mm × 76-mm PUF plug (*179*) has been widely used for indoor air sampling, as well as for personal exposure monitoring. It is the basis of EPA Method TO-10 for ambient air (*59*), EPA Method IP-8 for indoor air (*197*), World Health Organization IARC Method 24 for indoor air (*198*), and two standard methods published by the ASTM (*95*): ASTM 4861 for pesticides and PCBs in air, and ASTM 4947 for chlordane and heptachlor in indoor air. This sampler has been used for indoor, outdoor, and personal air monitoring in several large studies, including the EPA Nonoccupational Exposure Study (*199, 200*) and Agricultural Health Study (*201*). A fine-particle filter was used in the later study (*202*) and is recommended in ASTM 4861. The 2.5-μm cut-point *impactor* developed by Marple et al. (*203*) has been adapted to this sampler (*204*).

PUF has two distinct disadvantages, however. Because of its low density (typically 22 mg/cm^3), its retention capacity for SVOCs is substantially lower than that of granular sorbents, which are 10–15 times as dense. Consequently, breakthrough volumes are much smaller for PUF. For example, lower-chlorinated biphenyls (<3 Cl) and smaller PAHs (<3 rings) have been shown to be poorly retained by PUF when high-volume sampling is used (*178, 205, 206*). In general, PUF is not a good choice for SVOCs having vapor pressures greater than 10^{-3} to 10^{-4} kPa (depending on polarity). The penetration of PAHs through PUF as a function of volatility has been studied by You and Bidleman (*207*).

The range of applicability of PUF can be extended by combining it with a small quantity of a granular sorbent (e.g., 50–75 g) so as to keep the pressure drop minimized. "Sandwich" combinations of XAD-2, Tenax, or one of several other granular sorbents between two layers of PUF have been shown (*178, 208, 209*) to efficiently trap SVOCs with vapor pressures up to 10^{-3} kPa. Another potential disadvantage of PUF is its relatively poor ability to trap fine particles (<1 μm) compared to Florisil and XAD-2 (*210*). Whereas collocated PUF samplers with and without particle filters gave similar results for pesticides in residential indoor air (*211*), the possibilities of loss of particle-associated SVOCs when sampling with PUF dictates that it be used in combination with a fine-particle filter. The high static charge associated with XAD-2 makes it highly effective at trapping particles down to 0.1 μm in size (*210*). Consequently, the need for the use of particle filters may be obviated in the case of XAD-2 and similar porous polymer resins. Sandwich traps of PUF and XAD-2 should also be more effective at fine-particle entrapment than PUF alone.

Next to PUF, XAD-2 [a poly(styrene–divinylbenzene) resin] has seen the most use for SVOC air sampling. It has been most widely used for PAHs (*205, 212–218*) but has also been used for PCBs (*219–221*), PCDDs/PCDFs (*222*), and other SVOCs (*223–225*). It is specified as an alternative to PUF in EPA Methods TO-13 (*59*) and IP-7 (*197*) and is also used in several NIOSH and Occupational Safety and Health Administration methods for SVOCs (*88, 226*). XAD-2 is said to have better capacity for PAHs than Tenax-GC (*227*). A sandwich combination trap of XAD-2 and PUF is the basis of EPA Method CLP-2 for monitoring ambient air for SVOCs monitored at hazardous waste and Superfund sites (*228, 229*) and is being incorporated into the revised EPA TO methods for ambient air.

Other polymeric sorbents that have been used to collect various SVOCs from air are XAD-4 (*95, 230–232*), XAD-7 (*225*), Tenax-GC (*178, 212, 232–234*), Chromosorb 101 (*233*) and 102 (*178, 235*), and Porapak R (*178, 236*). There has been some limited use of inorganic sorbents, such as Florisil (*178, 234, 237*) and silica gel (*238, 239*). Graphitized polymer sorbents are generally not used for SVOCs because of the difficulties associated with recovery. Bidleman (*240*) published a good review of sorbents for SVOCs in 1985.

Collected compounds are most often recovered for analysis by Soxhlet extraction, although shaking or sonication with solvents and supercritical fluid extraction (*241–243*) have been used with varying success. Polyurethane foam may also be extracted by repeated compression under several exchanges of solvent, but this procedure is labor-intensive and can require more solvent than Soxhlet extraction. Only limited success has been reported (*233, 244*) with thermal desorption of SVOCs with vapor pressures down to 10^{-5} kPa from Tenax-GC.

Chemical Artifact Formation and Storage Stability

Aside from the particle filter volatilization artifact previously discussed, the principal concern has been with potential analyte loss or alteration due to chemical reactions during and after sampling. Because thermal desorption is rarely employed for SVOCs, artifact formation during analysis is generally of little concern. Oxidation by atmospheric ozone and nitration by nitrogen dioxide have been the subject of the most studies, particularly in the case of PAHs. For example, benzo[*a*]pyrene (BaP) on particulate matter is known to react with ozone to form BaP-4,5-oxide (*245*). Losses of 40–80% of BaP applied to particulate filters were typically observed when air containing 30 ppb or higher concentrations of O_3 was pulled across them for 4–24 h (*246–248*). Likewise, losses were found to occur on storage of spiked filters in the dark (*248*). Other PAHs are also particularly prone to epoxidation (e.g., benzo[*a*]anthracene and cyclopenta[*c,d*]pyrene). Nitration of PAHs by NO_2 and nitric acid (*249*) and sulfonation by SO_2 (*250*) also have been similarly observed. Perhaps 50% or more of the observed losses, however, may not occur with actual airborne particulate matter, but result because the spike analytes remain primarily on the surface of the particles where they are more vulnerable to reaction. In most studies, the PAHs were also applied at higher concentrations than those typical for ambient air.

Grosjean et al. (*251*) exposed BaP, perylene, and 1-nitropyrene spiked at 0.05, 0.25, and 4.1 $\mu g/cm^2$ on clean and particulate-loaded (ambient, diesel, and fly ash) GFFs and Teflon filters to 100 ppb concentrations of O_3, NO_2, and SO_2 for 3 h at face velocities corresponding to high-volume sampling. They saw no significant losses under any conditions. The principal reason for the apparent discrepancy between the work of Grosjean et al. and that cited previously is likely the result of the much lower spiking levels used by Grosjean et al. Whereas they spiked at levels 10–800 times higher than the typical BaP loadings found on Hi-Vol filter samples of ambient air (ca. 5 ng/cm^2), most other investigators used much higher PAH spiking levels and usually higher reactant gas concentrations.

Likewise, Coutant et al. (*252*) found no significant losses of BaP on ambient air particulate matter during *collocated sampling* with a traditional sampler and one fitted with an ozone denuder. BaP concentrations on QFFs used to sample 19–21 m^3 of ambient air fortified to contain 180 ppbv O_3 were 95–97% of those obtained in simultaneous, collocated, unfortified air samples collected with a sampler capable of removing 90–95% of the atmospheric O_3. Brorström et al. (*253*) performed similar collocated sampling in which they introduced NO_2 at 30–200 ppb into the inlet of one sampler but not the other. They reported 2–38% (means: 2–19%) lower concentrations of BaP and BaA when the filters were exposed to NO_2. On the other hand, Arey et al. (*254*) found that high-volume sampling for airborne PAHs during periods of high ambient NO_2 concentrations (ca. 160 ppb) did not apparently react with perdeuterated BaP, perylene, pyrene, or fluoranthene spiked on filters and backup vapor traps.

Relatively little has been published on chemical artifact formation associated with solid sorbents for SVOCs and NVOCs. Hanson et al. (*255*) reported the production of decomposition products from XAD-2 and Tenax-GC when stack gas effluent containing 460–810 ppm of NO_x was passed through them. Lindskog et al. (*256*) observed 20–50% losses of three PAHs that had been vapor-spiked to XAD-2 when air containing 100 ppb of NO_2 was pulled through it at 1 L/min for 6 h. At higher concentrations (500–800 ppb) losses were 62–92%, with anthracene being the most affected. In contrast, when PUF was studied, there were no losses of acenaphthene or fluorene, and only a 25% loss of anthracene at 800 ppb for 22 h (*256*). Similar studies with O_3 showed a 20% loss of anthracene after 6 h at 50 and 100 ppb, and SO_2 at 1000 ppb caused little or no losses of the PAHs (*256*). Parallel ambient air sampling with PUF and Tenax-GC by Ligocki and Pankow (*206*) gave much higher results for acenaphthylene and 15–40% larger values for other PAHs when PUF was used.

Storage stability studies on pesticides and PCBs spiked on PUF have shown (*95, 179*) generally good recoveries even after 30 days at ambient temperatures. However, Chuang et al. (*205*) found BaP to be less stable on PUF than on XAD-2. This finding may be due to the higher porosity of PUF, which could allow greater infusion of gaseous oxidants and conditions more favorable to evaporative losses during storage. In any case, filter and sorbent samples should always be stored under refrigeration (preferably frozen) until analyzed. Storage times should be limited to 2–4 weeks. It is also better to extract the sampling media before long-term storage.

Background contamination is another factor that should be considered in any good monitoring practice. All sampling media should be considered to have potentially interfering extractable organic chemicals until they have been checked. Hunt and Pangaro (*257*) investigated potential contamination of XAD-2 and -4, as well as Ambersorbs 340 and 348, and identified a number of contaminants. The PUF sold for air sampling is manufactured for upholstery and insulation use, and may contain many extraneous chemicals unless the supplier has taken adequate steps to clean it up. Sorbents should always be desorbed before use in a more rigorous manner than will be employed to recover the analytes sought in actual sampling.

Phase Distribution

Most SVOCs have vapor pressures that allow them to exist in the atmosphere distributed between the gaseous and particle-associated phases, and the equilibrium between phases will vary with ambient temperature, concentrations, particle size and type, and other atmospheric variables. Deposition rates for particle-associated SVOCs are substantially larger than those for the gas phase. Photochemical and oxidation reactions may also be enhanced by surface sorption. Therefore, knowledge of *phase distributions* is valuable in predicting the fate and long-range transport of SVOCs in the atmosphere.

Approximate equilibrium distributions between gas and particulate-associated (surface-sorbed) phases can be calculated from Junge's equation:

$$\Phi = c\Theta / (p_o + c\Phi) \quad (2)$$

where Φ is the ratio of the amount of organic compound on the particulate matter to that in both the gaseous and particulate-associated phases, Θ is the surface area of the particulate matter per unit volume of air, and c is a constant that depends on the molecular weight and heat of condensation (*7*). Figure 6 shows a plot of Φ versus Θ, with vapor pressure isobars superimposed. The Θ values for remote, rural, and urban atmospheres are indicated. Equation 2 would suggest, therefore, that organic chemicals with $p_o > 10^{-5}$ kPa would exist almost entirely in the gas phase and those with $p_o < 10^{-9}$ kPa would be nearly all in the particulate-associated phase in remote atmospheres.

When contaminated particles are emitted into the atmosphere from sources, partitioning of the surface-sorbed SVOCs into the gas phase will likely commence by

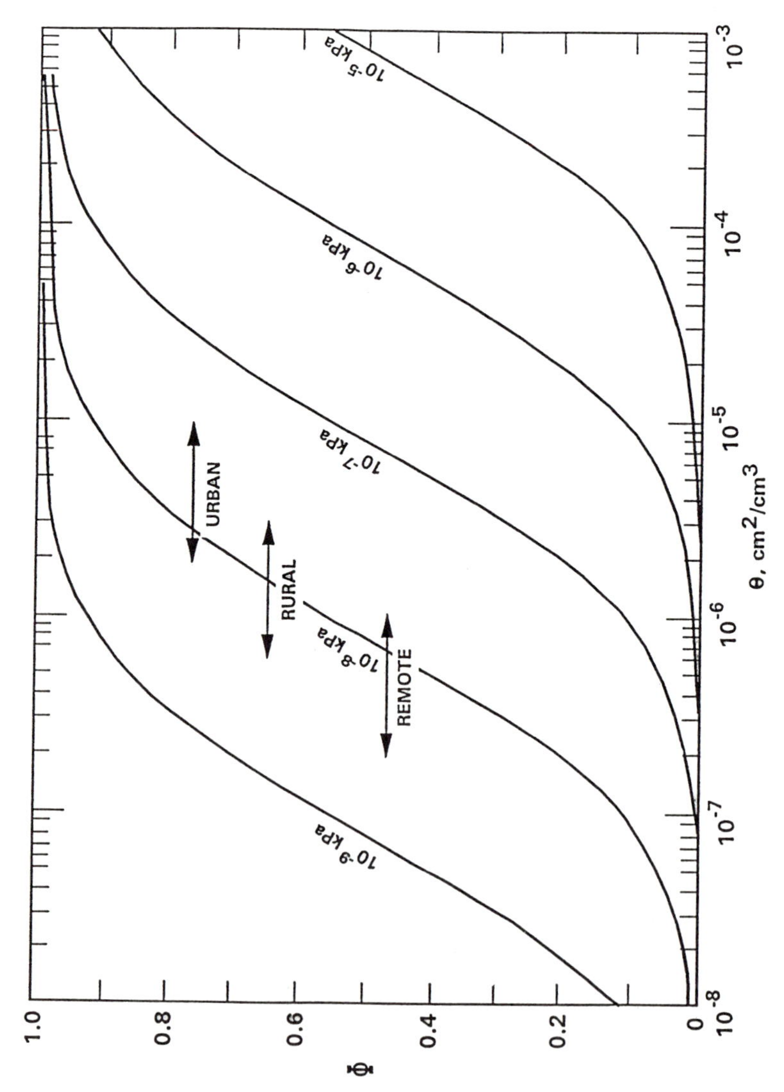

Figure 6. The fraction (Φ) of surface-sorbed organic chemicals on suspended airborne particulate matter as a function of surface density (Θ) and vapor pressure (p_o). (Adapted from reference 7.)

vaporization. The extent of loss of a given SVOC from the suspended particles in transit from a source to a sampler placement site will depend on its vapor pressure, on the distance transported, air temperature, and to a lesser degree on humidity, wind velocity, and particle size. Once collected onto a filter, volatilization is likely to occur at an accelerated rate, making it impossible to obtain accurate measurements of organic particulate loadings by filtration sampling. Consequently, efficient collection of most SVOCs requires a sampler equipped with a particle filter in series with a vapor trap capable of retaining both particles and gases.

When investigators have employed traditional samplers designed to collect gas phase and particulate-associated organic compounds from ambient air, compounds with $p_o > 10^{-8}$ kPa have been found primarily in the vapor traps of these samplers. For example, in 1972, the EPA used an air sampler with a backup vapor trap to sample for pesticides ($p_o = 5 \times 10^{-4}$ to 2×10^{-8} kPa) in three states (NY, TX, and FL) (*258*). Only rarely were any pesticides found on the particle filter. Billings and Bidleman (*212*) used a similar high-volume filter sampling system employing a PUF vapor trap to measure PCBs and pesticides in ambient air. They reported the following average percentages of each compound found in the filter: Aroclor 1016, <2%; Aroclor 1254, <6%; chlordane, <2%; and *p,p'*-DDT, 17%. Harvey and Steinhauer (*259*) employed a high-volume sampler to pull air at 620–680 L/min through a 10-cm GFF followed by a vapor trap containing distillation column packing (6.4-mm ceramic saddles) coated with OV-17 silicone oil to collect PCBs from ambient air over the Atlantic Ocean. Less than 1% of the PCBs were found on the particle filter. Jackson and Lewis (*260*) used a high-volume air sampler (225 L/min) that pulled air through a GFF followed by a PUF vapor trap to monitor the air in connection with the removal of PCB-contaminated soils in North Carolina. Samplers situated 30 m downwind of excavation activities showed that the majority (50–90%) of the Aroclor 1260 released into the air was collected by the PUF. Only in one case was more PCB (59%) found on the filter than in the vapor trap, and this resulted from excessive windblown dust created by a mechanical sweeper. In these studies, the samplers were turned off about 1 h after digging operations ceased. Had sampling continued for substantially longer periods, it is likely that even less PCB would have been detected on the filters. Recent evidence suggests that coplanar PCBs (nonsubstituted or monosubstituted with an ortho chlorine) may be associated to a greater degree with airborne particulate matter (*261*).

Cautreels and van Cauwenberghe (*234*) used a low-volume sampler (7 L/min) with a GFF and a Tenax–GC trap to study PAH phase distribution in the air in Belgium. They found that PAHs with $p_o > 10^{-7}$ kPa were found primarily in the vapor trap and those with $p_o < 10^{-8}$ kPa and below were mostly on the filter. Similar results were obtained by Yamasaki et al. (*166*) in Japan, Trane and Mikalsen in Norway (*183*), and by Keller and Bidleman (*184*) and Galasyn et al. (*262*) in the United States, all using high-volume samplers with GFFs backed with PUF traps. During summer months, Yamasaki found as much as 20% of the BaP ($p_o = 7.3 \times 10^{-10}$ kPa) in the vapor trap. Lewis et al. (*263*) reported finding 80–90% of PAHs with $p_o > 10^{-6}$ kPa on XAD-2 and the remainder on QFFs when sampling at 110 L/min for 24-h periods in Boston and Houston during February and April 1991. PAHs with p_o between 10^{-7} and 10^{-9} kPa were 70% on the particle filter.

Unfortunately, none of these studies could establish the real distributions of SVOCs in the ambient air. It is apparent that the filter and vapor trap should be extracted and analyzed together or the analytical results combined to avoid the risk of misinterpreting what is found on the two media as indicative of atmospheric phase distribution.

Gaseous compounds present in the atmosphere from fugitive emissions (or from evaporative desorption from suspended particulate matter) may, of course, be sorbed from the air onto airborne particles. The rate of sorption would be dependent on the frequency of collisions between suspended particles and the gaseous molecular species in the great air mass, but the efficiency of this process is questionable. The mechanics of atmospheric gas–particle interactions are complex and are governed by a host of chemical, physical, and environmental factors, including the number and size of particles, temperature, air mixing, photoreactivity, and chemical reactivity. Only a very small fraction of the collisions between gas molecules and particles are believed to lead to sorption. Highly carbonaceous material, such as that emitted from coal-burning stacks, will, of course, sorb organic compounds far more efficiently than highly siliceous particles, such as airborne sandy soil. Some clue as to the efficiency of the process may be found in the calculations of Judeikis and Siegel (*245*). Using a modified particle–gas collision equation containing both sorption and desorption parameters, they calculated that for conditions of sorption without reaction, only 10^{-7}% of a gas originally present at 1 ppm would be sorbed after several days' exposure to 100 μg/m^3 levels of particulate matter of the type normally found in the ambient atmosphere. Concentrations of SVOCs in the ambient atmosphere are normally at picogram per cubic meter or lower levels, making the probability of gas–particle collisions less likely and the potential particle-scavenging process even less efficient.

Whereas higher ambient air temperatures shift the phase equilibrium in favor of the gas phase, the effects of relative humidity (RH) are less well understood. Atmospheric moisture will compete with SVOCs for sorptive sites on airborne particles and may result in some desorption. However, the presence of water on the particles may also increase surface sorption through solubilization of the SVOC, especially in relatively water-soluble compounds. Of course, raindrops and fog are also known to scavenge SVOCs from air (*264–266*). Thibodeaux et al. (*267*) attempted to model the effects of atmospheric moisture on gas–particle partitioning using modifications of the Brunauer–Emmett–Teller (BET) isotherm. Their predictions suggested minimum particulate-associated SVOCs at 60% RH.

Although most particle-associated SVOCs are probably surface sorbates, some may be irreversibly sorbed onto carbonaceous matter or occluded inside particles. Only these SVOCs would be expected to be efficiently retained by a particle filter.

Gas–Particle Separators

In order to accurately determine the proportions of airborne SVOCs and NVOCs present in the gaseous and particle-associated phases at the time of sampling, a means must be employed to separate and maintain the integrities of the two phases. The first published attempt at this separation was by Bramesberger and Adams in 1965 (*268*). They designed a low-volume sampler (1 L/min) that directed the air first onto a rotating disk impactor (from which the collected particles were continuously swept into *n*-decane), then passed it through a midget impinger containing the same solvent. It is doubtful that this device cleanly separated the two phases, and its low flow meant it did not provide sufficient sensitivity for most ambient air monitoring applications.

Another approach that has been considered is to continuously move the filtered particulate matter out of the air sampling stream and trap the vapors downstream. Although this method has not been reported as a means of determining SVOC phase distributions, it has been used to monitor the inorganic composition of airborne particulate matter on a temporal basis (*269*). Electrostatic precipitators (ESPs) also have been used to collect airborne particles from air during high-volume sampling (*270, 271*). The potential for oxidation of the targeted organic chemicals by ozone produced by the electric field may limit this approach, as found by Kaupp and Umlauf (*270*), who obtained much lower values for particle-associated PAHs with an ESP than with a traditional filter sampler. Another study (*271*) using a massive air volume sampler (18,500 L/min), however, reported no evidence of oxidation artifacts.

The most successful approach to date at separating gas-phase from aerosol-phase organic compounds is based on the use of diffusion denuder separation technology. This separation technology is based on the fact that gases have diffusion coefficients that are generally orders of magnitude higher than those of particles. If air is drawn at a given flow rate through a channel of the proper width and length, particles will tend to pass straight through while gases will diffuse and collide with the walls (Figure 7). If the walls are coated with a sorbent (or reactive trapping medium), gases can be effectively removed from the air stream and the particles can be trapped downstream. If a traditional particle filter is used for the latter purpose, then a backup vapor trap will be required to collect SVOCs that are stripped from the particles during sampling.

The simplest *denuder* is a wall-coated open tube (Figure 7a). To effect an efficient (>90%) separation of SVOCs with typical diffusion coefficients of 0.06–0.07 cm^2/s from particles with diameters >0.01 μm using an open-tube denuder 1 m in length, air flow rates cannot exceed ca. 4 L/min (*272, 273*). Therefore, to achieve air sampling rates adequate for ambient air monitoring, several of these tubes must be bundled together (Figure 7b). Coutant et al. (*252*) used this approach to construct a denuder system consisting of seven 1.5-cm i.d. × 61-cm open tubes with the interior walls coated with Dow Corning high-vacuum silicone grease. This sampler was capable of separating more than 90% of gas-phase PAHs from particles at a flow rate

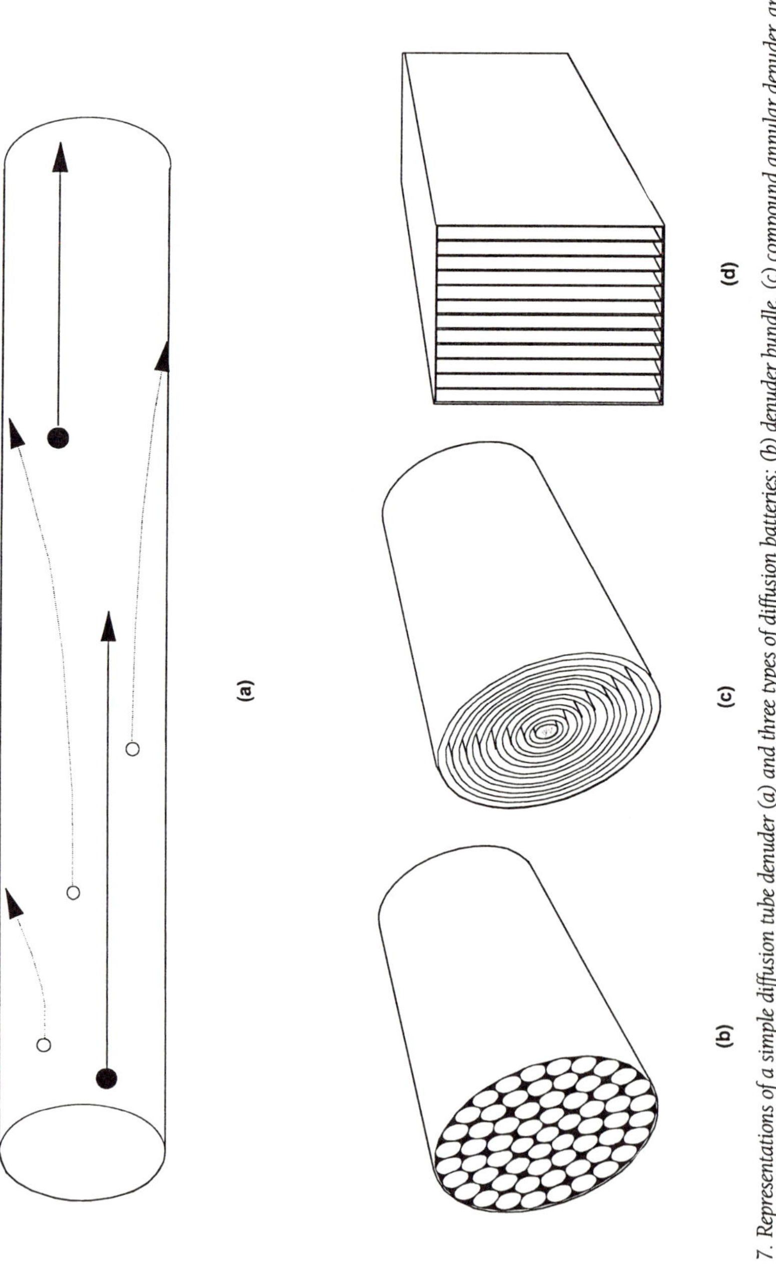

Figure 7. Representations of a simple diffusion tube denuder (a) and three types of diffusion batteries: (b) denuder bundle, (c) compound annular denuder, and (d) parallel plate denuder. In (a), the large solid circles represent particles; small open circles represent gas molecules.

of 15 L/min. The denuder assembly was designed to fit on the front of a PS-1 sampling module (*178*). The particulate phase was collected on a QFF placed behind the denuder tubes, and a backup PUF trap was used to collect PAHs that volatilized from the filter during sampling. Because it was not possible to recover the gas-phase PAHs for direct analysis, the sampler had to be collocated with a traditional sampler to determine the total airborne PAH concentration, C_t. Analysis of the filter and PUF trap behind the denuder bundle gave the particle-phase concentration, C_p; the gas-phase concentration, C_g, was inferred by difference, $C_t - C_p$. This approach is referred to as "denuder difference sampling", and is subject to subtraction errors, particularly if there are large differences between the two concentrations. Separate analysis of the PUF trap from the denuder sampler afforded an approximation of the volatilization artifact, A_v. However, this estimation may have been in excess of the real value of A_v, because C_g in the air passing through the particle filter was near zero in the denuder sampler, as opposed to relatively high values of C_g in a traditional sampler. Despite the limitations of the sampler and its low sampling rate, Coutant et al. (*252*) were able with 24-h sampling to make the first direct measurements of phase-distributed PAHs. They showed that the A_v for PAHs with p_o between 10^{-5} and 10^{-8} kPa ranged from 7 to 92%, depending on the ambient temperature. The median C_g for these PAHs, however, was determined by denuder difference to range from 26 to 43%.

A nested-tube diffusion battery of the type shown in Figure 7c was developed by Coutant et al. (*274*) and was capable of accommodating air sampling rates of up to 200 L/min. The high-efficiency compound annular denuder (CAD) was constructed from twelve 20.3-cm long nested aluminum cylinders with 1.6-mm internal annuli. The outside diameter was 8.2 cm, and like its predecessor, it was designed to attach to the PS-1 sampler. The inside walls were coated with silicone grease, which was found to be 92–98% efficient at removing gas-phase PAHs (*275*). Denuder difference field measurements with the high-volume CAD using QFFs and XAD-2 vapor traps (*275*) agreed well with the previous findings (*252*) on PAH phase distributions. Subsequent investigations with the CAD carried out in Boston, MA, and Houston, TX, by Lewis et al. (*263*) showed that PAHs with p_o between 10^{-4} and 10^{-6} kPa exhibited median A_v values of 17–44% at 7–19 °C, and denuder difference measurements indicated that 40–71% of these PAHs were originally in the gas phase (Table VIII). For PAHs with p_o between 10^{-7} and 10^{-9} kPa, A_v values were 9–15%, and 13–23% were originally in the gas phase. No volatilization artifact was measurable for PAHs with p_o below 10^{-9} kPa.

Lane et al. (*276*) also developed a CAD, but one limited to a much lower sampling rate (16.7 L/min). It consisted of six nested borosilicate glass cylinders, 60 cm long, coated with crushed Tenax-GC in SE-54 silicone GC column phase. [Coutant et al. (*275*), however, reported that SE-54 oxidized during sampling on their high-volume CAD.] The annular gap width was 2 mm. They, too, found that gas-phase SVOCs could not be efficiently recovered from the CAD and thus had to resort to denuder-difference measurements, which were achieved with a design that included two sampling lines, one with and one without a CAD. Both lines were backed up

Table VIII. Medians of Empirically Determined Volatilization Artifacts and Phase Distributions of Polycyclic Aromatic Hydrocarbons in Boston and Houston

Compound	*Vapor Pressure (kPa)*	*Concentration (ng/m³)*	*Percent of Total in Vapor Trap*[a]	A_V *(% of Total)*[b] *Median*	*Range*	*Percent Originally in Gas Phase*[c]
Fluorene	ca. 10^{-4}	4.5	95	44	11–68	52
Fluorenone	ca. 10^{-5}	1.5	84	17	8–67	71
Phenanthrene	2.2×10^{-5}	10.0	90	32	12–40	59
Anthracene	3.3×10^{-6}	0.6	83	44	19–47	40
Pyrene	3.1×10^{-6}	5.0	82	30	13–65	51
Fluoranthene	6.5×10^{-7}	3.1	80	24	3–72	56
Cyclopenta[*c*,*d*]pyrene	ca. 10^{-7}	0.2	29	15	0–36	16
Benz[*a*]anthracene	1.5×10^{-8}	0.4	27	9	9–34	13
Chrysene	1.7×10^{-9}	0.5	33	9	0–17	23

NOTE: Data were compiled from both Boston and Houston sites. Median temperature was 11 °C; temperature range was 7–19 °C. A_V is volatilization artifact.
[a]Determined by separately analyzing filter and vapor trap of PS-1 basic sampler.
[b]Determined by separately analyzing filter and vapor trap of denuder sampler and calculating artifact.
[c]Calculated from difference in total (vapor + particle) concentrations determined by traditional and denuder samplers.

with particle filters and sorbent beds. The CAD-equipped line also included fine-particle (<2.5 μm, 15 L/min) and coarse-particle (2.5–10 μm, 1.67 L/min) inlets behind the CAD, each backed up with sorbent traps. Field studies with this sampler showed that hexachlorobenzene ($p_o = 2.2 \times 10^{-6}$ kPa) and α- and γ-hexachlorocyclohexanes ($p_o = 7.4 \times 10^{-6}$ kPa) were 97–100% in the gas phase in the ambient air at remote locations in Ontario, Canada (*277*). For their field work, they used Florisil in the sorbent traps.

Subramanyam et al. (*278*) used the sampler developed by Lane et al. (*276*) with Tenax backup traps to determine PAH phase distributions in the air in Baton Rouge, LA. They reported that phenanthrene and fluoranthene were 98–99% in the gas phase.

A parallel-plate diffusion battery design (Figure 7d) was employed by Fitz (*172*) to eliminate the sorption artifact associated with the determination of volatilizable organic carbon by air filtration sampling. The denuder consisted of 15 quartz-fiber strips (Pallflex QAST), 25.4 cm long and 4.3 cm wide, separated by glass rods to provide 0.3-cm annuli. It was placed in the center of a 4.4-cm × 116.8-cm housing attached to two 47-mm Nucleopore filter packs containing the same filter material used in the denuder. The flow rate used was 15 L/min and the theoretical removal efficiency of the denuder was said to be 99.97% for anthracene. An identical parallel sampler without the denuder was operated simultaneously for denuder-difference comparison of volatilizable organic carbon. The denuder was reported to reduce the organic carbon on the back filter of the filter pack by more than a factor of four, to levels comparable to that of the filter blank.

Eatough et al. (*279*) constructed a high-volume parallel plate denuder similar to that of Fitz (*172*) to determine organic compounds associated with airborne fine particulate matter. This sampler consisted of 17 parallel strips of carbon-impregnated filters (CIFs), 4.5-cm × 58-cm long, separated by 2-mm annuli and was capable of operation at 200 L/min. Particles passing through the diffusion battery were trapped on a QFF backed up with a CIF to trap the volatilization artifact. The CIFs were thermally desorbed at 300 °C and the QFFs at 800 °C for total volatilizable carbon determinations. Because of the irreversible nature of the sorption of SVOCs to the CIFs, extraction and specific compound analysis was not possible. Field studies (*280*) in Canyonlands National Park, UT, showed that an average of 55% of particulate-associated carbon was lost during sampling. These results agree with previously reported findings for PAHs (*252, 263*).

A novel glass annular denuder was recently developed by Gundel et al. (*281*). It was made from a simple, commercially available annular denuder (URG 2000-30 B, University Research Glassware, Chapel Hill, NC) used for removal of inorganic acid gases, and similar to that published by Possanzini et al. (*282*). The Gundel denuder consisted of a single glass tube, 2.4-cm i.d. × 18.5 cm long, inside of which was positioned a 2.2-cm o.d. glass rod, leaving a 1-mm annulus. The sand-blasted walls of this annulus were coated with very finely ground (<0.5 μm) XAD-4. The major advantage of the denuder was the ability to recover the gas-phase SVOCs from the wall-coated sorbent by solvent extraction. When backed up with a filter and vapor trap, direct measurements of C_g and C_p were possible. However, the capacity of the denuder for PAHs was determined to be ca. 5 L/min for 3-h sampling periods, limiting its use for many airborne SVOCs.

Krieger and Hites (*283*) constructed a diffusion battery from a bundle of 120 short (25-cm) sections of fused silica capillary column (DB-1 WCOT, 0.32-mm i.d.) held together with epoxy resin that could be thermally desorbed for specific SVOC analysis. The system was of low capacity, accommodating flow rates of only 1.3–1.5 L/min. However, total transfer of collected gas-phase SVOCs to the GC–MS via a cryotrap provided detection limits comparable to those achievable with high-volume sampling and traditional solvent extraction. Particles passing through the denuder assembly were collected on a 4.7-cm QFF, which was followed by a 4.7-cm-diameter × 2-cm-long PUF plug. The PUF plug could not be thermally desorbed, and consequently comparable sensitivity was not obtainable for the determination of particle-associated SVOCs. Problems with cracking of the epoxy binding of the denuder bundle during thermal desorption at 220 °C were later identified (*284*). In recently published work, Krieger and Hites (*284*) have constructed similar denuders with 50–120 capillary tubes of either 0.32-mm DB-1 or 0.53-mm Quadrex 007-1 GC column sections cut to 30-cm lengths. These systems were extracted by either solvent elution or on-line supercritical fluid extraction procedures. Experiments performed with four or five denuders in series indicated an 80% collection efficiency for removing gas-phase PCBs. Although no plugging of the capillary tubes by large particles was observed in field evaluations, the denuder has not yet been characterized with respect to particle transmission efficiency.

A glass honeycomb denuder developed by Koutrakis et al. (*285*) is a variation of the multiple-tube diffusion battery (Figure 7b). Designed to collect inorganic acid or basic gases, the denuder assembly consists of 212 hexagonal glass tubes packed into a cylinder 4.7 cm i.d. × 3.8 cm long. The width of the diffusion channels was 0.2 mm. Sampling rates of 10 L/min were determined to be possible for efficient collection of HNO_3 and NH_3. To date, there has been no report of attempts to coat the denuder walls with a sorbent suitable for the collection of gas-phase SVOCs.

A laminar flow separator capable of separating a known fraction of the gas phase from the particulate phase has been recently designed by Turpin et al. (*286*). It is a low-volume sampler that pulls ambient air at ca. 1.5 L/min into a diffusion separator through which a core flow of clean, particle-free, ultrapure air is maintained at approximately the same rate. Gases, because of their much higher diffusivities, migrate into the core flow and are largely separated from the particles, which remain in the annular flow surrounding the core (Figure 8). Ambient air particles larger than 0.05 μm are said to be >95% excluded from the core flow, while most of the gaseous SVOCs diffuse into the core air stream, with the efficiency depending on the diffusion coefficient of the compound. The authors calculated a theoretical separation at 25 °C of 67% of the gas-phase phenanthrene ($p_o = 2.2 \times 10^{-5}$ kPa) from the annular flow to the core flow. Separation efficiencies would be higher for more volatile compounds, lower for less volatile compounds (e.g., 72% for naphthalene and 60% for BaP). The core flow is collected to obtain a direct measure of the gas phase, and the annular flow is separately collected to determine the combination of the aerosol phase and remaining gas phase. Mathematical corrections to the latter are necessary to estimate the concentrations of the particle-associated SVOCs. The authors used tandem PUF plugs, each 1.5-cm o.d. × 5.5-cm long, as trapping media and 60 mL of 5% methanol-modified supercritical CO_2 for extraction (60–70% efficiency). Extracts were concentrated to 1–2 mL for GC–MS analysis. The stated detection limits of 20–50 pg injected on column would translate to about 5–10 ng/m^3 for a 24-h air sample, not good enough to detect many SVOCs at ambient air concentrations.

Kaupp and Umlauf (*270*) used a five-stage, low-pressure impactor combined with a GFF to separate airborne particles from the gas phase and compared the results to those obtained by traditional filtration. The study was to test the theory that A_v would be less in the impactor. Both samplers were backed up with XAD-2 vapor traps. The impactor was operated at 153 L/min at a site collocated with the modified Hi-Vol operating at 300–350 L/min. Total and particulate concentrations of atmospheric chlorobenzenes (3 Cl–6 Cl), hexachlorocyclohexanes (HCHs), PCBs, chloroanisole, *p,p′*-DDT, *p,p′*-DDE [1,1-dichloro-2,2-bis(*p*-chlorophenyl)ethylene], and PAHs were measured for comparison. No chlorobenzenes (p_o = 10^{-2}–10^{-6} kPa), HCHs (10^{-6} kPa), or chloroanisole (ca. 10^{-2} kPa) were found in the particle phase by either sampler. BaP (7.3×10^{-10} kPa) and benzo[*g,h,i*]perylene (1.3×10^{-11} kPa) were found only on particles. The particle-associated fractions of PCBs (10^{-6}–10^{-8} kPa) ranged from <1% to 40% for the traditional sampler and <1% to 57% for the impactor. The hexachloro- and heptachlorobiphenyl congeners

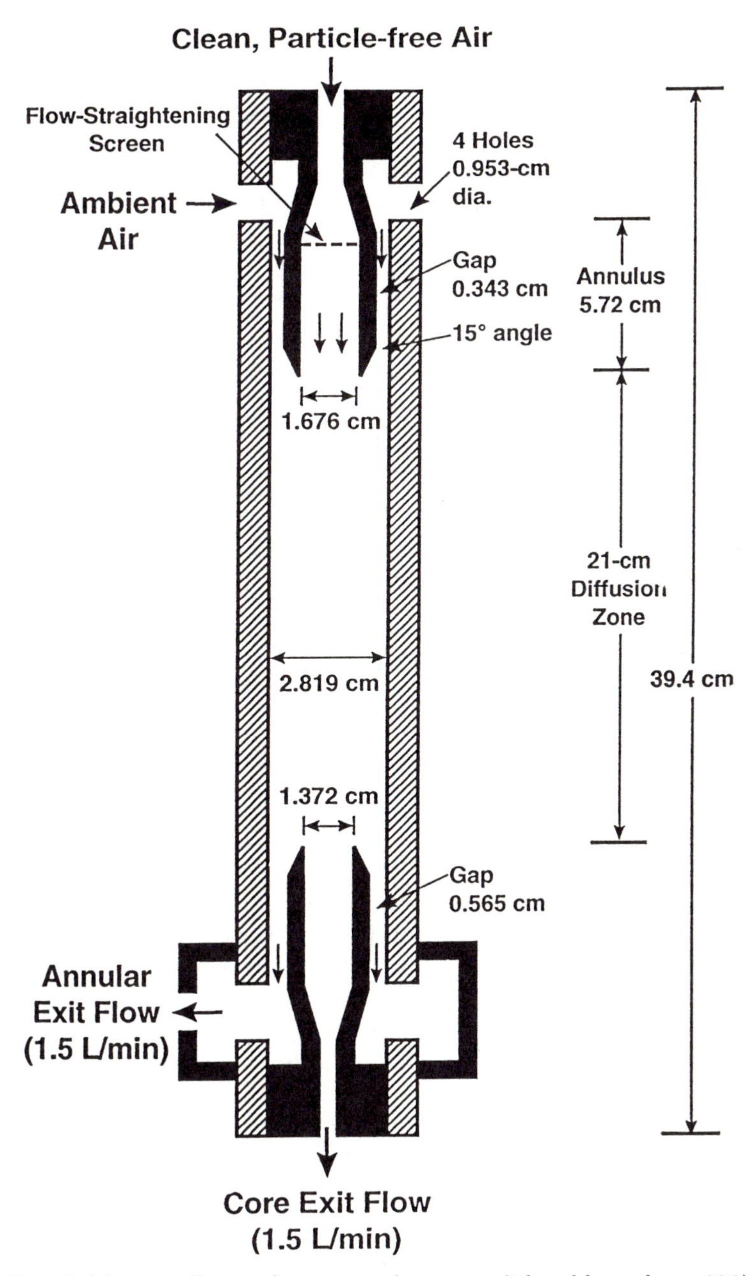

Figure 8. Schematic of laminar-flow gas–particle separator. (Adapted from reference 286.)

sought were found at an average 36% lower particle-phase concentration with the impactor. However, pyrene (3.1×10^{-6} kPa) and fluoranthene (6.5×10^{-7} kPa) showed a slightly lower degree of particle association with the impactor (21.2% vs. 25% and 13.4% vs. 15.6%, respectively). Mixed results were obtained for DDT (2.5×10^{-8} kPa) and DDE (ca. 10^{-7} kPa). Particulate DDT was lower for the impactor (19% vs. 32%), while DDE was higher (9.4% vs. 6.8%). From these data, it is difficult to conclude that volatilization artifacts are reduced when impactor sampling is used. Van Vaeck et al. (*287*) had previously reported that the high-volume cascade impactor (1,080 L/min) gave comparable results to high-volume filter sampling (200–400 L/min) for *n*-alkanes, carboxylic acids, and PAHs on collected airborne particulate matter.

The virtual impactor (*288*) is yet another approach recently taken by investigators interested in determining the phase distributions of organic chemicals. In a virtual impactor, air is drawn through an acceleration nozzle into a collection probe of slightly larger diameter. Particles larger than a certain cut point are collected in the probe, while gases and smaller particles are deflected away. The particles collected in the probe are then removed with a fraction of the total air flow passing through the sampler. Consequently, particles larger than the impactor's cut point can be effectively separated from the main air stream and separately trapped. A high-volume virtual impactor (HVVI) capable of separating gases and particles smaller than 0.1 μm from larger particles at 225 L/min has been developed by Sioutas et al. (*289, 290*). The fine particles (<0.1 μm) and about 80% of the gas phase are trapped in a combination particle filter–sorbent trap assembly, while the particles >0.1 μm and ca. 20% of the gas phase are similarly collected in another channel. The volatilization artifact may be reduced for the large particles because of the lower air flows (ca. 45 L/min) through this channel. While the HVVI effectively separates the gas phase from large particles, it does not separate gases from the fine particles, which by virtue of their larger surface area-to-mass ratios may contain the bulk of particulate organic compounds. The system is currently undergoing evaluation for SVOCs.

Despite recent progress in the design and development of gas–particle separators for airborne SVOCs, an efficient system has yet to be developed that will permit complete separation and recovery of phase-distributed organic compounds.

Modeling Phase Distribution

As previously discussed, Junge's equation may be used for general predictions of the approximate phase distributions of airborne organic chemicals. However, it does not take into account the effects of temperature and other atmospheric variables on phase distribution. It is based on the Langmuir isotherm:

$$\Theta_L = bP/(1 + bP) \tag{3}$$

where Θ_L can be taken as the fraction of airborne particle surface covered with sorbate, b is the ratio of the rate of adsorption to desorption, and P is the sorbate partial vapor pressure. The value of b is assumed to be independent of surface coverage and is temperature-dependent:

$$\ln b = a + E/RT \tag{4}$$

where a is a constant, E is the energy of sorption, R is the gas constant, and T is temperature in degrees Kelvin.

Coutant et al. (*252*) applied the more complex Dubinin–Radushkevich (D–R) isotherm (*291*) to estimate the atmospheric phase distributions of PAHs. The general form of the D–R isotherm has been previously given (*see* eq 1). Using this equation, w is taken as the mass fraction of the sorbate on the particles at partial pressure P, w_0 is the saturation capacity of the sorbate particles at equilibrium vapor pressure P_0, B is a constant that is characteristic of the particle surface, and β is the affinity coefficient for the sorbate, which can be defined in terms of its polarizability. The term $RT \ln (P/P_0)$ from eq 1 is the isosteric heat of sorption, which has units of energy per mole. The square of this term can be plotted against $\ln w$ to provide an isotherm for the concentration of an organic compound on airborne particulate matter at a given temperature. Coutant et al. (*292*) determined values of w_0 and B for a number of PAHs from plots of $\ln w$ vs. $[RT \ln (P/P_0)]^2$ using ambient air data collected over a two-year period in Columbus, OH. The data were collected using both open-tube and compound annular denuders (*252, 274*). Plots for individual PAHs yielded essentially constant values of B and values of w_0 that reflected the overall composition of the ambient PAHs. This research also showed that the volatilization artifact could be obtained roughly from the equation:

$$A_V(\%) = 76 + 7.6 \log p_o \pm 15 \tag{5}$$

where p_0 is the saturation vapor pressure of the PAH. Kelly et al. (*293*) compared predictions made with the D–R isotherm to ambient air measurements made in Boston, MA, and Houston, TX, with the high-volume CAD (*274*). Agreement between the model and field measurements was good for some PAHs and not for others. The model also performed better in Boston than in Houston. They concluded that the better predictions for Columbus were due to the more consistent airborne PAH composition there.

Yamasaki et al. (*166*) used an extension of the Langmuir equation to relate PAH phase distributions to total suspended particle (TSP) concentrations and temperature:

$$\log (C_g)(\text{TSP}) / C_p = -A / T + B \tag{6}$$

where A and B are constants for individual PAHs. Using PAH data collected over a 1-year period in Osaka, Japan, with a conventional Hi-Vol sampler fitted with PUF backup traps, they calculated values of A and B for several PAHs. However, they did not take into account volatilization artifacts, assuming that C_g corresponded to the amount found on the PUF and C_p to that on the GFF. Their data plots using eq 6 showed a direct relationship between C_g (or the fraction found in the vapor trap) and ambient temperature. Both the real atmospheric phase distributions and A_v will, of course, depend on temperature in the same way.

Bidleman et al. (*184, 294, 295*) has applied the Yamasaki equation to a number of SVOCs at locations all around the northern hemisphere and obtained results consistent with those reported by Yamasaki. They found, however, that ratios of the amounts (V) of airborne semivolatile organochlorine compounds collected on PUF to what was retained by the high-volume filters (F) were considerably larger than for PAHs with similar vapor pressures. This difference in behavior was ascribed to stronger sorption of PAHs to particulate matter, occlusion within the particles, or other possible factors (*295*). For organic compounds that are solids at ambient temperatures (as most SVOCs and NVOCs are), Bidleman (*295*) questions the use of the vapor pressure of the solid (p_S) as opposed to that of the subcooled liquid at 25 °C (p_L) for predicting airborne particle association and V/F ratios. The p_L values are much lower for compounds with high melting points and may be a better predictor of phase distributions (*207*). The two vapor pressures are related by the equation:

$$\ln (p_S / p_L) = -6.8(T_m - T) / T \quad (7)$$

where T and T_m are the ambient and melting point temperatures, respectively, in degrees Kelvin.

Rounds and Pankow (*296*) developed a rather complex "radial diffusion" model to describe gas–particle distribution kinetics both in the atmosphere and on filters. Their model considered airborne particles to be spheres with porous outer shells and nonporous cores. Thus they diverged from the traditional models that deal strictly with surface sorption. The model predicted rapid attainment of equilibrium for most compounds found in air. They cautioned that the model has not been validated for actual air sampling and should be used only to provide rough approximations.

Predicting the distributions of SVOCs between the gas and particulate phases by models is still an uncertain science. No model will be able to take into account nondesorbable compounds, such as those embedded within particles. Residence time of the particles in the atmosphere is another parameter that will be difficult to treat. Sampling artifacts are also very dependent on the amount of time particles remain in the air stream on filters after collection, as well as temporal changes in temperature and pollutant concentrations. Modeling will be most successful for steady-state atmospheres, remote from sources.

Pankow (*297*), Bidleman (*295*), and Gill et al. (*298*) have published comprehensive reviews on atmospheric phase distribution theory.

Real-Time Monitoring for SVOCs and NVOCs

Real-time monitoring is used less frequently for SVOCs and NVOCs than for VOCs because of their low volatilities and tendencies to be phase-distributed, which complicate sampling. Prager and Deblinger (*299*) developed a continuous detector for organophosphorus pesticides in 1967. Air was sampled at 300 L/min through a thermionic GC detector, which responded to concentrations of the target compounds down to 100 μg/m^3. Continuous measurements have also been achieved using a flame photometric detector, with only slightly better sensitivity (*300*). Neither of these systems was sensitive enough for ambient air monitoring. Real-time monitoring with an atmospheric pressure chemical ionization mass spectrometer (APCI–MS) has provided better detection limits for SVOCs, but is still inadequate for most ambient air monitoring applications. Lane et al. (*301*) determined the sensitivity of the Sciex TAGA APCI–MS for several PAHs in air. They observed detection limits of 50–250 ng/m^3 for gas-phase PAHs (2- to 4-ring) with real-time monitoring at 90 L/min. The detection limit achieved with the same instrument for the pesticide methomyl ($p_o = 4.5 \times 10^{-6}$ kPa) was 500 ng/m^3 (*302*).

Even less research has been directed at real-time measurement of organic aerosols. A newly developed monitor can, however, provide real-time detection and semiquantitative measurement of surface-associated PAHs. The Gossen 1000i PAH Ambient Analyzer (U.S. distributor: EcoChem Technologies, Inc., West Hills, CA) operates on the principle of photoelectric ionization (*303, 304*). It samples air at 4 L/min first through an electrostatic precipitator to remove charged particles, then through an ionization chamber, where the PAHs on carbonaceous particle surfaces are selectively ionized by 185-nm UV light. The positively charged particles are collected on a filter inside an aerosol electrometer for measurement. The electrometer output is proportional to the total PAH concentration. The monitor does not respond to gas-phase PAHs and has been shown to compare well with collocated indoor air sampling using a medium-volume CAD gas–particle separator (*305*). The detection limit was reported to be ca. 10 ng/m^3 total PAH (*305*).

Conclusions

The determination of organic chemicals in ambient, indoor, and personal respiratory air will remain a challenge to environmental and occupational monitoring communities for some time into the future. The current demands on environmental scientists are to provide demonstrated methods for the HAPs specified for control by the Clean Air Act Amendments of 1990. According to a recent EPA survey (*3, 4*), there are published methods that have been applied to ambient air measurement of 109, or 65%, of the 166 organic CAA HAPs. However, many of these methods have not been fully evaluated with respect to collection and recovery efficiencies, artifact problems, and field worthiness. Some do not have the required sensitivities to meet all the needs of ambient air monitoring. Others that may be adequate for ambient air monitoring are not amenable to indoor or nonoccupational personal exposure monitoring. Although no single method, or even several methods, will be sufficient to span the range of HAPs, there is a great need for more generic methods capable of covering multiple classes of chemicals.

EPA methods such as TO-13 and TO-14 have potential application for broad classes of SVOCs and VOCs, and together cover more of the CAA HAPs than any other two methods. Method TO-14 is limited to nonpolar VOCs, but may soon be replaced by a new method, TO-15, which will cover polar VOCs as well. Although TO-13 has the capability of collecting PAHs, pesticides, and many other SVOCs and NVOCs, it has only been validated for PAHs with three rings or more. It is also based on the use of a high-volume sampler without a size-selective inlet and cannot be used for indoor or personal exposure monitoring.

Currently, there are no commercially produced gas–particle separators capable of meeting the needs of ambient air monitoring. When using a traditional sampler for phase-distributed SVOCs, separate analysis of the particle filter and backup vapor trap will not provide direct information on the physical states of the compounds sought. Therefore, it is expedient to combine the two sampling media for extraction and/or analysis. Separate analyses will cost money, time, and method sensitivity, as well as open the door subsequently to erroneous interpretation of results.

Although it is good practice to use a second sorbent trap to backup the primary trap, the detection of a given analyte on that trap should only be taken as a qualitative indication that breakthrough has occurred during sampling. Adding the analytical results together can often lead to underestimation of air concentrations. Distributive volume sampling (use of two or more parallel sampling tubes to sample different air volumes) provides a better validation that breakthrough did not occur.

In general, VOCs and SVOCs cannot be readily determined in the same air sample. Even if the sampling system is capable of simultaneously collecting both, VOCs are usually present in ambient air at two to three orders of magnitude higher concentrations than SVOCs (and, of course, NVOCs). Therefore, if the sample volume is sufficient to detect the SVOCs, the capacity of the sorbent is likely to be exceeded for the VOCs. SVOCs usually require solvent desorption, which generally cannot be used for VOCs. If on-line supercritical fluid extraction is employed, the broad concentration ranges may exceed the linear range of the analytical detector.

Finally, no method should be assumed to perform adequately unless it has been validated (preferably by the user) under the conditions of its intended application. At a minimum, the sampling media should be spiked with the analytes of interest (or their isotopically labeled analogs) and subjected to the same or greater air flow rates and volumes that will be employed in the field. This dynamic retention test will usually provide a reasonable estimate of sampling efficiency. It is, of course, better to generate spiked atmospheres to be introduced into the sampler, but this procedure is difficult for less volatile compounds and with high-volume samplers. A simple low-volume vapor generator for SVOCs is described in ASTM Standard 4861 (95). With canisters and grab bags, breakthrough is not a problem, but wall losses may occur; therefore, storage stability and recovery tests are essential. Sorbents and filters also need to be evaluated for storage stability, as well as artifact formation during sampling. Sorbents and filters should be chilled or frozen immediately after sampling (including during transit to the laboratory) and be extracted as soon as possible after arrival at the laboratory. Extracts can usually be safely stored for extended periods (e.g., up to 1 year) at temperatures below −10 °C. Sampling media should not be stored for more than 30 days in a freezer unless kept under nitrogen. The use of isotopically labeled internal standards or other surrogates is very helpful in determining losses during storage, as well as during sampling and sample workup.

Quality assurance is always an essential element of any air monitoring program. At least 10% of all samples should be field blanks and controls. Air sampler flow rates should be calibrated and audited at the beginning and end of each sampling period. Accurate field and laboratory records should be kept and, if appropriate, chain-of-custody procedures strictly followed. Validation of analytical results by an external laboratory is always a good practice when feasible. Many important decisions have been made on the basis of air monitoring data. Unfortunately, not all of the data have been reliable.

Acknowledgments

The preparation of this document has been funded wholly or in part by the EPA. It has been subjected to the Agency's peer and administrative review, and it has been approved for publication as an EPA document. Mention of trade names or commercial products does not constitute endorsement or recommendation for use.

We thank John Clements (retired) and Lance Wallace of EPA for helpful advice and suggestions.

References

1. *Toxic Substances Control Act Chemical Substance Inventory: 1985 Edition;* U.S. Environmental Protection Agency: Washington, DC, 1985; Vols. I–V; EPA–560/7–85–002.
2. *The Clean Air Act* (as amended through December 31, 1990); 42 U.S. Code, 74.01–76.26; U.S. Government Printing Office: Washington, DC, 1991.
3. Kelly, T. J.; Mukund, R.; Gordon, S. M.; Hays, M. J. *Ambient Measurement Methods and Properties of the Clean Air Act Hazardous Air Pollutants;* U.S. Environmental Protection Agency: Research Triangle Park, NC, 1994; EPA–600/R–94–098.
4. Mukund, R.; Kelly, T. J.; Gordon, S. M.; Hays, M. J.; McClenny, W. A. *Environ. Sci. Technol.* **1995**, *29,* 183A–187A.
5. Lewis, R. G. *Proceedings of the 1986 EPA/APCA Symposium on Measurement of Toxic Air Pollutants;* Air & Waste Management Association: Pittsburgh, PA, 1986; Publication VIP–7, pp 134–145.
6. Clements, J. B.; Lewis, R. G. In *Principles of Environmental Sampling;* Keith, L. H., Ed.; ACS Professional Reference Book; American Chemical Society: Washington, DC, 1988; pp 287–296.
7. Junge, C. E. In *Fate of Pollutants in the Air and Water Environments;* Suffet, I. H., Ed.; Wiley–Interscience: New York, 1977; Advances in Environmental Science and Technology; Part 1, Vol. 8, pp 7–26.
8. Bloemen, H. J. Th.; Burn, J. In *Chemistry and Analysis of Volatile Organic Chemicals in the Environment;* Bloemen, H. J. Th.; Burn, J., Eds.; Blackie Academic & Professional: London, 1993; p xi.

9. Appel, B. R. In *Measurement Challenges in Atmospheric Chemistry;* Newman, L., Ed.; Advances in Chemistry 232; American Chemical Society: Washington, DC, 1993; pp 1–40.
10. Sheldon, L. S.; Sparacino, C. M.; Pellizzari, E. D. In *Indoor Air and Human Health;* Gammage, R. B.; Kaye, S. V.; Jacobs, V. A., Eds.; Lewis Publishers: Chelsea, MI, 1985; pp 335–349.
11. Winberry, W. T., Jr. *Am. Environ. Lab.* **1993,** *5,* 46–57.
12. Wallace, L. A. In *Handbook of Hazardous Materials;* Corn, M., Ed.; Academic: New York, 1993; pp 713–722.
13. Mackay, D.; Shiu, W. Y.; Ma, K. C. *Illustrated Handbook of Physical–Chemical Properties and Environmental Fate for Organic Chemicals;* Lewis Publishers: Chelsea, MI, 1992; Vols. I–III.
14. Howard, P. H. *Handbook of Environmental Fate and Exposure Data for Organic Chemicals;* Lewis Publishers: Chelsea, MI, 1990; Vol. I–III.
15. Verschueren, K. *Handbook of Environmental Data on Organic Chemicals;* Van Nostrand Reinhold: New York, 1977.
16. *The Pesticide Manual,* 9th ed.; Worthing, C. R.; Hance, R. J., Eds.; The British Crop Protection Council: Farnum, England, 1991.
17. Jones, D. L.; Bursey, J. *Simultaneous Control of PM–10 and Hazardous Air Pollutants II: Rationale for Selection of Hazardous Air Pollutants as Potential Particulate Matter;* U.S. Environmental Protection Agency: Research Triangle Park, NC, 1993; EPA/452/R–93/013.
18. Bidleman, T. F. *Anal. Chem.* **1984,** *56,* 2490–2496.
19. Otson, R.; Fellin, P. In *Gaseous Pollutants: Characterization and Cycling;* Nriagu, J. O., Ed.; Wiley & Sons: New York, 1992; pp 335–421.
20. Moschandreas, D. J.; Gordon, S. M. In *Organic Chemistry of the Atmosphere;* Hansen, L. D.; Eatough, D. J., Eds.; CRC Press: Boca Raton, FL, 1991; pp 121–153.
21. Lewis, R. G. *Chin. Chem. Soc. (Taiwan)* **1989,** *36,* 261–277.
22. De Bortoli, M.; Knoppel, H.; Mølhave, L.; Seifert, B.; Ullrich, D. *Interlaboratory Comparison of Passive Samplers for Organic Vapors with Respect to Their Applicability to Indoor Air Pollution Monitoring: A Pilot Study;* Office for Official Publications of the European Communities: Luxembourg City, Grand Duchy of Luxembourg, 1984; EUR–9450–EN.
23. Seifert, B.; Abraham, H. J. *Int. J. Environ. Anal. Chem.* **1983,** *13,* 237–254.
24. Weschler, C. J.; Shields, H. C. *Environ. Int.* **1989,** *15,* 593–604.
25. Brown, R. H.; Purnell, C. J. *J. Chromatogr. (Amsterdam)* **1979,** *178,* 79–90.
26. Krost, K. J.; Pellizzari, E. D.; Walburn, S. G.; Hubbard, S. A. *Anal. Chem.* **1982,** *54,* 810–817.
27. Wallace, L. A. *The Total Exposure Assessment Methodology (TEAM) Study: Summary and Analysis;* U.S. Environmental Protection Agency: Washington, DC, 1987; Vol. 1, EPA/600/6–87/002a.
28. Gordon, S. M. In *Advances in Air Sampling;* Lewis: Chelsea, MI, 1988; pp 133–142.
29. Rothweiler, H.; Wäger, P. A.; Schlatter, C. *Atmos. Environ.* **1991,** *25B,* 231–235.
30. Walling, J. F.; Berkley, R. E.; Swanson, D. H.; Toth, F. J. *Sampling Air for Gaseous Organic Compounds Using Solid Sorbents;* U.S. Environmental Protection Agency: Washington, DC, 1982; EPA/600/4–82–059.
31. Pellizzari, E. D.; Demian, B.; Krost, K. J. *Anal. Chem.* **1984,** *56,* 793–798.
32. Pellizzari, E. D.; Krost, K. J. *Anal. Chem.* **1984,** *56,* 1813–1819.
33. Walling, J. F.; Bumgarner, J. E.; Driscoll, D. J.; Morris, C. M.; Riley, A. E.; Wright, L. H. *Atmos. Environ.* **1986,** *20,* 51–57.
34. Hodgson, A. T.; Girman, J. R. In *Design and Protocol for Monitoring Indoor Air Quality;* Nagda, N. L.; Harper, J. P., Eds.; American Society for Testing Materials: Philadelphia, PA, 1987; pp 244–256.

35. Walling, J. F. *Atmos. Environ.* **1984,** *18,* 855–859.
36. McClenny, W. A.; Pleil, J. D.; Evans, G. F.; Oliver, K. D.; Holdren, M. W.; Winberry, W. T. *J. Air Waste Manage. Assoc.* **1991,** *41,* 1308–1318.
37. Sheldon, L.; Zelon, H.; Sickles, J.; Eaton, C.; Hartwell, T. *Indoor Air Quality in Public Buildings: Volume II;* U.S. Environmental Protection Agency: Research Triangle Park, NC, 1988; EPA/600/S6–88/009b.
38. Spicer, C. W.; Holdren, M. W.; Slivon, L. E.; Coutant, R. W.; Graves, M. E.; Shadwick, D. S.; McClenny, W. A.; Mulik, J. D.; Fitz-Simons, T. R. *Proceedings of the 1986 EPA/APCA Symposium on Measurement of Toxic Air Pollutants;* Air Pollution Control Association: Pittsburgh, PA, 1986; Publication VIP–7, pp 45–60.
39. De Bortoli, M.; Knöppel, H.; Pecchio, E.; Vissers, H. In *Indoor Air '87;* Seifert, B.; Esorn, H.; Fischer, M.; Rudën, H.; Wegner, J., Eds.; Institute for Water, Soil, and Air Hygiene: Berlin, Germany, 1987; Vol. 1, pp 139–143.
40. Cao, X.-L.; Hewitt, C. N. *Chemosphere* **1993,** *27,* 695–705.
41. McCaffrey, C. A.; MacLachlan, J.; Brookes, B. I. *Analyst (Cambridge, U.K.)* **1994,** *119,* 897–902.
42. Heavner, D. L.; Ogden, M. W.; Nelson, P. R. *Environ. Sci. Technol.* **1992,** *26,* 1737–1746.
43. Tsuchiya, Y. *Chemosphere* **1988,** *17,* 79–82.
44. Chan, C. C.; Vainer, L.; Martin, J. W.; Williams, D. T. *J. Air Waste Manage. Assoc.* **1990,** *40,* 62–67.
45. Sturges, W. T.; Elkins, J. W. *J. Chromatogr.* **1993,** *642,* 123–134.
46. Bishop, R. W.; Valis, R. J. *J. Chromatogr. Sci.* **1990,** *28,* 589–593.
47. Pollack, A. J.; Holdren, M. W.; McClenny, W. A. *J. Air Waste Manage. Assoc.* **1991,** *41,* 1213–1217.
48. Pollack, A. J.; Holdren, M. W. *Evaluation of an Automated Thermal Desorption System;* U.S. Environmental Protection Agency: Research Triangle Park, NC, 1991; final report by Battelle Columbus Operations under U.S. EPA Contract 68–D0–0007, Work Assignment 11.
49. Ciccioli, P.; Cecinato, A.; Brancaleoni, E.; Frattoni, M.; Liberti, A. *J. High Resolut. Chromatogr.* **1992,** *15,* 75–84.
50. O'Doherty, S. J.; Simmonds, P. G.; Nickless, G.; Betz, W. R. *J. Chromatogr.* **1993,** *630,* 265–274.
51. O'Doherty, S. J.; Simmonds, P. G.; Nickless, G. *J. Chromatogr.* **1993,** *657A,* 123–129.
52. Pollack, A. J.; Gordon, S. M.; Moschandreas, D. J. *Evaluation of Portable Multisorbent Air Samplers for Use with an Automated Multitube Analyzer;* U.S. Environmental Protection Agency: Research Triangle Park, NC, 1993; EPA/600/R–93/053.
53. Ciccioli, P.; Brancaleoni, E.; Cecinato, A.; Sparapani, R.; Frattoni, M. *J. Chromatogr.* **1993,** *643,* 55–69.
54. Kruschel, B. D.; Bell, R. W.; Chapman, R. E.; Spencer, M. J.; Smith, K. V. *J. High Resolut. Chromatogr.* **1994,** *17,* 187–190.
55. "Carbotrap—An Excellent Adsorbent for Sampling Many Airborne Contaminants"; *The Supelco Reporter* **1988,** *II*(3); Supelco: Bellefonte, PA.
56. Matisová, E.; Škrabáková, S. *J. Chromatogr.* **1995,** *707A,* 145–179.
57. Werner, M. D.; Winters, N. L. *Crit. Rev. Environ. Sci. Technol.* **1986,** *16,* 327–356.
58. Helmig, D.; Greenberg, J. P. *J. Chromatogr.* **1994,** *677A,* 123–132.
59. Winberry, W. T., Jr.; Murphy, N. T.; Riggin, R. M. *Methods for Determination of Toxic Organic Compounds in Air: EPA Methods;* Noyes Data Corporation: Park Ridge, NJ, 1990.
60. Pleil, J. D.; Oliver, K. D.; McClenny, W. A. *J. Air Pollut. Control Assoc.* **1987,** *37,* 244–248.
61. Janicki, W.; Wolska, L.; Wardencki, W.; Namieśnik, J. *J. Chromatogr.* **1993,** *654A,* 279–285.

62. McClenny, W. A.; Pleil, J. D.; Holdren, M. W.; Smith, R. N. *Anal. Chem.* **1984**, *56*, 2947–2951.
63. Kelly, T. J.; Callahan, P. J.; Pleil, J. D.; Evans, G. F. *Environ. Sci. Technol.* **1993**, *27*, 1146–1153.
64. Gordon, S. M.; Miller, M. *Analysis of Ambient Polar Volatile Organic Compounds Using Chemical Ionization–Ion Trap Detector;* U.S. Environmental Protection Agency: Research Triangle Park, NC, 1989; EPA/600/3–89/070.
65. Ogle, L. D.; Brymer, D. A.; Jones, C. J.; Nahas, P. A. *Proceedings of the 1992 EPA/A&WMA Symposium on Measurement of Toxic and Related Air Pollutants;* Air and Waste Management Association: Pittsburgh, PA, 1992; Publication VIP–25, pp 25–30.
66. Jesser, R.; Reiss, S. *Am. Environ. Lab.* **1995**, *7*(2), 16–17.
67. Reiss, S.; Jesser, R. In *Proceedings of the 1994 EPA/A&WMA International Symposium on Measurement of Toxic and Related Air Pollutants;* Air and Waste Management Association: Pittsburgh, PA, 1994; Publication VIP–39, p 911.
68. McClenny, W. A.; Gerald, N. O. *Am. Environ. Lab.* **1994**, *October/November,* 37–60.
69. McClenny, W. A. Presented at the 1994 EPA/A&WMA Symposium on Measurement of Toxic and Related Air Pollutants, Durham, NC, May 3–6, 1994.
70. Ryan, J. F.; Seeley, I.; Broadway, G. M.; McClenny, W. A. In *Proceedings of the 1993 EPA/A&WMA International Symposium on Field Screening Methods for Hazardous Wastes and Toxic Chemicals;* Air and Waste Management Association: Pittsburgh, PA, 1993; Publication VIP–33, pp 1049–1053.
71. Woolfenden, E. A.; Broadway, G. M.; Higham, P.; Seeley, I. *Am. Environ. Lab.* **1993**, *5*(6), 33–36.
72. Seeley, I. R.; Broadway, G. M.; Tipler, A. In *Proceedings of the 1993 EPA/A&WMA International Symposium on Field Screening Methods for Hazardous Wastes and Toxic Chemicals;* Air and Waste Management Association: Pittsburgh, PA, 1993; Publication VIP–33, pp 628–633.
73. Brixen, T. V.; Stewart, J. K. In *Proceedings of the 1993 EPA/A&WMA International Symposium on Field Screening Methods for Hazardous Wastes and Toxic Chemicals;* Air and Waste Management Association: Pittsburgh, PA, 1993; Publication VIP–33, pp 634–639.
74. Greenberg, J. P.; Lee, B.; Helmig, D.; Zimmerman, P. R. *J. Chromatogr.* **1994**, *676A*, 389–398.
75. Farmer, C. T.; Milne, P. J.; Riemer, D. D.; Zika, R. G. *Environ. Sci. Technol.* **1994**, *28*, 238–245.
76. Broadway, G. M.; Trewern, T. In *Proceedings of the 13th International Symposium on Capillary Chromatography;* Hüthig: Heidelberg, Germany, 1991; pp 310–320.
77. Holdren, M. W.; Smith, D. L.; Pollack, A. J.; Pate, A. D. *Analysis of Connecticut Database;* U.S. Environmental Protection Agency: Research Triangle Park, NC, 1993; final report by Battelle Columbus Operations under U.S. EPA Contract 68–D0–0007, Work Assignment 36.
78. Pollack, A. J.; Gordon, S. M. In *Proceedings of the 1993 EPA/A&WMA International Symposium on Field Screening Methods for Hazardous Wastes and Toxic Chemicals;* Air and Waste Management Association: Pittsburgh, PA, 1993; Publication VIP–33, pp 761–766.
79. Moschandreas, D. J.; Relwani, S. M. *J. Exposure Anal. Environ. Epidemiol.* **1991**, *1*, 357–367.
80. Moschandreas, D. J.; Akland, G. G.; Gordon, S. M. *J. Exposure Anal. Environ. Epidemiol.* **1994**, *4*, 395–407.
81. Shields, H. C.; Weschler, C. J. *J. Air Pollut. Control Assoc.* **1987**, *37*, 1039–1045.
82. Otson, R.; Fellin, P.; Barnett, S. E. *Proceedings of the 85th A&WMA Annual Meeting;* Air & Waste Management Association: Pittsburgh, PA, 1992; Paper 92–80.07.
83. Cao, X.-L.; Hewitt, C. N. *J. Chromatogr.* **1994**, *688A*, 368–374.

84. Lewis, R. G.; Mulik, J. D.; Coutant, R. W.; Wooten, G. W.; McMillin, C. R. *Anal. Chem.* **1985,** *57,* 214–219.
85. Coutant, R. W.; Lewis, R. G.; Mulik, J. D. *Anal. Chem.* **1986,** *58,* 445–448.
86. Chai, M.; Pawliszyn, J. *Environ. Sci. Technol.* **1995,** *29,* 693–701.
87. Tejada, S. B. *Int. J. Environ. Anal. Chem.* **1986,** *26,* 167–185.
88. *NIOSH Manual of Analytical Methods,* 2nd ed.; Department of Health and Human Services, National Institute for Occupational Safety and Health. U.S. Government Printing Office: Washington, DC, 1977–1982; Vol. 1–7.
89. Levin, J.-O.; Lindahl, R. *Analyst (Cambridge, U.K.)* **1994,** *119,* 79–83.
90. Otson, R.; Fellin, P.; Tran, Q.; Stoyanoff, R. *Analyst (Cambridge, U.K.)* **1993,** *118,* 1253–1259.
91. Mulik, J. D.; Lewis, R. G.; McClenny, W. A.; Williams, D. D. *Anal. Chem.* **1989,** *61,* 187–189.
92. Gersling, K. L.; Rappaport, S. M. *Environ. Int.* **1982,** *8,* 153–158.
93. Leaderer, B. P.; Zagraniski, R. T.; Berwick, M.; Stolwijk, J. A. J. *Am. J. Epidemiol.* **1986,** *124,* 275–289.
94. Daniel, M.; Sullivan, R.; Poslusny, M.; Krein, C. *Am. Environ. Lab.* **1993,** *5*(8), 24–25.
95. *Annual Book of ASTM Standards;* American Society for Testing and Materials: Philadelphia, PA, 1995; Vol. 11.03.
96. Arnts, R. R.; Tejada, S. B. *Environ. Sci. Technol.* **1989,** *23,* 1428–1430.
97. Grosjean, D. *Environ. Sci. Technol.* **1991,** *25,* 710–715.
98. Miksch, R. R.; Anthon, D. W.; Fanning, L. Z.; Hollowell, C. D.; Revzan, K.; Glanville, J. *Anal. Chem.* **1981,** *53,* 2118–2123.
99. Kelly, T. J.; Fortune, C. R. *Int. J. Environ. Anal. Chem.* **1994,** *54,* 249–263.
100. Kuwata, K.; Akiyama, Y.; Yamasaki, H.; Kuge, Y.; Kiso, Y. *Anal. Chem.* **1983,** *55,* 2199–2201.
101. Nishioka, M. G.; Burkholder, H. M. *Evaluation of an Anion Exchange Resin for Sampling Ambient Level Phenolic Compounds;* U.S. Environmental Protection Agency: Research Triangle Park, NC, 1990; final report by Battelle Columbus Operations under U.S. EPA Contract 68–02–4127, Work Assignment Nos. 69 and 80.
102. Rasmussen, R. A.; Khalil, M. A. K. In *Proceedings of the NATO Advanced Study Institute on Atmospheric Ozone: Its Variation and Human Influences;* Aikin, A. C., Ed.; U.S. Department of Transportation: Washington, DC, 1980; pp 209–231.
103. Arnst, R.; Meeks, S. *Atmos. Environ.* **1981,** *15,* 1643.
104. Evans, G. F.; Lumpkin, T. A.; Smith, D. L.; Somerville, M. C. *J. Air Waste Manage. Assoc.* **1992,** *42,* 1319–1323.
105. McClenny, W. A.; Lumpkin, T. A.; Pleil, J. D.; Oliver, K. D.; Bubacz, D. K.; Faircloth, J. W.; Daniels, W. H. *Proceedings of the 1986 EPA/APCA Symposium on Measurement of Toxic Air Pollutants;* Air Pollution Control Association: Pittsburgh, PA, 1986; Publication VIP–7, pp 402–418.
106. Pate, B.; Jayanty, R. K. M.; Peterson, M. R.; Evans, G. F. *J. Air Waste Manage. Assoc.* **1992,** *42,* 460–462.
107. Pleil, J. D.; Oliver, K. D.; McClenny, W. A. In *Indoor Air '87;* Seifert, B.; Esorn, H.; Fischer, M.; Rudën, H.; Wegner, J., Eds.; Institute for Water, Soil, and Air Hygiene: Berlin, Germany, 1987; Vol. 1, pp 164–168.
108. Pleil, J. D.; McClenny, W. A.; Holdren, M. W.; Pollack, A. J.; Oliver, K. D. *Atmos. Environ.* **1993,** *27A,* 739–747.
109. Raymer, J. H.; Thomas, K. W.; Cooper, S. D.; Whitaker, D. A.; Pellizzari, E. D. *J. Anal. Toxicol.* **1990,** *14,* 337–344.
110. Whitaker, D. A.; Fortmann, R. C.; Lindstrom, A. B. *J. Exposure Anal. Environ. Epidemiol.* **1995,** *5,* 89–100.
111. Oliver, K. D.; Pleil, J. D.; McClenny, W. A. *Atmos. Environ.* **1986,** *20,* 1403–1411.

112. Jayanty, R. K. M. *Atmos. Environ.* **1989,** *23,* 777–782.
113. Coutant, R. W.; McClenny, W. A. *Proceedings of the 1991 EPA/A&WMA Symposium on Measurement of Toxic and Related Air Pollutants;* Air and Waste Management Association: Pittsburgh, PA, 1991; Publication VIP–21, pp 382–388.
114. Coutant, R. W. *Theoretical Evaluation of Stability of Volatile Organic Chemicals and Polar Volatile Organic Chemicals in Canisters;* U.S. Environmental Protection Agency: Research Triangle Park, NC, 1992; EPA/600/R–92/055.
115. Schuyler, A.; Stauffer, J. W.; Loope, C. E.; Vargo, C. R. *Process Control Qual.* **1992,** *3,* 167–171.
116. Holdren, M. W.; Skarpness, B. O.; Strauss, W. J.; Keigley, G. W. *Effectiveness of Silcosteel Treatment of Canisters;* U.S. Environmental Protection Agency: Research Triangle Park, NC, 1994; final report by Battelle Columbus Operations under U.S. EPA Contract 68–D0–0007, Work Assignment 45, Task 3.
117. Shelow, D.; Silvis, P.; Schuyler, A.; Stauffer, J.; Pleil, J. D.; Holdren, M. W. *Proceedings of the 1994 EPA/A&WMA Symposium on Measurement of Toxic and Related Air Pollutants;* Air & Waste Management Association: Pittsburgh, PA, 1994; Publication VIP–39; p 904.
118. Shelow, D.; Schuyler, A. Presented at the 1995 EPA/A&WMA Symposium on Measurement of Toxic and Related Air Pollutants, Research Triangle Park, NC, May 16–18, 1995.
119. Skelding, T. *Survey of Portable Analyzers for the Measurement of Gaseous Fugitive Emissions;* U.S. Environmental Protection Agency: Research Triangle Park, NC, 1992; final report by Radian Corporation under U.S. EPA Contract 68–D1–0010.
120. Berkley, R. E.; Varns, J. L.; McClenny, W. A.; Fulcher, J. *Proceedings of the 1989 EPA/APCA Symposium on Measurement of Toxic and Related Air Pollutants;* Air Pollution Control Association: Pittsburgh, PA, 1989; Publication VIP–13, pp 13–18.
121. Berkley, R. E.; Varns, J. L.; Pleil, J. D. *Environ. Sci. Technol.* **1991,** *25,* 1439–1444.
122. Berkley, R. E.; Miller, M.; Chang, J. C.; Oliver, K. D.; Fortune, C.; Adams, J. *Field Screening Methods for Hazardous Wastes and Toxic Chemicals: Proceedings of the U.S. EPA/A&WMA International Symposium;* Air & Waste Management Association: Pittsburgh, PA, 1993; Publication VIP–33, pp 682–687.
123. Overton, E. B.; Grande, L. H.; Sherman, R. W.; Collard, E. S.; Steele, C. F. *Measurement of Toxic and Related Air Pollutants: Proceedings of the U.S. EPA/A&WMA International Symposium;* Air Pollution Control Association, Pittsburgh, PA, 1989; Publication VIP–13, pp 13–18.
124. Chang, J. S. C.; Gordon, S. M.; Berkley, R. E. *Measurement of Toxic and Related Air Pollutants: Proceedings of the U.S. EPA/A&WMA International Symposium;* Air & Waste Management Association: Pittsburgh, PA, 1990; Publication VIP–17, pp 830–835.
125. Carney, K. R.; Overton, E. B.; Mainga, A. M.; Steele, C. F. *Field Screening Methods for Hazardous Wastes and Toxic Chemicals: Proceedings of the U.S. EPA/A&WMA International Symposium;* Air & Waste Management Association: Pittsburgh, PA, 1993; Publication VIP–33, p 149.
126. Bruns, M. W.; Hammarstrand, K. G. *Proceedings of the 1994 EPA/A&WMA Symposium on Measurement of Toxic and Related Air Pollutants;* Air & Waste Management Association: Pittsburgh, PA, 1994; Publication VIP–39; pp 55–62.
127. McClenny, W. A.; Oliver, K. D.; Pleil, J. D. *Environ. Sci. Technol.* **1989,** *23,* 1373–1379.
128. Hill, H. H., Jr.; Siems, W. F.; St. Louis, R. H.; McMinn, D. G. *Anal. Chem.* **1990,** *62,* 1201A–1209A.
129. Karpas, Z. *Anal. Chem.* **1989,** *61,* 684–689.
130. Fetterolf, D. D.; Clark, T. D. *J. Forensic Sci.* **1993,** *38,* 28–39.
131. Fytche, L. M.; Hupé, M.; Kovar, J. B.; Pilon, P. *J. Forensic Sci.* **1992,** *37,* 1550–1566.
132. Limero, T. F.; Cross, J.; Brokenshire, J.; Hinton, M.; James, J. T. *Field Screening Methods for Hazardous Wastes and Toxic Chemicals: Proceedings of the U.S. EPA/A&WMA Interna-*

tional Symposium; Air & Waste Management Association: Pittsburgh, PA, 1993; Publication VIP–33, pp 809–822.

133. Edgerton, S. A.; Kenny, D. V.; Joseph, D. W. *Environ. Sci. Technol.* **1989,** *23,* 484–488.
134. Kelly, T. J.; Kenny, D. V. *Atmos. Environ.* **1991,** *25A,* 2155–2160.
135. Gordon, S. M.; Kenny, D. V.; Kelly, T. J. *J. Exposure Anal. Environ. Epidemiol.* **1992,** *Suppl. 1,* 41–54.
136. Cooks, R. G.; Glish, G. L.; McLuckey, S. A.; Kaiser, R. E. *Chem. Eng. News* **1991,** *69*(12), 26–41.
137. Kelley, P. E.; Hoekman, D.; Bradshaw, S. In *Proceedings of the 41st ASMS Conference on Mass Spectrometry and Allied Topics;* American Society for Mass Spectrometry: Santa Fe, NM, 1993; p 453a.
138. Kenny, D. V.; Callahan, P. J.; Gordon, S. M.; Stiller, S. W. *Rapid Commun. Mass Spectrom.* **1993,** *7,* 1086–1089.
139. Goeringer, D. E.; Asano, K. G.; McLuckey, S. A.; Hoekman, D.; Stiller, S. W. *Anal. Chem.* **1994,** *66,* 313–318.
140. Kelley, P. E. U.S. Patent 5 134 286, 1992.
141. Tabone, V.; Yelton, R.; Kelley, P. E.; Stiller, S. W. In *Proceedings of the 42nd ASMS Conference on Mass Spectrometry and Allied Topics;* American Society for Mass Spectrometry: Santa Fe, NM, 1994; p 181.
142. McLuckey, S. A.; Glish, G. L.; Asano, K. G.; Grant, B. C. *Anal. Chem.* **1988,** *60,* 2220–2227.
143. McLuckey, S. A.; Glish, G. L.; Asano, K. G. *Anal. Chim. Acta* **1989,** *225,* 25–35.
144. Gordon, S. M.; Chang, J. S. C.; Miller, M.; Pleil, J. D.; McClenny, W. A. *Measurement of Toxic and Related Air Pollutants: Proceedings of the U.S. EPA/A&WMA International Symposium;* Air and Waste Management Association: Pittsburgh, PA, 1991; Publication VIP–21, pp 596–601.
145. LaPack, M. A.; Tou, J. C.; Enke, C. G. *Anal. Chem.* **1990,** *62,* 1265–1271.
146. Wise, M. B.; Thompson, C. V.; Buchanan, M. V.; Merriweather, R.; Guerin, M. R. *Adv. Star Commun.* **1993,** *8,* 14–22.
147. Gordon, S. M.; Callahan, P. J.; Kenny, D. V. In *Proceedings of the 1994 EPA/A&WMA Symposium on Measurement of Toxic and Related Air Pollutants;* Air and Waste Management Association: Pittsburgh, PA, 1994; VIP–39; p 614.
148. Franzen, J.; Gabling, R.–H.; Preidel, M.; Schubert, M. *Field Screening Methods for Hazardous Wastes and Toxic Chemicals: Proceedings of the U.S. EPA/A&WMA International Symposium;* Air & Waste Management Association: Pittsburgh, PA, 1993; Publication VIP–33, pp 638–642.
149. Gordon, S. M.; Callahan, P. J.; Kenny, D. V.; Pleil, J. D. In *Field Screening Methods for Hazardous Wastes and Toxic Chemicals: Proceedings of an International Symposium;* Air & Waste Management Association: Pittsburgh, PA, 1995; Publication VIP–47; Vol. 1, pp 670–679.
150. Wise, M. B.; Thompson, C. V.; Guerin, M. R. In *Proceedings of the 42nd ASMS Conference on Mass Spectrometry and Allied Topics;* American Society for Mass Spectrometry: Santa Fe, NM, 1994; p 285.
151. Fortune, C. R. *Proceedings of the 84th A&WMA Annual Meeting;* Air & Waste Management Association: Pittsburgh, PA, 1991; Paper 91–68.13.
152. Penrose, W. R.; Buttner, W. J.; Stetter, J. R.; Findlay, M. W.; Tome, M. A.; Yue, C.; Stetter, T. *Field Screening Methods for Hazardous Wastes and Toxic Chemicals: Proceedings of the U.S. EPA/A&WMA International Symposium;* Air & Waste Management Association: Pittsburgh, PA, 1993; Publication VIP–33, pp 89–93.
153. Beyermann, K.; Eckrich, W. Z. *Anal. Chem.* **1974,** *269,* 279–284.
154. Lewis, R. G. In *Air Pollution from Pesticides and Agricultural Processes;* Lee, R. E., Jr., Ed.; CRC Press: Boca Raton, FL, 1976; pp 51–94.

155. DeRoos, F. L.; Tabor, J. E.; Miller, S. E.; Lewis, R. G.; Wilson, N. K. In *Proceedings 1986 APCA–EPA Symposium on Measurement of Toxic Air Pollutants;* Air & Waste Management Association: Pittsburgh, PA, 1986; APCA Publication VIP–7, pp 217–229.
156. Van Vaeck, L.; Van Cauwenberghe, K.; Janssens, J. *Atmos. Environ.* **1984,** *18,* 417–430.
157. Storey, J. M. E.; Luo, W.; Isabelle, L. M.; Pankow, J. F. *Environ. Sci. Technol.* **1995,** *29,* 2420–2428.
158. Appel, B. R.; Tokiwa, Y.; Kothny, E. L. *Atmos. Environ.* **1983,** *17,* 1787–1796.
159. Cotham, W. E.; Bidleman, T. F. *Environ. Sci. Technol.* **1992,** *26,* 469–478.
160. Davis, C. S.; Caton, R. B.; Guerin, W. C.; Tam, W. C. In *Polynuclear Aromatic Hydrocarbons: Mechanisms, Methods and Metabolism;* Cooke, M.; Dennis, A. J., Eds.; Battelle: Columbus, OH, 1984; pp 337–349.
161. Guerin, S. G.; Davis, C. S.; Caton, R. B. *Evaluation of Alternatives to Hi-Vol Sampling for Polynuclear Aromatic Hydrocarbons;* Concord Scientific Corporation: Downsville, Ont., Canada, 1984; Report CSC.J212.02 to the Ontario Ministry of the Environment.
162. Riggin, R. M.; Coutant, R. W. "Phase Distribution and Sample Integrity for PAHs in Ambient Air"; Presented at the Fourth Annual National Symposium on Recent Advances in the Measurement of Pollutants from Ambient Air and Stationary Sources, Raleigh, NC, May 8–10, 1984.
163. Rounds, S. A.; Tiffany, B. A.; Pankow, J. F. *Environ. Sci. Technol.* **1993,** *27,* 366–377.
164. Zhang, X.; McMurry, P. H. *Environ. Sci. Technol.* **1991,** *25,* 456–459.
165. Pupp, C.; Lao, R. C.; Murray, J. J.; Pottie, R. F. *Atmos. Environ.* **1974,** *8,* 915–925.
166. Yamasaki, H.; Kuwata, K.; Miyamoto, H. *Environ. Sci. Technol.* **1982,** *16,* 189–194.
167. Watts, R. R.; Hoffman, A. J.; Wilkins, M. C.; House, D. E.; Burton, R. M.; Brooks, L. R.; Warren, S. H. *J. Air Waste Manage. Assoc.* **1992,** *42,* 49–55.
168. de Raat, W. K.; Bakker, G. L.; de Meijere, F. A. *Atmos. Environ.* **1990,** *24A,* 2875–2887.
169. Gray, H. A.; Cass, G. R.; Huntzicker, J. J.; Heyerdahl, E. K.; Rau, J. A. *Sci. Total Environ.* **1984,** *36,* 17–25.
170. Otson, R.; Leach, J. M.; Chung, L. T. K. *Anal. Chem.* **1987,** *59,* 1701–1705.
171. Greaves, R. C.; Barkley, R. M.; Sievers, R. E. *Anal. Chem.* **1985,** *57,* 2807–2815.
172. Fitz, D. R. *Aerosol Sci. Technol.* **1990,** *12,* 142–148.
173. McDow, S. R.; Huntzicker, J. J. In *Sampling and Analysis of Airborne Pollutants;* Winegar, E. D.; Keith, L. H., Eds.; Lewis Publishers: Boca Raton, FL, 1993; pp 191–208.
174. Turpin, B. J.; Huntzicker, J. J.; Hering, S. V. *Atmos. Environ.* **1994,** *28,* 3061–3071.
175. Hart, K. M.; Pankow, J. F. *Environ. Sci. Technol.* **1994,** *28,* 655–661.
176. Appel, B. R.; Hoffer, E. M.; Kothny, E. L.; Wall, S. M.; Haik, M. *Environ. Sci. Technol.* **1979,** *13,* 98–104.
177. Lewis, R. G.; Brown, A. R.; Jackson, M. D. *Anal. Chem.* **1977,** *49,* 1668–1672.
178. Lewis, R. G.; Jackson, M. D. *Anal. Chem.* **1982,** *54,* 592–594.
179. Lewis, R. G.; MacLeod, K. E. *Anal. Chem.* **1982,** *54,* 310–315.
180. Bidleman, T. F.; Olney, C. E. *Bull. Environ. Contam. Toxicol.* **1974,** *11,* 442–450.
181. Lewis, R. G.; Martin, B. E.; Sgontz, D. L.; Howes, J. E., Jr. *Environ. Sci. Technol.* **1985,** *19,* 986–991.
182. MacLeod, K. E. *Environ. Sci. Technol.* **1981,** *15,* 926–928.
183. Trane, K. E.; Mikalsen, A. *Atmos. Environ.* **1981,** *15,* 909–918.
184. Keller, C. D.; Bidleman, T. F. *Atmos. Environ.* **1984,** *18,* 837–845.
185. Hunt, G. T.; Pangaro, N. In *Polynuclear Aromatic Hydrocarbons: Mechanisms, Methods and Metabolism;* Cooke, M.; Dennis, A. J., Eds.; Battelle: Columbus, OH, 1984; pp 583–608.
186. Brenner, K. S.; Mäder, H.; Steverle, H.; Heinrich, G.; Womann, H. *Bull. Environ. Contam. Toxicol.* **1984,** *33,* 153–162.
187. DeRoos, F. L.; Tabor, J. E.; Miller, S. E.; Watson, S. C.; Hatchel, J. A. *Evaluation of an EPA High-Volume Air Sampler for Polychlorinated Dibenzo-*p*-Dioxins and Polychlorinated Diben-*

zofurans; U.S. Environmental Protection Agency: Research Triangle Park, NC, 1987; EPA/600/4–86/037.
188. Eitzer, B. D.; Hites, R. A. *Environ. Sci. Technol.* **1989,** *23,* 1389–1395.
189. Harless, R. L.; Lewis, R. G.; McDaniel, D. D.; Gibson, J. F.; Dupuy, A. E. *Chemosphere* **1992,** *25,* 1317–1322.
190. Stratton, C. L.; Whitlock, S. A.; Allan, J. M. *A Method for the Sampling and Analysis of Polychlorinated Biphenyls (PCBs) in Ambient Air;* U.S. Environmental Protection Agency: Research Triangle Park, NC, 1978; EPA–600/4–78–048.
191. Hart, K. M.; Isabelle, L. M.; Pankow, J. F. *Environ. Sci. Technol.* **1992,** *26,* 1048–1052.
192. Hunt, G. T.; Maisel, B. E. *J. Air Waste Manage. Assoc.* **1991,** *42,* 672–680.
193. Marty, J. C.; Tissier, M. J.; Saliot, A. *Atmos. Environ.* **1984,** *18,* 2183–2190.
194. Hawthorne, S. B.; Miller, D. J.; Langenfeld, J. J.; Kreiger, M. S. *Environ. Sci. Technol.* **1992,** *26,* 2251–2262.
195. Erickson, M. D.; Michael, L. C.; Zweidinger, R. A.; Pellizzari, E. D. *Environ. Sci. Technol.* **1978,** *12,* 927–931.
196. Giam, C. S.; Atlas, E.; Chan, H. S.; Neff, G. S. *Atmos. Environ.* **1980,** *14,* 65–69.
197. Winberry, W. T., Jr.; Forchard, L.; Murphy, N. T.; Ceroli, A.; Phinnay, B.; Evans, A. *Methods for Determination of Indoor Air Pollutants: EPA Methods;* Noyes Data Corporation: Park Ridge, NJ, 1993.
198. Lewis, R. G. In *Environmental Carcinogens; Methods of Analysis and Exposure Measurement;* Seifert, B.; Von de Wiel, H. J.; Dodet, B.; O'Neill, I. K., Eds.; International Agency of Research on Cancer, World Health Organization: Lyon, France, 1993; Vol. 12 (Indoor Air), pp 353–376; IRAC Scientific Publication 109.
199. Hsu, J. P.; Wheeler, H. G.; Shattenberg, H. J., III; Camann, D. E.; Lewis, R. G.; Bond, A. E. *J. Chromatogr. Sci.* **1988,** *26,* 181–189.
200. Whitmore, R. W.; Immerman, F. W.; Camann, D. E.; Bond, A. E.; Lewis, R. G.; Schaum, J. L. *Arch. Environ. Contam. Toxicol.* **1994,** *26,* 47–59.
201. Geno, P. W.; Camann, D. E.; Villalobos, K.; Lewis, R. G. In *Proceedings of the 1993 U.S. EPA/A&WMA International Symposium;* Air & Waste Management Association: Pittsburgh, PA, 1993; Publication VIP–34, pp 698–705.
202. Lewis, R. G.; Fortmann; R. C.; Camann; D. E. *Arch. Environ. Contam. Toxicol.* **1994,** *26,* 37–46.
203. Marple, W. A.; Rubow, K. L.; Turner, W.; Spengler, J. D. *J. Air Pollut. Control Assoc.* **1987,** *37,* 1303–1307.
204. Lewis, R. G.; Camann, D. E.; Harding, H. J.; Stone, C. L. Presented at the Fourth International Aerosol Conference, American Association of Aerosol Research, Los Angeles, CA, August 29–September 2, 1994; paper 9C3.
205. Chuang, J. C.; Hannan, S. W.; Wilson, N. K. *Environ. Sci. Technol.* **1987,** *21,* 798–804.
206. Ligocki, M. P.; Pankow, J. F. *Anal. Chem.* **1985,** *57,* 1138–1144.
207. You, F.; Bidleman, T. F. *Environ. Sci. Technol.* **1984,** *18,* 330–333.
208. Zaranski, M. T.; Bidleman, T. F. *J. Chromatogr.* **1987,** *409,* 235–242.
209. Zaranski, M. T.; Patton, G. W.; McConnell, L. L.; Bidleman, T. F.; Mulik, J. D. *Anal. Chem.* **1991,** *63,* 1228–1232.
210. Kogan, V.; Kuhlman, M. R.; Coutant, R. W.; Lewis, R. G. *J. Air Waste Manage. Assoc.* **1993,** *43,* 1367–1373.
211. Camann, D. E.; Harding, H. J.; Lewis, R. G. In *Indoor Air;* Walkinshaw, D. S., Ed.; Canada Mortgage and Housing Corporation: Ottawa, Ont., Canada, 1990; Vol. 2, pp 621–626.
212. Billings, W. N.; Bidleman, T. F. *Atmos. Environ.* **1983,** *17,* 383–393.
213. Nishioka, M. G.; Lewtas, J. *Atmos. Environ.* **1992,** *26A,* 2077–2087.
214. Kaupp, H.; Umlauf, G. *Atmos. Environ.* **1992,** *13,* 2259–2267.
215. Hippelein, M.; Kaupp, H.; Dörr, G.; McLachlan, M. S. *Chemosphere* **1993,** *26,* 2255–2263.

216. Alfeim, I.; Lindskog, A. *Sci. Total Environ.* **1984**, *34,* 203–222.
217. Baker, J. E.; Eisenreich, S. J. *Environ. Sci. Technol.* **1990**, *24,* 342–352.
218. Chuang, J. C.; Mack, G. A.; Kuhlman, M. R.; Wilson, N. K. *Atmos. Environ.* **1991**, *25,* 369–380.
219. Doskey, P. V.; Andren, A. W. *Anal. Chim. Acta* **1979**, *110,* 129–137.
220. Andersson, K.; Levin, J.-O.; Nilsson, C.-A. *Chemosphere* **1981**, *10,* 137–142.
221. Manchester-Neesvig, J. B.; Andren, A. W. *Environ. Sci. Technol.* **1989**, *23,* 1138–1148.
222. McLachlan, M. S.; Hutzinger, O. *Organochlorine Compounds: Dioxin '90;* Hutzinger, O.; Fiedler, H., Eds.; Ecoinforma Press: Bayreuth, Germany, 1990; Vol. 1, pp 441–444.
223. Henriks-Eckerman, M.-J. *Chemosphere* **1990**, *21,* 889–904.
224. Turén, A.; Larsson, P. *Environ. Sci. Technol.* **1990**, *24,* 554–559.
225. Levin, J.-O.; Andersson, K.; Karlsson, R.-M. *J. Chromatogr.* **1988**, *454,* 121–128.
226. *OSHA Analytical Manual,* 2nd ed.; U.S. Department of Labor, Occupational Safety and Health Administration: Salt Lake City, UT, 1990; Parts I and II.
227. Lee, M. L.; Novotny, M. V.; Bartle, K. D. *Analytical Chemistry of Polycyclic Aromatic Compounds;* Academic: New York, 1981; Chapter 4.
228. Sullivan, R. J.; Zimmerman, M.; Pankas, S. M.; Gebhart, J. E. *Measurement of Toxic and Related Air Pollutants: Proceedings of the U.S. EPA/A&WMA International Symposium;* Air & Waste Management Association: Pittsburgh, PA, 1992; Publication VIP–25, pp 506–515.
229. *Statement of Work for Analysis of Ambient Air;* Contract Laboratory Program, U.S. Environmental Protection Agency: Research Triangle Park, NC, 1993; Rev. IAIR01.2.
230. Chuang, J. C.; Kuhlman, M. R.; Wilson, N. K. *Environ. Sci. Technol.* **1990**, *24,* 661–665.
231. Woodrow, J. E.; Seiber, J. N. *Anal. Chem.* **1978**, *50,* 1229–1231.
232. Umlauf, G.; Kaupp, H. *Chemosphere* **1993**, *27,* 1293–1296.
233. Eiceman, G. A.; Karasek, F. W. *J. Chromatogr.* **1980**, *200,* 115–124.
234. Cautreels, W.; van Cauwenberghe, K. *Atmos. Environ.* **1978**, *12,* 1133–1141.
235. Seiber, J. N.; Shafik, T. M.; Enos, H. F. In *Environmental Dynamics of Pesticides;* Hague, R.; Freed, V. H., Eds.; Plenum: New York, 1975; pp 17–43.
236. Jackson, M. D.; Lewis, R. G. *Bull. Environ. Contam. Toxicol.* **1979**, *21,* 202–205.
237. Giam, G. S.; Chan, H. S.; Neff, G. S. *Anal. Chem.* **1975**, *47,* 2319–2320.
238. Que Hee, S. S.; Sutherland, R. G.; Vetter, M. *Environ. Sci. Technol.* **1975**, *9,* 62–66.
239. Smith, R. M.; O'Keefe, P. W.; Hilker, D. R.; Aldous, K. M. *Anal. Chem.* **1986**, *58,* 2414–2420.
240. Bidleman, T. F. In *Trace Analysis;* Lawrence, J. F., Ed.; Academic: New York, 1985; Vol. 4, pp 51–100.
241. Wright, B. W.; Wright, C. W.; Gale, R. W.; Smith, R. D. *Anal. Chem.* **1987**, *59,* 38–44.
242. Raymer, J. H.; Pellizzari, E. D. *Anal. Chem.* **1987**, *59,* 1043–1048.
243. Hawthorne, S. B.; Krieger, M. S.; Miller, D. J. *Anal. Chem.* **1989**, *61,* 736–740.
244. Pankow, J. F.; Kristensen, T. J. *Anal. Chem.* **1983**, *55,* 2187–2192.
245. Judeikis, H. S.; Siegel, S. *Atmos. Environ.* **1973**, *7,* 619–631.
246. Seifert, B.; Peters, J. *Atmos. Environ.* **1980**, *14,* 117–119.
247. Van Vaeck, L.; Van Cauwenberghe, K. *Atmos. Environ.* **1984**, *18,* 323–328.
248. Lee, F. S.-C.; Pierson, W. R.; Ezike, J. In *Polynuclear Aromatic Hydrocarbons: Chemistry and Biological Effects;* Bjorseth, A.; Dennis, A. J., Eds.; Battelle: Columbus, OH, 1980; pp 543–563.
249. Pitts, J. N., Jr.; Van Cauwenberghe, K. A.; Grosjean, D.; Schmid, J. P.; Fritz, D. R.; Belser, W. L.; Knudson, G. B.; Hynds, P. M. *Science (Washington, D.C.)* **1978**, *202,* 515–519.
250. Tebbens, B. D. *Am. Ind. Hyg. Assoc. J.* **1966**, *15,* 415–422.
251. Grosjean, D.; Fung, K.; Harrison, J. *Environ. Sci. Technol.* **1983**, *17,* 673–679.

252. Coutant, R. W.; Brown, L.; Chuang, J. C.; Riggin, R. M.; Lewis, R. G. *Atmos. Environ.* **1988,** *22,* 403–409.
253. Brorström, E.; Grennfelt, P.; Lindskog, A. *Atmos. Environ.* **1983,** *17,* 601–605.
254. Arey, J.; Zelinska, B.; Atkinson, R.; Winer, A. M. *Environ. Sci. Technol.* **1988,** *22,* 457–462.
255. Hanson, R. L.; Clark, C. R.; Carpenter, R. L.; Hobbs, C. H. *Environ. Sci. Technol.* **1981,** *6,* 701–705.
256. Lindskog, A.; Brorström–Lundén, E.; Iverfeldt, A. In *Polynuclear Aromatic Hydrocarbons: Measurements, Means, and Metabolism;* Cooke, M.; Loening, K.; Merritt, J., Eds.; Battelle: Columbus, OH, 1987; pp 537–544.
257. Hunt, G.; Pangaro, N. *Anal. Chem.* **1982,** *54,* 369–372.
258. Compton, B. *Prog. Anal. Chem.* **1973,** *5,* 133–152.
259. Harvey, G. R.; Steinhauer, W. G. *Atmos. Environ.* **1974,** *8,* 777–782.
260. Jackson, M. D.; Lewis, R. G. *Sampling and Analysis of Toxic Organics in the Atmosphere;* Verner, S. S., Ed.; American Society for Testing and Materials: Philadelphia, PA, 1980; pp 36–47, ASTM Special Technical Publication 721.
261. Falconer, R. E.; Bidleman, T. F. *Atmos. Environ.* **1994,** *28,* 547–554.
262. Galasyn, J. F.; Hornig, J. F.; Soderberg, R. H. *J. Air Pollut. Control Assoc.* **1984,** *34,* 57–59.
263. Lewis, R. G.; Kelly, T. J.; Chuang, J. C.; Callahan, P. J.; Coutant, R. W. In *Critical Issues in the Global Environment: Papers from the 9th World Clean Air Congress;* Air & Waste Management Association: Pittsburgh, PA, 1992; Vol. 4, Paper IU–11E.02.
264. Ligocki, M. P.; Leuenberger, C.; Pankow, J. F. *Atmos. Environ.* **1985,** *19,* 1609–1617.
265. Ligocki, M. P.; Leuenberger, C.; Pankow, J. F. *Atmos. Environ.* **1985,** *19,* 1619–1626.
266. Glotfelty, D. E.; Seiber, J. N.; Liljedahl, L. A. *Nature (London)* **1987,** *325,* 602–605.
267. Thibobeaux, L. J.; Nadler, K. C.; Valsaraj, K. T.; Reible, D. D. *Atmos. Environ.* **1991,** *25A,* 1649–1656.
268. Bramesberger, W. L.; Adams, D. F. *J. Agric. Food Chem.* **1965,** *13,* 552–554.
269. Courtney, W. J.; Rheingrover, S.; Pilotte, J.; Kaufmann, H. C.; Cahill, T. A.; Nelson, J. W. *J. Air Pollut. Control Assoc.* **1978,** *28,* 224–228.
270. Kaupp, H.; Umlauf, G. *Atmos. Environ.* **1992,** *26A,* 2259–2267.
271. Jungers, R.; Burton, R.; Claxton, L.; Lewtas–Huisingh, J. In *Short-Term Bioassays in the Analysis of Complex Environmental Mixtures. II;* Waters, M. D.; Sandu, S. S.; Lewtas-Huisingh, J.; Claxton, L.; Nesnow, S., Eds.; Plenum: New York, 1981; pp 45–65.
272. Gormley, P.; Kennedy, K. *Proc. R. Ir. Acad.* **1949,** *52A,* 163–169.
273. Hinds, W. C. *Aerosol Technology;* John Wiley & Sons: New York, 1982; pp 133–152.
274. Coutant, R. W.; Callahan, P. J.; Kuhlman, M. R.; Lewis, R. G. *Atmos. Environ.* **1989,** *23,* 2205–2211.
275. Coutant, R. W.; Callahan, P. J.; Chuang, J. C.; Lewis, R. G. *Atmos. Environ.* **1992,** *26A,* 2831–2834.
276. Lane, D. A.; Johnson, N. D.; Barton, S. C.; Thomas, G. H. S.; Schroeder, W. H. *Environ. Sci. Technol.* **1988,** *22,* 941–947.
277. Lane, D. A.; Johnson, N. D.; Hanley, M.-J. J.; Schroeder, W. H.; Ord, D. T. *Environ. Sci. Technol.* **1992,** *26,* 126–133.
278. Subramanyam, V.; Valsaraj, K. T.; Thibodeaux, L. J.; Reible, D. D. *Atmos. Environ.* **1994,** *28,* 3083–3091.
279. Eatough, D. J.; Wadsworth, A.; Eatough, D. A.; Crawford, J. W.; Hansen, L. D.; Lewis, E. A. *Atmos. Environ.* **1993,** *27A,* 1213–1219.
280. Tang, H.; Lewis, E. A.; Eatough, D. J. *Atmos. Environ.* **1994,** *28,* 939–947.
281. Gundel, L. A.; Lee, V. C.; Mahanama, K. R. R.; Stevens, R. K.; Daisey, J. M. *Atmos. Environ.* **1995,** *29* (in press).

282. Possanzini, M.; Febo, A.; Liberti, A. *Atmos. Environ.* **1983,** *17,* 2605–2610.
283. Krieger, M. S.; Hites, R. A. *Environ. Sci. Technol.* **1992,** *26,* 1551–1555.
284. Krieger, M. S.; Hites, R. A. *Environ. Sci. Technol.* **1994,** *28,* 1129–1133.
285. Koutrakis, P.; Sioutas, C.; Ferguson, S. T.; Wolfson, J. M.; Mulik, J. D.; Burton, R. M. *Environ. Sci. Technol.* **1993,** *27,* 2497–2501.
286. Turpin, B. J.; Liu, S.-P.; Podolski, K. S.; Gomes, M. S. P.; Eisenreich, S. J.; McMurry, P. H. *Environ. Sci. Technol.* **1993,** *27,* 2441–2449.
287. Van Vaeck, L.; Broddin, G.; Cautreels, W.; Van Cauwenberghe, K. *Sci. Total Environ.* **1979,** *11,* 41–52.
288. Loo, B. W.; Jaklevic, J. M.; Goulding, F. S. In *Fine Particles: Aerosol Generation, Measurement, Sampling, and Analysis;* Liu, B. Y. U., Ed.; Academic: New York, 1976; pp 311–350.
289. Sioutas, C.; Koutrakis, P.; Burton, R. M. "Development of a New Semivolatile Organic Compound Sampler"; In *Measurement of Toxic and Related Air Pollutants;* Air & Waste Management Association: Pittsburgh, PA, 1994; p 424.
290. Sioutas, C.; Koutrakis, P.; Burton, R. M. *Part. Sci. Technol.* **1995,** *12,* 207–221.
291. Flood, E. A. In *The Solid–Gas Interface;* Flood, E. A., Ed.; Marcel Dekker: New York, 1967; Vol. 1, pp 11–73.
292. Coutant, R. W.; Callahan, P. J.; Chuang, J. C. *Annular Denuder Sampler for Phase-Distributed Semivolatile Organic Chemicals;* U.S. Environmental Protection Agency: Research Triangle Park, NC, 1989; EPA/600/3–89/029 (NTIS PB 89–169 858).
293. Kelly, T. J.; Chuang, J. C.; Callahan, P. J. *Research for Ambient Polar Volatile Organics and Semi-Volatile Phase-Distributed Organics Utilizing TAMS Sites;* U.S. Environmental Protection Agency: Research Triangle Park, NC, 1992; NTIS 92–164 979.
294. Bidleman, T. F.; Billings, W. N.; Foreman, W. T. *Environ. Sci. Technol.* **1986,** *20,* 1038–1043.
295. Bidleman, T. F. *Environ. Sci. Technol.* **1988,** *22,* 361–367.
296. Rounds, S. A.; Pankow, J. F. *Environ. Sci. Technol.* **1990,** *24,* 1378–1386.
297. Pankow, J. F. *Atmos. Environ.* **1987,** *21,* 2275–2283.
298. Gill, P. S.; Graedel, T. E.; Weschler, C. J. *Rev. Geophys. Space Phys.* **1983,** *21,* 903–920.
299. Prager, M. J.; Deblinger, B. *Environ. Sci. Technol.* **1967,** *1,* 1008–1013.
300. Yuen, D. Y.; Gavin, E. L.; Brand, F. *Advances in Instrumentation and Control: Proceedings of the ISA Conference;* Instrument Society of America: Research Triangle Park, NC, 1991; Vol. 46, pp 647–654.
301. Lane, D. A.; Sakuma, T.; Quan, E. S. K. In *Polynuclear Aromatic Hydrocarbons: Chemistry and Biological Effects;* Bjorseth, A.; Dennis, A. J., Eds.; Battelle: Columbus, OH, 1980; pp 199–214.
302. Williams, D. T.; Denley, H. V.; Lane, D. A.; Quan, E. S. K. *Am. Ind. Hyg. Assoc. J.* **1982,** *43,* 190–195.
303. Burtscher, H.; Scherrer, L.; Siegmann, H. C. *J. Appl. Phys.* **1982,** *53,* 3787–3791.
304. Niesser, R. *J. Aerosol. Sci.* **1986,** *17,* 705–714.
305. Wilson, N. K.; Barbour, R. K.; Chuang, J. C.; Mukund, R. *Polycyclic Aromat. Compd.* **1994,** *5,* 167–174.
306. Klein, A. W.; Harnish, M.; Porenski, H. J.; Schmidt-Bleck, F. *Chemosphere* **1981,** *10,* 153–207.
307. Mackay, D.; Patterson, S.; Cheung, B. *Chemosphere* **1985,** *15,* 1397–1400.

Chapter 24

Current and Emerging Sampling and Analytical Methods for Point Source and Non-Point Source Emission Measurements

Eric D. Winegar and Larry O. Edwards

Recent regulatory initiatives in the air quality arena have introduced a number of challenges to the environmental measurement community. These challenges include an increased number of target analytes (from the six criteria pollutants to 189 hazardous air pollutants of the Clean Air Act), more sources from which to obtain data (both smaller point sources and numerous non-point or area sources), lower emission levels to quantify (down to parts-per-billion concentrations), and a larger amount of data needed (for continuous compliance assessment and process monitoring). These challenges have encouraged the development of additions to the usual repertoire of source testing techniques. This chapter presents a summary of the most frequently used source testing methodologies along with discussions of non-point source approaches. In addition to the standard approaches, an introduction is presented to several emerging technologies that are used as a response to some of the current emission measurement problems.

THE ASSESSMENT OF AIR QUALITY in the United States in the past has been dependent on a number of federal reference methods supplemented by adaptations, modifications, and often special research methods. This situation continues today. However, recent regulatory initiatives have enhanced or added to that general list of methods that had its genesis more than 20 years ago. In addition, regulatory demands (e.g., the Clean Air Act Amendments of 1990) have presented several challenges to the sampling and analytical community to develop and validate methods that are appropriate for the wider range of chemical compounds for which measurement data must be obtained.

The purpose of this chapter is to present a survey of the more common sampling and analytical methods that are currently used for determining the emissions from various types of point and non-point sources. It is not intended to be a comprehensive examination of all sampling methods and issues surrounding source testing. The focus will be on the most frequently used federal reference methods; many are also the basis for modified sampling and analytical methodologies. In addition, a brief survey will be presented of some emerging technologies that are being increasingly used to address the demands of today's regulations. We begin with some background material on the evolution of the regulatory basis for these methods.

Air Quality Regulatory Development

Until several decades ago, air pollution had been perceived to be primarily due to the easily identified sources contributing dust and smoke to the air. About 40 years ago, the photochemical formation of air pollutants was first discovered, which led to a better understanding of the scope and complexity of air quality problems. The role of major gaseous pollutants was reasonably well understood by the time of the passage of the Clean Air Act of 1970. At the same time, the U.S. Environmental Protection Agency (EPA) was established, and it was their responsibility, as stated in the preamble, to "protect and enhance the quality of the nation's air resources so as to promote the public health and welfare and the productive capacity of its population...."

Based on the adverse health effects of air pollution, National Ambient Air Quality Standards (NAAQS) were established for the six "criteria pollutants" as follows: sulfur dioxide (SO_2), nitrogen oxides (NO_x), carbon monoxide (CO), lead (Pb), ozone (O_3), and particulate matter with an aerodynamic diameter of ≤ 10 µm (PM_{10}). In 1987, based upon a better understanding of the health effects of particles, PM_{10} replaced PM as a criteria pollutant. Safe atmospheric concentrations were established for each criteria pollutant. Nationwide monitoring programs were put in place, and areas that were out of compliance had to submit an attainment plan to the EPA. Volatile organic compounds (VOCs) came to refer to airborne organic substances that, through chemical reactions in the atmosphere, lead to the formation of ozone. A more recent development, once again based on health standards, has lead to the establishment of another category of toxic pollutants called "hazardous air pollutants" (HAPs). These HAPs are often generically referred to as "air toxics" and are toxic; they may or may not also promote the formation of ozone.

The establishment of standards and the need to monitor for criteria pollutants required the development of measurement methods that could provide an accurate and reproducible body of data in order to determine compliance status and from which policy decisions could be made (*1*). Two different types of measurement methods were required. One need was to monitor the ambient concentration of the criteria pollutants, but separate methods were needed to measure the emissions of these pollutants from point sources such as smoke stacks. Methods were developed and promulgated under several different EPA authorities including New Source Performance Standards (NSPS) in 1975, Prevention of Significant Deterioration (PSD), and National Emission Standards for Hazardous Air Pollutants (NESHAPs) in 1980. Later, measurement methods were promulgated under other authorities such as the Resource Conservation and Recovery Act (RCRA), resulting in the SW–846 collection of methods (2).

For HAPs, which can be either acutely toxic or carcinogenic or both, no ambient air quality standards have been established. Rather, safe exposure levels are determined through the use of risk assessments that determine concentrations at which there are negligible or acceptable risks to humans or the environment. Most of the time, measurement at the source is required to estimate the risk to nearby receptors.

The focus in the past on the criteria pollutants led to a narrow set of sampling and analytical methods directed to just those pollutants. The methods that were developed were limited and unable to cope with the substantially increased demands created by the Clean Air Act Amendments of 1990. In many cases, this sweeping legislation called for the measurement of HAPs for which there were no methods; thus, a new generation in sampling and analytical methodology was introduced. One of the centerpieces of Title III of the Clean Air Act Amendments of 1990 is a list of 189 hazardous substances, some of which are whole classes of compounds (e.g., arsenic compounds or coke oven emissions). These hazardous substances comprise seven general categories, listed as follows with a breakdown of the total number of compounds in each class:

1. Aliphatic and cyclic hydrocarbons (saturated and unsaturated, including chlorinated hydrocarbons) (30 chemicals)
2. Aromatic compounds (including aromatic hydrocarbons, phenolic compounds, and phthalates) (34 chemicals)
3. Nitrogenated organic compounds (including nitroaromatics) (49 chemicals)
4. Oxygenated organic compounds (including alcohols, aldehydes, α,β-unsaturated carbonyls, carboxylic acids, esters, ethers, ketones, and oxides) (35 chemicals)
5. Pesticides and herbicides (15 chemicals)
6. Inorganic chemicals (including metals) (24 chemicals)
7. Sulfates (2 chemicals)

The 189 HAPs are listed according to chemical class in Table I. This list was compiled by Congress from several sources: (1) the Emergency Planning and Community Right-to-Know Act of 1986 (SARA Section 313); (2) the Comprehensive Envi-

Table I. Hazardous Air Pollutants Listed Under Title III of CAAA

Chemical Classification	*Pollutant*
Aliphatic and cyclic hydrocarbons	
Saturated hydrocarbons	hexane
	2,2,4-trimethylpentane
Unsaturated hydrocarbons	1,3-butadiene
Saturated halogenated hydrocarbons	bromoform
	carbon tetrachloride
	chloroform
	1,2-dibromo-3-chloropropane
	ethyl chloride (chloroethane)
	ethylene dibromide (1,2-dibromoethane)
	ethylene dichloride (1,2-dichloroethane)
	ethylidene dichloride (1,1-dichloroethane)
	hexachloroethane
	methyl bromide (bromomethane)
	methyl chloride (chloromethane)
	methyl chloroform (1,1,1-trichloroethane)
	methyl iodide (iodomethane)
	methylene chloride (dichloromethane)
	propylene dichloride (1,2-dichloropropane)
	1,1,2,2-tetrachloroethane
	1,1,2-trichloroethane
Unsaturated halogenated hydrocarbons	allyl chloride
	chloroprene
	1,2-dichloropropene
	hexachlorobutadiene
	hexachlorocyclopentadiene
	tetrachloroethylene (perchloroethylene)
	trichloroethylene
	vinyl bromide
	vinyl chloride
	vinylidene chloride (1,1-dichloroethylene)
Aromatic compounds	
Aromatic hydrocarbons	benzene
	biphenyl
	catechol
	coke oven emissions
	cumene
	ethyl benzene
	naphthalene
	polycyclic organic matter
	styrene
	toluene
	xylenes (isomers and mixture)
	o-xylene
	m-xylene
	p-xylene

Table I. Hazardous Air Pollutants Listed Under Title III of CAAA—*Continued*

Chemical Classification	*Pollutant*
Aromatic compounds *(Continued)*	
Halogenated aromatic hydrocarbons	benzotrichloride
	benzyl chloride
	chlorobenzene
	1,4-dichlorobenzene
	hexachlorobenzene
	polychlorinated biphenyls (aroclors)
	2,3,7,8-tetrachlorodibenzo-*p*-dioxin
	1,2,4-trichlorobenzene
Phenolic compounds	cresols/cresylic acid (isomers and mixtures)
	o-cresol
	m-cresol
	p-cresol
	pentachlorophenol
	phenol
	2,4,5-trichlorophenol
	2,4,6-trichlorophenol
Phthalates	bis(2-ethylhexyl)phthalate (DEHP)
	dibutylphthalate
	dimethyl phthalate
	phthalic anhydride
Nitrogenated organic compounds	acetamide
	acetonitrile
	2-acetylaminofluorene
	acrylamide
	acrylonitrile
	4-aminobiphenyl
	aniline
	o-anisidine
	benzidine
	diazomethane
	3,3-dichlorobenzidene
	diethanolamine
	N,N-diethyl aniline (*N,N*-dimethylaniline)
	3,3-dimethoxybenzidine
	dimethyl aminoazobenzene
	3,3′-dimethyl benzidine
	dimethyl carbamoyl chloride
	dimethyl formamide
	1,1-dimethyl hydrazine
	4,6-dinitro-*o*-cresol, and salts
	2,4-dinitrophenol
	2,4-dinitrotoluene
	1,2-diphenylhydrazine
	ethyl carbamate (urethane)
	ethylene imine (aziridine)
	ethylene thiourea

Continued on next page

Table I. Hazardous Air Pollutants Listed Under Title III of CAAA—*Continued*

Chemical Classification	*Pollutant*
Nitrogenated organic compounds (*Continued*)	hexamethylene-1,6-diisocyanate
	hexamethylphosphoramide
	hydrazine
	methyl hydrazine
	methyl isocyanate
	4,4-methylene bis(2-chloroanilene)
	methylene diphenyl diisocyanate (MDI)
	4,4-methylenedianiline
	nitrobenzene
	4-nitrobiphenyl
	4-nitrophenol
	2-nitropropane
	N-nitroso-*N*-methylurea
	N-nitrosodimethylamine
	N-nitrosomorpholine
	pentachloronitrobenzene (quintobenzene)
	p-phenylenediamine
	1,2-propyleneimine (2-methyl aziridine)
	quinoline
	2,4-toluene diamine
	2,4-toluene diisocyanate
	o-toluidine
	triethylamine
Oxygenated organic compounds	
Alcohols	methanol
Aldehydes	acetaldehyde
	formaldehyde
	propionaldehyde
α,β-Unsaturated carbonyls	acrolein
Carboxylic acids	acrylic acid
	chloroacetic acid
Esters	ethyl acrylate
	methyl methacrylate
	vinyl acetate
Ethers	bis(chloromethyl)ether
	chloromethyl methyl ether
	dibenzofurans
	dichloroethyl ether (bis(2-chloroethyl)ether)
	1,4-dioxane (1,4-diethyleneoxide)
	glycol ethers
	methyl *tert*-butyl ether (MTBE)
Ketones	methyl ethyl ketone (2-butanone, MEK)
	methyl isobutyl ketone (MIBK)
Oxides	epichlorohydrin (1-chloro-2,3-epoxypropane)
	1,2-epoxybutane
	ethylene oxide

Table I. Hazardous Air Pollutants Listed Under Title III of CAAA—*Continued*

Chemical Classification	*Pollutant*
Oxygenated organic compounds *(Continued)*	
Oxides	propylene oxide
	styrene oxide
Other carbonyls and oxygenates	acetophenone
	caprolactam
	2-chloroacetophenone
	ethylene clycol
	hydroquinone
	isophorone
	maleic anhydride
	phosgene
	1,3-propane sultone
	β-propiolactone
	quinone
Pesticides and herbicides	captan
	carbaryl
	chloramben
	chlordane
	chlorobenzilate
	2,4-dichlorophenoxyacetic acid
	dichlorodiphenyldichloroethylene (DDE)
	dichlorvos
	heptachlor
	lindane (all isomers)
	methoxychlor
	parathion
	propoxur (baygon)
	toxaphene (chlorinated camphene)
	trifluralin
Inorganic compounds	antimony compounds
	arsenic compounds (including arsine)
	asbestos
	beryllium compounds
	cadmium compounds
	calcium cyanamide
	carbon disulfide
	carbonyl sulfide
	chlorine
	chromium compounds
	cobalt compounds
	cyanide compounds
	hydrochloric acid
	hydrogen fluoride (hydrofluoric acid)
	lead compounds
	manganese compounds
	mercury compounds

Continued on next page

Table I. Hazardous Air Pollutants Listed Under Title III of CAAA—*Continued*

Chemical Classification	*Pollutant*
Inorganic compounds *(Continued)*	mineral fibers (fine)
	nickel compounds
	phosphine
	phosphorus
	radionuclides (including radon)
	selenium compounds
	titanium tetrachloride
Sulfates	diethyl sulfate
	dimethyl sulfate

NOTE: CAAA is Clean Air Act Amendments of 1990.

ronmental Response, Compensation, and Liability Act of 1980 (CERCLA Section 104), which identified high-priority environmental contaminants by the Agency for Toxic Substances and Disease Registry; and (3) state or local air pollution control agency regulations that had established an "acceptable ambient concentration" or standard. EPA has stated its intention that validated methods will someday exist for these regulated compounds.

State Regulatory Developments

Several states have active air quality regulatory programs that parallel or are more stringent than the EPA's activities. The recent trend has shown a similar pathway of moving beyond criteria pollutants toward regulation of air toxics, from a command-and-control mind-set to a risk-based or market-based approach. The most notable example is the State of California, which is often at the forefront of regulatory development on air quality. One example of this development was the so-called California "Air Toxics Hot Spots" law of 1988 (AB2588) that required facilities to develop emission inventories for a large number of toxics and carcinogens (more than 300 substances). As in the case of the Clean Air Act, validated measurement methods did not exist for most of these compounds. Although most emissions also were able to be quantified by alternative approaches, such as mass balance or engineering calculations, often methods were extended beyond their original intended uses and beyond the limits of the method validation.

The net effect of both the federal and state actions is that the demands for measurement, particularly of HAPs, and from a wider array of sources for a larger list of pollutants, has presented challenges to the environmental measurement community. Because the measurement data are to be used in making risk management decisions and for enforcing regulatory standards, high-quality, reliable, comparable data are required.

Types of Sources

Emissions of volatile and semivolatile compounds and particulate matter occur from a multitude of sources. The most obvious emission source is a smoke stack or vent; these have come to be called *point sources.* These sources comprise a few square feet (a point) from which emissions emanate with an active, confined flow. Another type of source, and one that is less obvious and more difficult to measure, is a *non-point source,* or an area source. (In the Clean Air Act Amendments of 1990, particularly in Title III, all nonmajor sources are referred to as area sources. In this chapter, we instead use the more general definition of an area source, meaning emissions from an area larger than a few square feet.) Such a source may be, for example, a lagoon or pond, the surface of a landfill, a series of nonducted building vents, leaks through building windows, or seams. In short, it describes an emission from an area as distinct from a point source. There usually is no confined flow and no specific point of emission. Occasionally there may be an active flow, as from a dissolved air flotation tank, but most often the flow cannot be measured using standard techniques.

Because they are easy to identify, measure, and control, point sources have traditionally been the focus of regulatory abatement programs. Large stacks servicing boilers and furnaces are an easily recognized source of criteria pollutants such as NO_x, SO_2, particulate matter, and CO. Other obvious conventional point sources subject to easy identification, measurement, and control are industrial processes such as chemical process operation vents and stacks at refineries. More recently, with all the larger point sources identified and controlled, the focus now includes smaller, nonconventional point sources, such as dry cleaners, printing operations, automotive paint shops, oil tanker emissions, storage tanks, transfer operations, and other smaller sources. Previously, these minor sources have escaped scrutiny, but it is now known that a very large number of these small sources constitute a major source of emissions, especially for HAPs. Besides the obvious large targets, the Clean Air Act Amendments of 1990 target the great number of these minor sources, particularly in ozone nonattainment areas.

Non-point sources include those sources from which emissions occur but cannot be localized or ascribed to a particular point. Examples would include impoundments or aeration ponds at publicly owned treatment works (POTW), hazardous waste impoundments, service stations, fugitive emissions from many types of chemical and refinery processes, and any other types of emissions that do not emanate from a single, well-defined source. A third distinct type of source is mobile emissions, including exhaust and evaporative emissions from automobiles, trucks, trains, aircraft, recreational vehicles, or golf carts. It has been shown (3) that, in some areas, up to two-thirds of the air toxics in the urban ambient air derive from mobile sources. This type of source is difficult to measure and control. Most testing is done in laboratories on dynamometers using the Federal Test Procedure (4) in which primarily total hydrocarbon emissions, carbon monoxide, and NO_x are measured, although some recent programs have included several toxics, including specific hydrocarbon species, alcohols, and aldehydes (5). Emissions are greatly affected by many factors such as driving habits and new types of reformulated fuels, and much remains to be learned about actual mobile source emissions. A discussion of measurement methods for mobile sources is beyond the scope of this chapter.

Measurement of Emissions

In the measurement of emissions, two critical pieces of information are usually measured: the concentration of the pollutants (e.g., parts per million (ppmv), $\mu g/m^3$, or lb/ft^3), and the volumetric flow of the air stream containing the pollutants [e.g., cubic feet per minute (CFM) or m^3/min]. The emission rate is calculated as the product of these two factors, which results in a mass per unit time emission rate: $(lb/ft^3) \times (ft^3/h) = lb/h$. The emission calculation is contrasted to ambient air methods in which only a concentration value (e.g., ppmv) is determined.

For non-point sources, there is no active or well-characterized flow for the mass transport of the emitted compounds away from the source. Therefore, the second key term in emission calculations is absent. However, for this type of source, various methods, although often not simple, can be used to estimate a flow (e.g., tracers) and to measure the emission rate directly (e.g., the EPA isolation flux chamber). These approaches will be discussed, but first the major methods used to measure emissions from classical, confined point sources will be described.

Standard Point Source Methods

Most pollutant sampling falls into one of two general categories: gaseous and particulate matter. These categories are self-evident for many target compounds of interest; however, when choosing a particular method to collect a sample, the physical and molecular properties of the pollutant must be known. A generally accepted breakdown of categories is illustrated in Table II, in which definitions of volatile, semivolatile, and particulate matter are given. Semivolatile substances may exist as either a vapor or as a solid or both, depending upon physical conditions of the pollutant stream, primarily temperature. For example, naphthalene at room temperature is a solid, but it has a significant vapor pressure such that some molecules will be present in the gaseous state. Any complete measurement of naphthalene or other semivolatile compounds at room temperature must therefore account for both

Table II. Definitions of Chemical Volatility

Category	*Vapor Pressure at STP*	*Boiling Point*
Volatile	>0.1 mmHg	<100 °C
Semivolatile	0.1 to 10^{-7} mmHg	100–325 °C
Nonvolatile (particulates)	<10^{-7} mmHg	>325 °C

NOTE: STP is standard temperature and pressure.

phases. At elevated temperatures typical of combustion sources, the vapor pressure effect of the compound is increased. Nonvolatile chemicals include most metals and other substances with very low vapor pressures (using the 325 °C criterion), such as asbestos fibers and very heavy organic chemicals like dioxins and benzopyrenes.

When these categories are used to separate the Clean Air Act Amendment HAPs, 108 are volatile chemicals, 62 are semivolatile chemicals, and 19 are nonvolatile chemicals. For sampling purposes, either a gas, a solid, or both must be collected. Gases may be collected as whole samples (e.g., a sample of the vapor is drawn into an inert bag or canister), collected on sorbent materials such as activated charcoal, dissolved into liquids, or measured directly by spectrometry or chromatography. A few gases, such as formaldehyde, are sampled by in situ derivatization by selective compounds, and the results of the reaction are captured and analyzed.

Particulate matter is usually collected on filters, but it is considerably harder to obtain a representative particulate matter sample than it is to collect a representative gas sample. Gaseous pollutant streams are usually well mixed, and individual molecules can easily follow complex flow paths through valves or nozzles. Particles, which have mass and momentum of their own, cannot always follow gaseous, sharply curved flow lines, and great care must be taken to collect all and only those particles that were initially in the volume of gas sampled or metered. Because semivolatile compounds are usually wholly or partially in the condensed phase, a more elaborate particle sampling method, often in tandem with gaseous sampling methods, must be to used to collect them. When particles of only a certain size are to be measured—for example, the criteria pollutant PM_{10}—special devices are used. These methods are described later.

Isokinetic Sampling

Because of differential velocities from flow disturbances, particles may not be uniformly distributed in a gas stream. Therefore, to ensure that a particulate sample is representative, sampling is conducted at a regular array of locations covering the entire cross-sectional area of the duct or stack, and in such a way that the gas–particle mixture being sampled is drawn into the nozzle at the same velocity as the gas flowing in the duct or stack. The nozzle must be aligned directly into the stream flow. If the stream velocity at the nozzle is at exactly the same rate as the overall flow in the stack, there will be, in effect, no local flow disturbance, and the gas containing particles of all sizes and densities will enter the nozzle without curving from its linear trajectory. This condition is termed isokinetic (same velocity) sampling. If, for example, gas were entering the nozzle at a rate greater than it was flowing in the stack, it would curve at the last second as it was sucked into the nozzle. The heavier or more dense particles that were moving in that gas stream but slightly outside the cylinder defined by the nozzle would be unable, because of their own momentum, to curve sharply with the gas into the nozzle. Therefore, they would not be collected and would not subsequently impact onto the filter behind the nozzle. This condition is referred to as super-isokinetic sampling, and if it is greater than 10%, then the EPA defines the sample to be nonrepresentative of the particulate matter moving up the stack or along the duct, and the result is considered invalid. A similar discussion is appropriate for sub-isokinetic sampling.

The protocol for collecting an isokinetic sample is given in EPA Reference Method 5 (6). (*See* Table V, which contains references in the *Code of Federal Regulations* for all Federal Reference Methods.) It is required whenever particles or liquid droplets are entrained in a gas stream. Sampling for only gases does not require isokinetic conditions. EPA Methods 1–4 refer to procedures that are required before an EPA Method 5 test can be run. Table III lists some of the key physical parameters that Methods 1–4 cover. They determine the sampling location, the required number of sampling points, the approximate velocity and temperature, the moisture, and the molecular weight of the gas. All these are necessary to select the nozzle size and approximate sampling rate before beginning any Method 5 sampling. Together, EPA Methods 1–5 specify exactly how a particulate sample must be collected. EPA Method 5 is not trivial and requires special equipment (described in this section) and trained, experienced operators.

Another feature of EPA's Method 5 is that it collects the particulate matter on a hot filter. This feature is necessary because nearly all combustion off-gases contain a substantial amount of water vapor. If this water vapor were to condense to liquid

Table III. Key Flow and Physical Measurement Parameters and Corresponding EPA Methods

Parameter	*EPA Method*	*Method Description*
Stack dimensions and configuration (i.e., cyclonic vs. noncyclonic flow); sampling port placement	1	Sample and velocity traverses for stationary sources
	1A	Traverse points in small ducts
Velocity traverses	2	Determination of stack gas velocity and volumetric flow rate (type S pitot tube)
	2A	Direct measurement of gas volume through pipes and small ducts
	2B	Determination of exhaust gas volume flow rate from gasoline vapor incinerators
	2C	Flow rate determination in small ducts using a pitot tube
	2D	Flow rate determination in small ducts using a ratemeter
Molecular weight of gas	3	Gas analysis for CO_2, O_2, excess air, and dry molecular weight
	3A	Determination of O_2 and CO_2 concentrations in emissions from stationary sources (instrumental analyzer procedure)
Moisture content	4	Determination of moisture content in stack gases

water, it would affect the collection of the particulate matter and gases, and may run back out of the nozzle with sampled material dissolved or suspended in it. Therefore, the entire probe and filter are heated to about 120 °C to ensure that all water stays in its vapor form and will pass through the filter as a gas. The nozzle, which is in the stack, is at the same temperature as the stack gas. In fact, EPA defines particulate matter as that material that will not pass through a filter at 120 °C. Only after the particulate material has been filtered out of the heated stream are the sampled gases cooled and the water allowed to condense. Whenever a sample is collected from any combustion source, provisions must be made for the large amounts of water vapor nearly always present in such off-gases.

Because special conditions are often required to quantitatively collect certain substances that are associated, in whole or in part, with the particulate matter, many modifications of the basic EPA Method 5 have been developed. One modification requires that the condensate collected behind the hot filter be evaporated, the residue weighed, and the weight of that residue added to the mass collected on the filter; this fraction is called condensable particulate matter. In these modifications, the sampling is isokinetic, but special filtration, condensation, or reaction techniques are added to make sure that all of the target substance is collected. Modified Method 5 procedures exist for the collection of semivolatile dioxins, polynuclear aromatic hydrocarbons (PNAs), polychlorinated biphenyls (PCBs), metals that may sometimes be partially or wholly volatilized (such as mercury, lead, or arsenic), and metals in a particular oxidation state (such as hexavalent chromium). All of these examples involve substances of high toxicity and considerable environmental concern. The EPA has promulgated the detailed modifications in such methods as Method 0010 (for semivolatile chemicals) (2), Method 12 (lead) (6), and Method 29 for multiple metals (6). For example, to make sure that all of the heavy semivolatile organic material is collected, an organic sorbent trap (XAD) is inserted after the filter and in front of the impinger; for multiple metals, the impingers are filled with acids and/or oxidative solutions to ensure that all volatile metallic compounds are captured.

The main components of the sampling train are, for descriptive purposes, often divided into the hot section or front half (i.e., up to and including the hot filter) and the cold section or back half (traps and/or impingers where the volatile components, including water, are captured). Figure 1 shows the main components of a Method 5 sampling train, which consist of the following:

- *Probe nozzle*—Stainless steel or glass with a sharp, tapered (30 degrees) leading edge. The nozzle size depends upon the velocity of the air stream sampled, according to the isokinetic requirement. Glass nozzles are required for most metals sampling as well as certain other applications. Metal is not acceptable for organic sampling with Modified Method 5.
- *Heated probe liner*—Stainless steel, borosilicate, or quartz-glass tubing with a heating system capable of maintaining a temperature of 120 ±

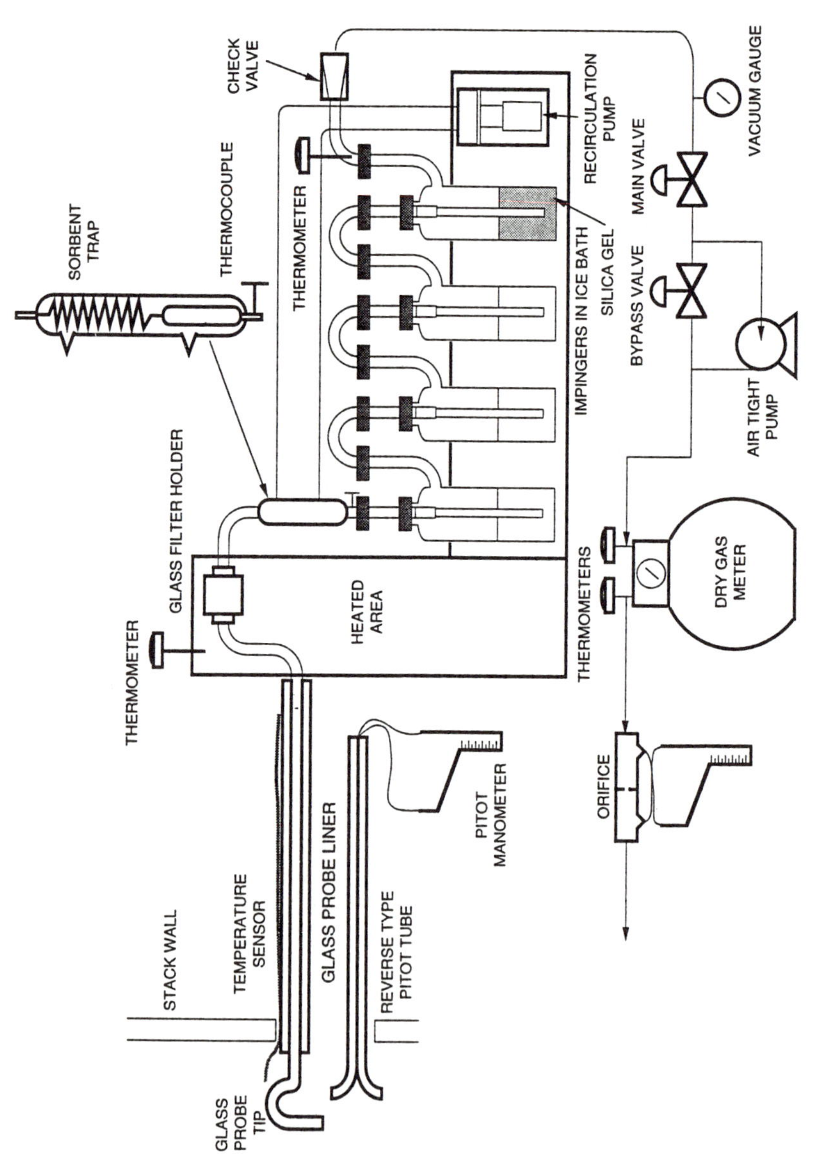

Figure 1. SW–846 Method 0010/Modified Method 5 sampling train.

14 °C. A thermocouple must also be present to document the probe temperature. Stainless steel cannot be used when sampling for metals.

- *Pitot tube*—Type S pitot tube with a known coefficient to measure the velocity pressure. It must be within 1 in. of the nozzle and is attached to a differential pressure gauge. A thermocouple is usually attached to the pitot tube assembly to measure the stack temperature.
- *Filter holder*—Borosilicate glass filter holder with a glass or Teflon frit (filter support) and an inert gasket (e.g., Teflon). Each method should be checked for specific filter requirements.
- *Filter heating system*—The chamber housing the filter and filter holder is heated. This system is often called the "hot box"; it maintains the temperature around the filter holder during sampling.
- *Organic sampling module*—This module most often comprises three sections: a gas conditioning section, a sorbent trap, and a condensate knockout trap. The gas conditioning system is a condenser that cools the gas from the heated filter unit to below 20 °C. The sorbent trap contains 20 g of XAD-2 and is maintained at 17 ± 2 °C. This unit is most often cooled by the circulating cooling ice water from the impinger bath. The sorbent module is maintained vertically so that condensed water can pass through it to the knockout impinger directly below. Figure 1 contains a schematic showing a cutout of the sorbent trap that is included when sampling for semivolatile organic chemicals in Modified Method 5.
- *Impinger train*—Four (or sometimes more) 500-mL impingers connected in series and immersed in an ice bath to condense water vapor for moisture determinations and to protect the downstream metering and pumping system. The first, second, and third impingers contain known amounts of water or an appropriate caustic trapping solution (e.g., as when the stack gas contains HCl), and the final contains silica gel desiccant.
- *Pump and metering system*—Pump, vacuum gauge, thermometers, and dry gas meter that are used to pull sample, at an adjustable flow, through the train and to measure its volume, temperature, and rate.

In addition to this equipment, some means of determining the gas density and barometric pressure are needed for accurate calculations of the volume sampled. During sampling, calculations based primarily on the stack velocity, as measured by the pitot tube located in close proximity to the nozzle, are made continuously, and the flow rate is adjusted at the pump to ensure that the sampling is isokinetic at all times.

The samples that are collected in this manner are subjected to gravimetric and general chemical analysis, such as atomic absorption for metals. Either separate portions of the trains (i.e., filter, sorbent, and impingers) can be analyzed separately, with the results combined, or they can be combined. For organic chemicals, the primary analysis relies on an extraction followed by full scan GC–MS, such as in the SW–846 Method 8270 analysis. For dioxins, furans, and polynuclear aromatic hydrocarbons, specialized analyses such as high-resolution GC–MS can be conducted.

One variation of Method 5 particulate sampling is to collect the particles in the stack at stack conditions. In this approach, EPA Method 17, a filter is located directly behind the nozzle and allowed to come up to stack temperature, which will probably be different than 250 °F. The filter holder is somewhat bulky and may cause flow disturbances in the region of the nozzle; in addition, EPA particulate sampling does not allow the sampling apparatus to subtend more than 5% of the cross-sectional area, and this restriction may disallow the use of Method 17. One advantage is that, with the filter in the stack, the probe does not need to be heated or cleaned. EPA Method 5 is always acceptable, but in some situations, Method 17 may be appropriate and captures the true in-stack particulate material (e.g., at 500 °F).

Multimetal Train/Method 0012

Multiple metals sampling evolved from the source assessment sampling system (SASS) (7), which provided a large body of data in one sampling run, including metals speciation, size fractionation, and a means to collect organic chemicals. However, the equipment for this method was cumbersome, and not frequently used, which led to the development of other approaches. The single-metal approach has relatively recently evolved into a more comprehensive method called the multimetals train. Two very similar methods are described in Method 0012 (2) and Method 29 (6). Prior to its development, methods were validated for only one metal at a time, which meant that multiple samplings had to be performed when data for several metals were required. The multimetals train combines several different modifications into one comprehensive method.

The multimetals train is used to determine emissions of metals from sources. It is based on Method 5 because most metals are associated with particulate matter, even at high temperatures, with some notable exceptions such as mercury. Therefore, sampling must be done isokinetically. This method is referred to as Method 0012 in SW–846, and it has been validated for antimony, arsenic, barium, beryllium, cadmium, chromium (total), lead, mercury, silver, and thallium. Hexavalent chromium, which accounts for much of the metals toxicity, has its own method that is described later.

As with Method 5, the equipment setup consists of two main sections: the heated front section, and the cooled back section. A major difference exists, however, in that a glass nozzle and a glass-lined probe are required in the front section. All other connections are to be of glass or Teflon. This requirement is to avoid the potential for contamination from any metal components in the sampling lines. The usual S-type pitot tube and thermocouple are attached for velocity and temperature measurements. Figure 2 illustrates the equipment setup.

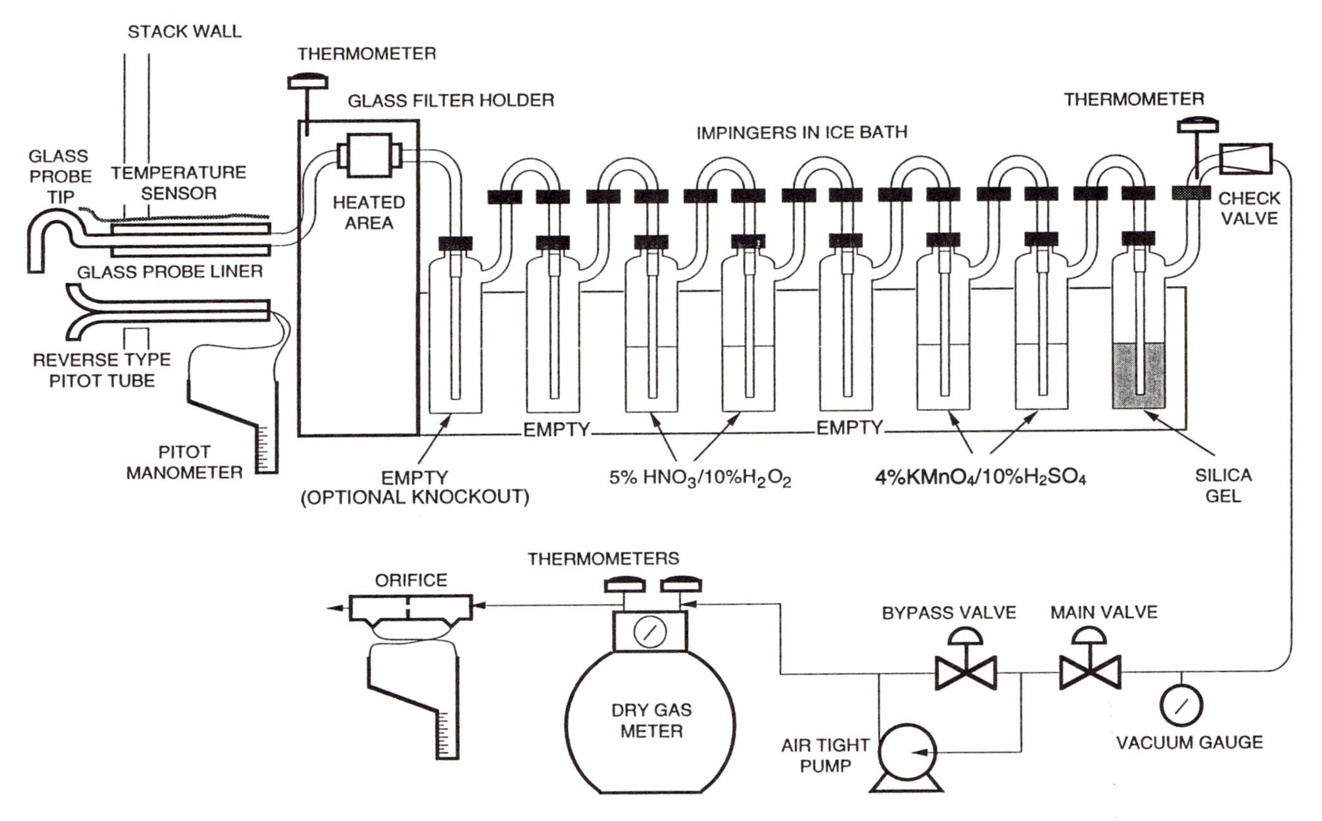

Figure 2. SW–846 Method 0012 multiple metal sampling train.

The heated (120 ± 14 °C) quartz filter is backed by a Teflon support and collects particulate matter, which is weighed and later analyzed for metals of interest. After the filter, the collected gas is drawn through several impingers in series that contain various chemical solutions. Up to eight impingers can be present, including empty knockout traps and the desiccant impinger for removing moisture prior to the pumping–metering system. The first two impingers contain 5% nitric acid (HNO_3) and 10% hydrogen peroxide (H_2O_2) solutions, which will oxidize and capture most metals. The last two impingers contain 4% potassium permanganate ($KMnO_4$) and 10% sulfuric acid (H_2SO_4) solution, which is directed toward capture of mercury.

After sampling, the glass probe and nozzle are rinsed with a nonmetallic brush and 0.1 N nitric acid for recovery of any attached particulate material. This rinsate is combined with the impinger solutions for analysis. The filter is extracted and analyzed separately.

These solutions may be analyzed by an array of metals methods given in SW–846, such as inductively coupled argon plasma emission spectroscopy (ICAP), atomic absorption (AA), and graphite furnace atomic absorption (GFAA). This choice is made depending upon the matrix of the sample and the sensitivity required.

Hexavalent Chromium

Another important Modified Method 5 procedure is used to sample for total and hexavalent chromium. Hexavalent chromium is of special interest because of its very high toxicity; the other forms of chromium, metallic and trivalent chromium, are much less toxic. These multiple oxidation states present an unusual sampling problem in that the metal must not change its oxidation state during the sampling, handling, and analytical process. This requirement is made more difficult by the fact that hexavalent chromium is very readily reduced to trivalent chromium by contact with most organic material and many other metals. Therefore a very special method has been developed by EPA and is published in a set of procedures describing the testing of boilers and industrial furnaces that are used to burn hazardous materials (8) (Table IV).

The method is, of course, isokinetic, but it seeks to stabilize the hexavalent chromium ions immediately after the sample enters the nozzle. Results indicated that, using the conventional Method 5 sampling method, some of the hexavalent chromium was being converted to trivalent chromium in the probe and on the filter. Chromium will remain in the hexavalent state if it is dissolved in a basic solution (e.g., NaOH). Thus, the first two impingers are filled with a strongly basic solution, but what is unusual is that some of that solution is pumped into the probe immediately behind the nozzle so that the gas–particle sample impinges into the basic solution as soon as possible (Figure 3). The solution in the probe is then sucked back into the first impinger and recirculated. Of course, there is no filter in the system. Because hexavalent chromium is so reactive, the entire train through the second

Table IV. Road Map of EPA Sampling Method Numbers

Method Number	*Source in Regulations*	*Designed Application*	*Comments*
EPA methods 1–29	40 CFR 60, Appendix A	NSPS	Originally intended as source methods for criteria pollutants; some additional methods for noncriteria pollutants have been added over the years.
EPA methods 101A, 104–107	40 CFR 61	NESHAPS methods	Based on risk analysis.Very difficult to develop. There will be no more.
EPA 200 series	40 CFR 51, Appendix M. Not attached to any regulations, they stand as guidelines. Source-specific SIP rules in 40 CFR 52. Mostly for criteria pollutants and CAA requirements	Methods to be required by SIPs under CAA Title V	201, 201A: PM_{10} 202: condensed PM 203: continuous opacity monitor 203A–C: visible emissions including fugitives 204A–F: VOC capture efficiency 205: dilution calibration
EPA 300 series	40 CFR 63, MACT standards	Methods to measure HAPs on sources requiring MACT under CAA Title III	Mostly under development, but many expected soon. If no method, use 301. 301: validation protocol 302: generic GC–MS (coming) 303: coke oven visible emissions 304: wastewater organics 305: HON wastewater VOCs 306: hex chromium (electroplating) 307: VOCs from degreasers
No numbers yet	40 CFR 64, Appendixes A–E	Enhanced monitoring for Title VII compliance. QA and QC information	Will include new generation of monitoring, intermittent monitoring, record keeping, and reporting. Each facility monitoring plan will be proposed by facility in Title V permit.
Method numbers 00XX and XXXX	Office of Solid Waste, 40 CFR 266. Generally not published in FR. Final BIF rules will appear here.	Test and analytical methods for solid and hazardous waste regulations	00XX (e.g., 0010) sampling methods in SW–846. XXXX (e.g., 8270) analytical methods in SW–846. Proposed BIF rules in FR 7/17/91.
EPA TO-XX	Compendium methods published in separate volume	Ambient air sampling methods with analytical methods	Sometimes used in source work (e.g., TO-14 for VOCs by GC). Sampling methods may not be appropriate for sources. Agency should approve before use as source method.
	40 CFR 60, General Provisions	Big picture of environmental source sampling	Information on CEMs, other useful tidbits; a road map to other parts of 40 CFR.

NOTE: CFR is *Code of Federal Regulations*; CAA is Clean Air Act; HAPs are hazardous air pollutants; QA is quality assurance; QC is quality control; TO is toxic organic; VOC is volatile organic compound; BIFs are boilers and industrial furnaces; FR is *Federal Register;* NSPS is new source performance standards; NESHAPS is national emission standards for hazardous air pollutants; SIP is state implementation plan; MACT is maximum achievable control technology; HON is hazardous organic NESHAPS; and CEMs are continuous emission monitors.

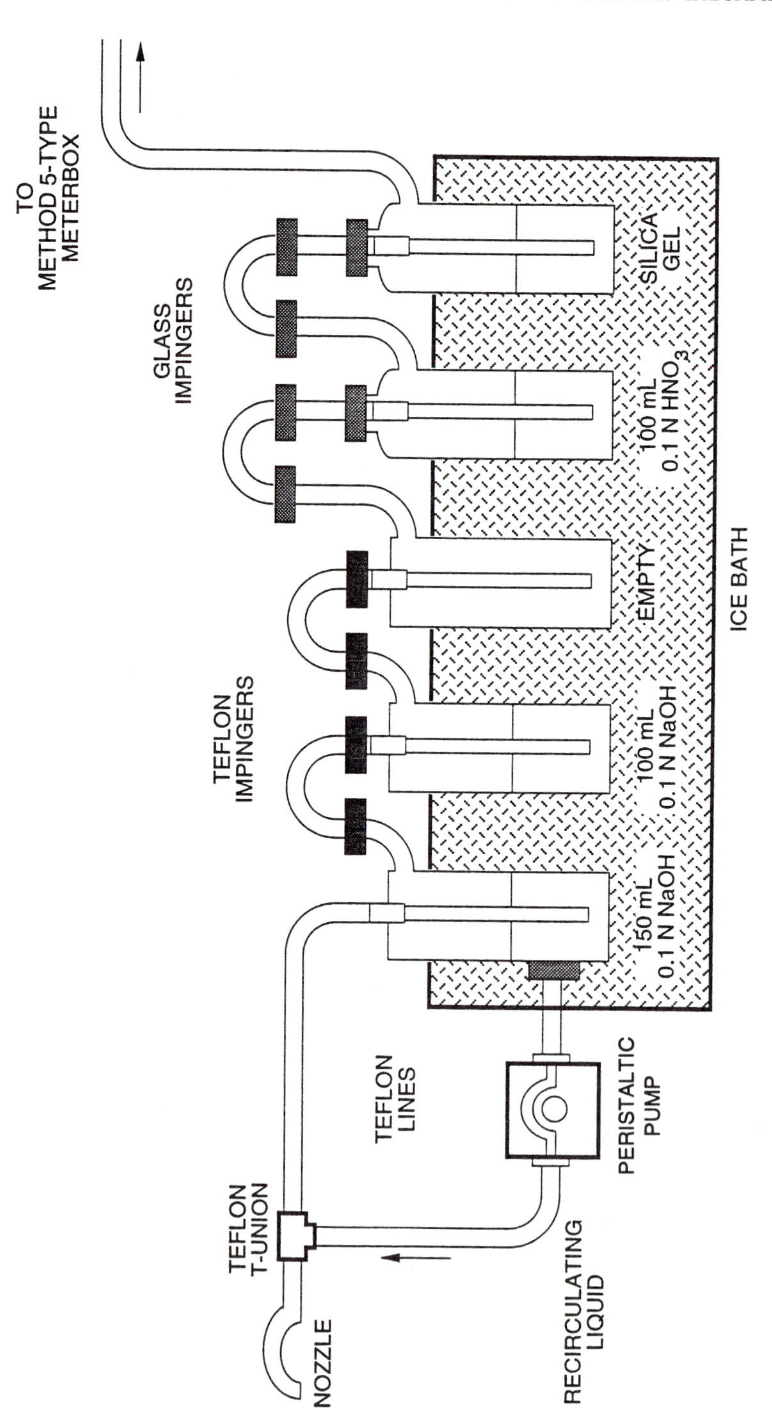

Figure 3. Hexavalent chromium sampling train.

impinger, to where all the chromium is removed, must be made of only Teflon or glass. The method also quantitatively collects chromium in its other oxidation states as well, and the sample may also be analyzed for total chromium using conventional analytical methods. The total particulate mass cannot be determined by this method.

One of the criteria pollutants is particulate material <10 µm in diameter, referred to as PM_{10}. For testing in a source, PM_{10} sampling is described in EPA Method 201A–CSR (constant sampling rate) (Table V). The modification to Method 5 for PM_{10} sampling consists of inserting a cyclone immediately behind the nozzle and operating the sampling train at a constant rate such that the cut point of the cyclone is 10 µm. As with Method 17, the cyclone must be allowed to come to stack temperature; if it is not fully in the stack, it must be heated to stack temperature. Because the sampling rate cannot be varied (i.e., the cut point of the cyclone would change), the method allows for ±20% deviation from isokinetic sampling; if the variation is greater, the method cannot be used (*see* Method 201). Practical problems with the method include the need for many closely spaced nozzles, a source with a nearly constant flow regime, and the need for a 6-in. port for certain sized nozzle–cyclone combinations.

To overcome the Method 201A problem of not being able to vary the flow through the cyclone, and therefore through the nozzle, EPA developed a somewhat complicated variation of Method 201A wherein a portion of the cleaned, sampled gas is recirculated to the sampling stream just behind the nozzle. By varying this recirculation rate and keeping the overall pump rate constant, the flow rate through the nozzle can be varied while keeping the flow through the cyclone constant. The procedure, Method 201–EGR (exhaust gas recirculation), requires special equipment and should only be run by experienced operators.

Of concern to many agencies is PM_{10} that may not be collected using Methods 201 or 201A. Some substances would pass wholly or in part through a filter at 250 °F, such as some medium-weight organic chemicals or semivolatile inorganic compounds (e.g., sulfuric acid). EPA has developed Method 202, which involves measuring the material that is captured in the chilled water impinger behind the filter. This method simply dries and heats (to 250 °F) the impinger catch and weighs the residue. This fraction is termed the condensable particulate material (CPM), and some agencies may require that the CPM be added to the Method 201/201A catch to define total PM_{10}. The method is not suitable if substances are present in the sampled gases that would not normally condense when injected into the atmosphere but would be captured by reactions in the aqueous impinger [e.g., ammonia, which will form salts such as $(NH_3)_2SO_4$, or hydrides].

Sampling methods have also been developed that separate particulate material into a series of size-fractionated catches to estimate the size distribution of the particulate material in the source. Most of these methods use a device that forces the sampled particle–gas stream to turn more and more sharply above a substrate surface such that particles of larger aerodynamic size cannot negotiate the turn with the gas and impact onto the substrate surface. The devices are called impactors, and the

Table V. Summary of EPA Emission Test Methods and Modifications

Method[a]	*Reference*		*Description*
	FR No.	*Date*	
			Part 60 Appendix A
1–8	42 FR 41754	08/18/77	Velocity, Orsat, PM, SO_2, NO_x, etc.
	43 FR 11984	03/23/78	Corrections and amendments to Methods 1–8
1/24	52 FR 34639	09/14/87	Technical corrections
	52 FR 42061	11/02/87	Corrections
2–25	55 FR 47471	11/14/90	Technical amendments
1	48 FR 45034	09/30/83	Reduction of number of traverse points
1	51 FR 20286	06/04/86	Alternative procedure for site selection
1A	54 FR 12621	03/28/89	Traverse points in small ducts
2A	48 FR 37592	08/18/83	Flow rate in small ducts (volume meters)
2B	48 FR 37594	08/18/83	Flow rate (stoichiometry)
2C	54 FR 12621	03/28/89	Flow rate in small ducts (standard pitot)
2D	54 FR 12621	03/28/89	Flow rate in small ducts (ratemeters)
2E (P)	56 FR 24468	05/30/91	Flow rate from landfill wells
2F	Tentative		3-D pitot for velocity
3	55 FR 05211	02/14/90	Molecular weight
3/3B	55 FR 18876	05/07/90	Method 3B applicability
3AA	51 FR 21164	06/11/86	Instrumental method for O_2 and CO_2
3B	55 FR 05211	02/14/90	Orsat for correction factors and excess air
3C (P)	56 FR 24468	05/30/91	Gas composition from landfill gases
3	48 FR 49458	10/25/83	Addition of QA–QC
4	48 FR 55670	12/14/83	Addition of QA–QC
5	48 FR 55670	12/14/83	Addition of QA–QC
5	45 FR 66752	10/07/80	Filter specification change
5	48 FR 39010	08/26/83	DGM revision
5	50 FR 01164	01/09/85	Incorporates DGM and probe cal procedures
5	52 FR 09657	03/26/87	Use of critical orifices as cal standard
5	52 FR 22888	06/16/87	Corrections
5A	47 FR 34137	08/06/82	PM from asphalt roofing (Proposed as Method 26)
5A	51 FR 32454	09/12/86	Addition of QA–QC
5B	51 FR 42839	11/26/86	Non-sulfuric acid PM
5C	Tentative		PM from small ducts
5D	49 FR 43847	10/31/84	PM from baghouses
5D	51 FR 32454	09/12/86	Addition of QA–QC
5E	50 FR 07701	02/25/85	PM from fiberglass plants
5F	51 FR 42839	11/26/86	PM from FCCU
5F	53 FR 29681	08/08/88	Barium titration procedure
5G	53 FR 05860	02/26/88	PM from woodstove (dilution tunnel)
5H	53 FR 05860	02/26/88	PM from woodstove (stack)
6	49 FR 26522	06/27/84	Addition of QA–QC
6	48 FR 39010	08/26/83	DGM revision
6	52 FR 41423	10/28/87	Use of critical orifices for FR/vol measure
6A	47 FR 54073	12/01/82	SO_2/CO_2
6B	47 FR 54073	12/01/82	Auto SO_2/CO_2
6A/B	49 FR 09684	03/14/84	Incorporates cal test changes
6A/B	51 FR 32454	09/12/86	Addition of QA–QC
6C	51 FR 21164	06/11/86	Instrumental method for SO_2
6C	52 FR 18797	05/27/87	Corrections
7	49 FR 26522	06/27/84	Addition of QA–QC

Table V. Summary of EPA Emission Test Methods and Modifications—*Continued*

Method[a]	*Reference*		*Description*
	FR No.	*Date*	
7A	48 FR 55072	12/08/83	Ion chromatograph NO_x analysis
7A	53 FR 20139	06/02/88	ANPRM
7A	55 FR 21752	05/29/90	Revisions
7B	50 FR 15893	04/23/85	UV NO_x analysis for nitric acid plants
7A/B	Tentative		High SO_2 interference
7C	49 FR 38232	09/27/84	Alkaline permanganate/colorimetric for NO_x
7D	49 FR 38232	09/27/84	Alkaline permanganate/IC for NO_x
7E	51 FR 21164	06/11/86	Instrumental method for NO_x
8	36 FR 24876	12/23/71	Sulfuric acid mist and SO_2
8	42 FR 41754	08/18/77	Addition of particulate and moisture
8	43 FR 11984	03/23/78	Miscellaneous corrections
9	39 FR 39872	11/12/74	Opacity
9A	46 FR 53144	10/28/81	Lidar opacity; called Alternative 1
10	39 FR 09319	03/08/78	CO
10	53 FR 41333	10/21/88	Alternative trap
10A	52 FR 30674	08/17/87	Colorimetric method for PS-4
10A	52 FR 33316	09/02/87	Correction notice
10B	53 FR 41333	10/21/88	GC method for PS-4
11	43 FR 01494	01/10/78	H_2S
12	47 FR 16564	04/16/82	Pb
12	49 FR 33842	08/24/84	Incorporates method of additions
13A	45 FR 41852	06/20/80	F (colorimetric method)
13B	45 FR 41852	06/20/80	F (SIE method)
13A/B	45 FR 85016	12/24/80	Corrections to Methods 13A and 13B
14	45 FR 44202	06/30/80	F from roof monitors
15	43 FR 10866	03/15/78	TRS from petroleum refineries
15	54 FR 46236	11/02/89	Revisions
15	54 FR 51550	12/15/89	Correction notice
15A	52 FR 20391	06/01/87	TRS alternative/oxidation
16	43 FR 07568	02/23/78	TRS from kraft pulp mills
16	43 FR 34784	08/87/78	Amendment to Method 16, H_2S loss after filters
16	44 FR 02578	01/12/79	Amendment to Method 16, SO_2 scrubber added
16	54 FR 46236	11/02/89	Revisions
16	55 FR 21752	05/29/90	Correction of figure (±10%)
16A	50 FR 09578	03/08/85	TRS alternative
16A	52 FR 36408	09/29/87	Cylinder gas analysis alternative method
16B	52 FR 36408	09/29/87	TRS alternative/GC analysis of SO_2
16A/B	53 FR 02914	02/02/88	Correction 16A/B
17	43 FR 07568	02/23/78	PM, in-stack
18	48 FR 48344	10/18/83	VOC, general GC method
18	49 FR 22608	05/30/84	Corrections to Method 18
18	52 FR 51105	02/19/87	Revisions to improve method
18	52 FR 10852	04/03/87	Corrections
18	57 FR 62608	12/31/92	Revisions to improve QA–QC
19	44 FR 33580	06/11/79	F-factor, coal sampling
19	52 FR 47826	12/16/87	Method 19A incorporated into Method 19
19	48 FR 49460	10/25/83	Corrections to F factor equations and F_c value
20	44 FR 52792	09/10/79	NO_x from gas turbines

Continued on next page

Table V. Summary of EPA Emission Test Methods and Modifications—*Continued*

Method[a]	*Reference*		*Description*
	FR No.	*Date*	
20	47 FR 30480	07/14/82	Corrections and amendments
20	51 FR 32454	09/12/86	Clarifications
21	48 FR 37598	08/18/83	VOC leaks
21	49 FR 56580	12/22/83	Corrections to Method 21
21	55 FR 25602	06/22/90	Clarifying revisions
22	47 FR 34137	08/06/82	Fugitive volatile emissions
22	48 FR 48360	10/18/83	Add smoke emission from flares
23	56 FR 5758	02/13/91	Dioxin/dibenzofuran
24	45 FR 65956	10/03/80	Solvent in surface coatings
24A	47 FR 50644	11/08/82	Solvent in ink (proposed as Method 29)
24	Tentative		Solvent in water-borne coatings
24	57 FR 30654	07/10/92	Multicomponent coatings
24	Tentative		Radiation-cured coatings
25	45 FR 65956	10/03/80	TGNMO
25	53 FR 04140	02/12/88	Revisions to improve method
25	53 FR 11590	04/07/88	Correction notice
25A	48 FR 37595	08/18/83	TOC/FID
25B	48 FR 37597	08/18/83	TOC/NDIR
25C (P)	56 FR 24468	05/30/91	VOCs from landfills
25D (P)	56 FR 33544	07/22/91	VOCs from TSDF (purge procedure)
25E (P)	56 FR 33555	07/22/91	VOCs from TSDF (vapor pressure procedure)
26	56 FR 5758	02/13/91	HCl
26	57 FR 24550	06/10/92	Corrections to Method 26
26 (P)	57 FR 62608	12/31/92	Expand Method 26 to HCl, halogens, and other hydrogen halides
26A (P)	57 FR 62608	12/31/92	Isokinetic HCl, halogens, and other hydrogen halides method
27	48 FR 37597	08/18/83	Tank truck leaks
28	53 FR 05860	02/26/88	Woodstove certification
28A	53 FR 05860	02/26/88	Air-to-fuel ratio
29	Tentative		Multiple metals
Part 60 Appendix B			
PS-1	48 FR 13322	03/30/83	Opacity
PS-1	Tentative		Revisions
PS-1	48 FR 23608	05/25/83	SO_2 and NO_x
PS-1–5	55 FR 47471	11/14/91	Technical amendments
PS-3	48 FR 23608	05/25/83	CO_2 and O_2
PS-4	50 FR 31700	08/05/85	CO
PS-4A	56 FR 5526	02/11/91	CO for MWC
PS-5	48 FR 32984	07/20/83	TRS
PS-6	53 FR 07514	03/09/88	Velocity and mass emission rate
PS-7	55 FR 40171	10/02/90	H_2S
Part 60 Appendix F			
Prc 1	52 FR 21003	06/04/87	Quality assurance for CEMs
Prc 1 (P)	54 FR 52207	12/20/89	Revision

Table V. Summary of EPA Emission Test Methods and Modifications—*Continued*

Method[a]	Reference FR No.	Reference Date	Description
			Part 60 Appendix J
App-J	55 FR 33925	08/20/90	Woodstove thermal efficiency
	48 FR 44700[b]	09/29/83	S-factor method for sulfuric acid plants
	48 FR 48669[b]	10/20/83	Corrections to S-factor publication
	49 FR 30672[b]	07/31/84	Add fuel analysis procedures for gas turbines
	51 FR 21762[b]	06/16/86	Alternative method for low-level concentrations
	54 FR 46234[b]	11/02/89	Miscellaneous revisions to Appendix A, 40 CFR Part 60
	55 FR 40171[b]	10/02/90	Monitoring revisions to Subpart J
			Part 60
	54 FR 06660	02/14/89	Test methods and procedures review (40 CFR 60)
	54 FR 21344	05/17/89	Correction notice
	54 FR 27015	06/27/89	Correction notice
			Part 61 Appendix B
101	47 FR 24703	06/08/82	Hg in air streams
101A	47 FR 24703	06/08/82	Hg in sewage sludge incinerators
101	49 FR 35768	09/12/84	Corrections to Methods 101 and 101A
102	47 FR 24703	06/08/82	Hg in H_2 streams
103	48 FR 55266	12/09/83	Revised Be screening method
104	48 FR 55268	12/09/83	Revised Be method
105	40 FR 48299	10/14/75	Hg in sewage sludge
105	49 FR 35768	09/12/84	Revised Hg in sewage sludge
106	47 FR 39168	09/07/82	Vinyl chloride
107	47 FR 39168	09/07/82	Vinyl chloride in process streams
107	52 FR 20397	06/01/87	Alternative calibration procedure
107A	47 FR 39485	09/08/82	Vinyl chloride in process streams
108	51 FR 28035	08/04/86	Inorganic arsenic
108A	51 FR 28035	08/04/86	Arsenic in ore samples
108B	55 FR 22026	05/31/90	Arsenic in ore alternative
108C	55 FR 22026	05/31/90	Arsenic in ore alternative
108B/C	55 FR 32913	08/31/90	Correction notice
111	50 FR 05197	02/06/85	Polonium-210
114 (P)	54 FR 09612	03/07/89	Monitoring of radionuclides
115 (P)	54 FR 09612	03/07/89	Radon-222
			Part 61
	53 FR 36972	09/23/88	Corrections
			Part 51 Appendix M
201	55 FR 14246	04/17/90	PM-10 (EGR procedure)
201A	55 FR 14246	04/17/90	PM-10 (CSR procedure)
201/A	55 FR 24687	06/18/90	Correction of equations
201	55 FR 37606	09/12/90	Correction of equations
202	56 FR 65433	12/17/91	Condensable PM
203 (P)	57 FR 46114	10/07/92	Transmissometer for compliance
203A (P)	58 FR 61640	11/22/93	Visible emissions (2–6-min avg)
203B	58 FR 61640	11/22/93	Visible emissions (time exception)
203C	58 FR 61640	11/22/93	Visible emissions (instantaneous)
204	Tentative		VOC capture efficiency

Continued on next page

Table V. Summary of EPA Emission Test Methods and Modifications—*Continued*

Method[a]	Reference		Description
	FR No.	*Date*	
204A	Tentative		VOC capture efficiency
204B	Tentative		VOC capture efficiency
204C	Tentative		VOC capture efficiency
204D	Tentative		VOC capture efficiency
204E	Tentative		VOC capture efficiency
204F	Tentative		VOC capture efficiency
205 (P)	59 FR 39501	08/03/94	Dilution calibration verification
206	Tentative		NH_3
			Part 63 Appendix A
301	57 FR 61970	12/29/92	Field data validation protocol
302	Tentative		Generic GC–MS procedure
303 (P)	57 FR 57534	12/04/92	Coke oven door emissions
304 (P)	57 FR 62608	12/31/92	Biodegradation rate
305 (P)	57 FR 62608	12/31/92	Compound specific liquid waste
306 (P)	58 FR 65768	12/16/93	Hexavalent chromium
306A (P)	58 FR 65768	12/16/93	Simplified chromium sampling
306B (P)	58 FR 65768	12/16/93	Surface tension of chromium suppressors
307 (P)	58 FR 65793	11/29/93	Solvent degreaser VOC
308 (P)	58 FR 66079	12/17/93	Methanol
309 (P)		06/06/94	Aerospace solvent recovery material balance
310	Tentative		Residual hexane in EPDM rubber
311 (P)	59 FR 62652	12/06/94	VOC HAPs for furniture industry
312	Tentative		Residual styrene in SBR rubber
313	Tentative		Residual styrene in SBR rubber

NOTES: FR is *Federal Register*. Tentative means under evaluation. DGM is dry gas meter; FCCU is fluid catalytic cracking unit; ANPRM is advance notice of proposed rule making; SIE is specific ion electrode; TRS is total reduced sulfur; F_c is volume of combustion components per unit of heat; TGNMO is total gaseous non-methane organic chemicals; TOC is total organic compound; TSDF is toxics storage and disposal facility; MWC is municipal waste combustor; EGR is exhaust gas recirculation; CSR is constant sampling rate; EPDM is ethylene–polypropylene–diene monomer; SBR is styrene–butadiene rubber; NDIR is nondispersive infrared spectrometry; VOC is volatile organic compound; and cal is calibration.

[a]Method numbers followed by (P) are proposals.
[b]Alternative and miscellaneous procedures.

SOURCE: Reproduced from reference 48.

substrate surface may be filter material, a thinly oiled film, or other suitable surfaces. Commercial devices separate the particulate material into 8 or 10 factions, each is weighed, and computer programs are used to "smooth" the particle size distribution curve. This type of sampling uses Method 5-type apparatus, but as with many other Method 5 modifications, it requires specialized, expensive equipment and experienced operators. As an alternative to impactors, cyclone stacks (up to five) are available that separate the particulate material into several fractions. The cut points of cyclones are not as sharp as are the cut points of impactor stages, but cyclones are useful if larger amounts of material are needed in each size range for analysis, because impactors are capable of collecting only submilligram quantities of particulate material in each stage.

Gas-Phase Sampling

The sampling of gases is much simpler than sampling for particulate matter. Because vapor samples are usually well mixed and no biases are introduced when a gas is pulled into a nozzle, sampling often consists of little more than pulling a slip stream of the gas through a tube (probe). Isokinetic sampling is normally not necessary, and a special nozzle directed into the stream flow is not required. However, several situations do require some special attention. If the gas stream contains particulate material of no interest, it should be filtered. If the gas stream is hot, allowances must be made for any condensing species. The most frequent condensate is water; most vapor collection devices cannot tolerate water. Therefore, if water vapor or droplets are present, EPA's volatile organic sampling train (VOST) should be used.

Volatile Organic Sampling Train

This method is described in SW–846, Method 0030 (Figure 4). In the procedure, a sample of the gas is obtained nonisokinetically, filtered (usually by a glass wool plug in the nozzle), kept above the condensation temperature of water, and delivered to a cooling coil condenser positioned above an organic sorbent trap filled with Tenax. The gas is cooled in the condenser, and the remaining vapor and condensed water are passed down through the Tenax trap where the organic constituents are quantitatively captured. The water collects in a vessel below the first Tenax trap, and the dried gas passes through a second trap filled with Tenax and charcoal to ensure capture of all volatile organic compounds. The volume of gas and condensed water are measured so that the quantity of gas sampled may be accurately determined.

The collected sorbent tubes are analyzed via thermal desorption and GC–MS. Care must be taken to prevent saturation of the analytical system. To avoid the problem, the analysis of screening samples or the use of distributive sampling is recommended. Distributive sampling refers to the use of several sample volumes that bracket the most probable one. In this way, an unknown concentration will more likely be appropriately captured.

Sampling Method for Volatile Organic Compounds

The proposed SW–846 Method 0031, sampling method for volatile organic compounds (SMVOC), is based on the VOST method but is designed to extend the applicability of the VOST-type sorbent approach to compounds having boiling points from −15 °C to 121 °C. [Method 0031 is available on the EPA on-line Technology Transfer Network (TTN)]. The primary change from the VOST approach is the configuration of the sorbent traps. For SMVOC, there will be three traps, two containing Tenax and the third containing Anasorb-747. This method is not suitable for polar or reactive species.

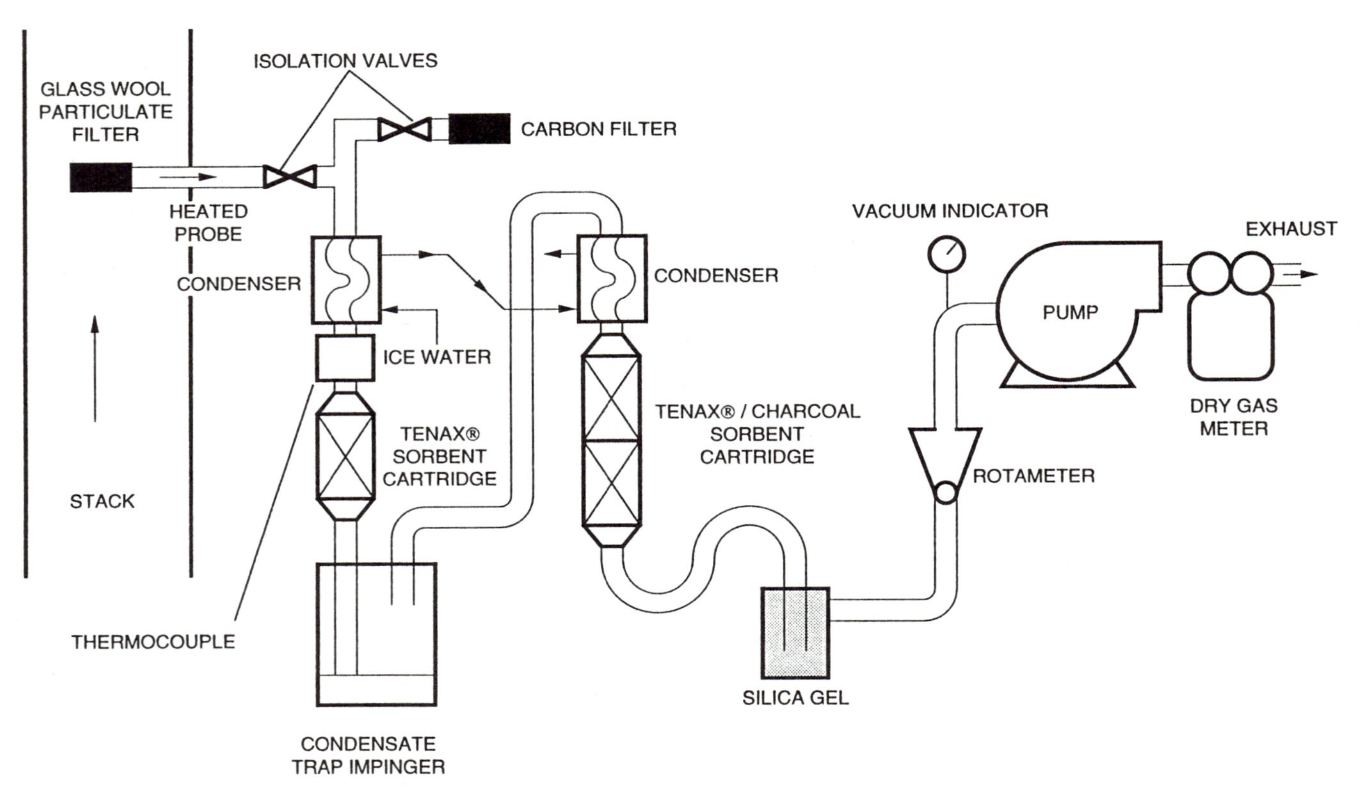

Figure 4. SW–846 Method 0030 volatile organic sampling train (VOST).

Formaldehyde

Another sampling method that is used for several specific compounds involves the technique of reactive capture. These compounds are usually unstable or easily lost or converted. With this approach, the target pollutant is passed through a medium or bubbled through a solution, and a specific quantitative reaction occurs. A derivative compound is formed that stays in the matrix and is easily measured, the quantity of which is directly related to the amount of the original pollutant. Reactive capture methods are used for ethylene oxide, formaldehyde and other aldehydes, and other compounds. We shall look more closely at the important compound formaldehyde as illustrative of reactive capture sampling.

The exact protocol is given in SW–846, Method 0011 (Figure 5). In this method, the gas containing formaldehyde is bubbled through an aqueous solution containing 2,4-dinitrophenylhydrazine (DNPH); this compound reacts selectively with the formaldehyde to form a readily soluble, stable derivative, hydrazone. If the sampled gas contained water vapor, it condenses into the impinger; particulate material will also be captured, but the reaction and derivative product are so specific to formaldehyde that the other substances create no interference. As with other methods, the volume of gas and amount of condensate are measured. The solution is then sent to a laboratory where it is analyzed by HPLC for the formaldehyde-derived hydrazone.

This methodology has been validated at many diverse sources for formaldehyde and acetaldehyde, and a modification using field extraction to enhance the stability of the acrolein–hydrazone derivative has been proposed with some limited validation data (9). Other attempts to extend the analyte list to higher aldehydes and ketones have had only limited success; their limited success suggests that either the hindered access to the carbonyl moiety by the bulky DNPH molecule restricts the reaction, or the kinetics of the derivatization reaction are too slow for the short period of time during which the gas being sampled traverses the length of the impinger and the target molecules are expected to diffuse into the bulk solution and react (*10, 11*). This question will require resolution as propionaldehyde is a Clean Air Act HAP and will need to be quantified for some sources.

Method 18

EPA Method 18 is used for the measurement of gaseous organic compounds by gas chromatography (6). This method can be considered a self-validating method because it requires method performance data for particular applications prior to full use. Direct on-line GC analysis is preferred, but this type of analysis is frequently impossible for reasons of safety or access. When direct analysis is performed, a dilution interface is often required to reduce the moisture percentage and reduce the pollutant concentration into the analytical range of the instrument. Alternatively, samples of the gas may be captured for later analysis. Several types of sampling media for this alternative are discussed, including glass sampling bulbs, evacuated

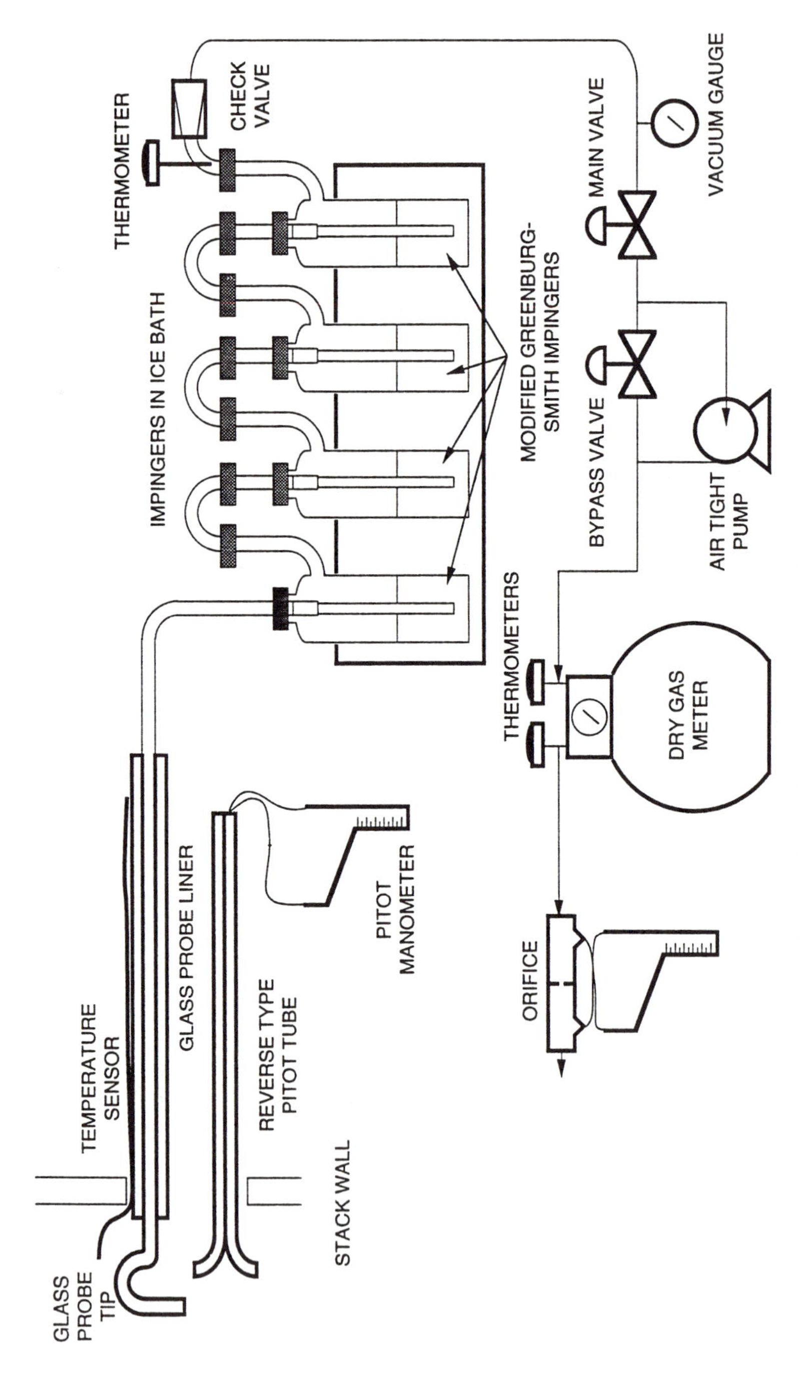

Figure 5. SW–846 Method 0011 formaldehyde sampling train.

stainless steel spheres [electropolished tanks, such as Summa-polished canisters are not mentioned specifically, but sometimes are used], and various sorbents. The Summa passivation process, consisting of a proprietary electropolishing technique (of Molectrics Corporation), is specified for use in several ambient air methods, including TO-3, TO-12, and TO-14. Recent clarification from the EPA on the use of Summa canisters states that Summa canisters cannot be used for Method 18 (*12*). The rationale for this prohibition is that the severe matrix in many sources (e.g., high moisture, acid gases, high temperature, or particulate matter) presents several potential problems to acceptable sample recovery (*13*). Some state or local jurisdictions may approve the use of canisters for specific situations, although this application is not specifically termed Method 18.

Method 18 is useful for situations in which a specific method has not been developed and the stack conditions are relatively mild. However, high temperatures, high moisture, or a corrosive matrix in a stack or vent may necessitate the use of other methods such as VOST.

Once the gas containing the pollutant of interest is cleaned of particulate matter and is cooled to ambient temperature, one of several sample media can be chosen. Most directly, a whole gas sample is captured in an inert container. Either Tedlar bags or evacuated Summa-polished, stainless steel canisters are used, although as mentioned previously, EPA will not accept canister data for compliance-related sampling. Tedlar bags are relatively cheap, are light weight, require no cleaning (they should not be reused), and are transparent so that the sample being collected may be observed. For example, if water condenses inside the bag, it may be necessary to use Method 0030 instead. However, because the bags are transparent, gases collected in them must be protected from light if they are photosensitive. Also, the gas must be drawn into the bag using some pumping apparatus, which sometimes can introduce the possibility of cross-contamination. Evacuated canisters, on the other hand, use the vacuum in the canister to draw the sample—no pump is required. Flow regulators are often used in conjunction with canisters to guarantee a uniform flow and at such a rate that the sampling will last for a desired period (a few minutes up to 24 h). In some cases, pumps are used with canisters, which will withstand pressures above atmospheric, if necessary. Canisters must be precleaned, but they are rugged and very easy to use. Both Tedlar bags and canisters have limitations on the types of gases they may be used for. For example, canisters should not be used for acid gases (e.g., HCl) for fear of corrosion or analyte loss, nor for any sulfur compounds (e.g., H_2S, mercaptans, and sulfides) because of irreversible loss on the metal surface.

Along with the choice of sampling media, a range of GC detectors can be used, as long as they are appropriate for the target species. The most frequent detectors in current use are the flame ionization, photoionization, and electron capture detectors. Although the MS detector should, in principle, function similarly to nonspecific detectors for detecting eluting species from the GC column, EPA is developing a MS-specific Method 18, which takes advantage of the additional information available from GC–MS (*14*).

Tedlar Bag Method

EPA has recently been developing an alternate method targeting Clean Air Act HAPs, many of which are polar or labile and not amenable to standard methods, including Method 18. The value of whole air samples versus samples from sorbents has been recognized, but stainless steel canisters are excluded because of the uncertainty in water condensation and degradation of the canister surface.

Method 0040, a sampling method for principal hazardous constituents from combustion sources using Tedlar bags, has only been recently proposed and open to public comment (*15*). This method is based on the approach by which the air stream is sampled through a heated probe into a Tedlar bag in an airtight chamber (*16*). The chamber, sometimes called a lung chamber sampler, is evacuated at a known rate, causing the stack gas to be pulled into the Tedlar bag at that same rate. Previous approaches had required a much higher sampling flow rate (1–1.5 L/min), whereas this approach allows a more manageable 50-mL/min flow rate. Various sizes of bags are available for sample collection.

The collected sample is analyzed according to appropriate analytical protocols. Because the Tedlar bag generally has a short hold time (24–72 h), it must be transported to the laboratory quickly for analysis, or analyzed on site.

Canister Sampling and New Sorbents

As may be inferred from previous discussion, there is a quiet debate on the usefulness and validity of volatile organic chemicals collected in Summa-polished canisters for certain types of stationary sources. Summa canisters have found favor in some quarters as an easy way to capture whole air samples of ambient air. Ambient Method TO-14 is the VOC method of choice for broad spectrum analyses for ambient air, which can be used for some relatively simple point and non-point sources. Some users are attempting to extend it to more "classical" sources, including combustion sources. However, there are serious doubts about the validity of some source data from samples collected in canisters. The key reasons are water vapor, particulate matter, and sorption or reactions with metal surfaces. A sample from a high-temperature and high-moisture source will undoubtedly condense once cooled. That condensate can be a sink for losses of VOCs or can accelerate other chemistry or surface reactions. Although this type of loss is not a problem for some sources, some regulators are reluctant to approve of canisters for certain types of source sampling. Further work is being done, mainly by users, to define the extent of canister use in traditional source testing situations. Today, they are used primarily for compounds and in situations in which satisfactory accuracy, precision, and recovery have been well demonstrated.

Besides whole air sampling techniques, a new generation in solid sorbents is being developed (*17*). The carbonized molecular sieve type of sorbent material, coupled with various other sorbent media, is expanding the range of volatilities that can be sampled, including use in moist air streams that have heretofore not been amenable to some sorbent types. For example, the alternative method to VOST (i.e., SMVOC) uses one of these sorbents, Anasorb-747.

Listing of EPA Regulatory Methods

The EPA has a range of reference methods that apply to particular regulatory scenarios. Table IV is a road map through these regulations [generally the *Code of Federal Regulations* (CFR)] and indicates where these methods have been published. Over the years, additions and modifications to these methods have been published, many of which are clarifications or changes in quality assurance procedures. Table V is a listing of those modifications as well as the original method references.

New Point Source Methods

Because no specific sampling or analytical methods exist for a large percentage of the CAA HAPs, the selection of "acceptable" sampling and analytical methods for the majority of these compounds could be tentatively made on the basis of chemical or physical similarity. However, three fundamental concerns must be satisfied prior to using a method to obtain critical compliance data (*18*):

1. Is the target analyte sampled quantitatively?
2. Does the laboratory sample preparation capture the collected analyte completely?
3. Is the analytical method capable of the required accuracy, precision, and detection limits for all the target analytes?

EPA's recognition of the timetable of the Clean Air Act enforcement provisions and its inability to develop and validate methods for all HAPs in that time frame led to the promulgation of a procedure for method validation. The result was EPA Method 301, which can be used to develop a procedure of equivalency to a Federal Reference Method for compounds that cannot be sampled under the standard methods (*19*). Method 301 states that the following procedures must be followed to assess bias and precision:

- Dynamic spiking of the target compounds into the field sampling equipment should be performed and carried through the entire sampling and analytical procedure for determination of bias.
- Multiple collocated samples must be collected for assessment of the method precision.
- Alternatively, a comparison with a previously established method may be made for assessment of bias.

Once the proposed method has gone through this procedure and results in an acceptable bias factor, the sampling and analytical data it produces can be used for compliance purposes. Two recent successful applications of this procedure were for polar organic species for which there were no appropriate sampling and analytical techniques (*20, 21*). EPA constantly monitors methods in use, and modifications such as those in Table V continue to be promulgated (*22, 23*).

New Developments in Source Monitoring

Evolution of Conventional Point Source Methods to Semicontinuous or Continuous Monitoring

Given the increasing demands on facilities to provide accurate data for emission inventories and the tough compliance requirements mandated by new regulations, the traditional point source measurement approaches are evolving from the usual one-point-in-time sampling into a new concept of semicontinuous or continuous monitoring. As contrasted with continuous monitoring, in which the instrumental response is essentially instantaneous, allowing continuous reporting of data, ceasing only for maintenance, semicontinuous monitoring in this context would be defined as a series of consecutive short-term monitoring events over an extended period of time. Semicontinuous monitoring approaches provide a much larger data set than conventional integrated methods allow that can be used for temporal profiling or other uses. The enhanced monitoring provisions (*24*) mandated by section 114 of the Clean Air Act are moving facilities toward a state of continuous compliance, and they provide impetus to have more data available instead of the periodic conventional snapshot source test. In addition, the high cost of collecting only a small snapshot of a facility's emissions and the frequent difficulties in obtaining the appropriate quality data are also leading to the examination of alternatives.

These pressures have encouraged developments in instrumental approaches that can provide continuous or nearly continuous data at a lower overall cost. Although the initial costs are invariably high, the cost per data point decreases substantially over the lifetime of the instrumentation. These developments parallel the movement by EPA toward continuous emission methods (CEMs) for criteria pollutants. Wet chemical methods exist for NO_x and other species, but these methods have for the most part been supplanted by CEMs. The field is also moving toward a more continuous measurement of the more difficult HAP organic species, which is providing the impetus for the development of newer technologies such as in-stack optical measurement methods and on-line GCs.

These techniques do not yet have the imprimatur of EPA approval, but as the methods are proven from experience or go through the Method 301 validation, they should gain acceptance. Most of these developments have focused on the volatile species. The semivolatile and nonvolatile species present greater challenges due to the need to sample isokinetically, the degree to which they are part of the particulate matter, and the relatively lower concentration compared to many VOCs. It is probable that speciated semivolatile measurements will, for the foreseeable future, continue to rely upon standard conventional methods such as Modified Method 5.

Physical Measurements

As with any point source method, any semicontinuous or continuous method must also measure both volumetric flow and concentration to determine emission rate (e.g., lb/h). The flow measurement requires several pieces of information: temperature, pressure, water concentration, and, depending on the measurement device, molecular weight of the gas stream.

Temperature is easily measured using thermocouples. Electronic pressure transducers must be used for continuous pressure readings. Moisture can be less easily measured using optical or electrode techniques. A recently developed instrument measures water using continuous automated mass balance (25). Flows with particulate matter can be monitored using ultrasonic means, and flows without particulate matter can be continuously monitored automatically using a flow tube calibrated for the size of the stack or vent to measure velocity and static pressures (Annubar Flow Measurement, Dieterich Standard, Boulder, CO).

EPA reference methods are starting to be automated, although the intent is to simplify the normal sampling. An automated method exists for Method 2 flow determinations (Auto-Probe 2000, United Sciences Inc., Gibsonia, PA) and for Modified Method 5 (Auto5 Automatic Stack Sampler, Graseby–Anderson, Smyrna, GA). Continuous particulate emission monitors have been around for some time (Triboflow 2604 and 2602, Auburn International, Danvers, MA; P-5A Particulate Monitor, Environmental Systems Corporation, Knoxville, TN) that may eventually replace the need for periodic Method 5 measurements, but none have gained EPA acceptance to date.

Chemical Measurements

Chemical species measurements are more difficult because of the diversity of the target species, the concentrations, and the current techniques available to measure them. A promising technique is open-path (e.g., across-stack) or extractive optical methods, primarily using Fourier transform infrared spectrometry (FTIR). UV spectroscopy is also used, but less frequently (26). One of the challenges to the FTIR approach is to overcome the interferences from ubiquitous atmospheric species, namely, water and carbon dioxide. Water and carbon dioxide strongly absorb IR at many transition bands of importance in the measurement of toxic compounds, forcing reliance on weaker bands. The net result is more difficult detection (i.e., higher detection limits) and identification (27).

The across-stack approach is limited by the short path length and the possible fouling of optical windows. The interferometer and detector must also be situated at an often inconvenient location, such as across the stack. Extractive sampling requires a sampling setup containing heated or conditioning lines to transport unmodified gases to an internal reflecting cell, called a white cell. This cell allows multiple internal reflections to obtain lower detection limits.

Real-Time Total PAH Monitor

Measurement of polycyclic aromatic hydrocarbon compounds (PAHs) is difficult because of the need to sample isokinetically and the relatively low concentration typical for this class of compounds. For the Method 5 approach, this low concentration requires a long sampling period of up to several hours, followed by a complicated extraction and analysis by GC–MS or HPLC. Therefore, collecting a greater number of data points for assessments of temporal processes is difficult.

An optical method exists for real-time speciation of PAHs, but the approach is not yet reduced to common practice (*28*). However, an instrument for semiquantitative real-time measurement of nonspeciated total PAH on fine airborne particles is currently available (PAS 1000e Stack Emission PAH Monitor, EcoChem Technologies, West Hills, CA). This PAH monitor operates by photoelectric ionization of the easily ionizable PAH molecules adsorbed on the surface of carbonaceous aerosols. Light at 185 nm selectively ionizes the PAH adsorbed on the aerosol surface. Gases and other particulate material remain un-ionized and do not interfere.

Although there are many PAH compounds, for a typical point source monitoring situation such as a combustion source, the PAH profile is relatively constant, which allows for a universal calibration. The detection limit is approximately 10 ng/m^3, which is sufficiently low for most applications.

This instrument has not yet gained widespread acceptance in the U.S. monitoring community because of its lack of speciation and its lack of official EPA recognition as a reference method. Several research groups are currently working to develop applications that should enhance its capability as an emerging point source monitoring method.

Ion Mobility Spectrometry

A promising technique for monitoring semivolatile compounds at low concentrations is the ion mobility spectrometer. This instrument operates on the principle of ionization of molecules by a ^{63}Ni source in a small chamber. The ions are then electrically injected into a drift region where they move under the influence of an electric field through a counterflow of clean buffer gas. This movement is dependent upon the molecules' mass, charge, and size, which control the time at which the molecules arrive at the Faraday plate detector. This drift time is characteristic of each type of molecule and is similar to a retention time in GC.

High-molecular-weight compounds, not typically amenable to rapid analysis, can be analyzed within seconds with this system. In addition, commercial point source monitors are available for acid gases such HCl and HF (from Environmental Technology Group, Atlanta, GA). Although this technology appears promising, several challenges exist, such as nonlinearity of response and interferences (*29*).

On-Line MS

The value of the unequivocal identification of organic molecules based on their mass spectra is widely recognized, given the popularity of GC–MS. However, GC–MS requires the coupling of two dimensions of separations: GC based on retention time, and MS based on the mass fragments. In some situations in which complex mixtures do not exist, however, the MS dimension is sufficient to provide adequate concentration data.

In this approach, the gas stream to be sampled is directed into the ionization chamber for direct ionization of the entire mixture (*30*). Software is used to deconvolute the resulting mix of mass–ion data. This sample is rapidly flushed out of the chamber, and a continuous stream of data is acquired. This principle has been used for years as a process monitoring technique, but additional sensitivity and selectivity are now needed. This approach continues to be developed further, although some commercial products are available.

On-Line GC

On-line GC is not new; process GC has existed for quite some time. However, the Title III monitoring requires lower detection limits for more compounds. Past instrumentation focused on higher concentrations. New sample introduction techniques to lower detection limits are the next wave of process GC instrument development, including more sensitive detectors and preconcentration techniques.

Parametric Monitoring

Parametric monitoring refers to the option of monitoring an alternate parameter as a surrogate for the target toxic species (*31*). For example, if it can be demonstrated that the concentration of carbon monoxide is proportional to the emission of a target toxic gas, then once that relationship is proven, only carbon monoxide need be monitored. The use of already established testing methods enhances the potential of this option. However, establishing the relationship between an easily measured parameter such as carbon monoxide and a difficult parameter such as a volatile organic combustion product has not been very successful in the past. Advances in this part of the science await developments in understanding the relationships between various parameters.

Non-Point Source Test Methods

Although the emissions from non-point sources were ignored or estimated for many years, the past ten years have witnessed the development of methods that were specifically designed for the special problems of non-point emissions. Chief among these methods is the EPA surface isolation flux chamber. Other methods are used less commonly but have their role in the armamentarium of available methods. Table VI presents a summary of non-point assessment techniques.

EPA Surface Isolation Flux Chamber

The surface isolation flux chamber is based on the principle of isolating a well-defined segment of the emitting surface under a sealed dome (chamber), flowing a known amount of clean air through the chamber, and sampling the chamber gases for the appropriate target compounds using standard techniques (*32, 33*). The most commonly employed EPA-approved chamber consists of a 16-in.-diameter stainless steel cylinder, capped with a Plexiglas dome. Ports for introduction of sweep air and for sampling are placed in the dome. A large port for eliminating excess sweep air is at the top of the chamber. Figure 6 shows the chamber and associated equipment.

The chamber operates by setting up an equilibrium of the emitting gases with the sweep air. At least five chamber-volume exchanges are required prior to collecting the sample. The volume of the chamber is approximately 30 L, and at 5 L/min, the equilibration time is about 30 min.

Table VI. Summary of Point and Non-Point Emission Rate Measurement Tools

Measurement Method	*Types of Applicable Sources*	*Limitations or Comments*
Point source methods (various)	All process vents	Requires measurable gas flow.
Surface isolation flux chamber	Active landfills	Small surface area measured.
	Inactive landfills	Small vents and surfaces can be measured.
	Surface impoundments	Equipment is floated on surface.
	Land treatment	Treatment cycle variabilities affect data collection design.
Headspace	Similar to flux chamber	Used for concentration screening measurements, mainly for relative comparisons.
Concentration profile technique	Surface impoundments, land treatment	Requires complex equipment; meteorological conditions must meet criteria; not suitable for small surface areas.
Transect technique	Landfills, surface impoundments, land treatment, irregular area sources unsuitable for flux chamber	Meteorological conditions must meet criteria; substantial effects from upwind sources
Tracers	Any liquid process	Sample- and labor-intensive.
Air monitoring/ modeling technologies	Most non-point sources	Meteorological data requirements often limiting factor; costly for both components.

SOURCE: Reproduced from reference 49.

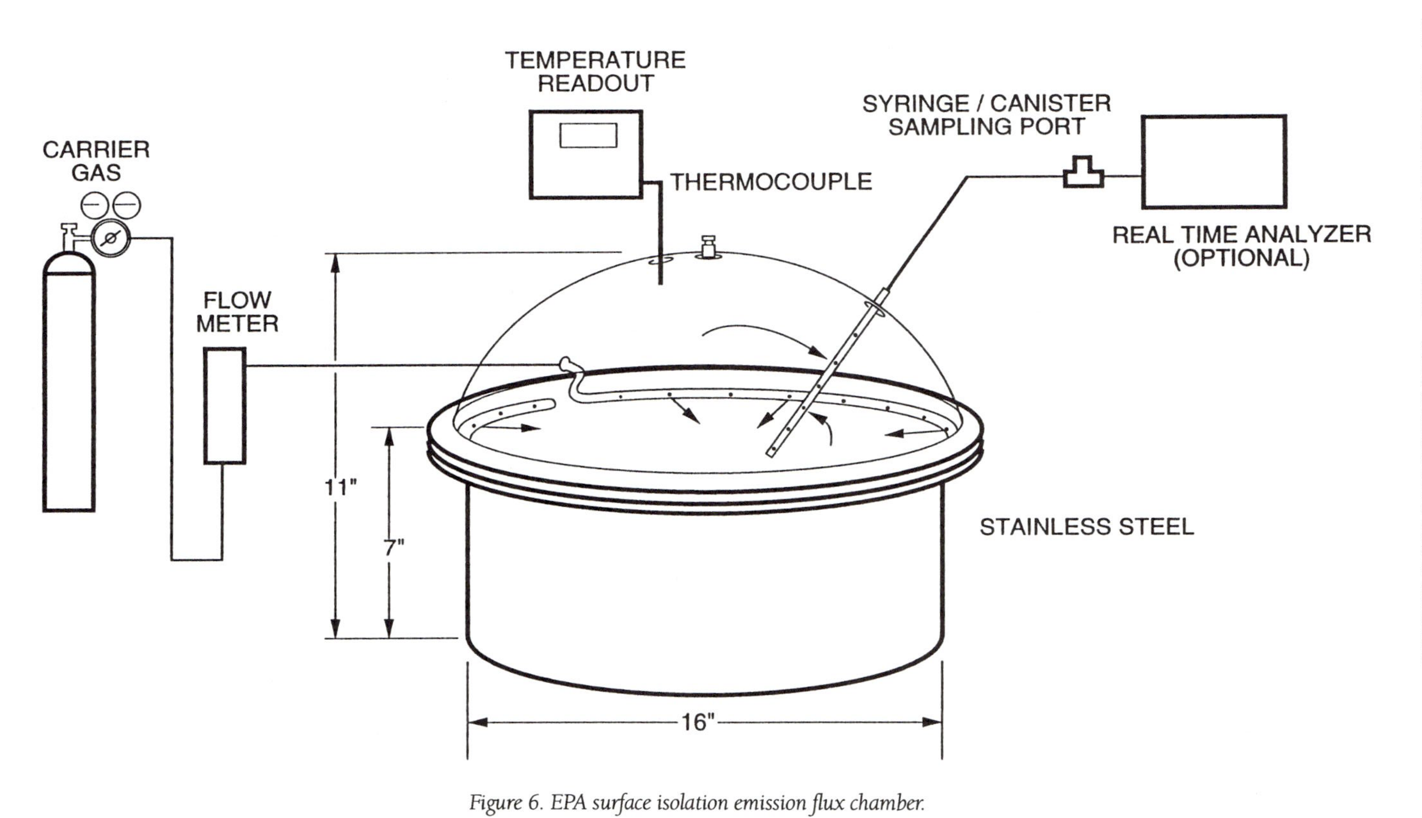

Figure 6. EPA surface isolation emission flux chamber.

The air in the chamber is sampled using a variety of methods, which are based mainly on the EPA Compendium Methods for ambient air (34). The ambient air methods are appropriate for this type of sampling instead of stationary source methods because of the relatively mild conditions that are similar to ambient conditions (e.g., no water condensation). The temperatures and humidity are usually close to ambient, although sampling on liquids, such as lagoons or aeration ponds, can produce high-moisture samples. On some soil surfaces, higher humidities would also be expected. Higher temperature surfaces (e.g., 80–90 °C) such as soot decanters from acetylene production or freshly laid asphalt can also be sampled as long as the appropriate sampling and analytical methods are used (35).

Table VII shows a list and description of the ambient methods that can be used for sampling from flux chambers. Some of the methods, such as TO-13, require modification because the maximum sampling rate that can be used without altering the flux conditions in the chamber is 2.5 L/min (36).

As in any emission determination, the flow and concentration parameters are required. The flow factor comes from the sweep air (5 L/min), and the concentration comes from the sampling data. This product is divided by the surface area of the chamber to yield an emission rate of mass per unit surface area per unit time (generally, $\mu g\ m^{-2}\ min^{-1}$).

The flux chamber is the easiest area source measurement technique, but it is limited to only small surfaces for each sample, although a multipoint design can be used to representatively sample larger areas. In addition, the flux chamber also has the potential to perturb the surface emission dynamics. For these reasons, alternative methods are available and should be considered in some situations.

Concentration Profile

Emissions measurements from larger scale surfaces can be accomplished using the concentration profile method (37). Figure 7 shows the experimental setup, in which sample collection points are set up at logarithmically spaced heights at a downwind location along the plume centerline. These sampling points are arrayed on a sampling mast, at the top of which is the wind direction sensor. Several parameters are measured at each height, including wind velocity, temperature, relative humidity, and volatile species concentration. The emission rate is obtained as the ratio of various parameters, but it depends upon the relative flux of water vapor or another tracer species to target chemical species. Because of this reliance on the estimated water vapor flux, this method is valid only in stable atmospheric conditions.

Transect Sampling

The principle behind the transect sampling is to capture the cross section of an emitted plume from an area source by using horizontal and vertical arrays of samplers. Three equally spaced air samplers are placed on a central 3.5-m mast. Wind direction and speed and temperature are monitored continuously. Two other 1.5-m

Table VII. EPA Compendium of Methods for the Determination of Toxic Organic Compounds in Ambient Air

Method	*Toxic Organic Analyte*	*Description*
TO–1	VOCs	Air sample is drawn through Tenax sample tube, heat desorbed into a cold trap, and subsequently analyzed by GC–MS.
TO–2	VOCs	Sampled through a carbon molecular sieve sample tube, heat desorbed into a cold trap; analyzed by GC–MS.
TO–3	VOCs	Sampled through a collection trap submerged in liquid argon; analyzed by GC using temperature programming techniques with FID and PID detection.
TO–4	Organochlorine pesticides and PCBs	Sampled on a glass fiber filter and polyurethane foam, recovered by solvent Soxhlet extraction; analyzed by GC–ECD.
TO–5	Aldehydes and ketones	Air is sampled through a midget impinger and then analyzed by HPLC. Up to 15 aldehydes and ketones are determined.
TO–6	Phosgene	Air sample is drawn through midget impinger. After collection the solution is reduced to dryness, dissolved in acetonitrile, and analyzed by HPLC.
TO–7	*N*-Nitrosodimethyl-amine	Sample is drawn through a sorbent cartridge, which is solvent extracted by elution with dichloromethane and analyzed by GC–MS.
TO–8	Phenols and methylphenols (cresols)	Air is drawn through two midget impingers using NaOH solution. Phenols are trapped as phenolates and determined by reverse-phase HPLC.
TO–9	Polychlorinated dibenzo-*p*-dioxins	Inlet filter and polyurethane foam cartridge air sample is benzene extracted and analyzed by GC–MS.
TO–10	Organochlorine pesticides	Sampled through a polyurethane foam cartridge, solvent extracted; analyzed by GC–ECD.
TO–11	Formaldehyde	Air is drawn through a coated silica gel cartridge to form a stable derivative of formaldehyde. The sample is solvent eluted and then determined by HPLC.
TO–12	Non-methane organic compounds	Air is drawn through a glass bead trap at –186 °C (liquid argon), helium flushed to remove methane, heated to 90 °C, and flushed into a FID giving total non-methane organic compounds.
TO–13	PNAs	Air sample is drawn through sorbent and polyurethane foam cartridges, solvent Soxhlet extracted, and then analyzed either by GC–FID, GC–MS, or HPLC.
TO–14	VOCs	Air sample is drawn into a passivated preevacuated canister. The sample is then concentrated in a cryogenic trap and analyzed by GC with one or more appropriate GC detectors including MS, ECD, ELCD, or FID.
TO–15	VOCs	Extension of TO–14, with a larger analyte list that contains additional compounds, including polar compounds. Designed to provide data for many Title III VOCs.

NOTES: TO means toxic organic; VOCs are volatile organic compounds; PCBs are polychlorinated biphenyls; FID is flame ionization detector; PNAs are polynuclear aromatic hydrocarbons; PID is photoionization detector; ECD is electron capture detector; and ELCD is electrolytic conductivity detector.

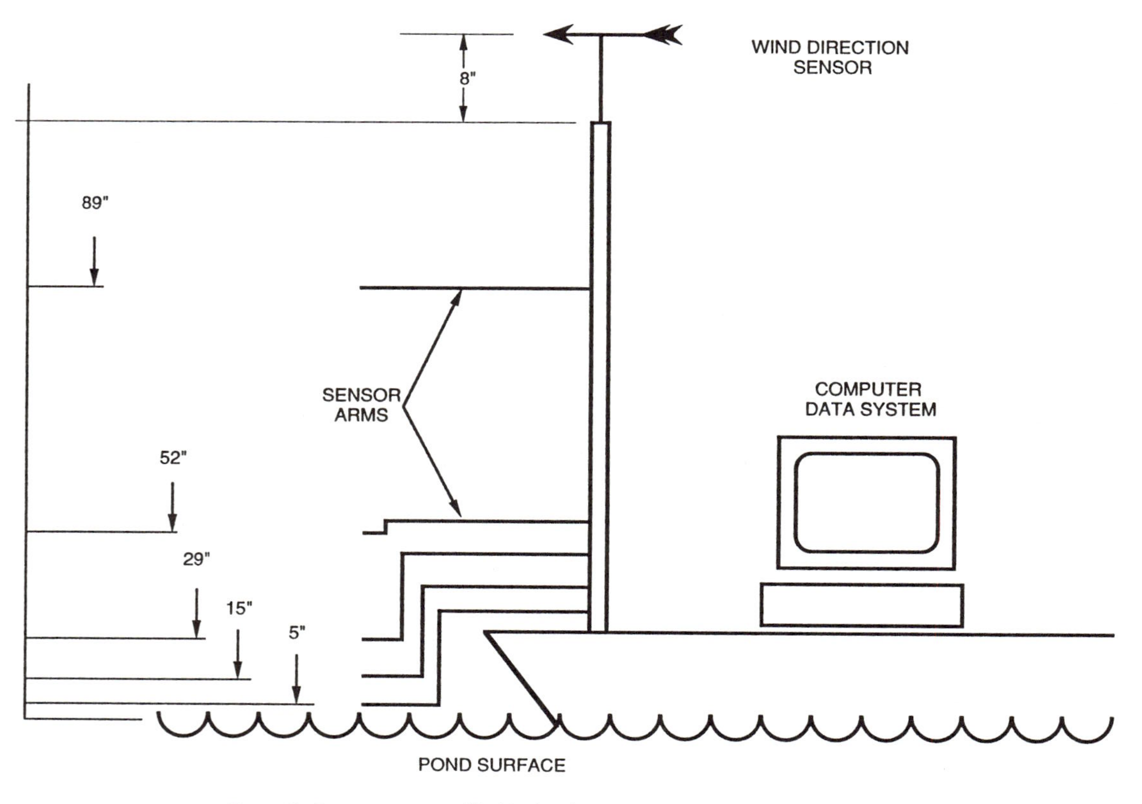

Figure 7. Concentration profile (C-P) technique. Mast sample collection system.

masts are placed on either side of the main mast at equal spacing, and one additional mast is placed downwind (Figure 8) (*38*). The masts are spaced to provide a two-dimensional spatial coverage of the plume, although the height of the plume is not captured.

Air samples are collected during appropriate meteorological conditions, although this method is still susceptible to variabilities in the wind direction and speed. If real-time analyzers are used, the entire network can be shifted to account for changes in the plume dispersion.

Tracer Flux

The tracer flux method relies on injecting a known amount of tracer gas into a liquid or gaseous process stream at the source. The tracer, often sulfur hexafluoride, is then monitored along with the target chemical species to determine the dilution factor. The resultant ratio of target compound to tracer species provides a measure of the flux of the target species (*39*). The advantage of this approach is that detailed meteorological data and complicated sampling setups are not required. On the other hand, a large number of samples typically must be collected in order to minimize uncertainties.

Method 21

Fugitive emissions of gaseous organic compounds from valves, flanges, pumps, and other gas and fluid handling equipment are determined using EPA Method 21 (*6*). Correlation equations have been derived from a "bagging" study in which components were covered with a Mylar tent, air flowed through the bag, and the inside air was sampled. A large number of these measurements were used to generate emission factors for the various components. Once the correlation equations have been established, testing of similar components by field screening instruments provides data for calculation of emissions from the emission factors. Refineries and chemical plants use these methods extensively to gather fugitive emission information. Correlation equations, relating parts-per-million values at the surface of the component to the mass leak rate (e.g., g/s), are available from the EPA and industry-specific trade groups.

Emerging Non-Point Source Methods

The EPA isolation flux chamber is the most commonly used technique to measure surface emissions. However, the flux chamber can alter the surface or the air flow over it, sampling is labor-intensive under some measurement conditions (e.g., monitoring water surfaces), and it measures only a small portion of the flux of an area, thus requiring extrapolation to the entire surface (*40*). In addition, highly time-

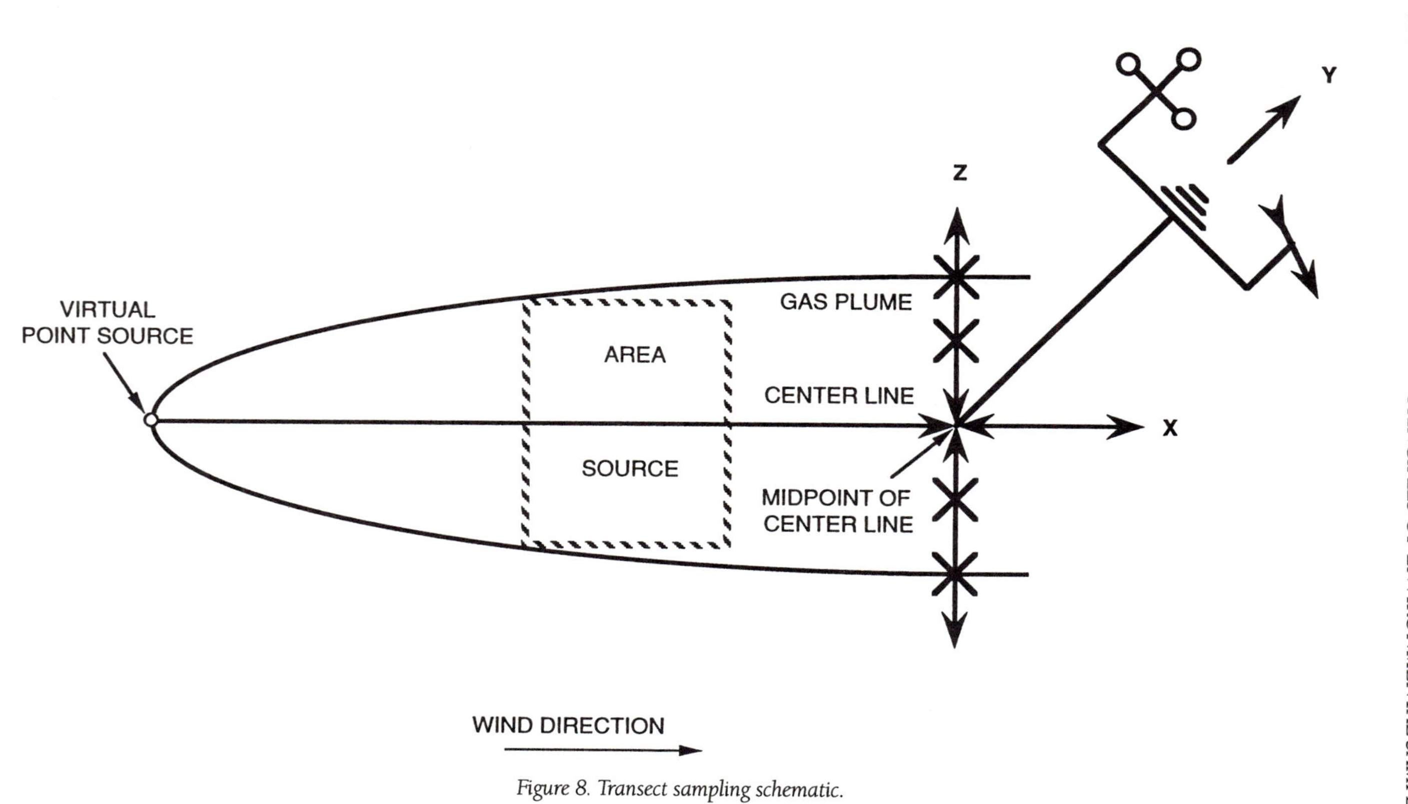

Figure 8. Transect sampling schematic.

resolved or long-term measurements over days or weeks are difficult to obtain as the flux chamber is essentially limited to a few grab samples or over a period of only several hours.

Alternatives have been in the research arena for years, but now they are starting to be used more widely as emission measurements of atmospherically important chemicals (e.g., methane and CO_2) from land surfaces such as tundra and peat bogs are being instigated (*41*). Because these alternatives evolved from a research perspective, they are often more complicated than conventional approaches but offer benefits that cannot be found in other methods. Two of the most commonly used approaches involve open-path techniques.

Open-Path and Extractive Optical Remote Sensing

Optical methods for atmospheric measurements have been in development stage for many years but now are gaining more acceptance as the costs have decreased and familiarity with the equipment and approach has increased. Several commercial instruments are now available (Midac Corporation, Los Angeles, CA; Environmental Technology Group, Atlanta, GA; Nicolet Corporation, Madison, WI). Optical methods for assessing non-point emissions rely mainly on the open-path technique of passing the light beam across the source and using modeling or on-site micrometeorological monitoring to obtain the flow portion of the equation (*42, 43*).

This approach is not without problems, given the large uncertainty of the meteorological data. A way around this difficulty is to use a tracer compound as an internal standard for determining the ratio of the unknown target concentrations with the known emission rate of the tracer. This method has shown fairly favorable comparison with the flux chamber for aeration ponds (*44*).

Micrometeorological Methods

Although research has been performed for many years on this general approach, recent years have seen increased use of micrometeorological techniques. These methods measure the mean flux of species from a large surface area (often acres in size) by using fast response chemical and meteorological sensors to capture the minute fluctuations that occur in the boundary layer of the atmosphere. The chemical flux is assumed to be proportional to the heat or water vapor flux, which are more easily measured on a high-frequency (>60 Hz) basis (*41*). Two main micrometeorological techniques are the aerodynamic gradient method and the eddy correlation method (*45*).

These methods place high demands on the data collection system, which may include high-frequency three-dimensional air movement, water vapor fluctuations, and fast chemical sensing. Because of these demands, such methods are limited to chemicals for which rapid sensors exist. A more accessible approach that modifies the general method by relaxing some of the data collection demands is the relaxed eddy correlation method. In this method, sample is collected in two directions: wind going up and wind going down. The sampling equipment is tied directly to wind sensors that control a rapid valving system to capture the short-term fluctuations of the air movement. Figure 9 shows a schematic of one such approach that was used to measure the emissions of pesticides from a large field (45). The agreement with mass balance measurements was good in this example.

The micrometeorological methods compete with the flux chamber and other indirect approaches for attention in measurement programs. Studies are under way to compare these approaches and to determine the limits of each method (46).

Analytical Issues

Analytical challenges from new regulations have arisen alongside sampling challenges, although the analytical challenges have been minimized because of how the analytical approach is often tied to the sampling approach. For example, a VOST sample is processed much differently than a Tedlar bag sample in order to introduce the sample into the analytical instrument; this processing difference results in a major disadvantage for the VOST sample due to the inability of analyzing multiple samples from the VOST tube. This limitation can be a real hindrance if prior information is not known about the expected concentrations.

Analytical methods have been previously developed for many of the newer target species. EPA's SW–846 methods contain the analytical protocols for a large portion of the new target analytes because many encompass the hazardous substance lists from other previous regulatory efforts. Some newer methods, such as the multimetals Method 0012, SMVOC (Method 0031), and the Tedlar bag method for volatile organic chemicals (Method 0040), include new analytical procedures. However, for most of the new target species, no validation of analytical protocols coupled with specific source testing methods exist.

In general, a few major analytical methods would seem to apply to the analysis of the new target compounds. For the most part, analytical laboratories are approaching the problem by using the volatility and polarity classifications to provide initial judgments about the appropriate analytical approach to use. For example, a large number of volatile organic species can be analyzed using the general Method 18 approach of self-validating the target analytes. This approach often requires additional up-front method development, but in the absence of other solutions, it will be required.

For a large number of the semivolatile organic chemicals, SW–846 Method 8270, which uses an extraction followed by GC–MS, has been judged by molecular properties alone to be an appropriate analysis method. In essence, the Modified Method 5 (SW–846 Method 0010) procedures are augmented to include additional new target analytes. In many cases, the main validation that is performed consists of a media spike and matrix spike. However, for full acceptability, the complete Method 301 validation procedure would need to be followed.

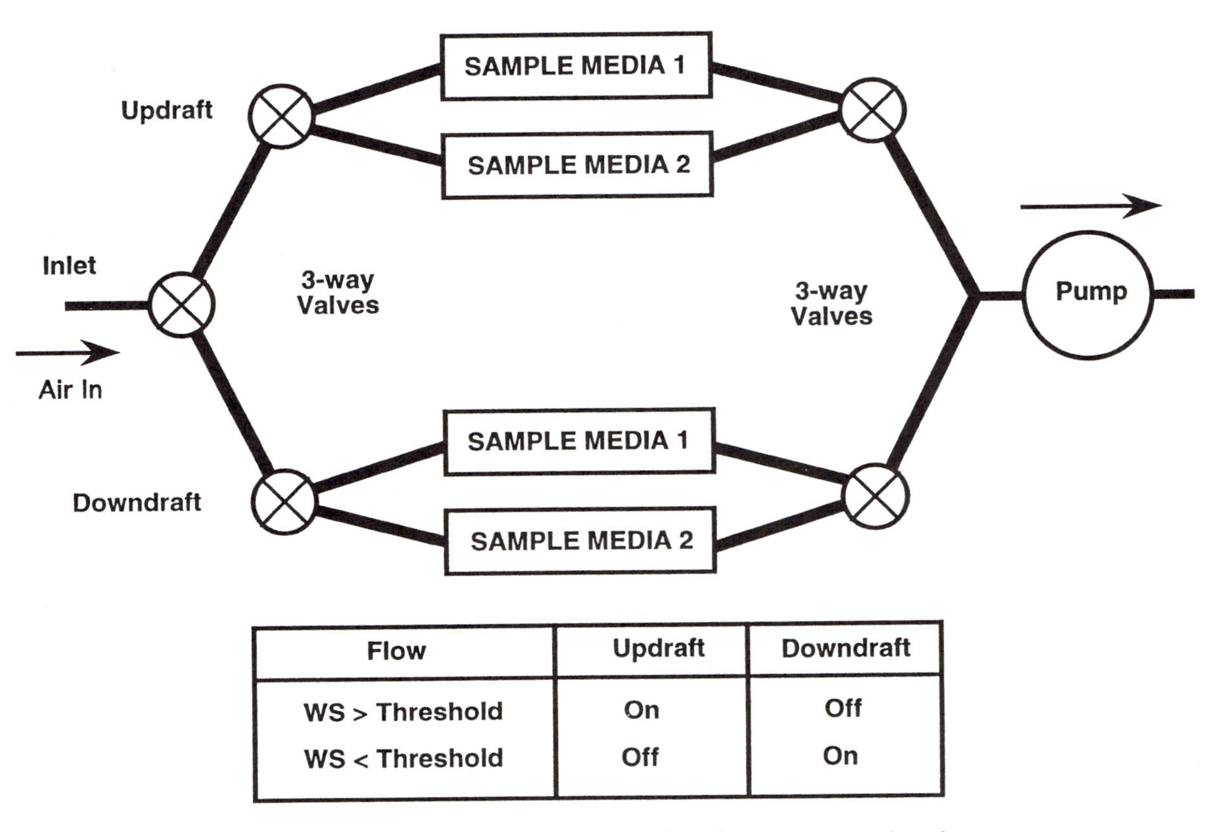

Flow	Updraft	Downdraft
WS > Threshold	On	Off
WS < Threshold	Off	On

Figure 9. Relaxed eddy correlation technique sampling schematic. WS is wind speed.

Quality Assurance

The most important aspect of any measurement program is the correct application of the data quality objective (DQO) process (47). This process is designed to force the discipline of examining the problem in a systematic fashion on a programmatic basis. From a methods perspective, the five main aspects of quality assurance that must be accounted for in any environmental measurement are precision, accuracy, representativeness, comparability, and completeness. For emission measurements, these quality assurance factors apply to both the flow rate determination and the chemical concentration determination. Indirectly related is the in-stack detection limit that is required for analyte detection, which is a function of the analytical detection limit and the volume of the emission stream collected. When the procedures are followed, the EPA reference methods cited will provide adequate quality for the majority of emission measurements.

Summary and Conclusions

The challenges to established sampling and analytical methods posed by new regulations are being met by both the EPA and industry. Both point and non-point sources are being scrutinized, with developments moving rapidly. Approaches are evolving from standard, traditional methods, in which one or only a few data points are collected for a snapshot in time, to greater number of data points to semicontinuous monitoring. These developments will add to the amount of data that are available for enforcement and policy decision making and lead to a better understanding of the temporal fluctuations of emitted pollutants.

References

1. *Fed. Regist.* **1970,** *36,* 24877.
2. *Test Methods for Evaluating Solid Waste, Physical/Chemical Methods, SW–846 Manual,* 3rd ed.; U.S. Government Printing Office: Washington, DC, 1986; Document 995–001–0000001; available from Superintendent of Documents.
3. Cooper, F. I. et al. *Proceedings of the Conference on Current Issues in Air Toxics;* Mother Lode Chapter of the Air and Waste Management Association: Sacramento, CA, 1991; p 214.
4. *Code of Federal Regulations,* Title 40, Part 86, Subpart B.
5. Siegl, W. O.; Richert, J. F. O.; Jensen, T. R.; Schuetzle, D.; Swarin, S. J.; Loo, J. F.; Prostack, A.; Nagy, D.; Schlenker, A. M. *Improved Emissions Speciation Methodology for Phase II of the Auto/Oil Air Quality Improvement Research Program—Hydrocarbons and Oxygenates;* Society of Automotive Engineers: Detroit, MI, 1993; SAE paper 930142.
6. *Code of Federal Regulations,* Title 40, Part 60, Appendix A.
7. Hamersma, J. W.; Reynolds, S. L.; Maddalone, R. F. *IERL–RTP Procedures Manual: Level I Environmental Assessment;* U.S. Environmental Protection Agency: Research Triangle Park, NC, 1976; 600/2–76–160a.
8. *Fed. Regist.* **1991,** *56,* 7134.

9. Freeman, R. R. *Proceedings of the Conference on Current Issues in Air Toxics;* Mother Lode Chapter of the Air and Waste Management Association: Sacramento, CA, 1992; p 48.
10. Steger, J. In *Air and Waste Management Association Annual Meeting Proceedings;* Air and Waste Management Association: Pittsburgh, PA, 1994; paper 94–TA27.12.
11. Serne, J. C.; White, J. O.; Burdette, J. W. In *Air and Waste Management Association Annual Meeting Proceedings;* Air and Waste Management Association: Pittsburgh, PA, 1993; paper 93–A763.
12. Dishakjian, R. *Technology Transfer Network, Emissions Measurement Test Information Center,* U.S. Environmental Protection Agency: Research Triangle Park, NC; notice (bulletin board communication) from January 1995.
13. Johnson, L. U.S. Environmental Protection Agency: Research Triangle Park, NC, personal communication, 1995.
14. Air and Waste Management Association. Workshop notes: Advanced Emission Measurement Workshop; U.S. Environmental Protection Agency, San Francisco, CA, 1993.
15. Johnson, L. D. U.S. Environmental Protection Agency, Research Triangle Park, NC, personal communication, 1995.
16. Pau, J. C.; Knoll, J. E.; Midgett, M. R. *J. Air Waste Assoc.* **1991,** *41,* 1095.
17. Betz, W. In *Sampling and Analysis of Airborne Pollutants;* Winegar, E. D.; Keith, L. H., Eds.: Lewis: Boca Raton, FL, 1993; pp 91–102.
18. Bursey, J. T.; Merrill, R. G.; Jones, D. L.; Moody, T. K.; Blackley, C. R.; Lynch, S. K.; Kuyendal, W. B. *1991 EPA/AWMA Symposium on Measurement of Toxic and Related Air Pollutants;* Air and Waste Management Association: Pittsburgh, PA, 1991; pp 1000–1005.
19. EPA Method 301, *Fed. Regist.* **1992,** *57,* 61970.
20. Crawford, R. J.; Lloyd, C. R.; Elam, D. L. Presented at the 87th Meeting of the Air and Waste Management Association, Cincinnati, OH, 1994; paper 94–TA27.02.
21. DeWees, W. G.; Steinsberger, S.; Buynitsky, W. D.; Rickman, E. E.; Jayanty, R. K. M.; Knoll, J. R.; Pau, J. C. Presented at the 86th Air and Waste Management Association National Meeting, Denver, CO, 1993; paper A1054.
22. Wilshire, F. W.; Johnson, L. D. *J. Air Waste Manage. Assoc.* **1993,** *43,* 117.
23. Wilshire, F. W.; Johnson, L. D.; Hinshaw, G. D. *Hazard. Waste Hazard. Mater.* **1994,** *11,* 277.
24. *Fed. Regist.* **1993,** *58,* 54648.
25. Gordon, C. L.; DeFriez, H. H.; Froberg, W. R. *Proceedings of 1991 International Symposium on Measurement of Toxic and Related Air Pollutants;* Air and Waste Management Association: Pittsburgh, PA, 1991; p 1021.
26. Spellicy, R. L. *Proceedings of "Current Issues in Air Toxics";* Winegar, E. D., Ed.; Mother Lode Chapter of the Air and Waste Management Association: Sacramento, CA, 1992.
27. Geyer, T. J.; Plummer, G. M.; Dunder, T. A.; Shanklin, A. A.; Gronsshandler, L. M.; Royals, P.; Blancshan, G. C.; Staughsbaugh, R.; Worthy, M. *Proceedings of the 1993 International Symposium on Measurement of Toxic and Related Air Pollutants;* Air and Waste Management Association: Pittsburgh, PA, 1993; p 423.
28. Thijssen, J. H.; Toqan, M. A.; Beer, J. M.; Sarofin, A. F. *Combust. Sci. Technol.* **1991,** *90,* 101–110.
29. Hill, H. H.; Siems, W. F.; St. Louis, R. H.; McMinn, D. G. *Anal. Chem.* **1990,** *62,* 1201A–1209A.
30. Kinner, L. L.; Plummer, G. M. *Proceedings of the 1993 International Symposium on Measurement of Toxic and Related Air Pollutants;* Air and Waste Management Association: Pittsburgh, PA, 1993; p 414.
31. Clean Air Act, Title V, Section 504(b), *Fed. Regist.* **1993,** *58,* 54648–54699.
32. Schmidt, C. E. In *Sampling and Analysis of Airborne Pollutants;* Winegar, E. D.; Keith, L. H., Eds.; Lewis: Boca Raton, FL, 1993; pp 39–55.

33. *Measurement of Gaseous Emission Rates from Land Surfaces Using an Emission Isolation Flux Chamber Users Guide;* EPA Contract 68–02–3889, Radian Corporation: Austin, TX, 1986.
34. Winberry, W. T.; Murphy, N. T.; Riggan, R. M. *Methods for Determination of Toxic Organic Compounds in Air;* Noyes: Park Ridge, NJ, 1990.
35. Winegar, E. D.; Schmidt, C. E. "Results of Field Assessment of Volatile and Semivolatile Organic Compounds from the Acetylene Unit Process Decanter, " report for confidential client, 1993.
36. Schmidt, C. E. consultant, Red Bluff, CA, personal communication, 1994.
37. Thibodeaux, L. G.; Parker, D. G.; Heck, M. M. *Measurements of Volatile Chemical Emissions from Wastewater Basins;* Hazardous Waste Engineering Research Laboratory, U.S. Environmental Protection Agency: Cincinnati, OH, 1982; EPA 600/5–2–82/095.
38. Schen, T. T.; Schmidt, C. E.; Card, T. R. *Assessment and Control of VOC Emissions from Waste Treatment and Disposal Facilities;* Van Nostrand Reinhold: New York, 1993.
39. Howard, T.; Lamb, B.; Bainsberger, W. L.; Zimmerman, P. *J. Air Waste Manage. Assoc.* **1992,** *42,* 1336–1344.
40. Aneja, V. P. *J. Air Waste Manage. Assoc.* **1994,** *44,* 977–982.
41. Delany, A. C. In *Measurement Challenges in Atmospheric Chemistry;* Newman, L., Ed.; Advances in Chemistry Series 232; American Chemical Society: Washington, DC, 1993; pp 91–100.
42. Minnich, T. R.; Scotto, R. L.; Leo, M. R.; Solinski, P. J. In *Sampling and Analysis of Airborne Pollutants;* Winegar, E. D.; Keith, L. H., Eds.; Lewis: Boca Raton, FL, 1993; pp 247–256.
43. Carter, R. E.; Lane, D. D.; Marotz, G. A.; Thomas, M. J.; Hudson, J. L. *Proceedings of 1992 International Symposium on Measurement of Toxic and Related Air Pollutants;* Air and Waste Management Association: Pittsburgh, PA, 1992; p 601.
44. Schmidt, C. E. consultant, Red Bluff, CA, personal communication, 1994.
45. Majewski, M. S.; Glofelty, D. E.; Pau, U. K. T.; Seiber, J. N. *Environ. Sci. Technol.* **1990,** *24,* 1490.
46. Aneja, V. P. *J. Air Waste Manage. Assoc.* **1994,** *44,* 977–982.
47. *Guidance for the Data Quality Objectives Process;* U.S. Environmental Protection Agency: Washington, DC, 1994; QA/G–4.
48. *Technology Transfer Network;* U.S. Environmental Protection Agency: Research Triangle Park, NC; notice (bulletin board communication).
49. Radian Corporation. *Estimation of Baseline Air Emissions at Superfund Sites;* Air/Superfund National Technical Guidance Study Series, Vol 2; Environmental Protection Agency: Austin, TX, 1990.

Chapter 25

Sampling for Exposure Assessment in the Workplace and Community

Strategies and Methods

C. Herndon Williams

New strategies and techniques have evolved to address exposure assessments in the industrial workplace and in the community. Historically, emphasis has been placed on occupational exposures, but there is now much more concern for and awareness of exposures occurring in the urban community and inside residences. Reliable measurements of exposure are a necessary element of any risk assessment. Strategies and methods are presented that are currently being used to conduct exposure assessments for chemical vapors and dust present in a broad concentration range from 0.1 parts per billion by volume to 100 parts per million. Most of the methods have been developed to assess the inhalation exposure potential through air measurements, but methods for dermal exposure and biological monitoring (total exposure) are also presented.

THE OBJECTIVE OF SAMPLING FOR EXPOSURE ASSESSMENT is to estimate the dose of a chemical received by a person over a specified period of time and under a representative set of conditions. Air sampling is used to estimate exposure by inhalation of airborne chemicals as particulate matter and/or vapor and by deposition of the chemical on the skin of the exposed person. Sampling of surfaces in contact with the skin is used to assess the potential for chemical exposure through dermal absorption. Both air and surface sampling will be addressed in this review, as well as the use of biological matrices (urine, blood, and exhaled air) in estimating exposure dose.

3152–4/96/0521$15.00/0

Estimates of inhalation exposure to chemicals with acute toxic effects are derived from measurements made over a short time span: from direct-reading instruments that make instantaneous (≤1-min response time) measurements of air concentrations to integrated samples collected over a 15–30-min period. These measurements of acute exposure are routinely made in the industrial workplace and at hazardous waste sites, but are also done in the community or in indoor air.

Estimates of inhalation exposure to chemicals with chronic health effects (including carcinogens) are derived from measurements of air concentrations integrated or time weighted over an 8-h work shift or a 24-h day. Measurements of 24-h integrated air concentrations can be repeated periodically (e.g., on an every 6th- or 12th-day frequency) to obtain an estimate of an annual mean air concentration. An annual mean concentration or dose can be used to model or predict long-term health effects.

Estimates of exposure are one element of data input into the procedure for risk assessment. The simplest approach to risk assessment involves the comparison of a measured exposure with a standard or guideline value established by a regulatory agency or professional society. An example of this type of assessment would be the comparison of a breathing-zone air concentration measured over an 8-h work shift with the relevant Occupational Safety and Health Administration (OSHA) permissible exposure level (PEL) (*1*). Such comparisons provide an indication of safe vs. unsafe conditions for the specific exposure scenario. Although this approach is widely used, it has the fundamental limitation that there is little consistency in the criteria being employed by various regulatory agencies and professional organizations in deriving air quality standards.

Estimates of exposure can also be used in quantitative risk assessments in which mathematical models are used to provide a numerical assessment of risk, usually in terms of excess deaths per million persons exposed. This approach has several limitations: (1) the models used are often very conservative and can greatly overestimate risk, (2) the models have generally only been developed for a chronic, cancer-death endpoint, and (3) there is little consensus on the best mathematical model to use.

The question of how best to use the data from an exposure assessment is currently an area of active discussion and research. Occupational exposure guidelines exist for about 1000 chemicals and are being continuously reviewed and developed by the American Conference of Governmental Industrial Hygienists (ACGIH). These guidelines, the threshold limit values (TLVs) (*2*) of the ACGIH, are based on both epidemiological data and animal studies and are the most consistent set of guidelines available to the practicing professional for occupational exposure assessment. The U.S. Environmental Protection Agency (EPA) is developing a similar set of inhalation exposure standards for continuous community exposure in the EPA's IRIS database (*3*), but inhalation standards exist for only about 30 chemicals.

The objective of this review is to provide an overview of some of the sampling approaches that can be used to estimate chemical exposures in the workplace and in the community. In the author's experience, some of the same monitoring approaches can be used for both applications, but they often require modification and validation when used in community exposure scenarios. The objectives of this review chapter can be summarized as follows:

- to provide an overview of sampling methods useful both in workplace monitoring and community exposure assessment;
- to describe some cost-effective monitoring strategies that have been used in both the workplace and the community; and
- to define field quality control (QC) procedures that have been used to validate nonroutine measurement methods.

This review will provide an overview of sampling methods, monitoring strategies, and field QC procedures used in methods validation. Analytical methods used for exposure assessments will be treated, but not in detail.

The scope of this review will address the following elements:

- The target chemicals are the 1000 air toxic chemicals for which OSHA has established an occupational PEL. These target chemicals include gases, vapors, aerosols, organic compounds, inorganic compounds, heavy metals, mineral acids, and odorants.
- Sampled media include air (breathing zone, ambient, and indoor air), surfaces, and biological matrices.
- Monitoring methods are capable of trace level (parts-per-billion) quantitation and chemically specific qualitative analysis.
- Sampling times extend from 15 min to 24 h.

As this list indicates, the scope is defined in terms of the target chemicals and chemical classes to be measured, the medium that is being sampled, the level of quantitation and qualitative analysis, and the duration of each sample. Not treated in this review is the area of real-time, direct-reading instrumentation or field monitoring methods that do not have the requisite sensitivity or specificity to address chronic health effects of individual chemicals. Some chemicals with which the author has some experience that are frequent targets for exposure assessments are benzene, ethylene oxide, formaldehyde, arsenic, mercury, and hexavalent chromium; frequently targeted chemical classes include aromatic hydrocarbons, chlorinated hydrocarbons, inorganic acids, chlorinated dioxins, volatile organic compounds (VOCs), and particulate matter with diameters ≤ 10 μm (PM_{10}). These chemicals usually have industrial or hazardous waste emission sources but present exposure hazards in the workplace and in the community.

Model of Total Human Exposure

Figure 1 shows a model for total human exposure and excretion in the workplace and in the community. The routes of exposure and excretion are essentially the same for both scenarios. However, one difference is the duration of the chemical exposure: 8–10 h/day for workplace exposure and up to 24 h/day for some population groups in the community (e.g., children). Permissible workplace chemical concentrations are usually higher than those for community exposure, in part because workplace exposures usually occur for only 8–10 h/day. The nonexposure period (14–16 h/day) allows chemicals in the body to begin to be metabolized and excreted with a chemically specific half-life that ranges from minutes to days. The nonexposure period is usually much shorter or zero for community exposures.

An assessment of total human exposure should address all of the relevant exposure pathways. For most industrial-source exposure scenarios, however, the major pathway is through inhalation, and most exposure standards have been developed for inhalation. The measurement technology and the risk assessment guidelines are much more highly developed for inhalation exposures. Methods and guidelines for dermal exposures and biological matrices are a very active area of research currently.

Elements of an Exposure Assessment Study

The elements of an exposure assessment study can be summarized as follows:

- objectives, and the intended use of data [data quality objectives (DQOs)]
- target chemicals and their physical state (solid, liquid, or vapor)
- routes of exposure to be evaluated
- exposure standards or guidelines
- sampling and analytical methods, and limits of detection
- sampling and study duration (acute, chronic, or both)
- representative activities or conditions during exposure
- quality assurance–quality control (QA–QC) strategy for field and laboratory.

One of the primary determinants in any study are the objectives of the study, that is, the intended use of the data. Some of the potential uses of exposure assessment data are as follows:

- risk assessment by comparing measured exposures with standards or guidelines established by regulatory agencies (e.g., OSHA) or professional organizations (e.g., ACGIH);

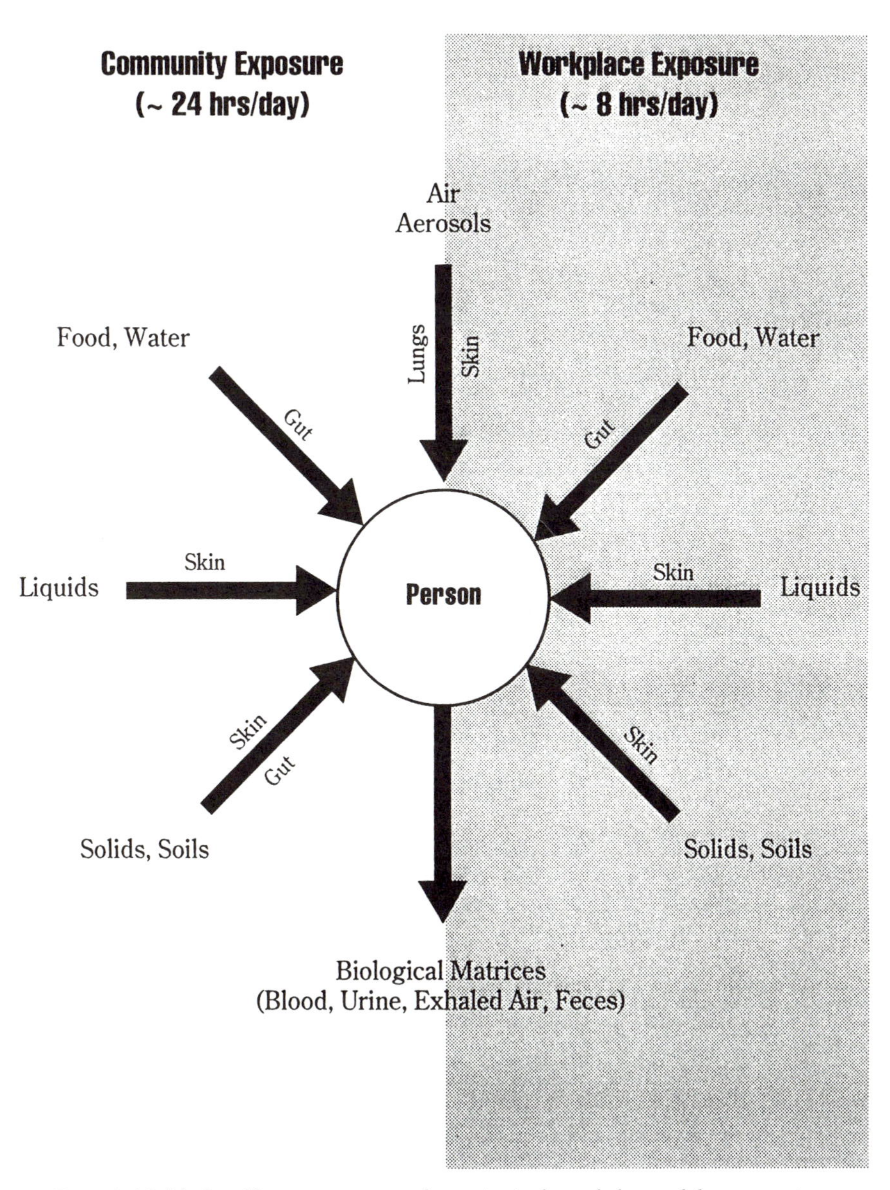

Figure 1. Model of total human exposure and excretion in the workplace and the community.

- risk assessment by using the measured exposures as input data for a quantitative risk assessment model;
- the establishment of source–receptor relationships;
- input to an epidemiological study;
- comparison with exposures predicted by a model (e.g., by air dispersion modeling);
- input to a reverse dispersion model to estimate a chemical emission rate; and
- helping to evaluate an emission control technology.

Major consideration should be given to the level of QA–QC that will be required to meet the DQOs.

Some of the possible exposure assessment measurement parameters are as follows:

- vapor concentration in breathing zone air (concentration × time = dose)
- particulate concentration in the breathing zone (concentration × time = dose)
- distribution of a target chemical between vapor and particulate matter
- concentration on the skin or on a surface in contact with the skin
- concentration in a biological matrix: exhaled air, urine, or blood
- concentration at the source or in a contaminated matrix (e.g., soil)
- time variability of the exposure concentration.

An exposure assessment will usually involve the measurement of several of these parameters depending on the objectives of the study. An exposure assessment would be incomplete if it did not include enough of these parameters to characterize the relevant exposure pathways (e.g., an inhalation exposure assessment that treated only exposures by vapor or particulate matter but not both). Similarly, an exposure assessment should normally include provision to estimate the time variability of the exposure concentration (e.g., day to day, or season to season).

Exposure Standards and Guidelines

A number of regulatory agencies and professional societies have established legal standards and advisory guidelines for use in defining a hazardous chemical exposure. The best known of these are the inhalation PEL standards legislated under OSHA and the equivalent TLVs established by ACGIH. Some of the other national entities that have established exposure guidelines are the National Institute of Occupational Safety and Health (NIOSH), the EPA, the American Industrial Hygiene Association (AIHA), and the National Academy of Sciences. One state agency with established air quality guidelines is the Texas Natural Resource Conservation Commission (TNRCC). A number of states other than Texas have also established their own set of workplace or, more commonly, ambient air guidelines.

An overview of all of the various chemical inhalation exposure standards and guidelines is given in Table I. Table I lists the exposure or testing duration, the defined acronym for the standard or guideline, and the exposure scenario. Generally, the OSHA PEL is the only legal standard in Table I; all other entries are advisory guidelines or recommendations.

In planning for an exposure assessment, all of the relevant exposure standards or guidelines for all of the proposed target chemicals should be researched and

Table I. Overview of Chemical Inhalation Exposure Standards and Guidelines

Exposure or Testing Time	*Relevant Inhalation Exposure Standards*		*Exposure Scenario*
	Organization	*Standard*	
Instantaneous (<1 min)	ACGIH OSHA	TLV (ceiling) PEL (ceiling)	Peak exposure to lethal chemicals
15 min	ACGIH OSHA NIOSH AIHA	TLV–STEL PEL–STEL REL–STEL WEEL–STEL	Acute exposure in the industrial workplace or at hazardous waste sites
30 min	NIOSH TNRCC	IDLH ESL (30 min)	Acute exposure in emergency escape Acute exposure in the community
1 h	AIHA	ERPG	Acute exposure in emergency response
8 h	ACGIH OSHA NIOSH AIHA EPA ACGIH	TLV–TWA PEL–TWA REL–TWA WEEL–TWA NCEL BEI	Chronic exposure in the industrial workplace or at hazardous waste sites
>24 h (continuously)	EPA TNRCC	RfC ESL (annual) Odor threshold	Chronic exposure in the community

NOTES: PEL is permissible exposure level by Occupational Safety and Health Administration (OSHA); REL is recommended exposure level by National Institute of Occupational Safety and Health (NIOSH); TLV is threshold limit value by American Conference of Governmental Industrial Hygienists (ACGIH); WEEL is workplace environmental exposure level by American Industrial Hygiene Association (AIHA); IDLH is immediately dangerous to life and health (NIOSH); STEL is short-term exposure limit (15 min); TNRCC is Texas Natural Resource Conservation Commission; BEI is biological exposure indices (ACGIH); ERPG is emergency response planning guideline (AIHA); TWA is time weighted average (8–10 h/day); ESL is (health) effects screening levels (TNRCC); RfC is reference concentration (EPA); and NCEL is new chemical exposure limit by EPA.

reviewed. When exposure standards or guidelines exist for target chemicals, then exposure measurement can be directly compared to achieve an evaluation of exposure risk.

In order to be able to make a valid comparison with an exposure standard or guideline, several criteria must be satisfied. First, the time period of the measured exposure must correspond to the exposure or testing time specified in the standard or guideline, as shown in Table I. The sensitivity of the monitoring method must also be high enough to address the standard or guideline. A minimum sensitivity would be a limit of detection (LOD) equal to 0.2 times the standard or guideline. Similarly, the specificity of the method must be sufficient such that the target chemical is resolved from any significant interference, artifact, or blank contamination.

Exposure Assessment Sampling Strategies

The elements of an exposure assessment sampling strategy are as follows:

- identification of representative individuals or groups whose exposure will be measured
- identification of target chemicals and exposure routes to be measured
- placement of samplers or sampling sites
- field QC
- relationship between samples
- monitoring method parameters
- level of measurement validation.

The first two elements in this list follow from the objectives of the study and are outside the scope of this review. The selection of exposure groups has been treated in several studies (4, 5).

The placement of samplers and the field QC selected for the sampling strategy are designed to create a matrix or relationship between the various exposure measurements. In a larger sense, this matrix or relationship helps to demonstrate that the source–receptor relationship is known and understood and that the measured exposures are credible. This approach is shown schematically in Figure 2, in which three sites were selected for exposure measurement to test three expected levels of exposure, ranging from no (or minimal) exposure for a control group to maximum exposure. The elements of field QC and their suggested minimum frequency are as follows: field duplicates, 20%; field blanks, 10%; field spikes, 10%; and control samples, 10%. The objectives of the field QC are to demonstrate precision, accuracy, blank level, and reasonable relationships, and the ability to detect the target chemicals in the actual sample matrix.

Figure 2. Sampling strategy depicting degrees of impact associated with placement of sampler.

The use of the strategies shown in Figure 2, along with the suggested elements and frequencies of field QC listed previously, can generate an exposure data set whose internal coherence can be evaluated in terms of accuracy, precision, and representativeness. One important element of this approach is that the matrix of measured exposure values can be compared internally for consistency and reasonableness and externally against standards and guidelines. The level of QC in the measurement is often designed to validate the measurement method as well as the data. The result can be an exposure data set that exhibits a high level of confidence and credibility.

In developing a sampling strategy for an exposure assessment a number of parameters need to be considered and maximized. A list of the most important monitoring method parameters includes

- sampled air matrix parameters
- target chemical scope
- sensitivity (LOD < exposure standard)
- specificity or resolution
- interferences and artifacts
- DQOs (accuracy and precision)
- level of validation: existing and required
- cost
- ruggedness.

The specification of these parameters will follow from the objectives and intended data use and will usually require a detailed knowledge of conditions that will be encountered in the exposure assessment. Sampled air matrix parameters are subject to a wide range of values, as shown in Table II. Selection of an optimized monitoring method will require a specification of the actual conditions to be encountered in the exposure assessment.

Table II. Sampled Air Matrix Parameters and Values

Parameter	*Value*
Temperature	Ambient (0–100 °F)
Pressure	Ambient (1 atm)
Relative humidity	20–100%
No. of chemicals/sample	5–150
Concentration range	0.1–10,000 ppbv
	0.2–20,000 $\mu g/m^3$
Time of sampling	15 min–24 h

NOTE: ppbv means parts per billion by volume.

There are many more target chemicals and exposure scenarios than there are monitoring methods to address them. Method development or modification is required for most nonroutine exposure assessments. Some sources of exposure assessment methods are OSHA, NIOSH, EPA, American Society for Testing and Materials (ASTM), and the chemical literature. These sources will give methods that have been validated for a well-defined scope of applicability. For example, OSHA and NIOSH have developed many methods for measuring chemical exposures in air in the industrial workplace. Some of these methods can be judiciously adapted or modified for measuring community exposure. For example, Radian has modified NIOSH Method 1501–Aromatic Hydrocarbons to measure benzene in ambient air to a limit of detection in the range 0.1–0.2 parts per billion by volume (ppbv). In all cases, the modified method must be validated for the new application.

Air Sampling Methods

Workplace

Industrial hygiene (IH) air sampling methods for personal (breathing zone) exposures in the industrial workplace have received attention since the passage of the OSHA Act in 1970. Both OSHA (6) and NIOSH (7) have published and updated validated methods for IH monitoring, developed to measure single chemicals (e.g., NIOSH Method 2538–Acetaldehyde) or small groups of chemicals in one chemical family (e.g., NIOSH Method 1501–Aromatic Hydrocarbons). Sampling media for these methods are often available from commercial sources. Some of the parameters for these IH methods are given in the left column of Table III.

The left column of Table III also gives an overview of the types of air sampling media that are often specified in OSHA or NIOSH methods. Treated sorbents and filters are used when the target chemical is reactive or unstable. A reagent chemical (derivatizing agent) is dosed on a sorbent or filter to react quantitatively with the target chemical and form a more stable derivative. For example, charcoal treated with hydrobromic acid (HBr) is used to sample for ethylene oxide so that the reactive eth-

Table III. Air Sampling Methods and Media in the Workplace and the Community

Workplace	*Community*
Air Sampling Methods	
15 min to 8 h per day sampling duration	30 min to 24 h per day sampling duration
Air volumes of 1–50 L from the worker's breathing zone or fixed sites in the workplace	Air volumes of 150 L to 150 m^3 from ambient air or indoor air
Address PEL–STEL and/or PEL–TWA	Address EPA–RfC or TNRCC–ESL
LODs of 0.1× PEL–TWA (0.01–1 ppmv)	LOD of 0.01–1 ppbv
Battery-powered air sampling pumps	Battery-powered pumps or AC
Commercial sampling pumps and media generally available	Some commercial instrumentation and sampling media available
Uses inexpensive, readily accessible technology in sampling and analytical approaches	Meteorological data and or directional samplers needed
Validated to satisfy NIOSH criterion of measured exposure within ±25 of the true value with 95% confidence	
Air Sampling Media	
Filters	Filters
Filter with a cyclone or impactor for size selection	Filters with dichotomous particle size selection
Filter + sorbent	Filter + sorbent
Treated sorbents and filters	Treated sorbents and filters
Sorbent (active or passive)	Metal canisters
Liquid-filled impingers (active or passive)	Diffusional denuders

NOTE: LOD means limit of detection. *See* Table I for other abbreviations.

ylene oxide is converted to bromoethanol. Liquid-filled impingers are used less frequently in workplace monitoring because of their lack of robustness, although spill-proof impingers and passive impingers have been developed recently and are commercially available.

Filter media available for workplace monitoring include mixed cellulose–ester (MCE), poly(vinyl chloride) (PVC), Teflon, and glass fiber. Sorbent media available for workplace monitoring include activated charcoal, silica gel, graphitized carbons, carbon molecular sieve, porous polymers (Tenax, XAD), polyurethane foam (PUF), graded sorbent multisection samplers, and hopcalite. MCE filters are widely used for sampling airborne particulate matter for metals and asbestos, whereas PVC and Teflon are often used for organic particulate matter. Glass fiber filters may be a good choice when a derivatizing agent will be dosed on the filter.

Activated coconut-shell charcoal has received the most application for organic vapors in IH monitoring, although a number of new synthetic carbon sorbents have been developed over the past 5 years. Some of these new sorbents are more amenable to thermal desorption of the target chemicals, replacing the older solvent-desorption approach. However, most OSHA or NIOSH methods are based on the use of solvents for desorption or digestion. Several commercial suppliers are very active in the development of new sorbents for IH monitoring. Silica gel is used for many polar compounds like alcohols. The porous polymers (e.g., Tenax or XAD) and PUF plugs are used for less volatile chemicals such as polychlorinated biphenyls (PCBs), polycyclic aromatic hydrocarbons, and pesticides.

Community

Air sampling methods for ambient air in the community (including indoor air) are available for only a relatively small number of toxic chemicals. Most of the validated methods were developed by EPA (8) for the National Ambient Air Quality Control (NAAQC) air pollutants and other chemicals involved in the photochemical formation of ozone (e.g., VOCs). This situation is likely to change as the result of the 1990 Clean Air Act Amendments, which defined a list of 189 hazardous air pollutants (HAPs). However, there is no current regulatory requirement to measure exposure to these chemicals in the ambient air.

The requirements for air sampling methods for the community are listed in the right column of Table III. Community air monitoring usually requires longer periods of sampling (24 h), higher sampled air volumes (150 L to 150 m^3), and lower LODs (0.01–1 ppbv) than most workplace air monitoring. These community requirements are also more appropriate for monitoring ambient air quality at a Superfund hazardous waste site where airborne chemical migration into populated areas could be a problem.

There is much less commercially available instrumentation and sampling media for community air monitoring. The right column of Table III also lists some of the air sampling media currently being used in ambient air monitoring. Ambient air monitoring equipment usually requires AC (alternating current) power, which limits the placement of samplers in remote areas. Ambient air sampling also requires wind direction data, obtained by the placement of a portable meteorological station at the sampling site or by the use of wind-directional samplers. Wind-directional samplers have a built-in wind vane that turns the sampler on only when the wind is from a prescribed quadrant.

It is sometimes advantageous to develop ambient air samplers that use modified IH methods to achieve the monitoring requirements given for community air sampling listed in Table III. Figure 3 shows a sampler developed by Radian that used a combination of filters, treated sorbents, and evacuated metal canisters to perform concurrent ambient air monitoring for four target chemical groups: (1) VOCs, (2) ethylene oxide and propylene oxide, (3) aliphatic and aromatic aldehydes from C_1 to C_8, and (4) hexavalent chromium [Cr(VI)]. Radian used a combination of modified OSHA, NIOSH, and EPA methods to achieve ppbv (or μg/m^3) limits of detection in 24-h integrated air samples for these four target chemical groups. The methods used for this application are listed in Table IV.

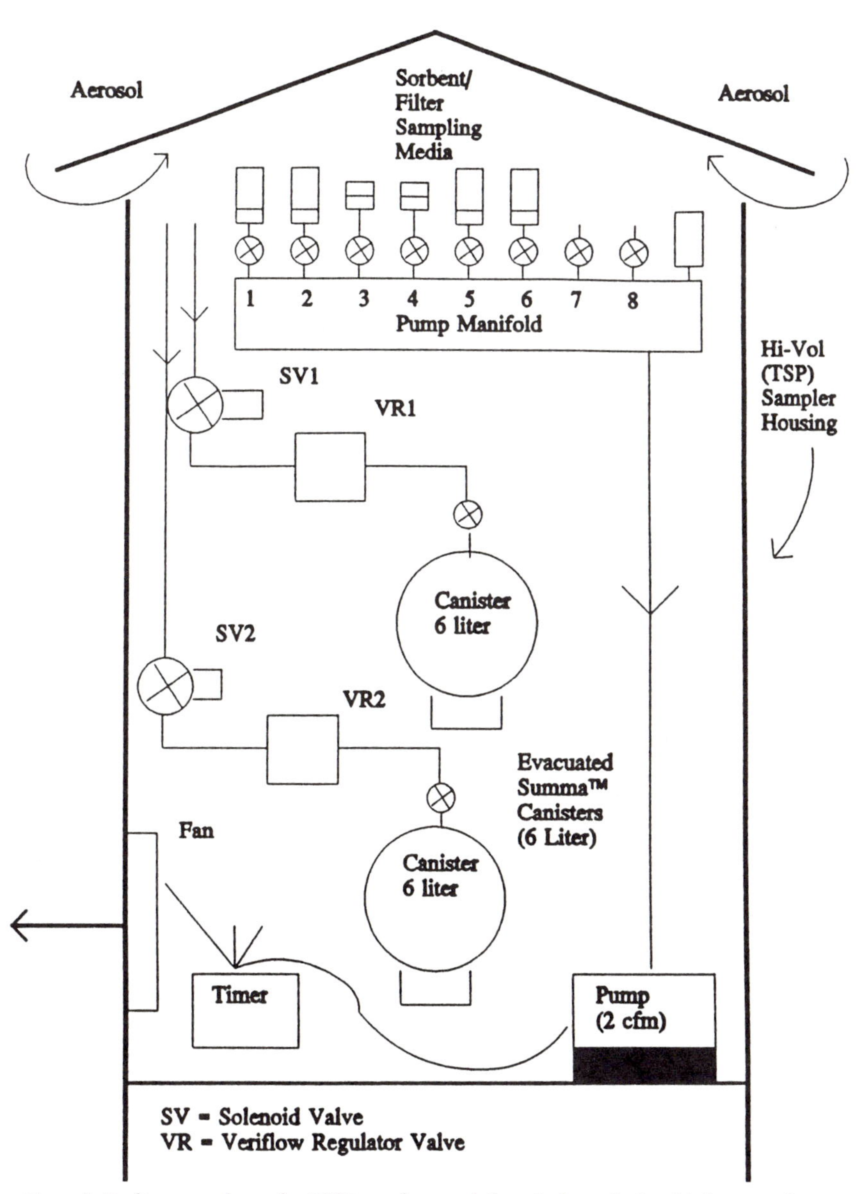

Figure 3. Radian aerosol sampler. TSP is total suspended particulate; cfm is cubic feet per minute.

Table IV. Target Chemical Groups, Monitoring Methods, and Approaches Used with Radian Aerosol Samples

Target Chemical Group	*Monitoring Method*	*Sampling and Analytical Approach*
VOCs: C_2 to C_{10}	EPA TO–14 (Radian modification)	Evacuated, Summa canister, cryogenic inlet, HRGC, multiple detectors (FID, PID, HSD).
Aldehydes: C_1 to C_8	EPA TO–11	Silica gel cartridge treated with 2,4-dinitrophenylhydrazine, solvent desorption, HPLC–UV.
Ethylene oxide, propylene oxide	OSHA–50 (Radian modification)	Activated charcoal sorbent treated with hydrogen bromide, solvent desorption, derivatization, GC–ECD.
Cr(VI) on particulate matter	NIOSH–7600 (Radian modification)	PVC membrane filter, acid digestion, colored complex formation, visible absorption spectrophotometry.

NOTES: HRGC is high-resolution gas chromatography; FID is flame ionization detector; PID is photoionization detector; HSD is halogen-specific detector; and ECD is electron capture detector.

Biological Sampling

Biological exposure indexes (BEIs) have been established for workplace exposure to about 30 industrial chemicals by the ACGIH (*9*). All but two of the chemicals having BEIs also have established values of the TLV that address inhalation exposure, but the BEIs allow an estimation of total exposure (i.e., including dermal and ingestion routes), as well as nonoccupational exposure.

The BEIs apply to both the target chemical and its metabolite in one or more of three biological matrices: exhaled air, urine, and blood. Table V gives the various sampling media and times specified by the ACGIH for measuring the target chemical in the biological matrix. A wide range of sampling times is evident: from the end of a breath of exhaled (alveolar) air, to the end of a work shift, to the end of the work week. Biological monitoring methods have been addressed (*10, 11*) in two recent books.

Biological monitoring has a number of limitations, which have resulted in a small number of chemicals for which BEIs have been established. There is generally a very limited amount of data that allow a correlation to be established between a concentration measured in a biological matrix and the occurrence of adverse health effects. Technical issues include knowledge about the routes of metabolism and identities of metabolites, and individual variability in the rates of metabolism and excretion. BEIs could play a very valuable role for those chemicals whose primary route of exposure is not inhalation, such as glycol ethers, dinitrotoluene, aniline, creosote, and isocyanates (*12*).

Table V. Biological Sampling Media and Sampling Times in the Workplace

Sampling Media	*Sampling Times*
Exhaled air for the target chemicals	End of exhaled (alveolar) breath; End of shift
Urine for the target chemical or metabolite	End of shift, and preshift End of work week
Blood for the target chemical	End of shift, and preshift End of work week

Dermal Sampling

Surface wipe sampling has been used by OSHA compliance officers as one method of evaluating workplace exposures. OSHA has developed a detailed protocol for collecting surface wipe samples (*13*), although OSHA has no standards or guideline for evaluating the measured surface concentrations. A recent survey (*14*) has concluded that there is generally no quantitative correlation between surface wipe concentrations and workplace air concentrations.

Surface wipes and dermal sampling have been extensively applied in a few applications: in pesticide applications, in health physics, and in cleanup of contamination by PCBs, chlorinated dioxins, and furans. Some of the dermal sampling methods that have been used in the workplace and the community are

- wet surface wipes (horizontal and vertical surfaces)
- vacuuming or brushing of surface dust
- skin wipe or wash
- patches on clothing, and gloves on hands
- fluorescent emissions from the target chemical or a surrogate tracer.

These applications are also limited by the establishment of only very few standards and guidelines. However, in this active area of research (*12, 14, 15*), models are being developed that would aid in the interpretation of measurements of skin and surface concentrations.

Analytical Methods

The primary focus of this review has been on sampling strategies and methods, but analytical methods are an integral part of the overall strategy. The analytical methods that have been applied to exposure assessments are

- gas chromatography (GC) with a variety of detectors: flame ionization (FID), photoionization (PID), electron capture (ECD), and halogen-specific (HSD)
- GC–mass spectrometry (MS)
- atomic absorption spectroscopy (AAS), inductively coupled argon plasma–atomic emission spectroscopy (ICAP–AES), and cold vapor atomic absorption spectroscopy (CVAAS)
- ion chromatography (IC)
- high-performance liquid chromatography (HPLC) with UV and fluorescent detection.

These methods have been used traditionally in environmental analysis and do not include some of the methods that have been developed more recently or that require costly instrumentation, such as Fourier transform infrared spectroscopy (FTIR) and ICAP–MS.

Some of the elements of analytical QC that would be considered or used in an exposure assessment study are instrument calibration (frequency, range, and stability); calibration or resolution check sample; system blank; matrix, reagent, or method blanks; duplicate analysis; limit of detection measurement; desorption recovery; storage stability of target chemical on sample matrix; vapor breakthrough on sorbents; and interferences and artifacts.

Using the traditional analytical methods listed previously offers several advantages. The methods are well characterized and the instrumentation is well developed and competitively priced. Thus the per-sample analytical costs are very competitive. The lower analytical costs make it feasible to incorporate high levels of field and laboratory QC to help validate novel monitoring methods, while at the same time achieving high levels of sensitivity and specificity. Overall, the choice of traditional methods can result in a known level of uncertainty and a high level of confidence in the measurement results. However, these traditional methods generally require the use of a remote laboratory and do not make use of newly developed analytical technology.

Conclusions

The conclusions from this review can be summarized as follows. (1) Some of the same monitoring methods can be applied in both the workplace and community. (2) Many methods exist that can be adapted to cost effectively perform novel exposure assessments. (3) Sampling strategies should incorporate several elements of field QC to validate measurements. (4) Methods can be combined to achieve a personal total body exposure assessment.

This chapter presents a number of methods and a sampling strategy that can be used to conduct a cost-effective total exposure assessment study in either the workplace or in the community. The approach makes full use of methods validated by OSHA, NIOSH, and EPA, but it modifies these methods as necessary to achieve the objectives of the study. The strategy uses high levels of field and laboratory QC and multiple sample sites or subjects to establish quantitative relationships between the measured exposure parameters. This protocol serves to validate the measurements by estimating their accuracy and precision. As importantly, these relationships demonstrate the representativeness of the exposure measurements by testing the knowledge of the source–receptor interaction. The result can be a credible, consistent set of exposure measurements.

References

1. *Code of Federal Regulations,* Title 29, Part 1910, § 1000, 1993; pp 6–19.
2. *ACGIH 1994–1995 Chemical Substance Threshold Limit Values (TLVs);* American Conference of Governmental Industrial Hygienists: Cincinnati, OH, 1994.
3. *EPA Integrated Risk Information System (IRIS),* an online database created by the U.S. Environmental Protection Agency and distributed as part of the National Library of Medicine's Toxicology Data Network (TOXNET), Bethesda, MD, 1994; (301) 496-6531.
4. *A Strategy for Occupational Exposure Assessment;* Hawkins, N. C.; Norwood, S. K.; Rock, J. C., Eds.; American Industrial Hygiene Association: Fairfax, VA, 1991.
5. *Human Exposure Assessment for Airborne Pollutants;* National Research Council, National Academy of Sciences: Washington, DC, 1991.
6. *OSHA Analytical Methods Manual,* 2nd ed.; U.S. Department of Labor, Occupational Safety and Health Association, Directorate of Technical Support, Salt Lake City Analytical Laboratory: Salt Lake City, UT, 1990.
7. *NIOSH Manual of Analytical Methods,* 3rd ed.; Eller, P. M., Ed.; U.S. Department of Health and Human Services, Centers for Disease Control and Prevention, National Institute of Occupational Safety and Health: Cincinnati, OH, 1994.
8. *Compendium of Methods for the Determination of Toxic Organic Compounds in Ambient Air;* U.S. Environmental Protection Agency, Environmental Monitoring Support Laboratory: Research Triangle Park, NC, 1984; EPA 600/4–84/041.
9. *Threshold Limit Values and Biological Exposure Indices, 1994–1995;* American Conference of Governmental Industrial Hygienists: Cincinnati, OH, 1994.
10. Baselt, R. C. *Biological Monitoring Methods for Industrial Chemicals,* 2nd ed.; PSG Publishing: Littleton, MA, 1988.
11. Lauwerys, R. R.; Hoet, P. *Industrial Chemical Exposure, Guidelines for Biological Monitoring,* 2nd ed.; Lewis Publishers and CRC Press: Boca Raton, FL, 1993.
12. Klinger, T. D.; McCaskle, T. *Am. Ind. Hyg. Assoc. J.* **1994,** *55,* 251.
13. *Sampling for Surface Contamination;* U.S. Department of Labor, Occupational Safety and Health Administration: Washington, DC, 1990; OSHA Instruction 2–2.2B.
14. Caplan, K. J. *Am. Ind. Hyg. Assoc. J.* **1993,** *54,* 70.
15. Leung, H. W.; Paustenbach, D. J. *Appl. Occup. Environ. Hyg.* **1994,** *9,* 187.

Chapter 26

Particulate Matter with Aerodynamic Diameters Smaller than 10 μm

Measurement Methods and Sampling Strategies

Judith C. Chow, John G. Watson, and Frank Divita, Jr.

Filter-based particle and gas sampling systems consist of more than the mechanical device used to acquire the sample. The laboratory analyses to be applied, the types of filters that are amenable to those analyses, the minimum deposits needed on these filters, the sampling hardware that extracts pollutants from the atmosphere onto the filters, and the procedures that ensure the accuracy, precision, and validity of the acquired atmospheric concentrations must all be considered. Methods to design these systems and examples of successful design are presented. Different systems are needed for determining compliance with air quality standards, source apportionment, and visibility studies. Systems can meet multiple purposes when they are properly designed.

PARTICLES SUSPENDED IN THE ATMOSPHERE are collected and analyzed for many different reasons. Particles are most often measured to (1) determine compliance with air quality standards, (2) identify chemical components that might be deleterious to public health, (3) evaluate the extent and causes of visibility impairment, and (4) apportion and quantify the chemical constituents of suspended particulate matter to their emitting sources. Measuring compliance with the National Ambient Air Quality Standards (NAAQS) for PM_{10} (particles with aerodynamic diameters < 10 μm) requires the use of sampling systems that have received

3152–4/96/0539$18.75/0

the U.S. Environmental Protection Agency's (EPA) designation as reference or equivalent instruments (*1*). This designation is specific for mass measurements, but not for particulate chemical composition. The types of samplers used, the filters on which they sample, and the analyses to be performed on each type of filter should be specified before a single sample is collected in an air pollution study. Many options are available for these specifications.

This chapter describes (1) the general requirements for a sampling system, including the components that make up these systems; (2) available filter-based particle and gas sampling systems; (3) commonly applied filter analysis methods; and (4) sampling and analysis strategies. PM_{10} sampling and analysis methods and strategies are emphasized here because of the importance of PM_{10} in current ambient air quality standards. The methods are general, however, and can be applied to other particle size fractions for a variety of measurement purposes. This article complements others that describe airborne particle sampling (*2, 3*) and analysis (*4–6*) methods in greater detail.

Aerosol Sampling and Analysis Requirements

Most commercially available sampling systems have been developed to determine compliance with the NAAQS for PM_{10}, which are 150 $\mu g/m^3$ for a 24-h average and 50 $\mu g/m^3$ for an annual arithmetic average (*7*). PM_{10} sampling and analysis requirements include (1) wind-tunnel testing of sampler inlets; (2) sampling efficiency and alkalinity of filter media; (3) stability of sample flow rates; and (4) precision of gravimetric analysis. Future revisions to the NAAQS may identify specific chemical components in particles and additional particle size fractions that adversely affect public health.

To evaluate visibility reduction, it is necessary to measure the mass and chemical composition of particles with aerodynamic diameters less than ~2.5 μm ($PM_{2.5}$) because particles in this range scatter light most efficiently (*8, 9*). The major constituents of $PM_{2.5}$ that contribute to light extinction are sulfates, nitrates, ammonium, organic carbon, elemental carbon, and crustal species (e.g., aluminum, silicon, iron, calcium, titanium) (*10*); each of these chemicals has a different effect on light extinction. Samples need to be taken over time periods corresponding to visible haze (daylight hours), thereby requiring multiple samples of less than 24-h duration.

Receptor models use the chemical and physical characteristics of gases and particles measured at source and receptor to both identify the presence of, and quantify the source contributions to, air quality at the receptor (*11–13*). To differentiate the contributions of one source from another, the chemical and physical characteristics must be such that (1) they are present in different proportions in different source emissions; (2) these proportions remain relatively constant for each source type; and (3) changes in these proportions between source and receptor are negligible or can be approximated. Source apportionment of PM_{10} uses software (*14*), applications and validation protocols (*15*), and model reconciliation protocols (*16*) prescribed by the EPA to focus emissions reduction strategies on the largest contributors.

To meet these different study objectives, samples must be obtained with (1) well-defined size fractions; (2) sampling surfaces that do not react with or add to the measured species; (3) filter media compatible with analysis methods; (4) sufficient aerosol deposits for the desired analyses without over- or underloading the filters; and (5) instruments that are available, cost-effective, and have established operating procedures. Several sampling systems have been constructed and applied using a combination of different sampling inlets, sampling surfaces, filter media, filter holders, and flow controllers.

Sampling inlets remove particles that exceed a specified aerodynamic particle diameter and operate on the principles of direct impaction, virtual impaction, cyclonic flow, selective filtration, and elutriation (*3*). Inlets are characterized by sampling effectiveness curves, measured in a wind tunnel, that show the fraction of spherical particles of unit density penetrating through the inlet to the filter surface (*17, 18*). Figure 1 shows an example of these curves for PM_{10} inlets. The 50% cut-point (d_{50}) is the diameter at which one-half of the suspended particles pass through the inlet. Acceptable size-selective inlets have sampling effectiveness curves that are independent of wind speed and wind direction. The d_{50} varies with flow rate through the inlet. Tested inlets are available with flow rates of ~1000 L/min for high-volume sampling (*19–21*), ~100 L/min for medium-volume sampling (*22, 23*), and ~10–20 L/min for low-volume sampling (*24–27*). Medium- and high-volume inlets are often used to collect samples in parallel on several substrates because flow rates can be kept high enough to obtain an adequate deposit for analysis.

After passing through the inlet, air is drawn through filters that retain the suspended particles. EPA filter requirements for PM_{10} sampling specify weight losses or gains due to mechanical or chemical instability of less than a 5-$\mu g/m^3$ equivalent, 0.3 μm dioctyl phthalate (DOP) sampling efficiency in excess of 99%, and alkalinity of less than 25 μeq/g to minimize absorption of sulfur dioxide and nitrogen oxides (*28*). These are the minimum requirements for samples that require chemical analyses. The most commonly used filter media for atmospheric particle and gas sampling are cellulose fiber, glass fiber, Teflon-coated glass fiber, Teflon membrane, etched polycarbonate membrane, quartz fiber, and nylon membrane. None of these materials is perfect for all purposes. Watson and Chow (*6*) summarize different filter media and their relevant physical and chemical characteristics. Lippmann (*29*) identifies several specific types of filter media and reports experimental results on their sampling efficiencies and pressure drops.

Filter blank levels must be low to permit aerosol chemical analysis. Even batches of ultrapure filters have been found to be contaminated, and a sample from each batch of filters (1 out of 50–100 filters) should be submitted to the intended chemical analyses prior to use in a field study. Elevated levels of lead, calcium, and nitrate have been found in batches of blank Teflon membrane filters. Filters may also become contaminated in the field or during handling by passive deposition before and after sampling. Dynamic field blanks can be placed in the field under situations

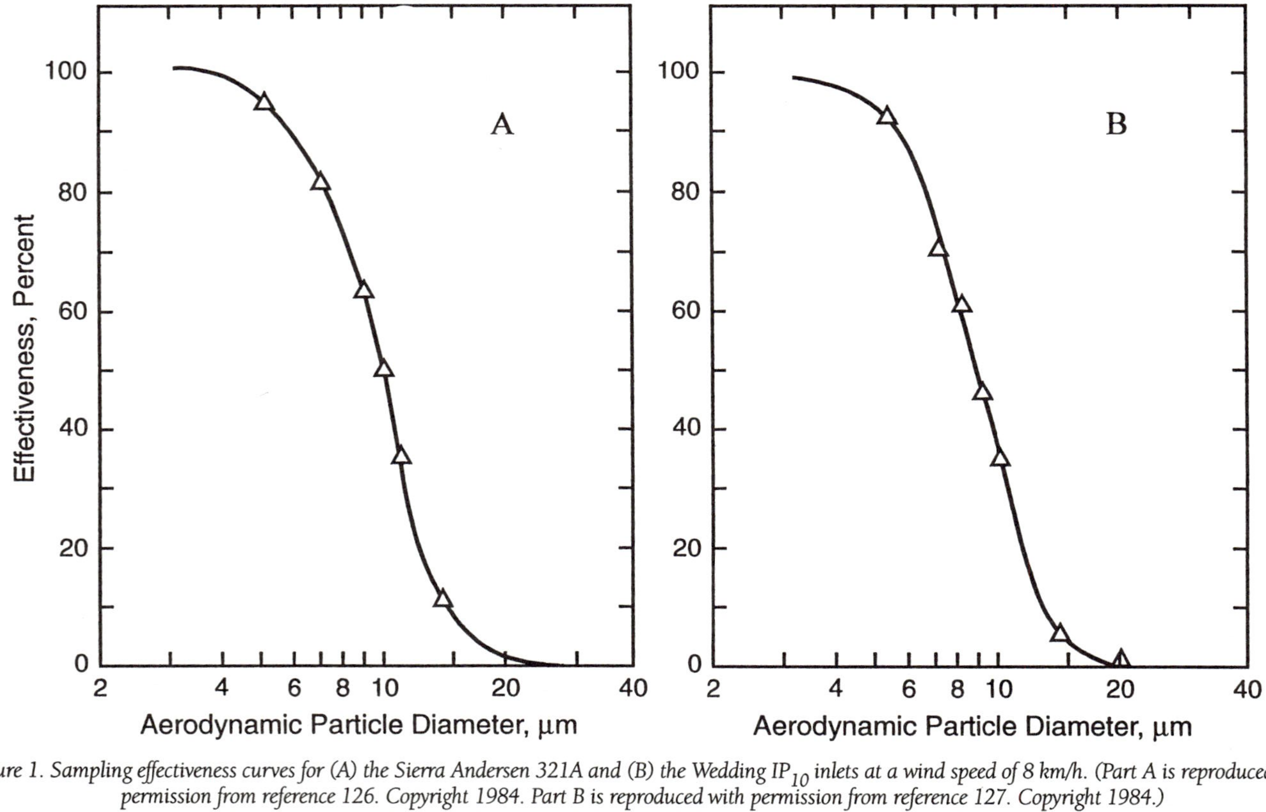

Figure 1. Sampling effectiveness curves for (A) the Sierra Andersen 321A and (B) the Wedding IP_{10} inlets at a wind speed of 8 km/h. (Part A is reproduced with permission from reference 126. Copyright 1984. Part B is reproduced with permission from reference 127. Copyright 1984.)

similar to that of the sampled filter. These filters are then analyzed so average blank levels can be subtracted from the chemical measurement.

Filters should be protected from contamination prior to, during, and after sampling by loading into and unloading from an appropriate filter holder in a clean environment. These holders should (1) accommodate commonly available air sampling filters [e.g., 37 or 47 mm diameter, 8 × 10 in. (20.3 × 25.4 cm) rectangles]; (2) provide a homogeneous deposit; (3) mate to the sampler and to the flow system without leaks; (4) be composed of inert materials that do not adsorb acidic gases or contaminate the filters; (5) have a low pressure drop across the empty holder; and (6) be easy to use, durable, and reasonably priced. Table I lists several of the characteristics associated with commonly used filter holders. Lippmann (*29*) provides a more comprehensive, but less current, summary of filter holder characteristics. In-line holders often concentrate the particles in the center of the substrate, and this uneven distribution will bias the results if analyses are performed on portions of the filter. Therefore, open-faced filter holders are a better choice for ambient aerosol sampling systems. All PM_{10} reference samplers use open-faced filter holders.

Air is passed through filters using a pump (*2*). The quantity of air per unit time must be precisely measured and controlled to determine particle concentrations and to maintain the size-selective properties of the sampling inlet. Manual volumetric, automatic mass, differential pressure volumetric, and critical orifice (or throat) volumetric flow control systems are commonly applied.

For manual volumetric flow control, a valve between the filter and pump is adjusted until the desired flow rate through the filter is obtained. In the absence of leaks, pump degradation, and filter loading, the pressure drop across the filter and the flow rate remain constant. Although the filter loads up during sampling, thereby increasing the resistance and reducing the flow, this loading does not ordinarily reduce the flow rate by more than 10% when ambient concentrations are less than ~200 $\mu g/m^3$. Automatic mass flow controllers detect the passage of air with a thermal anemometer, then adjust pump motor speeds to maintain this flow at a constant level. As their name implies, these controllers measure mass rather than volume, and Wedding (*30*) estimates potential differences in excess of 10% between mass and volumetric measurements of the same flow rates, depending on temperature and pressure variations. Mass flow controllers need to be recalibrated each season and for different elevations to compensate for changes in temperature and pressure. Many high-volume PM_{10} samplers use mass flow controllers.

A constant pressure can be maintained across an orifice by a diaphragm-controlled valve located between the filter and the orifice. As the pressure drop across the orifice increases because of filter loading, the diaphragm opens to bring it back into equilibrium. This differential pressure method is used on the Graseby–Andersen dichotomous sampler and on the sequential filter sampler for PM_{10} monitoring. A critical orifice or critical throat consists of an opening between the filter and the pump. The pressure downstream of the orifice must be less than 53% of the upstream pressure to maintain constant flow, thereby requiring large pumps and low flow rates (typically less than 20 L/min with commonly available pumps). The

Table I. Summary of Filter Holder Characteristics

Manufacturer	Manufacturer's Part Number	Filter Size (mm)	Materials of Construction			Adapters	Configuration		Approximate Cost/Unit ($)
			Base	Gasket/O-Ring	Support Grid		Open	In-Line	
Gelman Instrument Company	1107	25	Delrin	Viton O-ring	Stainless steel screen	Nylon	X		89
600 South Wagner Road	1209	25	Stainless steel	Viton O-ring	Stainless steel screen	Nylon		X	163
Ann Arbor, MI 48106	1220	47	Aluminum	Viton O-ring	Stainless steel screen	Nylon	X		269
(313) 665-0651	1235	47	Aluminum	Viton O-ring	Stainless steel screen	Polyethlyene		X	269
	2220	47	Stainless steel	Viton O-ring	Stainless steel screen	Polyethylene		X	289
	1109	25	Delrin	Viton O-ring	Stainless steel screen	Polyethylene		X	89
Millipore Corporation	SX0002500	25	Polypropylene	Silicone	Polypropylene	NA		X	8
397 Williams Street	XX3002500	25	Stainless steel	Teflon	Stainless steel screen	NA		X	123
Marlborough, MA 01752	XX3002514	25	Stainless steel	Teflon	Stainless steel screen	NA		X	166
(508) 624-8400	M000025A0	25	Polypropylene	NA	Cellulose pad	NA	X	X	107
	XX5004700	47	Aluminum & stainless steel	Teflon	Stainless steel screen	NA		X	530
	XX4304700	47	Glass-filled polystyrene	Silicone O-ring	Polystyrene	NA		X	105
	SX0004700	47	Polypropylene	Silicone O-ring	Polystyrene	NA		X	120
Nuclepore Corporation	420410	47	Polycarbonate	Ethylene, propylene	Polycarbonate	NA		X	29
7035 Commerce Circle	420500	47	Polycarbonate	Ethylene, propylene	Polycarbonate	NA	X		31
Pleasanton, CA 94566	421700	47	Stainless steel	Teflon	Stainless steel screen	NA		X	92
(800) 882-7711	420210	25	Polycarbonate	Ethylene, propylene	Polycarbonate	NA		X	12
	420300	25	Polycarbonate	Ethylene, propylene	Polycarbonate	NA	X		20
	421600	25	Stainless steel	Teflon	Stainless steel screen	NA		X	84
Graseby–Andersen, Inc.	273	47	Stainless steel	Viton O-ring	Stainless steel screen	NA		X	790
4801 Fulton	273–LI	47	Stainless steel	Viton O-ring	Stainless steel screen	NA		X	790
Industrial Boulevard	273–AL	47	Aluminum	Viton O-ring	Stainless steel screen	NA		X	790
Atlanta, GA 30336	273–O	47	Stainless steel	Viton O-ring	Stainless steel screen	NA	X		790
(404) 691-1910	273AL–O	47	Aluminum	Viton O-ring	Stainless steel screen	NA	X		790
	240–P	37	Polypropylene	NA	NA	NA	X		5
	G–300	8×10	Stainless steel	NA	NA	NA	X		95

Table I. Summary of Filter Holder Characteristics—*Continued*

Manufacturer	*Manufacturer's Part Number*	*Filter Size (mm)*	*Materials of Construction*			*Adapters*	*Configuration*		*Approximate Cost/Unit ($)*
			Base	*Gasket/O-Ring*	*Support Grid*		*Open*	*In-Line*	
BGI, Inc.	F–1	47	Aluminum	Silicone	Ni-plated brass	Ni-plated brass		X	215
58 Guinam Street	F–2	47	Aluminum	Silicone	Ni-plated brass	Ni-plated brass	X		215
Waltham, MA 02154	F–5/2	47	Teflon	Silicone	Teflon	NA		X	325
(617) 891-9380	F–7	47	Stainless steel	Silicone	Stainless steel screen	Stainless steel		X	350
	F–3	47	Aluminum	Silicone	Ni-plated brass	Ni-plated brass	X	X	275
Savillex Corporation 6133 Baker Road	0–473–4N 1/4-in. NPT	47	PFA Teflon	Viton O-ring	PFA Teflon	PFA Teflon	X		62
Minnetonka, MN 55345 (612) 935-4100	0–473–6N 3/8-in. NPT	47	PFA Teflon	Viton O-ring	PFA Teflon	PFA Teflon	X		62
	2–25–2 1/8-in. Tubing	47	PFA Teflon	Viton O-ring	PFA Teflon	PFA Teflon		X	43
	4–25–4 1/4-in. Tubing	47	PFA Teflon	Viton O-ring	PFA Teflon	PFA Teflon		X	44
University Research Glassware	URG–2000– 30F–2 stage	47	Teflon	Viton O-ring	Teflon-coated stainless steel double stage	Teflon		X	315
118 East Main Street P.O. Box 368	URG–2000– 30F–3 stage	47	Teflon	Viton O-ring	Teflon-coated stainless steel triple stage	Teflon		X	362
Carrboro, NC 27510 (919) 942-2753	URG–2000– 30F–4 stage	47	Teflon	Viton O-ring	Teflon-coated stainless steel four stage	Teflon		X	515

NOTE: NA means not applicable; PFA means perfluoralkoxy.

critical throat (*31*), also applied to PM_{10} high-volume samplers, diffuses the flow before and after the orifice to recover a portion of the energy lost by a conventional critical orifice.

Aerosol Sampling Systems

Hering (*32*), Perry (*33*), and Rubow and Furtado (*2*) describe commercially available systems for ambient aerosol sampling. High-volume PM_{10} samplers, such as those shown in Figures 2 and 3, are most commonly used because they are designated reference methods for PM_{10}. In these samplers, a low-pressure blower draws air through an 8 × 10 inch quartz fiber or glass fiber filter. The Wedding units use critical throat flow control, whereas the Graseby–Andersen units offer a choice between mass flow control or critical throat control. Procedures for these samplers are well established (e.g., *34, 35*). Frequent inlet cleaning is necessary for accurate particle size sampling by these units.

Figure 4 shows a low-volume dichotomous sampler with a virtual impactor for simultaneous $PM_{2.5}$ and PM_{10} measurements. A differential pressure regulator controls flow rates for this unit. Because of the virtual impactor, 10% of the $PM_{2.5}$ (fine) particles are sampled on the coarse particle (PM_{10} minus $PM_{2.5}$) filters, although this fraction can be accurately estimated and adjusted (*36*).

Figure 5 shows a sequential filter sampler (SFS) equipped with a medium-volume PM_{10} inlet. Up to 12 sampling ports divert flow from one channel to the next by means of a programmable timer that controls solenoid valves. Filters are loaded into open-faced 47-mm Nucleopore filter holders in a laboratory, and several filter media can sample simultaneously from different ports. Flows can be controlled manually or by differential pressure controllers.

Beta attenuation monitors (BAM) (e.g., *37–41*) attenuate beta rays (moderately high energy electrons) emitted by a radioactive source when they pass through particles deposited on a filter. Figure 6 shows an example of a BAM that qualifies as an equivalent method for PM_{10} compliance monitoring and can acquire continuous measurements over averaging times as short as 1 h. Beta-ray attenuation is related to the particulate mass collected on the filter. PM_{10} monitors draw a filter tape across the path between the beta emitter and a detector to measure blank attenuation, then across a sampling area where particles are collected on the tape, and finally through the detector area to measure the attenuation of the filter and the deposit.

The tapered element oscillating microbalance (TEOM) shown in Figure 7A (*42, 43*) also has equivalence status for PM_{10} compliance monitoring and can measure continuous hourly values. The TEOM mounts a small filter on the narrow end of a hollow tapered tube (Figure 7B). The wider end of the tube is fixed, while the narrow end oscillates in response to an applied electric field. Air is drawn into a low-volume inlet and through a filter and tapered tube leading to a flow controller. The oscillation frequency changes as the mass loading on the filter increases. PM_{10} concentrations are then calculated by comparing the measured frequency with standards.

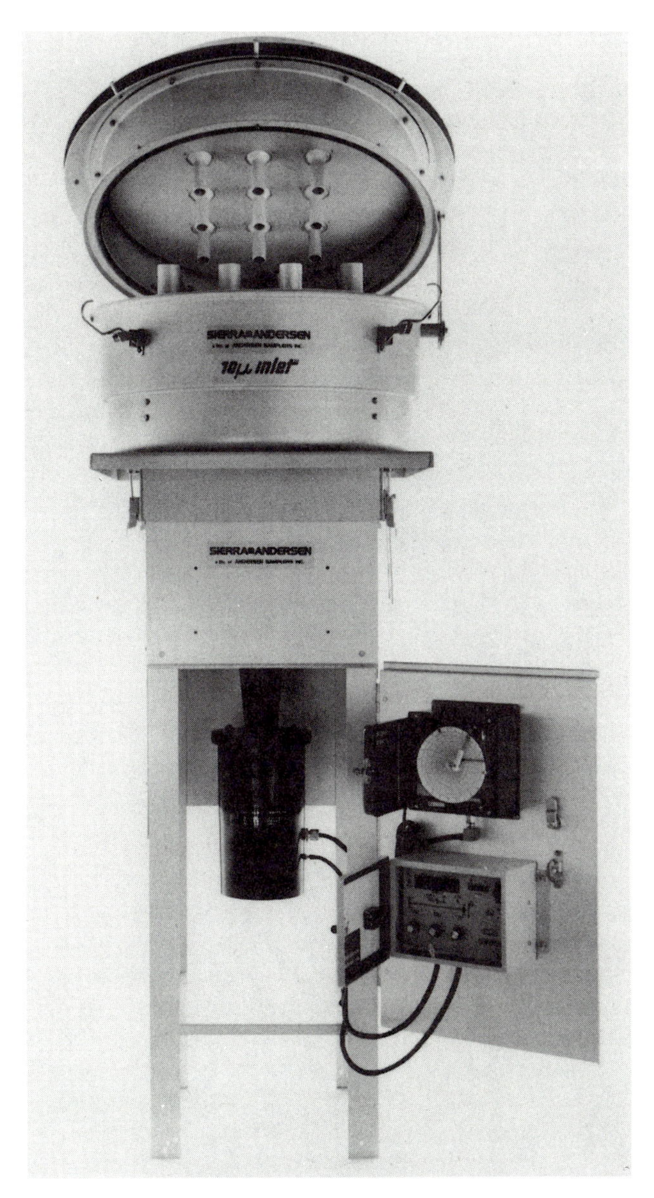

Figure 2. Graseby–Andersen high-volume sampler with PM_{10} *inlet (designation no. RFPS-1287-063 and RFPS-1287-064). (Courtesy of Graseby–Andersen Inc.)*

Figure 3. Wedding & Associates critical flow high-volume sampler with PM_{10} inlet (designation no. RFPS-1087-062). (Courtesy of Wedding & Associates.)

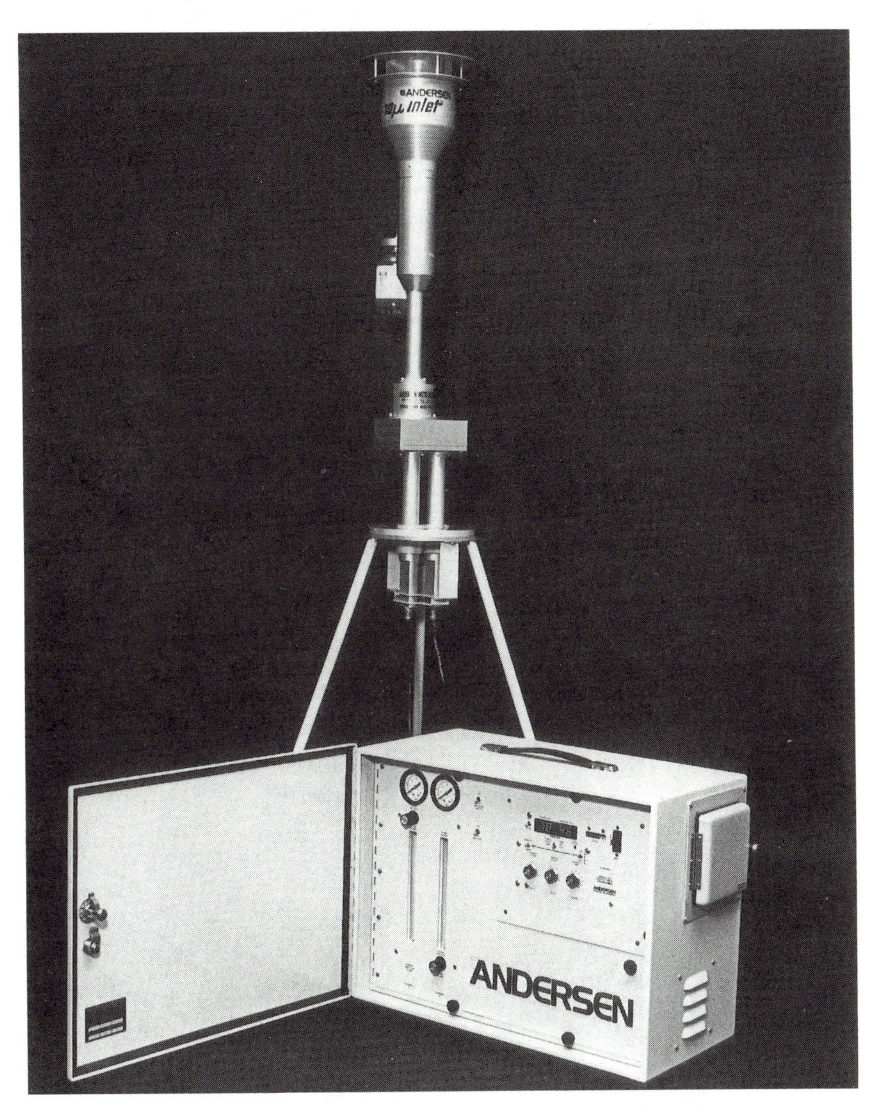

Figure 4. Graseby–Andersen low-volume PM_{10} dichotomous sampler (designation no. RFPS-0389-073). (Courtesy of Graseby–Andersen Inc.)

Figure 5. Sequential filter sampler (SFS) with a Graseby–Andersen SA-254 medium-volume PM_{10} inlet (designation no. RFPS-0389-071).

Figure 6. Graseby–Andersen beta attenuation particulate monitor (designation no. EQPM-0990-076). The inlet is identical to that on the dichotomous sampler in Figure 4. (Courtesy of Graseby–Andersen, Inc.)

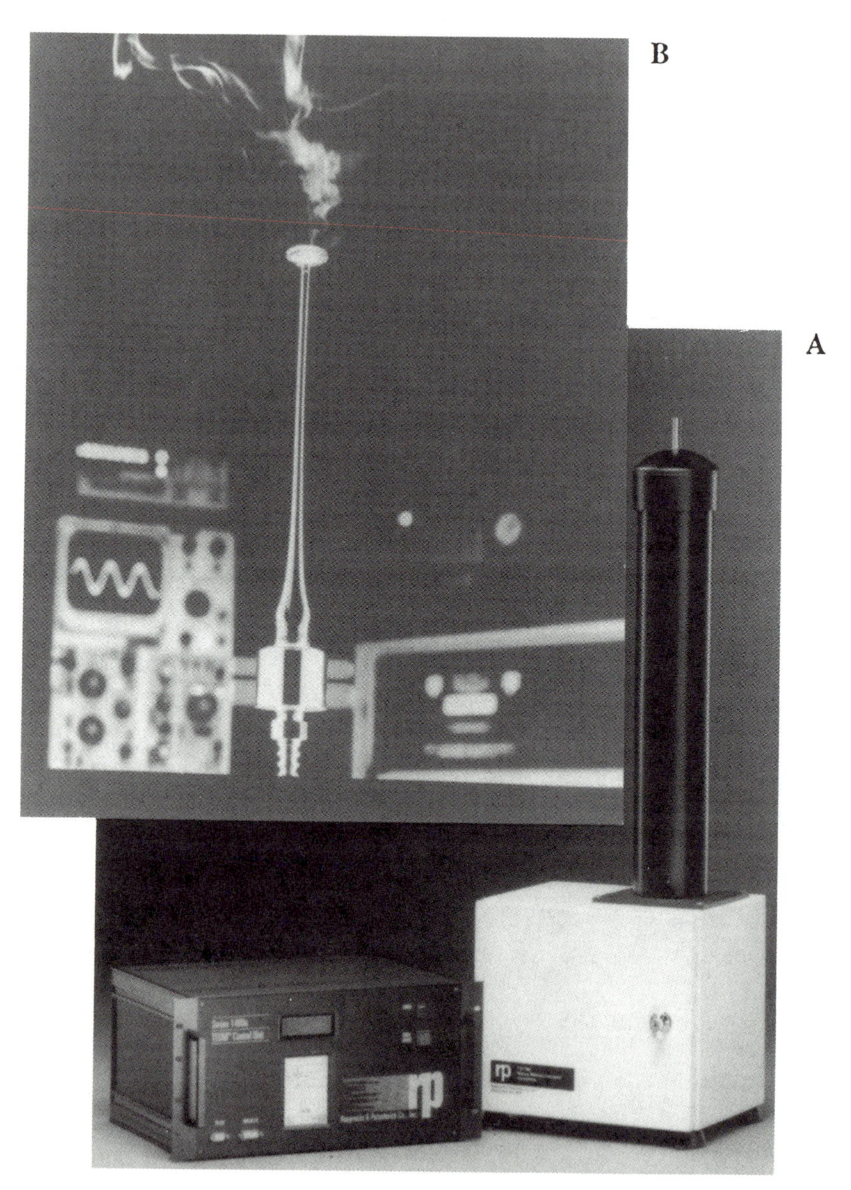

Figure 7. Rupprecht & Patashnick Co. tapered element oscillating microbalance series 1400A ambient particulate monitor (A) and tapered element (B) (see text). (Courtesy of Rupprecht & Patashnick Company.)

Figure 8 shows a minivolume portable survey sampler (AIRmetrics, Springfield, OR) that can be deployed in spatially dense PM_{10} sampling networks. This sampler does not have EPA reference or equivalent status, but it has been used in several source apportionment studies (e.g., *44*). These battery-powered units can be hung from power poles and building walls and do not require complicated sampler siting efforts. This inexpensive sampler contains an impactor inlet, removable filter holder, flow meter, timer, tubing and fittings, pump, and battery pack.

In addition to the PM_{10} reference and equivalent sampling systems, Table II lists a number of samplers that have been developed and deployed for specialized purposes. The denuders referred to in Table II remove gases from the sampling stream while allowing particles to pass through them. Sulfur dioxide, hydrochloric acid, nitric acid, ammonia, and other gases are absorbed on a basic or acidic coating applied to the inner surfaces of the denuder (*45–49*). Concentrations of these gases can be determined when denuders are washed with appropriate solvent. Cascade impactors partition particles into 5–10 smaller size ranges, from ~0.05 to ~15 μm, onto substrates suitable for chemical analysis (*50–53*).

Chemical Analysis Methods

A large variety of chemical analyses have been applied to suspended particulate samples, and this section only reviews those most commonly used for PM_{10} samples. These methods can be divided into four categories: (1) mass, (2) elements, (3) water-soluble ions, and (4) organic and elemental carbon. Some methods are nondestructive, and these are preferred because they preserve the filter for other uses. Methods that require destruction of the filter are best performed on a section of the filter to save a portion of the filter for other analyses or as a quality control check on the same analysis method. Table III identifies the elements and chemical compounds commonly found in air using these methods with typical detection limits.

Less common analytical methods, which are applied to a small number of specially taken samples, include isotopic abundances (*54–56*), mineral compounds (*57–59*), organic compounds (*60–67*), and single-particle characterization (*68*). The cited references provide information on sampling and analysis methods for these highly specialized methods.

Mass Measurement Methods

Mass is measured gravimetrically by weighing filters before and after sampling and subtracting the preweight from the postweight. PM_{10} reference methods require that filters be equilibrated for 24 h at a constant (within ±5%) relative humidity between 20 and 40% and at a constant (within ±3 °C) temperature between 15 and 30 °C to minimize the liquid water associated with soluble compounds and to minimize the loss of volatile species. Nominal relative humidities of 20–30% and temperatures of 15–20 °C preserve particle deposits during weighing. Balance sensitiv-

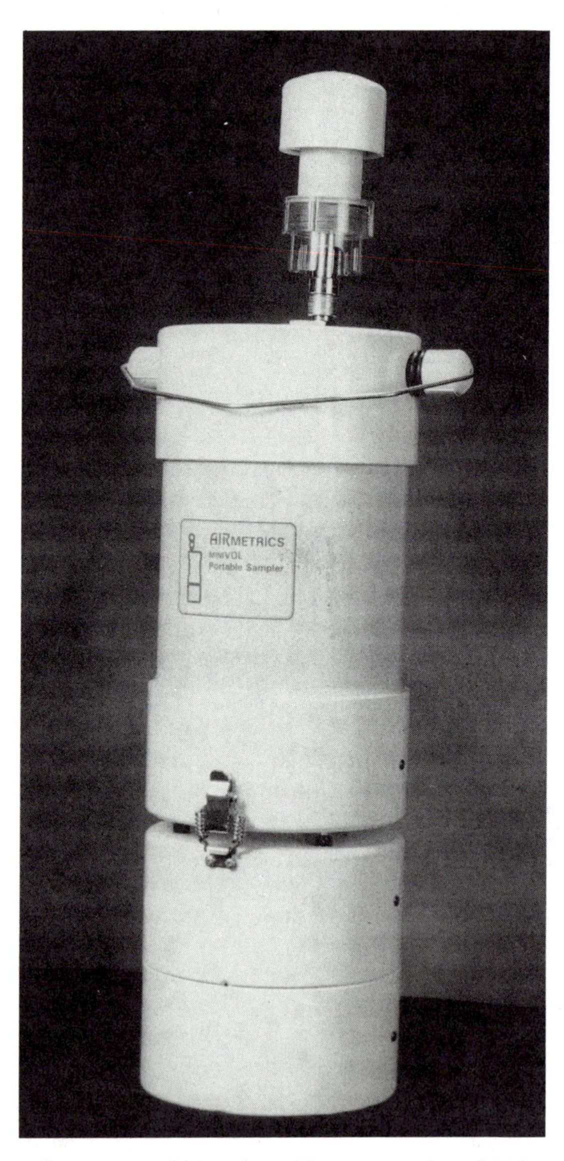

Figure 8. AIRmetrics battery-powered Minivol portable survey sampler with PM_{10} inlet. (Courtesy of AIRmetrics, Inc., Springfield, OR.)

Table II. Examples of Ambient Aerosol Sampling Systems Assembled for Major Studies

Sampling System	*Descriptive Reference*	*Inlet and Particle Size (μm)*	*Flow Rate (L/m)*	*Sampling Surface*	*Filter Media*	*Filter Holders*	*Features*
Western Region Air Quality Study (WRAQS) Sampler	Tombach et al. (*111*)	Aluminum high-volume impactor (PM_{15})	113 out of 1130	Aluminum and copper	47-mm Teflon membrane 47-mm quartz fiber	Nuclepore polycarbonate in-line	
		Steel medium-volume cyclone ($PM_{2.5}$)	113	Aluminum and copper	47-mm Teflon membrane 47-mm quartz fiber	Nuclepore polycarbonate in-line	
California Institute of Technology Sampler	Solomon et al. (*112, 113*)	Aluminum low-volume impactor (PM_{10})	16.7	Stainless steel and aluminum	47-mm Teflon membrane 47-mm quartz fiber	Gelman stainless steel in-line	
		Aluminum low-volume cyclone ($PM_{2.5}$)	22	Teflon-coated aluminum and glass	47-mm Teflon membrane 47-mm quartz fiber 47-mm nylon membrane	Gelman stainless steel in-line	
Size Classifying Isokinetic Sequential Aerosol Sampler (SCISAS)	Rogers et al. (*114*)	Aluminum high-volume impactor (PM_{15})	113 out of 1130	Aluminum and poly(vinyl chloride)	47-mm Teflon membrane 47-mm quartz fiber	Nuclepore polycarbonate open-face	Sequential sampling
		Steel medium-volume cyclone ($PM_{2.5}$)	113 out of 1130	Stainless steel and aluminum	47-mm Teflon membrane 47-mm quartz fiber	Nuclepore polycarbonate open-face	
Southern California Air Quality Study Sampler (SCAQS)	Fitz et al. (*115*)	Aluminum medium-volume impactor (PM_{10})	35 out of 113	Stainless steel and aluminum	47-mm Teflon membrane 47-mm quartz fiber	Gelman stainless steel in-line	Option to add 20-cm flow homogenizer

Continued on next page

Table II. Examples of Ambient Aerosol Sampling Systems Assembled for Major Studies—*Continued*

Sampling System	*Descriptive Reference*	*Inlet and Particle Size (µm)*	*Flow Rate (L/m)*	*Sampling Surface*	*Filter Media*	*Filter Holders*	*Features*
SCAQS *(Continued)*		Bendix 240 cyclone ($PM_{2.5}$)	35 out of 113	Teflon-coated aluminum and Teflon	47-mm Teflon membrane 47-mm quartz fiber 47-mm impregnated quartz fiber 47-mm nylon membrane 47-mm etched polycarbonate	Gelman stainless steel and Savillex PFA Teflon in-line	Option to add 20-cm flow homogenizer
Harvard/EPA Annular Denuder System (HEADS)	Koutrakis et al. (*116–118*)	Teflon-coated low-volume glass impactor ($PM_{2.5}$)	10	Glass	37-mm Teflon membrane 37-mm impregnated quartz fiber	Graseby–Andersen open-face ring	Includes sodium carbonate coated denuders to collect acidic gases (e.g., nitric acid, nitrous acid, sulfur dioxide, organic acids) and citric acid coated denuders to collect ammonia
Stacked Filter Unit (SFU)	Cahill et al. (*71*)	Large-pore etched polycarbonate filters ($PM_{2.0}$ to $PM_{3.0}$)	10	Polycarbonate	47-mm etched polycarbonate membrane 47-mm Teflon membrane	Nuclepore	Uses large-pore etched polycarbonate filters as $PM_{2.5}$ inlet
Interagency Monitoring of Protected Visual Environments (IMPROVE) Sampler	Eldred et al. (*119*)	Aluminum low-volume cyclone (PM_{10})	18	Aluminum	25-mm Teflon membrane 25-mm quartz fiber	Nuclepore polycarbonate open-face	Nitric acid denuders can be placed in inlet line
		Aluminum low-volume cyclone ($PM_{2.5}$)	24	Aluminum	25-mm Teflon membrane 25-mm quartz fiber 47-mm nylon membrane	Nuclepore polycarbonate open-face	

Table II. Examples of Ambient Aerosol Sampling Systems Assembled for Major Studies—*Continued*

Sampling System	*Descriptive Reference*	*Inlet and Particle Size (μm)*	*Flow Rate (L/m)*	*Sampling Surface*	*Filter Media*	*Filter Holders*	*Features*
Sequential Filter Sampler (SFS)	Watson et al. (*44*)	Aluminum medium-volume impactor (PM_{10})	20 out of 113	Aluminum	47-mm Teflon membrane 47-mm quartz fiber	Nuclepore polycarbonate open-face	Option to add nitric acid denuders in the sampling stream. Sequential sampling
		Steel medium-volume cyclone ($PM_{2.5}$)	20 out of 113	Teflon-coated aluminum	47-mm Teflon membrane 47-mm quartz fiber 47-mm nylon membrane 47-mm impregnated cellulose fiber	Nuclepore polycarbonate open-face	
New York University Medical Center/ Sequential Acid Aerosol Sampling System (NYUMC/SAASS)	Thurston et al. (*120*)	Teflon-coated glass low-volume impactor ($PM_{2.5}$)	4	Teflon-coated glass	37-mm Teflon membrane 37-mm nylon membrane	Graseby–Andersen open-face rings	Sequential sampling
California Acid Deposition Monitoring Program (CADMP) Dry Deposition Sampler	Chow et al. (*109*)	Aluminum medium-volume impactor (PM_{10})	20 out of 113	Aluminum	47-mm Teflon membrane 47-mm impregnated cellulose fiber	Savillex open-face	Includes nitric acid denuders. Sequential sampling
		Teflon-coated steel medium-volume cyclone ($PM_{2.5}$)	20 out of 113	PFA Teflon-coated aluminum	47-mm Teflon membrane 47-mm nylon membrane	Savillex PFA Teflon open-face	

Continued on next page

Table II. Examples of Ambient Aerosol Sampling Systems Assembled for Major Studies—*Continued*

Sampling System	*Descriptive Reference*	*Inlet and Particle Size (μm)*	*Flow Rate (L/m)*	*Sampling Surface*	*Filter Media*	*Filter Holders*	*Features*
Versatile Ambient Pollutant Sampler (VAPS)	Stevens et al. (*121*)	Teflon-coated aluminum low-volume elutriator (PM_{10}) and Teflon-coated aluminum low-volume virtual impactor ($PM_{2.5}$)	33	Teflon-coated aluminum	47-mm Teflon membrane 47-mm etched polycarbonate membrane 47-mm quartz fiber	University Research Glassware glass filter pack (Model 2000-30F)	Includes annular denuders to capture nitric acid, nitrous acid, and sulfur dioxide; and polyurethane foam to collect organic compounds
BYU Organic Sampling System (BOSS)	Eatough (*122*); Eatough et al. (*123,124*)	Teflon-coated aluminum medium-volume cyclone ($PM_{2.5}$)	140 L/m through inlet and 35 L/m per channel	Teflon-coated stainless steel	47-mm quartz fiber 47-mm activated charcoal impregnated filter (CIF)	University Research Glassware glass filter pack (Model 2000-30F)	A multichannel diffusion denuder sampler to determine semivolatile organic compounds
BYU Big BOSS	Tang et al. (*125*)	Aluminum high-volume virtual impactor with 2.5 mm, 0.8 mm, and 0.4 mm cutpoint	1130 L/m through inlet, with 11, 60, 93, and 200 L/m per channel	Teflon-coated stainless steel	47-mm quartz fiber 47-mm activated charcoal impregnated filter (CIF)	University Research Glass filter pack (Model 2000-30F)	A multichannel diffusion denuder sampler to determine semivolatile organic compounds

Table III. Analytical Measurement Specifications for Air Filter Samples

	Minimum Detection Limit (ng/m^3)								
Species	*XRF*[a]	*PIXE*[b]	*INAA*[b]	*ICP–AES*[b]	*AA Flame*[b]	*AA Furnace*[b]	*AC*[b]	*IC*[b]	*TOR*[b]
Ag	6	NA	0.12	1	4	0.005	NA	NA	NA
Al	5	12	24	20	30	0.01	NA	NA	NA
As	0.8	1	0.2	50	100	0.2	NA	NA	NA
Au	2	NA	NA	2.1	21	0.1	NA	NA	NA
Ba	25	NA	6	0.05	8	0.04	NA	NA	NA
Be	NA	NA	NA	0.06	2	0.05	NA	NA	NA
Br	0.5	1	0.4	NA	NA	NA	NA	NA	NA
Ca	2	4	94	0.04	1	0.05	NA	NA	NA
Cd	6	NA	4	0.4	1	0.003	NA	NA	NA
Ce	NA	NA	0.06	52	NA	NA	NA	NA	NA
Cl	5	8	5	NA	NA	NA	NA	NA	NA
Co	0.4	NA	0.02	1	6	0.02	NA	NA	NA
Cr	1	2	0.2	2	2	0.01	NA	NA	NA
Cs	NA	NA	0.03	NA	NA	NA	NA	NA	NA
Cu	0.5	1	30	0.3	4	0.02	NA	NA	NA
Eu	NA	NA	0.006	0.08	21	NA	NA	NA	NA
Fe	0.7	2	4	0.5	4	0.02	NA	NA	NA
Ga	0.9	1	0.5	42	52	NA	NA	NA	NA
Hf	NA	NA	0.01	16	2,000	NA	NA	NA	NA
Hg	1	NA	NA	26	500	21	NA	NA	NA
I	NA	NA	1	NA	NA	NA	NA	NA	NA
In	6	NA	0.006	63	31	NA	NA	NA	NA
K	3	5	24	NA	2	0.02	NA	NA	NA
La	30	NA	0.05	10	2,000	NA	NA	NA	NA
Mg	NA	20	300	0.02	0.3	0.004	NA	NA	NA
Mn	0.8	2	0.12	0.1	1	0.01	NA	NA	NA
Mo	1	5	NA	5	31	0.02	NA	NA	NA
Na	NA	60	2	NA	0.2	<0.05	NA	NA	NA
Ni	0.4	1	NA	2	5	0.1	NA	NA	NA
P	3	8	NA	50	100,000	40	NA	NA	NA
Pb	1	3	NA	10	10	0.05	NA	NA	NA
Pd	5	NA	NA	42	10	NA	NA	NA	NA
Rb	0.5	2	6	NA	NA	NA	NA	NA	NA
S	2	8	6000	10	NA	NA	NA	NA	NA
Sb	9	NA	0.06	31	31	0.2	NA	NA	NA
Sc	NA	NA	0.001	0.06	50	NA	NA	NA	NA
Se	0.6	1	0.06	25	100	0.5	NA	NA	NA
Si	3	9	NA	3	85	0.1	NA	NA	NA
Sm	NA	NA	0.01	52	2000	NA	NA	NA	NA
Sn	8	NA	NA	21	31	0.2	NA	NA	NA
Sr	0.5	2	18	0.03	4	0.2	NA	NA	NA
Ta	NA	NA	0.02	26	2000	NA	NA	NA	NA
Th	NA	NA	0.01	63	NA	NA	NA	NA	NA
Ti	2	3	65	0.3	95	NA	NA	NA	NA
Tl	1	NA	NA	42	21	0.1	NA	NA	NA
U	1	NA	NA	21	25,000	NA	NA	NA	NA
V	1	3	0.6	0.7	52	0.2	NA	NA	NA
W	NA	NA	0.2	31	1000	NA	NA	NA	NA

Continued on next page

Table III. Analytical Measurement Specifications for Air Filter Samples—*Continued*

	Minimum Detection Limit (ng/m^3)								
Species	*XRF*[a]	*PIXE*[b]	*INAA*[b]	*ICP–AES*[b]	*AA Flame*[b]	*AA Furnace*[b]	*AC*[b]	*IC*[b]	*TOR*[b]
Y	0.6	NA	NA	0.1	300	NA	NA	NA	NA
Zn	0.5	1	3	1	1	0.001	NA	NA	NA
Zr	0.8	3	NA	0.6	1000	NA	NA	NA	NA
Cl^-	NA	NA	NA	NA	NA	NA	NA	50	NA
NH_4^+	NA	NA	NA	NA	NA	NA	50	NA	NA
NO_3^-	NA	NA	NA	NA	NA	NA	NA	50	NA
SO_4^{2-}	NA	NA	NA	NA	NA	NA	NA	50	NA
EC	NA	NA	NA	NA	NA	NA	NA	NA	100
OC	NA	NA	NA	NA	NA	NA	NA	NA	100

NOTES: Minimum detection limit is three times the standard deviation of the blank for a filter of 1 mg/cm^2 area density. NA means not available. ICP–AES is inductively coupled plasma with atomic emission spectroscopy; AA is atomic absorption spectrophotometry; PIXE is proton induced X-ray emissions analysis; XRF is X-ray fluorescence analysis; INAA is instrumental neutron activation analysis; IC is ion chromatographic analysis; AC is automated colorimetric analysis; TOR is thermal optical reflectance analysis; OC is organic carbon; and EC is elemental carbon.

[a]Concentration is based on 13.8-cm^2 deposit area for a 47-mm filter substrate, with a nominal flow rate of 20 L/min for 24-h samples with 100-s radiation time.

[b]Concentration is based on the extraction of one-half of a 47-mm filter in 15 mL of deionized-distilled water, with a nominal flow rate of 20 L/min for 24-h samples.

SOURCE: PIXE data are from Eldred (72), INAA data are from Olmez (75); ICP–AES data are from Harman (*81*); AA furnace data are from Fernandez (77). For flame AA, data for Ba, Be, Ca, Co, K, and Na are from Harman (*81*);other flame AA data are from Fernandez (77).

ity must be equal to or better than 100 μg for high-volume samples, 10 μg for medium-volume samples, and 1 μg for low-volume samples of 24-h duration. Gloved hands and a clean work area are prerequisites for weighing filters intended for chemical analysis.

Elemental Analysis Methods

Photon-induced X-ray fluorescence (XRF) (*69, 70*) and proton-induced X-ray emission (PIXE) (*71, 72*) are the most commonly used methods for aerosol elemental analysis because they are nondestructive. The particle deposit is irradiated by high-energy X-rays (XRF) or protons (PIXE) that eject inner shell electrons from the atoms of each element in the sample. When a higher energy electron drops into the vacant lower energy orbital, a fluorescent X-ray photon is released. The energy of this photon is unique to each element, and the number of photons is proportional to the concentration of the element. Concentrations are quantified by comparing photon counts for a sample with those obtained from thin-film standards of known concentration. Lower energy X-rays from elements such as sodium, magnesium, aluminum, silicon, sulfur, phosphorus, and chlorine are easily absorbed by the filter and even by the particles in which they are contained. XRF and PIXE are most effectively applied to samples collected on membrane-type filters such as Teflon or polycarbonate membrane filter substrates. These membrane filters collect the deposit on their surfaces, which eliminates biases due to absorption of X-rays by the filter material. These filters also have a low areal density that minimizes the scatter of incident X-rays, and their inherent trace element content is very low. Glass and quartz fiber filters used for high-volume aerosol sampling do not exhibit these features. Particles are collected within the filter matrix, which attenuates the emitted X-rays. Impurities observed in various types of glass and quartz fiber filters include aluminum, silicon, sulfur, chlorine, potassium, calcium, iron, nickel, copper, zinc, rubidium, strontium, molybdenum, barium, and lead. Concentrations for aluminum, silicon, and phosphorus cannot be determined for quartz fiber filters because of the large silicon content of the filters.

Destructive methods for aerosol elemental analysis include (1) instrumental neutron activation analysis (INAA), (2) atomic absorption spectrophotometry (AAS), and (3) inductively coupled plasma with atomic emission spectroscopy (ICP–AES).

In INAA (*73–75*), a sample is inserted into the core of a nuclear reactor and irradiated for periods ranging from a few minutes to several hours. Several elements transform into radioactive isotopes under neutron bombardment. These isotopes are identified by their gamma rays. The number of gamma-ray photons counted by a germanium detector is proportional to the amount of the parent element present in the sample. A sample remains after INAA that could conceivably be submitted to other analyses. However, the intense neutron irradiation often pulverizes the filter, and the residual radioactivity needs to decay before the sample can be removed.

In AAS (*76, 77*), a few milliliters of the entirely extracted sample are injected into a flame where the elements are vaporized. A light beam with wavelengths specific to the elements being measured is directed through the flame to be detected by a monochromator. The light absorbed by the flame containing the sample extract is compared with the absorption from known standards to quantify the elemental concentrations. ICP–AES (*78–81*) is similar to AAS, except that the dissolved sample is injected into an atmosphere of argon gas seeded with free electrons induced by high voltage from a surrounding Tesla coil. High temperatures in the plasma raise valence electrons above their normally stable states. A photon of light unique to the element that was excited is emitted when these electrons return to their stable states.

Water-Soluble Ion Measurement Methods

Aerosol ions refer to chemical compounds that are soluble in water. Abundant aerosol ions include sulfate, nitrate, chloride, fluoride, phosphate, ammonium, potassium, and sodium. Magnesium, calcium, iron, and manganese may also have soluble fractions specific to emission sources and may participate in atmospheric aerosol chemistry. To measure aerosol ions, the filters are first extracted in distilled–deionized water and then filtered to remove undissolved particles. The major sampling requirement for water-soluble ion analysis is that the filter material be hydrophilic, allowing water to penetrate the filter and fully extract the desired chemical compounds. Small amounts of ethanol or other wetting agents can be added to the filter surface to aid the wetting of hydrophobic filter materials, but this procedure introduces the potential for sample contamination.

Simple ions, such as sodium, magnesium, potassium, and calcium, are quantified by AAS as described previously. Polyatomic ions such as sulfate, nitrate, ammonium, and phosphate must be quantified by other methods such as ion chromatography (IC) and automated colorimetry (AC). Simple ions, such as chloride and fluoride, may also be measured by IC and AC.

The operating principle for AAS was described in the preceding section. IC can be used for both anions (fluoride, phosphate, chloride, nitrate, or sulfate) and cations (potassium, ammonium, or sodium) with separate columns. Applied to aerosol samples, the anions are most commonly analyzed by IC, and the cations are most commonly analyzed by a combination of AAS and AC. In IC (*82–89*), the sample extract passes through an ion-exchange column to separate ions in time for individual quantification by an electroconductivity detector. Prior to detection, the column effluent enters a micromembrane suppressor where the chemical composition of the eluant is altered, resulting in a matrix of low conductivity. The ions are identified by their elution or retention times and are quantified by the conductivity peak area or peak height.

Ammonium is often measured by AC, which can also be used to measure sulfate, nitrate, and chloride. In AC, a peristaltic pump introduces air bubbles into the sample stream at known intervals to separate individual samples in a continuous stream. The stream is mixed with various reagents and subjected to appropriate reaction periods before submission to a colorimeter. The ion being measured usually reacts to form a colored liquid. The liquid absorbance is related to the amount of the ion in the sample by Beer's law. This absorbance is measured by a photomultiplier tube through an interference filter that is specific to the colors generated by the species being measured.

The methylthymol blue (MTB) method (*90*) is applied to analyze sulfate. Nitrate is reduced to nitrite, which reacts with sulfanilamide to form a diazo (*91*). This diazo is then reacted to an azo dye for colorimetric determination. Ammonium is measured with the indophenol method (*92*).

Carbon Measurement Methods

Although there are thousands of carbon-containing organic compounds in suspended particles, practical analyses are usually limited to the determination of total carbon and its organic and elemental components. Organic and elemental carbon are operationally defined by the measurement method (*93*). Elemental carbon is sometimes termed "soot", "graphitic carbon", or "black carbon". Light-absorbing carbon is a more useful concept than elemental carbon for visibility reduction. Several consistent but distinct fractions of carbon in both source and receptor samples are desired for source apportionment. There are many organic materials that absorb light (e.g., coffee, tar, asphalt, or motor oil), so "elemental carbon" is an imprecise name for light-absorbing carbon. Chow et al. (*94*) document several variations of the thermal (T), thermal optical reflectance (TOR), thermal optical transmission (TOT), and thermal manganese oxidation (TMO) methods for organic and elemental carbon. The TOR (*94, 95*) and TMO (*96, 97*) methods have been most commonly applied in aerosol studies. Sampling on quartz fiber filters, which can withstand high temperatures without melting or burning, is needed for thermal carbon analyses.

The TMO method uses manganese dioxide as an oxidizing agent in contact with the sample during the analysis. The sample is heated and the oxidized carbon leaves the combusted sample as carbon dioxide, which is converted to methane and quantified with a flame-ionization detector. Carbon evolving at 525 °C is classified as organic carbon, and carbon evolving at 850 °C is classified as elemental carbon.

The TOR method submits a sample to volatilization at temperatures ranging from ambient to 550 °C in a pure helium atmosphere, then to combustion at temperatures between 550 and 800 °C in a 2% oxygen–98% helium atmosphere with several temperature ramping steps. The carbon released at each temperature is converted to methane and quantified with a flame-ionization detector. The reflectance from the deposit side of the filter punch is monitored throughout the analysis. This reflectance usually decreases during volatilization in the helium atmosphere owing to the pyrolysis of organic material. When oxygen is added, the reflectance increases as the light-absorbing carbon is combusted and removed. Organic carbon is defined as that which evolves prior to reattainment of the original reflectance, and elemental carbon is defined as that which evolves after the original reflectance has been attained.

Sampling and Analysis Strategies

A sampling and analysis strategy must begin with a clear and complete statement of the study objectives. Compliance determination, source apportionment, and control strategy evaluation are the most common objectives for PM_{10} nonattainment areas. Compliance determination requires that PM_{10} mass be measured at least every sixth day using an EPA-designated reference or equivalent sampler. Adherence to compliance monitoring methods is not sufficient to provide samples amenable to chemical analysis. Source apportionment and control strategy evaluation require chemical speciation, so additional measures must be taken when these objectives are to be addressed.

Once study objectives are determined, the chemicals to be measured must be specified and their expected concentrations must be estimated. When source apportionment is an objective, it is desirable to obtain chemicals that are present in the sources suspected to contribute to PM_{10}. Table IV identifies several source types

Table IV. Typical Chemical Abundances in Source Emissions

Source Type	Dominant Particle Size	Chemical Abundances <0.1%	0.1–1%	1–10%	>10%
Geological material	Coarse	Cr, Zn, Rb, Sr, Zr	Cl^-, NO_3^-, SO_4^{2-}, NH_4^+, P, S, Cl, Ti, Mn, Ba, La	OC, EC, Al, K, Ca, K, Fe	Si
Motor vehicle	Fine	Cr, Ni, Y^+	NH_4^+, Si, Cl, Al, Si, P, Ca, Mn, Fe, Zn, Br, Pb	S, Cl^-, NO_3^-, SO_4^{2-}, NH_4^+	OC, EC
Vegetative burning	Fine	Ca, Mn, Fe, Zn, Br, Rb, Pb	NO_3^-, SO_4^{2-}, NH_4^+, Na^+	K^+, K, Cl, Cl^-	OC, EC
Residual oil combustion	Fine	K^+, OC, Cl, Ti, Cr, Co, Ga, Se	Na^+, NH_4^+, Zn Fe, Si	Ni, OC, EC, V	S, SO_4^{2-}
Incinerator		V, Mn, Cu, Ag, Sn	K^+, Al, Ti, Zn, Hg	NO_3^-, Na^+, EC, Si, S, Ca, Fe, Br, La, Pb	SO_4^{2-}, NH_4^+, OC, Cl
Coal-fired power plant	Fine	Cl, Cr, Mn, Ga, As, Se, Br, Rb, Zr	NH_4^+, P, K, Ti, V, Ni, Zn, Sr, Ba, Pb	SO_4^{2-}, OC, EC, Al, S, Ca, Fe	Si
Marine	Coarse	Ti, V, Ni, Sr, Zr, Pd, Ag, Sn, Sb, Pb	K, Ca, Fe, Cu, Zn, Ba, La, Al, Si	NO_3^-, SO_4^{2-}, OC, EC	Na^+, Na, Cl^-, Cl

NOTE: OC is organic carbon, and EC is elemental carbon.

commonly found in PM_{10} nonattainment areas, along with the types of chemicals known to be present in these source emissions. Table IV can be used in conjunction with Table II to determine which methods should be applied to obtain the needed measurements. Expected emissions cycles should be examined to determine sampling periods and durations. For example, residential wood burning usually contributes to samples taken during the nighttime, whereas agricultural burning usually contributes to samples taken during the daytime. Although these two source types may be indistinguishable based on their chemical profiles, their different diurnal cycles provide convincing evidence that one or the other is a major contributor when daytime and nighttime samples are available.

Next, sample durations and flow rates must be matched with lower quantifiable limits for the desired chemicals and compared against expected concentrations for these chemicals. Urban samples acquire adequate deposits for analysis with flow rates as low as ~20 L/min for sample durations as short as 4 h. Samples at nonurban sites may require >100-L/min flow rates for 24-h durations to obtain an adequate deposit. The analytical laboratory should be involved in the study design to ensure compatibility among sampling methods, analysis methods, and lower quantifiable limits. When these specifications are complete, one of the sampling systems described previously can be matched with appropriate filter media, analysis methods, and operating procedures.

Although this procedure is ideal, time, equipment, and financial constraints often require compromises. The following subsections provide guidance on what can be done with different sampling and analysis configurations in a stepwise fashion.

Analysis of Archived PM_{10} Filters

As noted earlier, samples taken without the intent of chemical analysis can rarely be used to provide defensible chemical concentrations for source apportionment or other purposes requiring precise chemical speciation. Chemical measurements are still useful in a semiquantitative or qualitative sense to identify, although not to quantify, the major source types contributing to high PM_{10} concentrations, and to examine intersite variability. Elemental, ion, and carbon analysis methods can be applied to these filters subject to the limitations stated previously. Archived filters should be reweighed prior to sectioning, and the reweight should be compared with the final weight that was taken immediately following sampling. This protocol will provide an approximation for the particulate matter (typically volatile organic compounds and nitrates) that has been lost during storage. Archived filters of greatest interest are those with PM_{10} concentrations in excess of 150 $\mu g/m^3$.

Filters from all sites within the air quality management area should be examined on a PM_{10} standard violation day, even though the violation may not be exceeded at every site. Differences in chemical content among sites, coupled with knowledge of emissions source locations, will assist in determining whether chemical contributions have a local or a regional effect on PM_{10}. If blank filters from lots corresponding to the samples have been archived, these should be submitted to the same analyses as the exposed filters. If blanks levels for a chemical exceed 30% of actual sample concentrations, the concentration of that species should not be used for interpretive purposes as it probably does not accurately represent the concentration in the deposited particles.

Planned High-Volume PM_{10} Sampling

If it is known that chemical analyses will be applied to some or all of the high-volume PM_{10} samples, several precautions should be taken. When procuring filters, minimally acceptable blank concentrations should be stated for each chemical to be quantified. One filter from each labeled batch of filters received from a vendor should be submitted to the same chemical analysis methods that might be applied to all filters. If the blank levels exceed 30% of the expected concentrations for any of the chemicals to be analyzed, then the batch of filters should be returned to the manufacturer. Specifying the maximum tolerable levels in the purchase agreement will provide an opportunity to obtain replacements at no additional cost. Filter holders that mate to the high-volume PM_{10} sampler should be obtained, and filters should be loaded and unloaded using gloved hands in a laboratory setting. Each filter should be folded in half with the exposed side inward and stored in a Ziplock bag. These bags should be placed in hanging folders with their corresponding data sheets and stored in a refrigerator or freezer after weighing.

High-Volume and Dichotomous PM_{10} Sampling

As noted previously, the quartz fiber filters used for high-volume sampling are especially poor for elemental measurements owing to the penetration of particles into the filter and high blank levels for many of the species to be measured. A dichotomous sampler operated alongside the high-volume PM_{10} sampler with Teflon membrane filters supplies two, 37-mm-diameter samples (one for $PM_{2.5}$ and one for coarse particles), which can be submitted to XRF or PIXE analysis for elements. Portions of the high-volume quartz fiber filter can be submitted to ion and carbon analyses. This procedure provides accurate measurements for receptor modeling in most cases, plus the additional information from two different particle sizes obtained from the dichotomous sampler. Because of the slight differences between the sampler PM_{10} cut points, the concentrations are not entirely comparable for the two samplers. A more accurate sampling method would be to collocate two dichotomous samples, one with Teflon membrane filters and one with quartz fiber filters. When all of the filters are submitted to gravimetric analysis, additional validation information is available by comparing the two PM_{10} concentrations. Because both sampling systems are reference samplers, compliance can be determined from either sampler. All filter media should undergo the acceptance testing, handling, and storage procedures described previously.

High-Volume or Dichotomous PM_{10} and Continuous PM_{10} Sampling

A continuous TEOM or BAM can be operated along with a high-volume or dichotomous PM_{10} sampling system. This configuration is most useful when short-term events, such as fires or windblown dust, are hypothesized to be major contributors to excessive PM_{10}. The dichotomous sampler using Teflon membrane filters is preferred because this combination allows particle size as well as mass and elements to be measured. When the TEOM is equipped with a bypass channel (*98*), a quartz fiber filter can be taken simultaneously with the hourly measurements to be analyzed for ions and carbon. If a high-volume PM_{10} sampler with a quartz fiber filter is used, the bypass channel should use a Teflon membrane filter for elemental analysis.

Sequential Filter Sampling

When high PM_{10} levels are suspected to result from multiday buildup from a variety of sources, it is desirable to have daily samples available that can be submitted to analysis. High-volume and dichotomous PM_{10} sampling can be applied to this task, but this schedule requires someone to change the filters at midnight every day, or several samplers and a timing mechanism to switch between them. Manpower and equipment costs can become prohibitive. Many sampling sites have limited space and cannot accommodate a large number of sampling systems. In this case, a sequential filter samplers (SFS) using Teflon membrane filters for mass and elemental analysis and quartz fiber filters for ion and carbon analysis is a good choice. The SFS can also be applied to situations in which more than one sample per day is needed to bracket emissions events with samples amenable to chemical analysis (*99–104*). In this case, two to six samples of 4- to 12-h duration are taken sequentially and analyzed separately for the desired chemical species.

Saturation Sampling

Sometimes, one or more source categories may be identified as major contributors to elevated PM_{10}, but the chemical profiles of specific emitters are too similar to differentiate them from each other. In this case, portable survey samplers using Teflon membrane filters can be located within and around the suspected emitters (*44, 105–107*). If the objective of the study is to characterize fugitive dust sources, mass and elemental analyses are sufficient to separate this source category from others by receptor modeling. Several studies have applied the survey sampler to characterize the contributions from residential wood combustion. In this case, collocated samplers with Teflon membrane and quartz fiber filters are required for full chemical speciation. The remaining geological source contributions can be used in spatial receptor models to identify the locations of specific emission sources (*108*).

Denuder and Impregnated Filter Sampling

When secondary ammonium sulfate and ammonium nitrate are major contributors, one or more sites should be operated to obtain precursor concentrations of nitric acid and ammonia gas (e.g., *109*). In the eastern United States, sulfuric acid and ammonium bisulfate are also important components. In this situation, denuder methods (*45–49*) can be applied to obtain accurate measures of the secondary aerosol and the precursor gases. These precursor gas measurements should be accompanied by collocated temperature and relative humidity measurements so that equilibrium receptor models (e.g., *110*) can be applied to determine whether the secondary particles are limited by emissions of ammonia or oxides of nitrogen.

Summary and Conclusions

Several sampling systems are available for PM_{10} compliance sampling. All systems are filter-based, although some can monitor mass as the deposit is accumulating. Each system consists of an inlet to determine the particle sizes collected, filter media onto which the particles are deposited, filter holders to position and protect the filters, and pumps with flow controllers to draw air through the filters and inlets. Several different filters are needed for chemical analysis, and the sampling must be matched with the analysis methods prior to monitoring. This matching ensures that filter blank levels, filter deposits, and sampling substrates will be amenable to the desired analytical methods.

No single sampling system can meet all needs, and it is often necessary to adapt existing sampling components to the specific situation being studied. Many examples of successful sampling systems that can be copied or modified to meet these specific needs have been published.

References

1. *Fed. Regist.* **1987**, *52*, 45684–45685.
2. Rubow, K. L.; Furtado, V. C. In *Air Sampling Instruments for Evaluation of Atmospheric Contaminants*, 7th ed.; Hering, S. V., Ed.; American Conference of Governmental Industrial Hygienists: Cincinnati, OH, 1989; pp 241–274.
3. Watson, J. G.; Chow, J. C. In *Aerosol Measurement: Principles, Techniques and Applications;* Willeke, K.; Baron, P. A., Eds.; Van Nostrand Reinhold: New York, 1993; pp 622–639.
4. Appel, B. R. In *Aerosol Measurement: Principles, Techniques and Applications*, Willeke, K.; Baron, P. A., Eds.; Van Nostrand Reinhold: New York, 1993; pp 233–259.
5. Chow, J. C.; Watson, J. G. *Guidelines for PM_{10} Sampling and Analysis Applicable to Receptor Modeling, Final Report;* prepared by Desert Research Institute, Reno, NV, for the Office of Air Quality Planning and Standards, U.S. Environmental Protection Agency: Research Triangle Park, NC, 1994; DRI Document 2625.1F.
6. Watson, J. G.; Chow, J. C. In *Sampling of Environmental Materials for Trace Analysis;* Markert, B., Ed.; VCH-Publisher: Weinheim, Germany, 1994; pp 83–175.
7. *Fed. Regist.* **1987**, *52*, 24634.
8. Friedlander, S. K. *Smoke, Dust, and Haze;* John Wiley & Sons: New York, 1977.
9. Hinds, W. C. *Aerosol Technology: Properties, Behavior, and Measurement of Airborne Particles;* John Wiley & Sons: New York, 1982.
10. Sloane, C. S.; Watson, J. G.; Chow, J. C.; Pritchett, L. C.; Richards, L. W. In *Transactions, Visibility and Fine Particles;* Mathai, C. V., Ed.; Air and Waste Management Association: Pittsburgh, PA, 1990; pp 384–393.
11. Watson, J. G. *J. Air Pollut. Control Assoc.* **1984**, *34*, 619–623.
12. Hopke, P. K. In *Chemical Analysis;* John Wiley & Sons: New York, 1985; p 76.
13. Hopke, P. K. In *Receptor Modeling for Air Quality Management;* Hopke, P. K., Ed.; Elsevier Science: Amsterdam, Netherlands, 1991; pp 1–10.
14. Watson, J. G.; Robinson, N. F.; Chow, J. C.; Henry, R. C.; Kim, B. M.; Pace, T. G.; Meyer, E. L.; Nguyen, Q. *Environ. Software (Southampton, U.K.)* **1990**, *5*, 38–49.
15. Watson, J. G.; Chow, J. C.; Pace, T. G. In *Data Handling in Science and Technology. Volume 7: Receptor Modeling for Air Quality Management;* Hopke, P. K., Ed.; Elsevier Science: New York, 1991; pp 83–116.
16. Pace, T. G. In *Receptor Modeling for Air Quality Management;* Hopke, P. K., Ed.; Elsevier Science: Amsterdam, Netherlands, 1991; pp 255–297.
17. Watson, J. G.; Chow, J. C.; Shah, J. J.; Pace, T. G. *J. Air Pollut. Control Assoc.* **1983**, *33*, 114–119.
18. Wedding, J. B.; Carney, T. C. *Atmos. Environ.* **1983**, *17*, 873–882.
19. Wedding, J. B.; McFarland, A. R.; Cermak, J. E. *Environ. Sci. Technol.* **1977**, *11*, 387–398.
20. McFarland, A. R.; Ortiz, C. A.; Rodes, C. E. In *Proceedings of the Technical Basis for a Size Specific Particulate Standard;* Air Pollution Control Association: Pittsburgh, PA, 1980; p 59.
21. Wedding, J. B.; Weigand, M. A. *Atmos. Environ.* **1985**, *19*, 535–538.

22. Olin, J. G.; Bohn, R. R. Presented at the 76th Annual Meeting of the Air Pollution Control Association, Atlanta, GA, 1983.
23. Wedding, J. B.; Weigand, M. A.; Ligotke, M. W.; Baumgardner, R. *Environ. Sci. Technol.* **1983,** *17*(7), 379–383.
24. McFarland, A. R.; Ortiz, C. A.; Bertch, R. W. *Environ. Sci. Technol.* **1978,** *12,* 679–682.
25. John, W.; Reischl, G. *J. Air Pollut. Control Assoc.* **1980,** *30,* 872–876.
26. John, W.; Hering, S.; Reischl, G.; Sasaki, G. V. *Atmos. Environ.* **1983,** *17,* 373–382.
27. John, W.; Wall, S. M.; Wesolowski, J. J. *Project Summary: Validation of Samplers for Inhaled Particulate Matter;* U.S. Environmental Protection Agency, Environmental Monitoring Systems Laboratory: Research Triangle Park, NC, 1983; EPA 602/S4–83–010.
28. *Fed. Regist.* **1987,** *52,* 24663–24666.
29. Lippmann, M. In *Air Sampling Instruments for Evaluation of Atmospheric Contaminants,* 7th ed.; Hering, S. V., Ed.; American Conference of Governmental Industrial Hygienists: Cincinnati, OH, 1989; pp 305–336.
30. Wedding, J. B. *Atmos. Environ.* **1985,** *19*(7), 1219–1222.
31. Wedding, J. B.; Weigand, M. A.; Kim, Y. J.; Swift, D. L.; Lodge, J. P. *J. Air Pollut. Control Assoc.* **1987,** *37,* 254–258.
32. Hering, S. V. In *Air Sampling Instruments for Evaluation of Atmospheric Contaminants,* 7th ed.; Hering, S. V., Ed.; American Conference of Governmental Industrial Hygienists: Cincinnati, OH, 1989; pp 337–385.
33. Perry, W. H. In *Air Sampling Instruments for Evaluation of Atmospheric Contaminants,* 7th ed.; Hering, S. V., Ed.; American Conference of Governmental Industrial Hygienists: Cincinnati, OH, 1989; pp 291–303.
34. *Quality Assurance Handbook for Air Pollution Measurement Systems;* U.S. Environmental Protection Agency, Office of Research and Development, Environmental Monitoring Systems Laboratory: Research Triangle Park, NC, 1989; Section 2.11.0; EPA 600/3–89–027a.
35. Watson, J. G.; Bowen, J. L.; Chow, J. C.; Rogers, C. F.; Ruby, M. G.; Rood, M. J.; Egami, R. T. In *Methods of Air Sampling and Analysis,* 3rd ed.; Lodge, J. P., Jr., Ed.; Lewis Publishers: Chelsea, MI, 1989; pp 427–439.
36. Evans, J. S.; Ryan, P. B. *Aerosol Sci. Technol.* **1983,** *2,* 531–536.
37. Lillienfeld, P.; Dulchinos, J. *Am. Ind. Hyg. Assoc. J.* **1972,** *33,* 136–145.
38. Husar, R. B. *Atmos. Environ.* **1974,** *8,* 183–188.
39. Lillienfeld, P. *Staub-Reinhalt. Luft* **1975,** *25,* 458.
40. Macias, E. S.; Husar, R. B. *Environ. Sci. Technol.* **1976,** *10,* 904–907.
41. Lillienfeld, P. Presented at the 72nd Annual Meeting of the Air Pollution Control Association, Cincinnati, OH, 1979.
42. Patashnick, H.; Rupprecht, E. G. *J. Air Waste Manage. Assoc.* **1991,** *41,* 1079–1083.
43. Rupprecht, E.; Meyer, M.; Patashnick, H. Presented at the European Aerosol Conference, Oxford, U.K., 1992.
44. Watson, J. G.; Chow, J. C.; Frazier, C. A.; Lowenthal, D. H. *Program Plan for an Imperial Valley/Mexicali* PM_{10} *Source Apportionment Study, Draft Report;* prepared by Desert Research Institute, Reno, NV, for Region IX, U.S. Environmental Protection Agency: San Francisco, CA, 1991; DRI Document 8623.1 D1.
45. Shaw, R. W.; Stevens, R. K.; Bowermaster, J.; Tesch, J. W.; Tew, E. *Atmos. Environ.* **1982,** *16,* 845–853.
46. Vossler, T. L.; Stevens, R. K.; Paur, R. J.; Baumgardner, R. E.; Bell, J. P. *Atmos. Environ.* **1988,** *22*(8), 1729–1736.
47. Appel, B. R.; Winer, A. M.; Tokiwa, Y.; Bierman, H. W. *Atmos. Environ.* **1990,** *24A,* 611–616.
48. Stevens, R. K.; Purdue, L. J.; Barnes, H. M.; Ward, R. P.; Baugh, J. O.; Bell, J. P.; Sauren, H.; Sickles, J. E., II; Hodson, L. L. In *Transactions, Visibility and Fine Particles;* Mathai, C. V., Ed.; Air and Waste Management Association: Pittsburgh, PA, 1990; pp 122–130.

49. Cheng, Y. S. In *Aerosol Measurement: Principles, Techniques and Applications;* Willeke, K.; Baron, P. A., Eds.; Van Nostrand Reinhold: New York, 1993; pp 427–451.
50. Berner, A.; Lurzer, C. H.; Pohl, L.; Preining, O.; Wagner, P. *Sci. Total Environ.* **1979,** *13,* 245–261.
51. Hering, S. V.; Flagan, R. C.; Friedlander, S. K. *Environ. Sci. Technol.* **1979,** *12,* 667–673.
52. Marple, V. A.; Liu, B. Y. H.; Kuhlmey, G. A. *J. Aerosol Sci.* **1981,** *12,* 333–337.
53. Raabe, O. G.; Braaten, D. A.; Axelbaum, R. L.; Teague, S. V.; Cahill, T. A. *J. Aerosol Sci.* **1988,** *19,* 183–195.
54. Jackson, M. L. *Spec. Pap.—Geol. Soc. Am.* **1981,** (186), 27–36.
55. Currie, L. A. In *Particulate Carbon: Atmospheric Life Cycle;* Wolff, G. T.; Klimisch, R. L., Eds.; Plenum: New York, 1982; pp 245–260.
56. Hirose, K.; Sugimura, Y. *Earth Planet. Sci. Lett.* **1984,** *70,* 110–114.
57. Davis, B. L. *Atmos. Environ.* **1978,** 12, 2403–2406.
58. Davis, B. L. *Atmos. Environ.* **1980,** *14,* 217–220.
59. Schipper, L. B., III; Chow, J. C.; Frazier, C. A. Presented at the 86th Annual Meeting of the Air and Waste Management Association, Denver, CO, 1993; Document 93–MP–6.03.
60. Appel, B. R.; Hoffer, E. M.; Haik, M.; Wall, S. M.; Kothny, E. L. *Characterization of Organic Particulate Matter;* California Air Resources Board: Sacramento, CA, 1977; Document ARB–R–5–682–77–72.
61. Simoneit, B. R. *Atmos. Environ.* **1984,** *18,* 51–67.
62. Flessel, P.; Wang, Y. Y.; Chang, K. I.; Wesolowski, J. J.; Guirguis, G. N.; Kim, I. S.; Levaggi, D.; Wayman, S. *J. Air Waste Manage. Assoc.* **1991,** *41,* 276–281.
63. Hildemann, L. M.; Markowski, G. R.; Cass, G. R. *Environ. Sci. Technol.* **1991,** *25,* 744–759.
64. Li, C. K.; Kamens, R. M. *Atmos. Environ.* **1993,** *27A*(4), 523–532.
65. Rogge, W. F.; Hildemann, L. M.; Mazurek, M. A.; Cass, G. R.; Simoneit, B. R. T. *Environ. Sci. Technol.* **1993,** *27*(4), 636–651.
66. Rogge, W. F.; Mazurek, M. A.; Hildemann, L. M.; Cass, G. R.; Simoneit, B. R. T. *Atmos. Environ.* **1993,** *27A*(8), 1309.
67. Rogge, W. F.; Hildemann, L. M.; Mazurek, M. A.; Cass, G. R.; Simoneit, B. R. T. *Atmos. Environ.* **1993,** *27,* 1892–1904.
68. Casuccio, G. S.; Schwoeble, A. J.; Henderson, B. C.; Lee, R. J.; Hopke, P. K.; Sverdrup, G. M. In *Transactions: Receptor Models in Air Resources Management;* Watson, J. G., Ed.; Air and Waste Management Association: Pittsburgh, PA, 1989; pp 39–58.
69. Dzubay, T. G.; Stevens, R. K. *Environ. Sci. Technol.* **1975,** *9,* 663–667.
70. Jaklevic, J. M.; Loo, B. W.; Goulding, F. S. In *X-Ray Fluorescence Analysis of Environmental Samples*, 2nd ed.; Dzubay, T. G., Ed.; Ann Arbor Science: Ann Arbor, MI, 1977; pp 3–18.
71. Cahill, T. A.; Eldred, R. A.; Feeney, P. J.; Beveridge, P. J.; Wilkinson, L. K. In *Transactions, Visibility and Fine Particles;* Mathai, C. V., Ed.; Air and Waste Management Association: Pittsburgh, PA, 1990; pp 213–218.
72. Eldred, R. A. Crocker Nuclear Laboratory, University of California, Davis, CA, personal communication, 1993.
73. Dams, R.; Robbins, J. A.; Rahn, K. A.; Winchester, J. W. *Anal. Chem.* **1970,** *42,* 861–867.
74. Zoller, W. H.; Gordon, G. E. *Anal. Chem.* **1970,** *42,* 257–265.
75. Olmez, I. In *Methods of Air Sampling and Analysis*, 3rd ed.; Lodge, J. P., Jr., Ed.; Lewis Publishers: Chelsea, MI, 1989; pp 143–150.
76. Ranweiler, L. E.; Moyers, J. L. *Environ. Sci. Technol.* **1974,** *8,* 152.
77. Fernandez, F. J. In *Methods of Air Sampling and Analysis*, 3rd ed.; Lodge, J. P., Jr., Ed.; Lewis Publishers: Chelsea, MI, 1989; pp 143–150.
78. Fassel, V. A.; Kniseley, R. N. *Anal. Chem.* **1974,** *46*(13), 1110A–1120A.

79. McQuaker, N. R.; Khickner, P. D.; Gok, N. C. *Anal. Chem.* **1979,** *51,* 888–895.
80. Lynch, A. J.; McQuaker, N. R.; Bicun, D. I. *J. Air Pollut. Control Assoc.* **1980,** *30,* 257–259.
81. Harman, J. N. In *Methods of Air Sampling and Analysis*, 3rd ed.; Lodge, J. P., Jr., Ed.; Lewis Publishers: Chelsea, MI, 1989; pp 88–92.
82. Small, H.; Stevens, T. S.; Bauman, W. C. *Anal. Chem.* **1975,** *47,* 1801–1809.
83. Mulik, J.; Puckett, R.; Williams, D.; Sawicki, E. *Environ. Sci. Technol.* **1976,** *9*(7), 653–663.
84. Mulik, J. D.; Puckett, R.; Sawicki, E.; Williams, D. In *Methods and Standards for Environmental Measurement;* National Bureau of Standards: Washington, DC, 1977; Document 464.
85. Butler, F. E.; Jungers, R. H.; Porter, L. F.; Riley, A. E.; Toth, F. J. In *Ion Chromatographic Analysis of Environmental Pollutants;* Sawicki, E.; Mulik, J. D.; Wittgenstein, E., Eds.; Ann Arbor Science: Ann Arbor, MI, 1978; pp 65–76.
86. Mueller, P. K.; Mendoza, B. V.; Collins, J. C.; Wilgus, E. S. In *Ion Chromatographic Analysis of Environmental Pollutants;* Sawicki, E.; Mulik, J. D.; Wittgenstein, E., Eds.; Ann Arbor Science: Ann Arbor, MI, 1978; pp 77–86.
87. Rich, W.; Tillotson, J. A.; Chang, R. C. In *Ion Chromatographic Analysis of Environmental Pollutants;* Sawicki, E.; Mulik, J. D.; Wittgenstein, E., Eds.; Ann Arbor Science: Ann Arbor, MI, 1978; pp 17–29.
88. Small, H. In *Ion Chromatographic Analysis of Environmental Pollutants;* Sawicki, E.; Mulik, J. D.; Wittgenstein, E., Eds.; Ann Arbor Science: Ann Arbor, MI, 1978; pp 11–21.
89. Appel, B. R.; Wehrmeister, W. J. In *Ion Chromatographic Analysis of Environmental Pollutants;* Mulik, J. D.; Sawicki, E., Eds.; Ann Arbor Science: Ann Arbor, MI, 1979; pp 223–233, Vol. 2.
90. *Standard Methods for the Examination of Water and Wastewater,* 14th ed.; American Public Health Association, American Water Works Association, Water Pollution Control Federation: Washington, DC, 1975; p 628.
91. *Standard Methods for the Examination of Water and Wastewater*, 15th ed.; American Public Health Association, American Water Works Association, Water Pollution Control Federation: Washington, DC, 1980.
92. *Methods for Chemical Analysis of Water and Wastes;* U.S. Environmental Protection Agency: Cincinnati, OH, 1979; EPA–600/4–79–020.
93. Grosjean, D. Presented at the Second Chemical Congress of the North American Continent, Las Vegas, NV, 1980.
94. Chow, J. C.; Watson, J. G.; Pritchett, L. C.; Pierson, W. R.; Frazier, C. A.; Purcell, R. G. *Atmos. Environ.* **1993,** *27A,* 1185–1201.
95. Huntzicker, J. J.; Johnson, R. L.; Shah, J. J.; Cary, R. A. In *Particulate Carbon: Atmospheric Life Cycle;* Wolff, G. T.; Klimisch, R. L., Eds.; Plenum: New York, 1982; pp 79–88.
96. Mueller, P. K.; Fung, K. K.; Heisler, S. L.; Grosjean, D.; Hidy, G. M. In *Particulate Carbon: Atmospheric Life Cycle;* Wolff, G. T.; Klimisch, R. L., Eds.; Plenum: New York, 1982; pp 343–370.
97. Fung, K. K. *Aerosol Sci. Technol.* **1990,** *12,* 122–127.
98. *Operating Manual: Partixol Model 2000 Air Sampler, Version 1.00;* Rupprecht & Patashnick Company: Albany, NY, 1994.
99. Watson, J. G.; Chow, J. C.; Richards, L. W.; Neff, W. D.; Andersen, S. R.; Dietrich, D. L.; Houck, J. E.; Olmez, I. *The 1987–88 Metro Denver Brown Cloud Study. Volume I: Program Plan;* Desert Research Institute: Reno, NV, 1988; prepared for the 1987–88 Metro Denver Brown Cloud Study, Denver, CO; DRI Document 8810.1 F1.
100. Watson, J. G.; Chow, J. C.; Richards, L. W.; Haase, D. L.; McDade, C.; Dietrich, D. L.; Moon, D.; Chinkin, L.; Sloane, C. *The 1989–90 Phoenix PM_{10} Study. Volume I: Program*

Plan; prepared by Desert Research Institute, Reno, NV, for Arizona Department of Environmental Quality: Phoenix, AZ, 1990; DRI Document 8931.2 F.

101. Watson, J. G.; Chow, J. C.; Richards, L. W.; Haase, D. L.; McDade, C.; Dietrich, D. L.; Moon, D.; Sloane, C. *The 1989–90 Phoenix Urban Haze Study. Volume II: The Apportionment of Light Extinction to Sources—Final Report;* prepared by Desert Research Institute, Reno, NV, for Arizona Department of Environmental Quality: Phoenix, AZ, 1991; DRI Document 8931.5 F1.
102. Chow, J. C.; Watson, J. G.; De Mandel, R.; Chan, W.; Cordova, J.; Fairley, D.; Fujita, E. M.; Levaggi, D.; Long, G.; Perardi, T.; Rothenberg, M. *Measurements and Modeling of PM_{10} in San Francisco's Bay Area. Volume I: Program Plan;* prepared by Desert Research Institute, Reno, NV, for Bay Area Air Quality Management District: San Francisco, CA, 1993; DRI Document 3654.1 D3.
103. Chow, J. C.; Watson, J. G.; Lu, Z.; Lowenthal, D. H.; Frazier, C. A.; Solomon, P. A.; Magliano, K. L. *Atmos. Environ.* **1996,** *30*(12), 2079–2112.
104. Chow, J. C.; Watson, J. G.; Lowenthal, D. H.; Countess, R. J. *Atmos. Environ.* **1996,** *30*(9), 1489–1499.
105. *Results from the Summer 1992 PM_{10} Saturation Monitoring Study in the Ashland, KY, Area;* prepared by TRC Environmental Corporation, Research Triangle Park, NC, for U.S. Environmental Protection Agency, Technical Support Division, Monitoring and Reports Branch, Monitoring Section: Research Triangle Park, NC, 1992.
106. *PM_{10} Saturation Sampler O&M/QA Manual;* Region X, U.S. Environmental Protection Agency: Research Triangle Park, NC, 1992.
107. Ringler, E. S.; Shrieves, V. X.; Berg, N. J. Presented at the 86th Annual Meeting of the Air and Waste Management Association, Denver, CO, 1993; paper 93–TA–28.04.
108. Lu, Z.; Chow, J. C.; Watson, J. G.; Frazier, C. A.; Pritchett, L.; Dippel, W.; Bates, B.; Jones, W.; Torres, G.; Fisher, R.; Lam, D. Presented at the 87th Annual Meeting of the Air and Waste Management Association, Cincinnati, OH, 1994; paper 94–WA195.02P.
109. Chow, J. C.; Watson, J. G.; Bowen, J. L.; Frazier, C. A.; Gertler, A. W.; Fung, K. K.; Landis, D.; Ashbaugh, L. L. In *Sampling and Analysis of Airborne Pollutants;* Winegar, E. D.; Keith, L. H., Eds.; Lewis Publishers: Chelsea, MI, 1993; pp 209–228.
110. Watson, J. G.; Chow, J. C.; Lurmann, F.; Musarra, S. *Air Waste* **1994,** *44,* 261–268.
111. Tombach, I. H.; Allard, D. W.; Drake, R. L.; Lewis, R. C. *Western Regional Air Quality Studies. Visibility and Air Quality Measurements: 1981–1982;* prepared by AeroVironment, Monrovia, CA, for Electric Power Research Institute: Palo Alto, CA, 1987; Document EA–4903.
112. Solomon, P. A.; Fall, T.; Salmon, L.; Lin, P.; Vasquez, F.; Cass, G. R. *Acquisition of Acid Vapor and Aerosol Concentration Data for Use in Dry Deposition Studies in the South Coast Air Basin;* Environmental Quality Laboratory, California Institute of Technology: Pasadena, CA, 1988; Vol. II.
113. Solomon, P. A.; Fall, T.; Salmon, L.; Cass, G. R.; Gray, H. A.; Davison, A. *J. Air Pollut. Control Assoc.* **1989,** *39,* 154–163.
114. Rogers, C. F.; Watson, J. G.; Mathai, C. V. *J. Air Pollut. Control Assoc.* **1989,** *39,* 1569–1576.
115. Fitz, D.; Chan, M.; Cass, G.; Lawson, D.; Ashbaugh, L. Presented at the 82nd Annual Meeting of the Air and Waste Management Association, Anaheim, CA, 1989.
116. Koutrakis, P.; Wolfson, J. M.; Slater, J. L.; Brauer, M.; Spengler, J. D. *Environ. Sci. Technol.* **1988,** *22,* 1463.
117. Brauer, M.; Koutrakis, P.; Wolfson, J. M.; Spengler, J. D. *Atmos. Environ.* **1989,** *23*(9), 1981–1986.
118. Koutrakis, P.; Thompson, K. M.; Wolfson, J. M.; Spengler, J. D.; Keeler, G. J.; Slater, J. L. *Atmos. Environ.* **1992,** *26A*(6), 987–995.

119. Eldred, R. A.; Cahill, T. A.; Wilkinson, L. K.; Feeney, P. J.; Chow, J. C.; Malm, W. C. In *Transactions: Visibility and Fine Particles;* Mathai, C. V., Ed.; Air and Waste Management Association: Pittsburgh, PA, 1990; pp 187–196.
120. Thurston, G. D.; Gorczynski, J. E., Jr.; Jaques, P.; Currie, J.; He, D. *J. Exposure Anal. Environ. Epidemiol.* **1992,** 2(4), 415–428.
121. Stevens, R. K.; Pinto, J.; Conner, T. L.; Willis, R.; Rasmussen, R. A.; Mamane, Y.; Casuccio, G.; Benes, I.; Lanicek, J.; Subri, P.; Novak, J.; Santroch, J. Presented at the 86th Annual Meeting of the Air and Waste Management Association, Denver, CO, 1993.
122. Eatough, D. J. "Determination of the Size Distribution and Chemical Composition of Fine Particulate Semi-Volatile Organic Compounds Using Diffusion Denuder Technology: Results of the 1992 Study in Azusa, California. " Department of Chemistry report, Brigham Young University, Provo, UT, 1993.
123. Eatough, D. J.; Wadsworth, A.; Eatough, D. A.; Crawford, J. W.; Hansen, L. D.; Lewis, E. A. *Atmos. Environ.* **1993,** *27A*(8), 1213–1219.
124. Eatough, D. J.; Tang, H.; Machir, J. Presented at the 86th Annual Meeting of the Air and Waste Management Association, Denver, CO, 1993; Paper 93–RA–110.04.
125. Tang, H.; Lewis, E. A.; Eatough, D. J.; Burton, R. M.; Farber, R. J. *Atmos. Environ.* **1994,** *28*(5), 939–950.
126. McFarland, A. R.; Ortiz, C. A. *Characterization of Sierra-Andersen Model 321a 10-μm Size Selective Inlet for Hi-Vol Samplers;* Texas A&M University: College Station, TX, 1984; Document 4716/01/02/84/ARM.
127. McFarland, A. R.; Ortiz, C. A. *Wind Tunnel Characterization of Wedding* IP_{10} *10-μm Inlet for Hi-Vol Samplers;* Texas A&M University: College Station, TX, 1984; Document 4716/01/06/84/ARM.

Chapter 27

Effects of Environmental Measurement Variability on Air Quality Decisions

John G. Watson

The variability, or uncertainty, of environmental measurements can have a major effect on the decisions made about an environmental issue. Alternatives are evaluated prior to making a decision by using combinations of measurements and models to quantify the ramifications of those alternatives. Both measurement and model uncertainty affect the decision-making process, but these uncertainties are seldom quantified and are rarely part of the decision. Uncertainty estimates are used in the decision-making process by ignoring uncertainty, defining margins of safety, or balancing costs against benefits.

AIR POLLUTION FROM ANTHROPOGENIC SOURCES is a fact of life today and has been for thousands of years. Air pollution probably came into being when the first fire was lit. As humans discovered new ways to apply the conversion of fuel into energy, energy into sustenance and leisure, and sustenance and leisure into more people, air pollution emissions increased. The first recorded decision to control air pollution emissions was issued in 13th century London as a royal proclamation prohibiting the use of coal (*1*). Though the "measurements" on which this decision were based (e.g., observations of black chimney plumes, reduced visibility, soot deposits on clothing, and sickness and death due to respiratory disease) were imprecise and variable, the regulatory decision was undoubtedly correct.

Most of the environmental decisions before the latter half of the 20th century were made without the benefit of extremely precise measurements. These decisions were fairly obvious, however, and did not require such precision. Today, this situation is no longer the case. The inherent variability of environmental measurements can make the difference between alternative outcomes to a decision-making process. This variability must therefore become an integral part of that decision making. The objectives of this chapter are to categorize the types of environmental decisions made with respect to air quality, specify the uses of air quality measurements in decision making, and summarize the uses of measurement variability in decision making. Examples specific to suspended particulate matter are supplied to clarify certain concepts. These examples are intended to be illustrative rather than comprehensive; full examination of past and present air quality decisions is beyond the scope of this chapter.

Types of Air Quality Decisions

Many approaches can be applied to the study of the environment, and these approaches often yield differing conclusions and alternatives for environmental control. The decision maker must select the best of these alternatives, and the "best decision" normally is defined as that which is the least disruptive and most cost-effective. The decision maker is often confronted with alternatives in four aspects of air quality: effects determination, identification of pollutants causing effects, source attribution, and emissions controls.

Effects Determination

The first air quality decision involves the perception that a problem exists and is primarily caused by an atmospheric constituent. An effect can be one of human respiratory distress, damage to materials or plants and wildlife, visibility impairment, or a perceived nuisance. These effects only become problems when they are defined as such by the general population in which they occur, as evidenced by the differences in environmental concern between developed and developing countries (2). Environmental observations are used to determine the existence of a problem (e.g., the number of sicknesses and deaths in a given period). These observations are highly variable. Exposure to atmospheric constituents is one of many possible causes in most cases.

Identification of Pollutants Causing Effects

In order to be classified as an air pollutant, an atmospheric constituent must have the chemical and physical properties and be present in sufficient quantities to cause an unacceptable effect. The decision to make such a classification is based on measurements that establish the cause and effect relationship, and these measurements exhibit large uncertainties. For example, the U.S. primary air quality standard for suspended particulate matter has been revised to consider size classification as opposed to total suspended particulate matter in response to more precise measurements of particle entry into the human lung (*3*). The original measurements on which the earlier standard was based (*4*) have been improved over the intervening time period (*5, 6*), and the decision regarding the tolerable physical properties and concentrations of suspended particulate matter is different.

Source Attribution

Excessive exposures to air pollutants can only be reduced by curtailing their emissions from anthropogenic sources. Because many such sources usually exist, decisions must be made with respect to the contributions from each one to ambient concentrations. These decisions are based on emissions, meteorological, and ambient air measurements, all of which have substantial measurement uncertainties. The relationships between emissions and receptors are embodied in *air quality models*, which introduce additional uncertainty because they cannot be perfect representations of reality.

Emissions Controls

Many alternatives exist with respect to reducing emissions from major contributors, and the appropriate decision, from a purely cost-effective standpoint, is that which achieves the greatest reduction in ambient concentrations per dollar invested in emissions controls. These decisions rely on measurements of control efficiency and the assumption that a reduction in emissions will be accompanied by a proportional reduction in ambient concentrations. Large uncertainties are associated with these measurements and assumptions. For example, even though road dust has been identified as a major contributor to suspended particulate matter in many urban areas, and street-sweeping programs have been implemented to reduce road dust, this control has not been shown to be very effective (*7, 8*). Nonlinear photochemical processes may cause oxidant and secondary particle levels to increase at certain locations in response to decreases in nitrogen oxide emissions (*9*).

Air Quality Models

Measurements are always used in models to arrive at a decision. The word *model* is used in its broadest sense to include all of the methods used to interpret environmental measurements. The most widely used air quality models for each of the decision categories described in the previous section are descriptive, covariational, classificatory, mechanistic, and inferential. The effects of environmental measurement uncertainty on decisions cannot be separated from the models in which those measurements are used.

Descriptive Models

Descriptive models summarize the spatial, temporal, and statistical distributions of individual observations. The model results show which air pollutants are reaching levels of concern and are the basis for defining the existence of a problem. Descriptive models require a large number of measurements in space and time. For example, an annual geometric average of total suspended particle concentration in excess of 75 $\mu g/m^3$ has been deemed unacceptable by national ambient air quality standards, and a model of one 24-h measurement taken every 6th day has been deemed adequate to determine whether or not this standard is violated in a given year (*10*).

Covariational Models

Covariational models calculate measures of association between two or more variables and are often used to assign effects to a pollutant species or a pollutant to an emissions source. The output data of these models, such as correlation coefficients, are high absolute values when variables change in the same manner over a period of time or over a geographical area; the output data of these models are low absolute values when covariation is lacking. By themselves, these models establish only whether the values for a set of variables change in the same way. Ware et al. (*11*), for example, pointed to many confounding variables not related to air pollution that could cause similar variabilities among morbidity, sulfur dioxide concentrations, and suspended particulate matter found to be statistically significant in several epidemiological studies.

Classificatory Models

Classificatory models define categories based on the achievement of certain measured values for a set of attributes. Each category has a specific set of decisions associated with the members of that category. The prevention of significant deterioration regulations provided by the Clean Air Act Amendments of 1977 (*12*), for example, automatically impose certain air quality modeling, monitoring, and emissions requirements on 28 stationary source types that have the potential to emit more than 100 tons/year of any criteria pollutant. Other source types may emit up to 250 tons/year without invoking this decision-making process. All proposed sources are classed based on emissions calculations derived from environmental measurements.

Mechanistic Models

Mechanistic models, also known as *source-oriented models*, contain mathematical descriptions of the interactions among variables derived from fundamental physical and chemical laws. They directly relate a cause to an effect and can be used to hypothesize the effects of alternative decisions. Air quality dispersion models that use meteorological and emissions data to calculate ambient concentrations (*13*) are the most common example of these models.

Inferential Models

Inferential models, also known as *receptor-oriented models*, can be derived either from fundamental physical and chemical laws or from empirical observations. Inferential models differ from mechanistic models in that their primary set of input data is that which is measured at the receptor, not at the source. The chemical mass balance receptor model (*14*), for example, is a U.S. Environmental Protection Agency (EPA) model that calculates contributions to ambient concentrations of suspended particulate matter from a large number of chemical species measured on the receptor sample and in the source emissions. These models are most often used to decide which source types are the major contributors to unacceptable pollutant levels.

Uses of Measurements in Models

All models are simplified descriptions of complex systems, and these simplifications are effected by making certain assumptions. A model is only valid to the extent that these assumptions are complied with in a specific application (*15*). This definition leads to two sources of uncertainty in a model result: *measurement uncertainty*, which results from the variability of the environmental measurements used as model input data, and *model uncertainty*, which is caused by deviations from the model assumptions (*16*). Either one of these uncertainties may be the dominant cause of uncertainty in the model result used for decision making. Environmental measurements are used not just as model inputs, but also to derive fixed constants, to test compliance with model assumptions, and to estimate measurement uncertainty.

Model Input Measurements

Every model requires some data on which to operate for the period of time being examined. For example, mechanistic air quality models may require boundary and initial conditions of the precursor, intermediate, and end-product species as well as three-dimensional wind fields and atmospheric stability estimates. The chemical mass balance receptor model requires ambient concentrations and source compositions for a number of chemical species. These variables are considered input data when measured for the time simulated.

Parameter Measurements

Parameters are constants that are not supplied to the model on a case-by-case basis, as are input measurements. Parameters are obtained by a theoretical calculation, by measurements made elsewhere and assumed to be appropriate for the place and time being studied, or by tacit assumption that the value of a variable is constant or negligible. Reaction rates, emissions rates, transformation rates, dispersion parameters, and source compositions are common parameters in mechanistic and inferential models. Values for these variables are rarely measured over the period of time being modeled because of lack of feasibility or excessive measurement costs. Parameters normally carry higher levels of variability than input data because parameters are not specific to each case being studied.

Model Validation Measurements

Additional measurements that may not appear in the model as input data or as parameters can be used to determine the extent of deviations from model assumptions. The use of measurements in this way is the first step in quantifying model uncertainty. For example, Gordon et al. (*17*) applied the chemical mass balance model to a subset of chemically speciated suspended particulate matter concentrations. They then calculated the concentrations of the remaining species and compared these calculations with their measured values to evaluate the validity of their source contribution estimates. In this way, they found that several key species must be included in the model to comply with the chemical mass balance model assumption of linearly independent source compositions.

Measurement Uncertainty Measurements

The magnitude and distribution of variability in the model input data, parameters, and validation measurements can only be determined by replicate measurements of the same variables. Most models assume that a single point measurement represents a spatial or temporal distribution, and an estimate of the variability over space or time is needed if an adequate value for measurement uncertainty associated with a model result is to be derived. Measurements "collocated" with a resolution finer than the temporal and spatial scales of the model can be used to estimate the uncertainty of input data and parameters. These uncertainties can also be estimated from periodic performance tests of each measurement method if analytical variability is larger than spatial or temporal variability (*18*).

Quantifying Uncertainty

Incorporating the uncertainty of environmental measurements into the decision-making process presupposes (1) that the uncertainty associated with an individual measurement can be quantified, (2) that these measurement uncertainties can be propagated through the models to provide a reasonable confidence level around the model result that is used to justify the decision, and (3) that these measurement uncertainties can be compared with model uncertainties to determine which is of greater concern. Several methods have been applied to the quantification of uncertainty, though these methods are largely unproven.

Uncertainty of an Individual Measurement

These uncertainties are ones of precision, accuracy, and validity (*19*). *Precision* is a measure of the variability of measurements of the same quantity by the same method. Precision can be determined from the standard deviation of repeated measurements and is usually assumed to follow a normal distribution. Precision is the easiest uncertainty to quantify because some form of performance testing or replicate analysis comprises the quality control of most measurement networks. Very few data bases for ambient or source measurements report these precisions, however. The sulfate regional experiment (*20*) data base, for example, has been widely used to decide when and where high levels of particulate sulfate occur and to test air quality models precisely because this data base is one of the few measurement programs with extensive quantification of measurement precision.

Accuracy is the difference between measured and referenced values, and this difference is expected to be within the precision interval for the measurement to be deemed accurate. Primary standards, reference materials, and equivalency protocols have been established by the National Bureau of Standards and the EPA for several pollutants measured at sources and in ambient air in order to quantify the accuracy of these measurements (*21*). Standardization methods are still inadequate to establish the equivalency of many measurements, however. Decisions regarding the attainment of the proposed national air quality standard for particulate matter having diameters $\leq 10\ \mu m$, for example, are still clouded by demonstrated differences among simultaneous measurements by different sampling systems (*22, 23*), even though each of these systems has passed prescribed equivalency tests.

Measurement validity consists of two components, method validity and sample validity. Method validity requires the identification of measurement method assumptions, the quantification of effects of deviations from those assumptions, the ascertainment that deviations are within reasonable tolerances for a specific application, and the creation of procedures to quantify and minimize those deviations in a specific application. Sample validity consists of procedures that identify deviations from measurement assumptions for each measurement. Each individual value needs to be designated as valid, valid but suspect, or invalid based on predefined criteria. The validity of a measurement is the most difficult uncertainty to determine and is often the cause of very high or low measurement values. For example, a recent examination of the highest inhalable particulate concentrations in EPA's inhalable particulate network (*24*) demonstrated that 10 out of the 50 samples selected for intensive study yielded mass concentrations that were inconsistent with the sample chemical composition and size distribution. Without these nonroutine measurements that allowed the measurement validity to be questioned, the highest concentrations might have been attributed to pollution sources instead of to anomalies in the measurement process.

Measurement Uncertainty in Models

As difficult as estimating the uncertainty of an individual measurement is, combining these estimates from a number of measurements that are acted upon by a decision-making model is even more challenging. Analytical and computationally intensive methods have been applied to this task.

If a model output can be expressed as some function of a set of environmental measurements, and if the errors in those measurements are random and normally distributed about the true value, then an analytical expression for the model output variable can be derived from the theory of maximum likelihood (*25*). For example, such expressions have been derived for the Gaussian plume (*16*) and the chemical mass balance (*26*) air quality models, which are commonly used to decide which particulate emitters are the major contributors to ambient concentrations. These analytical error propagation methods make numerous assumptions, however, and the range of practical applications for which they are valid remains to be determined.

Computationally intensive methods rely on repeated applications of the model to input data that have been altered in a way that mimics the uncertainty caused by the measurement process. The distribution of model results derived from these repeated applications defines the portion of their uncertainty caused by the variability of the environmental measurements. *Monte Carlo methods* (*27*) perturb all input data with random numbers drawn from an assumed error distribution. Though this method is very effective, a large number of trials are required to achieve statistical stability, and the error distributions of input data are often unknown. The *bootstrap method* (*28*) recalculates statistical parameters, such as those yielded by description, classification, and correlation models, from many randomly selected subsets of the model input data set. The fundamental assumption of this method is that measured data sets define their own distributions of random error. This method does not require an assumed distribution, though extension of this model to mechanistic and inferential models is not apparent, and many trials are still required for statistical stability. The Fourier amplitude sensitivity test (*29*) associates each uncertain model input variable with a specific frequency in a Fourier transform space of the model variables. The model equations are solved for discrete values of the Fourier transform variables. The uncertainty of the model result is a function of these Fourier coefficients. This method requires fewer computations than the Monte Carlo and bootstrap methods. Each of these computationally intensive methods is based on a number of assumptions that may not be complied with in actual practice, and the methods require substantial additional testing to define the limits of their application.

Model Uncertainty

Model uncertainty is nearly impossible to quantify with current technology. No air quality model is ever applied in a situation without some deviations from its basic assumptions. A testing protocol for identifying those deviations and determining their significance involves (1) explicitly listing those assumptions as part of the model derivation from basic principles, (2) analytically introducing deviations from those assumptions and quantifying the effects on model results, (3) performing controlled experiments to verify model components, (4) comparing model results with those of a more complete or more widely accepted model, (5) comparing model results with ambient measurements, and (6) introducing unique tracers into the modeled system (*30*). Each of these methods has been applied to each of the decision-making models described, though none of these applications has adequately quantified model uncertainty over the range of typical applications in which decisions are made.

Using Uncertainty in Making Decisions

As already noted, the variability of environmental measurements on which decisions are based is rarely quantified and reported. This measurement uncertainty is even more rarely propagated through decision-making models and weighed against a quantitative estimate of model uncertainty. Even if these indicators of measurement variability were available, how could they be used in the decision-making process? Quantitative uncertainty estimates are ignored, used to describe worst-case situations, or used to weigh costs against benefits.

Ignoring Uncertainty

When measurements and model outputs are calculated to two or more significant figures without the specification of uncertainty intervals, the implicit assumption is that these intervals do not exist. The decision is based only on the value at hand with no consideration of whether that value overestimates, underestimates, or is totally unrelated to the true value. Many day-to-day decisions are made by ignoring the effects of variability in environmental measurements.

Worst Case

When uncertainty is quantified, postulating the value for a measurement or model output that will result in the most deleterious outcome is possible. If this outcome is still acceptable, then a decision about a proposed action can be made with confidence. Conversely, if the best possible outcomes are unacceptable, then a negative decision about the proposed action can be easily justified. For example, national ambient air quality standards are set well below the level at which any deleterious effects to the most sensitive segments of the population have been observed. Most air quality modeling conducted prior to permitting a new source is performed under worst-case emissions and meteorological conditions by using models that are expected to overestimate rather than underestimate ambient concentrations. This use of uncertainty in decision making is adequate for the majority of day-to-day environmental decisions such as those involving emissions source locations and emission controls. The worst case is an improbable event, however, and decisions based on this analyses are often more restrictive than is necessary for adequate environmental protection.

Cost–Benefit Analysis

Analytical frameworks have been developed to integrate uncertainty estimates from measurements and models into the decision-making process. These frameworks also integrate the four categories of air quality decisions cited at the beginning of this chapter and attempt to balance the uncertainties associated with one category against those of the other categories. The outcomes of these frameworks are positive (i.e., benefits) and negative (i.e., costs) measures associated with each alternative decision. These frameworks explicitly separate costs and benefits because the relative weight attached to each is really a subjective judgement provided by the decision maker. These frameworks are crude at this time and are rarely used as a fundamental basis for decisions. They do provide a conceptual structure for integrating decision categories, models, and uncertainty estimates.

A *decision tree approach* (*31*) identifies all of the possible outcomes of a series of decisions and assigns a cost and a probability to each of these outcomes. Estimates of measurement and model uncertainty are part of the decision tree and are used to compute the potential environmental costs of delaying a decision as well as the potential benefits of uncertainty reduction. This approach is appealing because the value added to the decision-making process is determined by more precise environmental measurements. This model could be used to optimize monitoring networks by selecting only those observations, locations, and measurement methods that would make a difference in the selection of outcomes in the decision tree. These decision trees must be custom-designed for each environmental concern, and the cost of such a design may exceed the cost of an incorrect decision arrived at by a worst-case analysis.

A *risk assessment* truly integrates models used in each decision-making category. For example, a computer code for air quality risk assessment has been implemented with quantitative modules for emissions, transport, exposures, and effects (32). The output of these risk assessments is in terms of a probable number of deaths, sicknesses, or number of dollars spent. Alternative decisions regarding emissions controls, locations of emissions sources with respect to exposed populations, and the physical and chemical nature of emissions can be evaluated by comparing the outputs for each alternative. Uncertainty is incorporated both implicitly, by expressing results in terms of statistical probabilities rather than as direct outcomes, and explicitly, by propagating uncertainties from one model to another by using the methods described in this section.

Conclusions and Future Research

The variability of environmental measurements has an important effect on the decisions based on those measurements. The uncertainty that results from that variability should be a formal part of the decision-making process. This uncertainty will not be incorporated until the measurement uncertainties are quantified as part of the measurement process and then propagated through the models that are normally used as the justification for decisions. Methods have been proposed for estimating and propagating these uncertainties, but these methods are largely untested.

Further research is needed to determine the range of applicability of uncertainty estimation methods in practical situations. Additional research is needed to derive new methods for incorporation into measurement and modeling practice. Finally, frameworks need to be developed and tested that will allow the uncertainties associated with different decision categories to be balanced against each other for the purposes of specifying additional environmental measurements that will reduce the uncertainty of an environmental decision.

References

1. Halliday, E. C. WHO Monograph Series No. 46; World Health Organization: Geneva, Switzerland, 1961.
2. Watson, J. G.; Broten, A. R.; Smith, S. M.; Chow, J. C.; Shah, J. J. *Proc. 76th Annu. Meet. Air Pollut. Control. Assoc.* Atlanta, GA, 1983.
3. *Fed. Regist.* **1987**, *52(126)*, 24634–24749.
4. *Air Quality Criteria for Particulate Matter*; U.S. Department of Health, Education, and Welfare: Washington, DC, 1969.
5. *Air Quality Criteria for Particulate Matter and Sulfur Oxides*; U.S. Environmental Protection Agency: Research Triangle Park, NC, 1980.
6. Swift, D. L.; Proctor, D. F. *Atmos. Environ.* **1982**, *16*, 2279.
7. Portland Road Dust Demonstration Project; Oregon Department of Environmental Quality: Portland, OR, 1983.
8. Gatz, D. F.; Wiley, S. T.; Chu, L. C. *Characterization of Urban and Rural Inhalable Particulates*; Illinois Department of Energy and Natural Resources: Springfield, IL, 1983.

9. Roth, P. M.; Reynolds, S. D.; Tesche, T. W.; Gutfreund, P. D.; Seigneur, C. *Environ. Int.* **1983**, *9*, 549.
10. Akland, G. G. *J. Air Pollut. Control Assoc.* **1972**, 22, 264.
11. Ware, J. H.; Thibodeau, L. A.; Speizer, F. E.; Colome, S.; Ferris, B. G. *Environ. Health Perspect.* **1981**, *41*, 255.
12. *Prevention of Significant Deterioration Workshop Manual*; U.S. Environmental Protection Agency: Research Triangle Park, NC, 1980.
13. Turner, D. B. *J. Air Pollut. Control. Assoc.* **1979**, *29*, 940.
14. Williamson, H. J.; DuBose, D. A. *User's Manual for Chemical Mass Balance Model*; U.S. Environmental Protection Agency: Research Triangle Park, NC, 1983.
15. Goodall, D. W. In *Mathematical Models in Ecology, The Twelfth Symposium of the British Ecological Society*; Blackwell Scientific: Blackwell, England, 1972; p 173.
16. Freeman, D. L.; Egami, R. T.; Robinson, N. F.; Watson, J. G. *J. Air Pollut. Control Assoc.* **1986**, *36*, 246.
17. Gordon, G. E.; Zoller, W. H.; Kowalczyk, G. S.; Rheingrover, S. H. In *Atmospheric Aerosol: Source/Air Quality Relationships*; Macias, E. S.; Hopke, P. K., Eds.; ACS Symposium Series 167; American Chemical Society: Washington, DC, 1981; pp 51–74.
18. Watson, J. G.; Lioy, P. J.; Mueller, P. K. In *Air Sampling Instruments for Evaluation of Atmospheric Contaminants*, 6th ed.; Lioy, P. J., Ed.; American Conference of Governmental Industrial Hygienists: Cincinnati, OH.
19. Hidy, G. M. *Environ. Sci. Technol.* **1985**, *19*, 1032.
20. Mueller, P. K.; Hidy, G. M. *The Sulfate Regional Experiment: Report of Findings*; Electric Power Research Institute: Palo Alto, CA, 1983.
21. Greenberg, R. R. *Anal. Chem.* **1979**, *51*, 2004.
22. Rodes, C. E.; Holland, D. M.; Purdue, L. J.; Rehme, K. A. *J. Air Pollut. Control Assoc.* **1985**, *35*, 345.
23. Wedding, J. B.; Lodge, J. P.; Kim, Y. J. *J. Air Pollut. Control Assoc.* **1985**, *35*, 649.
24. Rogers, C. F.; Watson, J. G. *Proc. 77th Annu. Meet. Air Pollut. Control Assoc.* San Francisco, CA, 1984.
25. Bevington, P. R. *Data Reduction and Error Analysis for the Physical Sciences*; McGraw–Hill: New York, 1969.
26. Watson, J. G.; Cooper, J. A., Huntzicker, J. J. *Atmos. Environ.* **1984**, *18*, 1347.
27. Tiwari, J. L.; Hobbie, J. E. *Math. Biosci.* **1976**, *28*, 25.
28. Efron, B. *SIAM Rev.* **1979**, *21*, 460.
29. McRae, G. J.; Tilden, J. W.; Seinfeld, J. H. *Comput. Chem. Eng.* **1982**, *6*, 15.
30. Javitz, H. S.; Watson, J. G. In *Receptor Methods for Source Apportionment: Real World Issues and Applications*; Pace, T. G., Ed.; Air Pollution Control Association: Pittsburgh, PA, 1986; pp 161–174.
31. Balson, W. E.; Boyd, D. W.; North, D. W. *Acid Deposition: Decision Framework*; Electric Power Research Institute: Palo Alto, CA, 1982; Vol. 1.
32. Eschenroeder, A. Q.; Magil, G. C.; Woodruff, C. R. *Assessing the Health Risks of Airborne Carcinogens*; Electric Power Research Institute: Palo Alto, CA, 1985.

Chapter 28

Airborne Sampling and In Situ Measurement of Atmospheric Chemical Species

Roger L. Tanner

Common aspects of airborne sampling and analysis procedures are discussed whereby the composition of gaseous, aerosol, and condensed water droplet phases may be characterized in three dimensions in the boundary-layer atmosphere. The scope includes a review of techniques for isolating a particular phase during airborne sampling, techniques for continuous in situ measurements of trace gases and aerosols, and means for collection of condensed-phase samples. Time resolution considerations for airborne measurements are discussed, and some examples comparing real-time and integrative measurements are included.

SAMPLING AND IN SITU MEASUREMENT of chemical species in ambient air are important components in studies of atmospheric chemistry. Most major pollutant emission sources, and dry deposition sinks as well, are located at the Earth's surface. Therefore, the boundary layer of the atmosphere must be well-mixed vertically, or lacking that, airborne (or tower) measurements must be made to characterize the composition of gaseous, aerosol, and condensed water phases in three dimensions in the boundary-layer atmosphere. Special considerations apply in the sampling and in situ measurement of atmospheric constituents aloft. However, airborne sampling and analysis procedures have aspects in common, whether the focus of the measurement program is global tropospheric studies in remote locations, heterogeneous and homogeneous photochemical smog studies, or precipitation scavenging and acidic deposition studies. These common aspects are the central theme of this chapter, which is not designed as a comprehensive review of continuous

3152–4/96/0587$15.00/0

methods for the species of interest, but as a guide to the principles and application of these methods in the atmospheric environment.

Throughout this chapter, an in situ chemical measurement in the atmosphere is defined as follows: the collection of a representative sample of a definable atmospheric phase and the determination of a specific chemical moiety in that phase, in situ (in its original location), with definable precision and accuracy (*1*). This definition is required because most constituents of interest in the atmosphere are not inert ingredients sequestered in a single phase but reactive trace constituents that may be distributed between two or more phases depending on the physical and chemical properties of the atmosphere being sampled. The scope of this chapter includes discussion of the following: (a) techniques for isolating a particular phase during sampling from an airborne platform; (b) techniques for in situ, continuous measurement of trace gases including ozone, nitrogen oxides [e.g., NO_y, NO, NO_2, HNO_3, HONO, peroxyacetyl nitrates (PANs), and organic nitrates], sulfur dioxide, ammonia, nitric acid vapor, carbon monoxide, and hydrogen peroxide; (c) techniques for continuous measurement of aerosol species and size distributions; (d) means for collection of condensed aqueous-phase samples (e.g., cloud and rain liquid water, supercooled droplets, ice clouds, and snow); and (e) measurement of related physical parameters. Measurement time resolution questions will also be discussed in the context of what constitutes a "real-time" measurement, and some examples of comparisons of real-time and integrative measurements of the same atmospheric species will be given.

Separation of Phases

Air pollutants exist, usually at trace levels, in the atmosphere in one or more physical states depending on the presence or absence of condensed phases. That is, air pollutants may be present as gases or vapors, as surface or bulk constituents of aerosol particles and droplets, or as constituents of cloud liquid water (including supercooled water) droplets, cloud ice particles, and various forms of precipitation. Proper sampling requires that a single phase be sampled if inferences are to be made from the chemical composition of that sample. This requirement is particularly important when, under the atmospheric conditions sampled, a pollutant species may be distributed between more than one phase (e.g., nitrate and nitric acid between gas and aerosol phases depending on the temperature and the pH of the aerosol droplet) (2). A few representative methods for separation of phases during atmospheric sampling from an airborne platform are discussed in this section.

Cloud droplets, aerosols, and gaseous phases are usually separated on the basis of differences in aerodynamic properties. In the absence of a condensed aqueous phase, sampled air is simply filtered to remove aerosol particles by impaction, diffusion, or interception onto the filter surface, and the air stream is then analyzed for the gaseous constituent of interest. Several assumptions must be valid for this sampling procedure to yield accurate results. Gaseous constituents to be analyzed must not react irreversibly with the filter surface or with collected aerosol particles; otherwise negative errors will result. Collected aerosols must also not release or retain gaseous constituents upon changes in sampling conditions. Both of these problems have been observed in continuous and integrated filter analyses of nitric acid (*3, 4*) as a result of the phase equilibrium that exists between nitric acid vapor and particulate ammonium nitrate in ambient air (*1, 5*). The occurrence of phase equilibration during sampling may also affect the accuracy of chemical determinations done on the collected aerosol phase, but these determinations are integrative measurements and thus not the subject of this chapter.

In the presence of a cloud liquid water phase, most water-soluble aerosol particles are incorporated (scavenged) into the cloud water phase, which is a major sink process for removal of sulfate from the atmosphere whenever the scavenging clouds actually produce precipitation. This process can be observed directly in airborne sampling (*6*) by simultaneously recording aerosol light scattering with a nephelometer and cloud liquid water content with a heated wire probe (Johnson–Williams sensor or King probe). The cloud droplet size distribution (approximately Gaussian and centered about 10–20-μm diameter) overlaps the coarse-particle regime of ambient aerosols but may be separated from the fine, accumulation-mode aerosols by impaction techniques. Airborne sampling of gases and aerosols may be conducted continuously in the presence of cloud liquid water (i.e., in cloud interstitial air) by removing the cloud droplets with a centrifugal rotor (*7*), cyclone, or other impaction device and then directing the air to the continuous gaseous or aerosol instruments. This cloud-free air may also be filtered of any nonscavenged fine-mode aerosol at sampling rates as high as 1 m^3/min.

Sampling of Condensed Phases

Direct sampling for liquid water in clouds may be done from an aircraft with a multiple slotted-rod collector (*8*) inserted through the skin of the aircraft; the water is collected by gravity inside the aircraft and subsequently analyzed. Chemical measurements that can be performed continuously on collected clouds will be discussed briefly in the next section. Rainwater can also be collected aloft in a similar way, but because rain droplet size distributions are considerably larger (typically 0.1–2-mm geometric mean diameter), a single, large slotted rod is used for collection of rain droplets.

In practice, several problems are apparent when using these collection devices (*9*). The collection efficiency, even for a single rain or cloud droplet size distribution, depends on the physical location of the collector on the aircraft fuselage. Collection efficiency may also vary with aircraft speed and angle of attack because of physical distortion of the collector, which occurs because collectors are usually made of somewhat flexible plastic. Finally, although the rain collector is observed to collect very little cloud water, the converse is not true and cloud water samples collected from precipitating clouds provide data that are more difficult to interpret (*10*).

Collection of supercooled cloud water or precipitation presents few sampling difficulties because any available surface outside the slip stream of the aircraft will serve as a freezing site for the supercooled water. Sampling ambiguities result from melting the collected ice from the collection surface for subsequent chemical analysis. This result is caused by the fact that two phase transitions have already been performed on the sample (liquid-to-solid on the collection surface and solid-to-liquid during the melting), and such phase transitions may result in phase exclusion or chemical transformations of certain trace constituents. For example, the solubility of SO_2 in ice is a subject of active investigation (Lee, Y.-N., Brookhaven National Laboratory, unpublished data).

Collection of snow from airborne platforms is, on the other hand, very problematical at this time. Several types of cyclone-based devices have been investigated, but the performance of these devices is generally much poorer than predicted; thus, further investigation in this area is warranted.

Continuous Gas-Phase Techniques

Techniques for the continuous, in situ measurement of trace gases from an airborne platform are well-developed in some cases and still in the preliminary development stage for others. The commonly used technique for each gas considered (O_3, CO, NO_y, SO_2, NH_3, NO_2, PAN, and H_2O_2) is described in this section, and the discussion moves from well-established techniques to more recent, state-of-the-art innovations.

Ozone

Measurement of ozone may be made in real-time from airborne platforms at ambient air concentrations by using the ethylene chemiluminescence technique (*11*). Observed precision and accuracy of ±10% or better in the 20–200-ppb range are usually more than adequate for most applications, and the observed time resolution of 1–3 s is exceptionally good. In fact, a similar design, the nitric oxide chemiluminescence instrument for ozone, having a time resolution of about 0.1 s, has been used for aerial eddy correlation flux measurements (*12*). These types of measurements are of great interest in dry deposition studies.

Carbon Monoxide

CO is nearly always measured by using a nondispersive IR spectroscopic technique. IR light attenuation is compared with that observed in a reference cell containing a known amount of CO in excess of expected measured values. The limit of detection (LOD) for conventional instruments mounted on an aircraft is about 50 ppb, time resolution is 10 s, and precision and accuracy at >10 times the LOD are about ±10%. An instrument with improved performance characteristics for airborne use has been reported (*13*).

Nitrogen Oxides

Air free of aerosols and condensed-phase atmospheric water (clouds and precipitation) may be analyzed for a variety of nitrogen oxide and oxyacid species. Measurements of nitric oxide (NO) and other nitrogen oxides and oxyacids are most frequently made by using techniques based on the chemiluminescent reaction of ozone with NO (*14*). Chemiluminescence is produced when a portion of the excited-state NO_2 molecules formed in the O_3–NO reaction luminesce in the spectral region beyond ~600 nm. Instruments for NO_y based on ozone chemiluminescence consist of a chamber for mixing ambient air with excess ozone and a window for viewing the filtered chemiluminescence >600 nm with a red-sensitive photomultiplier tube. Nitrogen species other than NO are converted to NO by passage over a heated catalyst, which is usually Mo at 375 °C.

Commercial instruments having two parallel sampling paths are available for real-time analysis of NO and NO_y, but their LOD is only about 2 ppb. Airborne applications in nonurban regions in which frequently $[NO_y] < 2$ ppb require the modification of commercial instruments to improve the signal-to-noise ratio. Optimizing the chemiluminescence detector response has been achieved (*15, 16*) in several ways: improving light collection efficiency with polished, gold-coated reaction chambers; using low-noise, red-sensitive photomultiplier tubes; and increasing the intensity of chemiluminescent light by operating at high-volume flow rates under reduced chamber pressure in an enlarged chamber at higher ozone concentrations. Background ("zero-air" response) stability has been improved through the use of a prereactor, in which sampled ambient air is contacted with ozone prior to admission to the reaction chamber. This zero-air response compensates for the effects of quenching gases and spectrally interfering compounds in ambient air.

Determination of gaseous nitric acid concentrations may be made from airborne platforms by filter (*17*) or at the surface by diffusion-denuder (*18*) techniques with time resolution of 15–30 min. However, the only available real-time technique uses a two-channel chemiluminescence analyzer in which one sample stream passes through a nylon filter prior to the Mo catalyst. Nylon removes HNO_3 and other acidic gases, but transmits other nitrogen oxides and organonitrogen compounds with high efficiency, with the probable exception of HONO (*19*). Nitric acid is determined continuously from the difference in the two NO_y channels with and without the nylon filter. Serial sampling of HNO_3 by this method with a single-channel instrument is not appropriate because of serious adsorption effects exhibited by nitric acid vapor under ambient air sampling conditions.

Significant difficulties arise in the determination of NO_2 using the ozone chemiluminescence approach because the Mo catalyst nonspecifically converts several nitrogen oxides and oxyacids to NO in addition to NO_2. Two separate approaches to specifically convert NO_2 to NO have been taken. Ferrous sulfate has been used for NO_2 conversion to NO (*20*), but humidity-dependent adsorption effects have been reported for this catalyst when PAN is present (*21*). Photolytic conversion of NO_2 to NO has been reported for surface measurements of NO_2 (22), and this method is promising for future airborne applications.

An entirely different approach to real-time NO_2 determination based on observation of the surface chemiluminescence of NO_2 in the presence of an aqueous luminol solution has been reported (*23*). A commercial instrument based on this principle has also been introduced, and investigations now in progress will help to establish its suitability for airborne, real-time NO_2 measurements in the presence of other NO_y species.

Sulfur Dioxide

Most real-time measurements of SO_2 made from an aircraft have used a modification of a commercial flame photometric detector (FPD) (*24*), although a recent commercial version of a pulsed-fluorescent instrument for SO_2 is promising. The sensitivity of the commercial FPD must be enhanced by the addition of a known background of a sulfur compound, usually SF_6 (60 ppb) in the hydrogen fuel gas supply (*25*). The resulting LOD then approaches 0.2 ppb for SO_2 (10-s time constant), which is adequate for most ambient applications. Sulfur gases (mostly SO_2 in ambient air) are determined with the FPD after removal of aerosols by filtration onto materials that are SO_2-inert (polytetrafluoroethylene (Teflon) or acid-treated quartz).

In the airborne use of the real-time FPD for SO_2, controlling the mass flow of sampled air is essential. This controlling is done with a constant-pressure sampling manifold, or by inserting mass-flow controllers in the hydrogen and exhaust-gas lines (the exhaust-gas lines are placed downstream of a trap for removal of condensable water). The sensitivity and especially the baseline (zero-air) signal are quite sensitive to the H_2/O_2 ratio in the burner. Calibrations with an instrument modified as described are linear with <10-ppb full-scale sensitivity, and this modified instrument has been used successfully on many clear- and cloud interstitial-air sampling missions (*10*).

Ammonia

Methods for determination of ammonia (NH_3) in real-time from an aircraft platform are few in number and have been successfully used in only a few studies. A continuous method for NH_3 with a time resolution of a few minutes has been reported (*26*). This method is based on Venturi collection, derivatization to an isoindole, and determination by fluorescence in a flow-through fluorimeter. The LOD reported (0.3 ppb) is adequate for surface sampling (*27*) but was seldom exceeded in airborne missions with this apparatus. Use of an IR heterodyne radiometer for vertical profile measurements of NH_3 from a surface location has also been reported (*28*); this technique is complex and the instrumentation expensive, but it does have the advantage of being able to determine ammonia levels in the vertical direction for some cases. Ammonia levels, determined by a tungstic acid denuder with ozone chemiluminescence detection of thermally evolved NO_y (*29*), have been measured aloft (*30*) with <10-min sampling times, but real-time use is probably not possible.

Gaseous Peroxides

Measurement capabilities for gas-phase peroxides (e.g., H_2O_2, and methyl hydroperoxide and peracetic acid when present) have lagged behind developments in methods for peroxides in atmospheric liquid water samples, despite the ease with which H_2O_2, at least, can be scrubbed from air samples. This ease is due to the difficulties in collecting $H_2O_2(g)$ without generating "artifact" peroxide by aqueous-phase chemistry involving, for example, ozone and radical species. Recently, three methods have been demonstrated for artifact-free collection of H_2O_2 based on prompt derivatization and analysis (*31*), or ozone-removal techniques (*32, 33*), and H_2O_2 has been directly observed by a diode-laser absorbance technique (*34*) as well. Collected peroxide is determined by peroxidase-catalyzed *p*-hydroxy phenylacetic acid fluorescence dimer (*31, 32*) or hemin-catalyzed luminol (*33*) techniques. Airborne measurements of gaseous H_2O_2 have been reported for only one of these gas-phase techniques (*35*) by using continuous, in situ determination of collected peroxides with a time resolution of <2 min. An intercomparison of all gaseous peroxide methods currently under development was conducted near Los Angeles, CA, in August 1986, and the results of this study (to be reported at the 3rd International Conference on Carbonaceous Particles in the Atmosphere, Berkeley, CA, Oct 1987) should help establish the best method for use in future airborne studies.

Continuous Aerosol Techniques

Several techniques exist for the continuous, in situ determination of aerosol number, size distribution, and mass. (Techniques for mass determination are derived from light-scattering measurements with a nephelometer.) These techniques include the electrical aerosol analyzer; optical particle counters; optical particle probes (PMS probes) for fine aerosols, clouds, and rain droplet distributions; and even near-real-time mass distributions using impactor separation with a piezoelectric balance. Most of these techniques have received substantial use on aircraft platforms and have effective time resolutions for size and mass distribution determinations of a few minutes. The exceptions are the nephelometer from which fine mass concentrations may be determined with a time resolution of about 1 s, and the PMS probes, which measure number distributions each second, and provide statistically valid size distribution data for fine aerosols and cloud droplets when averaged for about 5 min. Time resolution of the PMS probes varies with number, concentration, and phase of particles present (and aircraft speed) and is in the range from 0.1 s in liquid clouds to as long as 30 min for dry particles in the free troposphere.

Aerosol sampling techniques must take into consideration the aerodynamic properties of the sampled particles. That is, the aerosol must be sampled isokinetically if the original aerosol size distribution is to be maintained. This requirement need not be rigorously adhered to if only fine aerosol (particles <2 μm in diameter) is being sampled, but is critical if coarse-particle or cloud droplet analyses are to be performed.

Chemically specific measurement methods that can achieve continuous, near-real-time determinations of aerosol species are, in general, not available. The only significant exception to this statement is a FPD for the determination of aerosol sulfur (i.e., mostly sulfate in ambient air). (The FPD application to the measurement of ambient SO_2 was discussed previously.) Aerosol sulfur concentrations are determined by the FPD after removal of gaseous sulfur compounds by passage through a diffusion-denuder tube coated with lead(II) oxide or another equivalent sulfur gas sink. By using sensitivity enhancement by SF_6 addition to the H_2 fuel and a constant-pressure inlet or mass-flow controls to reduce the sensitivity of the baseline to pressure changes, a real-time FPD detector for aerosol sulfur can be used in airborne measurements with the same sensitivity and calibration factors as the analogous FPD-based system described previously for SO_2.

Time Resolution Considerations

The *time resolution* of an in situ measurement is the time required for the real-time sensor to reach 90% of the final response to a step change in concentration of the measured species. This time response consists of the residence time of the sampled air in the instrument, the time required for the trace constituent to reequilibrate with the inlet and intrainstrumental flow lines, and the time response of the instrumental sensor itself. For trace reactive gas species, the reequilibration time is usually the determining factor in establishing the overall response time in airborne sampling. Typical response times for airborne instruments vary from about 1 s for ozone (ethylene chemiluminescence) and light scattering (nephelometer) to about 3 s for nitrogen oxides (ozone chemiluminescence), and as much as 10–20 s for SO_2 and aerosol sulfur (FPD technique). As noted, the chemiluminescent technique cannot be used in serial fashion with a nylon filter to determine nitric acid because the reequilibration time is excessively long for this highly reactive gas; a two-channel system is used instead. Step changes in nitric acid concentration require substantially longer time periods for reequilibration than observed with other nitrogen oxides (e.g., NO and NO_2).

Time resolution considerations are somewhat altered for integrated sampling techniques such as the tungstic acid technique for ammonia or the impinger–fluorescence analysis technique for peroxides. Here, a fixed amount of analyte is required for analysis; thus, lower air concentrations translate directly to longer time resolution. Other techniques such as the continuous concurrent scrubber technique for peroxides require an in situ derivatization procedure with mixing of reagents prior to determination; hence, the time resolution is determined by the relevant chemical and mixing kinetics. Thus, it is important in evaluating any method's time resolution for airborne application to establish the limiting factors: sample volume, reequilibration volume, or postcollection factors. Only then can the optimum choice of method for a particular application be made.

Comparison of Real-Time and Integrative Measurements

Comparison of data from real-time and integrative techniques employed on airborne platforms to measure trace species in the atmosphere are not numerous in the refereed literature. A few examples will be given in this section, although this list is not intended to be comprehensive.

Nitric acid data obtained by most published methods have been compared extensively with surface measurements (*3, 4, 36*), and new data from an intercomparison conducted in 1985 and reported at the 192nd meeting of the American Chemical Society (ACS) will be available in the literature. Airborne intercomparison data are much more rare, although one comparison of filter-pack (*17*) and real-time chemiluminescent HNO_3 techniques taken from the work of Kelly (*37*) has been reported (*1*). The agreement is good ($r^2 > 0.5$, the slope is about 0.75, and the intercept is not significantly different from zero), considering that half-hour averages of the real-time data are taken, the sampling lines are not identical for the two methods, and the filter-pack data are subject to some systematic artifacts (*3*). Precision and accuracy are estimated at ±25% on the basis on this study. Another report (*38*) of airborne use of this technique is less optimistic; further work in this area is strongly recommended.

Sulfur dioxide data have been obtained simultaneously by the FPD real-time method and by the impregnated filter approach as early as the mid-1970s, as part of the studies of SO_2-to-SO_4 oxidation in power-plant plumes. Filter-pack measurements of SO_2 and sulfate were used to deduce conversion rates and to deduce real-time data to define plume shapes and homogeneity of background

SO_2 levels. Systematic attempts to correlate integrated averages of real-time data across the plume with filter data were few and generally not successful because of sampling and instrumental difficulties. Correlation of surface measurements of SO_2 by the same techniques has also been problematical because most large data sets have been acquired with the commercial FPD, which is insufficiently sensitive for background ambient measurements, even in northeastern North America. One exception is the comparison of filter and real-time SO_2 data obtained over a full year of measurements at Whiteface Mountain, NY, by Kelly (*39*). In that comparison, real-time SO_2 = 0.75(filter SO_2) + 0.60 ppb. This regression equation value had an r^2 value of 0.67 for 44 sampling periods. This agreement is considered good given the range of ambient SO_2 concentrations and the large temporal extent of the measurements. Care must also be taken in SO_2 intercomparisons to accurately zero the FPD and to account for drifts in the zero signal due to changing sampling conditions (e.g., barometric pressure, CO_2 concentration, and absolute humidity).

FPD-determined surface aerosol sulfate data have also been compared with filter measurements (*40, 41*). When artifact-free quartz filters were used and the sulfate levels were well above the FPD limit of detection, the agreement was quite acceptable: The slope was 1.0 ± 20%, $r^2 > 0.9$, and the intercept was variable but <2-μg/m^3 of sulfate.

In summary, several techniques are available for real-time, in situ measurement of trace atmospheric species from airborne platforms. However, much work remains to be done to improve the techniques available, to introduce new techniques for use in aircraft (in particular, in situ or near-real-time methods for reactive hydrocarbons and aldehydes), and to extend the range of species for which adequate intercomparisons have been performed.

Acknowledgments

I acknowledge the many helpful discussions with Peter Daum and Thomas Kelly. This work was performed as part of the Processing of Emissions by Clouds and Precipitation (PRECP) program under the auspices of the National Acid Precipitation Assessment Program (NAPAP) and of the U.S. Department of Energy under contract no. DE–AC02–76CH00016.

References

1. Tanner, R. L. In *Chemistry of Acid Rain*; Johnson, R. W., Ed.; ACS Symposium Series 349; American Chemical Society: Washington, DC, 1987.
2. Stelson, A. W.; Friedlander, S. K.; Seinfeld, J. H. *Atmos. Environ.* **1979**, *13*, 369–371.
3. Spicer, C. W.; Howes, J. E., Jr.; Bishop, T. A.; Arnold, L. H.; Stevens, R. K. *Atmos. Environ.* **1982**, *16*, 1487–1500.

4. Forrest, J.; Spandau, D. J.; Tanner, R. L.; Newman, L. *Atmos. Environ.* **1982**, *16*, 1473–1485.
5. Tanner, R. L. *Atmos. Environ.* **1982**, *16*, 2935–2942.
6. ten Brink, H. M.; Schwartz, S. E.; Daum, P. H. *Atmos. Environ.* **1987**, *21*, in press.
7. Walters, P. T.; Moore, M. J.; Webb, A. H. *Atmos. Environ.* **1983**, *17*, 1083–1091.
8. Winters, W.; Hogan, A.; Mohnen, V.; Barnard, S. ASRC Publication No. 728; Atmospheric Sciences Research Center, State University of New York at Albany: Albany, NY, 1979.
9. Huebert, B. J.; Baumgardner, D. *Atmos. Environ.* **1985**, *19*, 843–846.
10. Daum, P. H.; Kelly, T. J.; Schwartz, S. E.; Newman, L. *Atmos. Environ.* **1984**, *18*, 2671–2684.
11. Nederbragt, G. W.; Van der Horst, A.; Van Duijn, J. *Nature (London)* **1965**, *206*, 87.
12. Lenschow, D. H.; Pearson, R.; Stankow, B. B. *J. Geophys. Res.* **1982**, *87*, 8833–8837.
13. Dickerson, R. R.; Delaney, A. C. *Anal. Chem.* **1987**, *59*, in press.
14. Fontijn, A.; Sabadell, A. J.; Ronco, R. J. *Anal. Chem.* **1970**, 42, 575–579.
15. Delaney, A. C.; Dickerson, R. R.; Melchoir, F. L., Jr.; Wartburg, A. F. *Rev. Sci. Instr.* **1982**, *53*, 1899–1902.
16. Tanner, R. L.; Daum, P. H.; Kelly, T. J. *Int. J. Environ. Anal. Chem.* **1983**, *13*, 323–335.
17. Daum, P. H.; Leahy, D. F. Report No. BNL-31381R2; Brookhaven National Laboratory: Upton, NY, 1985.
18. Shaw, R. W., Jr.; Stevens, R. K.; Bowermaster, J.; Tesch, J. W.; Tew, E. *Atmos. Environ.* **1982**, *16*, 845–853.
19. Sanhueza, E.; Plum, C. N.; Pitts, J. N., Jr. *Atmos. Environ.* **1984**, *18*, 1029–1031.
20. Kelly, T. J.; Stedman, D. H.; Ritter, J. A.; Harvey, R. B. *J. Geophys. Res.* **1980**, *85*, 7417–7425.
21. Tanner, R. L.; Lee, Y.-N.; Kelly, T. J.; Gaffney, J. S. Presented at the 25th Rocky Mountain Conference, Denver, CO, August 1983; Paper No. 102.
22. Kley, D.; Drummond, J. W.; McFarland, M.; Liu, S. C. *J. Geophys. Res.* **1981**, *86*, 3153–3161.
23. Wendell, G. J.; Stedman, D. H.; Cantrell, C. A.; Damrauer, L. D. *Anal. Chem.* **1983**, *55*, 937–940.
24. Garber, R. W.; Daum, P. H.; Doering, R. F.; D'Ottavio, T.; Tanner, R. L. *Atmos. Environ.* **1983**, *17*, 1381–1385.
25. D'Ottavio, T.; Garber, R.; Tanner, R.; Newman, L. *Atmos. Environ.* **1981**, *15*, 197–203.
26. Abbas, R.; Tanner, R. L. *Atmos. Environ.* **1981**, *15*, 277–281.
27. Kelly, T. J.; Tanner, R. L.; Newman, L.; Galvin, P. J.; Kadlacek, J. A. *Atmos. Environ.* **1984**, *18*, 2565–2576.
28. Hoell, J. M.; Harward, C. N.; Williams, B. S. *Geophys. Res. Lett.* **1980**, *7*, 313–316.
29. Braman, R. S.; Shelley, T. J.; McClenny, W. A. *Anal. Chem.* **1982**, *54*, 358–364.
30. LeBel, P. J.; Hoell, J. M.; Levine, J. S.; Vay, S. A. *Geophys. Res. Lett.* **1985**, *12*, 401–404.
31. Lazrus, A. L.; Kok, G. L.; Lind, J. A.; Gitlin, S. N.; Heikes, B. G.; Shetter, R. E. *Anal. Chem.* **1986**, *58*, 594–597.
32. Tanner, R. L.; Markovits, G. Y.; Ferreri, E. M.; Kelly, T. J. *Anal. Chem.* **1986**, *58*, 1857–1865.
33. Groblicki, P. J.; Ang, C. C. *Proceedings of a Symposium on Heterogeneous Processes in Source-Dominated Atmospheres*; Report LBL-20261; Lawrence Berkeley Laboratory: Berkeley, CA, 1985; pp 86–88.
34. Slemr, F.; Harris, G. W.; Hastie, D. R.; Mackay, G. I.; Schiff, H. I. *J. Geophys. Res.* **1986**, *91*, 5371–5378.
35. Heikes, B. G.; Kok, G. L.; Walega, J. G.; Lazrus, A. L. *J. Geophys. Res.* **1987**, *92*, 915–931.
36. Anlauf, K. G.; Fellin, P.; Wiebe, H. A.; Schiff, H. I.; Mackay, G. I.; Braman, R. S.; Gilbert, R. *Atmos. Environ.* **1985**, *19*, 325–333.

37. Kelly, T. J. Report No. BNL–38000; Brookhaven National Laboratory: Upton, NY, 1986.
38. Walega, J. G.; Stedman, D. H.; Shetter, R. E.; Mackay, G. I.; Iguchi, T.; Schiff, H. I. *Environ. Sci. Technol.* **1984**, *18*, 823– 826.
39. Kelly, T. J. Report No. BNL–37110; Brookhaven National Laboratory: Upton, NY, 1985.
40. Camp, D. C.; Stevens, R. K.; Cobourn, W. G.; Husar, R. B. *Atmos. Environ.* **1982**, *16*, 911–916.
41. Morandi, M.; Kneip, T.; Cobourn, J.; Husar, R.; Lioy, P. J. *Atmos. Environ.* **1983**, *17*, 843–848.

Chapter 29

Aerometric Measurement Requirements for Quantifying Dry Deposition

B. B. Hicks, T. P. Meyers, and D. D. Baldocchi

Dry deposition can sometimes be measured by intensive micrometeorological techniques, but these methods require either demanding chemical precision (e.g., for the gradient method) or rapid frequency response (e.g., for eddy correlation). For tower application, eddy correlation requires a frequency response of about 0.5 Hz in daytime, and about 5 Hz at night. For aircraft application, the usual specification is for 5–10-Hz response. For routine measurement programs, use is frequently made of less direct approaches, in which dry deposition is computed from atmospheric concentration and site-specific deposition velocity data. Available information limits such simplified approaches to only a few chemical species (e.g., O_3, SO_2, and HNO_3) and particulate sulfate, nitrate, and ammonium. In general, average concentration data can be used.

DRY DEPOSITION HAS DIFFERENT MEANINGS in different disciplines. To some workers, it is the particle fraction of the total exchange of airborne trace substances to the surface. To others, it is only the large-particle component of this transfer (i.e., the deposition of particles due to gravity). However, in most contemporary considerations, dry deposition involves both gas and particle exchange: *Dry deposition* is the aerodynamic exchange of trace gases and aerosols from the air to the surface as well as the gravitational settling of particles. In this context, dry deposition is viewed as a parallel to wet deposition, namely, as a mechanism by which airborne pollutants are transferred to the surface unaccompanied by hydrometeors.

In some locations and for some chemical species, dry deposition can greatly exceed wet deposition, and the opposite situation can also occur. In general, the uncertainty regarding dry deposition is a major impediment to the resolution of questions concerning the fate of emissions into the atmosphere, such as are of principal concern to the National Acid Precipitation Assessment Program (NAPAP).

Except in some special circumstances, existing technology cannot measure dry deposition rates of any trace atmospheric constituent in a routine way. Nor is there any generally recognized simple method by which dry deposition fluxes can be modeled. The difficulties involved have been reviewed extensively elsewhere (*1–3*). None of the special methods that have been used (including micrometeorology, snowpack accumulation, mass balance calculations, leaf washing, and throughfall studies) is appropriate for all species, nor is any method yet developed to the extent permitting application in other than intensive experiments. The main application of these methods is to develop relationships describing dry deposition, which can then be used to evaluate deposition either from other information generated by models or from measurements of concentrations and other variables at monitoring sites. No experimental methods are yet available to address the matter of dry deposition in highly complex terrain, although some techniques (e.g., throughfall approaches) can be used in circumstances more complex than can other methods. Micrometeorological methods are especially constrained to simple circumstances.

A key issue is the credibility of dry deposition measurements. In general, experimental methods fall into three convenient categories: measurements at the surface, measurements of the flux through the air to the surface, and estimates derived from knowledge of concentration and deposition velocity.

Measurements at the surface itself (e.g., throughfall and snowpack accumulation methods) provide the most direct quantifications of dry deposition, but are possible only for a few chemical species (mainly particulate) and for some surfaces.

For some chemical species, measurements of fluxes through the air to the surface (e.g., micrometeorological gradient and covariance methods) can be made accurately, but these methods are representative of surface values only if the surface is horizontally uniform and if conditions are not changing rapidly with time.

Estimates derived from knowledge of, or assumptions concerning, concentrations and deposition velocities are susceptible to errors arising from specification of both the appropriate deposition velocity (V_d) and the atmospheric concentration at some specified height ($[C]$), and also from errors associated with the failure of the assumption that $F = V_d[C]$, where F is the flux from the surface. Therefore, in this context, consideration should be given to several independent sources of error:

- errors arising from the fact that the deposition velocity concept is not always appropriate (e.g., in complex terrain, over patchy surfaces, or for pollutants potentially having a surface source such as NH_3 or NO_x);
- errors arising from the specification of $[C]$, either by measurement or modeling;
- errors arising from the specification of V_d, either by estimation from field data or by computation in transport models; and
- errors arising from the incorrect assumption that the surface is a perfect sink (i.e., $[C] \neq 0$ in the final receptor).

The present purpose of this chapter is not to examine the problems associated with dry deposition measurement, nor to review the possible techniques that offer promise at this time. These matters have been addressed in detail elsewhere (3). Instead, the purpose is to summarize the techniques presently used to estimate dry deposition from aerometric measurements. The intent is to examine the philosophies underlying the different networks presently being operated in North America, to explore the measurements being made in these networks, and to review the techniques by which the dry deposition data derived from these programs will be verified. The focus is on the in-air measurement and interpretation of concentrations, both on a routine monitoring basis and as a component of intensive research programs. The question of sampling requirements for liquid sampling (e.g., for throughfall studies) and for sampling of solids (e.g., for surface accumulation methods) will not be addressed. In all cases, however, a need arises for better chemical sensors suitable for use in intensive research applications, and for better understanding of the processes governing deposition and surface emission.

At present, application of the deposition velocity concept in measurement applications is limited to O_3, SO_2, HNO_3, and submicrometer sulfate and nitrate particles. For these species, there is need to consider the requirements of a routine measurement program based on network measurement of air concentrations. The following discussion concentrates on the methods by which air chemistry and critical supporting variables can be measured in order to estimate dry deposition of those chemical species for which current knowledge is best. Measurement of air concentrations of trace species for which deposition velocities are presently not well-known will not be considered in this chapter.

Needs for Intensive (Research) Measurement

The needs of a wide variety of measurement techniques have recently been summarized in the proceedings of a NAPAP workshop on dry deposition (3). In general, surface sampling methods provide detailed information on specific surfaces by overaveraging times that vary (according to the particular method) from hours to months. Methods based on micrometeorology (e.g., gradients and covariances) provide averages over larger spatial scales (e.g., of the order of 1 km for tower-based sensors, and 10–100 km for aircraft systems), but are typically limited to averaging times of about 1 h. In essence, these methods are designed to develop and "calibrate" simple inferential methods, by application of which larger areas and longer averaging times can be addressed.

All chemical species have well-recognized needs for sensors having rapid response. The application is in conjunction with "eddy correlation" measurements, in which vertical fluxes (F) are determined as the average covariance between concentration mixing ratios (X) and the vertical wind component (w).

$$F = D < w'X' > \quad (1)$$

where the primes denote deviations from mean values, and the angle brackets indicate time averaging. The symbol D denotes the density of air. The quantity F is necessarily statistical, and related statistical uncertainties due to sampling can be readily computed. In general, the turbulent structure of the atmosphere imposes a limit on the accuracy with which F can be determined. As a guiding rule, the standard deviation associated with any single evaluation of F is typically not less than 10% for a half-hour averaging period.

In practice, micrometeorological requirements for measuring eddy fluxes are somewhat flexible. Although eddy correlation is sometimes claimed to require exceedingly rapid response of the sensor system (often stated as many Hertz), in most field applications, 0.5-Hz response of the chemical sensor will permit adequate field measurement of the turbulent flux. The key consideration is that sufficient information is known about the high-frequency component of vertical turbulent exchange that corrections can be applied with confidence to account for the effects of some sensor response limitations (4). At night, however, considerably higher frequencies contribute to the turbulent exchange phenomenon and impose greater demands for high-frequency sensor performance. For aircraft operation, the central consideration is the speed of the aircraft; in most situations, 5–10-Hz response is adequate. Table I presents a summary of fast-response chemical sensors presently used in dry deposition studies.

All micrometeorological methods involve an assumption that the fluxes measured at the height of the turbulence sensors are the same as the fluxes to the underlying surface. To assure the validity of this assumption, measurements must be conducted only in conditions that are not changing with time, and at locations where the surface is horizontally uniform. These stationarity and "good-fetch" requirements are fundamental to all micrometeorological research and impose severe limitations on the general applicability of all micrometeorological techniques. In practice, the sensors must be mounted sufficiently high so

Table I. Summary of Some Chemical Sensors Used in Recent Studies of the Turbulent Exchange of Trace Gases

Chemical	*Sensor*	*Use*	*Reference*
SO_2	Flame photometry	EC	5
	UV absorption	EC	6
	Bubblers	G	7
NO	O_3 luminescence	EC	8
NO_2	O_3 luminescence	EC	9, 10
	Luminol	EC	11
O_3	NO luminescence	EC	12[a]
	NO luminescence	EC	13[b]
	Ethylene luminescence	EC	9, 14
	Bubblers	G	15
HNO_3	Nylon filters	G	16

NOTE: References are selected to identify typical applications of the devices, not the origin of the development. Abbreviations are as follows: EC indicates use in an eddy correlation study, and G indicates that the application involved the measurement of vertical gradients.
[a]Information is contained on towers only.
[b]Information is contained on aircraft only.

that no single surface element will unduly influence the flux being measured, and sufficiently low so that flux divergence terms can be ignored.

Special considerations arise in the case of particle fluxes. Some workers employ sensors capable of detailed resolution by particle size (*17*) but then confront problems concerning the counting statistics in each size range (*18*). Other workers (*19*, *20*) sample by using a broad particle size range, thus reducing the adverse effects of unfavorable counting statistics but forfeiting the capacity to study particle deposition as a function of particle size. At this time, an outstanding need exists for a sensing system that would provide chemical-species-specific outputs as a function of particle size and have about 1-Hz response.

In some situations, gradient methods may be used. In such cases, vertical differences in air concentration of the chemical species in question are measured and interpreted by using an atmospheric diffusivity (K) derived from other sources:

$$F = DK[d < X > /d(z - d)] \tag{2}$$

where z is the height above the ground, and d is the height of the relevant zero plane and corresponds to the effective source or sink height of the material being considered. In practice, the requirement for accurate diffusivity information is a severe limitation because of difficulties that arise over forests or any similar surface for which the precise distribution of sources or sinks is not well-known (*21*). These difficulties are sufficient that gradient methods are often viewed with substantially less favor than the more direct eddy correlation methods. However, for some chemical species for which fast-response sensors are not available, no alternative remains but to employ gradient methods in studies carefully designed to avoid such problems. Table I includes references to some representative gradient studies of dry deposition, both for chemical species that can also be measured by eddy correlation (e.g., O_3) and for species for which gradient techniques are at present the only viable option (e.g., especially HNO_3).

The ability to measure a difference in concentrations between two levels in the air does not necessarily mean that deposition fluxes can be determined. In general, the differences that must be measured are of the order of 0.2%–5% of the concentration, depending on the choice of heights above the surface and on the deposition velocity of the material in question. In order to address questions concerning the processes involved in the air–surface exchange phenomenon, these vertical differences must be resolved with a precision typically better than 10%, so that the precision required of the chemical sensor is likely to be in the range 0.02%–0.5%.

Both eddy correlation and gradient studies are sufficiently difficult that the methods are normally tested by comparison against the requirement for heat energy balance at the surface. Thus, measurements are normally made of the sensible and latent heat fluxes (H and L_wE, respectively) by applying equations 1 and 2 to measurements of temperature and humidity. In this way, an objective test of the validity of the micrometeorological technique is possible by verifying that $H + L_wE = R_n - S - G$, where R_n is the net radiation flux, S is the storage of heat associated primarily with changes of canopy biomass temperature with time, and G is the conduction of heat into the ground. The steps involved in such a verification have been discussed elsewhere (22).

"Eddy accumulation" is frequently advocated as a method for avoiding the complexity of a fast-response chemical sensor, while retaining the benefits of direct flux measurement associated with eddy correlation. In concept, eddy accumulation profits from the use of a fast-response sampling system to replace the usual fast-response air chemistry sensor: air is sampled and stored or measured at a rate proportional to w' with separate samplers for updraft and downdraft systems. The difference between values of $<X>$ determined independently for the updraft and downdraft samples can be easily translated into a value of the desired eddy flux F. (In simple concept, an eddy accumulation system requires some system to correct for the impossibility of determining $<w>$ with precision.)

In practice, the promise of the eddy accumulation concept has yet to be fulfilled. Hicks and McMillen (23) simulated the technique and demonstrated the great difficulties associated with several critical components of any eddy accumulation system. Speer et al. (*24*) presented results of field tests of a prototype eddy accumulation system that appears to have operated with insufficient precision to measure trace gas-exchange rates.

Businger (25) reviewed the use of micrometeorological methods as tools for investigating dry deposition, and especially for measuring deposition velocities. The reader is referred to reference 25 for details of the techniques summarized previously and for discussion of a variety of additional methods that are derived from the basic principles presented here. Wesely and Hart (26) presented a discussion of the influence of random noise on the interpretation of micro-meteorological data. In essence, pollutant sensors are inherently noisy; ensemble averages are typically required to reduce statistical uncertainty associated with eddy fluxes derived from their use.

Needs for Routine Measurement (Monitoring)

Routine monitoring of dry deposition is presently in its infancy, and little assurance has been given yet that the exploratory efforts presently underway will eventually prove to be successful. Most monitoring activities call for the measurement of the atmospheric concentrations of appropriate chemical species plus supporting information on meteorological and surface properties from which values of the relevant deposition velocities can be estimated. In this context, the surface flux can be inferred from four possibilities:

a. from direct measurements of both $[C]$ and V_d,
b. from measured V_d and values of $[C]$ derived from other sources,
c. from measured $[C]$ and values of V_d estimated from other information, or
d. from computed values of both V_d and $[C]$.

In practice, options a and b are impractical because direct measurements of V_d are not possible except in the unusual circumstances that permit application of one of the more direct methods of flux measurement. Option c is selected for application in the various monitoring programs presently underway or being initiated. These various programs differ mainly in the way in which they attempt to specify V_d. The "concentration monitoring" approach focuses attention on the need for accurate concentration data from which dry deposition can later be estimated by applying some value of V_d based on external information such as land-use categories. A recent modification calls for V_d to be derived by extrapolation from reference data obtained by using research-grade methods at a subset of locations. The research network operation of NAPAP is based on this "inferential" method, whereby a central array of core research establishment (CORE) stations provides reference data for a larger network of satellite stations. As will be discussed later in this chapter, an intermediate approach is being adopted for the dry deposition monitoring program presently being set up as part of the National Trends Network (NTN) of NAPAP. Option d reflects the approach adopted in numerical modeling schemes.

The need to specify time trends at specific locations may be satisfied by recording air concentrations alone, although some insecurity is related to the tacit assumption that V_d remains constant at each site. When spatial differences are an issue, the concentration monitoring approach requires augmentation to provide information on the spatial variability of the deposition velocity.

Specification of Atmospheric Concentration at Some Specified Height

For some chemical species, existing air pollution monitors offer a potential solution to the problem of routine measurement. This situation is especially the case for O_3, for which UV absorption methods are well-developed and for which alternative methods are not competitive. In this case, the ready availability of commercial sensors capable of yielding accurate hourly averages of ozone concentrations is of considerable benefit; ozone concentration records reveal consistent diurnal variability that imposes a need to consider hourly data more than in the case of pollutants such as SO_2 (27).

No monitor exists for either SO_2 or NO_x that has the sensitivity or freedom from drift necessary to permit continuous reliable operation at remote or rural background sites. Furthermore, no capability is available to monitor hourly fluctuations in HNO_3 concentration with assurity. In all of these cases, alternative methods for field measurement are presently being developed and tested. The following summary is intended to provide some guidance concerning several of the methods presently under development.

Filter Packs. Several different configurations of filter packs are now in routine use. Many share similar characteristics: a polytetrafluoroethylene (Teflon) or similar prefilter intended to remove particles from the airstream followed by a nylon filter to remove HNO_3, and a cellulose final filter previously doped with potassium or sodium carbonate or bicarbonate, and sometimes treated with glycerol, to ensure a moist surface. After an extended study of filter packs, a special subcommittee of the National Atmospheric Deposition Program (NADP) recommended in 1981 that several changes should be made to filter packs before widespread deployment in a monitoring program in the United States. To permit first-order calculation of deposition of sulfate and nitrate aerosol, the use of an open-face filter was severely criticized because the presence of large particles could cause great uncertainty in the computation of an appropriate deposition velocity. One recommendation was that some technique be used to prohibit large (i.e., >2-μm diameter) particles from entering the sampling system. To protect against liquefaction of the final doped filter in high humidities, either the sampled airstream should be heated slightly or the sampler should be turned off when the humidity exceeds some predetermined value. The construction of the filter pack was recommended to be molded tetrafluoroethylene (Teflon) rather than machined from a cast block. Finally, the flow rate and pressure drop data obtained at critical points of the sampler should be recorded for later scrutiny and for accurate computation of the volume sampled. Filter-pack units developed as a consequence of these recommendations are now available from commercial sources.

Bubblers. A variety of bubblers was also examined by the NADP committee. Although many bubblers were constructed to provide assured long-term reliability, they shared the common feature of a liquid collection medium in which chemical speciation can be changed substantially. Bubblers were not recommended for long-term routine monitoring in the United States because of the anticipation that evaporation would be rapid and that the matter of speciation would present problems.

Denuders. Recent attention has been focused on denuders of a variety of styles. Denuder difference methods have been successful in many situations, and the recent development of annular denuders has caused considerable excitement. Denuders are perceived by some workers to suffer from some potential problems regarding the efficiency with which they scavenge species such as HNO_3, especially in humid conditions. At this time, denuders of many varieties are being developed, and it can probably be predicted with confidence that the current developmental programs will eventually be successful. A special category of devices operates on the same principles as the denuders but is not reliant on the assumption of complete scavenging of the material being measured from the sampled airstream. Such devices also show considerable promise and appear to be reliable at their present stage of development.

Passive Monitors. For purposes of estimating dry deposition, passive monitoring devices have probably not been explored adequately. Such devices are an extension of the "lead candle" approach: a sensitized medium is contained in a protected environment, and a slow leak connects its sampling volume to the ambient atmosphere. This slow leak may be as simple as a hypodermic needle or a set of very small holes in an inert cover plate. In principle, the transfer of the material being measured to the measuring medium is controlled by molecular diffusivity and by the geometry of the exposure. In sufficiently controlled conditions, an accurate average concentration number should be obtained.

Guidelines adopted for chemical measurements made in the networks of the routine monitoring and research programs of NAPAP, with emphasis on the nested-network operation of the National Oceanic and Atmospheric Administration's (NOAA), Department of Energy's (DOE), and Environmental Protection Agency's (EPA) CORE research program (*28*), are presented in the appendix at the end of this chapter. Concentration monitoring measurement philosophies, containing various stages of augmentation to assist in specifying appropriate values of V_d, are being used in the programs of other agencies [e.g., the EPA, the

Electric Power Research Institute (EPRI), and the Atmospheric Environment Service of Canada (AESC)], and are also discussed in the appendix.

Derivation of Deposition Velocity

Rather complicated mathematical formalism lies behind the relatively simple concept of a deposition velocity. The overall behavior can be described in terms of a resistance model, in which the overall resistance to transfer (i.e., the reciprocal of the deposition velocity) is made up of individual resistances associated with specific transport processes combined in series and in parallel as in an electrical circuit.

Deposition velocities are known to vary with the surface area index, the texture of the surface, its biological properties, and the prevailing meteorological conditions. Surface wetness due to condensation (i.e., dew) is known to be an important property (*29*) that promotes the transfer of water-soluble trace gases. For the case of particles, the role of surface wetness might well be to increase the efficiency of retention of impinging particles through the generation of liquid bridges between the particles and the deposition substrate.

Most often, values of the deposition velocity are derived by using a multiple resistance transfer model, which is somewhat limited in its generality but can be applied with gradually improving confidence if circumstances are well-selected. At present, deposition velocity evaluation techniques are being developed both to interact with long-range transport models and to be driven by field data for application in measurement programs. All such methods for calculating deposition velocities rely on separate evaluation of the important contributing resistances, followed by a calculation of their combined effects. For both modeling and measurement applications, canopy resistances are assumed to be made up of individual subresistances, visualized in both series and parallel arrays to simulate the roles of different in-canopy biological and physical processes (*30, 31*). The techniques used to evaluate deposition velocities are not yet precise, and in some aspects they are grossly deficient. A recognized goal of ongoing research is to identify important processes and to formulate them in terms of variables suitable for modeling and routine measurement.

The resistance model provides a framework for coupling individual processes, some depending on the nature of the pollutant (e.g., chemical species and particle size) and some being surface-dependent. The limitations of the framework need to be remembered. Effects associated with particle sedimentation need to be considered very carefully, and perhaps most importantly, the method has difficulty separating emission from deposition. Suppose a diluted sulfur plume moves across a sulfur-rich swamp. As the air moves over the swamp, its concentrations of sulfur will increase as a result of the emissions from the surface. At some stage downwind, a measurement of sulfur species might inadvertently be coupled with some carefully parameterized deposition velocity, and a completely erroneous dry deposition estimate results. In all cases, care must be taken not to conclude that dry deposition is occurring when the concentrations that are measured in the air are in fact a consequence of local emissions from the surface.

Two alternative approaches are presently used to derive estimates of appropriate deposition velocities for interpreting field measurements of air concentrations. For application in numerical models, the surface is typically categorized in terms of its land use, on the basis of which estimates of appropriate surface resistances are computed by using available field observations as guidance (*32–34*). For field programs and for possible extension to routine monitoring applications, the surface can be considered as a composite of individual biological species, each having its own physiological behavior controlling such factors as stomatal resistance and cuticular uptake (*35, 36*). The focus of the following discussion is on the field experimental requirements.

The three major resistance components in the multiple resistance approach are an aerodynamic resistance (R_a), a near-surface boundary resistance (R_b), and a resistance representative of surface reactions and canopy uptake (R_c).

Aerodynamic resistance is fairly well-understood for uniform, relatively flat terrain. For rough surfaces, the formulation of R_a is affected by uncertainty about how to consider the consequences of source and sink distributions that differ from those for momentum. In daytime, values of R_a tend to be relatively low, and thus the precise value is often not important. (Exceptions are very reactive species such as HNO_3, HCl, and HF, which are believed to deposit efficiently to any surface with which they come in contact.) At night, R_a is usually much greater, except over extensive water surfaces.

Diffusive effects strongly influence the resistance associated with transfer across the boundary layer immediately adjacent to a surface. Thus, gas speciation and particle size and density become important factors. The results of modeling studies and theory tend to agree for simple surfaces. There remains considerable doubt about the best representation for forests and for other surfaces with large roughness elements. Problems appear to be greatest for the case of particle exchange.

Canopy resistances are usually controlled by biological factors, most of which vary with the state of the plant canopy, water stress, and soil conditions, as well as with radiation, humidity, temperature, and other meteorological factors. Any single evaluation should be viewed as only one representation of a widely varying phenomenon.

The NOAA, DOE, and EPA CORE research program (*28*) has developed the guidelines in the box on page 610 for measurements to be made in support of air concentration sampling for dry deposition measurement. These guidelines identify environmental factors that should be quantified to permit extrapolation from research-grade CORE sites to other locations where intensive research measurements are not made. Many of these variables are being measured at the regular dry deposition monitoring (concentration monitoring) sites presently being set up under EPA and EPRI programs.

Guidelines for Measurements Made in Support of Air Concentration Sampling for Dry Deposition Measurement

Meteorological Measurements

Variables. In addition to measurements of the atmospheric concentrations of the chemical species of interest, measurements are needed of the quantities presently used in dry deposition algorithms (*36*) for deposition velocities: wind speed, wind direction standard deviation, incoming shortwave solar radiation, air temperature, air humidity, surface wetness, and precipitation. Quantities that might prove necessary in the future include net radiation, wind direction, and precipitation chemistry.

Time Resolution. Time resolution should be sufficient to resolve features of the diurnal cycle, typically 1 h. (Some 15-min averages are recorded in current work and are used in the quality assurance and quality control procedures to derive 1-h averages.)

Site Quality. Site quality should be adequate to assure that critical measurements are representative of the local surroundings (*see* the previous paragraphs), and unaffected by local topographical features imposing variability in either surface heating (e.g., nearby parking lots or buildings), roughness, moisture exchange (i.e., evaporation rates, perhaps from a nearby water body or irrigated crop), or slope. The matter of slope variation is especially important. Because flow separation often occurs when a slope changes by about 15%, sites should be selected so that no nearby surfaces exceed a slope of more than 8° from the horizontal, nor should the local slope change by more than this amount.

Surface Measurements

Variables. Quantities presently used in V_d algorithms include snow cover, plant species distribution, leaf area index, and canopy height. Quantities that are expected to be needed in the future include leaf water potential and soil moisture.

Time Resolution. Time resolution should be that of the basic dry deposition requirement (e.g., 1 week for the regular monitoring programs of the NTN).

Sampling Protocols

The question of sampling time arises frequently. Some workers argue that short-term concentration data are required in order to evaluate dry deposition. The perceived problem is the strong diurnal cycle that doubtlessly exists in many locations of interest. Two choices have statistical validity: either consider short-term data so that the long-term average can be constructed by integrating through the diurnal cycle, or consider sampling periods long with respect to 1 day and apply appropriately corrected interpretation procedures. Tests conducted so far have demonstrated that day–night sampling might be beneficial for ozone but not for sulfur species (27).

Siting Requirements and Representativeness

The inferential method as is being used in the dry deposition research programs of NAPAP (CORE and CORE satellite programs) evaluates the average dry deposition flux to an area surrounding the point of observation, which is typically several hundred meters in diameter. Hence, the observation site should be located within an area that is both spatially homogeneous and representative of the larger region of interest in which the site is located. ("Homogeneity" includes both topographical features and the distribution of plant species and other surface elements. For example, proximity to a body of water or to an area of different roughness should be avoided.)

For purposes of estimating dry deposition at some selected location, the appropriate deposition velocity must be computed. The deposition velocity is usually controlled by surface factors more than atmospheric ones, and hence the requirements associated with the operation of a normal meteorological reporting station can be relaxed somewhat. The extent to which such relaxation is permissible remains to be assessed fully. Some sites have been set up in forest clearings, for example, so that tests can be made of this kind of meteorologically imperfect exposure. These tests have not yet been completed.

The indirect estimation of dry deposition from air concentration data as collected in some concentration monitoring programs is not necessarily so constrained, but the results are correspondingly susceptible to greater errors and uncertainties. In essence, the philosophy of such methods is to characterize the atmospheric chemistry of some area of interest, and then to compute dry deposition by using some estimated deposition velocity derived from consideration of such factors as land-use categories.

Data Verification

The regular dry deposition monitoring network presently being set up under sponsorship of NAPAP is a modified concentration monitoring network, in which air chemistry and selected meteorological variables are recorded such that dry deposition can be computed once quality-assured deposition velocity routines are available. The accuracy of the results obtained in such an operation is assured, at least in concept, by the use of verified deposition velocity algorithms driven by on-site measurements of critical controlling variables.

In parallel with this regular monitoring activity, a nested-network approach is being tested in which the inferential methods applied at a wide array of locations are tested and results verified at a subset of more intensive CORE research stations. In this network operation, the accuracy of routine measurements is assured by direct reference against higher grade measurements made at a subset of more intensive stations: the CORE stations of the nested-network array (*28*). The various pieces of these various networks and the goals of each component are summarized in the appendix at the end of this chapter.

The network presently in existence consists of stations and satellite stations. CORE stations, designed to provide reference "benchmarks" data, are located at Oak Ridge, TN (servicing southern forested surfaces); Argonne, IL (servicing midwestern cropland); and State College, PA (servicing eastern cropland). CORE satellite stations, used for extrapolating results obtained at CORE sites, are located at West Point, NY (variable surface); Whiteface Mountain, NY (sloping terrain); Champaign–Urbana, IL (cropland); Panola, GA (comparison site for calibrated watershed); Sequoia, CA (comparison site for throughfall and stemflow); Pawnee, CO (comparison site for micrometeorology, U.S. Forest Service); Borden, Ontario (comparison site for micrometeorology); and Shenandoah, VA (comparison site for throughfall, stemflow, and watershed). The CORE stations at Oak Ridge and State College, as well as the satellite stations at West Point and Whiteface Mountain, are also sites of the EPA Concentration Monitoring, which is a regular dry deposition monitoring network. This EPA network will soon expand to more than 30 locations.

Conclusions

Improvement in the state of the art of sampling atmospheric concentrations for deriving dry deposition information is urgently needed. In particular, the research programs are in need of improved sensors having fast response or high precision, and routine monitoring programs require relatively simple but reliable devices suitable for year-round deployment in the entire range of environmental conditions.

Sampling designed to provide information relative to dry deposition must contain strong meteorological and surface components because for many chemical species, the deposition velocity is as variable a quantity as the concentration.

No well-defined optimal method for sampling dry deposition has been presented. Instead, a range of experimental methods are still being tested and developed. In this presentation, the focus has been on those methods that rely on the measurement of concentrations in air. Consequently, little attention has been given to many alternative methods that are of obvious promise. A more complete examination of the various alternatives is available (3).

Abbreviations and Symbols

ADOM	Canadian Acid Deposition and Oxidant Model
AESC	Atmospheric Environment Service of Canada
C	concentration in air (kg/m^3)
CORE	Core research establishments (of the dry deposition nested-network research program)
d	zero-plane height (m)
D	air density (kg/m^3)
DOE	U.S. Department of Energy
E	evaporation rate ($kg/m^2/s$)
EPA	U.S. Environmental Protection Agency
EPRI	Electric Power Research Institute
F	flux from the surface ($kg/m^2/s$)
G	heat transfer into the ground (W/m^2)
K	eddy diffusivity (m^2/s)
L_wE	latent heat flux (W/m^2)
NADP	National Atmospheric Deposition Program
NAPAP	U.S. National Acid Precipitation Assessment Program
NTN	National Trends Network of NAPAP
NOAA	U.S. National Oceanic and Atmospheric Administration
PAN	peroxyacetyl nitrate
R_a	aerodynamic resistance (S/m)
R_b	quasilaminar boundary resistance (S/m)
R_c	canopy resistance (S/m)
R_n	net radiation (W/m^2)
RADM	Regional Acid Deposition Model
S	canopy heat storage rate (W/m^2)
V_d	deposition velocity (m/s)
VOCs	volatile organic compounds
w	vertical wind velocity (m/s)
X	concentration mixing ratio (kg/kg)
z	height above the ground (m)

Appendix—Summary of Existing Networks Providing Data on Dry Deposition in the United States

The existing program of the NAPAP is largely directed toward provision of routine dry deposition measurements. The overall approach has been to address the problem at three groups of sites constituting a nested network. These groups of sites contain several levels of complexity and have substantially different goals.

The EPA Routine Air Concentration Monitoring Dry Deposition Sites

An EPA NAPAP program is being initiated to monitor variables related to dry deposition at locations identified as requiring dry deposition data for modeling purposes. Presently there are five sites, but the program will expand to at least 30.

Weekly data are desired and are initially to be obtained by using day–night sampling. However, the equipment will apparently be capable of operation over shorter time intervals as well in order to test shorter term predictions of models. Similar concentration monitoring activities are being conducted by EPRI and AESC.

Goal. As a first step, the goal is to provide data for testing transport and deposition models, probably by using 12-h time resolution.

Chemical Focus. The focus is on species important in modeling such as SO_2, HNO_3, NO_2, H_2O_2, O_3, NH_3, volatile organic compounds (VOCs), and peroxy-acetyl nitrate (PAN), plus fine particulate SO_4^{2-}, NO_3^-, and NH_4^+. Coarse particles are not addressed.

Siting. Initial emphasis is on the Northeast, at locations required for testing the Regional Acid Deposition Model (RADM) and Canadian Acid Deposition and Oxidant Model (ADOM).

The NOAA Trial Program (CORE and Satellite Stations)

Methods for extrapolating from research-grade CORE stations to other locations of simpler measurement are being tested at an array of satellite sites associated with the CORE program. The intent is to develop methods suitable for routine measurement of dry deposition of a few key chemical species (especially SO_2 and HNO_3) in a manner compatible with the operating procedures of the existing wet deposition monitoring networks; the installation and operating costs should be no more than 5 times that of the wet deposition operation. Measurements of air chemistry and supporting atmospheric and surface data are obtained with equipment that is relatively inexpensive, rugged, and suitable for long-term operation. The necessary techniques for inferring dry deposition from such field data are still being developed fully, and present capabilities are such that only a limited number of chemical species and surface configurations can be addressed. Nevertheless, the approach appears sufficiently practical that some limited expansion is planned.

All sites are selected to be in regionally representative areas where appropriate deposition velocities can be estimated from field data by using current capabilities. The measurements focus on SO_2, HNO_3. and particulate anions and cations as reported by the NTN. Weekly averages of concentrations are measured with simple filter packs (initially), and supported by hourly meteorological and surface data for computing deposition velocities. The method relies on the ability to deduce appropriate deposition velocities provided measurements are made of controlling surface and atmospheric properties. The approach is known as the *inferential method*, in order to differentiate it from the more conventional *concentration monitoring approach*, which focuses on air quality alone, although often with supporting measurement of a selected set of meteorological variables. Estimated deposition fluxes derived at the satellite sites are referenced against more direct methods at the subset of special CORE sites.

Goal. As a first step, the goal is to develop a routine method for providing dry deposition data that would parallel the wet deposition data of NTN.

Chemical Focus. The measurements will ultimately focus on fine-particulate anions and cations as reported by NTN, plus SO_2 and HNO_3. In general, these are species for which inferential methods are appropriate. Coarse particles are not yet being considered.

Siting. The siting is a regionally representative subset of NTN.

CORE Network

The first goal of the NAPAP CORE sites is to provide reference data for assuring the quality of results obtained by using simpler methods elsewhere (e.g., at the CORE satellite stations). A second goal has been to develop and assess state-of-the-art methods for measuring dry deposition fluxes. Several methods are explored concurrently (e.g., eddy fluxes, gradients, foliar extraction, throughfall, and watersheds). For those cases where these methods can operate side by side, rigorous intercomparisons of the mature methods are used to assess the reliability with which dry deposition fluxes can be measured. Strengthened by such assessments, the different methods can be separately applied in regions of their special applicabilities (e.g., hillsides, flat terrain, canopy, or open areas).

However, these special methods are usually more complicated than is feasible for routine application (e.g., in a monitoring network) and are hence applied regularly, not necessarily continuously, but normally, in short intensive case studies.

The DOE, NOAA, and EPA NAPAP network of CORE research stations is intended to provide the basis for the deposition velocity parameterizations to be used in routine monitoring operations, and also to provide benchmark verifications of routine methods. This network is presently made up of only three stations. All chemical, meteorological, and surface data are recorded hourly or more frequently and augmented by routine application of more direct dry deposition measurement methods (e.g., eddy correlation for 1 week every month) to provide the desired CORE site comparisons of both deposition velocities and dry deposition fluxes.

Goal. The goal is to provide detailed data on dry deposition at a limited number of regionally representative sites, and to test and improve methods used in routine measurement programs.

Chemical Focus. The emphasis is on species for which more direct measurement methods are appropriate (e.g., SO_2, O_3, NO_2, HNO_3, and fine particles).

Siting. Emphasis is on regionally representative sites having uniform surface and having existing research capabilities for application of different measurement methods.

Acknowledgments

This work was carried out under the sponsorship of the U.S. Department of Energy and the National Oceanic and Atmospheric Administration (NOAA) as a contribution to the National Acid Precipitation Assessment Program (NAPAP) (Task Group II: Atmospheric Chemistry).

References

1. Hicks, B. B. *Water, Air, Soil Pollut.* **1986**, *30*, 75–90.
2. Voldner, E. S.; Barrie, L. A.; Sorois, A. *Atmos. Environ.* **1986**, *20*, 2101–2123.
3. Hicks, B. B.; Wesely, M. L.; Lindberg, S. L.; Bromberg, S. M. *Proceedings of NAPAP Workshop on Dry Deposition, Harpers Ferry, VA, March 25-27, 1986*; NAPAP Report; National Technical Information System: Oak Ridge, TN, 1987.
4. Hicks, B. B. *Boundary-Layer Meteorol.* **1972**, 3, 214–228.
5. Galbally, I. E.; Garland, J. A.; Wilson, M. J. G. *Nature (London)* **1979**, *280*, 49–50.
6. Nestlen, M. G. *Proceedings of the NATO Advanced Research Workshop on Acid Deposition Processes at High Elevation Sites, Edinburgh, Scotland, September 8-12,* **1986**, 4, 429–444.
7. Shepherd, J. G. *Atmos. Environ.* **1974**, *8*, 69–74.
8. Delany, A. C.; Fitzjarrald, D. R.; Lenschow, D. H.; Pearson, R., Jr.; Wendel, G. J.; Woodruff, B. *J. Atmos. Chem.*, in press.

9. Wesely, M. L.; Cook, D. R.; Williams, R. M. *Boundary-Layer Meteorol.* **1981**, *20*, 459–471.
10. Wesely, M. L.; Eastman, J. A.; Stedman, D. H.; Yalvac, E. D. *Atmos. Environ.* **1982**, *16*, 815–820.
11. Hicks, B. B.; Matt, D. R. *J. Atmos. Chem.*, in press.
12. Eastman, J. A.; Stedman, D. H. *Atmos. Environ.* **1977**, *11*, 1209–1211.
13. Lenschow, D. H.; Delany, A. C.; Stankov, B. B.; Stedman, D. H. *Boundary-Layer Meteorol.* **1980**, *19*, 249–265.
14. Wesely, M. L.; Eastman, J. A.; Cook, D. R.; Hicks, B. B. *Boundary-Layer Meteorol.* **1978**, *15*, 361–373.
15. Galbally, I. E.; Roy, E. R. *Proceedings, Quadrennial International Ozone Symposium, August 4–9, 1980, Boulder, CO*; National Center for Atmospheric Research: Boulder, CO, 1980; pp 431–438.
16. Huebert, B. J.; Robert, C. H. *J. Geophys. Res.* **1985**, *D1(90)*, 2085–2090.
17. Sievering, H. In *Precipitation Scavenging, Dry Deposition, and Resuspension*; Pruppacher, H. R.; Semonin, R. G.; Slinn, W. G. N., Eds.; Elsevier: New York, 1983; pp 963–978.
18. Fairall, C. W. *Atmos. Environ.* **1984**, *18*, 1329–1337.
19. Wesely, M. L.; Cook, D. R.; Hart, R. L.; Hicks, B. B.; Durham, J. L.; Speer, R. E.; Stedman, D. H.; Trapp, R. J. In *Precipitation Scavenging, Dry Deposition, and Resuspension*; Pruppacher, H. R.; Semonin, R. G.; Slinn, W. G. N., Eds.; Elsevier: New York, 1983; pp 785–793.
20. Wesely, M. L.; Cook, D. R.; Hart, R. L.; Speer, R. E. *J. Geophys. Res.* **1985**, *90*, 2131–2143.
21. Hicks, B. B. In *The Forest Atmosphere Interaction*; Hutchison, B. A.; Hicks, B. B., Eds.; Reidel: Boston, MA, 1985; pp 631–644.
22. Hicks, B. B.; Wesely, M. L.; Coulter, R. L.; Hart, R. L.; Durham, J. L.; Speer, R.; Stedman, D. H. *Boundary-Layer Meteorol.* **1986**, *34*, 103–121.
23. Hicks, B. B.; McMillen, R. T. *J. Clim. Appl. Meteorol.* **1984**, 23, 637–643.
24. Speer, R. E.; Peterson, K. A.; Ellestad, T. G.; Durham, J. L. *J. Geophys. Res.* **1985**, *90*, 2119–2122.
25. Businger, J. A. *J. Clim. Appl. Meteorol.* **1986**, *25*, 1100–1124.
26. Wesely, M. L.; Hart, R. L. In *The Forest Atmosphere Interaction*, Hutchison, B. A.; Hicks, B. B., Eds.; Reidel: Boston, MA, 1985; pp 591–612.
27. Meyers, T. P.; Yuen, S. *J. Geophys. Res.* **1987**, *D6(92)*, 6705–6712.
28. Hales, J. M.; Hicks, B. B.; Miller, J. M. *Bull. Am. Meteorol. Soc.* **1987**, *68*, 216–225.
29. Brimblecombe, P. *Tellus* **1978**, *30*, 151–157.
30. O'Dell, R. A.; Taheri, M.; Kabel, R. L. *J. Air Pollut. Control. Assoc.* **1977**, *27*, 1104–1109.
31. Wesely, M. L.; Hicks, B. B. Ibid., 1110–1116.
32. Sheih, C. M.; Wesely, M. L.; Hicks, B. B. *Atmos. Environ.* **1979**, *13*, 1361–1368.
33. Walcek, C. J.; Brost, R. A.; Chang, J. S. *Atmos. Environ.* **1986**, *20*, 949–964.
34. Voldner, E. S.; Sirois, A. *Water, Air, Soil Pollut.* **1986**, *30*, 179–186.
35. Baldocchi, D. D.; Hicks, B. B.; Camara, P. *Atmos. Environ.* **1987**, *21*, 91–101.
36. Hicks, B. B.; Baldocchi, D. D.; Hosker, R. P., Jr.; Hutchison, B. A.; Matt, D. R.; McMillen, R. T.; Satterfield, L. C. "On the Use of Monitored Air Concentrations to Infer Dry Deposition"; NOAA Technical Memorandum; National Oceanic and Atmospheric Administration: Boulder, CO, 1985; pp 1–65; ERL ARL–141.

Sampling Biota

Chapter 30

Sampling Aquatic Biological Matrices

J. A. Rogalla

Biological samples are collected for chemical and ecological analyses for research, to satisfy regulatory requirements, or to meet other environmental program objectives. In any sampling program, sample representativeness must be addressed through careful planning and data interpretation. Adapting the data quality objectives process to biological studies can help ensure that objectives are clearly defined, input from technical experts is obtained, and appropriate sampling and analytical procedures are selected. Statistical sampling approaches are often necessary or useful to plan sampling activities. Various types of sampling equipment and procedures are available to use in different situations; knowledge of site conditions, study objectives, and potential limitations is needed to help select the proper equipment for the job. Sample preservation is also an important element to consider to minimize interactions, chemical or physical changes, or degradation of the samples collected.

SAMPLES OF AQUATIC BIOLOGICAL ORGANISMS, from plankton to fish, are collected and analyzed for a variety of environmental programs. The results are used, for example, to support decisions related to hazardous waste investigations and impacts, to verify compliance with permit requirements, to satisfy research objectives, or to assemble reference collections. Both the sample design and sample collection techniques are critical elements of a successful sampling program, and both are addressed in this chapter. Planning and sample collection for various types of organisms must account for the fact that most species occur only in specialized habitats or conditions, and most organisms are mobile and thus able to move to more favorable conditions or avoid capture by sampling devices. One more complicating factor is that distributions are often patchy, a condition that increases variability and makes it more difficult to characterize the populations. These qualities make collecting truly representative biological samples a challenging task.

Planning and conducting a biological sampling program involves: (1) clearly defining the program objectives and expected data uses; (2) developing a reasonable and feasible sampling plan; (3) identifying the proper sampling and analytical techniques; (4) collecting, preserving, and analyzing the samples according to the plan; and (5) interpreting the results and determining whether representative and sufficient data have been collected to meet program objectives. If not, additional samples may be required, or the expectations for data use may be modified because of limitations in data quantity or quality.

This chapter focuses on the unique conditions and specific requirements or limitations surrounding biological sample collection; it is not meant to be a comprehensive sampling guide. Common sampling techniques and equipment, preservation requirements, and tips to avoid or limit common problems are described for plankton, bacteria, benthos, periphyton, and fish. Much of the sampling information provided is summarized from *Standard Methods for the Examination of Water and Wastewater (17th ed.)* (*1*); additional information was obtained from U.S. Environmental Protection Agency (EPA) guidance (*2, 3*), textbooks (*4–7*), and journal articles. These and other references cited should be used to obtain detailed descriptions of equipment and procedures and to evaluate other sampling alternatives.

Why Collect Biological Samples?

Biological samples are collected for both chemical and ecological analysis, and the results are used to meet a variety of environmental program objectives. These programs include:

- *Applied and pure research.* Biological research has long been conducted to advance the general understanding of biological populations and interactions, chemical composition, growth, and other facets of individuals or populations. The research may focus on ecological or chemical analysis, or a combination of both.
- *Reference collections.* Reference collections are assembled for use in validating taxonomic identifications, comparison populations between reference and study areas, and for educational use.
- *Ecological risk assessment.* Ecological risk assessment is becoming more common as measuring or calculating impacts on natural populations exposed to hazardous waste is needed to establish cleanup goals for remedial actions. For these studies, both ecological and chemical analyses may be needed to properly assess impacts (*8*).
- *Bioaccumulation studies.* Pollutants may be metabolized by the biota and concentrated in each successive step in the food chain. The concentrations that accumulate can be harmful to the aquatic organisms,

and fish consumed by humans may contain toxic levels of these chemicals.

- *Baseline and monitoring studies.* Plankton and fish are often sensitive to changes in their chemical or physical environments and are used to identify impacts from releases of toxic chemicals.
- *Permit requirements.* Biological sample collection may be required by Resource Conservation and Recovery Act (RCRA) or National Pollutant Discharge Elimination System (NPDES) (*9*) permits to monitor conditions or indicate whether a facility is in compliance with permit requirements. Often upstream and downstream populations are assayed or samples are collected to determine if there has been a release that could influence nearby or downstream populations.

The type of analysis required for biological samples varies with program objectives but falls into two general categories: ecological and chemical. Ecological analysis refers to identification and/or enumeration (counting) of organisms to determine species presence or absence, community composition, and diversity. Chemical analysis refers to the quantitative analysis of the chemicals or nutrients present in the organisms or selected organs (for fish).

If samples are used to document permit compliance, chemicals and water quality parameters are generally specified in the permit. Metabolism or growth studies may be conducted to calculate productivity and biomass. For most sampling programs, water or sediment samples are collected concurrently to provide information for comparison or correlation. For fish, organs (brain or liver) or tissues (muscle or fatty tissues) are often analyzed to determine whether bioaccumulation of toxic chemicals has occurred.

Applying the Data Quality Objectives Process

Data quality objectives (DQOs) are statements about the quality and quantity of data needed to support decisions and meet program objectives. Data quality objectives are an extension or adaptation of the scientific method, in which a hypothesis is proposed, the data needed to test it are identified, and data interpretation and hypothesis testing procedures are defined. The DQO process has been formally presented by the EPA (*10*) as a way to focus planning activities and ensure that all aspects of program planning and data use are addressed before data are collected. This process was originally developed for planning Comprehensive Environmental Response, Compensation, and Liability Act (Superfund) investigations, but it is easily adapted to the biological sampling programs described in this chapter.

The DQO process includes seven elements that should be addressed when a sampling program is developed. Several iterations of this process are often necessary to collect sufficient data to support the decision or complete the hypothesis testing. The data collected for each iteration are used to plan the next one. The seven steps are:

1. State the problem.
2. Identify the decision.
3. Identify inputs to the decision.
4. Define the study boundaries.
5. Develop a decision rule.
6. Specify limits on decision errors.
7. Optimize the design for obtaining data.

Detailed discussions and examples for each element are presented in the current version of the EPA DQO Guidance (*10*); application of several of these elements to biological sampling programs is described here.

Stating the problem concisely is a key to designing a successful sampling program. Biological sampling can be very expensive and time-consuming, and because data are often highly variable it is often difficult to collect sufficient data to make a decision or solve the problem with a desired level of confidence. These constraints may necessitate separating the problem statement, and therefore the sampling program, into several discrete steps. The decision or problem should be stated in question form. An example of a problem statement is "Have organic chemicals discharged from an industrial outfall reduced species numbers and diversity in downstream populations?" The decision to be made for a sampling program to address this problem statement will be a "yes/no" determination about whether there is a difference between upstream and downstream populations.

Identifying the inputs to the decision, or the data to be collected, usually requires participation by specialists from several biological disciplines, chemists, surface water hydrologists, statisticians, regulators, and others who will be involved in decision making. Interactive discussions will help identify related parameters that should be measured and the degree of variability likely to be encountered. Related parameters that could be important in evaluating biological results include general water quality parameters and the chemical composition of water and sediment, observations of the physical features of the sampling locations, weather, season, and time of day.

For the example given, the data needed as inputs to the decision probably would include ecological samples or observations of the species and numbers of organisms present upstream and downstream of the outfall; observations of streambed features; chemical concentrations of organic compounds in the organisms, water, and sediment from the outfall locations; and locations where organisms are collected or observed.

The interdisciplinary interaction needed to identify the inputs is also needed to define the study boundaries in time and space. The occurrence and distribution of organisms are greatly affected by substrate, season, life cycle, and other factors, and these in turn influence when and where samples should be collected. These factors are critical in determining whether representative data are collected and can be used to test the hypothesis or make the decision. If experts familiar with both the study objectives and the effects of these factors are not involved, unrepresentative or inaccurate data are likely to be collected.

For the example problem, influences on the study boundaries include stream flow velocity and characteristics (e.g., turbulent or laminar flow, eddies, or stream confluences), depth, light penetration, bottom composition (e.g., rocky, sandy, or silty) and features (e.g., pools and riffles, vegetation, or submerged logs), and whether the organisms are sedentary. Seasonality and life-cycle stages will also be important considerations for sampling whether the long-term effects of the discharges will be monitored or a single sampling event is expected to provide the data for decision making.

Developing a decision rule and specifying limits on decision errors are two of the most challenging and critical steps in the DQO process. Decision rules are "if...then" statements about actions to be taken depending on the data and outcome of the hypothesis testing. Limits on decision errors are usually defined in terms of the degree of statistical uncertainty associated with the decision and data sufficiency parameters for decision making. A decision rule for the example could be: "If releases of organic compounds from the wastewater outfall have caused decreases in species diversity and abundance in downstream populations, then the concentrations of these compounds in the wastewater must be reduced or eliminated."

Biological data are usually highly variable, and characterizing that variability must be one of the primary considerations in designing the sampling program. The statistical tests or process that will be used to quantify the variability and determine if additional samples are needed must be addressed in the decision error step. Power analysis and various types of comparison tests (*t*-test, analysis of variance, or multivariate procedures) are statistical procedures that can be used depending on the sampling design and distribution of the data. Textbooks and journal articles describe statistical tests that can be used for these purposes (*11–15*), and these or other sources should be referenced at this point in the planning stage and during data analysis. Consulting these sources will help identify appropriate sampling designs and ensure that the sampling design produces data that can be evaluated using the selected procedures. Even if a quantitative statistical analysis will not be performed, decision rules and decision error should be described as unambiguously as possible to avoid later indecision or misunderstanding about interpretation of the data.

Optimizing the sampling design is the final step in the DQO process and is the culmination of all the previous steps. The sampling design must identify the number and type of biological and chemical samples and locations, analytical methods, sampling duration and frequency, time of day, season, equipment, and related parameters to be observed or measured. Sample representativeness and variability will be the most critical and difficult factors to assess; therefore, the degree of replication, field observations to be recorded, training, standard operating procedures, and participation by involved technical experts should all be addressed in the sampling plan.

Biological sampling is often conducted using a statistical sampling design because quantitative inferences will be made about the populations sampled. Various types of random or systematic sampling designs are commonly used, such as cluster sampling, stratified random sampling, and grid sampling (*11*). These designs should be adapted to the field conditions by statisticians and knowledgeable experts. Nonrandom or directed sampling is used for some types of programs, such as when monitoring discharge points, pollutant sources, or sampling within water treatment systems. The sampling frequency, parameters, numbers, and locations for these types of programs are often defined in the regulations or permit governing the program. Field conditions, statistical considerations, and program objectives will also influence sampling locations for nonrandom sampling.

Biological Sampling Equipment and Procedures

Sample collection procedures must be selected considering both the organisms to be collected and the conditions that will be encountered. For instance, equipment used to collect benthos samples from a shallow stream will be different from that used to collect samples from a lake that is 30 ft deep. Safety and site accessibility are also factors to consider in the type of equipment used. Sometimes live samples are needed; this requirement will influence sample collection and handling. Whether these or any other factors are relevant should be identified and addressed during sample planning.

Equipment and procedures for collecting plankton, bacteria, periphyton, benthos, and fish are described in the following sections. For each type of sampling, general objectives and uses for the data, equipment, preservation, and special considerations or problems are discussed. Detailed descriptions of sampling equipment, procedures, preservation, and analytical methods are provided in method references and textbooks cited and should be consulted for detailed planning and procedures.

Plankton

Data Collection Objectives and Limitations. *Plankton* are free-floating plants (phytoplankton) and animals (zooplankton) that occur in most types of aquatic environments. They are often used as water quality indicators because they are ubiquitous and respond quickly to environmental changes (*16, 17*); therefore, they are good indicators of current local conditions. This information can be used to establish baseline conditions and monitor the community for shifts in diversity or composition. These shifts could represent impacts from exposure to releases of toxic chemicals or other changes in the water body. Concurrent collection and analysis of water or sediment samples is recommended to support data interpretation. Phytoplankton samples are also collected and analyzed for nutrients, to estimate "standing crop" (biomass) for a body of water, or analyzed for specific chemicals. Plankton sampling in any environment is usually conducted using a statistical random sampling design (random or stratified random within sites), which allows making inferences or predictions about the population that is represented. Replicating samples collected during a single event or replicating the entire sampling program over a period of time (e.g., seasonally or annually) is important in attempting to quantify variability.

A primary limitation on using plankton as indicators is that they often respond to factors such as food resources, nutrients, and predation in addition to the chemical or ecological response under evaluation. It is very difficult or impossible to adequately quantify all factors and the influence each has on species composition or abundance, and only broad inferences about the environment are valid. A second limitation for using plankton as water quality indicators is the potential for patchy distribution, both horizontally and vertically, and subsequent difficulty in obtaining representative samples. Patchy distribution results in either over- or underestimates of the population and pollutant exposure. Specific problems to be considered for plankton sampling include the following:

- If samples are collected from the wrong place or at the wrong time, few or no organisms may be obtained. Conversely, if samples are only collected from a few densely populated locations, the true conditions may not be represented, and incorrect or misleading results may be obtained.

- Plankton collected from fast-moving water will not represent the location from which they were collected. In these conditions, the duration and source of any exposure to pollutants or changing conditions cannot be easily or accurately characterized because the plankton are largely transient populations.

- Zooplankton can occur at different depths or zones within the water column depending on the current food supply and environmental conditions. Some zooplankton exhibit diurnal patterns, occur in narrow temperature or salinity ranges, or have seasonal population patterns that must be considered in the sampling design.

- Lakes, ponds, or slow-moving rivers or streams are the best environments to characterize using plankton data because it is easier to collect representative samples, and the plankton collected are less likely to be transient or temporary populations. Samples should be collected at multiple depths when temperature, nutrient, or chemical stratification is known or expected to occur. Multiple locations within the water body should also be sampled because variability in horizontal distribution also occurs.

- When sampling a stream or river, the sampling design should include transect sampling, or at least samples collected from near each bank because lateral mixing of the water may vary, causing uneven plankton distribution. Microhabitats caused by high or low flow rates, inflow from tributary streams, eddies, and laminar or turbulent flow can also occur, influencing plankton distribution and compromising the ability to collect representative samples. This phenomenon is especially true near a bend or tributary confluence where water flow is greatly disrupted.

Sample Preservation. Plankton sample preservation depends upon the organisms sampled and the measurements or analyses to be performed. It is important to plan for the total number of sample containers, preservatives, and total volume of sample (water or organisms) to collect for all chemical analyses. Large volumes of water may need to be collected and filtered to obtain sufficient plankton mass to meet the sensitivity requirements for chemical analyses. For chemical or nutrient analyses, samples are preserved as designated for the analytical method or class of analytes; the analytical method text contains this information, or information may be requested from the laboratory that will be performing the analyses. Laboratories performing chemical analyses often provide sample containers and preservatives. Samples may or may not need to be filtered or otherwise concentrated, and may require preservation with acid to $pH < 2$, shipping and storage at 4 °C, or storage in amber glass bottles or minimal exposure to light. If designated for the analytical method, filters can be frozen until sample preparation and analysis.

Plankton samples collected for ecological analyses such as identification and enumeration are preserved differently than those collected for chemical analyses. Phytoplankton samples collected for ecological analysis are most commonly preserved in Lugol's solution, a mixture of glacial acetic acid and potassium iodide (*1*); buffered formaldehyde is added for long-term storage. All samples should be stored in the dark. Other preservatives, such as Merthiolate and glutaraldehyde, for example, are also acceptable or useful under certain conditions, as described in *Standard Methods for the Examination of Water and Wastewater* (*1*). Zooplankton are usually preserved in 70% ethanol or 5% buffered Formalin solutions; as for phytoplankton, other preservatives may be used for specific applications or conditions (*1*).

Stains of various kinds may be added to the preservation solution to facilitate identification by highlighting physical features of the organisms. Most preservatives disrupt certain types of cells or cause loss or degradation of delicate features such as flagella, or other physical features needed to identify the organisms (*18, 19*). Therefore, it is important to analyze the samples as soon as possible, or evaluate the various preservatives available to identify the most appropriate one for the type of samples and analyses. Lugol's solution is the least damaging to samples. If live samples are being collected, refrigeration or a system that maintains ambient conditions is required, followed by immediate analysis or transfer to a long-term maintenance system as soon as possible.

Sampling Equipment, Procedures, and Considerations. Four types of sampling equipment are commonly used to collect phytoplankton and zooplankton samples: tow nets of various types; Van Dorn, Kemmerer, or Niskin bottle-type samplers; diaphragm or peristaltic pumps; and trap samplers. Sampling objectives, conditions, and collection efficiency should be considered when using each type of equipment.

Tow nets are usually constructed of fine mesh cloth with a rigid frame that holds the top of the net open, and a removable collection vessel at the bottom (Figure 1). Varying mesh sizes are available to collect different sizes of plankton. Nets are best used to sample low-density populations of larger sized zooplankton where qualitative to semiquantitative measurements or large biomass samples are needed; if quantitative data are needed, bottle-type samplers or pumps should be used. Vertical, horizontal, or oblique tows can be performed, and different equipment configurations can be used to ensure that the proper net orientation is maintained. Vertical tows are conducted from a stationary boat or sampling platform, whereas oblique or horizontal tows are conducted from a moving boat.

Whenever a tow net is used, some measure of the volume of water filtered to collect the plankton sample is needed to report biomass or the number of organisms on a volumetric basis. A rotameter may be attached to the net for a semiquantitative estimate, or a more qualitative estimate may be calculated using the tow length and area of the net opening. For any net, the calculated volume of water filtered to collect the plankton samples will be a maximum estimate because of the potential for the nets to become clogged and reduce the volume of water that actually flows through the net.

Special problems or considerations when using tow nets include the following:

- The nets can become clogged, reducing the volume of water that is filtered through the nets. Thus they are recommended for sampling low-density plankton populations. As nets become clogged, plankton may actually be backwashed or diverted from the net.
- Filtration efficiency for nets can be less than 50%. Low efficiencies indicate that significant turbulence or diversion of water at the mouth of the net occurs, and plankton are either diverted from the net or can avoid capture (for larger zooplankton), reducing the representativeness or completeness of the samples collected. The ratio of filtering area to the opening should be at least 3:1 to ensure good filtration characteristics.
- The tow speed for manual casts should be constant and slow (around 0.5 m/s) to avoid turbulence or diversion effects caused by higher speeds.

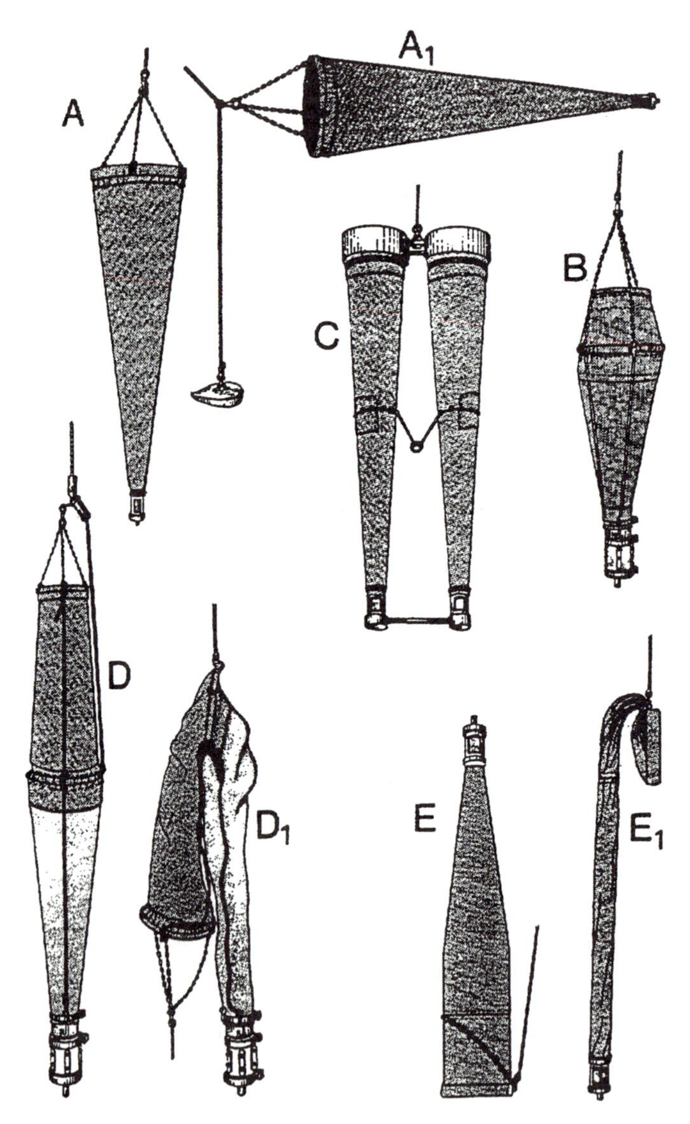

Figure 1. Commonly used plankton sampling nets: (A) simple conical tow net rigged for vertical tows (A_1 is rigged for horizontal or oblique tows); (B) Wisconsin (Birge) tow net with truncated cone to improve filtration efficiency; (C) Bongo net, which can be fitted with flow meters and opening/closing mechanisms; (D) Wisconsin net fitted with messenger-activated closing mechanism [D is open, D_1 is closed]; (E) free-fall net [E is open, E_1 is closed]. (Reproduced with permission from reference 1. Copyright 1989 American Public Health Association, American Water Works Association, and Water Environment Federation.)

- Difficulty in maintaining the desired tow configuration (e.g., vertical or horizontal) may occur if there are strong currents or winds at the sampling site. Weighting the net, sampling under optimal weather conditions, and other equipment designs can reduce these problems.

Van Dorn, Kemmerer, and *Niskin samplers* are bottle-type samplers consisting of hollow plastic or metal cylinders with end caps that are "tripped" to close off the ends of the sampler and isolate the sample (Figure 2). The sampler is lowered to the desired depth, and a weight, or messenger, is released to trip a spring mechanism and close the end caps. Van Dorn samplers are preferred because the end caps are pulled aside from the central cylinder, and little or no flow diversion or impedance occurs during sample collection. In deep water, Niskin samplers can be configured in series to collect simultaneous samples at selected depths.

Samples collected in bottle-type samplers provide quantitative data and are recommended for most types of plankton sampling. However, larger zooplankton may still be able to avoid capture, and results for these organisms should be considered to be semiquantitative. Different size classes of plankton can be segregated after collection by filtration through nets of differing mesh sizes. There are several considerations or limitations for using bottle-type samplers:

- Samplers made of linear polyethylene (LPE) or poly(vinyl chloride) (PVC) are usually preferred because metal samplers may contaminate samples and interfere with chemical analyses or productivity measurements.
- The sampler should be slowly lowered, not dropped, to the desired sampling depth because of the turbulence that occurs. Turbulence can disrupt the normal distribution of organisms and compromise the representativeness of the sample.
- The trigger mechanisms are delicate and easily damaged, so rough handling should be avoided. In addition, multiple attempts may be required if the mechanism does not trip both end caps of the sampler.
- The sample volumes needed for various analyses will dictate the size of the sampler or the number of casts needed to collect sufficient sample. Bottle-type samplers are available in volumes up to 10 L. For phytoplankton in oligotrophic waters, up to 6 L of water is recommended to yield sufficient sample; 0.5–1 L is recommended for eutrophic waters. For zooplankton, 5–10 L is recommended. Filtration or screening through nets is necessary to concentrate and collect both types of plankton.
- Larger zooplankton species can avoid capture in bottle-type samplers, and pumps or traps are recommended to collect quantitative samples of these organisms.

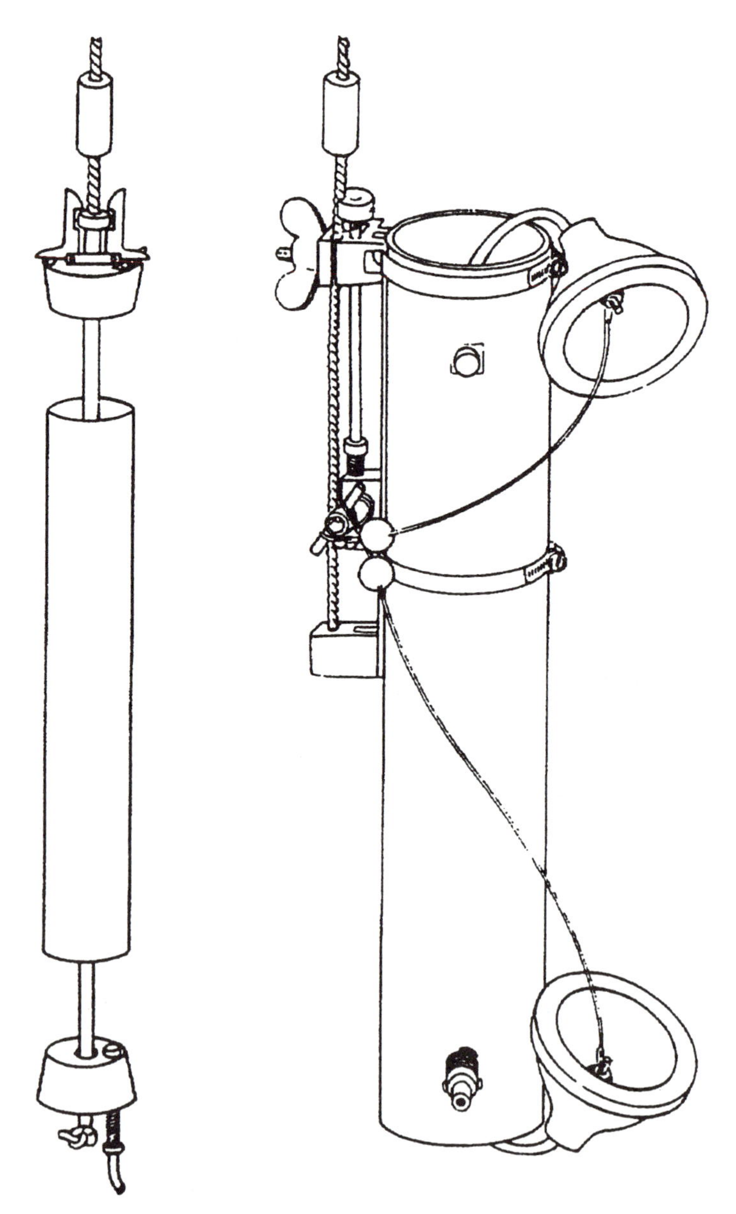

Figure 2. Kemmerer (left) and Van Dorn (right) samplers, common bottle-type samplers. (Reproduced with permission from reference 1. Copyright 1989 American Public Health Association, American Water Works Association, and Water Environment Federation.)

Peristaltic or *diaphragm pumps* can be used to quickly collect large volumes of water that can be filtered to obtain plankton samples. Centrifugal pumps can also be used but have impellers that can damage organisms; the peristaltic or diaphragm pumps are less damaging (*20*). Weighted hose or tubing is lowered to the desired depth, and the water is then pumped to the surface and collected in appropriate containers. The samples can also be directly filtered. Pumps are desirable because they can be used to collect integrated depth samples, or to collect samples from a specified depth. Considerations or limitations for pumps include the following:

- If a centrifugal pump is used, withdraw the sample before it flows past the impeller.
- If organochlorine analyses will be performed, use tetrafluoroethylene (TFE) tubing to avoid contamination problems.
- Larger zooplankton can avoid capture at the hose opening, so only semiquantitative samples are obtained.

Plankton traps are used to collect quantitative samples of large zooplankton. These traps are metal or clear acrylic boxes with a collection net and cup attached to one side to concentrate the zooplankton collected. They range from 10 to 30 L in capacity, and some have spring-loaded covers that isolate the sample, similar to bottle-type samplers. The Schindler–Patalas trap (*21*) is desired because it can be lowered into the water with little disturbance. Considerations and limitations for trap samplers include the following:

- Smaller size traps are unsuitable for oligotrophic conditions because insufficient numbers of zooplankton will be collected.
- Multiple samples may be needed to collect sufficient numbers of organisms when low-density populations are sampled.
- Metal traps cannot be used when samples will be analyzed for metals or used in productivity studies.

Periphyton

Data Collection Objectives and Limitations. *Periphyton* refers to attached protozoa, filamentous bacteria, rotifers, algae, and the unattached organisms that live among them (*22*). Periphyton occurs on submerged stones, sticks, plants, and other surfaces (e.g., pilings or debris). Periphyton organisms are used as indicators of pollution or other environmental changes. Because these organisms are stationary, they reflect changes occurring at the immediate location, and in rivers or streams they are especially good indicators directly downstream of a pollution source. Sampling designs usually include both upstream and downstream samples to provide direct comparison of affected and unaffected populations. Species diversity and numbers are often monitored because shifts in species composition can also reflect influences from pollution sources. Because natural substrates are often limited or highly variable at the desired sampling location, it is difficult to collect quantitative samples of periphyton. Therefore, artificial substrates are usually used to provide uniform surfaces, adequate surface area, and orientation within the water column (23).

Limitations on the use and representativeness of periphyton data are related to the density of organisms and the availability of substrate at the sampling or monitoring location.

The following limitations or considerations should be addressed when designing periphyton sampling programs and interpreting the data:

- No artificial substrate will duplicate natural growth, and preferential colonization of some substrates may occur. Therefore, natural substrates in the immediate vicinity of the sample location should be closely observed and used to at least qualitatively assess the representativeness of the samples collected.
- Glass slide samplers are most typically used because the materials are easily available and provide uniform and consistent samples. However, they do not mimic natural conditions. This factor must be considered during data interpretation, and observation of natural substrates in the vicinity of these types of samplers can provide valuable qualitative information.
- Variability of periphyton samples can be around 10–25% between replicate samples (*1*). Therefore, as many replicates as possible should be collected for each type of analysis to characterize variability and increase the ability to identify differences or changes. Replicates also ensure that sufficient sample volumes will be obtained to perform the needed analyses.
- The orientation and depth of the artificial substrates should be selected to mimic natural conditions and encourage colonization.
- To assess a pollutant source or disruption, upstream and downstream locations in rivers, and concentric configurations around potential sources of disruption in lakes should be used. Intensive sampling programs and replication are recommended to collect sufficient data to characterize temporal and spatial variability.

Sample Preservation. Samples collected for identification and counting should be preserved in Formalin or Merthiolate solutions. Preserve slides intact in storage containers, or scrape the periphyton into the containers. Chlorophyll samples should be placed in acetone and frozen with Freon or carbon dioxide, and stored on dry ice until analysis. All samples should be stored in the dark. Periphyton collected for chemical analyses should be preserved and stored as directed for the analytical method.

Sampling Equipment, Procedures, and Considerations. Periphyton samples are collected by scraping natural substrates, or providing artificial substrates for colonization. Samples collected by scraping natural substrates are qualitative, and it is often difficult to collect sufficient volumes of periphyton for analysis. Various scraping tools are available to collect samples from a variety of irregular surfaces, but success in using these devices to collect sufficient quantities of periphyton is limited.

Artificial substrates are most frequently used because they can be placed at a variety of locations and depths, and many replicates can be collected. Ceramic or clay plates can be used as substrates, but glass slides are most commonly used. Specially designed frames are usually used to hold replicate slides at the proper orientation and depth (Figure 3). The slides or plates are submerged for a designated growth period. A 1- to 2-week growth period is recommended during the summer, and longer periods are needed during the winter; the optimal growth period will vary depending on season, temperature, and nutrient availability. Limitations or considerations for using artificial substrates include the following:

- The same material should be used throughout the study to avoid introducing variability related to preferential colonization by different species.
- If the growth period is too long, sloughing may occur, and representativeness will be compromised. Variability between replicate samples will also increase.

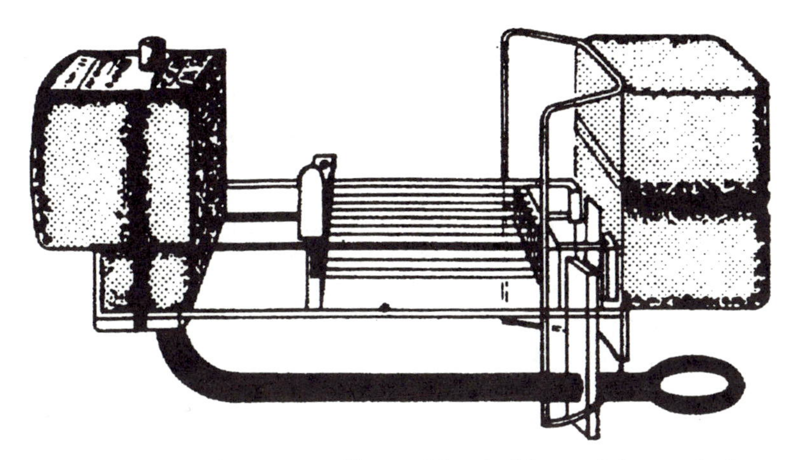

Figure 3. Periphyton sampler consisting of floating slide rack of clear vinyl plastic and polystyrene, used in streams and lakes. (Adapted from reference 1.)

- Grazing by macroinvertebrates may occur, also compromising the representativeness of samples. If grazing becomes a problem, increasing the substrate surface area and reducing the growth period to 7–10 days should minimize the impacts.

Bacteria

Data Collection Objectives and Limitations. Bacteria are single-celled organisms present in nearly all environments. Aquatic bacteria are collected most frequently as water quality indicators in water supply reservoirs, groundwater, rivers, water treatment systems, and discharge and receiving waters from wastewater facilities. The most common indicators are coliform bacteria; their presence in numbers above acceptable levels or standards indicates contamination from fecal material and the possibility that pathogens are also present. As for other biological analyses, bacteriological sample results should be interpreted along with other water quality parameters. Quantitative samples are needed to assess and document whether water quality criteria are satisfied, and replicate samples collected over a period of time are usually recommended to provide a more accurate representation of the conditions and to characterize the variability in the system. Limitations or considerations for collection and use of bacteria samples include the following:

- Samples should be collected from representative locations within the water source or treatment system, or at the discharge point and points upstream and downstream from it. Often sampling locations and frequency are designated in the regulations or discharge permits.
- Proper preservation and prompt analysis are important to reduce potential for changes in the samples that could invalidate results.

Sample Preservation. Bacteria samples are almost always analyzed by culturing the bacteria and counting the number of colonies that grow. The number of bacteria present in the original sample is estimated from these counts. Therefore, preservation and storage to maintain the samples in a viable condition is critical to obtain accurate data. If samples cannot be processed within 1 h after collection, they must be stored on ice in the dark for transport to the laboratory, and holding time between sample collection and analysis should not exceed 24 h.

For samples collected from water or wastewater treatment systems, preserving the samples to maintain the condition of the water at the time of sampling is necessary. Dehalogenating agents such as sodium thiosulfate must be added to stop bactericidal action in drinking water samples, and chelating agents such as ethylenediaminetetraacetic acid (EDTA) are added to wastewater or other samples that may contain metals that inhibit bacterial growth. The sampling locations and conditions will dictate whether these preservatives are needed and should be addressed in the sampling plan.

Sampling Equipment, Procedures, and Considerations. Bacteria samples are usually collected directly into the sample containers. Sterilized glass bottles are used for manual sample collection. For collecting samples at depth in a water body, special samplers such as a ZoBell JZ sampler are used (*24*). This type of sampler has a sealed sterile glass container that is mechanically opened and filled using a tripping mechanism similar to the bottle-type samplers described for plankton sampling. Limitations and considerations for bacteria sampling include the following:

- Because bacteria are easily transferred in the environment, precautions are needed to minimize sample container contamination and sample cross-contamination. Sampling containers must be kept closed until the time of sample collection and should be sealed immediately after being filled.
- If contamination of the tap or sampling port is possible, then the surface should be disinfected using sodium hypochlorite or other accepted disinfecting agent prior to sample collection.
- Samples collected from water systems or wells should be collected after running water from the tap or port for several minutes to ensure that the service line is purged, and a representative sample of the system will be collected.
- Samples of water supply reservoirs, rivers, or lakes should not be collected from too near the banks or too far from the withdrawal point to ensure representativeness. Sample locations for monitoring programs should be established and used consistently.

Benthos

Data Collection Objectives and Limitations. *Benthos* refers to the population of insect larvae, mollusks, sponges, nematodes, macrocrustaceans, and other sediment-dwelling invertebrates. They are sometimes defined as being retained in a U.S. standard no. 30 sieve (0.595-mm openings); however, for comparison with some historical data, a No. 18 sieve (1.0 mm) is sometimes used to collect samples (*1*). These organisms are very sensitive to pollutants, changes in organic loading, and substrate disruption and are good indicators of effects from these changes. Their use as indicators is also desirable because individuals in most species are relatively stationary and reflect local conditions. Changes in population density, species composition and diversity, and accumulation of toxic chemicals can occur (*25*); therefore, benthos samples are collected for both ecological and chemical analyses. Water and sediment samples collected from the same locations are strongly recommended to support data interpretation. In addition to analysis for specific chemical pollutants, dissolved oxygen, total organic carbon, and general water quality parameters should be considered. Sediment or substrate samples should also be analyzed for particle size distribution and composition.

Benthos sampling is usually a two-step process. Sediment is collected, and then the organisms are separated from the sediment and any rocks, sticks, or debris. Separation is accomplished by *sieving* (pouring the sediment into another container and then adding ambient water and gently swirling to remove silt and light sediment before pouring through a sieve or mesh) or *elutriation* (adding a solution that separates organisms from sediment based on density differences). Various elutriation solutions and equipment are available (26) to separate different types of organisms.

Although benthic organisms are highly sensitive and representative of localized conditions, several significant limitations or considerations must be addressed during sample planning and data interpretation:

- Many benthic organisms live in a very limited range of sediment conditions or microhabitats. Because most studies involve comparisons of unaffected and potentially affected populations, it is critical that field observations to identify comparable conditions in both affected and unaffected areas are made prior to selecting sampling locations. The factors that create microhabitats are related, but particular consideration for several of these may be critical to the success of the program. In lakes, comparability will be primarily influenced by depth, oxygen or temperature stratification, light penetration, water circulation or currents, and sediment composition (e.g., sand, gravel, or silt). In streams or rivers, it will be primarily influenced by depth, type of flow regime (e.g., turbulent, eddy, or laminar), gradient, the presence of riffles and pools, and sediment composition. Snags in lakes or streams also are important habitats for benthic organisms. Other factors may also need to be identified and considered for a particular situation.
- Although many organisms do not move far, some organisms (particularly in rivers) normally drift to downstream locations. Drifting organisms will not represent the location they are collected from. Drifting usually occurs at night, so arrangements for night sampling will be needed to collect some samples.
- When sampling streams or rivers, one should start with the farthest downstream location and proceed to upstream locations. This protocol minimizes influence from sediment washed downstream or drifting organisms released when the bottom is disturbed.
- Sediment samples collected for physical or chemical analysis should be taken from the upper few centimeters where most organisms live.
- Seasonal fluctuations in populations and numbers of individuals are common, and sampling schedules and data interpretation must account for these variations. Life cycles for many organisms are short, and in some cases, sampling conducted within days or even hours

will give vastly different results because of an insect hatch or some other dramatic natural occurrence. Multiple samples collected over a period of time are usually recommended to sufficiently characterize the populations.

- Critical times for sampling stream-dwelling benthic organisms are periods of high temperature and low flow. For estuarine, lacustrine, or marine environments, periods of maximum stratification and poor vertical mixing are critical. During these periods, the organisms will be the most stressed and susceptible to other disruptions because they will be closer to the extremes of the conditions they need to survive. When limited funds or other situations occur, sampling during these periods should be given priority.
- Sample collection usually involves collecting sediment, followed by separating the organisms by sieving or manually picking the organisms. This process can be tedious and fairly complicated, and attention to detail and consistency will be important to ensure complete and consistent sampling.

Sample Preservation. As for other organisms, benthic organisms collected for chemical analysis should be preserved as required for the analytical method. Samples collected for identification and counting should be transferred to storage containers and preserved in 10% buffered Formalin or 70% ethanol. Ethanol-preserved samples should not be filled more than half full with organisms to ensure adequate preservation. Organisms with calcareous shells or exoskeletons (e.g., mussels, snails, or crayfish) should be preserved in ethanol.

Sampling Procedures, Equipment, and Considerations. Selecting equipment for benthos sampling will depend upon whether the samples need to be quantitative or qualitative; whether the sample locations are from a small stream, lake or open water, or river; and the consistency of the substrate (rocky, sandy, or silty). The most commonly used types of samplers are grab samplers (usually dredges of some type), Surber square foot samplers, core samplers, drift samplers, emergence traps, and artificial substrates.

Other types of samplers such as suction samplers or tow nets are available, but they are not as widely used. Method references, aquatic biology textbooks, and manufacturers' instructions should be consulted for detailed descriptions of equipment and procedures.

Grab samplers are among the most commonly used benthic sampling devices. Ponar, Petersen, and Eckman dredges (Figures 4–6, respectively) are typical grab sampling devices, but other configurations are available (*1*). The three dredges mentioned all have a spring-loaded jaw mechanism that is tripped from the surface by a messenger. Specific advantages, conditions, and considerations for using dredges include the following:

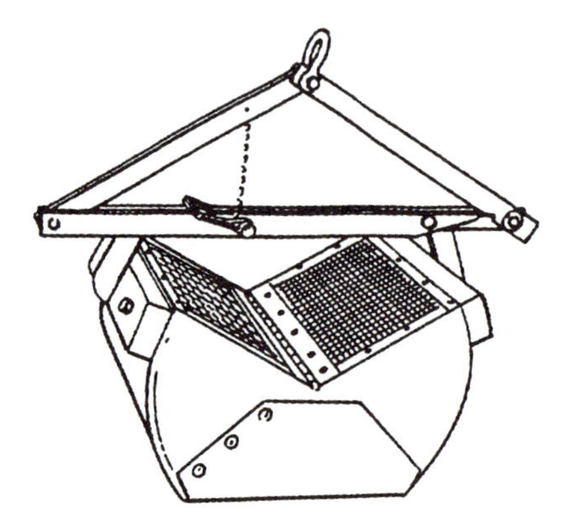

Figure 4. Ponar grab sampler used for benthos and sediment sampling. (Reproduced with permission from reference 1. Copyright 1989 American Public Health Association, American Water Works Association, and Water Environment Federation.)

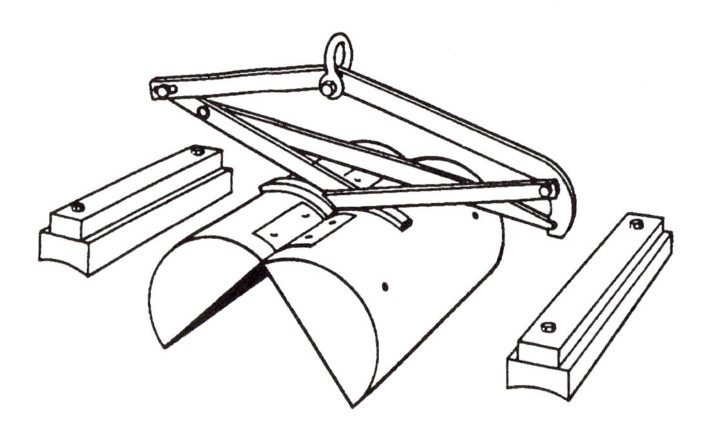

Figure 5. Petersen grab sampler used for benthos and sediment sampling. (Reproduced with permission from reference 1. Copyright 1989 American Public Health Association, American Water Works Association, and Water Environment Federation.)

- Sample loss or recovery is nearly always a concern when dredge sampling. Although quantitative samples are collected, some assessment of percent recovery is needed.
- Sticks, rocks, or other debris may prevent the jaws from closing completely, and some or all of the sample may be lost as the dredge is raised to the surface.
- Dredges should be lowered and raised at slow, constant speeds; they should never be dropped. This procedure minimizes sample loss and reduces the pressure wave that precedes the sampler at the sediment

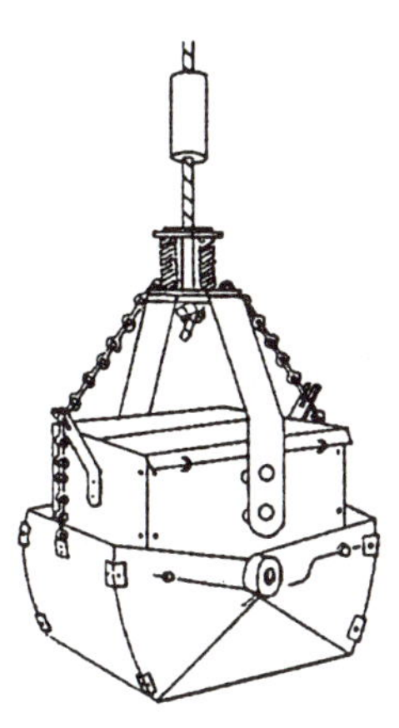

Figure 6. Eckman dredge used for benthos and sediment sampling. (Reproduced with permission from reference 1. Copyright 1989 American Public Health Association, American Water Works Association, and Water Environment Federation.)

surface. A pressure wave can wash away organisms and disturb loose sediments. This precaution is especially important for heavier dredges such as the Ponar and Petersen dredges. Using a constant speed also allows better assessment of when the bottom has been reached and helps ensure that the sampler is resting vertically on the bottom.

- Sometimes the messenger fails to trip the release mechanism, and several attempts may be needed to recover a sample. Bottom disturbance from multiple casts is likely in these cases; the cable or rope should be taut before releasing the messenger to ensure that it will have sufficient momentum to trip the release mechanism.
- Undisturbed samples are unlikely to be collected. If the actual location and distribution of organisms within the sediment are desired, some other device should be designed or a core sampler should be used.
- Ponar and Petersen dredges are used for similar conditions, which are deep water or strong currents. The Ponar dredge has a screened top that can reduce sample loss. Both of these devices are heavy, and a winch and cable setup is best for lowering and recovering the dredges. They are most effective for use in sand, gravel, or cobble bottoms.
- Safety considerations are important, especially for the heavier Ponar and Petersen dredges because of the potential for back injuries caused by lifting, and the danger of pinching or crushing fingers if the jaws are accidentally released.

- Eckman dredges are smaller than the Ponar or Petersen dredges and are best suited for use in shallower lakes and slow-moving rivers with little or no current. Some models have a screen that can be placed over the top to prevent sample loss. They work well in silt, muck, or sludge, but not in rocky or sandy bottoms.
- The dredges (especially Eckman) can tip, or rest unevenly on the bottom, resulting in sample loss or nonrepresentative samples. In soft sediments, they can sink too far, also causing sample loss and increasing disturbance of the sediment. A stabilizing platform or legs extending from each corner can be constructed to reduce or eliminate both of these problems (27). Any stabilizing modifications should be constructed of lightweight materials to prevent the dredge from becoming an anchor (27).

Surber square foot samplers are used to quantitatively sample shallow flowing streams less than 30 cm (1 ft) deep. They consist of two square foot frames fixed perpendicular to each other (Figure 7). The vertical frame has a net attached that extends downstream. "Wings" extending from the top of the vertical frame to the far edge of the horizontal frame help increase sample efficiency and recovery by directing the disturbed organisms into the net. To collect samples, the frame is anchored into the stream bottom, and the area inside is stirred to release any organisms present. Rocks or sticks should be turned over and examined, and any clinging or attached organisms should be dislodged and directed into the net. Larger organisms may need to be hand picked. The organisms are washed into the net by the stream flow and are collected by inverting the net into a bucket or other collection container. Special considerations for using these samplers include the following:

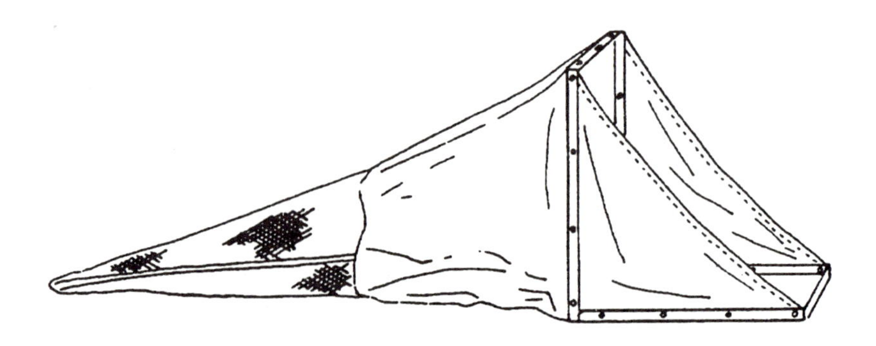

Figure 7. Surber square foot sampler used for macroinvertebrate sampling in shallow streams. (Reproduced with permission from reference 1. Copyright 1989 American Public Health Association, American Water Works Association, and Water Environment Federation.)

- Mesh size is important because it determines the size of the organisms that will be retained. A mesh size too large will not capture small instars or other organisms, but a mesh size too small can clog, causing backwashing and sample loss. The most common mesh size is nine threads per centimeter.
- The horizontal frame must be carefully anchored into the bottom. Gravel, rocks, or other materials can be used to fill gaps around the frame to prevent organisms from escaping or washing out of the sampler. An extension can be added to the back edge or the entire bottom frame to help secure the sampler in rocky or hard bottoms.
- Care must be taken not to disturb the site or upstream areas that could disturb organisms that might wash into the net at the sampling location.
- Sample replication is important to characterize sampling variability and variations in the distribution of sampled populations.

Core samplers are used in shallow streams for surface sediment sampling and in deeper streams or open water for core samples. As surface sediment samplers, they are best used to sample dense populations. Both manually and mechanically driven models are available. Core samplers provide a quantitative sample, and intact sediment cores can be collected for laboratory studies of sediment–water interface reactions, pollutant transport, or metabolism studies. Considerations for using core samplers include the following:

- As surface sediment samplers, they must be used in conjunction with sieves or nets to recover the samples.
- Cores recovered from depth may require extensive processing.
- Most cores have relatively small surface areas, so many samples may need to be recovered from a small area to obtain sufficient numbers of organisms to adequately represent the area.

Drift nets are used to quantitatively collect drifting organisms that are migrating or have been dislodged from upstream locations. These nets have a rigid frame opening attached to a closed-end net (Figure 8). Most have mesh sizes equivalent to a no. 30 screen. Drift nets are anchored into the stream bottom with the opening facing upstream, and the anchoring poles are height-adjusted to sample specific depths in the water column. The nets are set in flowing water at designated locations and time periods (usually 3 h). Considerations for drift net sampling include the following:

- Stream flow must be measured in conjunction with sample collection. Capture results are usually reported as (number of organisms per 24 h)/(m^3/24 h) (28).

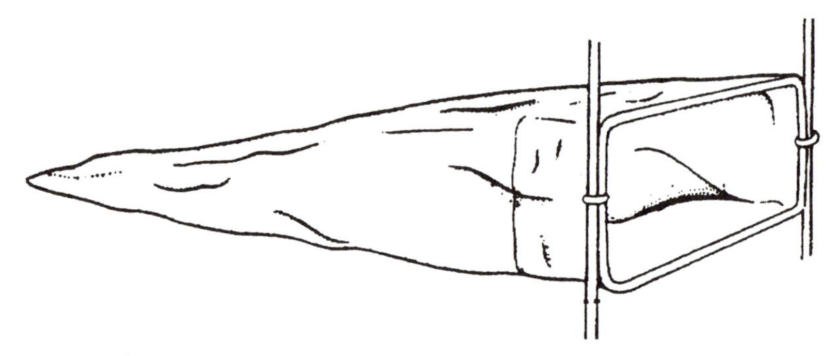

Figure 8. Drift net sampler used for macroinvertebrate sampling in shallow streams. (Reproduced with permission from reference 1. Copyright 1989 American Public Health Association, American Water Works Association, and Water Environment Federation.)

- Optimum sampling times to capture migrating organisms are from dusk to 1 A.M., but times should be consistent for all locations.
- The nets should be frequently checked to remove captured organisms and debris. This procedure will avoid clogging and backwash or diversion at the mouth of the net, improving sampling efficiency and representativeness.

Artificial substrates can be used to provide a qualitative assessment of the types of benthic organisms present at a location. They provide uniform substrates for colonization, similar to those described for periphyton sampling. Artificial substrates are usually colonized by insect larvae, crustaceans, sponges, and sometimes worms and mollusks. Because they are artificial, preferential colonization by organisms that either typically occur on similarly textured surfaces or under similar temperature, light, nutrient, and flow conditions will occur, and samples may not accurately represent natural communities. Representativeness should be assessed by observing the surrounding natural sediments and bottom conditions and comparing the observations with the samples that are obtained. Considerations and limitations for artificial substrate sampling include the following:

- Multiple plate samplers and basket samplers filled with rocks or similar material are the most common artificial substrates (*29–31*).
- Artificial substrates should be kept in place for 6 weeks.
- If the sampling locations are in areas accessible to the public or in waterways heavily used for recreation, then security and tampering may become an issue and substrates should be frequently checked. Subsurface floats are available to make the samplers less obvious or visible, and therefore less prone to tampering.

- The equipment should be properly secured to the bottom using rods driven into the bottom, anchors, and floats.
- The samplers should not be placed on the bottom at shallow sampling locations. The bottom of the sampler should be 1 ft above the sediment.
- Artificial substrates are subject to loss or burial from flooding or may be stranded if water levels decline.

Fish

Sampling Data Collection Objectives and Limitations. *Fish* are at or near the top of aquatic food webs and therefore can have a significant impact on other aquatic populations. Fish are also caught for human consumption, and bioaccumulation of toxic chemicals and the associated potential for human health impacts from consuming contaminated fish is often a primary objective in fish studies. The species of fish present, their numbers, and their distribution are all indicators of water quality and the general health of the ecosystem. Fish populations respond to changes in oxygen, light penetration, food supplies, temperature, pH, salinity, and pollutants. Considerations and limitations for fish sampling include the following:

- Regulations for collecting and disposing of fish samples must be understood and complied with. Special permits or licenses and notification of local officials are almost always required. Collection programs may also be highly visible to the public, and community notification or community relations programs may be needed. Permits and information identifying the group or study should be posted or readily available during field sampling activities.
- A site survey should be conducted to identify vegetation patterns, sediment types (e.g., rocky, sandy, or silty), lake inlets and outlets, potential point and non-point sources of pollutants (if the study is planned to evaluate pollutant effects), bottom contours, man-made or natural features such as waterfalls or weirs on streams, and other features that could influence where fish feed and gather to reproduce.
- Using glasses with polarized lenses will greatly improve visibility and the ability to identify fish and bottom features during visual surveys. Snorkeling or scuba equipment may be useful to obtain accurate observations of fish and habitat in clear water. (Special training and certification are needed for use of scuba equipment.)
- Water and equipment safety must be addressed in the sampling plan and staff training.

- Quantitative samples may be difficult to obtain because fish are highly mobile, and multiple sampling events or replicates may be needed to obtain representative, quantitative samples.
- Seasonal life cycles and migration patterns can greatly affect fish presence in a given water body and must be considered during sampling plan development and data interpretation.
- For nonrandom sampling, hydroacoustic equipment can be used to help locate fish. (For less sophisticated operations, a fish locator or depth sounder may suffice.) Several transects of the lake or reservoir should be cruised at a constant speed, and the locations should be recorded on a map for future reference and interpretation. This method may be used as the sole qualitative assessment of fish populations.

Sample Preservation. Fish collected for identification are usually frozen as soon as possible after collection or are identified in the field. If long-term preservation other than freezing is needed for example specimens, 10% Formalin is used; however, this technique is only recommended if other alternatives are not feasible. Also, shipping restrictions apply for specimens preserved using these fixatives. Fish less than 10 cm long can be preserved in the 10% Formalin solution without opening the visceral cavity. Larger specimens require injecting 10% Formalin into the visceral cavity or slitting the body cavity along the right side. For specimens longer than approximately 25 cm, pure Formalin is injected into the dorsal muscle mass. Specimens fixed in this way should be transferred to a 70% ethanol solution within two weeks for longer term storage. Whole fish, organ, or tissue samples taken for pathology should be preserved in buffered Formalin for at least 24 h before additional processing. Standard references (*1, 3*) should be consulted for additional fish preservation details.

If parasites are collected from the fish, they should be preserved as directed for those species. Fish scales can be collected and stored in fish-scale envelopes. If chemical analyses will be performed, the whole fish or selected organs, usually fillets, brain, liver, or fatty tissue, are collected and frozen using dry ice (*3*). For tissue analysis a minimum of approximately 100 g will be needed. If fish are wrapped in foil for shipment, the shiny side should not be allowed to contact the sample because slip agents used on the foil may cause contamination. It is usually preferable to have dissection performed at a clean staging area or in the laboratory by a qualified biologist or technician. If dissection is performed in the field, the equipment (e.g., surface or knives) should be rinsed with acetone followed by a distilled water rinse. If chemical analyses will be performed, additional decontamination steps may be required; these potential requirements should be evaluated by consulting with the laboratory.

Sampling Equipment, Procedures, and Considerations. The equipment and procedures used to collect fish samples will vary depending on (1) whether quantitative or qualitative data are needed, (2) whether selective or nonselective sampling is planned, (3) access, and (4) the water body being sampled. Regulatory policies and community interaction may also influence fish sampling and methods. Further information about fish sampling is provided in fisheries manuals and textbooks (32–34). The most commonly used equipment and procedures include angling or set lines; seines, gill nets, trawls, and fish traps; hydroacoustic methods; plankton nets for drifting fish eggs or fry; electrofishing; and ichthyocides. The operation of equipment for some of these methods requires extensive training for proper use and data interpretation, and specialists should be included as project team members or subcontractors to ensure that field activities are conducted safely and properly. Many of these procedures have some significant limitations that will affect both sampling success and data interpretation.

Angling is perhaps the most enjoyable way to sample fish (at least for many), but the success of the sampling effort is largely dependent on the skill of the angler and is generally selective for "sport" species. However, this technique may be the best option when the objective is to obtain a few samples for chemical analysis and other alternatives are not feasible. Angling may be done from a stationary position or by trolling from a boat. Samples can be quantified as catch per unit effort. Set lines (a line with a series of hooks) can also be used to obtain a small number of samples. These can be set at varying depths for a selected length of time (usually overnight). Samples collected using these techniques will be qualitative.

Seines, or *fish traps,* are probably the best option for qualitative to quantitative sampling where live capture or release are desired, such as for surveys or tagging. Trawls or gill nets may also be used, but all or most of the fish will be killed. Various types of these nets can be used for different conditions. Mesh sizes vary, so selective sizes of fish can be caught. All types of nets should be checked frequently to collect the fish and avoid oversampling or damage to the samples. Considerations or limitations for this type of equipment include the following:

- Most seines and fish traps will capture fish live so that catch and release studies can be performed, and only fish required for analysis need to be killed.
- Traps and nets are often selective (based on mesh size and other features), and placement is critical to obtaining representative samples. Anchoring nets or traps at proper depths and orientation will be critical to the success of the sampling program, and prior observations and knowledge of the sampling location conditions are required.
- Seines of various mesh sizes are used for qualitative shoreline sampling, and short passes are recommended to avoid losses.
- For large-scale studies in commercial fisheries, contracting with local fishermen to collect samples is often done; however, data are often unreliable for population-type studies and extra effort is recommended to evaluate conditions and sample representativeness. This

option is feasible if the sampling objective is to obtain fish for studies of impacts from human consumption.

- Gill nets can be used to selectively capture fish of specified sizes. Fish caught in gill nets will be killed. Recovery of these nets is critical because a lost net will continue to capture and kill fish; any lost gill nets should be reported to local fisheries officials.
- Nets or traps may be subject to vandalism or damage by boats. They should be clearly posted, and constant surveillance may be necessary in some areas.

Plankton nets or *traps,* or *bottle-type samplers* are used to collect fish eggs or small fry in the same manner described for plankton samples, and the same considerations and limitations apply.

Hydroacoustic methods use equipment to generate sound waves that travel through the water and reflect back from submerged or swimming objects or the bottom to a transducer mounted to a boat or fixed location. The signal is converted to produce a paper tape, video, or other display mechanism and provide a visual representation of the area scanned. Both vertical and side scanning may be conducted, depending on the objectives of the study. References available for acoustic methods should be consulted to aid in equipment selection and data interpretation (*35, 36*). Considerations or limitations for hydroacoustic techniques include the following:

- Fish have high acoustic impedance and produce strong signals. The trace is usually boomerang- or chevron-shaped, with the curved side up. A school of fish will blend together and show up as a cloud, complicating interpretation.
- Gas bubbles, floating objects, or density differences associated with a thermocline may cause interferences.
- The acoustic traces provide accurate representation of the bottom profile and distribution of fish in the water body.

Electrofishing is used in freshwater to obtain quantitative samples. This technique is useful in areas where use of traps or other types of nets is not feasible. These procedures are usually effective and allow most or all fish present to be collected. Electrofishing uses either alternating- or direct-current power sources. Alternating currents stun or kill fish, whereas direct currents induce a polarizing effect, attracting fish to one pole where they can be collected. Considerations and limitations for these procedures include the following:

- Electrofishing is useful in areas with uneven bottoms, in fast-flowing water, or when other obstructions limit use of nets or traps. Stunned fish may be swept downstream if there is a strong current, and provision for downstream collection may be necessary. It is not efficient in deep water or for bottom-dwelling species.

- Electrofishing is nonselective and will affect all fish in the area, although larger fish are more easily stunned than smaller fish.
- Direct-current devices are effective in turbid water or where heavy vegetation is present.
- Alternating-current devices are more likely to kill fish.
- Water hardness or ionic strength can affect the electrofishing success. In dilute, low-conductivity waters (such as alpine or bedrock lakes), electrofishing may not be feasible, or salt may be added to increase conductivity.
- Portable and boat-mounted electrofishing apparatus are available. Most electrofishing is conducted by experts trained in proper use of the equipment.

Ichthyocides are chemicals that act as a metabolic poison or anesthetic and are the least selective method of collecting fish. Several chemicals have been approved for use as ichthyocides by the EPA; however, rotenone is the most commonly used. They can provide the most accurate quantitative assessment of the fish population in terms of the percentage of the total population collected, but there are serious drawbacks to their use. Limitations and considerations include the following:

- Ichthyocides can have a slow reaction time and lack of sufficient killing power unless applied at high doses.
- The chemicals can have toxic effects to the users and other organisms in the lake or stream.
- Ichthyocides are highly controversial and usually generate negative publicity, so unless the objective is to eliminate or nearly eliminate fish from the target water body, ichthyocides are usually not feasible.

Summary

This chapter provides guidance and references for planning and implementing a successful biological sampling program. As discussed, planning this type of program must first include a clear definition of the problem or question to be addressed. The DQO process provides a useful framework to define the problem and to ensure that specialists in various disciplines are involved as necessary and that expectations for use of the data are realistic. The ultimate goal of any program is for the samples and data collected to be sufficient for use as planned to address the problem. The types of sampling equipment and sample preservation described are those most commonly used for biological sample collection. However, the proper equipment and preservatives may vary and must be carefully selected considering the conditions that are likely to be encountered and the limitations noted for each type of equipment or procedure. The guidance provided in this chapter is meant to direct program planners to possible sampling alternatives and alert them to potential issues or limitations that may be encountered.

References

1. *Standard Methods for the Examination of Water and Wastewater,* 17th ed.; Clesceri, L. S.; Greensburg, A. E.; Trussell, R. R., Eds.; American Public Health Association: Washington, DC, 1989.
2. *Biological Field and Laboratory Methods for Measuring the Quality of Surface Waters and Effluents;* U.S. Environmental Protection Agency: Washington, DC, 1973; EPA 670/4–73–001.
3. *A Compendium of Superfund Field Operations Methods;* U.S. Environmental Protection Agency: Washington, DC, 1987; EPA 540/P–87/001.
4. Bower, J. E.; Zar, J. H. *Field and Laboratory Methods for General Ecology;* William C. Brown: Dubuque, IA, 1977.
5. *Laboratory Manual of General Ecology;* William C. Brown: Dubuque, IA, 1967.
6. *A Manual on Methods for the Assessment of Secondary Productivity in Fresh Waters;* Downing, J. A.; Rigler, F. H., Eds.; IBP Handbook 17; Blackwell Scientific Publications: Oxford, England, 1984.
7. Wetzel, R. G.; Likens, G. E. *Limnological Analyses;* Springer-Verlag: Berlin, Germany, 1991.
8. *Framework for Ecological Risk Assessment;* Risk Assessment Forum, U.S. Environmental Protection Agency: Washington, DC, 1992; EPA 630/R–92/001.
9. *Code of Federal Regulations,* Title 40, Parts 122 and 264; Office of the Federal Register, National Archives and Records Administration; revised as of July 1993.
10. *Data Quality Objectives Process for Superfund, Interim Final Guidance;* Office of Emergency and Remedial Response, U.S. Environmental Protection Agency: Washington, DC, 1993; EPA 540/G–93/071.
11. Gilbert, R. *Statistical Methods for Environmental Pollution Monitoring;* Van Nostrand Reinhold: New York, 1987.
12. Albert, R.; Horwitz, W. In *Principles of Environmental Sampling;* Keith, L. H., Ed.; American Chemical Society: Washington, DC, 1988; pp 337–353.
13. Bourke, J. B.; Spittler, T. D.; Young, S. J. In *Principles of Environmental Sampling;* Keith, L. H., Ed.; American Chemical Society: Washington, DC, 1988; pp 355–361.
14. Garner, F. C.; Stapanian, M. A.; Williams, L. R. In *Principles of Environmental Sampling;* Keith, L. H., Ed.; American Chemical Society: Washington, DC, 1988; pp 363–374.
15. Hirsch, R. M.; Alexander, R. B.; Smith, R. A. *Water Resour. Res.* **1991,** *27,* 803–813.
16. Palmer, C. M. *Bull. N.Y. Acad. Sci.* **1963,** *108,* 389.
17. Gannon, J. E.; Stemberger, R. S. *Trans. Am. Microsc. Soc.* **1978,** *97,* 16.
18. Paerl, H. W. *Limnol. Oceanogr.* **1984,** *29,* 417.
19. Silver, M. W.; Davoll, P. J. *Limnol. Oceanogr.* **1978,** *23,* 362.
20. Beers. J. R. In *Phytoplankton Manual;* Sornia, A., Ed.; United Nations Educational, Scientific, and Cultural Organisation: Paris, 1978.
21. Schindler, D. W. *J. Fish. Res. Board Can.* **1969,** *26,* 1948.
22. Roll, H. *Arch. Hydrobiol.* **1939,** *35,* 39.
23. Sladeckova, A. *Bot. Rev.* **1962,** *28,* 286.

24. ZoBell, C. E. *J. Mar. Res.* **1941,** *4,* 173.
25. Barton, D. R. *J. Great Lakes Res.* **1989,** *15*(4), 611–622.
26. Robinson, S. M. C.; Chandler, R. A. *Limnol. Oceanogr.* **1993,** *38*(5).
27. Blomqvist, S. *Hydrobiologia* **1990,** *206,* 245–254.
28. Waters, T. F. *Annu. Rev. Entomol.* **1972,** *17,* 253.
29. Mason, W. T., Jr.; Weber, C. I.; Lewis, P. A.; Julian, E. C. *Freshwater Biol.* **1973,** *3,* 409.
30. Beak, T. W.; Griffing, T. C.; Appleby, G. In *Proceedings Biological Methods for the Assessment of Water Quality;* American Society for Testing and Materials: Philadelphia, PA, 1974.
31. Fullner, R. W. *J. Water Pollut. Control Fed.* **1971,** *43,* 494.
32. *Fisheries Techniques;* Nielsen, L.; Johnson, D. L., Eds.; American Fisheries Society: Bethesda, MD, 1983.
33. Everhardt, W. H.; Youngs, W. D. *Principles of Fishery Science;* Cornell University: Ithaca, NY, 1981.
34. Kushlan, J. A. *Trans. Am. Fish Soc.* **1974,** *103,* 348.
35. *Symposium on Fishery Acoustics;* Nakken, O.; Venema, S. C., Eds.; Fishery Report 300, Food and Agriculture Organisation of the United Nations Unipub.: Ann Arbor, MI, 1983.
36. Burczynski, J. *Introduction to the Use of Sonar Systems for Estimating Fish Biomass;* FAO Fishery Technical Paper 191, Food and Agriculture Organisation of the United Nations: Rome, Italy, 1979.

Chapter 31

Coping with Sampling Variability in Biota

Percentiles and Other Strategies

Richard Albert and William Horwitz

Principles of good sampling should be followed regardless of the population sampled. Especially important is the requirement that the sample taken be representative. A complication arising in sampling biota is the large variability encountered and the corresponding uncertainties in the estimates. Also, the distribution of values (e.g., mercury concentrations in fish) may not be normal (i.e., bell-shaped, or Gaussian). To reduce the uncertainties, such strategies may be resorted to as increasing the number of units sampled or breaking down the population into distinct subgroups. A naked, unqualified estimate of an average is useless. Associated with any average should be an indication of how the individual values are distributed. For biota, the various percentile points can be reported along with the confidence limits derived from the binomial distribution. The final results of sampling biota may well be an average plus key percentiles of the distribution values.

STATISTICAL MANIPULATION ALONE cannot reduce variability to manageable proportions; therefore, nonstatistical strategies must also be brought to bear. The more that is known about the target biota (i.e., plants and animals) before sampling, the more readily can the effects of variability be mitigated and biased estimates be guarded against. The emphasis of this chapter is on coping statistically with the inherently large variability encountered in sampling biota.

Example 1. The contamination of the bluefish by polychlorinated biphenyls (PCBs) off the Atlantic Coast of the United States is investigated. To check each bluefish is an absurd impossibility, so only some are checked, and the few that are caught will somehow represent all the bluefish. On the basis of the PCB concentrations measured in the bluefish sample, what can be said about the PCB concentrations in all Atlantic bluefish?

Example 2. The amount of aflatoxin in the U.S. peanut crop is investigated. The investigator goes to a nearby grocery store and buys all the peanuts therein. The aflatoxin content of each peanut is meticulously measured. From these measurements, what can be said about the contamination of the peanut crop?

Example 3. Having heard the famous rhyme about purple cows, an investigator wants to find out if any purple cows really do exist. While traveling through a farm district, the investigator spots 20 cows, but none are purple. What can be concluded about the number of purple cows in existence?

These three scenarios share something in common: Data are desired about all members of a certain large group (i.e., the population is Atlantic bluefish, U.S. peanuts, or cows), but only a minuscule fraction of the large group is inspected. From this fraction, information must be extrapolated to the entire population. The goal is to make estimates and inferences that are valid for the entire population, although data are contained on only a sample of the population. The fundamental question is this: What can be asserted, and how certain can the investigator be in these assertions? This chapter focuses on the certainty, or lack thereof, in estimating specific statistical quantities associated with various biota.

In assessing certainty or uncertainty (i.e., in trying to quantitate confidence or diffidence), a fundamental role is played by the variability of the population. If all peanuts were identical, then the sampling of just one peanut would suffice to monitor the entire peanut crop. Alas, in the real world, the pervasive existence of variability of all sorts compels us to sample more than just one unit: more than one bluefish, more than one peanut, and more than one cow.

Confronting variability and taming it so that valid estimates can be made with adequate precision is the burden of the working statistician. By clever use of the variability found in the sample itself, the statistician can typically associate with an estimate a *confidence interval*, which is a range of values that can be declared with a specified degree of confidence to contain the correct or true value for the population. One quality control manual defines confidence interval as "the interval between two values, known as *confidence limits* [italics added], within which it may be asserted, with a specified degree of confidence, that the true population value lies." The more variable the population, the broader are these confidence intervals.

Assuming the 20-cow sample to be representative, the statistician can be 95% confident that the actual percentage of cows that are purple is somewhere between 0.0% and 13.9% ($1 - (1 - 0.139)^{20} = 0.95$). For a 40-cow negative sample, the 95% confidence interval would be from 0.0% to 7.2% (1 − (1 −

$0.072)^{40} = 0.95$). A user of statistics should be critical and alert in making decisions based on sample estimates and corresponding confidence intervals. These statistical artifacts are only as sound as the assumptions and procedures that went into creating them. The validity of the confidence interval is inextricably linked to the validity of the sampling; coordination and quality control are called for throughout the whole complex integrated process that begins with a question and ends with a confidence interval.

In the following portions of this chapter, some mathematical "cookbook recipes" will be presented, but the essential points to remember are as follows:

Point 1. Sampling should not be undertaken until the questions to be answered have been determined and properly framed.

Point 2. The individuals included in the sample must have been chosen at random, or, more generally, by probability sampling from a population that is well-defined. If this procedure is not done, experienced statisticians have warned that the investigator can be 100% confident that the final confidence interval estimates will be catastrophically wrong.

Point 3. The distribution of values in a population is often Gaussian (i.e., "bell-shaped", or "normal"), so that the estimates of the average and of the standard deviation are all that is needed to characterize the entire distribution. Confidence intervals are readily computed from a sample of the distribution if the parent population is Gaussian.

Point 4. Non-Gaussian populations can be sampled by empirical estimates based on samples of the 10-, 25-, 50-, 75-, and 90-percentile points. Confidence intervals for these percentile-point estimates are much broader (i.e., less precise) than corresponding confidence intervals of percentile-point estimates for Gaussian populations. (The P-*percentile point* is that value at or below which *P* percent of all population values lie.)

Proper Sampling

The order of evolution is from raw data to information, from information to knowledge, and from knowledge to wisdom. The crucial first step is to intelligently gather the appropriate raw data. What is appropriate is determined by what sort of knowledge is sought. A campaign must be mapped out where all available resources are inventoried, and the scope of the questions is guided in part by the extent of these resources. After careful framing of the questions, the relevant population must be sought. Here, as elsewhere, the more that is known, the more that can be learned. For example, in a real-life bluefish survey, PCBs were known a priori to concentrate in the larger adult fish. Such knowledge served to direct the sampling procedure.

After question posing and population identifying, the actual sample is selected (e.g., the specific individuals, the specific bluefish, the specific peanuts, or the specific cows). Such selection must be done in a probabilistic way (i.e., every individual in the population has a prior known probability of being included in the sample). Only with such a probabilistic selection can certain potential biases be avoided.

The final phase in getting the raw data is to reduce the sample to manageable portions, aliquots, or pieces so that the chemical or other measurements can be made. The homogeneity of the test portion at this stage is crucial in minimizing the overall variability or uncertainty in the ultimate estimates of the average and associated confidence intervals. Obviously, cows can just be looked at for being purple, but a fish has to be skinned, deboned, and subjected to extraction of its fatty portion in order to estimate PCB content.

Further details on these consecutive links in the chain leading from the problems (i.e., the questions) to the answers (i.e., the estimates of averages and confidence intervals) are provided in the rest of this chapter.

Proper Questions

The toil and expense of a survey can be justified only if the questions answered have a value. Vague, unstructured exploratory surveys are wasteful. A need to answer certain questions must be present. Then the price of getting the answers should be looked at. The cost per datum can be disconcertingly high.

The routine analysis of aflatoxin concentration in peanuts at the parts-per-billion level costs perhaps $100 per data point. The moral of the story is to make sure the questions are worth asking in terms of dollars and cents and resources.

Questions should be couched in quantitative terms. In the real-life bluefish survey, the preliminary work established over two dozen questions: "What is the average PCB concentration in bluefish?" and "Are there significant differences in PCB concentrations based on geographical, sex, or seasonal differences?" Even "significant" should be assigned a more quantitatively precise meaning. Otherwise, the investigator would be violating the principle of posing only those questions that allow definite, clear-cut answers.

Basically, at this stage the investigator should decide what should be measured and how accurately it should be measured.

Proper Population

Coupled with the issue of what is measured is the issue of what it should be measured in. Needs include what is to be included as a potential candidate to be sampled, and, equally as important, what is not a candidate to be sampled.

Basically the right answer should not be obtained to a wrong question. Are PCB raw data desired from all bluefish at all levels in all regions? Definitely not. The purpose of one recent real-life bluefish survey was to assess human exposure to PCBs; accordingly, the field workers for this survey sampled only from those bluefish that were actually "landed" (i.e., brought to an Atlantic Ocean dock for sale or personal use). Only those bluefish entering the human food chain were of interest.

Proper Selection and the Importance of Randomness

Sufficiently large sample sizes can provide very "tight" estimates of averages (i.e., estimates bracketed by very narrow 95% confidence limits). The smaller the confidence interval, the more precise (and expensive) the estimated average. However, the danger always lurks that a precise estimate of the wrong average may be obtained. The persistent difference between the true or population average and the average obtained from repetitions (actual or potential) is a manifestation of bias. (In other words, the difference between the true population average and the estimate from the particular survey of the average may be called the *error of the estimate. Bias*, on the other hand, is a theoretical construct, and is equal to the algebraic average of such errors obtained from an indefinitely large number of repetitions of the survey.)

The sovereign remedy against bias is proper sample selection by a random process. The use of randomness helps preclude the presence of a persistent bias. A random sampling scheme can help ensure that in the long run, representative, nonbiased samples are obtained. In a random sampling scheme, every member of a population—every bluefish, every peanut, or every cow—has an equal probability of being included as part of the sample. In a random sampling scheme, the proportion of "positives" selected would in the long run be equal to the actual proportion of positives in the population (i.e., a representative sample would be achieved).

Knowledge is power, and foreknowledge of certain features in the target population can help reduce the variability in the sample, which in turn can lead to more precise estimates (i.e., to narrower confidence intervals). In particular, if before a survey is taken, a certain subpopulation (e.g., positives) is known to constitute a certain fraction of the target population, and the quantity to be estimated is especially variable among that subpopulation, then the investigator could randomly sample more frequently among the more variable population (i.e., larger random samples could be taken from the more variable subpopulations). Trivial recipes exist for combining the averages and variances of the nonoverlapping subpopulations (strata) into an overall average with an associated confidence interval surrounding this average. The resulting estimates are more precise but are also more liable to bias should the investigator be mistaken in the preliminary information on the population breakdown into strata.

Whether or not the investigator has prior information on the population subgroups, sampling should be done via random selection within each subpopulation or within the total population. However, in addition to knowing how to sample, the investigator needs to know how much to sample. Formulas exist to calculate the number (N) of items or individuals to be included. This number is directly proportional to the variability of the population expressed as the variance, but how can the investigator know this variability before the survey? Experience with previous similar surveys or results from a crude preliminary survey may give an adequate estimate of the variance, which in turn will help to estimate the sample size in the full-blown survey. The required sample size is also inversely proportional to the square of the a priori desired width of the confidence interval.

This section can be summarized as follows: Sampling must be random, either overall over a population or within its well-defined subpopulations, in order to ensure representativeness and lack of bias.

Final Link: Proper Test Sample and Measurement

A chain, being only as strong as its weakest link, is a good metaphor to describe the progression in estimating the relevant characteristics of a population. The investigator starts off by framing the questions in an answerable format and by defining in detail the population for which answers are desired. A random-sampling plan is then prepared that includes specifications as to how many individuals are to be included in the sample and from which strata these individuals are to be selected. Careful design and control up to this point help to cope with the intrinsic variability of the population.

The researcher is now faced with a hopefully representative sample and must make measurements on the individuals or units in the sample. How is the PCB content of a sample fish measured? How can the researcher ascertain the amount of aflatoxin in a mound of peanuts? [In fact, especially dramatic contributions to variability arising from sample inhomogeneity have been uncovered in the measurement of aflatoxin in peanuts (*1*).] The analytes whose concentrations are to be analytically measured are not uniformly distributed throughout the individual. This within-sample unit variability contributes its undesirable share to the width of confidence intervals obtained.

Homogenization or common sense can help remove this variability contribution. The U.S. Department of Agriculture, in its inspection of certain antibiotic residues in cattle, examines just the liver instead of the total carcass. Putting a liver into a high-speed blender is easier than putting in a whole cow. Similarly, often only after a fish has been gutted or fileted, certain chemical measurements are taken. The overall survey scheme can focus on just that portion of the individual where the analyte of interest is usually found.

Even with the best homogenization and the best selectivity of sites to remove within-sample unit variability, the analytical method itself has an irreducible, inevitable variability. Horwitz (2) has found that this variability, expressed as a relative standard deviation [RSD, where %RSD = 100 × (standard deviation/concentration)], often may be empirically given by the following equation:

$$\text{RSD}_{\text{method}} = 2^{1 - 0.5\log_{10}C} \tag{1}$$

where the concentration C is expressed as a decimal fraction (1.0 = 100% and 0.000001 = 1 ppm), and the $\text{RSD}_{\text{method}}$ is expressed as a percentage. For biota, this analytical source of variation would be expected to be negligible compared with those large variations from all the other sources such as sampling.

But what about bias? The analytical method may be adequately precise, but what if it is persistently off the mark? A classic horror story of bias in analytical methods comes from the nonbiotic realm of mineral composition. The details can be found in an article by Abbey (3). The author's conclusion is worth quoting: "There is no such thing as a bad method—only bad analysts who fail to allow for its limitations."

The lesson to be learned here is that even in the final stage, due care must be taken to eliminate avoidable variability and bias.

Variance and Confidence Intervals

Any distribution may be characterized by a variance, even if that variance equals infinity for certain U-shaped distributions. *Variance* is a measure of the variability in a population and can be used to compute confidence intervals about estimates of population averages. Distributions in biota might be expected to have especially large variances and hence especially large confidence intervals (i.e., estimates of averages would tend not to be very precise). However imprecise an estimate, at least statistical computations will alert the investigator to the magnitude of the imprecision. This section applies to all sampling, including sampling biota. However, the expected large variance and the possibility of non-Gaussian distributions in biota make more relevant the section on non-Gaussian distributions.

Measures of Variability

Consider a set of values, such as 244 PCB concentrations from 244 bluefish. The most convenient single statistic to characterize this set of values is the average, $\bar{X}$, given by $\bar{X} = (\Sigma X_i)/n$, where X_i is the PCB concentration measured in the ith bluefish, and n is the number of values (here $n = 244$).

A supplement to the average would be some measure of the spread or of the variability of the values. A simple measure of spread is the *range*, which is equal to the highest value minus the lowest value:

$$\text{range} = X_{(n)} - X_{(1)} = X_{\text{highest}} - X_{\text{lowest}} \tag{2}$$

where the integer subscripts in parentheses conventionally indicate the rank of the value when the values are arranged in increasing order from lowest to highest. Thus, $X_{(1)}$ is the lowest PCB value, and $X_{(244)}$ is the highest PCB value.

A statistic that is more stable than the range is the variance (V) of the set of values, which is simply the average of the squared deviation from the average of a set of values:

$$V = \frac{\Sigma(X_i - \bar{X})^2}{n} \tag{3}$$

Usually, however, the variance of the set of values is not what is wanted, but rather the variance of the population from which these values are merely samples. No matter what the population, the estimate of the population variance is conveniently obtained via multiplication of the sample variance by $n/(n - 1)$:

$$\text{population variance} = \frac{\Sigma(X_i - \bar{X})^2}{n - 1} \tag{4}$$

The average of a sample is a good estimate of the corresponding average of the population, but the variance of a sample must be multiplied by the $n/(n - 1)$ factor to get a good estimate of the population variance. For $n = 100$, no palpable differences between the population variance and the sample variance exist.

Other measures of variability come immediately to mind, such as the range divided by n, the average magnitude of deviation from the average, and the 75-percentile point minus the 25-percentile point. However, the variance remains the statistic of choice to indicate the variability. The prime reason is that variances are additive. This additivity means that the total variance can be expressed as a sum of the variance of assorted factors (e.g., within individual, within geographical region, and among geographical regions).

Moreover, the square root of the variance, which is the *standard deviation*, is the crucial quantity used in calculating confidence intervals. In Gaussian distributions, knowledge of the average and the variance (or equivalently, knowledge of the average and the square root of the variance) is enough to describe the whole distribution. In particular, with Gaussian distributions, the investigator can state a priori what percentage of the values in a population lie at or below such and such a number of standard deviations away from the average (*see* Table I in the appendix at the end of this chapter).

Once and For All: Standard Deviation Versus Standard Error

"The *standard error* [italics added] is the standard deviation of the average and is equal to the standard deviation based on the individual values divided by the square root of the number of individual values involved." This definition is the recipe for the standard error. A typical question is "But do you use the standard error or the standard deviation in calculating the confidence intervals?" The correct answer is yes. Simply put, when confidence intervals are desired for single values in a population, the standard deviation (as calculated from the sample by using the $n/(n-1)$ factor) should be used; when confidence intervals for averages are desired, the standard error (i.e., the standard deviation based on the sample divided by the square root of the number of individuals in the sample) should be used.

Qualitatively, taking averages leads to a tighter distribution: In any set of values being averaged, even an extreme value is likely to be neutralized and driven toward the population average. Nature is kind in that this tightening—this reduction in variance—is proportional to $1/n$, or for standard deviation, proportional to $1/n^{1/2}$.

The Magnificent Gaussian Distribution

Gaussian distributions, or at least near-Gaussian distributions, occur naturally. These distributions arise when many small independent fluctuations in components tend to cancel each other out to yield a stable average. Little subcomponents that add up and vary are what generate a Gaussian distribution. A more classic example is the distribution of IQs among a fixed population, where a wide variety of factors (e.g., genetic, economic, and social) combine to generate a final IQ. (IQ is a measure of how well one does on an IQ test.) Such small independently contributing components that fluctuate exactly match the model of a Gaussian distribution.

Consider the distribution of analytical results from an interlaboratory study of a given method applied to a given concentration of analyte in a given material. The roster of small, independently contributing factors might include sampling time, background noise, ambient temperature, and inhomogeneity.

Statisticians relish the Gaussian distribution because it is easy to handle: Knowledge of the average and standard deviation completely specifies the whole distribution. As Table II in the appendix at the end of this chapter shows, the location of every percentile point is known a priori. The fraction of the population that falls between any two specified values is known also.

Besides being handy, the assumption of a Gaussian distribution is often realistic. Moreover, rigorous statistical demonstration shows that even if a population is not Gaussian, its averages tend to become Gaussian-like. This point shows the true magnificence of the Gaussian distribution. The larger the sample size, the more Gaussian-like is the population of averages. Therefore, statisticians agonize over large sample sizes for two reasons: (a) the population gets more Gaussian-like, and the validity of the confidence-interval formulas increases; and (b) the standard deviation is diminished by a factor of $1/n^{1/2}$, which leads to tighter estimates (i.e., narrower 95% confidence intervals).

Another reason for advocating a large n is that the 1.96 in the 95% confidence interval formula

$$X \pm (1.96S/n^{1/2}) \tag{5}$$

is, strictly speaking, valid only for infinite n ($n = 30$ is acceptable), and the coefficient before the standard deviation (S), denoted $t_{0.025,n}$, is a function of n. For $n = 2$, the t coefficient is 12.71, and for $n = 30$, the t-coefficient is 2.05. Remembering 1.96 is easier than looking up the appropriate value for the t-coefficient, but the looking up should be done even if the refinement is only minor.

Confidence Intervals for More General Distributions

Gaussian distributions apply to continuous-valued variables, such as PCB or aflatoxin concentration. Obviously, an indefinitely large number of distributions for continuous values can be contrived, and many such distributions are well-known and fully described in the literature. Among such candidate distributions are the log-normal distribution, Pearson-types I–IV, the F-distribution, and the t-distribution ($t_{0.025,n}$ is a value in this distribution category). If the investigator knows a priori what distribution the population of values will have, then specialized tables can be looked at to calculate appropriate confidence intervals. Many times, Gaussian is a good guess, for which researchers should be thankful.

Biota distributions often fall into the dreaded "none-of-the-above" category: neither fish nor fowl. Particularly pernicious would be a bimodal distribution consisting of the superposition of two Gaussian distributions, each having its own average and its own standard deviation. Nevertheless, a confidence interval formula exists that is applicable to any continuous variable distribution, even a bimodal one. The price paid for the generality and the universal applicability of these so-called *Chebyshev intervals* is that the confidence intervals are quite broad.

The Chebyshev confidence interval is given by the following equation:

$$X \pm (ZS/n^{1/2}) = 100 \times [1 - (1/Z^2)]\% \text{ confidence interval} \quad (6)$$

where X and the standard deviation S are calculated from a large sample, and Z is any positive value greater than one. The price of ignorance is explicit: for $Z = 1.96$ with a Gaussian distribution, a 95% confidence interval is obtained; for an unknown distribution, for $Z = 1.96$, at least a 74% confidence interval is obtained, but the researcher does not know how much more confident he or she can be. To construct a 95% confidence interval about an average in the most general case, use $X \pm 4.47\ S/n^{1/2}$.

Even if the true population distribution is not known, this interval will be at least a 95% confidence interval. Had the population been Gaussian, this interval would be a 99.9+% confidence interval. Therefore, the way to tighten the interval—the way to get a more precise estimate—is to take a larger number of samples.

Confidence Intervals for Percentile Estimates

By applying due care and attention to details, the researcher can get a serviceable estimate of the average and the standard deviation. In fact, with sufficient skill and extraordinarily large samples (i.e., large n), very good estimates can be made of the population average. But often after all that labor, the mere average is not enough.

The researcher must be aware of how misleading an average alone can be, because the average tells nothing about the underlying spread of values. Think about the old joke about the statistician who drowned in the river whose average depth was 5 in.; the statistician just happened to be trying to walk on water 10 ft deep. The moral is this: Often the overall distribution is a crucial feature to assess. Especially in regulatory situations, great importance is attached to knowing what percentage of the population has values lying above the maximum level permitted by law. In short, the task of the statistician can also include that of estimating the distribution.

This section is devoted to a very simple, straightforward procedure for estimating the population distribution: reporting for the sample the 10-, 25-, 50-, 75-, and 90-percentile points, and then reporting the confidence interval for each of these percentile points. The appendix at the end of this chapter presents the confidence intervals for these percentile points as a function of sample size (n) in Table III. Statisticians sometimes use *box plots*, which are geographical depictions of percentiles.

Customer's Favorite: Two-Item Sample

Cost considerations lead users of statistics to minimize the sample size. Two-item samples are the ideal. (Even the most cost-conscious manager would grant that a one-item sample is too small, because at least two items are needed to estimate variability.) This section serves to show the quantitative consequences of picking just two individuals.

If the investigator is estimating some continuous quantity such as PCB concentration, then the 95% confidence interval is

$$[(X_1 + X_2)/2] \pm (12.706) \times \sqrt{(X_1 - X_2)^2/(2)(2)} \qquad (7)$$

$$[(X_1 + X_2)/2] \pm [6.353 \times (X_1 - X_2)] \qquad (8)$$

(For the connoisseur: The number 12.706 is the t-value for $n = 2$ at the 95% confidence interval, and the estimated variance from two values is $\Delta^2/2$, where $\Delta = X_1 - X_2$.)

Besides estimating continuous values, often the problem of estimating fractions is confronted (e.g., what fraction of the cow population consists of purple cows?). With fractions like this, any particular individual either belongs or does not belong to the class under investigation.

The crucial trick that provides a mechanism to convert continuous data to fractional or proportional type data is this: The statistician should mentally break down the population of values (e.g., actual PCB concentrations) into those values at or below the 90-percentile point and those values above the 90-percentile point. By definition, the probability of any value being at or below the *P*-percentile point is *P*% (or *P*/100 when probabilities are expressed as a decimal fraction). Therefore, if any item (e.g., a fish) is taken at random, the probability is 0.1 that its value (e.g., PCB concentration) is at or below the 10-percentile point. The probability is 0.9 that the value for any item is at or below the 90-percentile point. Associated with this probability is the probability that the value is above the 90-percentile point. This probability is 0.1 ($= 1 - 0.9$) because any value must be in one of the two classes, and therefore, the total probability must equal one.

For any two-item sample, the probability that both values are at or below the 90-percentile point is $0.9 \times 0.9 = 0.81$. The probability that both are above the 90-percentile point is $0.1 \times 0.1 = 0.01$. The probability that one value is at or below this point and that the other is above this point is $(0.9 \times 0.1) + (0.1 \times 0.9) = 0.18$. All possibilities have been exhausted, which is indicated by the fact that $0.81 + 0.01 + 0.18 = 1.00$.

For the 97.47-percentile point, the probability is $0.9747 \times 0.9747 = 0.9500$ that both item values lie below this point. This value of 0.9500 is also the probability that the higher of the two values is at or below the 97.47-percentile point.

Therefore, for the higher of the two-item values, the investigator can be 100% certain that it lies at or below the 100-percentile point of the population, 95% certain that it lies at or below the 97.47-percentile point of the population, 81% certain that it lies at or below the 90-percentile point of the population, and only 1% certain that it lies at or below the 10-percentile point of the population.

Before undertaking sampling, the researcher should know that the higher (or highest) value in the sample will be equal to or less than the 100-percentile point of the population (i.e., the population's maximum value). With only two items, the researcher can be 95% confident that the higher value is at or below the 97.47-percentile point. Not much improvement is obtained in going from 100% to 97.47%. But what would the researcher expect with only two items?

Bigger Samples

A more meaningful sample size would be 28 items, for then the researcher could be 95% confident that the highest value in the sample lies above the 90-percentile point of the population. (For the connoisseur, $0.9^{28.4} = 0.05$.)

For any either/or, yes/no, belongs/doesn't belong variable, the characteristic distribution to be used is known unequivocally as a *binomial distribution*. Use of the binomial distribution is not a mere approximation as is the case so often with the Gaussian distribution. The shape of the distribution depends only on two parameters: f, which is the fraction of the population that consists of "yes" items, and n, which is the number of items in the sample. By using conventional statistical techniques that are simple and obvious but tedious extensions of the work done for the two-item sample, the confidence intervals for percentile estimates can be obtained. Table III in the appendix at the end of this chapter reveals the results of some of these extensions. Specifically, for the indicated sample size (n) and the indicated sample percentile, the 95% and 99% confidence intervals are given for the true population fraction (f).

The following tactics are recommended when the population of values is so non-Gaussian that knowledge of the average and standard deviation is not enough to characterize the distribution of values. First, the sample values should be arranged from smallest to largest. Then the values for the (n/10)th, (n/4)th, (2n/4)th, (3n/4)th, and (9n/10)th items in this series should be reported. These five values are, respectively, the 10-, 25-, 50-, 75-, and 90-percentile points of the sample. These five values serve as estimates of the corresponding percentiles in the population. Next, after the computations are done, the confidence intervals for the estimates should be reported. For example, with a sample size (n) of 20, the 10th smallest value should be reported as the estimate of the 50-percentile point. (For the connoisseur: A subtle distinction, of no relevance for this exposition, exists between the *median*, which has 50% of the values below it and 50% of the values above it, and the *50-percentile point*, which has 50% of the values at or below it.) This estimate for the 50-percentile point could in reality (i.e., in the population) be reported as anywhere from the 27.2-percentile point to the 72.8-percentile point, with a confidence of 95%. These confidence intervals for percentile estimates are painfully broad, but they do become narrower with increasing sample size. These intervals are the best that can be done if no knowledge of the population distribution is available. These intervals have the virtue of being applicable to any distribution; they have the vice of being broad.

A fruitful comparison can often be made between the observed sample percentile points and the percentile points predicted on the basis of a Gaussian distribution having the average and standard deviation calculated on the basis of the sample. Large discrepancies reinforce the fears that the population is not Gaussian. Concordance among the percentile estimates, on the other hand, might indicate that the population is Gaussian after all. (*See* Table III in the appendix at the end of this chapter.)

Conclusion

The four key points are as follows: (1) frame the questions properly, (2) obtain representative samples, (3) estimate confidence intervals about estimates of population average under the assumption of a Gaussian distribution for the population of averages, and (4) report certain percentile points and their confidence intervals especially if the population has an unknown or non-Gaussian distribution.

If the sample is representative, the variability in the sample will reflect the variability in the population. Large variability leads to large confidence intervals encompassing the average. This problem is often remedied by adequate sample size. If the values are not part of a Gaussian population, estimates of the percentile points may suffice to answer relevant environmental and regulatory questions. Huff (4) stated this point:

> *By the time the data have been filtered through layers of statistical manipulation and reduced to a decimal-pointed average, the result begins to take on an aura of conviction that a close look at the sampling would deny... To be worth much, a report based on sampling must use a representative sample...from which every source of bias has been removed... Even if you can't find a source of demonstrable bias, allow yourself some degree of skepticism about the results.*

The investigator never knows for certain and can only play the odds. Statistics helps the investigator play these odds in a rational way.

Abbreviations and Symbols

Δ	difference between two values
Σ	summation
C	concentration
f	fraction of a population
n	sample size
N	number of items or individuals
P	percent of population values
PCBs	polychlorinated biphenyls
RSD	relative standard deviation
S	standard deviation ($S = V^{1/2}$)
t	Student's t-value, usually accompanied by subscripts indicating the probability level and number of degrees of freedom involved
V	variance
$\bar{X}$	average
$X_{(i)}$	the ith value when the n values of X are arranged in increasing order from lowest to highest
Z	a value, X, expressed in standard deviation units ($Z = X/S$)

Appendix

Table I. Confidence Intervals for Gaussian and Chebyshev Conditions

Z	*Gaussian Conditions*	*Chebyshev Conditions*
0.1	8.0	NA[a]
0.2	15.9	NA
0.3	23.6	NA
0.5	38.3	NA
1.0	68.3	NA
1.1	72.9	17.4
1.2	77.0	30.6
1.3	80.6	40.8
1.4	83.9	49.0
1.5	86.6	55.6
1.645	90.0	63.0
1.8	92.8	69.1
1.96	95.0	74.0
2.0	95.5	75.0
2.1	96.4	77.1
2.5	98.8	84.0
3.0	99.7	88.9
4.0	100.0	93.8
5.0	100.0	96.0

NOTE: The confidence intervals are presented as the percentage of values between $X \pm Z \times S$, where X is the average, Z is a value expressed in standard deviation units, and S is the standard deviation. These are confidence limits for single values.
[a]NA indicates not applicable.

Table II. Percentiles for a Gaussian Curve

Z	*Percentile*
−1.282	10
−0.674	25
0.000	50
0.674	75
1.282	90

Table III. Sample Size Dependence of Confidence Intervals for Percentile Estimates

n	Sample Percentile	95% Confidence Interval		99% Confidence Interval	
		Lower Value	Upper Value	Lower Value	Upper Value
10	10	0.3	44.5	0.1	54.4
10	50	18.7	81.3	12.8	87.2
10	90	55.5	99.8	45.6	100.0
20	10	1.2	31.7	0.5	38.7
20	25	8.7	49.1	5.8	56.0
20	50	27.2	72.8	21.8	78.2
20	75	50.9	91.3	44.0	94.2
20	90	68.3	98.8	61.3	99.5
40	10	2.8	23.7	1.7	28.3
40	25	12.7	41.2	10.0	46.1
40	50	33.8	66.2	29.5	70.5
40	75	58.8	87.3	53.9	90.0
40	90	76.3	97.2	71.7	98.3
50	10	3.3	21.8	2.2	25.8
50	50	35.5	64.5	31.6	68.5
50	90	78.2	96.7	74.2	97.8
60	10	3.8	20.5	2.6	24.1
60	25	14.7	37.9	12.3	41.8
60	50	36.8	63.2	33.1	66.9
60	75	62.1	85.3	58.2	87.8
60	90	79.5	96.2	75.9	97.4
80	10	4.4	18.8	3.3	21.7
80	25	16.0	35.9	13.7	39.4
80	50	38.6	61.4	35.4	64.7
80	75	64.1	84.0	60.7	85.3
80	90	81.2	95.6	78.3	96.7
100	10	4.9	17.6	3.8	20.2
100	25	16.9	34.7	14.8	37.7
100	50	39.8	60.2	36.9	63.1
100	75	65.3	83.1	62.3	85.2
100	90	82.4	95.1	79.8	96.2

References

1. Whitaker, T. B. *J. Am. Oil Chem. Soc.* **1972**, *49*, 590– 592.
2. Horwitz, W. *Anal. Chem.* **1982**, *54*, 67A–76A.
3. Abbey, S. *Anal. Chem.* **1981**, *53*, 528A–534A.
4. Huff, D. *How To Lie with Statistics*; Norton: New York, 1954; p 18.

Bibliography

Bennet, C.; Frankin, N. *Statistical Analysis in Chemistry and the Chemical Industry*; Wiley: New York, 1966.

Cochran, W. G. *Sampling Techniques*; Wiley: New York, 1977.
Huff, D. *How To Lie with Statistics*; Norton: New York, 1954.
Huntsberger, D.; Billingsley, P. *Elements of Statistical Inferences*; Allyn and Bacon: Boston, 1973.
Langley, R. *Practical Statistics Simply Explained*; Dover: New York, 1970.
Zervos, C.; Fringer, J. *J. Toxicol. Clinical Toxicol.* **1984**, *21* (special symposium issue on "Exposure Assessment: Problems and Prospects").

Chapter 32

Sample Size

Relation to Analytical and Quality Assurance and Quality Control Requirements

John B. Bourke, Terry D. Spittler, and Susan J. Young

Biota present unique challenges to valid sampling because of the vast size differences between species, variations within a study population, and tissue differentiation. Several considerations dictate the number of specimens involved and place limits because of growth stage, habitat, and availability. Subsample size selection must consider the limitations of the analytical method, sensitivity requirements, and statistical requirements. Variation in sampling and subsampling methods for a study of grapes is discussed. No differences were found over a three-fold subsample size range. Preliminary results from a study of subsample homogeneity over the range of 2–100 g are presented. These field-treated produce reflect nonuniform pesticide residue distributions requiring careful preparation.

Preliminary Considerations

Samples are taken for a number of reasons, including monitoring, regulatory activities, quality control, scientific study, disaster assessment, or just idle curiosity. Whatever the reason, this first collection of any material for the analysis of one or more analytes is considered the *primary sample*. The method of collection and sample size are described by the sampling plan, which in turn is designed to ensure that this primary sample will be representative of the whole. In most instances, this primary sample will be sent to the laboratory or some intermediate location for subdivision into manageable-sized secondary samples.

No one set procedure or sample size is appropriate for all sampling events. Many factors must be considered before sampling; the most important factor is the objective of the study for which the sample is being taken. The size and method of sampling for monitoring studies are considerably different than those employed when attempting to describe some cause-and-effect relationship as is usually the case in either applied or basic research projects. The precision and accuracy required by the experimental protocol has an influence on sample frequency and size. The degree of validation needed to ensure the integrity of results will also influence sample frequency and replication. Data destined for litigation are produced under conditions more stringent with regard to sample identity and custody then are results collected for incorporation into a scientific data base.

Therefore, before a sample is taken, a sampling plan or protocol must be developed; the purpose for undertaking the study will form the basis of this plan. If, for instance, monitoring is the objective and the investigator wishes to know only if some analyte is present, a single grab sample will be all that is required, and the sample may be relatively small. However, the concentration of analyte found is representative of the sample only and not the environment from which it is taken.

In other chapters included in this volume, problems were presented regarding sampling and the determination of sample size necessary to ensure that a primary sample is representative when studying analytes in water and air. Some of the same sampling philosophy applies with respect to the sampling of plants, animals, microorganisms, or other parts of the biota. Obviously, when sampling biota, a number of factors must be included such as purpose of the study, homogeneity of the matrix to be sampled, concentration of the analyte, efficiency of the methods of extracting and concentrating, and sensitivity of the method to be employed. Nature of the organism and size of the population under consideration as well as availability and cost of the material to be studied are factors more unique to biosamples. In addition, analyte physical characteristics such as phase, volatility, oxidation state, chemical properties, and biological activity will have pronounced influences on sampling, and these characteristics specify when, where, and how much material must be removed from the environment to ensure sample validity.

Distribution

When the homogeneity of the matrix to be sampled is referred to, not only the matrix itself must be considered, but also the distribution of the analyte throughout that matrix. For instance, if investigators wish to determine the residues of a pesticide applied to plants in a field, they need to know something about how the pesticide was applied. If the application was done by an accurately calibrated and designed ground boom sprayer, where care was taken to avoid skips and oversprays, then the number of sampling points required to describe the area will be determined by the number and uniformity of the plant population. If, however, the pesticide was applied by a sprayer having poor nozzle design and positioning or with improper control over swath width and placement, a nonuniform distribution of pesticide will result, and considerably more sampling sites will be required to adequately assess the coverage. The resulting primary sample may then become large. Although this composite sample may accurately describe the average analyte concentration on the plants, it will not tell us anything about the variability. If that profile is desired, produce from a large number of sampling points must be kept separate and analyzed independently.

This problem can be illustrated by visualizing the results of spring-applied nitrogen to winter wheat. If each pass across the field is separated by a distance greater than the application swath width, alternating light and dark green streaks will be seen. These streaks are the result of an uneven distribution of nitrogen. A single sample from such a field will represent only that part of the field from which it was taken. Walking down the field parallel to the direction of application and taking a number of samples will result in analytical data representative of that particular swath only. Walking across the swath paths, however, and taking samples at a spacing unequal to the swath width will better represent the average analyte content of the field. Keeping the original objectives in mind for the sampling is important. If application uniformity is the subject of interest, each sample must be analyzed individually, and the sample size will approach that of a subsample. If the mean application is the subject of interest, for example, as in yield studies, then all the individual samples can be composited, the composite made as homogeneous as possible, and a subsample taken for analysis. Duplicate subsamples in this case tell something about the homogeneity of the composite, and the total composite sample size will be dependent on the total number of field samples taken.

Sample Size

The size of each sample will depend on a number of factors but should be kept at a minimum. For biological materials, the limiting factor may often be the availability of substrate. The collection of uncommon or even rare and endangered organisms often limits the sample to the material that suggested the need for analysis (i.e., a dead organism). In another biological material, the ability to capture organisms may be impaired by the population's habitat, size, or life cycle. The cost of some biotic materials may make the acquisition of large samples impossible; thus, resources can limit sample size. In these cases, the data may represent only the individual, primary sample and not the extant population. Extrapolations from these results to describe the world at large may be tenuous at best. When working with larger organisms, one must decide if it is the average occurrence within a population that is desired or the probability of a single organism containing the analyte. In the case of average occurrence, many organisms may be composited to give a single sample, and in the case of the probability of a single organism, each sample will, by definition, be derived from a single organism.

The concentration of the analyte may have a significant bearing on the size of the sample. If the analyte is in a very low concentration, the sample may have to be rather large to allow for the extraction and concentration of analyte. Occasionally, the entire primary sample may have to be used as the laboratory sample, and extraction carried out on the whole sample as obtained. The size of each unit within a sample also affects primary sample size. A primary sample of peppercorns will be much smaller than an equivalent sample of oranges, which will in turn be smaller than one sample comprised of watermelons. In regard to the distribution of analyte at very low concentrations within a sample, one orange in a boxcar represents 1 ppm, and one orange in 1000 boxcars represents 1 ppb. Such visualizations put in perspective some of the problems of sampling and analysis.

So far, the pulling or taking of the primary sample has been discussed. Such a sample may be made up of snakes or snails or even puppy dog tails. Whatever the composition, the primary sample must be of sufficient size to be representative of the system or population that the sampling plan wished to address and is generally far larger than can be used for individual chemical or physical analyses. These samples must then be reduced to replicates of a test size that can be analyzed in the laboratory. The manner in which this reduction takes place is critical to the eventual validity of the data.

Size Reduction

Air and water samples, once taken, are and generally remain homogeneous. Although this homogeneity can also be true of biological fluids, it rarely holds for other materials. Reduction can therefore proceed in two ways: nonhomogeneously or homogeneously.

Nonhomogeneous Reduction

First, a selected portion of the sample is isolated for analysis because that portion is either of special interest, is the known accumulation site, is the regulated commodity, or represents a limitation in the method. Arguments can be made that this separation process is merely generating more primary samples and does not really constitute a reduction. In instances where this takes place directly on-site, as in picking of fruit or foliage or in removing blood or tissue specimens from captured and released animals, the argument is probably valid. Other times, the segregation of the primary sample into differentiated tissues is an intermediate step performed either in the laboratory or at selected facilities. Fat is removed from fish samples for analysis of lipophilic chlorinated pesticides and polychlorinated biphenyls (PCBs); food commodities are reduced to the edible components for regulatory or nutritional testing. Frequently, a collection of excised parts will subsequently undergo a second or homogeneous reduction process.

Homogeneous Reduction

The sample (primary or secondary) can be made homogeneous by grinding, milling, blending, chopping, mixing, or any of a number of physical processes such that further subsampling by techniques of dividing, riffling, or aliquot taking will produce uniform portions of a size amenable to the analytical method and representative of the whole. The final subsample is called a test sample and its size will be primarily determined by the concentration of the analyte, the sensitivity of the method, and the capability of the analytical equipment. At any of these steps in which a solution is produced, the increment becomes homogeneous, and further reductions by aliquot are redundant and for convenience only. At this stage, replicate test samples are produced and checked, and storage and recovery spike material is prepared and incorporated into the analytical scheme. That topic has been thoroughly covered in prior chapters.

Case History

An interesting series of studies illustrating these concepts was conducted in cooperation with the Geneva laboratories over 25 years ago and pertained to the selection of grape berries for surface stripping of pesticides by solvent for residue determination (*1–3*). Analytical subsamples consisting of as many as 900 grape berries were stripped of DDT (1,1′-(2,2,2-trichloroethylidene)bis[4-chlorobenzene]) by tumbling in solvent. The average DDT residues for each sample size were essentially equal at two different residue levels for samples of 300 berries or more. The 300 count was thus retained as the standard sample size. Standardizing on a berry count rather than a weight was done because the increase in berry size by growth during the residue-weathering study's time frame resulted in a smaller arithmetic change in surface area. Had weight been the constant, progressive sample changes would have been much greater because of the geometric surface-to-mass relationship.

The selection method for the 300 berries was found to be critical. One experiment involved the designation of 20 consecutive vines from a 70-vine replicate plot. Three clusters were randomly picked from each vine, and five individual berries were snipped from random sections of each cluster; the total was 300 berries. Residues were found to be 9.2 ppm. When five berries were selected from each of 15 clusters derived from only four random vines in the row, the average residue was 28% lower (6.6 versus 9.2 ppm). As a check, all berries from all clusters on four random vines were separated and randomized; 300 berries randomly selected and analyzed from these were found to have a DDT concentration of 9.2 ppm. The efficient first method duplicated the very thorough but time-consuming third method. The four-vine shortcut employed in the second trial gave inadequate representative samples.

Analysis of the surface of the grapes only and the separation of the berries from row, vine, and cluster stems constitute nonhomogeneous reductions. Yet, the final objective was still the production of analytical samples representative of the entire crop.

Current Study

The limits of homogeneous reduction is the topic of a current cooperative study between the Food and Drug Administration and Cornell University, and some preliminary illustrative data are available (Young, S. J.; Newell, D. F.; and Spittler, T. D.; in press). Most applied pesticides, with the possible exception of some having systemic mechanisms, are unevenly distributed in and on the target. In these experiments, methoxychlor (1,1′-(2,2,2-trichloroethylidene)bis[4-methoxybenzene]) is used as a foliar spray after the heads of cabbage are well-formed. Consequently, distribution on the leaf surfaces is not uniform and obviously heavier on the outer layers of the head. With the current trends toward microanalysis and solid-phase cartridge extraction, and the high costs of solvent purchase and disposal, a small subsample (e.g., 5.0 g) is more desirable than a large one (e.g., 500 g). But at what point do the fewer and fewer chopped pieces that make up size-reduced subsamples become nonrepresentative of the primary sample and give unacceptable replicate variation?

A large primary sample of methoxychlor-treated cabbage (ca. 10 kg) was thoroughly chopped, and 12 replicates of each of six subsample weights were obtained in a randomized manner. The subsamples were 100, 50, 25, 10, 5, and 2 g. Because this study was to test the representativeness of subsample size and not the miniaturization of the method, appropriate amounts of chopped, untreated check cabbage were added to bring the total test sample weight of each subsample to 100 g. Analytical results for this first series are shown in Table I along with the respective coefficients of variance (CV).

Although results for other commodities (e.g., green beans and apples) and full statistical analyses are incomplete, inspection reveals that at least in this instance, homogeneous reduction to 2 g is reasonable and representative of the primary sample. This statement does not mean that an analytical method based upon 2 g of extractable produce will give results of either this precision or

Table I. Means and Coefficients of Variation for Decreasing Subsample Weights

Sample Weight (g)	*Mean (ppm)*		*CV*[a]	
	Area	*Peak Height*	*Area*	*Peak Height*
100	8.882	8.874	4.638	4.421
50	9.066	9.048	3.251	3.146
25	9.062	9.063	3.650	3.703
10	9.399	9.401	5.116	4.978
5	9.525	9.510	4.784	4.915
2	9.883	9.899	6.011	5.725
	Overall mean[b] 9.296		Overall CV[b] 5.800	

[a]The abbreviation CV denotes the coefficients of variation.
[b]The overall value was calculated from 144 determinations.

accuracy, but rather that 2 g is a valid subsample size when employed with validated analytical procedures.

Size reduction is a determination that must still be made on the basis of each new set of circumstances, but eventually a reliable body of literature will accumulate to provide guidance and insight in establishing workable experimental designs.

References

1. Taschenberg, E. F.; Avens, A. W. *J. Econ. Entom.* **1960**, *53(2)*, 269–276.
2. Taschenberg, E. F.; Avens, A. W. *J. Econ. Entom.* **1960**, *53(3)*, 441–445.
3. Taschenberg, E. F.; Avens, A. W.; Parsons, G. M.; Gibbs, S. D. *J. Econ. Entom.* **1963**, *56(4)*, 431–438.

Chapter 33

Composite Sampling for Environmental Monitoring

Forest C. Garner, Martin A. Stapanian, and Llewellyn R. Williams

Guidance for selecting a plan to composite environmental or biological samples is provided in the form of models, equations, tables, and criteria. Composite sampling procedures can increase sensitivity, reduce sampling variance, and dramatically reduce analytical costs, depending on the exact nature of the samples, the analytical method, and the objectives of the study. The process of taking random grab samples and individually analyzing each sample for elements, compounds, and organisms of concern is very common in environmental, biological, and other monitoring programs. However, the process of combining aliquots from separate samples and analyzing this pooled sample is sometimes beneficial. The researcher must consider detection limits, probability of analyte occurrence, criterion level, sample size, aliquot size, analyte stability, the number of samples, analysis cost, sampling cost, monetary resources, biological interactions, chemical interactions, and other factors in order to make a wise decision of whether to composite or not, and how many sample aliquots to composite.

IN TYPICAL ENVIRONMENTAL, BIOLOGICAL, and other monitoring programs, each monitored site or individual is sampled, and each sample is analyzed individually. A chemical or biological measurement is thus obtained for each sample. This procedure is the optimal plan when a measurement is needed for every sample. However, combining aliquots from two or more samples and analyzing this pooled sample is advantageous at times. Such *composite sampling plans* can provide various advantages but should be used only when the researcher fully understands all aspects of the plan of choice. Composite sampling, as discussed in this chapter, applies to each of the following situations:

3152–4/96/0679$15.00/0

1. when samples taken from varying locations or individuals need to be analyzed to determine if the component of interest is present or exceeds criterion limits in any of the samples;
2. when aliquots of extracts from various samples composited for analysis need to be analyzed to determine whether the component of interest is present or exceeds criterion limits in any of the samples;
3. when representativeness of samples taken from a single site, waste pile, product lot, household, community, or population need to be improved by reducing intersample variance effects;
4. when representativeness of random aliquots removed from a potentially heterogeneous sample needs to be ensured by reducing the effect of variance between aliquots;
5. when the material available for analysis in samples of necessarily limited size, such as blood samples, needs to be increased to achieve analytical performance goals; and
6. when the confidentiality of the individual donors of samples needs to be ensured.

Applications of composite sampling have involved sampling bales of wool (*1-3*); estimating plankton in freshwater (*4-6*); and determining pyrethrin levels in fruit (*7*), fat content in milk samples (*8*), pesticide levels in drinking water (*9*), and pH in soil samples (*10*). Sobel and Oroll (*11*) listed many industrial applications. Garrett and Sinding-Larsen (*12*) presented applications in geochemistry and remote sensing. Much of the pioneering theory of composite sampling was provided by Dorfman (*13*). Watson (*2*) derived equations for relative testing costs for some types of composite sampling plans. Watson discussed the problems of detection and false detection of factors and optimum group size in composite sampling. Connolly and O'Connor (*8*) compared random and composite sampling methods. Brown and Fisher (*14*) considered the problem of estimating the mean of the characteristic of interest from composited samples.

Potential Advantages

Composite sampling can be used to reduce the cost of classifying a large number of samples. Suppose the objective of the monitoring program is to identify samples of hazardous material, diseased individuals, or defective units of product when the a priori probability of hazard, disease, or defect is believed to be very low. For example, when checking human blood samples for syphilis, analysts may reduce cost by pooling aliquots from pairs of samples. On the rare occasion when a composite sample tests positive, the two original samples can be found and tested to identify the diseased individual. Over a large number of samples, the cost of such a plan would be little more than half the cost of analyzing every sample. Costs might be reduced further by pooling three or more samples. However, if too many samples are pooled, then expected cost may rise, and other limitations may be reached.

In any monitoring program, sample analyses may occasionally indicate the presence of analyte when none is actually present. Such *false positives* may be caused by random errors, instrument anomalies, or laboratory contamination. A false positive is less likely to occur in a composite sampling plan for classification than in ordinary sampling plans. If a composite tests positive, then each of the contributing samples is tested individually. A false positive will only occur in the unlikely event that both the composite and an individual sample both falsely test positive. Thus, composite sampling may also provide an added benefit by reducing the number of false positive results.

Composite sampling may also be used to reduce intersample variance due to heterogeneity of the sampled material. This reduction can be useful when the objective is to characterize a potentially heterogeneous material by estimating the mean concentration of analyte or total amount of analyte present in the material. For example, a hazardous waste site might be characterized by chemically analyzing 10 surface soil samples and averaging the result. However, the same value might be obtained by pooling the 10 samples and performing just one analysis. By pooling 10 sample aliquots and homogenizing the composite, the variance due to sample differences (site heterogeneity) is reduced by a factor of 10. If the variability between samples is much larger than the variability within samples, then one analysis of a composite of 10 samples will yield an estimate of the average concentration of the site that is of similar quality (equal mean and nearly equal variance) to that of averaging 10 sample analyses. However, the same amount of information is not obtained because no variance estimate is possible from one analysis and no unusual samples can be identified. Therefore, this composite sample plan is recommended only when funds are limited and analytical costs are much larger than sampling costs.

Composite sampling can also be used, with certain limitations, to increase the amount of material available for analysis. This procedure could potentially increase method sensitivity and achieve lower method detection limits, which would reduce the rate of false negatives. Additionally, composite samples may be used to ensure confidentiality of the donor. This purpose may be beneficial when social concerns may discourage certain classes of individuals from participating in a random survey because of fear of identification; composite sampling may help to preserve randomness and anonymity.

Potential Limitations

When the objective of the monitoring program is classification, sample compositing may result in dilution of the analyte to a level below the detection limit of the analyte. This dilution would result in an error in decision making known as a *false negative*. Suppose that the detection limit of a particular analyte in drinking water is 0.005 mg/L, and samples from five wells having true concentrations of 0.003, 0.015, 0.0000, and 0.001 mg/L are composited. Further, suppose that the criterion level for this analyte is 0.005 mg/L. In this example, the true concentration of the composite would be 0.004 mg/L. Therefore, the analyte may not be detected in the composite. Such a situation is unacceptable because one well has levels of the analyte 3 times the criterion level. This problem can occur when the number of samples composited exceeds the ratio of the criterion level to the detection limit. Care should be taken so that sample dilution does not substantially reduce the ability of the researcher to identify target analytes in the composite. Many definitions of detection limit exist, and the researcher must be certain that the detection limit used implies a high probability of detection in the sample matrices of interest.

If sampling costs are greater than analytical costs, then analyzing each sample individually may be more cost-effective. Garrett and Sinding-Larsen (*12*) presented statistical and cost-based approaches to optimal composite sampling plans that allow for nontrivial sampling costs.

When considering multiple analytes in a composite, information regarding the relationships among analytes in individual samples will be lost. The experimenter cannot be sure from which sample each analyte came or in what proportion with other analytes. Further, analytes or organisms in separate samples may be mutually destructive; therefore, after some time, the composite may not be representative of any of the individual samples.

Composite Sampling for Classification

Suppose the experimental objective is to classify each of N units into exactly one of two categories. These discrete categories might be detected and nondetected, diseased and nondiseased, or defective and nondefective. Each unit could be tested separately, which would require N tests. Alternatively, a group of units could be tested simultaneously, and the group would be classified by the test into one of the two categories. Such a composite sampling plan is appropriate for discrete measurement methods.

Other studies (*2*, *11*, *13*, *15*) have considered the case where the number of defective units is a binomial variable. These studies used the expected number of tests for the entire population to determine the efficiency of composite sampling. In Dorfman's study (*13*), blood samples were pooled into groups. When a group was identified as defective, each blood sample in that group was tested individually. Li (*16*) proposed a multicycle version of Dorfman's method. Watson (*2*) derived equations for relative testing costs for discrete models and considered instances of zero and nonzero error variances. Watson discussed the problems of detection and false detection of factors and optimum group size. Hwang (*3*) proposed a method of composite sampling that does not require the assumption of binomial distribution and applies to the cases where either the distribution or an upper bound of the number of defective units is known. Hwang's algorithm for finding the optimum number of units to group into a composite sample that minimizes the total number of tests is also relatively efficient and simple to use. Mack and Robinson (*17*) researched the advantages of composite sampling in a national survey of pesticides in human adipose tissues and derived an approximate model of composite sample concentration from a model of individual sample concentration.

Now suppose that the objective of a sampling plan is to determine if the concentration of an analyte or organism exceeds a criterion level. For example, the criterion level may be the maximum concentration of iron deemed to be safe for drinking water. If each of N samples was tested individually, then N tests would be required. As for a discrete measurement method, use of a composite sampling plan for comparing continuous measurements to a criterion level may reduce the number of tests required. When determining the number of samples that can be pooled into the composite, researchers must consider the detection limit of the analyte. If the criterion level is near the detection limit of the analyte, then the number of samples that can be pooled into a composite is limited.

Thus far we have considered the measurement of only one analyte that will yield univariate data. The discrete and continuous measurement models can be extended to situations in which more than one analyte, organism, or attribute is measured. In such cases, a battery of tests is performed simultaneously. When a positive result occurs (discrete measurement model) or the level exceeds criterion (continuous measurement model) for any single test in the battery, then the n units are tested individually. If the multiple analytes are expected to be highly correlated, then a univariate approach may be useful.

In all of the classification models, the original samples are assumed to be homogeneous with respect to the analytes or organisms of interest.

Models for Discrete Distributions

The probability that the analyte or organism of interest occurs in a sample (p) may be estimated by the proportion of positive samples in a similar study, by presampling, or by any other reasonable procedure.

If the analyte or organism of interest is not found in the composite of n pooled samples, where n is the number of samples aliquoted and combined into each composite, then only one analysis needs to be performed. If the analyte or organism of interest is found in the composite, then $n + 1$ analyses must be performed. Hence, the total number of analyses (M) needed is 1 if the test of the composite is negative or $n + 1$ if the test of the composite is positive.

In order to design the sampling program that minimizes cost, the expected number of analyses, $E(M)$, must be calculated as follows:

$$E(M) = q^n + (n + 1)(1 - q^n) \quad (1)$$

where q is the probability that the analyte or organism of interest does not occur.

The relative cost factor (RCF) is the ratio of the expected number of analyses calculated by using composite samples to the number of samples that would have to be analyzed if samples were not composited. Therefore,

$$\mathrm{RCF} = \frac{NE(M)}{nN} = \frac{E(M)}{n} = \frac{q^n + (n+1)(1-q^n)}{n} = 1 + \frac{1}{n} - q^n \qquad (2)$$

This result was also found by Dorfman (*13*). The optimal composite sampling plan is that which has minimum cost relative to a traditional plan in which all samples are analyzed. Table I provides the value of n that minimizes the RCF for a given value of q.

The model described by equation 2 (model 1) can be generalized to accommodate cases where more than one analyte or organism is tested in each composite sample. The resulting multivariate model (model 2) assumes independence among the various analytes.

Model 2 is a simple extension of model 1. In equation 2, the expected number of analyses is simply the product of the expected number of analyses for each of the individual analytes or organisms over the number of analytes or organisms of interest (m). Therefore, the relative cost factor is

$$\mathrm{RCF} = \frac{E(M)}{n} = \frac{Q^n + (n+1)(1-Q^n)}{n} = 1 + \frac{1}{n} - Q^n \qquad (3)$$

where n is the number of samples aliquoted and combined into each composite, and Q is the product of q_i values, where $i = 1$ to m.

Table I. Optimal Values of *n* for Given Values of *q* or *Q*

p	q *or* Q	*Optimal Value of* n	*Expected Savings*[a] *(%)*
0.0001	0.9999	101	98.0
0.0002	0.9998	71	97.2
0.0004	0.9996	51	96.0
0.0008	0.9992	36	94.4
0.0016	0.9984	26	92.1
0.0032	0.9968	18	88.8
0.0064	0.9936	13	84.3
0.0128	0.9872	9	77.9
0.0256	0.9744	7	69.1
0.0512	0.9488	5	56.9
0.1024	0.8976	4	39.9
0.2048	0.7952	3	17.0
0.4096	0.5904	1	0.0

[a] Expected savings are the percent reduction of analytical costs expected when a composite sampling plan is used instead of when each sample is analyzed individually.

A limitation of model 2 is the assumption that the analytes of interest are not correlated. Extreme caution, therefore, should be exercised when this multivariate procedure is used. The optimal value of n can be obtained from Table I.

Models for Continuous Distributions

A model based upon a continuous distribution of measurements is appropriate when the concentration of the analyte or organism is measured and compared to a criterion level, rather than when the presence or absence of the analyte or organism is merely detected or recorded.

The criterion level for the composite sample should be determined so that the composite sample concentration will exceed the criterion level whenever any one sample exceeds the criterion level of the analyte or organism in the sample (c'). Therefore, the composite sample criterion (c) can be calculated as

$$c = \frac{c'}{n} \tag{4}$$

Let q be the probability that $y < c$, where y is the concentration of the analyte or organism in the composite of n sample aliquots of equal mass or volume. The value of q can be estimated by integrating the probability density function of the composite sample results. This probability density function may be estimated by using some knowledge of the density function of results for individual samples, or by using the central limit theorem to justify normality of y. If the composite sample analysis yields a result less than c, then each of the n samples must have concentrations less than c'. If the composite sample analysis yields a result in excess of c, then each of the n samples must be analyzed to determine which samples contributed the analyte. If M is the number of analyses performed for each group of n samples, then $M = 1$ if $y \leq c$, or $M = n + 1$ if $y > c$. Also,

$$\text{RCF} = \frac{E(M)}{n} = \frac{q + (n+1)(1-q)}{n} = 1 + \frac{1}{n} - q \tag{5}$$

The minimum RCF can be found by substituting various values of n. Restrictions on n do exist, however. If the method detection limit is D, then

$$n \leq \frac{c'}{D} \tag{6}$$

Model 3, which is described by equation 5, can be generalized to another model (model 4), where $m > 1$ analytes or organisms are measured. If aliquots are taken from n samples and combined for each composite, then the RCF is calculated according to equation 7.

$$\mathrm{RCF} = \frac{Q + (n+1)(1-Q)}{n} = 1 + \frac{1}{n} - Q \tag{7}$$

The probabilities q_i are estimated by integrating an estimated probability density function of the composite sample results. This density function may be estimated by using some knowledge of the probability density of the individual sample results or by using the central limit theorem assuming approximate normality of the composite sample results for each analyte. These steps can be simplified by assuming that the m analyte concentrations are independent, but this assumption may be violated in many applications. Such estimation is very involved mathematically. Therefore, a mathematical statistician should be consulted to ensure that good estimates are obtained for the probabilities q_i and the RCF. Such estimates could be used to obtain an optimal value of n, but a table of optimal values such as those in Table I, which applies to models 1 and 2 only, cannot be derived without specific knowledge of the statistical distribution of analyte concentrations.

The number of samples aliquoted and composited is restricted by detection limit considerations. The restriction is that n must be less than or equal to the minimum of the ratios of c'_i to D_i, where c'_i and D_i are the criterion level and detection limit, respectively, of the ith analyte from $i = 1$ to m.

Reducing Variance

The difference between the variances from individual and composite analyses can be illustrated with an example. If k samples are randomly taken from a hazardous waste site and each is quantitatively analyzed, then the mean of the k results is an estimator of the average concentration of the site. The variance of this estimator ($\sigma^2{}_M$) can be calculated as follows:

$$\sigma^2{}_M = \frac{\sigma^2{}_E + \sigma^2{}_H}{k} \tag{8}$$

where $\sigma^2{}_H$ is the variance between samples due to site heterogeneity, and $\sigma^2{}_E$ is the random error variance of the analytical method.

If, however, equal aliquots are taken from each of the k samples, and these aliquots are pooled and homogenized, then one analysis of this composite sample will also be an estimator of the average concentration of the site. The variance of this estimator ($\sigma^2{}_C$) is

$$\sigma^2{}_C = \sigma^2{}_E + \frac{\sigma^2{}_H}{k} \tag{9}$$

By comparing the estimator obtained by averaging k analytical results to that obtained by analyzing a composite of k samples, $\sigma^2{}_M$ and $\sigma^2{}_C$ will be very nearly equal if σ^2_H is much greater than σ^2_E. On many occasions, the variance between samples at different locations of a waste site, or that between aliquots of a single sample, greatly exceeds the random error variance of the analytical method. In such cases, using a composite sample approach to estimate the average concentration at the waste site is cost-effective. Rohde (*15*) developed distributional properties of the analytical result of the composite sample. Rohde considered complex situations in composite sampling, such as the presence of more than one variance component.

Increasing Sensitivity

Composite sampling may also be useful in increasing analytical sensitivity and potentially reducing the rate of false negatives. Occasionally, an analytical method will require a larger sample than is possible from one individual. For example, if a method requires 2 L of blood to achieve the required limit of detection, then no living human could be adequately sampled. However, if four 0.5-L samples could be taken from four similar subjects, perhaps from the same household, then the desired sensitivity could be reached. Analysis of the composite sample would yield a result appropriate for the household as a whole, but not necessarily for each included individual. This type of plan is recommended only when the level of concern is below the limit of detection of practical individual samples, and a larger amount of material is needed for each analysis than is possible from a single individual.

Achieving Confidentiality

Occasionally, obtaining a random sample of individuals for testing may be difficult because of personal concerns about being identified as diseased. In such cases, composite sampling may be a useful technique when the objective of the study is to estimate the proportion of individuals that are diseased but not to identify these individuals. This objective may be beneficial when social concerns may discourage part of the population from participating in a random survey because of fear of identification. In such cases, composite sampling may help preserve randomness through assured anonymity.

For example, suppose that one wishes to estimate the proportion of individuals exposed to the acquired immune deficiency syndrome (AIDS) virus. Certain individuals, for social reasons, may be less willing to volunteer for individual testing; thus, a biased sample may result. By pooling groups of $n > 2$ samples for each test, no single individual can be identified as an AIDS carrier, and a less biased or an unbiased sample may result. The proportion of individuals that would have tested positive (p) can be estimated from the observed proportion of groups that test positive (P) by

$$p = 1 - (1 - P)^{1/n} \quad (10)$$

This estimator may also be useful to obtain a cost-effective estimate of the proportion of individuals possessing a characteristic (especially when sampling costs are much less than analytical costs) even when confidentiality is not a specified requirement.

Discussion

The prudent use of composite sampling can dramatically reduce analytical costs of classification experiments, reduce intersample variance in estimation experiments, improve sensitivity when individual sample size is limited, or improve the randomness of surveys through assured confidentiality. These benefits can be realized only under certain objectives and experimental conditions, and generally only after careful consideration on the part of the experimenter.

In classification experiments, where individuals are classified into exactly one of two groups, using composite sampling is typically cost-effective when there is a low probability of detecting analyte (models 1 and 2) or when determining analyte above the criterion limit (models 3 and 4). The optimal number, n, of samples to composite for models 1 and 2 is shown in Table I. Finding the optimal value of n for models 3 and 4 is more difficult because the solution must account for the exact statistical distribution of analyte concentrations. The experimenter must be careful that n does not exceed the ratio of the criterion level to the method detection limit.

In experiments where the objective is to estimate the mean concentration of analyte over a region or throughout a population, composite sampling will reduce the intersample variance. The result will be a more cost-effective estimate of the mean concentration. This procedure is especially useful when the cost of analysis substantially exceeds the cost of sampling. This procedure should not be used when an analytical result is needed for each sample.

Occasionally, achieving desired analytical performance, such as detection limits, with individual samples will be impossible because of sample size limitations. Using composite sampling to accumulate sufficient material may be beneficial to achieve the desired analytical performance. The experimenter must be cautioned that concentration estimates will not be available for each individual. Conclusions must be drawn for groups of individuals.

In random surveys, such as one to estimate the proportion of a population having a disease, composite sampling may be used to ensure anonymity of the sample donors. This use could help to achieve randomness of the sample when individual concerns of identification could cause a biased sample.

In any composite sampling plan, interactions among analytes must be carefully considered. Care must be taken to ensure that analytes or organisms from different samples will not be mutually destructive nor will create analytical interferences. If such problems are suspected, then the relevant procedures must be modified to eliminate them or render them irrelevant. Otherwise, composite sampling may create a material that is not representative of any of the original samples.

Abbreviations and Symbols

σ^2_C	variance of measurements of composite samples
σ^2_E	random error variance component
σ^2_H	heterogeneity variance component
σ^2_M	variance of the mean of k analyses
c	criterion level for a composite sample
c'	criterion level for a single sample
c'_i	criterion level for the ith analyte
D	method detection limit
D_i	method detection limit of the ith analyte
$E(M)$	expected value (mean) of the variable M
k	number of samples to be analyzed
M	number of analyses necessary to classify n composited samples
m	number of analytes measured in each analysis
N	total number of samples
n	number of samples composited
P	observed proportion of composites that test positive
p	probability of presence of analyte in one sample
Q	probability of simultaneous absence of all analytes
q	probability of absence of analyte in one sample
q_i	probability of absence of the ith analyte
RCF	relative cost factor, which is the expected cost of a composite sampling plan expressed as a proportion of the cost of a conventional sampling plan
X_i	concentration of analyte in the ith sample
y	concentration of analyte in the composite sample

Acknowledgment

The research described in this chapter was funded wholly or in part by the U.S. Environmental Protection Agency through contract number 68-03-3249 to Lockheed Engineering and Management Services Company, Inc. This chapter has not been subjected to Agency review and does not necessarily reflect the views of the Agency; therefore, no official endorsement should be inferred.

References

1. Cameron, J. M. *Biometrics* **1951**, 7, 83–96.
2. Watson, G. S. *Technometrics* **1961**, 3, 371–388.
3. Hwang, F. K. *J. Am. Stat. Assoc.* **1972**, 67, 605–608.
4. Cassie, R. M. In *Secondary Productivity of Fresh Water*; Edmondson, W. T.; Winberg, G. G., Eds.; International Biological Programme Handbook No. 17; Blackwell Scientific Publications: Oxford, England, 1971; pp 174–209.
5. Hrbacek, J. Ibid., pp 14–16.
6. Heyman, U.; Eckbohm, G.; Blomquist, P.; Grundstrom, R. *Water Res.* **1983**, *16*, 1367–1370.
7. Ryan, J. J.; Pilon, J. C.; Leduc, R. *J. Assoc. Off. Anal. Chem.* **1982**, 65, 904–908.
8. Connolly, J.; O'Connor, F. *Ir. J. Agric. Res.* **1982**, *20*, 35–51.
9. Bruchet, A.; Cognet, L.; Mallevialle, J. *Water Res.* **1984**, *18*, 1401–1409.
10. Baker, A. S.; Kuo, S.; Chae, Y. M. *Soil Sci. Soc. Am. J.* **1981**, 45, 828–830.
11. Sobel, M.; Oroll, P. A. *Bell Syst. Tech. J.* **1959**, *38*, 1179–1252.
12. Garrett, R. G.; Sinding-Larsen, R. *J. Geochem. Explor.* **1984**, *21*, 421–435.
13. Dorfman, R. *Ann. Math. Stat.* **1943**, *14*, 436–440.
14. Brown, G. H.; Fisher, N. I. *Technometrics* **1972**, *14*, 663–668.
15. Rohde, C. A. *Biometrics* **1976**, 32, 273–282.
16. Li, C. H. *J. Am. Stat. Assoc.* **1962**, 57, 455–477.
17. Mack, G. A.; Robinson, P. E. In *Environmental Applications of Chemometrics*; Breen, J. J.; Robinson, P. E., Eds.; ACS Symposium Series 292; American Chemical Society: Washington, DC, 1985; pp 174–183.

Sampling Solids and Hazardous Wastes

Chapter 34

Volatile Organic Compounds in Soil

Accurate and Representative Analysis

James S. Smith, Leslie Eng, Joseph Comeau, Candice Rose, Robert M. Schulte, Michael J. Barcelona, Kris Klopp, Mary J. Pilgrim, Marty Minnich, Stan Feenstra, Michael J. Urban, Michael B. Moore, Michael P. Maskarinec, Robert Siegrist, Jerry Parr, and Roger E. Claff

In 1989 a more accurate method for sampling and analysis of volatile organic compounds (VOCs) in solid matrices was introduced to the environmental community. This chapter is a comprehensive update on the use of methanol as a field preservative for VOCs in solids. This review will include but will not be limited to the following: (1) field and laboratory comparison with standard U.S. Environmental Protection Agency (EPA) VOC sampling procedures, (2) determination of dense, nonaqueous-phase liquids (DNAPLs) in solid matrices, (3) determination of VOCs in concrete, and (4) obtaining a detection limit of 5 μg/kg or less by several methods. The VOC data from soils preserved in the field with methanol are orders of magnitude more accurate, and therefore more representative of actual field conditions, than are data obtained using currently approved methods. These results strongly recommend that methanol preservation of VOCs in soil samples be used whenever accurate results are desired.

Some environmental scientists are genuinely concerned that past and current methodologies used for sampling of soil for the determination of volatile organic compounds (VOCs) lead to inaccurate analytical results. The bias of the analytical results can be several orders of magnitude low, depending on the volatility of the chemical or chemicals of concern and the delay between sampling and analysis. For example, losses of VOCs during sampling of a typical perceived hazardous waste site using current, accepted sample collection technology cause analytical results that may be biased low and therefore not representative of actual site conditions. As a result, the inaccurate results may be used to select a "no remediation" alternative for the site. This decision may be totally inappropriate given the actual concentrations of VOCs present on the site.

Another major problem with current VOC soil sampling and analytical methodologies is that they result in poor efficiency of remediation efforts. Because of inaccurate data, remediation efforts have removed VOCs in excess of the *calculated* masses in the soil, yet the remediation has not removed the VOCs *actually present,* because the original determinations were grossly biased. This results in residual levels of VOCs in high enough concentrations to still pose considerable risk at sites that have been reclassified as "remediated".

Furthermore, there is great concern about the detection of dense, nonaqueous-phase liquids (DNAPLs) at some sites for source identification and the resulting remediation technology necessary to deal with them. DNAPL samples collected unpreserved can lose volatile components critical to accurate fingerprinting of the substance.

Despite all of these extremely important reasons and our desire to protect our environment at the most reasonable cost, sampling methodologies currently in use do not address the low-sampling bias. The methodology discussed in this chapter does address this bias, and it has been available for years. The authors suggest that this superior methodology be universally approved and used.

Historical Background

The U.S. Environmental Protection Agency (EPA) has issued a symposium summary (*1*) of the National Symposium on Measuring and Interpreting VOCs in Soils: State of the Art and Research Needs. The symposium was held in Las Vegas, NV, in January 1993. In that report, the EPA summarized all of the variables that affect or bias VOC results. However, the summary stated that

> decision-making needs and information adequacy must be considered within the overall perspective of the characterization and assessment process. It can be argued that due to a high degree of uncertainty present in exposure scenarios and health effects (i.e., orders of magnitude) coupled with potentially great spatial variability, *striving for accurate and precise quantitation of VOCs in discrete soil samples is unfounded and unnecessary* [emphasis added].

This statement is contrary to EPA drinking water maximum contaminant levels (MCLs) and cleanup levels specified in Superfund Records of Decision (RODs), as well as the Clean Water Act. It is clear that accurate and precise measurements of VOCs are extremely important for good environmental stewardship.

As a particularly telling example, in 1985 the U.S. Department of Justice filed suit against Raymark et al. (*United States of America v. Raymark et al.*, U.S. District Court, Eastern District, PA, Civil Action No. 85–3073). The United States relied on results from soil samples preserved with methanol in the field as the definitive analytical data for this lawsuit. The VOC of concern was trichloroethene (TCE). The question to be decided in this case was whether Raymark et al. were responsible for contaminated soil and groundwater. The Raymark site was located in an area where it had already been determined that another company had previously contaminated the groundwater with TCE, and the plume of contamination ran beneath the Raymark facility. In other words, was the TCE contamination of the soil at the Raymark facility present because of contamination by Raymark, or had TCE volatilized from the groundwater plume and collected in the soil gas above the groundwater table?

Because the solubility of TCE in water is 1100 mg/L, the government had to prove that there was evidence of a DNAPL in the soil at the Raymark site in order to show conclusively that TCE releases originated at the Raymark facility. The incriminating testimony at trial related to a soil sample that was collected using a split spoon from underneath the building housing a TCE vapor degreaser. This soil sample was split into three subsamples that were prepared by vertically cutting the split spoon sample of soil. The first subsample (no. 1 in Table I) was collected in a 4-oz jar by the contractor for EPA Region III. The second subsample (no. 2 in Table I) was collected by the environmental engineering firm retained by Raymark and was packed into a 40-mL volatile organic analysis (VOA) vial with as little headspace as possible and capped with a septum screw cap. The final subsample (no. 3 in Table I) was collected by the environmental consultant for the U.S. Department of Justice. This third subsample was the most disturbed; it was pushed down the split spoon into a jar containing 250 mL of methanol. The third subsample was collected approximately 3–5 min after the opening of the split spoon sampler.

Analytical results for the three subsamples are provided in Table I.

The expert retained on behalf of the U.S. Department of Justice testified during trial that the analytical result achieved for the methanol-preserved sample was both credible and accurate. Furthermore, it was this expert's opinion that the TCE results for the other two samples (nos. 1 and 2 in Table I) were biased low because of TCE losses by volatilization caused by sample handling in the field and in the laboratory.

Table I. Concentration of TCE in Soil

Subsample No.	*Sampler*	*TCE Concentration (mg/kg)*
1	EPA Region III	1.0
2	Contract laboratory	0.06
3	Methanol preserved	3300

NOTE: TCE is trichloroethene.

A witness for the United States from the EPA Region III laboratory concurred with the evaluation of the data performed by the expert for the U.S. Department of Justice. The Region III laboratory witness stated that the methanol-preserved sample was "immediately taken into the methanol and therefore stabilized as far as future or additional losses of volatile organics." Therefore, as early as 1987, expert scientists from both the EPA and other agencies concluded that more accurate results for VOCs in soil samples can be obtained by disturbing the sample as little as possible during collection and preserving it in the field by immersion into methanol.

Given the opinions expressed by expert testimony for the United States in *United States of America v. Raymark et al.*, the use of methanol in the field to preserve VOCs in soil samples by Royal Nadeau of the EPA Environmental Response Team in Edison, NJ (2), and the comparisons of methanol-preserved versus unpreserved analytical results for many samples prepared and analyzed under controlled conditions, both published and unpublished, it is expected that sampling protocols would have been changed to reflect the improved technology.

Urban et al. (3) presented such data at the 5th Annual Waste Testing and Quality Assurance Symposium (Washington, DC, July 1989). That study and the sampling methodology were published in the proceedings of that meeting as well as in an ASTM book (4). Table II shows the effect of the sampling protocols on the accuracy of VOC results. The results in Table II that have a superscript *a* indicate that TCE was detected but was below the limit of quantitation by the laboratory in the affected samples; the results that have a superscript *b* indicate that TCE was not detected. The associated numerical value is the detection limit (i.e., TCE was not detected. The indicated value is the detection limit).

Table II. Comparison of Method Results

Sample ID	*Field Methanol Immersion/Extraction*	*Laboratory Methanol Extraction*	*Purge and Trap*
A (2.0–4.0′)	124	2.7	—
A (6.0–7.8′)	79.9	5.5	—
B (1.95–4.0′)	212	47.2	—
D (1.0–3.0′)	5.31	0.27	—
E (3.0–5.0′)	0.81	0.32	—
I (1.5–3.5′)	0.66	—	0.015[a]
I (3.5–5.5′)	0.28	—	0.035
I (5.5–7.0′)	0.096[a]	—	0.065[a]
M (2.0–4.0′)	0.34	—	0.015[a]
H (2.0–3.5′)	0.11[a]	—	0.025[b]

NOTE: Values are in milligrams per kilogram. Sample IDs are shown in feet (e.g., sample A (2.0–4.0′) was dug 2–4 ft down from soil surface level).

[a]TCE was detected but was below the limit of quantitation by the laboratory in the affected samples.

[b]TCE was not detected above this numerical value.

SOURCE: Data are from references 3 and 4.

The methanol preservation method was also the method of choice for Siegrist and Jenssen (5), who demonstrated that low detection limits could be achieved using this methodology.

However, the current EPA-approved sampling protocol for VOCs in soil (6), which was approved in 1986 and was still in use in 1994, specifies that soil be placed into a 40-mL VOA vial, packed with limited headspace, and capped with a septum screw cap. Methanol field preservation for VOC analysis of soils has not yet been incorporated into approved EPA sampling methodology.

After all of the compelling scientific evidence, why is the approved method for sampling VOCs in soil unaltered? Barry Lesnik, the national organic program manager for the Resource Conservation and Recovery Act (RCRA) Program with the EPA's Office of Solid Waste (OSW), Washington, DC, and E. William Loy, a retired EPA chemist, state (7) that the methanol preservation methodology is not considered a viable option for VOC samples because it raises five major issues in the following categories:

- *Technical*—the availability of better or more appropriate methodology;
- *Policy*—EPA's commitment to reduction of the use of hazardous solvents in environmental methods;
- *Health*—the potential of exposure of sampling or laboratory personnel to significant quantities of methanol vapor;
- *Regulatory*—the potential to cause waste disposal problems; and
- *Transportation*—methanol-preserved samples may not be shipped by all shippers.

With regard to the technical issue, the current approach recommended by OSW (8) for preparation of solid samples for VOC analysis is a closed-system, purge-and-trap method (Method 5035) (9), which EPA believes has superseded the methanol "preservation" technique by better technologies offering its advantages but without its inherent drawbacks.

In response to the five concerns of Lesnik and Loy with respect to field preservation of soil samples with methanol, the authors offer the following rebuttal.

- *Technical*—Based on the experience of the authors, there has not been a "better or more appropriate" sampling methodology available at this time. Field GC methods are good, but they cannot be applied to soil DNAPL detection because the utilization of a headspace sample over a soil–water mixture has an upper bound for quantification. Also, water-soluble substances give a poor response.
- *Policy*—The authors do not disagree with the reduction in the use of solvents in environmental analytical methods if and only if accuracy

and precision are not sacrificed. This criterion apparently has been the EPA policy all along. New methods have been approved only after rigorous testing in order that the data can meet criteria for accuracy and precision.

- *Health*—Methanol has been and still is used in the environmental analytical laboratory as the EPA-recommended solvent for VOC standards and the medium level methodology for soil preparation for VOC analysis in both Contract Laboratory Program (CLP) protocols (*10*) and SW–846 methods (*6*). Also, since the use of methanol in the field by the EPA Environmental Response Team (*2*), numerous sampling groups represented by the authors of this chapter have used this sampling methodology without any reported or known toxic exposures to field personnel.
- *Regulatory*—It is not an imposition of unnecessary regulatory requirements for the EPA to approve a sampling protocol that obtained results far superior in accuracy to the presently approved methodology. The EPA has never hesitated to revise existing regulations or implement new regulations when better technology becomes available. It is unclear why use of the field methanol preservation methodology should be an exception.
- *Transportation*—Methanol-preserved soil samples can be shipped from most locations if the samples are properly packed and labeled according to Department of Transportation regulations. This procedure involves some inconvenience to field crews that is similar to the shipment of dioxin samples. The cost is minimal compared to the cost of sample collection and analysis.

With regard to Method 5035, Lesnik wrote in the February/March 1994 issue of *Environmental Lab* (*8*) that actual recoveries for VOCs ranged from 50 to 60%. The sampling procedure remained the same as that currently approved by EPA. The change is that the sample aliquot is not subsampled in the laboratory. Lesnik (*8*) writes the following about the sampling procedure:

> The sample is collected in a fritted 40-mL VOA vial fitted with a Teflon septum-seal cap and a magnetic stirring bar. (A core-sampling device with a diameter slightly less than that of the neck of the VOA vial is commercially available.) Vials may be taken to the field preweighed or may be weighed on site with a portable balance or scale. Samples should be screened to determine the optimum analytical sample size. The purging medium, including preservative, if desired, may be added at the time of sample collection. Immediately prior to analysis, water (if not added at time of collection), surrogate standards, and internal standards are added to the sample vial without breaking the hermetic seal.

The slurry is then preheated to 40 °C, then purged by passing an inert gas through the bottom of the vial while agitating the slurry with the magnetic stirring bar.

Although Method 5035 represents some improvement over other, currently approved methods, it still presents us with significant deficiencies:

- The sample size is small and thus may not be representative of field conditions.
- The use of water as the extractant will not permit determination of whether a DNAPL is present or absent.
- Lack of preservation (since preservation is optional and unspecified) can cause major bias when analyzing volatile hydrocarbons.

With regard to the last point, the underground storage tank (UST) programs and Jackson's data (*11*) can be considered. The EPA holding time for soil samples collected for VOC analysis is 14 days from the date of sample collection. VOCs include, inter alia, benzene, toluene, ethylbenzene, and xylenes (known collectively as "BTEX"). Jackson's findings demonstrate that orders of magnitude losses can occur over a 14-day time period when no field methanol preservation is employed, as is shown in Table III.

Other authors (*11–13*) have made similar observations. In particular, Hewitt (*12*) reported that no losses in the concentrations of benzene, toluene, trichloroethene, and *trans*-1, 2-dichloroethene occurred over a 98-day holding period when samples preserved with methanol in the field were stored at 22 °C. Yet, in certain soil samples stored in VOA vials, losses of benzene and toluene occurred at a 4 °C storage temperature within a 14-day holding time without preservation, as shown in Table IV.

Experimental Methods

Is the field methanol preservation methodology worth pursuing? The advantages have been documented.

Table III. Summary of Average BTEX and TPH Levels Measured over a 14-Day Period

Day	*Benzene (µg/kg)*	*Toluene (µg/kg)*	*Ethylbenzene (µg/kg)*	*Xylenes (µg/kg)*	*TPH (mg/kg)*
1	19,000	120,000	59,000	260,000	1,500
7	990	23,000	31,000	150,000	730
14	1,000	730	5,300	32,000	250

NOTE: BTEX refers collectively to benzene, toluene, ethylbenzene, and xylenes. TPH means total petroleum hydrocarbons.

Table IV. Mean Analyte Concentrations for Soil Subsamples Stored in VOA Vials with Purge-and-Trap Adaptor Cap at 4 °C

Compound	*Day 0*	*Day 4*	*Day 7*	*Day 14*
trans-1,2-Dichloroethene	9.52 ± 0.46	8.96 ± 0.23	9.59 ± 0.32	9.2 ± 1.2
Benzene	12.7 ± 0.35	12.7 ± 0.20	7.41 ± 1.8	ND
Trichloroethene	13.8 ± 0.70	12.6 ± 0.15	12.4 ± 0.67	11.9 ± 1.5
Toluene	33.1 ± 0.47	31.9 ± 0.97	26.5 ± 3.8	10.0 ± 2.5

NOTE: Values are in nanograms per gram. VOA means volatile organic analysis. ND means not detected.

SOURCE: Reproduced with permission from reference 12. Copyright 1994 American Society for Testing and Materials.

- More accurate analytical results are obtained using methanol preservation than any other soil sampling methodology.
- Methanol-preserved samples are more representative of true site conditions. In addition, the sample size can vary from 5 g to 1 lb or more depending on the field circumstances because sample size is not limited to a 40-mL vial. In most cases, a larger sample is more representative of field conditions.
- This method is the only documented method that provides a preservative that prevents biodegradation and lengthens holding times without adding bias to the analytical results.
- This method is the only documented method that can be successfully adapted to concrete borings, rock borings, and large sample sizes.
- This method is the only documented method that can provide proof of whether a DNAPL is present on the site.

Robert Schulte of the Department of Natural Resources and Environmental Control (DNREC) of the state of Delaware states emphatically:

> Accuracy is not the only purpose of the methanol field preservation method; the method eliminates missed analytical holding times, curtails qualification of data from low internal standard areas, increases the purge efficiency, adapts to more aggressive sampling and analysis plans [that] may put a strain on laboratory operations, and increases the solubility of volatiles, especially DNAPLs.

The state of Delaware DNREC Superfund Branch (*14*) mandates the use of the methanol preservation sampling method. Using the methanol preservation method for soil samples, the state of Delaware investigated a site that had previously been investigated by EPA and found by EPA to be uncontaminated by VOCs in the soil.

All of the soil samples collected for EPA were analyzed by the EPA CLP protocols, and no VOCs were detected above 10 μg/kg. Using field methanol preservation of soil samples, the Delaware DNREC found 10–1000 μg/kg of toluene, xylene, trichloroethene, *cis*-1, 2-dichloroethene, and ethylbenzene. Groundwater monitoring wells were subsequently installed. As a result, vinyl chloride, chloroethane, *cis*-1, 2-dichloroethene, and trichloroethene were detected in the groundwater above the MCLs for drinking water.

Karl Johnson of the Johnson Company (Montpelier, VT) located a concrete manhole that had been removed from a dry cleaning facility. This concrete structure had been outdoors exposed to the VT weather for 2–3 years on the back lot of the property. A concrete core was taken out of the bottom of the manhole, broken with a hammer, and placed into methanol. The chunks of concrete selected were as large as would fit into a 500-mL amber glass bottle containing 250 mL of methanol and CLP medium-level surrogates with a Teflon-lined lid. After 48 h, the concrete sample was analyzed; tetrachloroethene (PCE) was detected at 500 μg/kg. Analyzed after 1 week, the concrete sample result was 5000 μg/kg. Another analysis was conducted 1 year later, and the PCE result was still 5000 μg/kg. Because concrete is porous, one may use this technique to find the source of a chlorinated solvent release from sumps and floors made of concrete.

The Wells G & H Superfund site is located in the City of Woburn, MA. It has been on the National Priority List (NPL) since 1982. Beginning in 1992, the practice of field preservation of samples of unconsolidated deposits in methanol was compared to the standard EPA sample protocol on collocated samples. The methanol-preserved samples provided higher analytical concentrations of VOCs in all but one sample. The magnitude of difference between each comparative pair of samples was highly variable, but concentrations as high as 2000 μg/kg PCE were detected in methanol-preserved soil samples for which the equivalent non-methanol-preserved sample yielded a “nondetect” result at a reported detection limit of 5 μg/kg. These comparative results are given in Table V (*15*).

One of the major objections to the field methanol preservation methodology is that the resulting analytical detection limits are considerably higher than the reported detection limits for the approved EPA methods for VOCs in soil. The methanol method with 250 g of soil in 250 mL of methanol and modern mass spectrometers can give a detection limit near 50 μg/kg. The EPA CLP target VOC analytes are reported with a method detection limit (MDL) of 5 μg/kg.

Joe Comeau of Inchcape/Aquatec Testing Services in Burlington, VT, has evaluated MDLs for methanol extracts of VOCs eluting after bromochloromethane on a GC–MS using a packed GC column. Vacuum diversion is used prior to the elution of the internal standard, bromochloromethane. The MDLs for the EPA target compound list (TCL) ranged from 4 to 8 μg/kg when 2.5 mL of the methanol extract was added to 25 mL of pure water and the VOCs were concentrated by purge-and-trap methodology. Other techniques, such as direct single- or multiple-ion monitoring MS detection or injection of methanol extract directly into a GC, can decrease the detection limit for specific analytes. Also, a hexadecane extraction of the

Table V. Comparison of Results of Methanol-Preserved Versus Unpreserved Samples of Four Analytes

	Tetrachloroethene		*Trichloroethene*		*1,1,1-Trichloroethane*		*cis-1,2-Dichloroethene*	
Sample ID	*Unpreserved*	*Methanol Preserved*	*Unpreserved*	*Methanol Preserved*	*Unpreserved*	*Methanol Preserved*	*Unpreserved*	*Methanol Preserved*
B-1 (0–2.5′)	17	330	2.0[a]	43[a]	2.0[a]	43[a]	2.0[a]	43[a]
B-1 (4.5–6.5′)	2.1[a]	120	2.1[a]	45[a]	2.1[a]	45[a]	2.1[a]	45[a]
B-2 (0.05–2.5′)	120	120	4.0[a]	41[a]	4.0[a]	41[a]	4.0[a]	41[a]
B-2 (3.0–4.5′)	2.0[a]	44[a]	2.0[a]	44[a]	2.0[a]	44[a]	2.0[a]	44[a]
B-3 (0.05–2.5′)	64	1400	3.8[a]	89[a]	3.8[a]	89[a]	5.4	290
B-3 (2.5–4.5′)	2.0[a]	42[a]	2.0[a]	42[a]	2.0[a]	42[a]	2.0[a]	42[a]
B-4 (0.05–2.5′)	2600	59,000	86[a]	1500	86[a]	4600	86[a]	360[a]
B-5 (0.05–2.5′)	5.4	82	2.2[a]	47[a]	2.2[a]	47[a]	2.2[a]	47[a]
B-5 (4.5–6.5′)	20	580	3.8[a]	47[a]	3.8[a]	47[a]	3.8[a]	47[a]
B-6 (0.05–2.5′)	2.0[a]	51	2.0[a]	46[a]	2.0[a]	46[a]	2.0[a]	46[a]
B-7 (0.05–2.5′)	37	1700	3.0[a]	280[a]	8.0	12,000	3.0[a]	280[a]
B-7 (2.5–3.7′)	1.9[a]	48[a]	1.9[a]	48[a]	1.9[a]	48[a]	1.9[a]	48[a]

NOTE: Values are in micrograms per kilogram. Sample IDs refer to distance (in ft) below soil surface level at which sample was dug.

[a]These values were not detected; the number indicates the detection limit.

SOURCE: Data are from reference 15.

methanol could be used to concentrate the sample before analysis and improve detection limits even further. However, Schulte of Delaware DNREC points out that soil cleanup criteria are not below 0.5 mg/kg (*16*), and the methanol-preserved samples meet that detection level easily.

Conclusions

The authors of this chapter are or have been actively involved with the solutions to the problems involved with the sampling of VOCs in soil. Many have at one time or another used the methanol preservation technology. The states of Wisconsin and Alaska are using methanol preservation with their respective UST programs. The state of Delaware makes the methanol preservation of soil samples mandatory in its Superfund program. As we have shown, certain regions within EPA are accepting data from field methanol-preserved soil samples.

In addition, the large amount of work developed by the American Petroleum Institute (API) on this topic (*17*) needs to be carefully considered. API recommends the use of field methanol-preserved soil samples for petroleum hydrocarbons.

Simply stated, if we in the United States have determined that VOCs are to be remediated when they are released to the soil, then we must insist that accurate measurements be a primary goal of any environmental investigation. Currently, the field methanol preservation of soil samples is the most accurate, practical, and precise of the methods available. Because we cannot afford to put technology to work to improve our environment on the basis of spurious data, the use of field methanol preservation for the sampling of VOCs in soil must be considered in any quality assurance project plan. The claim that accurate measurements of VOCs in soil are unimportant is in direct conflict with available toxicity data for these compounds. None of the purported disadvantages of methanol described in this chapter has been observed in the field or laboratory. Methanol field preservation cannot be rejected based on scientific merit, as illustrated here. Based on the body of data collected, there appears to be no technical reason for continued EPA rejection of this proven methodology.

References

1. *Measuring and Interpreting VOCs in Soils: State of the Art and Research Needs;* U.S. Environmental Protection Agency: Washington, DC, 1993.
2. Urban, M. J., Envirotech Research, Edison, NJ, private communication.
3. Urban, M. J.; Smith, J. S.; Schultz, E. K.; Dickinson, R. K. *Proceedings of the 5th Annual Waste Testing and Quality Assurance Symposium;* U.S. Environmental Protection Agency: Washington, DC, 1989; pp II87–II101.
4. Urban, M. J.; Smith, J. S.; Schultz, E. K.; Dickinson, R. K. *Waste Testing and Quality Assurance: Third Volume;* Tatsch, C. E., Ed.; American Society for Testing and Materials: Philadelphia, PA, 1991; ASTM STP 1075.

5. Siegrist, R. L.; Jenssen, P. D. *Environ. Sci. Technol.* **1990**, *24*, 1387–1392.
6. *Test Methods for Evaluating Solid Waste, Physical/Chemical Methods (SW–846)*, 3rd ed.; U.S. Environmental Protection Agency: Washington, DC, 1986; pp 4-1 and 4-2.
7. Lesnik, B.; Loy, E. W. *Environ. Lab.* **1995**, *August/September*, 33–38.
8. Lesnik, B. *Environ. Lab.* **1994**, *February/March*, 10. Extract reprinted with permission from Douglas Publications, Inc.
9. *Test Methods for Evaluating Solid Waste, Physical/Chemical Methods (SW–846)*, 3rd ed.; Proposed Update 111, Method 5035, U.S. Environmental Protection Agency: Washington, DC, 1995; Revision 0.
10. *Statement of Work for Organic Analysis, Multi-Media, Multi-Concentration, OLM03.0;* Revision OLM03.1; U.S. Environmental Protection Agency, Contract Laboratory Program: Washington, DC, 1994.
11. Jackson, J.; Thomey, N.; Dietlein, L. F. *Proceedings of the 5th National Outdoor Action Conference on Aquifer Restoration, Well Water Monitoring, and Geophysical Methods;* National Well Water Association: Dublin, OH, 1991; pp 567–576.
12. Hewitt, A. D. *Proceedings of Volatile Organic Compounds (VOCs) in the Environment;* American Society for Testing and Materials: Philadelphia, PA, 1994.
13. King, P. In *Measuring and Interpreting VOCs in Soils: State of the Art and Research Needs;* U.S. Environmental Protection Agency: Washington, DC, 1993; Tab 19.
14. *Standard Operating Procedures for Chemical Analytical Programs Under the Hazardous Substance Cleanup Act;* Delaware Department of Natural Resources and Environmental Control, Superfund Branch: Dover, DE, 1992; Revised 1994, 1995, 1996.
15. *Pre-Design Work Plan, Remedial Design and Remedial Action for the Unconsolidated Deposits beneath the UniFirst Property;* Wells G & H Site: Woburn, MA, 1992; Vol. I.
16. *Screening Value Guidances;* Delaware Department of Natural Resource and Environmental Control: Dover, DE, 1996.
17. *Sampling and Analysis of Gasoline Range Organics in Soil, Health and Environmental Sciences;* API Publication No. 4516; American Petroleum Institute: Washington, DC, 1991.

Chapter 35

Sampling and Analysis for Mixed Chemical and Radiological Parameters

Matt Stinchfield

Sampling and analysis of environmental samples that may possess both chemical and radiological contamination offers some of the greatest challenges possible in the field of environmental sampling. Mixed chemical and radiological samples are subject to the typical problems of sampling design, matrix interference, sample heterogeneity, quality control, and method selection, as well as additional issues related to safety and health, sample shipping and handling, methods selection, and laboratory notification, licensing, and selection. General definitions and sources of official guidance are included in this chapter, but the emphasis is on practical decision making backed up by sound scientific practice.

THE HISTORICAL SEPARATION between the fields of hazardous substances and radioactive materials has rarely been bridged by institutions or individuals. Each discipline possesses experts with specific knowledge, experience, instrumentation, and equipment necessary to extract, produce, monitor, and clean up those substances with which they are most experienced. The development of regulations, personnel training, and analytical techniques for chemical substances has evolved almost independently from those of the radiochemical industry.

Perhaps this dichotomous evolution arose because of the national security aspects of nuclear research and development versus the commercial and public aspects of the chemical industries. Then again, this separation could be due to fundamentally different modes of action, environmental fate, epidemiology, and measurement techniques for chemical substances versus radioactive materials. Upon first inspection, chemical pollutants seem more common and widespread than radioactive materials. To someone with the traditional environmental sampling and analysis perspective, radiochemistry may seem an exotic subset of these activities.

Yet, everything in our world is composed, to some degree, of both hazardous chemicals and radioactive materials. Each day we are exposed to atmospheric pollutants from automobiles (chemical contaminants) and naturally occurring radiation from radon, potassium in food, radium in drinking water, and cosmic rays in the atmosphere (radioactive contaminants) (*1*). Because all matter is both chemical and radioactive, when is any environmental matrix considered a mixed chemical and radiological substance? To those entities possessing waste materials regulated concurrently by the U.S. Environmental Protection Agency (EPA) and the Nuclear Regulatory Commission (NRC), the level of concern is usually focused on the point at which a *mixed waste* is formed (*2, 3*). But the mixed substances sampled for in the environment are not limited to mixed wastes. It is a very broad category that spans from common drinking water samples to fish near a defense plant to uranium mill by-products. In this chapter the term "nonradioactive" will sometimes be used to refer to materials for which we are only concerned with traditional chemical analysis. This term is used despite the fact that all matter is radioactive to some extent.

If a mixed chemical and radiological (MCR) sample is defined as one for which both chemical and radiological parameters are measured, then we will have carved out the broadest possible definition. A slightly less broad definition would be one in which the samples are anticipated to comprise elevated levels of both chemical and radiological contaminants. However, we cannot know with certainty the concentration of all of the analytes of interest in our samples. Additionally, in certain cases, there may be no statutory level of concern that one can refer to; site-specific determinations of the status of elevated levels must be made.

MCR samples may be obtained from any environmental medium. These include air and gases, atmospheric particulates, soil, sediments, groundwater, surface water, solvents, process materials, drummed waste, waste piles, and biota. Even recycled steel, artillery ranges, and human bone tissue may require sampling.

Even if an MCR sample analyzed in a radiochemical laboratory is found to be at background levels, the fact remains that the collection, submission, analysis, and interpretation of that sample requires considerations that are above and beyond those required for simple chemical analysis. It is those differences and supplemental considerations that this chapter focuses on. The typical reader of this chapter is expected to be principally versed in chemical sampling and analysis, with radiochemical sampling and analysis as an added interest. There are, however, a growing number of sampling specialists with a specific radiochemistry background, as both the U.S. Department of Energy (DOE) and the U.S. Department of Defense (DOD) have begun the enormous task of characterizing the mixed waste sites of the nuclear defense complex.

Those familiar with the sampling and analysis of MCR substances will appreciate both the depth of the subject and the scarcity of standardized approaches or methods associated with mixed substances. For this reason, this chapter is presented as an overview of what to consider when planning for an MCR sampling event, more than it is geared toward providing specific methodologies for a certain kind of sampling.

Historical Perspective

As previously mentioned, the chemical and radiological industries have historically evolved along separate, but parallel, courses. The rudimentary differences between the two industries include regulations and enforcement, personnel training and experience, analytical methods, environmental fate of substances, safety and health, storage and disposal, public perception, and government posture.

The following discussion is not intended to belabor the differences inherent in each industry; rather it is for the purpose of historical perspective. It is probably safe to say that the two domains are converging in the late 20th century. This convergence is largely fueled by increased activity in the characterization and restoration of nuclear weapons plants, abandoned uranium mills, and petroleum refinery wastes. The success of any MCR substance sampling event will depend largely not on the differences between the two industries, but rather on the integration of experiences from both.

More frequently than ever, environmental professionals are being compelled to look closely at the dissimilar aspects of chemical and radiological sampling and analysis. This trend must extend to beyond the boundaries of the United States, as the political openness of Eastern Europe, Australia, and some Asian countries is revealing a growing list of mixed chemical and radiological predicaments.

As a point of departure for integrated sample design activities, let us examine three areas where profound differences in the two industries exist: regulations, personnel, and methods.

Regulatory Framework

In the United States there has been considerable regulation of both chemical and radioactive substances, but there is very little overlap. (Great Britain is an example of a country that has combined regulations for chemical and radioactive substances.) A discussion of regulations is pertinent, as common questions posed of site investigators often rely on comparison of site data to regulatory criteria. Common questions include

- Is this material a waste?
- Does this sample comprise elevated levels of some substance?
- Does this material pose unacceptable risk to human health or the environment?

Most often we obtain the standard for evaluating results from federal regulations. Beyond the regulations there is a complex set of guidance documents, regulatory interpretations, and legal precedents that are often useful but not always easily accessible. Within this context should lie an undercarriage of reproducible and defensible scientific practices.

The principal regulations applying to hazardous chemicals and radioactive materials in the environment are tabulated in Table I. Each listing shows the key term that is defined by the regulation and the citation for the regulations.

These regulations have only marginal commonality. However, three notable overlaps are worth mentioning. The hazardous materials standards, under the U.S. Department of Transportation, which apply to the packaging, labeling, and shipping of hazardous materials, deal with both chemical and radiological hazards. The principal concern here is what type of risks any substance poses in transportation (4). The federal drinking water standards include maximum contaminant levels (MCLs) for both chemical and radiological constituents. Drinking water compliance samples under the Safe Drinking Water Act (SDWA) must be tested for gross alpha and beta radiation and radium (5). Finally, there is some mention of mixed wastes in the Land Disposal Restrictions (LDR) component of the EPA's hazardous waste code [Resource Conservation and Recovery Act (RCRA)]. But for the most part, RCRA specifically exempts radioactive waste (5).

Personnel Training and Experience

The chemical and radiological industries have staffed themselves in analogous but dissimilar ways. They are analogous in that they each have scientists, engineers, and safety and health professionals, but different in that each of these personnel tends to come from different educational and experiential backgrounds.

Table I. Principal Regulations Applied to Chemical and Radioactive Substances

Substance	*Reference*
Chemical substances	
Hazardous waste	40 CFR § 260–268 (5)
Hazardous substance	40 CFR § 300–302
Toxic substance	40 CFR § 761–766
Toxic and hazardous substance	29 CFR § 1910.1200 (6)
Hazardous material	49 CFR § 170–181 (4)
Radioactive substances	
High-level waste; transuranic waste	10 CFR § 60 (7)
	40 CFR § 191
	DOE 5820.2 A (8)
Low-level waste; naturally occurring radioactivity	10 CFR § 61
	40 CFR § 193
	DOE 5820.2 A
Hazardous material	49 CFR § 170–181

NOTE: CFR is *Code of Federal Regulations;* DOE is U.S. Department of Energy.

The scientists of the chemical sampling and analysis industry are most often of a natural science background such as chemistry, geology, hydrology, biology, or ecology: subjects in which the diversity, order, and disorder of natural systems is taught. These individuals are typically well-suited to conduct environmental sampling and field work. Scientists in the radioactive materials sampling and analysis process tend to come more from the chemistry, math, nuclear engineering, and physics subject areas. In recent years, a sizable contingent from robotics, computer science, and mining fields have entered the radioactive materials workforce.

Traditionally, engineers in the environmental field have earned civil or chemical engineering degrees. In recent years, many institutions have offered specific environmental engineering programs, with an emphasis on managing contaminants in the environment. The engineers in radiochemistry are often nuclear engineers. Although they may be highly knowledgeable in nuclear materials extraction, purification, materials handling, and production, they may not have been exposed to the engineering aspects that relate to radioactive contamination in the environment.

Safety and health professionals are perhaps the most distinct and different of these examples. In the chemical world we rely on *industrial hygienists* (IH) to monitor, record, and offer advice on hazardous chemicals that may be exposing individuals. The IH is usually focused on measuring atmospheric contaminants that pose an inhalation or skin absorption risk. The techniques employed usually involve air sampling with sorbent media or filters or real-time electronic monitoring. Results are then used as the basis for personal protective equipment (PPE) recommendations.

The radiological health professional is termed a *health physicist* (HP). The HP conducts monitoring for various types of radiation, tracks worker exposures, and recommends strategies to reduce exposure. The instrumentation and calculations associated with health physics measurements are so distinct from those of the IH that rarely is one person found who is knowledgeable in both fields. These training differences then become an important factor in staffing MCR sampling events, as the talents of a variety of scientists, engineers, and safety professionals are usually required.

Analytical Techniques and Methods Sources

The instrumentation required and the methods used for quantifying nonradioactive chemical substances are well-known and somewhat standardized. Techniques for analyzing for metals include atomic absorption and inductively coupled plasma emission spectroscopy. A wide range of chromatography techniques have been extensively used for organic chemicals, including gas chromatography and high-performance liquid chromatography. Light spectroscopy is commonly used for the measurement of petroleum hydrocarbons. Ion-specific electrodes and ion chromatography are often employed for measurement of specific anions or cations. A variety of wet techniques can of course be used for determining contaminant concentrations, although most environmental laboratories are very production-oriented, and automating laboratory analyses is often a high priority.

In the radiochemistry field the analytical devices are almost completely different from the chemical arena. Samples may be measured for bulk properties such as gross alpha radiation and gross beta radiation, or for specific isotopes. Spectrometers are used to classify radioactive emissions according to their isotopic composition. With liquid scintillation, sample constituents irradiate luminescing compounds in a solvent cocktail. In neutron activation analysis, samples are irradiated, emit gamma emissions, and are speciated and quantified by a gamma spectrometer. Additionally, a myriad of wet methods are used to isolate specific isotopes, clean up samples, concentrate analytes, and eliminate interferences. Many of these techniques are not standardized, and expert radiochemists frequently devise improvements (*9, 10*). The common analytical instruments and methods sources for chemical and radiochemical analysis are summarized in Tables II and III, respectively.

As one can see from the differences in required instrumentation and methods, relatively few laboratories are completely prepared to analyze MCR samples. Indeed, although environmental laboratories for hazardous chemical samples probably number in the thousands, only about a dozen laboratories (commercial) are properly tooled and licensed for mixed substance analysis. The principal limiting factor is handling the radioactive samples properly. Some samples can only be managed at government laboratories in which sample preparation and analysis are conducted in isolated environments called hot cells.

Finding Common Ground

These differences may confound the sample planner, but incorporating these different realities into a single system is essential to a successful data-gathering event. An

Table II. Common Analytical Techniques for Chemical and Radioactive Substances

Chemical	*Radiological*
Inductively coupled plasma emission spectroscopy	Alpha and gamma spectrometry
Atomic absorption	Alpha and beta liquid scintillation
Gas chromatography	Neutron activation analysis
IR and UV–visible spectroscopy	Gross alpha and gross beta counting
Ion-specific electrode	Wet methods
Wet methods	

Table III. Representative Methods References for Chemical and Radioactive Substances

Chemical	*Radiological*
EPA SW–846 (*11*)	HASL manual (*15*)
Standard water and wastewater methods (*12*)	DOE methods compendium (*13*)
NIOSH methods (*14*)	DOE site-specific SOPs
EPA SDWA methods (*5*)	EPA SW–846 (*11*)
Good science and research methods	Good science and research methods (*16, 17*)

NOTE: SW is solid waste; NIOSH is National Institute of Occupational Health and Safety; SDWA is Safe Drinking Water Act, HASL is Health and Safety Laboratory, and SOPs are standard operating procedures.

MCR sampling event has the same goal common to separate chemical or radiological environmental sampling efforts: to obtain data of known, acceptable, and defensible quality within time and budget constraints.

Both the EPA (*18*) and the DOE (*19*) have outlined quality assurance (QA) criteria. EPA's criteria tend to focus on good laboratory practices supporting data quality, whereas the DOE's criteria are much broader, incorporating QA program requirements, monitoring, and the quality of facility design and construction. The American Society of Mechanical Engineers (ASME) has proffered NQA-1, *Quality Assurance Program Requirements for Nuclear Facilities* (*20*), which has been actively referenced by the DOE.

The EPA's works contain a number of helpful strategies. One is the so-called PARCC list of data quality indicators (DQIs): precision, accuracy (bias), representativeness, completeness, and comparability (*18*). However, these factors constitute only part of the combined technical and philosophical aspects of sample planning.

A data quality system that has illuminated some, but perplexed others, is EPA's data quality objectives (DQOs). Although DQOs may be defined in many ways, quite simply they constitute statements identifying

1. the decision type(s);
2. the intended data uses, users, and needs; and
3. the components of a data collection program that will meet the data needs while providing answers to support the decision type.

DQO development allows the scientist to "backward-chain" from the final use of the data, through the analysis and sampling aspects, to the point of sampling design. The DQOs may comprise quantitative and qualitative statements regarding the degree of uncertainty that the project can tolerate. The DQO process applies directly to MCR or straight radiochemical sampling and analysis because the DQOs are based on the scientific method. Especially when there is a lack of standardized methods for sampling or analysis, the sample planner may reliably fall back on good science (*21*).

Four principal steps in data collection have been described (22). They include (1) sample design, (2) sample collection, (3) analysis, and (4) data reporting, validation, and interpretation. Here again we have an essential similarity between chemical and radiological approaches. The first and last of these steps are the most philosophical, whereas the second and third are the most procedural.

For mixed substances, sample design must consider worker safety and health, laboratory selection and licensing, methods selection, quality assurance–quality control (QA–QC), and data validation. For those procedural steps such as sampling technique, analytical method, and safety and health approach, standard operating procedures (SOPs) help the sample planner achieve the goal of known and acceptable data quality.

Sampling Design Elements

In Chapter 2, Barcelona (22) outlines a series of sampling design elements. For the topic in this chapter, some of these have been consolidated, and other categories have been added. Through close examination of each of these design elements, special considerations that relate to MCR sampling events can be identified. If all of these intricate design elements are considered when planning an MCR sampling event, the project should proceed with relative ease and a high probability of success. Because of the cost of mixed substance sampling and analysis, careful sample planning and a written QA project plan generally pay large dividends through the avoidance of incorrectly obtained samples, reduced worker and analyst exposure, and correct analyte selection.

Program Purpose

Identifying the decision type is the first step. Examples include identifying worst-case concentrations, potential concentration ranges, background levels, or elevated levels, or determining if a defined waste or substance is present above statutory limits. The sampling program development process is more challenging for MCR events because the sampler (1) may not want to be close to the worst-case sample, (2) may have difficulty identifying background concentrations, or (3) may be unable to find an appropriate statutory limit for guidance. Also, traditional analytical QC may not be possible because of reduced aliquot sizes.

As an example, one might consider the sampling of 20 drums of unknown waste at a former defense site. Exterior readings with a radiation survey instrument show alpha and gamma radiation at 100× background. Although the statistical sample planner may want all of the drums sampled, the financial controller may only have funding to sample one-fourth, and the HP may think only one drum should be sampled so that the sampler can get in and out as quickly as possible. Furthermore, the HP may say an on-site analysis technique will eliminate the need to send hot samples off-site, but the method is not in anyone's compendium of approved methods. So begins the negotiation of data representativeness versus cost versus data acceptance versus worker safety.

Once the terms have been tentatively agreed on, establishing the level of data quality required and the terms of collecting that data can be discussed. Currently, radiochemistry has no equivalent to the EPA's Contract Laboratory Program (CLP) commonly cited sources for chemical analysis and data deliverables. Radiochemistry data validation is project-specific, laboratory-specific, and customer-specific. Radiochemistry laboratory managers currently spend a great deal of time educating data users about the data meaning and the relative quality of the data after the samples have already been analyzed. Unfortunately, much of this tutelage is after the analytical process is complete, not in the sample planning step.

The design of the data collection program must include a description of the analytes of interest, the sampling techniques, the analytical techniques, laboratory selection and licensing criteria, sample locations (fixed or flexible), QC analytical criteria and samples, data validation processes, sample packaging and handling considerations, what field screening or field analyses will be conducted, and a worker safety outline.

Analytes of Interest

The sampling plan must define what analytes are to be determined and which data quality parameters are expected of each analyte. The most common types of analytes for MCR samples are shown in Table IV. Some of these categories may contain hundreds of specific analytes.

Terminology and Definitions

A brief discussion of nomenclature is appropriate at this time. *Radiation* is the liberation of energy from matter, which may be a result of subatomic, atomic, or molecular interactions. We speak of *ionizing radiation* and *nonionizing radiation.* Examples of ionizing radiation include hospital X-ray machine emissions and energy radiated from a sample containing radium. Nonionizing radiation includes laser, microwave, and radio-frequency radiation. The scope of this chapter includes only ionizing radiation.

There are four basic forms of ionizing radiation. These are alpha (α), beta (β), gamma (γ), and neutron (n) radiation. An *alpha particle* is a charged free helium nucleus that comprises two protons and two neutrons. *Alpha radiation* is the weakest form of ionizing radiation, easily stopped by unbroken skin or clothing, but alpha particles are of great concern if ingested or inhaled. A *beta particle* is a free electron, which is more energetic than an alpha particle. *Beta radiation* can penetrate unshielded bodies and release energy into tissues. Thankfully, this form of radiation is less commonly encountered than alpha radiation. *Gamma radiation* is typically described as a particlelike wave of energy. It is highly energetic and can pass through persons and unshielded walls. *Neutron radiation* is propagated during nuclear reactions and by electronic devices and is generally not under consideration for environmental measurements.

Table IV. Common Analyte Categories of Interest

Chemical	*Radiological*
Heavy metals	Transuranic metals
Toxic anions (e.g., cyanide and nitrate)	Uranium decay products
Volatile and semivolatile organic compounds	Isotopic inventory
Polyaromatic hydrocarbons	Radioactive gases
Pesticides and herbicides	Gross radiation
Petroleum hydrocarbons	Surface dose rates
Explosives	Particulates
Ignitability, corrosivity, and reactivity	Decay heat (calculation)
Free liquids	

The *curie* (Ci) describes the inherent level of activity in the sample, similar to the concentration of the radioisotope(s). Most analytical results will be in terms of picocuries per unit mass or volume (e.g., pCi/g or pCi/L). One nanocurie (1×10^{-9} Ci) is equivalent to 37 atomic disintegrations per second, regardless of the radioisotope.

However, not all of the radioactivity in a sample is radiated outward, and that which is emitted from the sample is highest near the sample, with energy decreasing proportionally to the square of the distance. The energy radiated outward is often measured with a Geiger counter. Units of measure are either *rads* per hour (rad/h) or counts per minute (cpm). By knowing the efficiency of the measurement device and the specific isotope, one can calculate disintegrations per minute (dpm) or curies. The number of rads expresses the amount of ionizing radiation energy in the space where the measurement is taken.

Yet another measurement of radioactive energy is devised to record the energy impacting a human body. These measurements are most commonly obtained with dosimeters that are worn on certain parts of the body. Later, the dosimeters are read or analyzed, and the exposure is put in terms of *rems* per unit of time. The units of measure are most commonly expressed as millirems per quarter. Thus, of the radiation inherent in a sample, a portion radiates outward, and of that portion, a fraction exposes the sampler and analyst.

Sampling Technique, Analytical Methods, and Laboratory Selection

These factors might seem to form an unlikely grouping, but all are of great importance in the long-range reliability and acceptability of data generated from an MCR sampling event. Any demonstrable oversight, error, or poor selection of scientific methods will render the data vulnerable to invalidation. MCR sample planning, sampling, and analysis are costly and sometimes risky ventures. To be forced to repeat these activities wastes valuable resources and damages credibility.

Sampling Technique and Method Selection. Sample collection, radiation surveying, handling, labeling, and packaging are the first steps in data collection and should follow standard procedures. The sample planner should work with the sampling personnel in advance of sample collection to ensure mutual understanding of the steps to be taken to achieve accurate sample representativeness and worker safety. Samplers and analysts must be specifically trained in both chemical and radiochemical safety. Although these issues are not mutually exclusive, they often require enhanced communication.

Sampling protocols should be written down as SOPs. This written record serves to aid the field sampling personnel during sampling, document the procedures used, and defend against scrutiny later. Records of nonstandard procedures must be maintained in the project logbook or file. Sampling SOPs should provide sufficient detail needed to replicate sampling procedures and quality by all sampling personnel. SOPs often include equipment specifications and problem-debugging information. Without SOPs significant random and systematic sampling errors may be unnecessarily introduced into the total error equation as the unguided sampling technician often alters the prescribed technique for reasons of expediency or ignorance. The techniques selected for sampling must be compliant with the DQOs, agency overseers, and project resource limitations.

Analytical Methods Selection. Test method selection is principally a component of the sample planning stage. Methods should be agreed on prior to sample collection, as the method chosen will dictate sample quantity, container type, preservation, and shipping requirements. Method selection will be based on several factors. These include analytes of interest, ability to overcome matrix interferences, ability to achieve desired detection level, maintaining analyst exposure as low as reasonably achievable (*ALARA*), reliability and standardization of the methods, and consistency with the DQOs. Knowledge of potential matrix interferences often comes from industry experts, methods compendia, and research articles. Detection levels, particularly for chemical parameters, may not be achievable in samples with high radioactivity, as sample size may be restricted to an absolute minimum. One laboratory representative spoke of a 200-mg sample that was submitted for EPA Method 8270, which normally requires 1000 times that sample mass.

Laboratory Selection Criteria. In general, when planning for an MCR sampling event, one should select a laboratory that is permitted and accredited for both chemical and radiological analysis. Finding such a laboratory may be a difficult task, depending on the nature and volume of samples to be collected. One cannot simply send splits of the samples to two different laboratories (one for chemical and the other for radiological) because to do so would expose laboratory analysts and equipment to hazards for which they may not be trained, licensed, or monitored.

The licensing for chemical analysis typically involves state licensing board certification. The laboratory should be licensed in the same state as the site. Some reciprocity agreements exist between states, as well. Laboratories performing Superfund work may be CLP participants, and there are a number of independent accreditation organizations. The American Association of Laboratory Accreditation (AALA) is perhaps the most widely known independent accrediting organization. There has been interest expressed on behalf of the laboratory community to move toward national accreditation, but as of this writing, no national program has been established. Nearly all laboratories also participate in periodic analysis of performance evaluation (PE) samples from EPA's Environmental Monitoring and Systems Laboratory (EMSL) in Cincinnati, OH. Laboratories submit their results on standardized synthetic water pollution and water supply samples. They are scored against other laboratories in the program and issued acceptable/unacceptable results. DOE's Environmental Measurements Laboratory (EML) facility in New York, NY, also provides PE samples, as do some independent commercial firms.

Radiochemistry laboratories are also subject to licensing and accreditation. Licensing occurs through the state-level NRC equivalent body or the regional NRC office. For NRC delegation to the state in which the laboratory is located (a so-called *agreement state*), licensing will be on the state level; otherwise the regional NRC office will be responsible. Currently, about two-thirds of the states are agreement states. The NRC license is essential to the radiochemistry laboratory. Specific terms stated in the license will include maximum gross activity, maximum quantities of specific isotopes, source and sample security issues, worker training, spill or theft contingencies, exposure monitoring, leak testing, and decontamination.

Other programs in which radiochemistry laboratories may participate include AALA accreditation, EMSL Las Vegas (EMSL-LV) PE samples, and PE samples from private audit support firms. Both chemical and radiological laboratories may be Army Corps of Engineers (ACE) validated. ACE validation involves both an audit and successful analysis of Corps-supplied PE samples. ACE validation is only available to laboratories that have performed or are currently performing Corps work. The EPA has two guidances that may not specifically apply to MCR laboratories but that add another layer of laboratory defensibility. These are known as the Good Laboratory Practices (GLP) and the Good Automated Laboratory Practices (GALP). Laboratories that are performing work directly for the EPA may be required to adhere to GLP and GALP. Particularly valuable to today's mixed chemical and radiological analysis, the GALP guidance describes quality and defensibility issues for laboratory information management systems, mass spectrometer (MS) and ICAP data systems, automated sample preparation sequences, and other automated electronic data systems. Radiochemistry QA programs should also conform to ASME *NQA-1*.

Other issues in selecting a laboratory include analysis capabilities, work schedules, appropriate instrumentation and personnel inventories, and successful history in performing analyses similar to those of the upcoming project. Numerous QA–QC indicators can also be reviewed. Among these are a record of method detection levels at or below project criteria or regulatory thresholds. The laboratory should maintain and be able to demonstrate tolerances (acceptance criteria) for data evaluation and validation, should be able to supply necessary data deliverables, and should routinely perform sufficient QC to document results at the time samples are analyzed for the client.

Sampling Location

In addition to concerns over sampling personnel exposure to radiation, sampling location considerations for an MCR sampling event will contain the same elements as for any environmental sampling exercise. One must plan for the location of the sampling station, depth, position, and frequency. Procedures to ensure minimal sample disturbance and personnel exposure, as well as maximum representativeness, must be stated in a written plan prior to the sampling event.

Minimizing area disturbance and avoiding public visibility may also be important to speeding the sampling process. Even if the event is designed to characterize background levels of naturally occurring radioactive materials, uninformed parties may take offense if they find out that MCRs are being evaluated.

Often, real-time monitoring or field screening may be used to increase data volume and reduce the requirement for off-site sample shipment. These techniques are discussed in an upcoming section of this chapter.

QA–QC and Data Validation

It would certainly be convenient if one could simply adopt a standardized approach to QA and data validation or MCR results, but such is not the case. Few standardized data validation protocols exist in the environmental radiochemistry field. There is no current analogue to the CLP in radiochemistry.

The most reliable substitute for the dearth of validation standards is preplanning a project-specific QA plan. Such a plan will rely on stating desired method detection levels, instrument calibration techniques, laboratory control standard tolerances, matrix spike duplicate acceptance criteria, and corrective action triggers and responses. As with any discussion of environmental data in the late 20th century, the focus must be upon sound scientific practice. If one is attempting to compare different analytical methods, the sampling recovery rate of a new kind of sampling device, or the precision and bias of an environmental measurement, one will rely on experienced mathematicians to perform conventional and innovative statistical assessments.

The most important point with these project-specific assurances is that the approaches to be used on a certain MCR data collection event be accepted by the principal intended data users prior to commencing the sampling. This requirement may, of course, involve some elementary training in statistics for the data users. This type of training is advisable anyway because radiochemistry data are generally reported with associated counting error, something uncommon to nonradioactive chemical data users.

As with any sampling program, one will specify certain QC samples for both the field and the laboratory. Certain types of QC samples are not common in an MCR sampling event. These may include field matrix spikes (requires source material), field splits (in samples with high activity), and some isotopic laboratory standards. Alpha spectrometers are calibrated against a single isotope standard using a spike in the sample. Results are then background and recovery corrected. Gamma spectrometers are calibrated against a multi-isotope standard spanning a wide range of energies. As expected, these calibrants constitute a mixed waste.

The most fundamental QC measurements that are made in a radiochemistry laboratory are background radiation and counting efficiency. Counting efficiency expresses the percent efficiency of the radiation detector component to measure the radiation emitted from the sample. This measurement is done by exposing the detector to a standard source of known activity. Counting error is calculated using Gaussian statistics, and this uncertainty is typically reported with radiochemistry results. [Although Gaussian statistics are the convention, it has been suggested that because the data set involves a small number of counts, especially with background samples, Poisson statistics may be more appropriate. Gaussian statistics may underestimate the errors associated with counting low-activity samples (*10*).]

Internal QC checks for gross alpha, gross beta, and liquid scintillation counting may include instrument blanks (background counts), method blanks, laboratory control standards, and matrix spike–matrix spike duplicates. There are differences between laboratories in this regard, and as always, the laboratory should be actively involved in the specification of methods and QA–QC criteria during the sample planning process.

The results from radiochemical analysis are considerably different from those of conventional chemical analysis. In chemical analysis results are most often reported as mass of contaminant per unit mass or volume (e.g., [benzene] = 1.2 μg/L). Radioisotope concentrations are most frequently reported in units of radioactivity per unit mass or volume (e.g., 1.2×10^{-3} pCi/L as ^{228}Ra). The standard unit is the curie (Ci), with the pico (1×10^{-12}) prefix being most common in environmental measurements.

Sample Collection, Packaging, Storage, and Transport

Another important reason for thoughtful sample planning is to guide the complex procedures related to the field activities of sampling, packaging, and shipping. Sample collection may be considered to comprise the following elements: sampling apparatus, sample representativeness, and contamination avoidance. Packaging involves preservation, container selection, holding time, and half-life. Additionally, MCR samples have special transportation considerations.

Sampling Apparatus. The tenet for reducing radiation exposure among workers is "time, distance, and shielding". Likewise, sampling apparatus for MCR substances should involve reduced time at the location, increased distance from the substance, and worker shielding to the greatest extent reasonably achievable. These principles apply to reducing exposure in the laboratory as well. As one might imagine, expediting sample collection and analysis time to reduce exposure may cause a loss in sample representativeness and reproducibility, and other errors.

Some mixed waste sampling activities involve waste materials that are so radioactive as to prohibit workers from getting close enough to collect a sample. The high-level liquid waste tanks at the Hanford Reservation in Washington state and other DOE sites exemplify this dilemma. Radiation in the void space above the waste may be lethal to humans, exceeding 10,000 rad/h. Innovative sampling techniques must be validated, as with analytical methods. SOPs form the basis for gaining reproducible sampling results and serve as the foundation of data acceptance and validation activities.

For very radioactive sampling events the two approaches that are most often used are robotic sampling and remote on-site analytical testing (*23–25*). There are many challenges ahead for both of the areas. In robotics, issues of mobility and dexterity are complicated by the effects of ionizing radiation on video components, electronic circuitry, and radio communications. Remote, automated analysis will probably be the principal technique of the future, as sampling and traditional analysis exposes the worker to health risks and creates secondary wastes from personal protective equipment (PPE), sampling equipment, and laboratory wastes. DOE currently has several projects under way to automate labor-intensive (exposure-intensive) extraction procedures for isotopic analysis sample preparation (*25, 26*). Other bodies, such as the American Society for Testing and Materials (ASTM), have made available standard sampling methodologies for some conditions (*27*).

Sample Contamination and Representativeness. Sample contamination reduction involves the recognition, measurement of the effects, and control of contamination. Typically the sources of contamination involve equipment, handling, subsampling, preservatives, containers, ambient contamination, glassware, reagents, and analytical equipment carryover (*28*). Contamination of the sample may involve either an increase or decrease in the concentration of key analytes. Because one does not collect radioactive soil samples with radioactive trowels, there seems to be little additional concern about that source of contamination. However, with radioactive samples, there are other types of contamination or sample impairment.

One type of sampling error could occur if, while tritium-contaminated groundwater is being sampled, a small amount of equipment rinsate (low-tritium contaminated water) gets into the sample, effectively lowering the result. Conversely, if one were sampling sediments or ice cores that predated the nuclear era (ca. 1950), tritium concentrations would be expected to be lower than "young" water found on Earth in the 1990s. Because of elevated tritium levels in the world's young water supply, rinse water from today getting into samples older than 1950 could result in artificially high results (*29*).

Sample contamination may also occur in the laboratory. The reader is surely familiar with the effects of opening highly contaminated volatile organic samples in a laboratory: the volatile constituents may be detected in the results of laboratory blanks and low-level samples submitted in the same or different batches. With highly radioactive samples the same risks exist. Radioactive contamination in a laboratory can elevate background measurements, causing increased detection levels. Dilution of "hot" samples must occur in a dedicated "hot cell" to avoid lab contamination, just as high-level volatile organic compound samples should be segregated from low-level samples. Contaminants may be in any physical state and may in fact decay from one state to another, such as krypton and radon gases giving rise to solid daughter products.

Shortly after the failure of the Chernobyl nuclear reactor in 1986, radiochemistry laboratories throughout the United States (and doubtless, the world) observed elevated background levels of radiation in their laboratories. If nothing else, this event is a testament to the sensitivity of these measurements and to the effects of elevated background radiation. Background radiation will vary from site to site and laboratory to laboratory, depending upon elevation, local geology, and potentially upon proximity to nuclear test areas and facilities. Background levels of uranium have been used to help define uranium releases from a mill site (*30*).

With nonradioactive chemical samples, exposure of the sampler and the chemist to the hazards of the sample is usually controlled through a series of safe work practices, engineering controls, and PPE. A belief among those in the hazardous chemicals industry is that sufficient controls and PPE will reduce any exposure to acceptable risk. This belief is not always the case with radioactive samples, as substances such as plutonium are so extremely toxic, or cesium, so energetically radioactive, that no PPE or control technique may lower risk enough to justify having people even get close to the samples. This concern has fostered inventive research into sampling and analysis techniques that allow in situ measurements, remote handling, and automation of sample preparation (*22, 24*). Some on-site analytical techniques themselves involve additional emitted radiation. These include portable X-ray fluorescence (XRF) spectrometers and neutron activation analyzers (NAA). The sampler and health physics staff must also consider the safety issues associated with these sources (*31*).

Sample size is another consideration. Highly radioactive samples may cause undue exposure to samplers and analysts. However, submitting a very small sample will result in raising the detection levels of the nonradioactive constituents, often to detection levels above the concentration of the decision point. Standard QC such as matrix spike–matrix spike duplicates may be impractical or impossible with small-volume, high-activity samples.

The activity of the sample may not be correlated with the nonradioactive constituents, either, so small samples may actually cause a representativeness problem for nonradioactive analytes. The EPA produced a position paper that acknowledges this problem and offers several solutions (*32*). The use of surrogate waste samples, reduced sample volumes, and remote or field-screening-type analyses are all possible remedies to this issue.

Preservation, Filtration, and Packaging. These aspects of the MCR sample collection event tend to be similar to those for nonradioactive chemical sampling, with the exception of sample packaging and shipping. Parr et al. (*33*) advocate a thorough understanding of the physical, chemical, and biological transformations that the sample may undergo. From this understanding, one may select appropriate preservation techniques (*33*) and establish acceptable holding times. For radiochemical analysis, in the absence of a national methods compendium, it is advisable to obtain specific preservation and filtration recommendations from the radiochemistry laboratory during the sampling planning step. Some time-consuming sample preparation procedures, such as subsampling fish for fatty tissue samples or sieving debris samples, could expose the sampler to elevated levels of radiation.

Packaging and shipping radioactive samples is often a real challenge. Samples must be packaged in accordance with Department of Transportation (DOT) regulations (*34*), as with any sample. Radioactive samples are not exempt from waste regulations, as their chemical counterparts are under EPA. Sample containers must be surveyed at the surface and at a 1-m distance from the container for gross radiation (i.e., alpha, beta, and gamma). According to the exterior and interior activity levels, the sample will be given one of three DOT radioactive materials shipping descriptions and labels. Generally speaking, the dosage at the surface of the container may not exceed 1.0 mrem/h. Many commercial couriers or transporters are reluctant to carry radioactive materials, even at very low activities. Instances of couriers not willing to accept properly packed radioactive materials are fairly common. Unfortunately, this practice may lead MCR sampling personnel to improperly label packaging to ensure shipment.

Some samples are too "hot" to leave the site, thus forcing on-site analysis. The sample exit criteria must be determined on a site-by-site basis, further complicating matters. Currently no straightforward statutory limit defines what is and is not radioactive, so any levels of radiation above background will usually bring the sample under some NRC regulation, DOE administrative order, or DOT packaging and shipping requirements. Radioactive samples are very difficult to dispose of, and so are often returned to the site. This practice may cause more field services expense and can elevate worker exposure levels.

Holding Time and Half-Life

Holding time is a common consideration for chemical analysis. Biological, chemical, and physical transformations occur that can lower analyte concentrations or produce spurious secondary contaminants. With radioactive samples it may also be necessary to consider half-life. Simply put, *half-life* is the time that it takes for one-half of the atoms of a particular radionuclide to disintegrate into something else.

Radon, for example, has a half-life of about 3.8 days. If one were to collect a Summa canister full of air from a site and ship it cross-country to a laboratory for radon analysis, the sample would be expected to have a lower result than the original site conditions because of the rapid decay of this species. In this case there are two alternatives: (1) measure the longer lasting decay products (daughters) of the radon in the laboratory and back-calculate, or (2) measure radon in a real-time sense using a portable working level monitor on the site.

Field Screening and Field Determinations

Screening and analysis of samples directly in the field is an excellent way to avoid many of the sample packaging, shipping, and licensing requirements for radioactive samples. Indeed, if there were sufficient variety in the technologies available for on-site analysis, there would doubtless be a profound shift toward heavy reliance on field analytical methods. The state of the art, however, can only support certain types of field screening at this time.

Despite the limitations of field methods available, field screening and on-site analysis are beneficial in many ways. First, unstable samples can be promptly analyzed. Field analysis will simplify sample collection, packaging, preservation, and shipping requirements. Although site exit criteria may prevent larger samples from leaving the site, on-site analysis may allow larger samples. The direct benefits are in lowered detection levels and easier sample disposal. Not only does one reduce the generation of radioactive waste in this way, but one also avoids significant concerns of container surveying, labeling, and shipping with common carriers. On-site techniques also foster the collection of a higher volume of data from more locations. These additional data beneficially impact the statistics of sampling.

Safety and Health

All of the traditional safety and health aspects of an environmental sampling project will be encountered in the MCR sampling event, as well as radiation-specific aspects. PPE may or may not be beneficial in a radioactive environment. Alpha radiation is particularly stoppable with disposable coveralls, rubber gloves, and a high-efficiency particulate air (HEPA) respirator. Higher energy beta particles and gamma radiation, however, will pass right through most of these systems.

Work practices designed to reduce exposure must be developed and maintained. As previously mentioned, time, distance, and shielding should all be incorporated into the MCR sampling event. Sampling event planning will reduce time spent around radioactive materials. Specialized sampling strategies and equipment will help increase the distance between the sampler and the sample. Except in the case of remotely handled high-energy waste materials, PPE and physical barriers can usually provide the necessary shielding.

Personnel who are sampling or analyzing radioactive materials will normally be monitored for bodily exposure by several means. With dosimetry, small devices that record exposure to various types of radiation are worn. The most common of these are the thermoluminescent detector (TLD) badge and the pocket or pen dosimeter. The TLD badge is submitted periodically (weekly, monthly, quarterly, or annually) to a laboratory for developing. The report will describe the estimated dosage, in millirems per time period, to which the individual was exposed. The pocket dosimeter relies on the static charge caused when radiation impacts a metal foil. Greater doses of radiation yield greater deflection of a small piece of metal foil inside the device. The advantage of the pocket dosimeter is that it can be checked frequently by the wearer. These units have questionable accuracy, however, because they can receive regular static electric charges, or be discharged accidentally by physical disturbance, such as by dropping.

Real-time monitoring for worker protection will usually include a radiation survey meter (RSM). RSMs may have various probes and recording devices attached, giving them more versatility. Alpha, beta, and gamma probes are available.

Last, the health physicist (HP) may require a certain frequency of biological sampling to assess worker exposure. Most often such testing involves submission of urine samples for radioisotope determinations, but blood analysis for white blood count is also employed.

Summary of Critical Considerations

Because the MCR sampling and characterization field is an outgrowth of two historically separate industries, much information needs to be exchanged and many new methods need to be written and approved. Mixed substances fall under fundamentally different disciplines and regulations.

In sample planning several issues require extra attention. These include early involvement of the laboratory in the specification of methods and QC. Laboratory licensure, license compliance status, and possession of other permits is critical, both from a data defensibility standpoint and in terms of protecting one's own liability in the involvement of an MCR project. Safety and health of the sampling staff and the chemists in the laboratory may require safety training, site monitoring, worker surveillance, and protective equipment that is unfamiliar to even the experienced chemical substance handler.

Three principal areas will benefit greatly from standard written procedures, whether project-specific or adopted on a larger scale. The first of these is sampling technique and methodology. For persons working in this field outside of the DOE, few SOPs are available in the way of radioactive materials sampling. Sampling and field screening techniques are subject to validation by comparative testing, field ruggedness testing, and cost and time evaluation.

Analytical techniques are another area requiring standardization. Although SOPs and guidance documents exist for many radiochemistry methods, the industry is behind the solid waste and wastewater laboratory industry. There is no broadly accepted compendium of methods for radiochemists and mixed waste analysts equivalent to EPA's SW–846. Radiochemical analyses often depend on analyst experience, laboratory-specific protocols, and good science. However, whenever there is a reliance on "good science", there is a potential for charlatans to be performing "bad science", and clients unfamiliar with radiochemical analyses may not recognize the difference.

Finally, because worker safety and health around MCR substances is more complex, there must be renewed emphasis on safety training, monitoring, personal protection, safe work practices (such as time, distance, and shielding), and documented safety protocols. Bridging the historical gap between IHs and HPs is a vital component of building a safe workplace for MCR sampling and analysis personnel.

Conclusion

Substances that are both nonradioactive and radioactive in composition are abundant. As investigators of the environment, persons conducting mixed substance sampling are required to assimilate a wide variety of scientific, regulatory, social, economic, and political sensitivities. Scientific integrity is the greatest tool in obtaining credible data. When standardized methods are lacking, and even when they seem to prevail, intimate knowledge of the chemical, radiochemical, and ecological transformations of hazardous substances will form the basis for reliable sample planning, sampling, analysis, and interpretation of environmental data. Of these, the requirement for environmental investigators to be accurate, swift, thrifty, prudent, and defensible bears most heavily on the sample planning stage. Surely, sample planning in the realm of mixed chemical and radiological hazards requires multidisciplinary teams that embody a wide range of technical expertise and innovation.

Acknowledgments

This topic is one of the most controversial subjects in the field of environmental sampling and analysis. Even for the author, who possesses several years of experience in mixed waste sampling at abandoned uranium mines and mills, much remains to be defined in this subject area. My gracious appreciation goes to radiochemistry experts Ellen LaRiviere, Michael Neary, and Jeff Tye, and to friends and colleagues at the U.S. Department of Energy who offered their experience and perspective.

References

1. Murray, R. L. *Understanding Radioactive Waste*, 3rd ed.; Powell, J. A., Ed.; Battelle: Columbus, OH, 1989.
2. Gershey, E. L.; Klein, R. C.; Party, E; Wilkerson, A. *Low-Level Radioactive Waste: From Cradle to Grave*; Van Nostrand Reinhold: New York, 1990.
3. Parker, F. L. In *Low-Level Radioactive Waste Regulation: Science, Politics and Fear*; Burns, M. E., Ed.; Lewis Publishers: Chelsea, MI, 1988; pp 85–94.
4. *Code of Federal Regulations*, Title 49; U.S. Office of the Federal Register, National Archives and Records Administration: Washington, DC, 1993.
5. *Code of Federal Regulations*, Title 40; U.S. Office of the Federal Register, National Archives and Records Administration: Washington, DC, 1993.
6. *Code of Federal Regulations*, Title 29; U.S. Office of the Federal Register, National Archives and Records Administration: Washington, DC, 1993.
7. *Code of Federal Regulations*, Title 10; U.S. Office of the Federal Register, National Archives and Records Administration: Washington, DC, 1993.
8. *Radioactive Waste Management*; U.S. Department of Energy: Washington, DC, 1988; DOE/5820.2A.
9. *Mixed Waste Integrated Program: Technology Summary*; U.S. Department of Energy, Office of Environmental Management, Office of Technology Development, National Technical Information Service: Springfield, VA, 1993; DOE/EM–0125P.
10. Neary, M., NUS-Halliburton, Pittsburgh, PA, personal communication, 1994.
11. *Test Methods for Evaluating Solid Waste*, 3rd ed.; U.S. Environmental Protection Agency: Washington, DC, 1982; SW–846.
12. *Standard Methods for the Examination of Water and Wastewater*, 18th ed.; American Public Health Association: Washington, DC, 1992.
13. *DOE Methods Compendium Database*; Los Alamos National Laboratories, U.S. Department of Energy: Los Alamos, NM, 1994.

14. *NIOSH Manual of Analytical Methods*, 3rd ed.; Eller, P. M., Ed.; U.S. Department of Health and Human Services: Cincinnati, OH, 1984; DHHS 84–100.
15. *EML Procedures Manual*, 27th ed.; Environmental Measurements Laboratory, U.S. Department of Energy: New York, 1990; Vol. 1, HASL–300.
16. Das, H. A.; Faanhof, A.; van der Sloot, H. A. *Environmental Radioanalysis;* Elsevier: Amsterdam, Netherlands, 1983.
17. Parry, S. J. *Activation Spectrometry in Chemical Analysis;* John Wiley & Sons: New York, 1991; Vol. 119.
18. *Interim Guidelines and Specifications for Preparing Quality Assurance Project Plans;* Office of Monitoring Systems and Quality Assurance, Office of Research and Development, U.S. Environmental Protection Agency: Washington, DC, 1980; QAMS–005/80.
19. *DOE Order 5700.6B;* U.S. Department of Energy: Washington, DC, 1986; DOE/EH–1.
20. *Quality Assurance Program Requirements for Nuclear Facilities;* American Society of Mechanical Engineers: Fairfield, NJ, 1989; NQA–1.
21. Taylor, J. K. In *Principles of Environmental Sampling;* Keith, L. H., Ed.; American Chemical Society: Washington, DC, 1988; pp 101–107.
22. Barcelona, M. J. Chapter 2 of this book.
23. *Underground Storage Tank Integrated Demonstration: Technology Summary;* U.S. Department of Energy, Office of Environmental Management, Office of Technology Development, National Technical Information Service: Springfield, VA, 1993; DOE/EM–0122P.
24. *Technology Catalogue;* U.S. Department of Energy, Office of Environmental Management, Office of Technology Development, National Technical Information Service: Springfield, VA, 1994; DOE/EM–0138P.
25. Hollen, R. H. *Environ. Test. Anal.* **1994,** *3*(2), 36–39.
26. *Interim Mixed Waste Inventory Report: Waste Streams, Treatment Capacities and Technologies;* U.S. Department of Energy: Washington, DC, 1993; DOE/NBM–1100.
27. *Annual Book of ASTM Standards;* American Society for Testing and Materials: Philadelphia, PA, 1986; Vol. 11.03, Section II, D4096–82.
28. Lewis, D. L. In *Principles of Environmental Sampling;* Keith, L. H., Ed.; American Chemical Society: Washington, DC, 1988; pp 119–144.
29. Pearson, F. J.; Loosli, H. H.; Balderer, W. In *Applied Isotope Hydrogeology: A Case Study in Northern Switzerland;* Elsevier: Amsterdam, Netherlands, 1991.
30. Black, S. C. In *Principles of Environmental Sampling;* Keith, L. H., Ed.; American Chemical Society: Washington, DC, 1988; pp 109–117.
31. Jensen, J. C.; Danforth, N. P.; Stinchfield, M. R. A. Poster presented at 1991 Colorado Hazardous Waste Management Society Conference, Denver, CO, October 1991.
32. *Clarification of RCRA Hazardous Waste Testing Requirements for Mixed Waste;* draft guidance; U.S. Environmental Protection Agency: Washington, DC, 1992.
33. Parr, J.; Bollinger, M.; Callaway, O.; Carlberg, K. In *Principles of Environmental Sampling;* Keith, L. H., Ed.; American Chemical Society: Washington, DC, 1988; pp 221–230.
34. *A Review of the Department of Transportation Regulations for Transportation of Radioactive Materials;* U.S. Department of Transportation, Research and Special Programs Administration, U.S. Government Printing Office: Washington, DC, 1983.

Chapter 36

Integration of Immunoassay Field Analytical Techniques into Sampling Plans

Kevin J. Nesbitt and Kevin R. Carter

The advent of high-performance immunoassay field screening tools now allows project managers to effectively delineate contamination at a site in a flexible yet structured fashion. A major limiting factor in employing this technology to its fullest advantage has been the continued use of the traditional sampling strategies. The full integration of immunoassay techniques requires a more innovative sampling approach. Utilization of standard sampling strategies (e.g., random, biased, or stratified) employed in an iterative fashion will address these issues. A sampling strategy emphasizing the benefits of field testing using immunoassay will allow samples to be collected and analyzed accurately, more efficiently, and at a lower cost.

MANY QUESTIONS AND CONCERNS have been raised involving the time taken to complete site assessment and remediation projects and, occasionally, the quality upon completion of these projects. The concern raised by the current pace and cost of site investigation and remediation efforts, especially those associated with the Superfund Program, has prompted the U.S. Environmental Protection Agency (EPA) to evaluate methods that could improve the situation. One method of work that has been well-received is field testing using immunoassay methods. Because of the drive to improve the process and the proven validity of immunoassay technology, the EPA Office of Solid Waste has recently evaluated several immunoassay methods and approved several for inclusion in the solid waste methods guidance manual, SW–846. This acceptance has allowed site investigation personnel to incorporate immunoassay methods into sampling efforts when appropriate.

Environmental assessment and remediation efforts are heavily dependent on the analytical results obtained during the course of the project. The project analytical work is driven and controlled by the sampling plan, the key document for this portion of the project. As such, the plan should be designed for maximum effectiveness and data value. The perception that the relationship between field or laboratory immunoassays and standard laboratory analytical methods is mutually exclusive, rather than complementary, is a common one. In reality, immunoassays allow for more effective use of all available analytical resources. The use of effective sampling plan design, employment of immunoassay, and competent selection of samples for laboratory analysis allows for a controlled, effective utilization of the sampling and analysis resources. This sequencing provides several benefits. Using inexpensive, accurate, real-time sample analysis to direct the final laboratory confirmatory analysis decreases costs, increases efficiency, and increases data quality.

Both the method of analysis and individual samples must be considered resources for the purpose of investigating the site. Most project managers view labor and equipment as resources that should not be squandered; who would waste a day's utilization of either? A sample that is not used effectively has also been wasted. According to the EPA Environmental Monitoring System Laboratory Office of Research and Development's *Soil Sampling Quality Assurance User's Guide* (*1*),

> [t]he course of action often substituted for the difficult value judgment is to adopt as a guiding principle the concept that one should always strive to achieve the highest power and level of confidence (or the lowest probability of error) possible with existing resources. The resulting data are then used as the basis for making decisions with the assumption that this guiding principle gives the best possible result. Obviously, such an approach will rarely, if ever, be cost-effective. There are two types of errors: the data may be much better than required, which indicates resources have been wasted, or the data may not be of adequate quality, thereby resulting in decisions of doubtful validity.

Immunoassay is a recently introduced technology that can effectively shorten the timeline of a project, as well as increase data quality. Immunoassay may be used either in the field or the laboratory depending upon the requirements of the sampling effort. The challenge is the functional integration into project sampling plans, in accord with specific site requirements, regulatory guidelines, and the laboratory analytical resources available.

Background of Immunoassay Techniques

Immunoassay is a century-old medical technology, which has been successfully adapted for use in the environmental field. Currently used extensively as a critical medical diagnostic tool, the technology inexpensively provides fast, accurate information concerning a variety of individual analytes or classes of analyte.

Immunoassay technology requires three basic components: antibodies, analyte enzyme conjugate compounds, and target analytes. These items interact to form measurable complexes. The field immunoassay requires 20–30 min per batch of samples, allowing 4–10 samples to be tested per hour. The results show good correlation with conventional instrumental methods, with virtually no false negatives.

Immunoassays available for field and laboratory screening include polychlorinated biphenyls (PCBs) in soil, oils, and on surfaces; petroleum fuels in soil and water; polyaromatic hydrocarbons (PAHs) in soil; pentachlorophenol in soil and water; benzene in water; and mercury in soil and water.

Regulatory Status of Immunoassay Methods

The validity of immunoassay as an environmental analytical technique has led to its acceptance by the EPA as an analytical technology suitable for certain types of environmental contaminants (Table I). Several states have included immunoassay methods in their state-specific field screening or site assessment manuals and underground storage tank guidance manuals.

RCRA and CERCLA Analytical Requirements

Contaminant analysis is driven by the commonly accepted assumption that Resource Conservation and Recovery Act (RCRA) and Comprehensive Environmental Response, Compensation, and Liability Act (CERCLA) *require* the use of laboratory analytical methods listed in SW–846. In reality, a large degree of freedom actually exists in sampling and analytical methodology allowed by most programs.

For example, the requirements for CERCLA removal actions, as listed in 40 *Code of Federal Regulations* (CFR), section 300.415 (a) 4(ii), state that "if environmental samples are to be collected, the lead agency shall develop sampling and analysis plans that shall provide a process for obtaining data of sufficient quality and quantity to satisfy data needs" (2). As stated in 40 CFR, section 300.430, for remedial investigation, feasibility study, and selection of remedy, "site-specific data needs...should reflect the scope and complexity of the site problems being addressed" (3).

RCRA is the major regulatory driver of sampling efforts. Whereas the analytical methods for the identification and characterization of RCRA hazardous waste are prescribed, the investigation of RCRA Solid Waste Management Units (SWMUs) allows more flexibility in the analytical methods used. Specifically, the new subpart S regulations have several important objectives including flexibility in approach to remediation and investigation, minimization of procedural delays, and streamlining facility investigations (4).

Table I. Immunoassay Application Information

Immunoassay	*Corresponding EPA Method (SW–846)*	*Contaminant*		*Applications*	*Cost per Sample Analyzed ($)*
		Compound	*MDL*		
PCB RIS[c] soil, liquid, waste, and wipe tests	Method 4020	PCBs	0.5 ppm in soil 5 ppm in oil 10 μg on surfaces	Transformer or capacitor spills, electrical equipment maintenance, natural gas pipeline condensate cleanup, and ship maintenance and salvage	12–42
PETRO RIS[c] soil and water tests	Method 4030	Petroleum fuels	10 ppm in oil 150 ppb in water	Underground storage tank leaks, vehicle maintenance, fuels refining, and distribution and storage sites	12–35
PAH RIS soil test	Method 4035	Total PAHs	1 ppm	Manufactured gas plants, wood-treating facilities, steel-coking operations, and burn pits	16–56
PENTA RIS[c] soil and water tests	Method 4010	Pentachlorophenol	0.5 ppm in soil 5 ppb in water	Wood-treating plants	16–56
Benzene RIS[c] water test		Benzene	5 ppb	Gasoline spills, leaking USTs or ASTs, manufactured gas plants, and petroleum refineries	30
BiMelyze Mercury soil and water tests	Method 4500	Mercury	1 ppm in soil 10 ppb in water	Natural gas metering stations and chloralkali plants	40

NOTE: Applications listed refer to all project phases: assessment, remediation, and closure. MDL means minimum detection level; PCB means polychlorinated biphenyl; PAH means polyaromatic hydrocarbon; ppm means parts per million; ppb means parts per billion; UST means underground storage tank; and AST means above-ground storage tank.

Although some states have specified individual methods for the most common analyte classes, most site assessment programs allow a reasonable amount of flexibility in analytical methods. There are typically two generic guidelines: the method must be appropriate for the contaminant and the site, and the data must be used appropriately. These measures ensure that the data derived from a sampling effort are of appropriate quality and are compatible with previously gathered data, as well as data generated by other methods of analysis.

Integration of Immunoassay into Sampling Plans

Immunoassay will alter the sampling plan by allowing the sampler to alter the course of the effort while it is under way. During a preliminary site assessment the immediate return of the results gives the sampler the opportunity to focus on areas of contamination quickly, thereby utilizing the available time and resources most effectively. The question is, How does immunoassay change the sampling plan?

Immunoassay provides a means of receiving instant, accurate data. This information empowers the sampler to make on-site decisions regarding sample locations. The impact of this capability can have far-reaching results. The ability of the sampler to instantly adapt operations based on incoming data increases the effectiveness of his or her efforts.

Historical Sampling Strategies

Several common sampling strategies have been used for environmental assessment. *Random sampling* involves the selection of sampling point by chance, without taking into account sources of contamination, for example. However, although random sampling is a low-bias method, it does present the risk of failing to identify contamination (*1*). *Judgmental sampling* is the targeting of samples toward potential sources of contamination or along presumed routes of contaminant movement. Judgmental sampling allows for the introduction of bias into the sampling effort. Allowing sources of contamination to drive the sampling effort can hamper the ability to gain a good representation of contamination across the entire site. *Systematic sampling* uses grids to select sampling points. Systematic sampling has almost no inherent bias, yet may require large numbers of samples to be effective. All of these methods, like any others, have benefits and disadvantages. Immunoassay has the capability of maximizing the benefits and minimizing the disadvantages. Overall, this benefit is accomplished through increased numbers of samples analyzed at reduced costs levels. The impact of immunoassay analysis on sampling efforts is discussed in the following section.

Representative Example of Immunoassay Integration

Based upon previous project experience, the most effective sampling strategy has been a combination of judgmental and systematic sampling. The effective combination results from targeting the known or suspect contaminated areas with judgmental sampling. The immunoassay can be used to assess the contamination levels and extent quickly, thereby eliminating the obviously contaminated areas initially. The disadvantage of judgmental sampling, as mentioned previously, is the potential for introduction of bias into the sampling effort. An effective method of minimizing this bias is the concurrent use of a systematic sampling grid.

The systematic grid serves two important purposes in the sampling effort: the reduction of introduced bias and more complete coverage of the area of concern. This complete coverage will address the following issues:

- Has the site been adequately assessed to prevent unknown contaminated areas from being missed?
- Have soil homogeneity issues been adequately compensated for?

The most important disadvantage of systematic grid sampling is the potential requirement for large numbers of samples to be collected and analyzed. The relatively low cost of immunoassay methods reduces the problem, as does the immediate nature of the analytical results.

The illustration in Figure 1 offers a basic comparison of one of the most common sampling strategies against a combination of judgmental and systematic strategies. The common strategy uses judgmental samples to assess the contamination. Because the samples are usually collected and sent to an off-site lab, no analytical

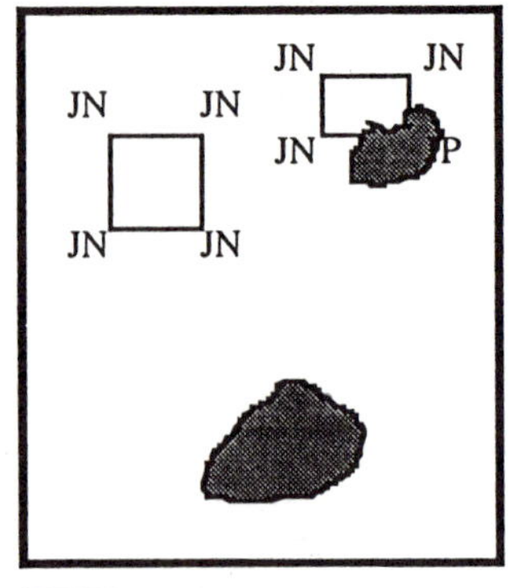

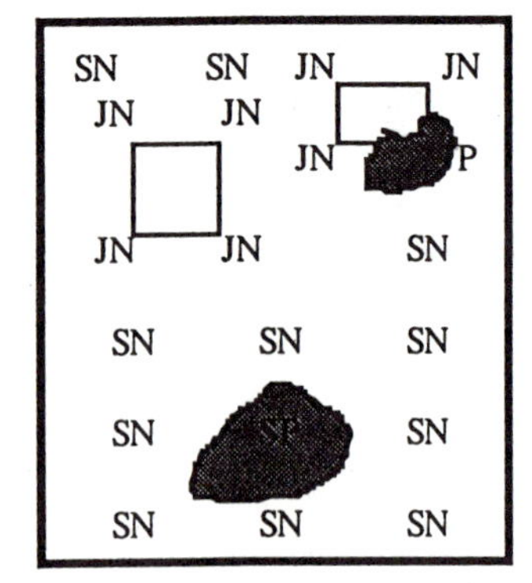

Suspected point source

Actual contamination

J - judgmental sampling point
S - systematic sampling point

N - negative sample
P - positive sample

Figure 1. Comparison of sampling strategies.

information is gathered during the sampling mobilization. Furthermore, the higher cost of the individual sample analyses may preclude the collection of additional samples on a systematic grid basis. Because of the exclusive use of judgmental sampling, a large area of contamination has not been detected.

In the immunoassay judgmental–systematic strategy the site has been more completely characterized. The judgmental samples that have tested positive have been surrounded by a localized grid and more clearly defined. In addition, a systematic sample has preliminarily identified an additional contaminated point and has been more comprehensively assessed with further samples. Additional benefit can be gained by completing this level of assessment during one mobilization. This protocol has the potential of saving substantial remobilization and equipment downtime expenses.

Benefits

Every sampling plan is site-specific, and analytical options will vary accordingly. The influences of the implementation of immunoassay methods can vary from minor to extensive. Clearly, all of these methods can be used separately or in tandem depending upon the requirements of the site. Immunoassay affords the following benefits to the sampling management team:

- The reduced costs allow for more samples to be collected and analyzed, effectively reducing bias and allowing for smaller sampling intervals.
- The rapid sample analysis allows flexibility during the effort.
- The field results create the ability to direct sampling for laboratory testing.
- The decreased timeline for sampling saves money and allows for additional technical effort if required.

The benefits of using immunoassay can be substantial, especially when considered cumulatively. A number of general considerations should be explored prior to the use of immunoassay. Competent planning will maximize the benefits achieved during any sampling effort, including those employing immunoassay. The ensuing section discusses some of the general issues relevant to employing immunoassay.

General Site-Specific Considerations

The inclusion of immunoassay into an integrated sampling effort involves several steps. First, available data and information should be reviewed to determine the best use of immunoassay. Second, the basic strategy should be identified. As a general guide, the best sampling strategy is a combination of judgmental and systematic sampling, as discussed previously.

The next step is the assessment of the total number of samples required and the available analytical resources. The type and capability of analytical methods available will depend upon the requirements of the site. The *Soil Sampling Quality Assurance User's Guide* (*1*) states that

> [t]he use of various levels of analyses should be considered when allocating resources for the remedial investigation/feasibility study. The cost of field methods (i.e., DQO level 1 and 2) time-to-data availability is usually considerably less than the cost of a DQO level 5 analysis. By coordinating levels 1 and 2 with the laboratory methods (i.e., level 3), a higher quality data set can be developed in less time and for less cost.

Additional efficiencies can be obtained by using the immunoassay to support and enhance the laboratory analyses and conversely by using the laboratory data to check the immunoassay.

The best use of immunoassay is a combination of using the data to both assess the site directly and to position the laboratory samples for maximum effectiveness. The appropriate data quality objective (DQO) level or levels should be identified. The application of an immunoassay-based method should be clear; one should know whether it will be used as the primary analytical method in a screening mode, as the primary analytical method, as the principal contamination assessment tool, or as the method of selecting sampling locations for the laboratory samples. Obviously, this assortment of options will vary in appropriateness based on the situation at the site.

After the initial information and resources have been reviewed, the sampling strategy will be finalized. The basic approach is to identify the potential sources of contamination (hot spots) and surround them using a grid system. Based upon the results of the immunoassay, those points with contamination are further surrounded by a systematic grid. Sample points that would likely confirm a contaminated area may not require analysis; however, those points that enter new areas must be sampled. This expanding grid allows contamination to be quickly and clearly defined. The final aspect of the sampling effort is the use of a systematic grid to assess the remainder of the site. This procedure will provide a complete, representative picture of the site.

The placement of laboratory-analyzed samples should be considered carefully to provide maximum data value. Is the sampling effort being completed to delineate contamination? If so, primarily contaminated samples identified by immunoassay should be selected for laboratory analysis. The laboratory analyses will provide information regarding the actual amounts of contamination and verification of the contaminant identity. If the sampling effort is being completed to investigate the success of remedial effort or for closure, samples identified as clean using the immunoassay should be chosen for laboratory testing by another method to confirm the results of the immunoassay.

Statistical Selection of Sampling Locations

The statistical selection of sampling locations is one of the more difficult tasks in the design of a sampling plan. Although a detailed discussion of the topic is beyond the scope of this chapter, there are two basic statistical approaches in sampling. The first, systematic grid sampling, may require large numbers of samples to adequately assess the contamination. The second approach is geostatistical sampling. Geostatistics is used primarily when soil homogeneity may be a concern Because contaminants may behave differently in various soils, in order to achieve a representative picture of the site, this variability must be addressed. Additionally, potential matrix effects may influence the analytical method being used. Although immunoassay methods are not commonly disrupted by soil matrix, soil variability should be addressed through analysis of collocated samples.

Conclusion

The benefits of immunoassay methods as environmental tools are clear. A more comprehensive delineation of contamination can be achieved via the use of immunoassay for a given cost. A more comprehensive site assessment can significantly reduce the cost of waste treatment or removal by providing a more concise picture of the extent of contamination. The reduced cost per sample allows for increases in sampling points at a site without an overall cost penalty. The rapid analysis on-site allows decreasing mobilization costs and a tremendous flexibility in the sampling effort. The benefits are clear; the challenge is the integration of the technology into the sampling plan.

The ability of immunoassay to enhance a sampling and analysis effort is a major step forward regarding contaminated site assessments. The ability to conserve funds and strengthen the information gathered will help increase the pace and quality of site investigations and remediations.

References

1. Barth, D. S.; Mason, B. J.; Starks, T. H.; Brown, K. W. *Soil Sampling Quality Assurance User's Guide,* 2nd ed.; Environmental Monitoring Systems Laboratory, Office of Research and Development, U.S. Environmental Protection Agency: Las Vegas, NV, 1989.
2. *Code of Federal Regulations,* Title 40; Subchapter J (7-1-93 ed.), p 163.
3. *Code of Federal Regulations,* Title 40; Subchapter J (7-1-93 ed.), p 175.
4. *Code of Federal Regulations,* Title 40; Chapter 1 (7-1-93 ed.), p 50.

Chapter 37

Preservation Techniques for Samples of Solids, Sludges, and Nonaqueous Liquids

Larry I. Bone

Although preservation techniques for solid, sludge, or nonaqueous samples have not been established, a few practices such as minimizing holding time, refrigeration, sealing sample bottles, and minimizing headspace are usually helpful. If samples are to be extracted prior to analysis, good preservation practice might involve putting the sample into the extraction medium in the field. Field preservation of samples in methanol prior to a volatile priority pollutant analysis that involves a methanol extract (SW-846 Method 5030) is probably a reasonable practice. The methanol helps preserve the sample and allows the laboratory selection of a more representative subsample. Selection of subsamples for actual analysis is quite critical for heterogeneous materials.

GOOD, RELIABLE ANALYTICAL DATA are very difficult to obtain from solids, sludges, and, to some extent, nonaqueous liquids. Sampling, preservation, and analytical procedures are better understood and are much more reliable for water and air than for solids. Not only are the methods for water and air better, but samples of these media are easier to spike, split, and dilute. The nonhomogeneous nature of the types of samples that are the subject of this section is at the heart of many problems.

Not only do scientists not know how to get good analytical results for solids and sludges, they are not even sure how to define the extent of the sample to be analyzed. For example, do those responsible for such investigations want to find the average concentration of some species over the entire solid, the worst-

3152–4/96/0737$15.00/0

case concentrations, the best-case concentrations, or something in-between? If investigators want an average over the entire sample, how large would a soil sample be? The entire Earth? Of course not. But where does the sample stop? An additional problem with heterogeneous solids and sludges is that discontinuities or variable contamination can often be seen. Everyone who has ever collected solid samples for chemical analysis has succumbed to the temptation to include "that little green spot" in the portion of the solid chosen for analysis. This inclusion immediately establishes a sample bias and suggests that a prescribed technique must be used regardless of what is seen if the investigator really wants an unbiased, but not necessarily representative, result.

This discussion has quite a bit to do with sample preservation. Before the proper technique can be chosen for any step in the sampling sequence, the investigator must somehow describe the universe to be sampled and decide what question the analytical results are expected to answer. Sometimes for hazard evaluations or plume definition in soils, a worst-case analysis is desired. That is, if a seam is present in the waste or soil, the scientist wants to analyze that seam rather than obtain an average result over the entire sample. In many cases, however, investigators may want the analytical result to be an average over the entire sample. In either case, deciding who is to make that decision is important: the sampler in the field or someone in the laboratory? If someone in the laboratory is chosen, the investigator may want to send an undisturbed sample to the laboratory without preservation. If the sampler in the field is chosen, the investigator will want to subdivide the sample in the field and preserve it. Often, samples are subdivided in the field and again in the lab without any real coordination. The result is that no one knows what kind of bias might be built into the results.

Prior Planning Is Essential

All phases of any sampling exercise need to be thoroughly preplanned. The goal of the sampling exercise must first be defined, and then its extent must be decided upon. The decision must be based on an honest assessment of the real question that the analytical results are supposed to answer. Data gathered simply for the data's sake are seldom of much real value.

The sampling plan must include all phases of the sampling exercise including containers, labels, field logs, sampling devices, blanks, splits, spikes, size of samples, composites, preservation techniques, chain-of-custody procedures, transportation, sample preparation in the field and in the lab, analytical procedures, and reporting. The sampling plan is probably best prepared by involving people from every step in the process, from the field to the laboratory, because decisions to be made on all of the various elements of the plan are interrelated. The plan should, however, have enough flexibility to allow for changes based on information collected in the field. No matter how much thought is given to possible conditions that may be encountered during the field work, unexpected conditions usually arise. The need for field splits, spikes, or composites will have a bearing on preservation methods. Preplanning is particularly important for preservation because preservatives may be put into the sample containers prior to taking them to the field. Other supplies for preservation such as ice, coolers, and sealing tape must also be available.

Standard Preservation Techniques

No prescribed preservation techniques for solids, sludges, or nonaqueous samples exist for either organic or inorganic analysis. A few helpful practices are widely accepted. These practices are to seal containers, minimize headspace, refrigerate samples during storage and transportation, and analyze the samples as soon as possible.

The need to analyze samples of this type as soon as possible cannot be over-emphasized because there is usually no way to preserve them. If the analysis requires an extraction or digestion, carrying out this step as soon as possible is usually acceptable. The sample can then be held for the standard holding time specified by the method. Even if solid samples are analyzed quickly, collecting, transporting, and storing them in an undisturbed condition is usually best if possible. For core samples from waste site investigations, either the entire core or at least a large portion of it is shipped to the lab to be sampled just prior to workup and analysis. These cores can be shipped either in the sampler, sealed in wax, wrapped in foil, or at least sealed in bottles. A 1-pt wide-mouth bottle is particularly useful. A portion of a core can be cut and trimmed so that it nearly fills the bottle. When the sample is to be analyzed, the analyst bisects the core lengthwise and takes the analytical sample from the center of the core. The analyst is to select the sample uniformly over the length of the split core if an average result is desired. Occasionally, however, a worst-case analysis is desired. In this case, the analyst selects a spot or a seam where the sample appears to be the most contaminated.

A few difficulties are involved in handling core samples this way. If a contract laboratory is being used for the analysis, the investigator has very little input into the choice of the actual sample to be tested; the investigator is at the mercy of the analyst. Human nature is such that the analyst may very well bias the sample if the sample is visibly heterogeneous. Of course, so would the person who is asking the analytical question, but at least the investigator would be aware of any bias he or she introduced. Consequently, selection of subsamples should be documented (i.e., a description of the subsample selected should be prepared by a trained observer).

The reason samples are preserved is to prevent any chemical change that might take place between the time the sample is taken and the time it is analyzed. The most frequent causes of these types of changes are volatilization, biodegradation, and oxidation–reduction. Both volatilization and biodegradation can be reduced by storing and transporting the samples at a reduced temperature. The standard ice temperature of about 4 °C is probably the most reasonable to use. Lower temperatures could reduce biological degradation and might reduce volatilization. However, freezing water-containing samples might fracture the sample or cause a slightly immiscible phase to separate and ultimately result in the release of volatile compounds. This opinion is based on the fact that lowering the temperature on some liquid-phase-packed chromatographic columns to dry-ice temperature will decrease retention times. This decrease probably results from solidification of the liquid phase.

Sealing samples and minimizing headspace is always a good idea. Anaerobic samples should not be exposed to air during storage and transport if a possibility of aerobic biodegradation or chemical oxidation exists. Sealing and reducing headspace will minimize the loss of volatile compounds.

Preservation of Samples of Volatile Organic Compounds

The most difficult problem that I have experienced in analyzing solid waste or contaminated soil samples is obtaining reliable reproducible analytical results for volatile organic compounds. In order to characterize wastes and contaminated soils at the Petro-Processors of Louisiana Superfund site, we have been attempting to run standard priority pollutant analysis for the volatile compounds on solid samples. An example of an actual blind split analysis that we carried out probably best illustrates the problem as well as suggesting a preservation technique that might be helpful.

During the field investigation of the Petro-Processors site, we wanted to check the validity of the analytical results received from a contract laboratory by running a blind split in our own Dow Louisiana Division environmental laboratory. I carefully chose a sandy sample from the bottom of a contaminated pond because the sample was easier to homogenize than most of the clay or silt samples we encountered, and the sample was also easy to split. In addition, the sample contained enough contamination to be well within the range of reliable analytical work.

The contract laboratory and the Dow laboratory ran metals and acid and base-neutral extracts by standard priority pollutant protocols and obtained good agreement. However, very poor agreement was obtained on the volatile organic compounds, as can be seen in Table I.

The contract laboratory analyzed for volatile compounds by placing a few tenths of a gram of the sample into 20 mL of water in a 25-mL vial. Volatile compounds were then sparged from the vial onto the 2,4-diphenyl-*p*-phenylene oxide resin (Tenax) adsorption column of a commercial purge-and-trap apparatus from which the samples were thermally desorbed and analyzed by gas

Table I. Volatile Analysis of a Soil Sample

Compound	*Contract Laboratory*	*In-House Laboratory*
Carbon tetrachloride	0.10	1853
1,2-Dichlorobenzene	0.10	7
1,4-Dichlorobenzene	0.10	2
1,3-Dichlorobenzene	0.10	4
1,2-Dichloroethane	0.10	525
1,1-Dichloroethane	0.10	4
1,2-Dichloropropane	0.10	664
Methylene chloride	0.10	8.3
1,1,2,2-Tetrachloroethane	80	1630
Tetrachloroethene	84	664
Toluene	13.5	2.0
1,1,2-Trichloroethane	92	3097
Trichloroethene	1.9	1740

NOTE: All results are in milligrams per kilogram.
SOURCE: Adapted from reference 1.

chromatography–mass spectrometry (GC–MS) Method 624 (2). The Dow laboratory extracted 4 g of the sample into 20 mL of methanol and then put 50–250 μL of the extract into water; this method is similar to SW–846 Method 5030 (3). The Dow samples were screened by purge-and-trap analysis on a gas chromatograph with Hall and photoionization detectors in series. (A Hall detector is an electrolytic conductivity cell.) The water-diluted methanol extract was also analyzed by purge-and-trap Method 624 (2). The contract laboratory reanalyzed their sample by using the Dow extraction method. In this later method, they used purge-and-trap gas chromatography–flame ionization detection (GC–FID) to analyze for the major purgeable compounds. They found higher concentrations than previously, but the results were still not in very good agreement with the Dow results.

We concluded from the results of Table I that the principal reason for the gross differences in the analytical results lies in the laboratory selection of the subsample to be analyzed from this heterogeneous solid. Selection of a larger sample for extraction followed by a split of the more homogeneous extract gives considerably more reliable results. Had the sample been even more heterogeneous, as most of ours were, the results probably would have been more divergent. Of course, even selection of a 4-g sample will give widely scattered results from a very heterogeneous sample. The lesson is that the full sampling process, completely through the laboratory subsampling, must be planned such that the sample selected is large enough to be representative of the universe from which the results are desired.

We also concluded that the base-neutral and acid extract results were a much better measure of the level of soil contamination at the Petro-Processors site than the volatile compounds regardless of the method used for the volatile analysis. In fact, when the U.S. Environmental Protection Agency subsequently analyzed soil samples from the site, they chose to only analyze for base-neutral and acid samples for exactly the same reason.

The experience just described also contains a plausible suggestion for preserving solid samples for volatile analysis. If the sample is to be extracted into methanol, the methanol might as well be added in the field to preserve the sample. A small wide-mouth bottle could be filled about one-half to two-thirds full of methanol in the laboratory prior to going into the field. The bottle and the methanol can then be preweighed so that the weight of sample to be added later can be calculated. In the field, a carefully selected, representative sample of about one-fourth the weight of the methanol is added to the sample bottle. The bottle is sealed, cooled to 4 °C, and transported to the laboratory. The filled bottle is again weighed in the laboratory to calculate the weight of the sample added. The laboratory can also finish the extraction by shaking, tumbling, or sonification. An appropriate ratio of methanol extract to water can be selected to achieve analytical sensitivity in the desired range.

As SW-846 Method 5030 (3) suggests, poly(ethylene glycol) probably could be used in place of distilled-in-glass methanol. In addition, tetraglyme might also be acceptable. Other cases may exist where a solvent to be used in a method might be added in the field to aid in preservation. Another example may be the *n*-hexadecane method for total volatile content described in SW-846 Method 8240 (3).

References

1. *Remedial Planning Activities Report*; Petro-Processors, Inc., NPC Services, Inc.: Baton Rouge, LA, 1985.
2. *Methods for Organic Chemical Analysis of Municipal and Industrial Wastewater*; U.S. Environmental Protection Agency. U.S. Government Printing Office: Washington, DC, 1982; EPA-600/4-82-057.
3. *Test Methods for Evaluating Solid Waste*; U.S. Environmental Protection Agency: Washington, DC, 1986; SW-846.

Chapter 38

Sampling and Analysis of Hazardous and Industrial Wastes

Special Quality Assurance and Quality Control Considerations

Larry P. Jackson

The design and conduct of sampling operations for industrial and hazardous wastes must involve the consideration of a series of related problems that are new to the technical community. Among these problems are a changing list of analytes by methods undergoing constant revision, the use of testing protocols and resulting data for nontechnical purposes, and the application of state-of-the-art methods for analytes and matrices in the absence of validation data. This chapter presents a brief discussion of each of these problems and suggests a strategy for dealing with them. The objective is to provide an approach that will allow the results from a sampling and analysis program to meet changing regulatory needs, be technically valid, and provide a means of documenting waste characteristics in a consistent manner so that the value of the data base is retained as regulatory requirements change.

PLANNING AND EXECUTION of a quality assurance (QA) and quality control (QC) program to support an environmental sampling and analysis project of industrial waste streams is complicated by problems arising from an inconsistent maze of regulatory requirements. Foremost among these problems are the rapid rate of introduction of new regulations; varying requirements of the regulatory programs among local, state, and federal agencies; and technical limitations of the methods required to sample and analyze the wastes.

These problems are compounded by the lack of guidance and assistance available from the agencies to resolve the problems prior to the submission of environmental permit applications or actual data. This lack of guidance results in the waste of countless hours of labor and many thousands of dollars for industry and regulators alike at a time when permit applications may take 18–24 months for review and when economically and environmentally important projects await a permit before implementation.

This chapter examines the Resource Conservation and Recovery Act of 1976 (RCRA), its subsequent amendments, and the supporting body of environmental regulations for QA and QC implications. This chapter presents examples of QA and QC problems, suggests limited solutions that may save time and money over the life of a project, and proposes a strategy for addressing the majority of the problems.

Institutional Problems

Regulatory Programs

Stated in simple terms, RCRA provides for the identification of hazardous and nonhazardous waste products and establishes guidelines for their disposal. RCRA was the last of the suite of major environmental regulations that grew out of the National Environmental Policy Act of 1972. RCRA was preceded by the Clean Air Act and Clean Water Act. Many of the technologies that grew out of the requirements of these first acts produce wastes that are controlled under RCRA. Examples of these types of technologies are smokestack particulate control, flue-gas desulfurization, and a host of wastewater treatment technologies. All of these processes produce a solid or sludge material that must be disposed. The 1984 Hazardous and Solid Waste Amendments (HSWA) to RCRA now require a minimum level of treatment for many wastes prior to disposal.

Each of the major acts is supported by a body of regulations written and maintained by separate offices within the U.S. Environmental Protection Agency (EPA). The actual implementation of the regulations is conducted by the regional EPA offices or state agencies that have applied for and been granted primacy by the EPA. In some of these states, the requirements are more strict than the federal standards. At each level, the interpretation of the regulations is left to local authorities. The lack of EPA accepted and published methods of analysis, accompanying method validation procedures, and sampling guidelines for many of the unusual matrices encountered in waste streams has resulted in many different local applications of the regulations. This nonuniform application results in procedures acceptable in one state or region not being accepted in another without some degree of modification. For each application of a sampling plan and its attendant QA and QC section, lengthy negotiations with local regulatory authorities must be completed before samples and data are obtained.

In some cases, the initial data submitted as part of a permit application are deemed inadequate upon review and thus require additional sampling and analysis.

The previous points make it appear that the fragmented regulatory structure and lack of published standards have an adverse impact only on industry. This situation is not the case. The public interest in a clean environment is also poorly served. The regulatory community, at all levels, is greatly overextended to issue and monitor permits. The level of technical sophistication required to conduct an intelligent sampling and analysis plan in the myriad of settings encountered in today's industrial and municipal settings is very high. What appear to be unnecessarily strict QA and QC requirements on sampling and analytical plans is an effort to protect the public interest in a clean environment in the absence of guidance. No matter how well-intentioned the applicant and regulator both may be, mistakes have been and will continue to be made. Potentially harmful situations will go undetected. Although good QA and QC standards will not eliminate mistakes, they will decrease the number. The costs, both in terms of time and money, of preparing and implementing an environmental sampling and analysis plan will be reduced.

Solutions

A variety of actions can be taken to address these types of problems. None will work overnight and all will take a commitment of time and resources on the part of private sector parties. The private sector and the regulatory community will have to minimize the adversarial nature of their relationship. Development of technically sound guidelines or standards will require presentation of sufficient supporting material or data to prove the validity of the approach sufficient to gain regulatory acceptance. Basically, the private sector must assume some of the burden of developing QA and QC guidelines for regulatory applications. This task can be achieved in three ways.

Trade associations such as the Chemical Manufacturers Association, Inc.; American Iron and Steel Institute; and American Textile Manufacturers Institute, Inc. should assemble the sampling practices and analytical methods used by their industry along with the QA and QC procedures used to ensure data quality. The methods, practices, and procedures should be written in a form suitable for publication. Descriptions of the intended applications as well as limitations or exclusions (where known) should be included. Data supporting the reliability of the intended application should be provided. The data should be of sufficient quality and quantity to withstand rigorous peer review. Once prepared, the documents should be submitted to EPA for review, potential modification, and eventual approval and inclusion in the list of approved procedures.

An alternative way to accomplish much the same thing is for individual companies or trade associations to join and participate in nationally recognized standard writing organizations such as the American Society for Testing and

Materials (ASTM). The ASTM is actively involved in writing standards for all types of applications and has developed procedures for ensuring the objectivity and technical validity of the standards that they publish. The EPA has a long history of accepting ASTM methods for regulatory application. In most cases where ASTM methods exist when regulations are initially proposed, the ASTM methods are included in the list of accepted methods.

Today, the regulatory process is being driven by timetables established by Congress. Regulations are being written in areas where no standard methods exist. The EPA is very active within the ASTM and seeks objective technical input at every opportunity to aid in developing good methods for regulatory application. This input includes sampling and analytical methods, validation procedures, and QA and QC requirements. Although the time frame of EPA's need for methods frequently is shorter than ASTM's procedural requirements for adoption, the EPA uses the technical information obtained by their participation in the process to develop their proposed methods, which are published for public comment. With more active participation from the public sector, ASTM methods can be developed faster and the necessary methods made available to meet the congressionally driven timetable.

In those cases where the EPA has not formally accepted the trade association or ASTM methods for regulatory use, the permit holder or applicant can still submit the method with its supporting documentation to local regulatory authorities in support of their use of the method. The method will be evaluated and accepted or rejected on the basis of its merits where no accepted method exists. In cases where EPA may have specified a method, the private sector party will have to submit additional data proving equivalency or establishing the lack of validity of the existing method for the particular application being considered. This process is not used as often as it should be to challenge poor practices. The private sector petitioners must provide sufficient data to support their applications, and the regulator must be receptive to a well-developed technical presentation. This approach has not met with much success in the past because many members of the regulated community and public interest groups have lobbied extensively from opposite points of view to influence the regulatory process. Both groups are beginning to accept the value of quality data to support their respective positions, and now is the time to increase the use of this approach.

Technical Problems

The past few years have brought many changes into the regulation of waste disposal practices. The Hazardous and Solid Waste Amendments of 1984 require the banning of the land disposal of certain hazardous wastes. The regulations governing a first group of these wastes went into effect on November 8, 1986, and these regulations include a broad range of new analytical requirements that exemplify the type of technical problems that must be addressed in a sampling and analysis project at an industrial waste site. The new regulations expand the list of analytes considered in making the hazardous–nonhazardous decision and introduce a new waste analysis test. Few of the existing environmental monitoring plans in place at industrial plants or proposed in pending permit applications cover these requirements.

Volatile and semivolatile organic compounds have been added to the list of analytes of concern. The list will continue to grow as the EPA gathers more data on the degree of risk associated with other compounds, and as the EPA can establish levels of these materials that are protective of the environment. The tests required to establish if the wastes are hazardous lack sufficient analytical sensitivity to reliably measure at the level deemed protective of the environment. The regulatory threshold value will decrease as the analytical methods become more sensitive and reliable. Procedures on sampling wastes are recognized within the regulations as a source of an unknown level of error. Sample holding-time limits have been established for the sample after collection and at an intermediate point in the analytical scheme. Reanalysis by the method of standard addition is required for each matrix if spike recoveries do not meet specified standards or if the measured value for the individual analyte falls within 20% of the regulatory threshold.

Before the technical details of a sampling and analysis plan are discussed, two institutional issues must be acknowledged. First, the costs associated with the analytical program establishing the environmentally safe performance of an industrial facility have increased from 10- to 100-fold in the last 15 years. The QA and QC share of the analytical budget is rising and will continue to rise as long as the regulatory requirements are vague. Second, the vagueness of the regulatory requirements and the known deficiencies of many of the required methods may allow for unintentional or intentional bias to influence the outcome of the sampling and analytical program used to establish the degree of hazard for any waste. It is hoped that both public interest groups and industry will realize that the costs for sound, objective decision making are worthwhile and that both groups are vulnerable to the adage, "Pay me now or pay me later."

Traditional QA and QC programs have a single goal: to ensure that a representative sample is available for analysis and that the analytical results are as accurate as the methods allow. Two ancillary goals should be added for QA and QC programs in support of hazardous waste determination. (1) The program should be flexible and anticipate the changes in regulatory direction to minimize permit modifications. (2) The program should be technically rigorous so that the measured values are of adequate precision and accuracy to withstand review on their own merits despite the lack of regulatory QA and QC procedures.

Modification of Quality Assurance and Quality Control Programs

Many QA and QC programs currently approved for use in regulatory settings do not address all the requirements of the new regulations. As they are modified, the QA and QC programs should be expanded to include the new types of samples and analytes covered in the new regulations and those that can be anticipated in the next 3–5 years. New analytes for regulation are likely to be chosen from Appendix VIII of the RCRA regulations. Careful review of existing data on starting materials, process chemistry, and actual waste stream composition should identify all compounds currently listed, and plans can be made at this time to include them as they're added in the future.

Two other changes can be made to existing QA and QC plans. Regulations require the taking of a sufficient number of samples to allow some preliminary testing to be done to decide on the proper protocol for hazardous determination. This determination can be done in the field as the samples are collected, and as a result, the actual number of samples collected and returned to the off-site laboratory can be reduced. The sampling plan and its attendant QA and QC section will have to be modified to document this determination, but the plan will be cost-effective if the total number of samples handled from the field to the laboratory is reduced significantly. Complete directions for analysis by the laboratory may allow a single sample to serve for the analysis of volatile and semivolatile organic compounds as well as inorganic compounds. This plan can also lead to significant reduction in sample acquisition and handling costs. Laboratory documentation practices will have to be instituted to ensure proper sample handling, but these costs should be less than working with multiple samples. In addition, biologically active wastes such as wastewater treatment sludges may contain both biologically resistant and nonresistant compounds. Modern sampling plans should contain directions on how to handle the samples and how long to store them to prevent alteration prior to analysis. The QA and QC plan should contain provisions for on-site analysis or field-spiking procedures to detect any alteration in the samples over time.

At present, the regulations do not require that field spikes be run on waste samples nor that laboratory spikes be run on the waste sample prior to conducting the *toxicity characteristic leaching procedure* (TCLP), which is the new test proposed by the EPA to determine if a waste is hazardous by the characteristic of toxicity. The absence of spiking procedures from the regulatory process is a recognition that nothing is known about how to do this test reliably. This absence of spiking procedures leads to questions of how to defend the value of the TCLP as a regulatory tool if spikes demonstrate that samples are not stable over the prescribed holding period or that analyte recoveries may be very low for certain types of compounds. This situation says nothing about being able to document that the performance of the TCLP is sufficiently precise on real-world samples to warrant its regulatory application. If any of these circumstances could be proven to exist, monumental problems would be caused for the EPA in promoting their whole regulatory scheme at a time when Congress demands regulation of wastes.

The fact that these types of QA and QC procedures are not required by regulation does not in any way relieve the sampling and analytical plan from considering them. These procedures are critical parts of any good sampling and analysis plan and must be included. Multimillion dollar decisions and permits to continue to operate depend on their results. Project sponsors must prove that the procedures proposed in the plan and implemented in the field and laboratory perform as anticipated regardless of whether the regulatory process requires the proof. This proof will require considerable effort on the part of the project sponsor and will constitute a major research project on its own. To make matters worse, proof will be required for each unique sample matrix. Effort of this type and expense may be beyond the resources of all but the largest companies. This type of project will require the commitment of industry through their trade associations. As the methods are developed and validated, they can be used to evaluate proposed regulations and improve the quality of data generated in the regulatory process. If done properly, the methods can be submitted to the EPA for inclusion in the list of accepted procedures as discussed previously.

The major questions to be addressed in this type of research project are as follows:

- How should spikes be added?
- How is the spiking level determined to ensure that appropriate levels of analytes will be found in the TCLP leachate?
- How is uniform distribution of the spike in the bulk sample achieved?
- How long are the spike samples aged prior to analysis?
- How is it determined that the sample-spike mixture is chemically passive so that observed values will be a measure of the sampling and analysis steps and not a measure of sample-spike alteration?

Interferences

As the number of analytes increases and the regulatory thresholds are lowered to near or below reliable quantitation limits, the importance of sample matrix effects will increase. The regulations currently imply that the analytical methods contain adequate discussion of interferences, but this implication is incorrect. The discussion of interferences is limited and often is concerned with those interferences present in relatively simple samples. Nowhere is this situation better illustrated then by EPA's own special analytical services (SAS) contracts issued as part of their contract laboratory program (CLP), which supports the Superfund activity. The SAS program was developed because the analytical complexity of many samples was beyond the capabilities of all but the most sophisticated laboratories. Special procedures are used with the permission of the EPA in these laboratories to analyze difficult samples. These procedures are frequently arrived at by verbal agreement between EPA personnel and the analyst. The procedures must be fully documented, but full validation is not required. This reasonable approach is denied to participants in the private sector when faced with RCRA-related problems.

Interferences are frequently considered to be compounds whose presence obscures the measurement of the analyte of interest by the introduction of an unrelated analytical signal where the analyte is measured. Interferences can also suppress the signal of interest to a level that the analyte cannot be accurately measured. These problems are sample- and method-specific and will have to be solved to the extent possible by existing procedures. The easiest solution may be the substitution of another approved method. The best example is the choice of standard flame atomic absorption spectrometry, graphite furnace atomic absorption spectrometry, or inductively coupled plasma spectrometry for the analysis of inorganic elements. These methods are all accepted methods and can be selected to avoid the most serious interferences.

Analysis of Organic Analytes

The analysis of organic analytes is not as easy. The number of broad-range analytical methods approved for general use is limited. Currently, combined gas chromatography–mass spectrometry (GC–MS) is the only universal method that can measure all volatile and semivolatile organic compounds. GC with specific-element detectors has limited utility in that it is most sensitive for compounds containing specific elements such as chlorine, sulfur, or nitrogen. These methods suffer from the same type of interferences as the inorganic methods, but they are not as interchangeable. In addition, these methods are generally more expensive. These methods suffer from another limitation: They are so sensitive that they are subject to detector overload causing a protective instrument shutdown or prolonged loss of measurement sensitivity. If the sample injected into the instrument has a single component in high concentration, the entire analytical procedure may be invalidated.

Validation of Screening Methods

Sampling and analysis plans will have to take advantage of the screening tests allowed under the regulations to establish that specific analytes are not present at regulatory threshold levels in the waste sample. Then a method can be selected from the approved measurement methods for the remaining material. Each of the screening methods will have to be validated because very few are approved for use at this time. As with the spiking method problem, the validation process requires considerable resources but may offer the only alternative to accurate measurement of many analytes in the presence of high concentrations of other compounds. The EPA is helping in this regard by publishing a general method validation guideline, and many of the new and reapproved methods are being published with matrix evaluation and method validation procedures. As with other discretionary issues, the private sector will have to negotiate with local regulatory authorities for the application of these new procedures. These procedures offer an opportunity to effect considerable cost savings over time once they gain widespread approval.

Abbreviations

ASTM	American Society for Testing and Materials
CLP	contract laboratory program
EPA	U.S. Environmental Protection Agency
GC–MS	gas chromatography–mass spectrometry
HSWA	1984 Hazardous and Solid Waste Amendments
QA	quality assurance
QC	quality control
RCRA	Resource Conservation and Recovery Act
SAS	special analytical services
TCLP	toxicity characteristic leaching procedure

Chapter 39

Cost-Effective Sampling for Spatially Distributed Phenomena

Leon E. Borgman, Ken Gerow, and George T. Flatman

Various measures of sampling plan cost and loss are developed and analyzed as they relate to a variety of multidisciplinary sampling techniques. The sampling choices examined include methods from design-based sampling, model-based sampling, and geostatistics. Graphs and tables are presented that allow the comparison of the various choices relative to a specified site and set of decision circumstances. The strengths and weaknesses of each choice are discussed, and special assumptions relative to prior knowledge are outlined. Examples are developed to illustrate the procedures.

THE EFFICIENT USE OF SUPPLEMENTARY INFORMATION is critically important in designing sampling plans that will produce the maximum amount of information for the money spent in data collection. The intricate interplay of subjectivity versus objectivity, generality of information gained versus narrowly focused estimation goals, the use to which the ultimate data are to be primarily applied, and other such considerations will be examined in the context of a sequence of simple examples. Because this presentation is directed to environmental specialists rather than statisticians, the theoretical basis of each example is sketched out briefly so the reader can gain some perspective on the conditions under which each method is appropriate.

Finally, various general considerations will be outlined and discussed to help the environmental scientist make considered judgments in this rather confusing area of selecting the best sampling plan for a particular problem. The emphasis will be placed on making estimates with controlled errors in all the subregions of an area,

3152–4/96/0753$16.50/0

rather than simply estimating the overall average for the whole area. The distribution of contamination over the space is generally more useful in remediation than any single average value for the area.

There are advocates for each particular sampling approach who know their methodology in great detail, but who are somewhat sketchy about other approaches to sampling. Some of these investigators rather stridently condemn methods other than their own and have an almost evangelical belief in their own techniques. We have encountered this enthusiasm in statisticians working with classical random sampling methods, geostatisticians using kriging and variograms, and statisticians in other specializations. Of course, most investigators maintain a more balanced and tolerant view. The purpose here is not to single out any group, but rather to demonstrate that each method is appropriate for its range of circumstances and within its frame of reference. More importantly, we hope to guide the reader to a reasonable understanding and appreciation for the various considerations that come into the process of selecting a sampling methodology, so that when controversy arises there is some basis for accommodation and adjustment.

Numerical Measures of Sample Plan Cost Effectiveness

Two different numerical measures of sample plan cost effectiveness are proposed. Each has been found useful under some circumstances. These are the CPE (cost per error), defined as the total cost of sample collection and laboratory analysis divided by a selected measure of estimate error, and the EPC (error per cost), which is the reciprocal of the CPE. The CPE is useful when a fixed allowable error is mandated, and the sample plan that minimizes the CPE will give the lowest cost. The EPC is appropriate when a fixed budget has been allocated for the data collection and analysis, and the sample plan having the lowest EPC will give the smallest standard error for the estimate.

A sample plan has a CPE and EPC rating in each subregion for which an estimate is to be prepared. The conservative extreme (i.e., the maximum value) of the rating over all estimates serves as an overall rating number for the sample plan, in general.

These ratings can take a number of functional forms depending on the application. Three particular cases will be discussed here. These are (1) one sample and one analysis, (2) one sample and several analyses, and (3) several samples and one analysis. Let $n_{sample,i}$ be the number of samples collected from the subregion i, and $C_{sample,i}$ be the collection cost per sample in that subregion. Similarly, let $n_{analysis,i}$ be the number of laboratory analyses performed for the samples from the subregion i, and $C_{analysis,i}$ be the cost per analysis of samples from that subregion. The root-mean-square error for the difference between the subregion actual average value and the estimate of the subregion i average value will be denoted by e_i. The two overall ratings are

$$\mathrm{CPE} = \frac{\sum_i \left(n_{\mathrm{sample},i} C_{\mathrm{sample},i} + n_{\mathrm{analysis},i} C_{\mathrm{analysis},i} \right)}{\sqrt{\mathrm{Max}_i \left(e_i^2 \right)}} \tag{1}$$

and

$$\mathrm{EPC} = \frac{\sqrt{\mathrm{Max}_i \left(e_i^2 \right)}}{\sum_i \left(n_{\mathrm{sample},i} C_{\mathrm{sample},i} + n_{\mathrm{analysis},i} C_{\mathrm{analysis},i} \right)} \tag{2}$$

where $\sqrt{\mathrm{Max}_i(e_i^2)}$ is the largest (over all i) observed e_i. The actual computation of e_i for each subregion will depend on the sampling and estimation procedures employed. For example, in geostatistics the spatial covariance function is modified to add a "nugget effect" consistent with the analysis error in estimating the value at each location. This modified covariance incorporates both the laboratory analysis error and the spatial variability into the estimation of the subregion average.

If each field sample unit is split into several separate laboratory units for analysis, then n_{analysis} will be larger than n_{sample}. This splitting is particularly appropriate if the cost of sample collection is high (e.g., deep core drilling) and the analysis error is substantial enough to benefit from averaging the determinations from several splits. Conversely, if the cost of sample collection in the field is relatively small compared with C_{analysis}, it may be desirable to combine, or composite, all the samples from a subregion into one or a few samples, and then to make careful determinations on the reduced number of samples. There are many variations on these choices, one of which may be optimal in a particular circumstance.

Design-Based, Model-Based, and Kriging-Based Sampling and Estimation

In this section, the pros and cons of imposing various model assumptions (and acting on them) in a sampling situation are considered. It is beyond the scope of this chapter to engage in a full discussion of this topic; detailed discussion of different approaches to a single data set will illustrate the important aspects of the general issue.

First a fictitious population is introduced, one with a specified structure. Then, four approaches to sampling from it (with a given estimation goal) are delineated; each successive method involves more sophisticated modeling. The four methods are contrasted and compared with respect to their costs and benefits.

The Example Population

As a means of motivating subsequent discussion, a simple example with known structure will be introduced. The locations are assumed to lie along a straight line of length 1000 m. A graph of the data for the example is shown in Figure 1.

Statistically the data were developed as a single realization of a stochastic process with structure given by

$$Y(x) = 5 + 0.1x + \varepsilon(x) \tag{3}$$

where $\varepsilon(x)$ is a stationary random function with expectation zero and covariance function, $C(h)$, given by

$$C(h) = \begin{cases} 6 \times (1 - |h| / 600) & \text{if } |h| < 60 \\ 0 & \text{otherwise} \end{cases} \tag{4}$$

where h is the distance between locations. However, in the following discussion, it will be assumed that the values in Figure 1 are not known and that a sampling plan is to be selected to estimate the average y-value for the intervals of x specified by (0, 500 m) and (500 m, 1000 m). The 95% confidence interval for each estimate is supposed to be ±1 part per million (ppm). It is also assumed that preliminary information suggests that the variance is about 6.5 ppm^2 in each subregion. This sampling (and estimation) goal will be used to clarify the properties of different sampling procedures described in subsequent sections.

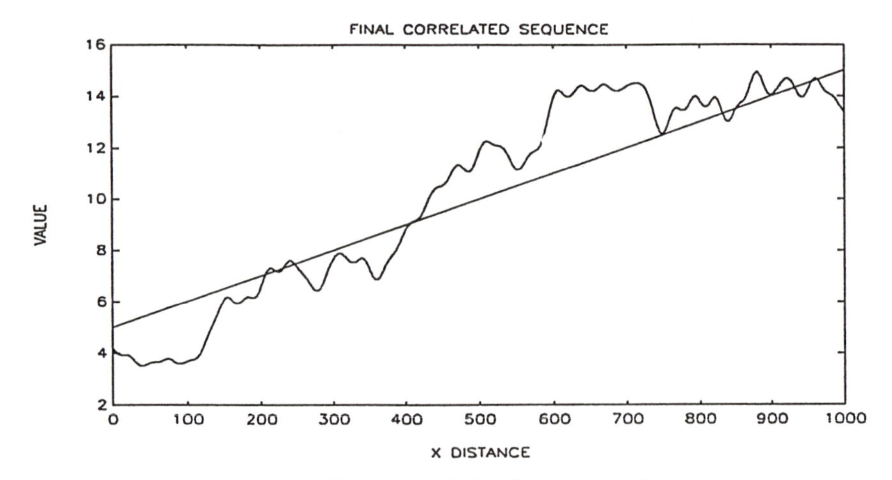

Figure 1. Fictitious population along an interval.

Method 1: Simple Random Sampling in Each Subregion

The two subregions (0, 500 m) and (500 m, 1000 m) are each sampled independently via simple random sampling. After n sample units are measured, the average of the measurements is used as an estimate of the subregion average. What sample size should be used to attain the target accuracy? This question will be answered after a brief review of the rationale of the technique.

The subregion is considered as consisting of a totality of N sample units (or "bags of dirt"), from which n units are to be selected for analysis. The selection technique is designed to imitate what one would get if all of the N bags of material were placed in a large container and mixed thoroughly so the bags end up being positioned randomly with respect to each other, with n bags selected from the mixed population of N bags in such a way as to make all bags equally likely to be included in the sample. The mixing process visualized purposefully does not use any spatial interrelationships that might have been present. It just leaves a big finite population of N items (bags) $(y_1, y_2,..., y_N)$ with population mean (μ) and variance (σ^2)

$$\mu = \frac{1}{N}\sum_{i=1}^{N} y_i$$
$$\sigma^2 = \frac{1}{N}\sum_{i=1}^{N}(y_i - \mu)^2 \qquad (5)$$

The sample mean ($\bar{y}$) of a sample of size n

$$\bar{y} = \frac{1}{n}\sum_{j=1}^{n} y_j \qquad (6)$$

is an estimate of μ.

What is the variance of the sample mean under the mixing assumption that each bag in the finite population is equally likely to be included in the sample? The derivation is basically an exercise in combinatorial algebra. There are $K = N!/[n!(N-n)!]$ different samples of size n that can be drawn from the finite population, and each sample is equally likely. Hence, the probability that a particular sample is drawn is $1/K$. Let these possible samples be enumerated with $k = 1, 2,..., K$, and let y_{jk} represent the jth value in the kth possible sample. The mean ($\mu_{\bar{y}}$) and variance of the mean ($\sigma_{\bar{y}}^2$) relative to the collection of all possible samples of size n are

$$\mu_{\bar{y}} = \sum_{k=1}^{K}\left(\frac{1}{n}\sum_{j=1}^{n} y_{jk}\right)/K$$
$$\sigma_{\bar{y}}^2 = \sum_{k=1}^{K}\left(\frac{1}{n}\sum_{j=1}^{n} y_{jk} - \mu_{\bar{x}}\right)^2/K \quad (7)$$

After some algebra, recognizing that a single individual sample unit, y_i, can be present in $(N!)/[(n-1)!(N-n-1)!]$ of the K samples, the formulas for the mean and variance of $\bar{y}$ are found to be

$$\mu_{\bar{y}} = \mu$$
$$\sigma_{\bar{y}}^2 = \frac{\sigma^2}{n}\left(\frac{N-n}{N-1}\right) \quad (8)$$

If the population size is much larger than the sample size ($N >> n$), then the variance of the sample mean reduces to the familiar σ^2/n. For much of what follows, $N >> n$ will be assumed.

It is assumed that, from preliminary information, the population variance is approximately 6.5. With the additional assumption of normality, the 95% confidence interval for μ is $\bar{y} \pm 1.96\sigma/\sqrt{n}$. Hence, the accuracy constraint is satisfied if

$$1.96\sigma/\sqrt{n} = 1.0$$
$$n = (1.96)^2\sigma^2 = (1.96)^2(6.5) = 25 \quad (9)$$

About 25 random sampling locations are needed in each interval, or a total of 50 sample units, overall.

Several comments can be made about the simple random sampling framework. The method involves very few subjective assumptions (it *does* assume normality of the sample means, which is at least approximately satisfied if the sample size is large enough). In some applications, the definition of the number of subregions and subregion boundaries can be quite subjective. Any spatial correlation that is present in the actual in situ samples is purposefully not used; the assumptions of spatial stationarity and intercorrelation are not required and are redundant in the framework of simple random sampling.

On the downside, if spatial correlation is present, this method will not take advantage of it. In that event, one can presumably get a more precise estimate (i.e., smaller variance) with more sophisticated techniques.

In sum, the population is thought of as a finite, fixed collection of objects, and the only randomness results from the selection of a sample from the set of all possible samples. In particular, the only realization of the stationary random function $\varepsilon(x)$ in eq 3 that is of interest is the realization actually present.

A disadvantage is that nonrandom samples selected for other purposes cannot be used to increase the estimate accuracy. Only those sample locations picked by a mechanical random method can legitimately be used in the estimator. This limitation also causes some logistical problems because each sample unit requires the movement to a new location that may not always be geographically convenient or even accessible.

The true mean of the population is the arithmetic average of all the bags of material. Hence, μ is the actual value that would be obtained if the sample were enlarged to be the full population. The sample, as a fraction of the total population, is an estimate of this actual value.

This procedure is different from many estimation procedures in which the sample mean is an estimate of the "expected value" relative to some probability law. The expected value, or population mean, refers in these other estimation procedures to a probabilistic average over an ensemble of possible realizations. A model illustrating this situation will be covered next. The spatial average over a geographical region has its own sampling variation and may be different from the expected value relative to a probability law because the actual population presented by nature is thought of as just one realization of a random process.

Interestingly enough, geostatistical procedures also directly estimate the block average that would be obtained if one dug up all the material and processed it totally. Thus, finite population random sampling and certain geostatistical procedures share this common property of seeking to estimate the actual spatial average of the in situ material.

The difference between the two approaches is ultimately philosophical. From the geostatistical perspective, the actual spatial average is thought of as a random variable; in the classical statistical school of thought, it is considered an unknown constant. Matheron (*1*) names the two approaches "transitive methods" and "intrinsic theory". He points out that the differences are epistemological and that similar results (from a practical viewpoint) are obtained from both approaches.

Method 2: Model-Based Sampling with Independent Error

Method 1 had the following model assumptions: no trend exists in y over the interval, and no correlation exists among the sampling errors. We will now assume a trend. Suppose that it is believed that the value of y appears to increase linearly as x increases, so that

$$y(x) \approx a + bx \tag{10}$$

Suppose further that supplemental information gives good physical reasons to believe that this relation is dependable and certain enough to be used in planning the sampling.

An estimate of the subregion average could be produced by making a straight line fit to the sample values by linear regression. Let $\hat{y}(x) = \hat{a} + \hat{b}x$ be the estimate from linear regression. For our example population, an estimate of the averages could be obtained by averaging this line over each interval. The notation < > will be used to denote a space average (here over an interval).

$$\langle \hat{y} \rangle_1 = \frac{1}{500} \int_0^{500} \hat{y}(x)\,dx$$
$$\langle \hat{y} \rangle_2 = \frac{1}{500} \int_{500}^{1000} \hat{y}(x)\,dx \qquad (11)$$

The subscripts indicate which of the two intervals are involved.

How large are the samples required for the target confidence interval size (±1 ppm), and what are the advantages and disadvantages of this approach? The determination of the necessary sample size requires making further assumptions about the underlying model structure. The usual linear model is

$$y(x_i) = a + bx_i + \varepsilon_i$$
$$E(\varepsilon_i) = 0 \quad \mathrm{Var}(\varepsilon_i) = \sigma^2 \qquad (12)$$

where $E(.)$ denotes expected value, a and b are regarded as constants, and the ε_i values are taken as independent. The locations at which values are measured are (x_1, $x_2, x_3, \ldots, x_n$), where some of the n sample locations are contained in each subregion. In other words, we will be using information from subregion 1 to improve the estimate for subregion 2, and vice versa. For the purposes of the following discussion, these assumptions will be accepted as valid. The integrals in eq 11 can be evaluated, for $j = 1, 2$, to get

$$\langle \hat{y} \rangle_j = \sum_{i=1}^{n} A_{ij} y_i$$
$$A_{ij} = \frac{1}{n} + \frac{\left(\langle x \rangle_j - \bar{x}\right)\left(x_i - \bar{x}\right)}{S_{xx}} \qquad (13)$$
$$S_{xx} = \sum_{i=1}^{n} (x_i - \bar{x})^2$$

and in this example, $< x >_1 = 250$ and $< x >_2 = 750$. The space average estimate can be written in terms of the y_i as

$$\langle \hat{y} \rangle_j = \hat{a} + \langle x \rangle_j \hat{b}$$

$$\hat{b} = \sum_{i=1}^{n} \frac{(x_i - \bar{x})}{S_{xx}} y_i \qquad (14)$$

$$\hat{a} = \bar{y} - \hat{b}\bar{x}$$

From standard regression theory, these are unbiased estimates in the sense that they have the same expected value as the expected value of the actual subregion space-integrated means given by

$$\langle y \rangle_j = a + \langle x \rangle_j b + \langle \varepsilon \rangle_j \quad j = 1,2 \qquad (15)$$

Two types of averages are involved here: the integral over x and the probabilistic or ensemble average. The difference (Δ_j) between the estimate of the strata space averages and the actual space averages is then

$$\Delta_j = \langle y \rangle_j - \langle \hat{y} \rangle_j$$
$$= (a - \hat{a}) + \langle x \rangle_j (b - \hat{b}) + \langle \varepsilon \rangle_j \qquad (16)$$

From eq 14, the variance of these differences under the assumption of independent errors is

$$\mathrm{Var}(\Delta_j) = \sum_{i=1}^{n} A_{ij}^2 \mathrm{Var}(y_i)$$
$$= \left[\frac{1}{n} + \frac{\left(\langle x \rangle_j - \bar{x} \right)^2}{S_{xx}} \right] \sigma^2 \qquad (17)$$

In order to compare the required sample size in this case with that in the simple random sample, choices have to be introduced for the x_i. Let $x_i = 1000i/(n + 1)$. This relationship spaces the observations evenly over the 1000 m and gives, with the use of the summation identities

$$\sum_{k=1}^{n} k = \frac{n(n+1)}{2}$$

$$\sum_{k=1}^{n} k^2 = \frac{n(n+1)(2n+1)}{6} \qquad (18)$$

and

$$\bar{x} = 500$$

$$S_{xx} = \frac{250{,}000}{3}\left[\frac{n(n-1)}{(n+1)}\right] \tag{19}$$

$\mathrm{Var}(\Delta_j)$ (for $j = 1, 2$) reduces to

$$\mathrm{Var}(\Delta_j) = \left[\frac{1}{n} + \frac{3(n+1)}{4n(n-1)}\right]\sigma^2 \tag{20}$$

$\mathrm{Var}(\Delta_j)$ is the same for both $j = 1$ and $j = 2$; this equality is a peculiarity of this example (exactly two subregions, and precisely splitting the region of interest into equal halves); this feature is not generally present. A tabulation of $1.96 \times \sqrt{\mathrm{Var}(\Delta_j)}$ (i.e., the 95% deviation for Δ) versus n is given in Table I. Some of the sample will be taken in each region; here, n is the *total* sample size. Visible in Table I is that the relationship between size of interval and n is not linear: if one doubles the sample size, one does not cut the length of the interval in half.

Method 3: Model-Based Sampling with Correlated Error

Once trend modeling assumptions have been introduced, it is natural to consider if other assumptions might be appropriate. A natural structure to introduce is that of correlated error. Suppose, for example, that prior information suggests that the covariance function for the error, $\varepsilon(x)$, in the model

$$y(x) = a + bx + \varepsilon(x) \tag{21}$$

Table I. 95% Confidence Deviation Assuming Error Independence

Sample Size	*95% Confidence Interval Deviation*
10	2.188
20	1.511
30	1.225
40	1.057
41	1.043
42	1.031
43	1.018
44	1.006
45	0.995

is approximated by

$$C(h) = 6.5 \times \left(1 - \frac{|h|}{h_0}\right) \tag{22}$$

for $|h| < h_0$, and 0 otherwise, where h_0 is the distance beyond which covariance is assumed to be negligible. The covariance function in eq 4 was used to generate the population. For the purpose of analyzing the sample from the population, we've estimated the variance to be 6.5, and h_0 (true value is 600) remains an unknown. The case of independent errors (method 2) can be thought of mathematically as a special case of the current model, with correlations equal to 0.

Several approaches can be considered. One could just estimate a and b by least-square formulas and use eq 14 as before, with the understanding that the variance of Δ_j in eq 16 now depends on the covariance function. Another option is to use the generalized linear model with correlated error. A third choice is to use a technique from geostatistics called "universal kriging" with an imposed assumption of linear trend. In the following, the first (method 3) and third (method 4) approaches will be examined, to find the average of $y(x)$ for x in (0, 500 m).

If eq 14 is combined with eq 16 for $j = 1$, one gets

$$\Delta_1 = \langle y \rangle_1 - \sum_{i=1}^{n} A_{i1} y_i \tag{23}$$

The weights A_{ij} are the same as for the uncorrelated error case just considered. The variance of eq 23 can be written as

$$\mathrm{Var}_{\mathrm{corr}}(\Delta_1) = \mathrm{Var}\big(\langle \varepsilon \rangle_1\big) + \mathrm{Var}\left(\sum_{i=1}^{n} A_{i1} y_i\right) - 2 \times \mathrm{Cov}\left(\langle \varepsilon \rangle_1, \sum_{i=1}^{n} A_{i1} y_i\right) \tag{24}$$

where the subscript "corr" reminds us that we are now assuming correlated errors. This expression can be further expanded to

$$\mathrm{Var}_{\mathrm{corr}}(\Delta_1) = \mathrm{Var}\big(\langle \varepsilon \rangle_1\big) + \sum_{i=1}^{n}\sum_{k=1}^{n} A_{i1} A_{k1} \mathrm{Cov}(\varepsilon_{i1}, \varepsilon_{k1}) - 2 \times \sum_{i=1}^{n} A_{i1} \mathrm{Cov}\big(\langle \varepsilon \rangle_1, \varepsilon_{i1}\big) \tag{25}$$

The variances and covariances in eq 25 can be expressed in terms of $C(h)$, with C_{BB} denoting $\mathrm{Var}(<\varepsilon>_1)$, C_{ik} representing $\mathrm{Cov}(\varepsilon_i, \varepsilon_k)$, and C_{Bi} standing for $\mathrm{Cov}(<\varepsilon>_1, \varepsilon_{i1})$, as

$$
\begin{aligned}
C_{BB} &= \left(\frac{1}{500}\right)^2 \int_0^{500}\int_0^{500} C(u-w)\,\mathrm{d}u\,\mathrm{d}w \\
C_{ik} &= C(x_i - x_k) \\
C_{Bi} &= \frac{1}{500}\int_0^{500} C(u-x_i)\,\mathrm{d}u
\end{aligned} \tag{26}
$$

where u and w are dummy variables of integration. Using eq 26, the expression for $\mathrm{Var}_{\mathrm{corr}}(\Delta_1)$ simplifies to

$$
\mathrm{Var}_{\mathrm{corr}}(\Delta_1) = C_{BB} + \sum_{i=1}^{n}\sum_{k=1}^{n} A_{i1}A_{k1}C_{ik} - 2\times\sum_{i=1}^{n} A_{i1}C_{Bi} \tag{27}
$$

This variance is for the least-squares regression case with correlated error. Table II compares sample size requirements for the regression method with universal kriging (discussed next). The first three methods (simple random sampling, uncorrelated-errors regression, and regression with correlated errors) are a set of nested models. Successively, each technique assumes more structure in the data: first a straight-line trend in the mean, and then nonindependent observations. In each step, one can gain precision in one's estimates by successfully incorporating existing structure. Caution is warranted, however, in that if one has incorrectly specified that structure, one's estimates will likely be biased.

Method 4: Universal Kriging with a Linear Trend

The universal kriging approach is also used for the case of correlated errors, with an additional twist: The coefficients A_{ij} are redefined to take full advantage of the presumed correlation structure in producing an estimator with smallest possible mean square error. We will here relabel the coefficients $A_{\mathrm{krig}(i,j)}$. Universal kriging determines the coefficients A_{i1} by minimizing

$$
E\left\{\left[\langle y(x)\rangle_1 - \sum_{i=1}^{n} A_{\mathrm{krig}(i1)}\,y(x_i)\right]^2\right\} \tag{28}
$$

where

$$
\begin{aligned}
\langle y(x)\rangle_1 &= a + b\langle x\rangle_1 + \langle \varepsilon(x)\rangle \\
y(x_i) &= a + bx_i + \varepsilon_i
\end{aligned} \tag{29}
$$

Table II. Correlated Error Sample Sizes for Regression and Kriging

h_0	Var_{corr} Sample Size	Var_{krig} Sample Size
10.0	37	37
100.0	16	9
200.0	12	8
300.0	8	7
400.0	6	6
500.0	6	5
600.0	6	5

NOTE: h_0 is the distance beyond which we assume negligible correlation between two points.

In order to ensure unbiasedness, two constraints are imposed on the coefficients:

$$\sum_{i=1}^{n} A_{\text{krig}(i1)} = 1.0$$
$$\sum_{i=1}^{n} A_{\text{krig}(i1)} x_i = \langle x \rangle_1 \tag{30}$$

The first constraint ensures that the weights will create a true weighted average of the observations; the second constraint ensures that the weighted observations are "balanced" (in a special sense) in the interval. After further computation (relatively sophisticated mathematics are involved; interested readers are referred to the Appendix), $\text{Var}_{\text{krig}}(\Delta_1)$ can be written in the same form as eq 25, replacing the A_{ij} terms by $A_{\text{krig}(ij)}$. The calculation procedures for $A_{\text{krig}(ij)}$ are shown in the Appendix.

Sample Sizes for Correlated Errors

Sample sizes for least-square regression estimates with correlated errors and for universal kriging can be approximated, respectively, by $1.96\sqrt{\text{Var}_{\text{corr}}(\Delta_1)}$ and $1.96\sqrt{\text{Var}_{\text{krig}}(\Delta_1)}$. The values of total sample size that give 95% confidence deviations of 1 ppm are shown in Table II, for a variety of hypothesized h_0 values.

For very small and very large h_0 values, there is almost no difference in the required sample sizes. The optimality of the kriging in minimizing errors reduces the sample sizes somewhat, as compared with the least-square regression, but the differences are not impressively large. The interesting aspect of Table II is that the required sample sizes decrease substantially as h_0 increases, regardless of which method is used. If the correlation scale is on the order of 300.0 or more, and the assumptions of the model are acceptable, then a sample of eight observations evenly spaced along the 1000.0-m line will give the same precision of estimate as the 50 observations of the earlier simple random sampling plan. Model assumptions take the place of data. This substitution, of course, is at the expense of a substantial increase in subjectivity, and a loss of objectivity. If it is cheap to collect and analyze samples, then the larger sample of 50 observations is probably preferable. However, if either sample collection or analysis is expensive, then the decrease to a sample of eight observations is very appealing.

Cost Effectiveness of Sampling Plans

The balance between greater objectivity (larger sample size and cost) and greater subjectivity (smaller sample and cost) is at the heart of the determination of cost effectiveness for sample plans. The cost per sample collection from the field and the cost per laboratory analysis of that data all have to be considered. Ultimately, the values from the laboratory analysis have to be representative of the in situ values in the field. Two possible errors are involved here, each with its own variance.

Even if the laboratory analysis gave the correct value of the sample, the average of the samples may not give the true average over the in situ region in the field. This difference between the sample value and the in situ region in the field that it is supposed to represent is called the fundamental error by Gy (2) and Pitard (3). No amount of sample splitting or analytical accuracy in the laboratory can reduce the fundamental error. The sample size calculations previously discussed here were for this fundamental error. The conclusion is that model assumptions, if correct, can allow smaller samples to get the same fundamental error.

The error in laboratory determination of the value for a sample is additionally added to the fundamental error to get the estimate of the in situ field values. Fortunately, the laboratory errors can be mitigated by averaging of multiple splits from the sample, or studies of procedural precision in the analytical techniques with chemical standards. These comments are not intended to minimize the importance of making the laboratory errors small. All types of error arising in the collection and handling of the samples are important. Having recognized this fact, it is important to appreciate that if a large fundamental error is present, it may be a waste of time to spend money making other errors small. If the fundamental error is large, then the values from the analysis will simply not represent what is in the field. Gy's fundamental error is examined from the viewpoint of statistics in Borgman et al. (4). Cressie (5) and Isaaks and Srivastava (6) provide excellent discussions of kriging. Cressie's study (5) is relatively theoretical and is written from a classical statistical perspective (largely), whereas the Isaaks and Srivastava's book (6) is applied and is written from a geostatistics perspective.

Stratified Random Sampling

In the preceding discussions of design-based, model-based, and kriging-based sampling and estimation, the study region was divided into two subregions, and thus the term "strata" may have seemed applicable. Stratified sampling methodology has been developed for times when one wishes to average over the subregions to perform a single estimation for the entire region. In the preceding section, separate estimates were required for each subregion; reserving the word *strata* (and methodology associated therewith) for "averaging over subregions" avoids confusion. If the two subregions in the preceding discussion had been further subdivided, with these subdivisions to be used in forming the subregion estimates, then some form of stratified sampling would have been appropriate for each subregion.

Most discussions of stratified sampling assume that the strata already have been identified; the discussion then focuses on optimal use of the strata. Choosing strata is not an easy task; it is rather like a much more difficult version of choosing bins for a histogram. With histograms, it is well understood that varying the number and width of bins (and choosing rules for dealing with ties) can dramatically alter the histogrammatic representation of the data. Each user must try different combinations of options and subjectively choose the combination that seems to best represent the data.

Here, choice of number and size (and shape) of strata will impact the ultimate cost-effectiveness of the resulting sampling. One should choose various combinations of number, size, and shape, and then assess them for their predicted cost effectiveness (a potentially tedious task). Another order of complexity is added by having to choose a variable over which to do the stratification. In a landscape setting, for instance, one could consider cover type, elevation, or distance from roads or water. Fortunately, some help is available for this problem. Bloch and Segal (7) consider the use of contingency-table cell aggregation, classification trees, and logistic regression as tools in strata definition.

Cost-Effective Stratified Random Sampling

The question of cost-effective sampling can be conveniently viewed from two perspectives. One perspective applies when one has a fixed budget and wishes to maximize the information value of the sampling effort. In the other, a specified level of precision is desired, and one wishes to determine the cost of achieving that precision. We will discuss cost-effective stratified random sampling from both perspectives.

The examination of cost-effective stratified random sampling does beg the question, "How do we optimally choose the strata in the first place?" For our purposes here, we restrict attention to the problem of devising a sampling scheme to estimate μ_y, the population mean of some variable Y over a finite population.

Fixed Budget and Optimal Precision

The optimal allocation of sampling effort for stratified random sampling, given (1) a fixed total budget, (2) known sampling (and analysis) costs for each stratum, and (3) known (or estimated) variances (in situ and analysis combined) for each stratum, is well known and documented in several standard sources (e.g., references 8 and 9). For the practitioner, however, several important comments are worth making. We work here with the formulation in Thompson (9). Suppose we have L strata, labeled $i = 1, 2, \ldots, L$. The total population size is $N = N_1 + N_2 + \ldots + N_L$. The variance in stratum i is denoted by σ_i^2, and the standard deviation by σ_i.

Let the total cost be $c = c_o + c_1 n_1 + c_2 n_2 + \ldots + c_L n_L$, where c_o is the fixed overhead cost; n_i is the sample size in stratum i; $i = 1,2, \ldots, L$ (there are L strata total); and c_i ($i = 1,2, \ldots, L$) is sampling (and analysis) cost of each sample in stratum i. In this setting, then, the sampling budget is $c_s = c - c_o$.

The optimal allocation of the sample n ($= \Sigma n_i$) is determined by

$$n_i = \frac{c_s N_i \sigma_i / \sqrt{c_i}}{\sum_{k=1}^{L} N_k \sigma_k \sqrt{c_k}} \tag{31}$$

In the denominator of eq 31, $\sqrt{c_k}$ appears as a multiplier, not a divisor. A derivation of this formula is found in reference 8 (p 109). Features of this formulation include the following:

1. The stratum sample sizes n_i are proportional to $N_i \sigma_i / \sqrt{c_i}$: that is, increased effort is put into larger strata, strata with higher variances, and strata with *lower* costs. Of course, all three of these factors interact in the sense that, for instance, high variance can offset lower cost in a given stratum.
2. If the costs of sampling, c_i, are all equal to, for example, c^*, then the formula reduces to the formula for optimal allocation for minimizing the variance:

$$n_i = \frac{\frac{c_s}{c^*} N_i \sigma_i}{\sum_{k=1}^{L} N_k \sigma_k} \tag{32}$$

3. If the variances are all equal, but the sampling costs are not, then the formula simplifies to

$$n_i = \frac{c_s N_i / \sqrt{c_i}}{\sum_{k=1}^{L} N_k \sqrt{c_k}} \tag{33}$$

4. If both variances and costs are assumed to be equal, then the formulation reduces to *proportional allocation*:

$$n_i = \frac{\frac{c_s}{c^*} N_i}{\sum_{k=1}^{L} N_k} = \frac{c_s N_i}{c^* N} \tag{34}$$

5. In its full-blown formulation (optimal allocation for differences in variance and sampling cost), the standard deviations σ_i themselves are not important; only the *relative* sizes of the standard deviations are important. This formulation is useful because, invariably, one does not know the variances and must apply the formula with some form of best guess (either from other data or perhaps from expert opinion). For purposes of solving the allocation problem, one could, for instance, choose the stratum that one believes to have smallest internal variance and set the standard deviation for that stratum to be 1.0. All the other standard deviations would follow, based on the assessment of variability relative to that baseline stratum.

This allocation scheme does not depend on a choice of total sample size, n. Indeed, it *generates* the total sample size, given a configuration for number of strata L, stratum population sizes (N_i), relative stratum standard deviations (σ_i), stratum sampling costs (c_i), and the allowable total sampling cost, c. We illustrate the effects of varying costs and (relative) standard deviations with the following.

Example of Optimal Allocation for Stratified Random Sampling

Suppose we have a finite population comprising three strata with the following information regarding that population:

1. The stratum population sizes are $N_1 = 2000$, $N_2 = 1500$, and $N_3 = 900$, for a total population size of $N = 4400$.
2. The variances are unknown, but practical experience suggests that the standard deviation among stratum 3 units is about twice that of stratum 1, while stratum 2 has standard deviation approximately 1.3 times that of stratum 1. Thus we have $\sigma_1 = 1$, $\sigma_2 = 1.3$, and $\sigma_3 = 2$.
3. The costs of sampling and analyzing units in strata 1 and 2 are equal ($c_1 = c_2 = \$10$), but because of difficult sampling logistics, the sampling costs in stratum 3 are $c_3 = \$20$.

4. The budget for the study is $10,000, with $2,000 allocated for fixed costs. Thus the sampling budget is c_s = $8,000.

With this information, and eq 31, we calculate that the optimal sample sizes in each stratum are

$$n_1 = \frac{[(1)(2000)(8000)]/\sqrt{10}}{[(1)(2000\sqrt{10})+(1.3)(1500\sqrt{10})+(2)(900\sqrt{20})]} = 246.32$$

$$n_2 = \frac{[(1.3)(1500)(8000)]/\sqrt{10}}{[(1)(2000\sqrt{10})+(1.3)(1500\sqrt{10})+(2)(900\sqrt{20})]} = 240.16 \qquad (35)$$

$$n_3 = \frac{[(2)(900)(8000)]/\sqrt{20}}{[(1)(2000\sqrt{10})+(1.3)(1500\sqrt{10})+(2)(900\sqrt{20})]} = 156.76$$

These, of course, would be rounded to whole numbers, yielding $n_1 = 246$, $n_2 = 240$, and $n_3 = 157$. We see that for strata 1 and 2, the stratum population sizes balanced the standard deviations, yielding very similar sample sizes. Stratum 3 was downweighted, despite its large variance, because obtaining samples is expensive there.

We point out that the sampling fractions, $f_i = n_i/N_i$, are f_1 =12.3%, f_2 = 16.0%, and f_3 = 17.4%. Thus the cheapest and least variable stratum (stratum 1) had the least intensive sampling, and despite having the smallest actual sample size, the most variable stratum (stratum 3) was sampled most intensively.

Variance of Sample Mean Under Stratified Sampling

We now turn to the variance of the estimator of the population mean to illustrate cost effectiveness. The variance of $\bar{y}_i$ (the estimator for the mean of each stratum) is

$$\mathrm{Var}(\bar{y}_i) = (1 - f_i)\frac{\sigma_i^2}{n_i} \qquad (36)$$

where f_i is the sampling fraction n_i/N_i in stratum i. As one samples more intensively in a given stratum, the variance $\bar{y}_i$ decreases, not only as n_i increases (the denominator term) but also as n_i increases relative to the stratum population size N_i. The overall estimate of the population mean is

$$\bar{y} = \frac{1}{N}\sum_{i=1}^{L} N_i \bar{y}_i \qquad (37)$$

which has variance

$$\mathrm{Var}(\bar{y}) = \sum_{i=1}^{L} \left(\frac{N_i}{N}\right)^2 (1 - f_i)\frac{\sigma_i^2}{n_i} \tag{38}$$

under the assumption that sampling is independent from stratum to stratum, and that N and the N_i values are known.

The formula simplifies when the sampling fractions are very small (as a general rule, e.g., when all are less than 1 or 2%). In that case the term $(1 - f_i)$ is negligible and may be ignored. Then

$$\mathrm{Var}(\bar{y}) = \sum_{i=1}^{L} \left(\frac{N_i}{N}\right)^2 \frac{\sigma_i^2}{n_i} \tag{39}$$

We shall see in the next section that, when this simplification is appropriate, the problem of sample size selection to achieve a desired precision is also greatly simplified.

Returning to the example, we now calculate the variance; of course, now the *actual*, not just the *relative*, standard deviations are required. For convenience, we will use as values $\sigma_1 = 100$, $\sigma_2 = 130$, and $\sigma_3 = 200$. Thus we have

$$\begin{aligned} \mathrm{Var}(\bar{y}) &= \sum_{i=1}^{L} \left(\frac{N_i}{N}\right)^2 (1 - f_i)\frac{\sigma_i^2}{n_i} \\ &= \left(\frac{2000}{4400}\right)^2 \frac{(1-0.12)100^2}{246} + \left(\frac{1500}{4400}\right)^2 \frac{(1-0.16)130^2}{240} \\ &\quad + \left(\frac{900}{4400}\right)^2 \frac{(1-0.174)200^2}{157} \\ &= 7.37 + 6.87 + 8.80 = 23.04 \end{aligned} \tag{40}$$

Cost Estimation for Desired Precision

Suppose we intend to make a 95% confidence interval for the population mean and that we desire that interval to have a prespecified width; that is, we wish to dictate the precision of our estimate. The algebraic interplay between the variance of $\bar{y}$ and the allocation formula is not straightforward; in particular, there is no explicit formula for sample size calculations (except for one simple case, which we will return to). We can, however, do the required calculations in a sequence of simple steps, and arrive at the answer. We illustrate the method using the example population.

For our example, suppose we wish a 95% confidence interval to be of total width 10 (units are the same as whatever it is we are measuring). If normality of the estimator is assumed, a 95% confidence interval is

$$\bar{y} \pm 1.96[\mathrm{SE}(\bar{y})] \tag{41}$$

where $\mathrm{SE}(\bar{y})$ is $\sqrt{\mathrm{Var}(\bar{y})}$, and therefore we may deduce that we need

$$\begin{gathered} 1.96[\mathrm{SE}(\bar{y})] = 5 \\ \Leftrightarrow \\ \mathrm{SE}(\bar{y}) = \frac{5}{1.96} = 2.55 \end{gathered} \tag{42}$$

A standard error of 2.55 corresponds to a variance of 6.5025. We have already seen from the example that a total investment of $8000 in sampling effort yielded a sample size of $n = 246 + 240 + 157 = 643$, and a variance of 23.04. We clearly need to allocate more money into the sampling effort than $8000. A plot of the relationship between variance and expenditure is shown in Figure 2. Based on the plot of Figure 2 alone, an expenditure of just over $20,000 is necessary. In fact, an expenditure of $20,500 produced a variance of 6.5 (with sample sizes of $n_1 = 628$, $n_2 = 612$, and $n_3 = 400$, for a total of 1640).

The relative sample sizes have not changed; n_1 for instance represents 628/1640 = 38.3% of the entire sample. This feature (constancy of relative sample sizes) is characteristic of the allocation formula for stratified random sampling. We can take

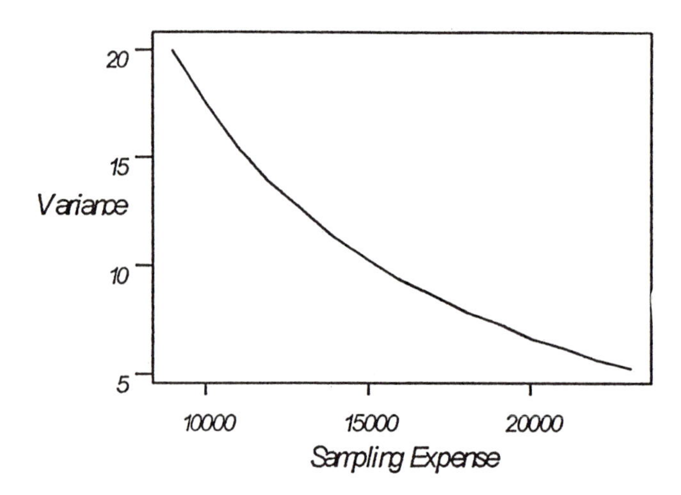

Figure 2. Relationship between variance and expense.

advantage of this feature to relatively easily deduce required sampling effort (with some specified precision as a goal) when the stratum population sizes are extremely large relative to the sample sizes (the aforementioned tiny sampling fractions).

The following situation is realistic. Suppose the samples are relatively small plots (square meters, for instance) and the strata are very large (counties in a state, for example). In that case, the estimate of the grand mean over the entire state would be

$$\bar{y} = \sum_{i=1}^{L} \frac{A_i}{A} \bar{y}_i \tag{43}$$

where A_i/A is the relative area of the ith stratum.

The solution to the allocation question, relative to desired precision, is an easy two-step process. Calculate the variance for an arbitrary expenditure. Then the desired variance is some multiple (e.g., m) of that variance. Simply divide the expenditures by m. We illustrate by an example.

Let the areas of three strata be $A_1 = 2000$ units, $A_2 = 1500$ units, and $A_3 = 900$ (so $A = 4400$). With relative standard deviations $\sigma_1 = 1$, $\sigma_2 = 1.3$, and $\sigma_3 = 2$, and sampling costs $c_1 = \$10$, $c_2 = \$10$, and $c_3 = \$20$, the optimal allocation problem yields the sample sizes $n_1 = 246$, $n_2 = 240$, and $n_3 = 157$, as before. The variance is different, however, because we are assuming very large populations of possible samples in each area. Specifically,

$$\begin{aligned} \mathrm{Var}(\bar{y}) &= \sum_{i=1}^{L} \left(\frac{A_i}{A}\right)^2 \frac{\sigma_i^2}{n_i} \\ &= \left(\frac{2000}{4400}\right)^2 \frac{100^2}{246} + \left(\frac{1500}{4400}\right)^2 \frac{130^2}{240} + \left(\frac{900}{4400}\right)^2 \frac{200^2}{157} \\ &= 8.40 + 8.18 + 10.66 \\ &= 27.24 \end{aligned} \tag{44}$$

This variance is larger than the first use of these numbers (23.04) in eq 40. There, the numbers 2000, 1500, and 900 were declared to be the actual stratum population sizes. As such, the sample sizes represented a substantial portion of the entire population; the finite population correction $(1 - f_i)$ appropriately reduced the variance from 27.24.

For sake of comparison to the earlier example, suppose that we want a 95% confidence interval with total width 10, so we desire the variance to be 6.5025. Now, 6.5025 = 27.24/4.19, so we estimate that we need to invest 4.19 × ($8000) = $33,520. In fact, $33,300 produced an estimated variance of 6.5; the discrepancy is just round-off error.

Summary and Conclusions

The following is a summary of the major conclusions of this study.

- Ultimately, the selection between sampling procedures comes down to a study of the cost effectiveness of each approach relative to the particular applications and purposes of the data collection. Either the smallest error bar width per unit of cost or the smallest cost per preselected error bar width can be used as measures of cost effectiveness.
- Cost effectiveness is also closely related to the balance between objectivity and subjectivity, relative to statistical relations taken as present in the in situ data values. Ideally, the same degree of subjectivity accepted in the engineering and other decision processes of the application planning will be permitted for the sampling plan selection.
- Sampling procedures, such as simple random or other design-based sample plans, may eliminate the use of data collected nonrandomly and prevent the introduction of other types of tangible and intangible prior information. This choice may force more data to be collected in order to achieve the required estimate accuracies, and, thus, increase the cost of sampling.
- Conversely, very subjective sampling procedures may introduce assumptions based more on "wishful thinking" than hard information and may lead to erroneous claims of greater accuracy for estimates than actually exists.
- The best choice lies in a careful balance between the objectivity and subjectivity permitted in the assumptions. This caveat is no different than for other engineering planning and problems. The most cost-effective sampling plan requires the maximum use of all reliable prior and supplementary information, while avoiding erroneous and biased input to the planning. Of course, avoiding such input is easier to say than to do. Critical judgment is an important prerequisite to designing the best sample plan.
- The careful inclusion of valid prior information in a model-based or geostatistical plan can greatly reduce the size of the sample required for a given estimate accuracy. However, the prior information must be at least approximately correct!
- If sample collection and analysis is cheap, then the increased objectivity of design-based sampling is probably the preferred approach.
- Much supplementary and prior information can be introduced into a design-based sample plan through the careful selection of strata. Much

can be gained in the assignment of sample size in each strata with consideration of various optimality relations to decrease variances.

- If either data collection or data analysis or both are substantially expensive, then the cost may be greatly reduced by incorporating (valid) supplementary information into design-based or geostatistical procedures. The reduction in sample sizes may be dramatic.
- In all error analysis, and the decision process based on the estimates made from the data, it is important to focus on the difference between the in situ field values or averages, and the estimates from the data. It is too easy to quote the error between the samples and the analysis results, and to forget that the true value in the samples may already differ greatly from the in situ average in the field (the fundamental error in Gy's terminology, or the kriging error in geostatistics).
- Honest and well-intentioned professionals may certainly disagree on the best sampling plan in a selected application. Having one, clear-cut solution that is best for all circumstances would be ideal. However, the real choice comes down to a balance between the money and the resources available, and the desired accuracy of the estimates.
- With a tolerant and nondogmatic willingness to compromise and to recognize the basis of competing procedures, an efficient use of the budget can, perhaps, be found. After all, if needlessly conservative sample plans are insisted on, then the extra cost of these may eliminate the remediation of other areas that also need to be studied as hazardous to human health. Conversely, if unreliable assumptions are introduced to cut costs, wrong decisions may be made on many of the areas investigated. Clearly, a careful, noncapricious, prudent compromise between procedures, based on the best available information, is required, if the correct decisions are to be made with minimal cost.

Appendix: Universal Kriging with a Linear Trend

The universal kriging approach determines the coefficients $A_{\mathrm{krig}(i1)}$ by minimizing

$$\mathrm{Var}_{\mathrm{krig}}(\Delta_1) = E\left[\langle y(x)\rangle_1 - \sum_{i=1}^{n} A_{\mathrm{krig}(i1)} y(x_i)\right]^2 \qquad (45)$$

where

$$\begin{aligned} \langle y(x) \rangle_1 &= a + b\langle x \rangle_1 + \langle \varepsilon(x) \rangle \\ y(x_i) &= a + bx_i + \varepsilon_i \end{aligned} \tag{46}$$

The constraint is also imposed that

$$E\left[\langle y(x) \rangle_1 - \sum_{i=1}^{n} A_{\text{krig}(i1)} y(x_i) \right] = 0 \tag{47}$$

This constraint provides a condition of unbiasedness. Equation 47 can be manipulated to the form

$$\begin{aligned} 0 &= E\left[a + b\langle x \rangle_1 + \langle \varepsilon(x) \rangle - \sum_{i=1}^{n} A_{\text{krig}(i1)} (a + bx_i + \varepsilon_i) \right] \\ &= a\left[1.0 - \sum_{i=1}^{n} A_{\text{krig}(i1)} \right] + b\left[\langle x \rangle_1 - \sum_{i=1}^{n} A_{\text{krig}(i1)} x_i \right] \end{aligned} \tag{48}$$

Consequently, unbiasedness can be imposed if the following equalities are required to hold:

$$\begin{aligned} \sum_{i=1}^{n} A_{\text{krig}(i1)} &= 1.0 \\ \sum_{i=1}^{n} A_{\text{krig}(i1)} x_i &= \langle x \rangle_1 \end{aligned} \tag{49}$$

With these constraints, one can manipulate eq 45 to the form

$$\text{Var}_{\text{krig}}(\Delta_1) = C_{BB} + \sum_{i=1}^{n} \sum_{k=1}^{n} A_{\text{krig}(i1)} A_{\text{krig}(k1)} C_{ik} - 2 \times \sum_{i=1}^{n} A_{\text{krig}(i1)} C_{Bi} \tag{50}$$

where C_{BB}, C_{ik}, and C_{Bi} are defined in eq 26. The criteria for universal kriging for this case, then, is to minimize eq 50 subject to the constraints in eq 49. This procedure can be done with Lagrangian multipliers by minimizing

$$\mathrm{Var}_{\mathrm{krig}}(\Delta_1) = C_{BB} + \sum_{i=1}^{n}\sum_{k=1}^{n} A_{\mathrm{krig}(i1)} A_{\mathrm{krig}(k1)} C_{ik} - 2 \times \sum_{i=1}^{n} A_{\mathrm{krig}(i1)} C_{Bi} + \lambda_1 \left(\sum_{i=1}^{n} A_{\mathrm{krig}(i1)} - 1.0 \right) + \lambda_2 \left(\sum_{i=1}^{n} A_{\mathrm{krig}(i1)} x_i \right) \tag{51}$$

with calculus by differentiating eq 51 with respect to $A_{\mathrm{krig}(i1)}$, λ_1, and λ_2, setting the derivatives all to 0, and augmenting the resulting set of equations with the two equations in eq 49.

This system of equations can be written in matrix form with the following definitions. Let $\mathbf{C}$ be the matrix whose (i,k) element is C_{ik}. Define $\mathbf{C}_B$ as the column vector whose ith element is C_{Bi}, and let $\mathbf{A}$ be the column vector whose ith component is $A_{\mathrm{krig}(i1)}$. Also, let $\mathbf{1}$ be a column vector of ones and $\mathbf{x}$ be the column vector whose ith component is x_i. The matrix equation whose solution gives the $A_{\mathrm{krig}(i1)}$ is

$$\begin{pmatrix} \mathbf{C} & \mathbf{1} & \mathbf{x} \\ \mathbf{1}^T & 0 & 0 \\ \mathbf{x}^T & 0 & 0 \end{pmatrix} \begin{pmatrix} \mathbf{A} \\ \lambda_1 \\ \lambda_2 \end{pmatrix} = \begin{pmatrix} \mathbf{C}_B \\ 1.0 \\ \langle x \rangle_1 \end{pmatrix} \tag{52}$$

In terms of these matrices and vectors, eq 51 can be expressed as

$$\mathrm{Var}_{\mathrm{krig}}(\Delta_1) = \mathbf{C}_{BB} + \mathbf{A}^T\mathbf{C}\mathbf{A} - 2\mathbf{A}^T\mathbf{C}_B \tag{53}$$

The top equality in eq 49 is

$$\mathbf{C}\mathbf{A} + \lambda_1\mathbf{1} + \lambda_2\mathbf{x} = \mathbf{C}_B \tag{54}$$

If eq 54 is multiplied by $\mathbf{A}^T$, and the constraints in eq 49 are used, then one develops the formula

$$\mathbf{A}^T\mathbf{C}\mathbf{A} = \mathbf{A}^T\mathbf{C}_B - \lambda_1 - \lambda_2\langle x \rangle_1 \tag{55}$$

This expression can be substituted into eq 53 to get the error expression

$$\mathrm{Var}_{\mathrm{krig}}(\Delta_1) = C_{BB} - \mathbf{A}^T\mathbf{C}_B - \lambda_1 - \lambda_2\langle x \rangle_1 \tag{56}$$

References

1. Matheron, G. *Cah. Cent. Morphol. Math. Fontainebleau* **1971**, 5, 210.

2. Gy, P. M. *Sampling of Particulate Materials Theory and Practice;* Elsevier Scientific: Amsterdam, Netherlands, 1982; p 431.
3. Pitard, F. F. *Pierre Gy's Sampling Theory and Sampling Practice Volume I, Heterogeneity and Sampling;* CRC Press: Boca Raton, FL, 1989; p 214.
4. Chapter 11 in this book.
5. Cressie, N. *Statistics for Spatial Data;* John Wiley & Sons: New York, 1991; p 900.
6. Isaaks, E. H.; Srivastava, R. M. *An Introduction to Applied Geostatistics;* Oxford University: New York, 1989; p 561.
7. Bloch, D. A.; Segal, M. R. *J. Am. Stat. Assoc.* **1989,** *84*(408), 897–905.
8. Cochran, W. G. *Sampling Techniques;* John Wiley & Sons: New York, 1977; p 428.
9. Thompson, S. K. *Sampling;* John Wiley & Sons: New York, 1992; p 343.

Chapter 40

Geostatistical Sampling Designs for Hazardous Waste Sites

George T. Flatman and Angelo A. Yfantis

This chapter discusses field sampling design for environmental sites and hazardous waste sites with respect to random variable sampling theory, Gy's sampling theory, and geostatistical (kriging) sampling theory. The literature often presents these sampling methods as an adversarial "either/or" philosophy; this chapter emphasizes when each should be used with a cooperative "both/and" philosophy. The intrasample variances, biases, or correlations must be taken care of by the use of Gy's sampling theory for both independent random variable sampling and analysis and correlated random variable sampling and analysis. The deciding factors in the choice of sampling design and analysis are not just intersample variances, biases, or correlations but also the discreteness of the waste under investigation, remediation as a unit, and the relative cost of samples versus the cost of remediation.

ENVIRONMENTAL SAMPLING is a multidisciplinary science. It requires chemists, media experts, risk assessors, and even statisticians. The sampling design is an integral part of the experimental design and data analysis, and most importantly, the data analysis cannot recover more information than the samples contain. Thus the statistician needs to be on the project from its inception. Optimal environmental sampling requires consideration of at least three branches of statistics. Classical random variable statistics (*1*) are needed in quality assurance (QA) and in the analysis of data that are reasonably independent (little or no process, spatial, or chronological correlation). Gy's theory of sampling (2) is needed for the definition of correctness for the "field sample" [determination of amount (mass or volume) sampled] and any samples taken in heterogeneous media (almost all environmental samples). Geostatistics, and its most used form, kriging (3), is needed for field sites with a spatial structure. The choice of sampling designs—

3152–4/96/0779$15.75/0

when to use classical random design or kriging's regular grid design—is a difficult decision. Even statisticians differ on such a question. This chapter discusses the statistical rules that enter into the decision. The decision depends on specifics of the site and remediation plan as well as statistical aspects. For example, Gy's theory must be used to take a correct sample for either random variable statistics (sampling or analysis) or geostatistics (sampling or analysis).

When I discussed the role of statistics in sampling design with a manager of a chemical laboratory, the manager confided in me that his statistician's recommendations were always illogical and irrational and contradicted common sense. We did not have time to discuss specifics, but I suspect the advice he received was also poor statistically because it confused the use of random variable statistics with the use of spatial statistics. If the correct branch of statistics has been chosen, statistical requirements can be explained from statistical theory in a logical and reasonable manner that does not defy common sense. It is important in a multidisciplinary project for all to be comfortable with the soundness of the decisions. Statisticians should be asked to explain the statistical requirements they recommend until all feel comfortable with the design.

Random Variable Statistics

A random variable has both magnitude and probability. It may come from a symmetric distribution such as normal or uniform, or from a skewed distribution such as lognormal or Poisson. Chemical environmental data sets are often assumed lognormal, and radioactive data sets are often assumed Poisson. Because both distributions are positively skewed, the estimate of the mean based on few samples has a higher probability of being underestimated than the mean of a normal distribution or any symmetric distribution with a strong central tendency. Random errors as monitored by QA are often assumed normal. The branch of statistics that deals with random variables gives us the statistical inferences that have tools for QA. Random variables provide measures of central tendency (such as mean, median, and mode), dispersion (such as range and standard deviation), and statistical inference (such as confidence intervals, prediction intervals, and tolerance intervals).

The mean and standard deviation are the statistics usually sought by a sampling campaign; they are sufficient statistics (i.e., completely define the distribution) for the normal distribution. The mean of any distribution becomes normal as the number of samples, n, becomes large. This property justifies the use of confidence intervals for the mean if, and only if, n is large enough ($n > 16$ for a symmetric distribution, and $n > 50$ for a skewed distribution). However, if the number of samples is much fewer than 50 samples from the typical environmental distribution in a confidence interval, then these limits are not to be trusted. Either knowledge of the distribution or transformation to normality is required for statistical inference about the variable, its distribution, or future samples. A listing of means and standard deviations or intervals, without investigating the distribution, is misleading and has the potential of inviting wrong decisions because the readers will assume normality. Nonparametric intervals and tests are available, but they lack power. For example, the critical values for one-sided intervals for probabilities (1 – α) of 0.95 and 0.99 using the Tchebycheff inequality are 4.472 (square root of 20) and 10.000 instead of the standard normal distribution values of 1.64 and 2.33. Most regulators will cringe at 4 or 10 in a compliance hypothesis test. Another consideration is that random variable sampling design requires rigorous definitions of the *population* and *sampling unit,* so that the design can give each sampling unit an equal probability of being chosen. This requirement will be discussed further.

Population Defined

In environmental samples, population is not as obvious or as well-defined a term as it is in statistical textbooks (e.g., all the cards in a deck, or the two sides of a coin). In site evaluation, the most obvious population is the waste site as a whole, but the usual site has more than one population of interest. It may have population(s) of plume(s) and background population(s). The population of interest is the population(s) of the plume(s). Waste plumes seldom honor property boundaries or travel in politically defined shapes such as city blocks. Thus the populations of interest are the plume(s) and the background, not a mixture of these. To average all the samples from the site would give an estimate of a mean from a mixture of populations, a "fruit salad" of plume(s) and background(s). If the location and extent of the plume or background are not known, but a map of mean contours (isopleths or isarithmic lines) is wanted for multiple remediations, then this situation would require geostatistical sampling and analysis. If the waste to be evaluated is well-defined and confined, such as liquid waste stored in 55-gallon drums or a waste pile on a tarp that will be disposed of as a unit, then the population of interest is the drum or pile and therefore classical statistics (a mean value) will be adequate for the decision.

Sampling Unit Defined

For textbook statistics, a sampling unit is a draw of a card or a flip of a coin, but for an environmental sampling the unit is complicated by natural variation (e.g., media heterogeneity or pollutant characteristics) and sampling tool variation and biases (4). In laboratory QA the unit may be the contents of every *i*th vial in the queue of the analyzing instrument. At an environmental site, the "sampling unit" is ambiguously used to refer to both the sample and the sample support. The sample is much smaller in volume or mass than the sample support, but if it is representative, it has approximately the same concentrations of the pollutant or the same values of some measured characteristic. The sample, simple or composite, is a small critical mass that is taken from the sample support for measurement. The sample support is the larger volume or mass of in situ media that is to be represented by the measure of the sample. The sampling support is often the same volume as the remediation unit. These two units are determined by the goal of the sampling campaign or the remediation option(s), but they must meet the requirements of Gy's theory of sampling and geostatistics, which are each discussed in subsequent sections of this chapter. The extractable mass or volume (field sample) cannot be dictated by the size of the sampling tool or the size of the official container. It should be determined by the heterogeneity of the media in accord with Gy's theory. Differing amounts of media of interest, because they are ambiguously called "sample", should be identified by size and use. The analysis sample (i.e., aliquot or split), used in its entirety by the chemist for analysis, has a mass less than that of the preparation sample, which has a mass less than or equal to that of the field sample. Each change of scale or reduction of sample mass must pass Gy's requirements (*see* the subsection ***Analytical Error***). The name of the sample is unimportant, but the change of mass is important. Any change in volume (mass) must be checked using a monogram made up for the current site. Extraction(s) for the field sample from the in situ sample support (i.e., sampling unit) must satisfy both Gy's theory requirements and geostatistical requirements.

Dealing with Correlation in Practice

In theory, the difference between an independent random variable and a random variable correlated in time or space is clear, but this difference is not so clear in practice. In practice, most environmental samples are correlated in either time or space, and possibly in both time and space, yet a random sampling or analysis is done. Even the analyses of the samples in the queue of a mass spectrometer (MS) are correlated somewhat in time, but this correlation is weak enough and the QA samples are spaced far enough apart that the correlation can be ignored. Correlation in space or time can be taken into account by slightly more complicated formulas in random variable statistics; Gilbert (5) gives relevant sediment and groundwater examples of how correlated sample units require more samples to be taken (larger n) than if the observations were independent. The critical criterion for using a spatial sampling and data analysis is the management decision or need to see a contour (isopleth) map of the pollutant location as well as concentration (these are kriging results) in place of a list or histogram of chemical analyses with a confidence interval about an estimate of some mean (random variable output).

Pierre Gy's Sampling Theory

Pierre Gy is a mining engineer and Francis Pitard is a chemist. Both men have had brilliant careers in process and mining quality control. Pitard has written a two-volume work (2) that captures and communicates their experiences in the sampling of heterogeneous media. These volumes are valuable for environmental sampling of soils or sediments. Pitard organizes the taking of "correct" samples with correct sampling tools, according to seven "errors". The emphasis on correct samples and tools is analogous to the emphasis from the U.S. Environmental Protection Agency (EPA) on representative samples. Because of the potential of one, some, or all seven of these errors to erode the correctness or the representativeness of an environmental sample, this chapter will refer to them as "variances" to stress their additivity for a component-of-variance model. "Variance" emphasizes the intrinsic nature of these errors or biases in heterogeneous material sampling, in contrast to the negative connotations of these terms in the vernacular ("error" as a careless mistake; "bias" as an intentional dishonesty). *Variance, error,* and *bias* are technical terms that describe differing problems with different solutions. An "error variance" is often thought of as symmetric with a mean (expectation) of zero and as reducible by taking more samples; a "bias variance" is one-sided (e.g., always too high or too low) and is reducible not by taking more samples but only through a correct sampling design. The symmetry or one-sidedness must be carefully thought out and often field-tested for all potential variance in any sampling design and QA plan.

This theory sounds like any QA plan talking about errors, but it refers to a different type of error and needs to be discussed in its own part of the QA plan. Specifically, it deals with intrasample error (errors within the sample) rather than intersample error (errors between samples). The various components of variance of this sampling theory sound trivially obvious when pointed out, but they are easily overlooked in the stress of formulating a QA or sampling plan. Leaving them out can be disastrous for QA and data quality objectives. Even though these sources of variation sometimes are obvious and trivial, they must be taken into account in every environmental sampling plan.

The Fundamental Error

This component of variance is a natural property of heterogeneous material. It is not an error in the sense of an avoidable mistake; however, if the sample planner does not take it into consideration it will generate unnecessary (avoidable) variance in the laboratory analyses. The variance is caused by the range of particle sizes in the medium and the fact that often only certain sized particles contain the pollutant of interest. This situation is illustrated in Figure 1; the shaded or lined particles are assumed to contain or carry the pollutant, and the other particles are the heterogeneous medium. Thus the chemical analysis depends on two values: the number of solid particles (percentage composition), and their concentration. This dependence adds another variance term or component of variance (percentage composition) to the analytic variance. The magnitude of this error is small in a fine or homogeneous soil or sediment but becomes larger as the medium becomes more heterogeneous in particle size and particle affinity for the pollutant of interest. This fundamental component of variance can be reduced by increasing the mass of the sample or by reducing the particle size of the sampling material by appropriate digestion.

To maintain the original level of accuracy, the sample material must always be reduced in maximum particle size before being reduced in mass or volume (split or aliquot). The mass of a sample required for a given relative variance [relative standard deviation (RSD) squared] can be read from Pitard's nomograms as a function of

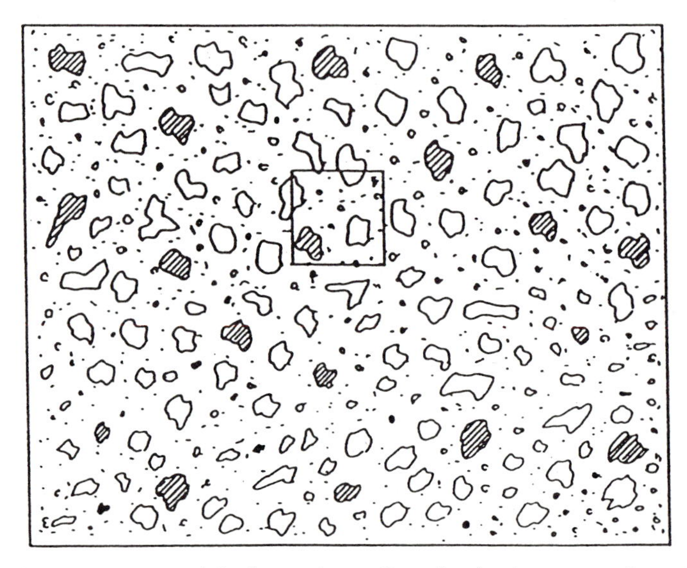

Figure 1. Heterogeneous material: fundamental error. (Reproduced with permission from reference 2, Vol 1. Copyright 1989 CRC Press.)

various physical properties, the most important one being maximum particle size of the medium (6). This relationship will be directly applicable to waste monitoring if the pollutants of interest are heavy metals, but the application to volatile chemicals or semivolatile chemicals remains to be developed. The EPA has a very readable document on this subject that presents an example nomogram for soil properties (7). The extension of Gy's theory to volatile chemicals and semivolatile chemicals is a very important but as yet undeveloped part of environmental sampling.

Grouping and Segregation Error

There is potential for this variance in any heterogeneous media. The grouping and segregation error develops through movement of samples through processing, handling, shipping, or mixing. The heterogeneity may be in density or size (also adhesion, cohesion, magnetism, affinity for moisture, and angle of repose of crystalline structure) so that the particles come together by groups during any movement or vibration. Figure 2 illustrates this type of error for the pile at the end of a conveyer belt. If the black particles contain the pollutant of interest, then a sample from the right side of the pile will be biased high and a sample from the left side will be biased low. In taking a sample of a waste stream or pile, the potential variance can be minimized by sampling along the gradient of grouping and segregation. For soil, gravel, or sediment being carried on a conveyor belt, the gradient of grouping and segregation would be across the belt orthogonal to the direction of motion, and thus a correct sample would be a rectangular (not a trapezoidal) section oriented across the belt. Sampling a pile, a truck, or a railroad car of waste in a correct manner is

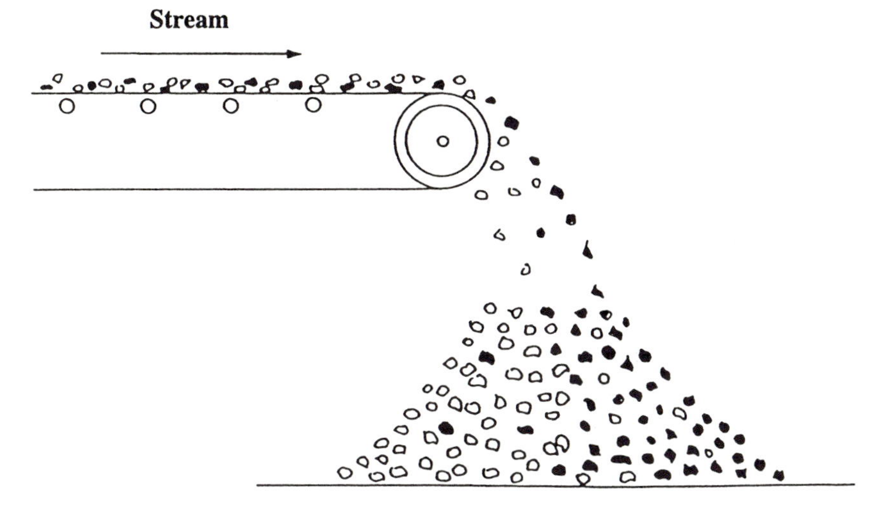

Figure 2. Grouping and segregation error. (Reproduced with permission from reference 2, Vol. 1. Copyright 1989 CRC Press.)

very difficult because of this component of variance. The correct time to sample is before the pile is built or the truck or railroad car is loaded. In sample preparation, Pitard suggests that the pouring of the well-mixed material from the V-blender, especially if the particulate material is allowed free fall of any distance, can undo (defeat) the blending (8). Aliquoting increases this error. The general rule is that as aliquot size decreases, the variance increases. Theoretically, as the size of the aliquot approaches the size of the grains of the sample, this error grows larger without bounds. The corollary to this theorem is the fact that the chemist, aliquoting to get the relatively small amount of material (analytical sample) actually required for the analysis, can turn the analytic equipment into a random number generator if the sample material has not been ground to the required fineness and aliquoted correctly.

Spatial and Periodic Errors

These error sources could be periodic and/or spatial structures on the scale of the extracted sample or the sample support (the in situ area or volume represented by the sample). If they were of a larger scale they would be studied by a time series analysis or a geostatistical analysis, but they are not of interest, and the decision statistic is the mean of the unit and not the means of the subunits. In the preceding discussion of classical statistics, the 55-gallon drum was assigned to a classical statistical analysis instead of a geostatistical analysis, even though there may have been a structure in concentration in the vertical dimension of the drum. No one wants a contour map of the concentration of pollutant inside of a drum because the drum will be remediated (disposed of) as a unit. However, this gradient cannot be ignored; instead it must be representatively sampled by sampling each layer proportional to its volume. This sampling is accomplished by the choice of sampling tools. To minimize the microspatial variance, a "composite liquid waste sampler" (COLIWASA) must be used. The name of the sampling tool tells an important principle. Compositing is an important tool in random variable statistics to save chemical analysis costs, but in spatial statistics it is used to ensure that the sample is representative of the in situ sample support. Subsample compositing is physically doing the same thing that statistical averaging does to the numerical values of replicate samples, except compositing loses the information about the variance or standard deviation, with the benefit of saving the cost of $(n - 1)$ chemical analyses. These are two quite different and important uses of compositing.

Increment Delimitation and Extraction Errors

These two variances arise from the interaction of a sampling tool with the heterogeneity of the media sampled. The circles in Figures 3 and 4 can represent the cutting edge of a plugging or coring device descending on the media to take a soil or sediment plug or core. In Figure 3, taking the shaded area of the larger particles would be the correct sample, but if the larger particles are hard compared to the softer interstitial material, the tool will not cut through the harder particles to give the desired correct sample. Rather, the large hard particles will be pushed out of the sample if their centers of gravity lie outside the corer, as illustrated by the white par-

Figure 3. Increment delimitation error. (Reproduced with permission from reference 2, Vol. 2. Copyright 1989 CRC Press.)

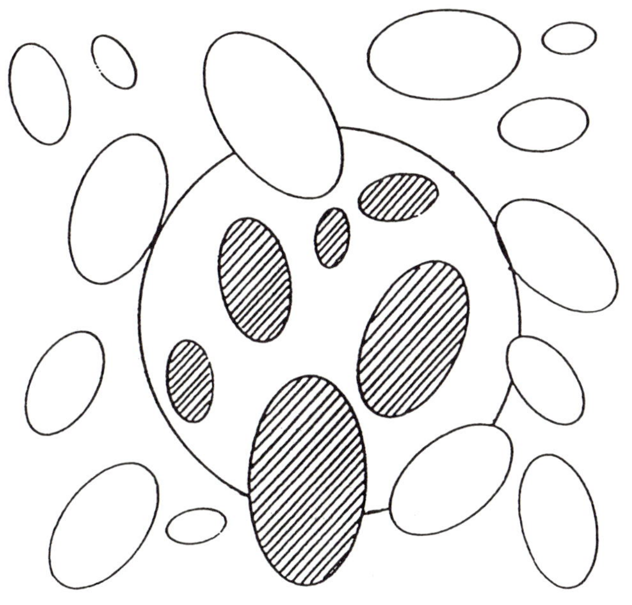

Figure 4. Increment extraction error. (Reproduced with permission from reference 2, Vol. 2. Copyright 1989 CRC Press.)

ticles in Figure 4. If their centers of gravity fall within the corer as illustrated by the shaded particles in Figure 4, then the particles in their entirety will be included in the sample. Either case is incorrect, but the two cases tend to average out. It is important to distinguish these two concepts: (1) the *delimitation error* is the variation caused by the inability to cut through all the heterogeneous media and take the part included in the circle of the coring or plugging device, and (2) the *extraction error* is the variation caused by taking or pushing out of the way the whole hard particle as a function of whether its center of gravity falls in or out of the circle of the corer or plugger. If the cylinder could be cut out exactly by a laser and then taken out intact by levitation as in science fiction, these two errors could be avoided. Today's solution to these problems is to have a corer or plugger that is at least two or three times the diameter of the largest particle size.

Analytical Error

The EPA and the American Chemical Society have published many excellent papers, proceedings, and books on this interdisciplinary subject. Therefore, to avoid duplication, we wish to speak only to the chemist's method of abstracting a much smaller sample (analytical sample) from the prepared sample. This step, because of the smallness of the mass of the analytical sample compared to the mass of the sample from which it comes, is the sample most apt to incur an unacceptable magnitude of Gy's fundamental error. If the analytical sample is taken by sticking a spatula randomly into the top of the material in the bottle and taking out the desired amount, such a sample is a grab sample and not an aliquot or split; the chemical analysis is very apt to give a value that is incorrect for Gy's theory and unrepresentative for regulatory use.

For an example of the grinding and splitting or aliquoting needed to acquire a correct and representative analytical sample, the critical path (A→B→C→D→E) should be traced through Figure 5, a nomogram adapted from references 2 (Vol. 1) and 7. (Grinding cannot be done, however, for volatile and semivolatile pollutants or to the media for a leach test.) First the nomogram must be made for the specific site (e.g., particle sizes and particle characteristics). The horizontal or *x*-axis is the sample weight in grams, and the vertical or *y*-axis is the RSD of the fundamental error; both axes are in log scale. In the center of the nomogram is a family of linear graphs that introduces the third variable, maximum particle size. Each particle size has its own line, and each line represents one and only one particle size.

The two ways to reduce the *y*-axis intercept, the RSD of the fundamental error, are: (1) to take a line with smaller particle size from the family of graphs, or (2) to take a larger weight of sample on the *x*-axis. First, in the family of linear graphs, the top line of the family represents the largest particle size, namely 75 mm, and intercepts the largest RSD on the *y*-axis. The next lower line is 25.4 mm, and so on down to the line with the lowest RSD, which is for a particle size of 0.2 mm. The 0.2-mm line is probably representative of QA internal standards, in contrast with Superfund's definition of soil as <2 mm and the definition from the Resource Conservation and Recovery Act (RCRA) of soil as <9 mm. These disparities in sizes might explain some of the bench chemist's problems with increasing variance or RSD (e.g.,

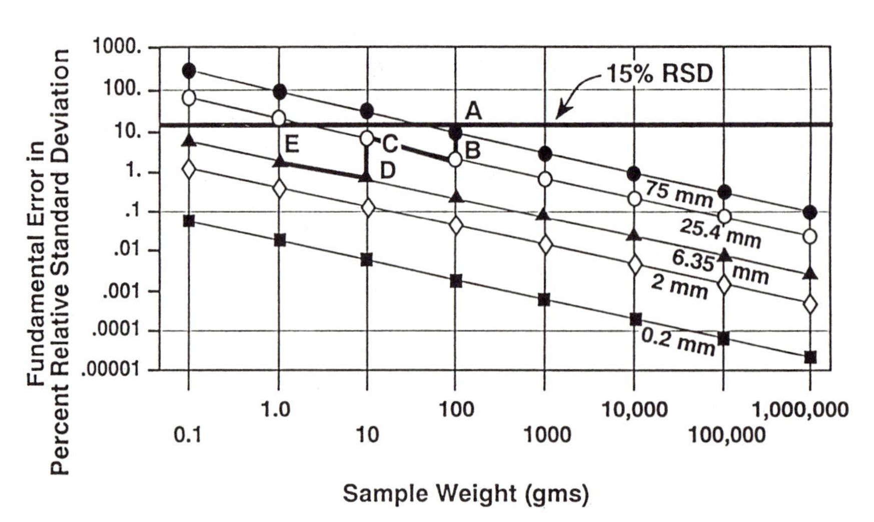

Figure 5. Maximum particle size: preparation error. (Adapted from reference 2, Vol. 1, and reference 7.)

square root of relative variance) as samples come from QA internal standards, Superfund samples, and RCRA samples. Second, each graph has a negative slope, which shows that as the mass of sample on the horizontal axis for a given particle size increases, the relative variance intercepted on the *y*-axis decreases. The horizontal line labeled 15% RSD represents the target accuracy or maximum acceptable RSD. If the maximum particle size of the material of interest is measured empirically to be 75 mm, and the pollutant of interest is one that can be pulverized or grown without loss, such as a heavy metal (Pb), then from the intersection of the horizontal line of maximum RSD = 15% and the downward sloping line of particle size 75 mm, the necessary minimum sample weight can be read on the horizontal or *x*-axis as 100 g. Thus the field technician or scientist must take a sample or composite subsamples so that a field sample of 100 g or more is obtained.

If the chemist is going to take an aliquot of 1 g for the analysis (analytical sample), then the preparation procedure must follow a path such as A→B→C→D→E in Figure 5. To maintain the accuracy of the 100 g of field sample whose maximum particle size is 75 mm, the digestion process must first grind and then split. Grinding reduces particle size and splitting reduces the mass of the sample. Grinding is going down on the nomogram from A to B representing pulverizing from a maximum particle size of 75 mm to a particle size of 25.4 mm, and aliquoting or splitting is moving to the left along the 25.4-mm line on the nomogram from B to C, representing aliquoting or splitting the sample of 100 g to a sample of 10 g. The new critical mass due to the particle reduction or the location of C on the new smaller particle line is the last integer weight tick line that intersects the new particle line just below the 15% relative error line. The amount of information in the 100 g of material of maximum particle size 25.4 mm at B appears to have an order of magnitude (axes in log scale) decrease in RSD. This apparent decrease is not true, because variance of an extracted sample is not reduced by grinding, or information is not created by digestion, but it does mean that now we can split the sample mass down to the new critical mass (10 g) on the current line (25.4-mm line) and still have the original RSD of the 100-g sample, namely 15%. A nomogram path has no lower RSD than its highest point (in this example, point A). Again, more digestion moves the sample from C on the 25.4-mm line to D on the 6.35-mm line. No information is created by grinding, but now the information in 100 g of 75-mm particles, namely 15% RSD, can be carried by a new critical mass of only 1 g as splitting or aliquoting moves us along the 6.35-mm line to E.

The process makes sense if the would-be user remembers that grinding reduces the critical mass needed to carry the same RSD and that the aliquoting or splitting removes only the unneeded mass. One might well ask, "Why the broken path? Wouldn't it be easier to grind all the way in one step and then split?" Yes, it would be simpler, but it would require the grinding of a larger mass of sample; the stepwise path minimizes the mass of material digested. In the interest of minimizing grinding or preparation, Pitard suggests sieving the material so the part less than the new maximum particle size falls through, and then grinding only the part that did not fall through, remembering to recombine the two.

This process sounds a little complicated because it is complicated, but with particle size analyses of the media of interest and with statistically guided preparation (pulverizing and splitting or aliquoting), a correct and representative analytical sample can be prepared for the chemical analysis.

Spatial Variable Statistics

The old adage that a chain is as strong as its weakest link implies that the prudent blacksmith will strengthen the weakest link and try to make all links equally strong. The application to environmental sampling is that error variances are a chain: the analytical variance, the sampling and handling variance, and the field variance are links. The goal of quality improvement is to make the sum of the variances as small as possible, and the cost-effective way to minimize this sum is to spend more resources on the variance link that is improved most cheaply. Because of diminishing returns in variance reduction, the optimal variance to reduce is often the biggest one. The field sampling variance is often the appropriate link or variance to reduce. Variance reduction is most obviously accomplished by taking more samples, but if sampling or analytic costs are high, increasing samples may be too expensive. In many cases, the field sampling variance is economically reduced by going from a random to a spatial variable sampling design.

The term *geostatistics* was coined by Matheron (9) to describe the study of regionalized or spatially correlated variables. In the past 20 years, the geostatistical literature has grown enormously, and many significant developments in theory and methodology have been presented. The practice of geostatistics has also spread from its original applications in the mining industry to such fields as soil science, forestry, meteorology, and environmental science.

The geostatistical methods described in this chapter, namely semivariograms and ordinary kriging, represent two of the approaches available to us, and we selected them primarily to illustrate geostatistical concepts and their implications for sampling programs. A discussion of the pros and cons of alternate approaches, such as generalized covariance and universal kriging, is beyond the scope of this chapter. More extensive treatments of the subject can be found in references 3 and 10.

Random or Spatial Variables

Most field sampling plans are based on random variable statistics and assume that the sample observations are independent and identically distributed (IID). However, field samples are usually spatially correlated. *Correlation* is a statistical measurement of the intuitive physical fact that samples taken close together are more similar in value than samples taken farther apart. Neglecting this correlation can make the statistics, tests, and sampling procedures that assume independence (IID) inappropriate (*11, 12*); using this correlation makes the statistics, tests, and sampling procedures of spatial statistics more appropriate and powerful. A truly random variable is completely described by its probability distribution. Samples are used to estimate this distribution and to estimate statistical descriptors such as mean, median, and standard deviation. In addition, spatial variables must be described by a measure of the correlation between each value and the values at nearby locations. Samples can be used to estimate the spatial correlation function and are frequently used to estimate localized mean values for remediation units or exposure units.

Localized mean estimates are often displayed in the form of isopleths or contour maps. A practical rule for the investigator is that if a contour map is a desired or even a plausible end product of a proposed study, geostatistical methods should be considered.

The implications for the design of a sampling program can be significant. Although random sampling is appropriate for random variables, Olea (*13*) demonstrated that the most effective sampling pattern for local estimation of spatial variables is the regular grid. Yfantis (*14*) evaluated triangular, square, and hexagonal grids. Also, geostatistical studies commonly use a multiphase approach, and the first sampling phase is oriented primarily toward estimating the spatial correlation (*15*).

Semivariograms for Quantifying Spatial Correlation

One way in which spatial correlation can be measured and displayed is by a *semivariogram*, or graph of the type shown in Figure 6. The dots are the empirical semivariogram representing experimental values computed from sample data; the fitted curve is a theoretical semivariogram or an estimation of a spatial correlation function

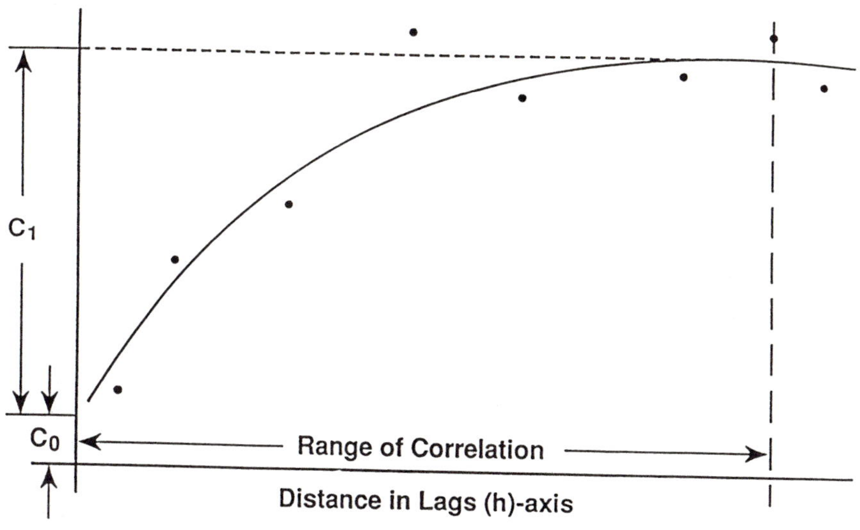

Figure 6. A typical semivariogram. (Reproduced from reference 16. Copyright 1988 American Chemical Society.)

assumed to be characteristic of the sampled area. The horizontal axis, called the *lag axis*, is the distance between points in linear units such as meters or kilometers; the vertical axis, called the *gamma axis*, is the variance of differences in pollution units squared, such as parts per million squared. The experimental points are computed by averaging data grouped into distance class intervals. Variance is a function of lag. The rising nature of the points and curve follows the principle of sampling that states the variance or difference between observations increases as the distance between their locations increases.

Sill and Range of Correlation. Figure 6 is typical of many semivariograms of chemical concentrations in the environment; the rise in variance has an upper bound known as the *sill*. When the variance reaches the sill, sample locations are far enough apart to make the samples independent. The distance on the lag axis at which the semivariogram's curve reaches the sill is the *range of correlation*. This distance is important to the sampling plan, the estimation of pollution over the area under investigation, and the interpolation error. The range of correlation explains a practical relationship between spatial variables and random variables; *random variables* are field samples that are farther apart than the range of correlation, and *spatial variables* are field samples closer together than the range of correlation. This range of correlation is important for choosing the correct analysis; if a classical random variable statistic is wanted, such as the mean or variance, then one type of sampling design that would ensure spatial independence of the samples would be any systematic random design requiring that all samples are at least the range of correlation apart (*17*). If a contour map of pollution isopleths or interpolation variance is wanted, then as the sampling locations get closer together, the local interpolation error decreases. Depending on the information wanted and the spacing of the sample locations, either random or spatial variance statistical analysis can be used on field samples.

Variance Model. In Figure 6, on the vertical axis of the fitted model the variance has two components, C_0 and C_1. The C_1 component of the variance is the measure of structural variation and has the characteristic of increasing variance between sample observations as the distance between sample locations increases. The C_0 component of the variance combines random variance factors, such as sampling and analytical error, along with any unmeasured spatial variance that may exist at distances smaller than the sampling interval; C_0 is constant for all lags. The relationship of C_0 to the need for compositing samples and the relationship of C_1 to the distance between sample locations will be discussed in a later section.

Anisotropy and Directional Semivariograms. The variance structure, as measured by the semivariogram, is often different in the range of correlation in different directions. This condition is called *anisotropy* and must be measured by directional semivariograms. Directional semivariograms are computed experimentally by grouping sample pairs into directional classes, or windows, as well as into distance classes. The directional ranges of correlation can change the geometry of the sampling grid and the orientation of the grid. Often, not enough preliminary data are available to compute directional semivariograms, and thus the sampling design must work with only an omnidirectional range of correlation. However, an omnidirectional range of correlation and a sampling design from it honor the variance–covariance structure more than conventional random variable methods that consider only a scalar variance.

Kriging for Surface Estimation

Kriging is a linear-weighted average interpolation technique used in geostatistics to estimate unknown points or blocks from surrounding sample data. By assuming that the spatial correlation function inferred from the experimental semivariogram is representative of the points to be estimated as well as those sampled, the interpolation error (kriging error or kriging standard deviation) associated with any estimate that is a linear-weighted average of sample values can be computed. The kriging algorithm computes the set of sample weights that minimize the interpolation error.

Kriging software usually offers both punctual and block output options. *Punctual kriging* treats the input values as located at points and output estimates as values at points. *Block kriging* estimates the output for an area or volume (called block) by averaging multiple points estimated over that area or volume. This difference is determined in the sampling and becomes important in the data analysis (*see* the subsection ***Sample Support and Estimation Blocks***).

Kriging has a number of characteristics of a desirable estimation method: sample weights can be adjusted for anisotropy; samples in correlated clusters can be down-weighted; the degree of smoothing increases as the random component (C_0) of the semivariogram model increases; and, when the semivariogram model is completely random ($C_1 = 0$), the kriging estimator becomes the sample mean, as in independent random sample statistics.

Spatial Outliers

Spatial outliers can be found by examining a geographical plot of the data; they may fit into a random variable histogram of all the data very well. In other words, a *spatial outlier* is a sample value that does not agree in magnitude with the values of its neighboring samples, especially the samples within a range of correlation. For example, a high (polluted) value in a low (background) neighborhood might be a spatial outlier but not a random variable outlier because the high value agrees with other polluted values. Once these outliers are identified, their location descriptions should be looked up in the sampling diary. If they are obviously from different sources that do not have the same correlation structure, they should be excluded from the semivariogram evaluation. The question of whether to include a spatial outlier in the final local estimate of concentration must be answered on a case-by-case basis. This matter involves the investigator's judgment, just as in the case of random variables.

The following discussion exemplifies an analysis of spatial outliers. Investigating the data from a city-wide sampling campaign for Pb, exploratory data analysts showed an empirical semivariogram with a range of correlation of at least 6 miles and two hot spots that were one order of magnitude higher in concentration than the rest of the data. The data set was printed out on a geographical plot that showed the two hot spots to be in sharp contradiction to their individual local neighborhoods, that is, every neighboring point and every point one neighbor out was at least one order of magnitude lower in concentration. The geographical map that identified the freeway system and the data showed that both points seemed very close to the freeways. In checking the sample log book this conclusion was confirmed; one of the aberrant samples was taken under a freeway overpass and the other at a freeway on-ramp. Freeway Pb is said to have a range of about 500 feet. Thus, because the two points represented a different source of Pb and had a much shorter range, they were excluded from the semivariogram computations. However, what was to be done with them in the kriging and mapping? If they were included in the kriging, they would spread their high values over circular areas of 6 miles in radius. This representation would be grossly untrue because the outliers are known to have a different source and a shorter range of correlation. The mapping would show a large area needing remediation that, in fact, did not need remediation. Nevertheless, the values had been found, and users of the map (risk assessors) needed to know of the hot spots. The compromise was to krige and contour the Pb concentrations of the other samples onto a kriging map and then just print the magnitude of such outliers at their respective locations on the map.

Spatial Soil Sampling

The growing number and complexity of toxic chemicals and hazardous waste sites call for a new statistical technique for monitoring with more efficient sampling designs and more precise data analysis. Geostatistics is a promising tool for these needs. This section traces the logic sequence of geostatistical analysis and then draws together the implications of geostatistical sampling design for soil pollution monitoring. Geostatistical sampling design has at least two phases: (1) the survey or the preliminary sampling to find the extent of the plume and to estimate a semivariogram, and (2) the census to take as many samples as needed to estimate the surface within the desired accuracy as calculated from the semivariogram model.

Sample Support and Estimation Blocks

The basic assumption of geostatistical sampling is to define and assign area or volume to all inputs and outputs. In monitoring for environmental protection, the spatial quantities to be defined and assigned are the *sampling unit* (area or volume), the *remediation unit,* and the *exposure unit.* Geostatisticians call the sampling unit the sample support. The sampling unit or support is ambiguous: it is used to refer to both the amount of medium extracted for the sample and the in situ area or volume represented by the sample. The context usually identifies whether the extracted support or the in situ support is meant. The remediation unit is determined by the method of remediation, and the exposure unit is determined by the risk assessor. For example, an appropriate remediation block might be a volume 250 ft long, 16 ft wide, and 0.5 ft thick, because this amount was the minimum volume to move economically. The shape is dictated by the up and back pass of a bulldozer with an 8-ft blade that scrapes up one truckload of contaminated soil. Sample unit, remediation unit, and exposure unit need to be defined (*18*) and then incorporated by a geostatistician into the sampling plan.

The critical mass of a correct sample should be calculated as previously explained (*see* the subsection ***Analytical Error***). The spatial variance of the sampling unit should be measured by taking "too many" equally spaced samples in several units in an exploratory sampling trip to the site. If the sampling unit is larger in spatial variance (a large spatial variance can be encountered in a small area), then the field samples design will have to use composite samples. In spatial compositing the geometry as well as the mass of the subsamples (samples to be composited) is important. The general rule is that subsamples should be equally spaced on the sampling unit. For example, if four subsamples can be afforded, then one should be taken from each quarter of the in situ support. Each subsample for the compositing should be a correct sample (*see* the subsection ***Analytical Error***). All samples for all analyses, even eye-balling, should have the same representativeness, which for composite samples means the same number of subsamples. The composite field sample, just like any other average, has a variance divided by the number of subsamples. *Homoscedasticity* (equality of variances) is a requirement for every data analysis even when eye-balling the data. If the quantity to be estimated (e.g., remediation or exposure unit) equals the sampling unit, punctual kriging analysis may be used because there is no change of scale or support. If the desired area or volume of estimation is larger than the sampling unit, block kriging will have to be used.

Survey or Semivariogram Sampling

In a multiphase sampling program using spatial statistics, the primary goal in the initial exploratory sampling is the collection of enough data to compute an empirical semivariogram and to determine the extent of the plume. These goals may conflict if limited resources are available. Widely spaced samples are needed to define extent, and closely spaced samples are often needed for semivariogram analysis. Approaches to this problem include regular grids (i.e., radial, square, or rectangular), transects, and combinations.

Burgess et al. (*19*) suggested transect sampling for variogram input, and this idea led to very good variograms in agricultural applications. However, in pollution monitoring, transects alone have given very noisy variograms. This result is probably due to intrinsic noise in pollution data, which is often highly skewed and contains high coefficients of variation. A combination exploratory grid, consisting of a grid of square sampling units having extended transects in the directions of the major axis and minor axis of the estimated plume (*20*), is illustrated in Figure 7. Prior information may be used to select the best grid orientation. For example, if the plume to be investigated was made by aerial deposition from an identifiable source, then wind roses can be examined for wind direction and magnitude, and topographic maps can be examined for natural barriers. Only the relatively regular grid concept is important in Figure 7; the orientation is site-specific.

If the extent of the plume must be found, and funds are limited, then the transect samples should be variably spaced closer together at the grid center and farther apart at the grid extremes. The purpose of this sampling is to capture the correlation structure of the plume. Inhabited areas have a high occurrence of disturbed sampling sites and local pollution from secondary sources, which are only stochastic noise to the semivariogram's calculation. Therefore, this noise should be avoided by this sampling. For example, aerially deposited smelter Pb should not be mixed with auto Pb by taking samples along the freeways. The samples from the semivariogram sampling can be pooled with the secondary mapping samples if they have the same support.

However, the semivariogram sampling often is the sampling that tests for the need for more compositing. If the support is changed between the samplings and we wish to pool the samples for analysis, then the change in support must be corrected before pooling. The sampling team must be aware of the need to keep all samples on the same support. When compositing, the same number and mass of

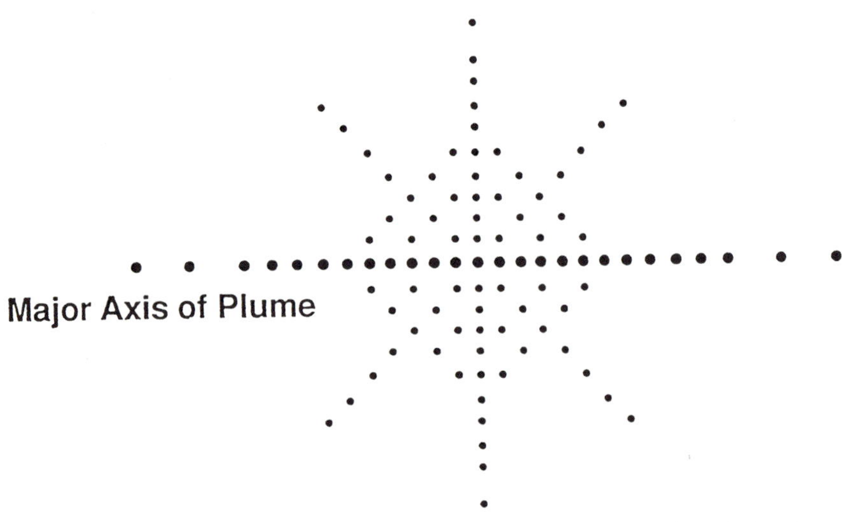

Figure 7. An exploratory grid design. (Reproduced from reference 16. Copyright 1988 American Chemical Society.)

subsamples and the same spacing or geometry must be maintained. When the sampled locations must move from the regular grid to avoid cultural improvements or natural barriers, then the spatial analysis program is corrected for this movement by the true coordinates of the new sample locations; however, no easy method is available for the program to correct for change of support. If the microvariation could be sampled and the support established before the semivariogram sampling, then a complex statistical problem could be avoided in the pooling of samples for the spatial analysis.

Some samples should be taken close together (in the scale of the sampling unit) to determine the need for composite samples. This sampling can be combined with field duplicates for quality analysis and control. Gy's fundamental error and compositing become more important as coring volumes decrease. These microvariation samples should be taken at a distance of a few multiples of the core's diameter apart. The distance between sample locations or grid unit's length needs to be estimated from the sample unit of interest (e.g., residential yard, city block, or square mile section) and the desired output unit (e.g., remediation unit, that is, the minimum volume of surface soil to be removed). The optimum exploratory sampling distance is a proper fraction of these measurements, but it is often determined by money available for sampling.

Census or Sampling for Map Making

In spatial statistics, the goal of secondary sampling is to uniformly cover the area in question with a density of samples sufficient to contour the plume with an acceptable error of interpolation. This sample coverage is accomplished by using the directional semivariograms to determine the orientation, shape, and size of the grid cell. Independent random variable statistics, in which the number of samples is computed, differs from spatial statistics, in which orientation, shape, and size of the grid are calculated and the number of samples is determined from the number of grid cells needed to cover the area.

If the directional semivariograms have a marked difference in their respective ranges of correlation, then the optimum cell geometry is not a square but a rectangle with the longer side in the direction of the longer range of correlation, and the ratio of the sides should be the ratio of the ranges of correlation. Thus the grid cell sides are of equal correlation or kriging (interpolation) variances rather than equal distance. This characteristic will save a lot of samples while retaining the same accuracy in both directions.

Boundary. For secondary sampling, the extent of the sampling grid must first be chosen. The sampling grid must extend beyond the suspected plume or area in question. The area in question must be bounded by sampling locations to avoid extrapolation in the kriging estimation algorithm for contouring. Extrapolation, which is estimating a value from data on only one side of the location of the point to be estimated, is likely to lead to unrealistically high or low values. If an action level has been set and a part of the plume has been adequately proven to be above or below the acting level, then that part of the plume need not be resampled. The sampling may be guided more by population areas or critical receptors than by the actual plume. The goals of the sampling must be written, and the areas of interest, action levels, and action areas (sampling unit, remediation unit, and exposure unit) must be defined before the optimum grid design can be made.

Compositing Samples Reduces Nugget. The next step in secondary sampling is choosing the sample support (*21*). If a residential yard is the sampling unit, then the ideal sampling process would be to take the entire yard, blend it to homogeneity, and remove the appropriate number of aliquots or splits to meet the volume needed by the laboratory for analysis. However, because few residents would donate their whole yard to science, and laboratory mixing equipment such as V-blenders or ball mills cannot homogenize so large a volume, this sampling unit must be represented by a few symmetrically laid out subsamples composited together. The number of subsamples is a compromise between the size of the microvariance and the amount of time and money allowed for the digestion of the subsamples. The subsamples are laid out symmetrically because a structural or spatial correlation may exist.

The mixing of the subsamples to achieve homogeneity is essential for compositing. If the medium is water, then the task is relatively easy; for soils or sediments, the task is difficult. Aliquots or splits should be taken after the mixing to make the final sample more representative. If a large nugget (e.g., $C_0 > 0.3$ relative variance) persists after Gy's critical mass calculations and compositing within the support, then the relative sizes of the field sampling and the laboratory analysis errors must be identified. The analysis of some pollutants has an analytical error that overwhelms the field sampling error and accounts for approximately all the semivariogram nugget.

The minimum volume at each step and especially the aliquot used by the chemical analyst in the lab must exceed the critical mass referred to in Gy's theory (*see* the section **Pierre Gy's Sampling Theory**).

Grid Unit Length or Distance Between Sample Locations. The range of correlations, the nugget (C_0), and the sampling budget determine the *grid unit length*, or the distance between sample locations. This length determination was discussed in mathematical detail by Yfantis et al. (*14*). Figure 8 shows the graphs of interpolation variance as a function of the ratio of grid spacing to range of correlation for a family of semivariograms. The model variograms each have relative C_1 and C_0 so that their sum equals 100%. The variograms differ only in the fraction of the sill ($C_0 + C_1$) represented by the nugget component (C_0) and the structure (C_1). If the semivariogram has a big nugget like the top graph of $C_1 = 10\%$ and $C_0 = 90\%$, then diminishing returns (the curve has less rapid vertical drop and becomes more horizontal) start and increase if the sample distance is less than two-thirds of the range of correlation. For a very low nugget, such as the lowest graph ($C_1 = 100\%$ and C_0

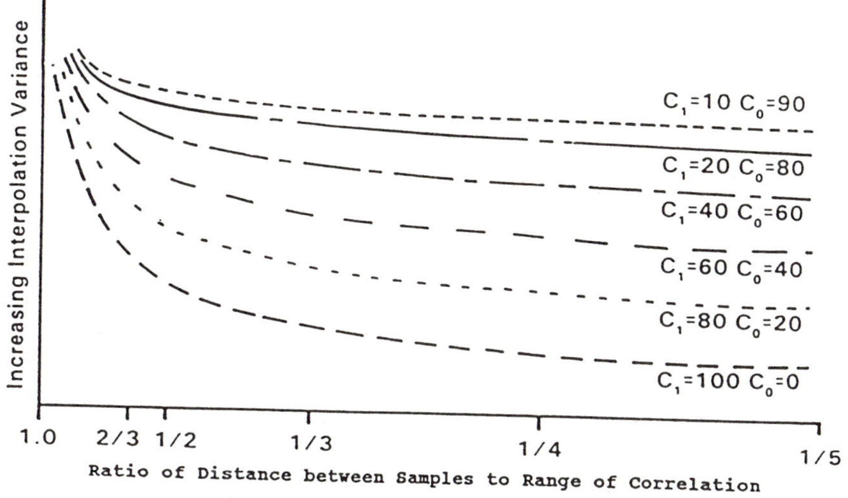

Figure 8. Diminishing information for additional samples. (Reproduced from reference 16. Copyright 1988 American Chemical Society.)

= 0%), diminishing returns do not start and increase until the sampling distance is less than one-half of the range of correlation. The general rule is that for smaller nuggets (C_0), the distance between sampling points on the sampling grid gets smaller. The grid should be laid out with no vertices unsampled. If this design exceeds budget, then the whole grid size should be adjusted, not just certain vertices left unsampled as in systematic random sampling.

Some real-world examples can clarify how the magnitude of the nugget (C_0) and the range of correlation determine the optimum cell size or distance between samples. One Pb smelter had a nugget of about 40% and a range of correlation of 3200 ft. In Figure 8, the family of diminishing return curves and the graph (for C_0 = 40%) indicates by observation and judgment that the point of diminishing returns is between one-third (33.3%) and one-fourth (25%) of the range of correlation, or 29% for the sake of argument. The sampling distance should not be less than 29% × 3200 ft, or 928 ft. Expressed as a function of money, the sampling distance should be the shortest affordable distance in keeping with the toxicity of the pollutant, but not less than 928 ft between samples. In contrast, a second Pb smelter had a semivariogram with a nugget of zero (0) and a range of correlation of 2400 ft. In Figure 8, the curve of diminishing returns for $C_0 = 0$ indicates by observation and judgment that the point of diminishing returns is between one-fourth (25%) and one-fifth (20%) of the range of correlation, or 22.5% for the sake of argument. In this case, the sampling distance should not be less than 22.5% × 2400 ft, or 540 ft. Expressed as a function of money, the sampling distance should be the shortest affordable distance in keeping with the toxicity of the pollutant, but not less than 540 ft. If the funding is adequate and the pollutants are of extreme toxicity, then the distance indicated by the point of diminishing returns should be used in minimum interpolation variance. If there is less money and the pollutant is less toxic, then a longer distance should be used for the grid cell's side. The directional semivariograms should orient the sampling grid.

The Pb smelters mentioned previously worked well with an east–west and north–south grid because the plume was formed by 80 years of aerial deposition. A third set of data, dioxin along a highway, gave a readable semivariogram in a direction of 13 degrees from east to west. This discovery took much searching because we started with the default directions (0, 45, 90, 135 degrees) of the semivariogram software; these default semivariograms showed no structure [pure nugget semivariograms (C_0 = 100%)]. After we discovered the semivariogram at 13 degrees the reason became obvious, because the road that was the transport of the pollutant ran at that angle and so should any sampling grid.

In the field, some vertices cannot be sampled because of man-made improvements or natural barriers, but these vertices must be sampled as closely as possible, and the actual coordinates should be used in the spatial analysis program.

Grid Orientation and Shape Versus Anisotropy. If the ranges of correlation are extremely different on the directional semivariograms, then the correlation structure is anisotropic. Optimum sampling patterns reflect this anisotropy. For example, the sides of a rectangular grid would be in the same ratio as the ranges of correlation for the corresponding directional semivariograms. This ratio was explained in detail by David (22), and a sampling design for logarithmic anisotropy was derived by Barnes (23). Anisotropy is a frequent occurrence, but often the semivariogram sampling gathers too few samples to measure it. Thus, more samples may be used cost-effectively in the semivariogram sampling in order to save samples in the larger census (or mapping) sampling by identifying and taking advantage of anisotropy.

Use of the triangular grid as opposed to the rectangular grid has been discussed (*13, 14*). If the nugget is large ($C_0 >> C_1$), little is gained by the triangular grid. Also, the triangular grid makes taking advantage of anisotropy more difficult. If a triangular grid is chosen, a theodolite, which is a surveying instrument, is not needed in the field; instead every other row of samples must be offset by one-half of a grid length. In practice, this action is easier than it sounds and almost as easy as the traditional square grid.

Beyond Anisotropy

Numerous additional geostatistical considerations affect environmental sampling. These considerations include spatial drift or trend, multivariate analysis, mixed or overlapping populations, concentration-dependent variances, and specification of confidence limits. Geostatistical techniques have been developed over the years to deal with these various problems, but an adequate discussion is beyond the scope of this chapter.

Acknowledgments

The EPA, through its Office of Research and Development, funded and performed the research described here.

References

1. Gilbert, R. O. *Statistical Methods for Environmental Pollution Monitoring;* Van Nostrand Reinhold: New York, 1987.
2. Pitard, F. F. *Pierre Gy's Sampling Theory and Sampling Practice;* CRC Press: Boca Raton, FL, 1989; Vol. 1 & 2.
3. Isaaks, E. H.; Srivastava, R. M. *An Introduction to Applied Geostatistics;* Oxford University: New York, 1990; pp 1–592.
4. Pitard, F. F. In *Pierre Gy's Sampling Theory and Sampling Practice;* CRC Press: Boca Raton, FL, 1989; Vol. 2, p 36.
5. Gilbert, R. O. In *Statistical Methods for Environmental Pollution Monitoring;* Van Nostrand Reinhold: New York, 1987; pp 35–42.
6. Pitard, F. F. In *Pierre Gy's Sampling Theory and Sampling Practice;* CRC Press: Boca Raton, FL, 1989; Vol. 1, pp 169–183.
7. *Preparation of Soil Sampling Protocols: Sampling Techniques and Strategies;* Center for Environmental Research Information: Cincinnati, OH, 1992; pp A1–A16; EPA/600/R–92/128.
8. Pitard, F. F. In *Pierre Gy's Sampling Theory and Sampling Practice;* CRC Press; Boca Raton, FL, 1989; Vol. 1, p 190.
9. Matheron, G. *Econ. Geol.* **1963,** *58,* 1246–1266.
10. Journel, A. G. *Geostatistics for the Environmental Sciences;* Stanford University: Stanford, CA, 1986.
11. Palmer, M. W. *Vegetation (Dordrecht, Netherlands)* **1988,** *75,* 91–102.
12. Cliff, A. D.; Ord, J. K. *Spatial Processes: Models and Applications;* Pion: London, 1981; pp 1–266.
13. Olea, R. A. *Math. Geol.* **1984,** *16*(4), 369–392.
14. Yfantis, E. A.; Flatman, G. T.; Behar, J. V. *Math. Geol.* **1987,** *19*(3), 183–205.
15. Flatman, G. T.; Yfantis, E. A. *Environ. Monit. Assess.* **1984,** *4,* 335–349.
16. Flatman, G. T.; Englund, E. J.; Yfantis, A. A. In *Principles of Environmental Sampling;* Keith, L. H., Ed.; ACS Professional Reference Book; American Chemical Society: Washington, DC, 1988; pp 73–84.
17. Borgman, L. E.; Quimby, W. F. In *Principles of Environmental Sampling;* Keith, L. H., Ed.; ACS Professional Reference Book; American Chemical Society: Washington, DC, 1988; pp 25–43.
18. Neptune, D.; Brantly, E. P.; Messner, M. J.; Michael, D. I. *Hazard. Mater. Control* **1990,** *May/June,* 19–25.
19. Burgess, T. M.; Webster, R.; McBratney, A. B. *J. Soil Sci.* **1981,** *32,* 643–659.
20. Starks, T. H.; Brown, K. W.; Fisher, N. J. In *Quality Control in Remedial Site Investigation;* American Society for Testing and Materials: Philadelphia, PA, 1986; Vol. 5, pp 57–66; ASTM STP925.
21. Starks, T. H. *Math. Geol.* **1986,** *18*(6), 529–537.
22. David, M. *Geostatistical Ore Reserve Estimation;* Elsevier Scientific: Amsterdam, 1977.
23. Barnes, M. G. *Statistical Design and Analysis in the Cleanup of Environmental Radionuclide and Other Spatial Phenomena;* TRAN STAT (Statistics for Environmental Studies) No. 13, Battelle Memorial Institute: Richland, WA, 1980; pp 1–21.

Glossary and Indexes

Glossary

Absorption – The penetration of one substance into the inner structure of another.

Accuracy – Difference between measured and referenced values.

Active sampling – A means of collecting an airborne substance that employs a mechanical device such as a pump or vacuum-assisted critical orifice to draw air into or through the sampling device.

Additive interferences – Interferences caused by sample constituents that generate a signal that adds to the analyte signal.

Adsorption – The adherence of contaminants to a material or to the micropores on the material surface.

Aerial photography – A method of characterizing a stream or land surface according to its appearance in either the visible or infrared spectrum. The camera can be mounted either in an aircraft or in a satellite.

Agreement state – A state that has been delegated the administration of federal Nuclear Regulatory Commission authority.

Air quality models – Models that relate emissions and receptors.

ALARA – Acronym for "as low as reasonably achievable", a philosophy for reducing hazardous exposures to the lowest practical level.

Alpha error – *See false positive.*

Alpha particle – A charged particle consisting of two protons and two neutrons, essentially a free helium nucleus, emitted from the nucleus of an atom.

Alpha radiation – Low-energy radiation that can be stopped by unbroken skin or clothing and that poses a substantial respiratory hazard.

Anisotropy – Condition in which the variance structure, as measured by the semivariogram, is often different in the range of correlation in different directions. In anisotropy, a spatial autocorrelation structure changes semivariogram values by direction as well as by distance apart or lags; in isotropy, spatial correlation varies only by distance apart or lags.

Audit sample – Prepared reference sample inserted into the sample processing procedure as close to the beginning as possible.

Automatic sampler – A device that can be programmed to collect samples for environmental analysis.

Background control – A sample collected from a location not known to contain the substance of interest.

Beta particle – A free electron emitted from the nucleus of an atom.

Beta error – *See false negative.*

Beta radiation – Moderate energy radiation that can pass through the skin and expose bodily tissues.

Bias – The nearness of a value to a "true" or assigned value. It is the systematic or persistent distortion of a measurement process that results in measurements consistently higher (or lower) than the "true" value. Systematic errors from interference, contamination, or operational errors contribute to bias.

Binomial distribution – Distribution of a variable that can be classified into one of two categories.

Biodegradation – Decomposition of a substance by microorganisms.

Blank – A sample matrix that has no, or unmeasurable, amounts of the chemical of interest. Blanks are named according to their use (e.g., field blank or method blank).

Blank sample – A clean sample or a sample of matrix processed to measure artifacts in the measurement process.

Blind sample – A subsample submitted for analysis with a composition and identity known to the submitter but unknown to the analyst.

Block kriging – Any type of kriging (simple, ordinary, or universal) that assumes the input is a set of points but gives the output as a block estimate (point values integrated over the block).

Bootstrap method – Computationally intensive measurement of uncertainty that involves recalculation of statistical parameters, such as those yielded by description, classification, and correlation models, from many randomly selected subsets of the model input data set.

Bottom core sampling – A technique for sampling the surface and near surface of the bottoms of streams, lakes, etc. A pipelike device, driven vertically into the bottom, then secured at both ends and withdrawn, is normally used.

Bottom grab sampling – A technique for grab sampling the surface and near surface of the bottoms of streams, lakes, etc. One of several dredgelike devices is normally used.

Box plots – Geographical depictions of percentiles.

Calibration check sample – A reference material used to check that the calibration is acceptable.

Case law – Rules of law established in judicial opinions as opposed to legislation.

Cation exchange capacity (CEC) – Test performed on the fine-grained (<2 mm) fraction of the soil.

Classificatory models – Models that define categories based on the achievement of certain measured values for a set of attributes. Each category has a specific set of decisions associated with the members of that category.

Collocated sample – One of two or more independent samples collected so that each is equally representative for a given variable at a common space and time.

Collocated sampling – The simultaneous collection of two or more samples by samplers placed side by side (often misnamed "co-located sampling"). The samplers must be placed close enough to each other to ensure that comparable samples are collected but be sufficiently separated to prevent oversampling.

Comparability – Confidence with which one data set can be compared to another.

Competent evidence – Information considered reliable.

Completeness – A measure of the amount of valid data obtained from a measurement system compared to the amount that was expected to be obtained under correct or normal conditions.

Compliance monitoring sample – A sample taken by the facility to demonstrate compliance with a regulatory requirement.

Composite sample – A mixture of samples collected from more than one sampling location or at the same location more than one time.

Composite sampling – A technique of sampling in which a series of discrete, single samples are combined as a function of either time, flow, or mass transfer. The pooled sample is analyzed to characterize an environmental matrix.

Concentration monitoring approach – Deposition velocity estimation method that focuses on air quality alone, although often with supporting measurement of a selected set of meteorological variables.

Conditional simulation – Statistical simulation conditioned on a relatively sparse data set generated to produce alternate, equally likely scenarios of what the spatial and temporal distribution of values may be. This method allows the consequences of various random field assumptions to be studied in the computer before actual field data are collected.

Confidence interval – Range of values that can be declared with a specified degree of confidence to contain the correct or true value for the population.

Confidence level – The probability of meeting a defined value given replicate measurements (e.g., a 95% confidence level implies that a stated value will be obtained in at least 95 out of 100 replicate measurements).

Contamination – Something inadvertently added to the sample during the sampling and analytical process.

Continual sampler – Sampler that withdraws a sample constantly and accumulates the withdrawn volume for collection at a later time.

Continuous monitoring – The measurement of air concentrations in real or near-real time. Such monitoring is usually automated and combines the collection of the sample with immediate or near-instantaneous analysis.

Control – Type of sample against which the results of a procedure are judged. A control can be a sample against which experimental samples are compared to determine anomalous appearance of a substance, or a sample used in quality control procedures to control instrument performance.

Control charts – Charts used to provide the most effective mechanism for interpreting blank results. In the control mode, control charts can be used to detect changes in the average background contamination of a stable system.

Correctness – A property of a sampling protocol that is necessary for unbiased determinations.

Correlation – A statistical cross-product to measure if two or more variables change together. There is simple correlation for two variables or multiple correlation for more. The range of the index is 1 to –1. It is 1 for synchronized change in the same direction, 0 for no synchronization, and –1 for synchronized change with opposite directions.

Covariational models – Models that calculate measures of association between two or more variables and are often used to assign effects to a pollutant, or a pollutant to an emissions source.

Criteria air pollutants – The original six pollutants regulated under the 1970 Clean Air Act to meet human health standards: lead, ozone, carbon monoxide, nitrogen dioxide, sulfur dioxide, and particulates.

Critical concentration – Determined on the first day of a holding time study, it is the concentration below which there is only a 5% chance that a measured analyte concentration would be observed.

Cryogenic sampling – The collection of substances from air by condensation in a cryogenic trap. Usually used in near real-time or sequential monitoring. Also called *freeze-trapping*.

Curie (Ci) – The unit of measure for activity, equal to exactly 3.7×10^{10} disintegrations per second.

Data quality – The degree to which data conform to a specified set of criteria. It is a measure of the error associated with a particular data set.

Data quality objectives (DQOs) – Qualitative and quantitative statements that specify the quality of data needed from a particular data collection activity.

Data validation – An independent systematic process for reviewing a body of data against a set of criteria to provide assurance that the level of quality of the data is known and documented.

Decision tree – Way of presenting a classification rule converted from a training set.

Denuder – A device designed to collect or remove gases from an air sampling stream while permitting the passage of particles. The gases are removed by diffusion from the airstream to a collecting surface or secondary airstream.

Descriptive models – Models that summarize the spatial, temporal, and statistical distributions of individual observations.

Discharge monitoring report – A compliance monitoring sample under the federal water pollution program, usually required on a monthly basis, that is made under oath.

Discovery – In civil litigation, exchange of information between the parties before trial.

Discrete sample – An individual sample collected within a short period of time and deposited in an individual container. It represents the conditions of the source at the time the sample was collected but does not necessarily represent the source at any other time. It is also known as a grab sample.

DQO-PRO – A series of computer programs used to calculate numbers of samples needed based on specific objectives such as confidence levels, and acceptable rates of false positive and false negative conclusions in the data.

DQO process – A seven-step planning procedure, based on the scientific method, that helps identify the type, quality, and quantity of data needed to satisfy the data user's needs.

Dry deposition – Aerodynamic exchange of trace gases and aerosols from the air to the surface as well as the gravitational settling of particles.

Electronic controller – The "brain" of the automatic sampler. It can be programmed to run any sampling routine.

Engineering judgment – Estimate of the characteristics of a waste made on the basis of knowledge of the processes that produce the waste.

Equipment blanks – Special type of field blank used primarily as a qualitative check for contamination rather than as a quantitative measure.

Error of the estimate – Difference between the true population average and the estimate from the particular survey of the average.

Estimation block – In situ volume represented by the estimated value.

Expert system – A computer program that emulates the logic of a human expert in a particular knowledge domain and that is used in making decisions.

Exposure unit – The smallest area or volume of the site that can have its own risk assessment. As an applied rule the sampling unit is smaller than or equal in area to the remediation unit, which often is smaller in area than the exposure unit. It is chosen by the risk assessor and the investigation team.

False negative – Error of concluding that an analyte is not present in the media sampled when it is present (also known as type II or beta errors).

False positive – Error of concluding that an analyte is present in the media sampled when it is not present (also known as type I or alpha errors).

Field blank – Blank used to provide information about contaminants that may be introduced during sample collection, storage, and transport.

Field duplicate – Two samples taken from and representative of the same population and carried through all steps of sampling and analysis in an identical manner.

Fifty-percentile point – Value that has 50% of the total values at or below it.

Flow-proportional composite samples – A series of samples of equal volume at equal flow volume intervals that are collected and deposited in one container, or samples that are collected at equal time intervals with the volume of each sample made proportional to the flow volume during the corresponding sampling period.

Freeze-trapping – *See cryogenic sampling.*

Fundamental error – This component of variance is a natural property of all heterogeneous materials and represents the difference between the value in the sample and the in situ region in the field that it is supposed to represent. A major focus of Gy's sampling theory, the fundamental error is associated with the constitution heterogeneity of the material and can never be eliminated.

Gamma axis – Squared variance of differences in pollution units, such as parts per million squared, used as the vertical axis for the semivariogram plot in geostatistics.

Gamma radiation – High-energy radiation that can pass through the body and unshielded walls and can travel significant distances.

Gamma ray – A wave of energy with some particlelike properties emitted from the nucleus of an atom.

Gaussian distribution – Distribution of values that arises when many small independent fluctuations in components tend to cancel each other out to yield a stable average.

Geostatistical estimation – Process of using space–time data, together with the assumptions and structural characterization of the random field, to estimate parameters or characteristics (properties) related to the random field.

Geostatistical sampling – Sampling procedures that are guided by the assumed properties of the random field and prior, or early, estimates of the covariance or variogram functions.

Geostatistics – Statistics used to describe the study of regionalized or spatially correlated variables. This branch of statistics answers the problems of earth mineral measurement (mining), natural resource inventory (forestry, ecology, and agriculture), and environmental protection. It has many old (tessellation and inverse distance) and new methods (theory of chaos and fractals). Kriging is its best known method.

Good data – Data that can be used with the specified level of confidence to make an inference about the environmental population from which samples were collected and analyzed.

Grab sample – A grab, or discrete, sample is one that is collected at a specific time and at a specific location.

Grid unit length – Distance between sample locations.

Gross sample (bulk sample) – Pool of two increments that is reduced or prepared as subsamples for analysis.

Grouping and segregation error – As described in Gy's theory of sampling heterogeneous media, it is the measurement inaccuracy caused when the mineralized or polluted particles sink, float, or shift in transportation or vibration, making further sampling biased because of the clustered distribution of the particles of interest.

Half-life – The time required for a radionuclide to lose one-half of its activity by decay.

Health physicist – An individual who specializes in the recognition, evaluation, and control of occupational exposures to radiation hazards.

Heteroscedastic – When the standard deviation of each component of the measurement error is allowed to depend on concentration.

Heteroscedastic curve (envelope of uncertainty) – Curve that relates interlaboratory coefficients of variation to concentration.

Homoscedasticity – The equality of variances across groups, such as treatments in an analysis of variance.

Hot cell – A shielded environment used for manipulating radioactive materials.

Hot spot – A localized area of contamination.

Hybrid plan – Sampling plan that incorporates elements of intuitive sampling plans and statistically based plans.

Impactor – A device for collecting airborne particulate matter in which the air being sampled is impacted or impinged against a surface.

Imprecision – The degree to which data from replicate measurements disagree. Random errors contribute to imprecision.

Incremental delimitation error – As described in Gy's sampling theory of heterogeneous materials, it is the measurement inaccuracy that deals with the definition and delimitation of the sample. The sample should include only the part of border particles that reside within the volume to be sampled, but this condition cannot be satisfied because the boundary of the sampler is not coterminous with the boundary of the medium particles.

Incremental extraction error – As described in Gy's sampling theory of heterogeneous materials, it is the measurement inaccuracy that deals with the integrity of the border of the sample as it is extracted. Particles on the border of the extraction tool should be partially included, but because of their hardness or the softness of interstitial material, they are either totally excluded or totally included.

Industrial hygienist – An individual who specializes in the recognition, evaluation, and control of occupational exposures to chemical and physical safety hazards.

Inferential method – Deposition velocity estimation method that relies on measurements made of controlling surface and atmospheric properties.

Inferential models (receptor-oriented models) – Models that can be derived either from fundamental physical and chemical laws or from empirical observations. These models differ from mechanistic models in that their primary set of input data is that which is measured at the receptor, not at the source.

Instantaneous sampling – The collecting of a sample over a very brief period of time (typically less than 5 min) to obtain a measure of the content of the sample at that instant in time. Also known as *grab sampling.*

Integrated sampling – The sampling of airborne substances over an interval of time to obtain an average concentration for that time interval (e.g., 24 h).

Interferences – Compounds whose presence obscures the measurement of the analyte of interest by the introduction of an unrelated analytical signal where the analyte is measured.

Interlaboratory precision (reproducibility) – Variation associated with two or more laboratories or organizations using the same measurement method.

Intralaboratory precision (repeatability) – Variation associated with a single laboratory or organization.

Intuitive sampling plan – Sampling plan based upon judgment, often by technical experts.

Ionizing radiation – Radiation composed of particles or waves that are themselves ionized or that are able to ionize other atoms by reaction with them (e.g., alpha, beta, gamma, and neutron radiation).

Isokinetic sampling – Method of obtaining a representative sample of an effluent stream containing particulate matter.

Kriging – Linear-weighted average interpolation technique used in geostatistics to estimate unknown points or blocks from surrounding sample data. It uses spatial correlation structure to compute the weights in order to improve the surface estimates at the unsampled points.

Laboratory control standard – A sample of media containing a known, validated concentration of the analyzed substance.

Lag axis – Distance between points in linear units such as meters or kilometers. The horizontal axis for the semivariogram plot in geostatistics is the lag axis.

Leaching – The process by which soluble contaminants are dissolved from the matrix of a material by a liquid, or compounds previously sorbed to the matrix are released.

Local control site – Control site that is near in time and space to the sample of interest.

Lot – A unit of material.

Low-level bias – Variability in data usually caused from sources of contamination in the sampling or measurement system.

Matched pair – Two observations, one with treatment and the other serving as control.

Matched-matrix field blank – Most common type of field blank, where the blank simulates the sample matrix.

Matched-pair Wilcoxon test – Test used to determine whether a treatment effect is significant compared to another observation that is similar in all ways with the first except that the treatment was not received.

Material evidence – Information that is an issue in a legal proceeding.

Matrix – The physical type of an environment sample (e.g., ambient air, sandy soil, drinking water, or fish tissue).

Matrix control (field spike) – Control used to estimate the magnitude of interferences caused by a complex sample matrix.

Measurement process – Process that includes the sampling design phase, sampling implementation phase, and analytical phase.

Measurement uncertainty – Uncertainty that results from the variability of the environmental measurements used as model input data.

Mechanistic models (source-oriented models) – Models that contain mathematical descriptions of the interactions among variables derived from fundamental physical and chemical laws.

Median – Value that has 50% of the values below it and 50% of the values above it.

Method blank (reagent blank) – Analyte-free media carried through the sample preparation and analytical procedures.

Mixed waste – A waste material that simultaneously meets the definition of a hazardous waste (chemical) and a radioactive waste.

Model uncertainty – Uncertainty that is caused by deviations from the model assumptions.

Monte Carlo method – Computationally intensive measurement of uncertainty that involves perturbation of all input data with random numbers drawn from an assumed error distribution.

Multiplicative interferences – Interferences that either increase or decrease the analyte signal by some factor without generating a signal of their own.

Neutron radiation – High-energy radiation that is formed by electronic devices and nuclear reactions.

Nonionizing radiation – Radiation that is neither itself ionized nor causes other atoms to become ionized by reaction with them (e.g., microwave, radio-frequency waves, and laser light).

Nonparametric procedures – Statistical techniques that can be applied without concern for the actual distribution of the underlying population from which the data were collected.

Nonvolatile organic chemicals – Organic compounds with saturation vapor pressures less than 10^{-8} kPa at 25 °C.

NORM – Acronym for "naturally occurring radioactive material".

Normal-level bias – Variability in data usually caused by operational or procedural errors in the collection, preparation, or analysis of a sample.

***P*-percentile point** – Value at or below which *P* percent of all population values lie.

Parameters – Constants that are not supplied to a model on a case-by-case basis, as are input measurements. These constants are obtained by a theoretical calculation, by measurements made elsewhere and assumed to be appropriate for the place and time being studied, or by tacit assumption that the value of a variable is constant or negligible.

Passive sampling – A means of collecting an airborne substance that depends on gaseous diffusion, gravity, or other unassisted means (such as particle deposition) to bring the sample to the collection surface or sorbent.

PCBs – Polychlorinated biphenyls manufactured as Aroclor mixtures.

Performance evaluation sample – A sample whose composition is unknown to the analyst.

Periodic error – As described in Gy's theory of sampling heterogeneous media, it refers to the within-sample time variations.

Phase distribution – The distribution of a chemical compound between the gaseous and condensed or particle-sorbed states in the atmosphere.

Photodegradation – The decomposition of a substance by ultraviolet light.

Plot – A geographical unit.

PM_{10} – Particulate matter having aerodynamic diameter smaller than 10 μm.

Polar organic compounds – Organic compounds that may exhibit a relatively high degree of internal polarization or may be readily ionized. They typically contain oxygen, nitrogen, sulfur, or other heteroatoms, and are more water-soluble than their nonpolar counterparts. They are usually more reactive and more strongly bind to surfaces and sorbents than nonpolar compounds.

Population – The set of all possible samples for a characteristic of interest. Using the word "population" ambiguously (e.g., for a mixture of populations) can lead to poor or wrong remediation decisions.

ppbv – A unit of measure of the concentration of a compound in the gaseous or vapor phase in air expressed as parts of the compound per billion (10^9) parts of air, both by volume.

Practical reporting time – As developed in holding time studies, it refers to the day when there is a 15% risk that a measured analyte concentration will be below the critical concentration.

Precision – Measure of the variability of measurements of the same quantity by the same method.

Preparation error (analytical error) – As described in Gy's sampling theory for heterogeneous media, it is the measurement inaccuracy that gives a mathematical relationship among several media parameters, maximum particle size, and the critical mass or volume of sample needed to be chemically analyzed for a specified accuracy such as relative standard deviation.

Primary chemical constituents – Category of chemical constituents that may represent known chemical species that have been identified in the sampling matrix or are required to be determined by regulation.

Primary sample – First collection of any material for the analysis of one or more analytes.

Primary sampling unit – Time period or volume of waste for which a composite estimate is desired. The unit may be a time period if the waste is a flowing stream, or a volume if the waste is an impoundment.

Protocol – Thorough, written description of the detailed steps and procedures involved in the collection of samples.

Punctual kriging – Any type of kriging (simple, ordinary, or universal) that assumes the input is a set of points and gives output as points.

Quality assurance (QA) – A system of activities that ensures the producer or user of a product, service, or data, that defined standards of quality with a stated level of confidence are met.

Quality control (QC) – A system of activities that controls the quality of a product, service, or data so that it meets the needs of users.

Quality control sample – An uncontaminated sample matrix spiked with known amounts of analytes from a source independent from the calibration standard.

rad – The unit of measure for absorbed dose, equal to 1×10^{-2} J/kg.

Radiation – Energy traveling through space in the form of particles, waves, or photons.

Random errors – Errors that vary unpredictably in both magnitude and direction but that can be estimated by the use of standard statistical techniques.

Random field sampling (geostatistical sampling) – Method in which the goal is to arrange sampling so as to make every individual of the total population equally likely to be selected in the sample.

Random variable – A characteristic of interest whose value is or has a component of chance. In environmental sampling the characteristic may be independent or correlated in time or space.

Range – Spread of values calculated by subtracting the lowest value from the highest value.

Range of correlation – Distance on the lag axis at which the semivariogram's curve reaches the sill. In the field it is the distance that defines nearest neighbors, neighborhoods, and representativeness. In kriging it is used for the weighting of coefficients to estimate the desired characteristic surface and the surface of errors of estimation.

Reactive sampling – The collection of compound from air or detection of a chemical in air by reacting it with a chemical reagent (e.g., derivatization).

Reagent blank (method blank) – Blank that contains any reagents used in the sample preparation and analysis procedure.

Relative standard deviation – Estimate of the average error in the measurement due to unassignable causes and usually expressed as a percentage of the average sample concentration.

Relevant evidence – Information that makes an inference more likely.

rem – The unit of measure for dose equivalency.

Remediation unit – The minimum area or volume of the site that can be remediated economically. It is chosen by the investigation and remediation team with regard to proximity to people, future use, and toxicity of pollutants.

Replicate measurement – Subsequent measurement on a unit that is not influenced by previous measurements on the unit; it is a measurement statistically independent of other measurements.

Replicate samples – Two or more samples representing the same population, time, and place, which are independently carried through all steps of the sampling and measurement process in an identical manner.

Representative sample – A sample that closely resembles, or is a subset of, the population being measured. A representative environmental sample is one that is collected and handled in a manner that both preserves its original physical form and chemical composition and prevents changes in the concentration of the materials to be analyzed or the introduction of outside contamination.

Representativeness – Correspondence between the analytical result and the actual environmental quality or the condition experienced by a contaminant receptor. It is a measure of how well a set of sample measurements yields information concerning the population.

Response – A numerical value obtained from execution of a protocol.

Routine-duplicate pair – Sample pair consisting of a duplicate sample collected coincidentally with the routine sample taken at some location.

Sample collection – Contact of the sample with the sampling device and its materials of construction.

Sample delivery system – A component of an automatic sampler. It is simply a pump designed to transfer the sample from the liquid source to the bottles for storage.

Sample storage system – A component of an automatic sampler. Generally, it consists of bottles that are placed in the base of the portable samplers or inside the refrigeration compartment of refrigerated models.

Sample transport line – A component of an automatic sampler. Generally, it is a plastic tube connected to the sample intake. The water or wastewater sample travels through the transport line until it reaches the sample containers for storage.

Sampling – Attempt to choose and extract a representative portion of a physical system from its surroundings.

Sampling equipment blank – A clean sample collected in a sample container with the sample collection device and returned to the laboratory as a sample.

Sampling media blank (trip blank) – Blank consisting of the sampling media used for collection of field samples.

Sampling unit – A volume of the medium (soil or sediment) to be represented by the sample. If there is a strong spatial correlation structure, the in situ unit is much larger than the extracted sample. The unit must be chosen by the site investigation team considering the future use of the site, proximity to people, and the toxicity of the pollutants. The sampling unit is called *sample support* in mining literature.

Scientifically true value – The true value (as compared with the *statistically true value*).

Secondary chemical constituents – Category of chemical constituents that may include transformation products of primary chemical species, environmental variables needed to characterize conditions or meet criteria for representativeness, or other chemical species that may be indicators of sample integrity.

Selectivity – *See specificity.*

Semivariogram – Graph used in geostatistics in which spatial correlation can be measured and displayed. It is the variance of pairs of points at each unit of distance apart (called lags) and is plotted as a function of lags.

Semivolatile organic chemicals – Organic compounds with saturation vapor pressures that fall between 10^{-2} and 10^{-8} kPa at 25 °C.

Sensitivity – The sensitivity of a protocol is a ratio that compares the effect on the response of changes in the property of interest with the standard deviation of the random error.

Sequential composite sample – A series of composite samples collected during time or flow subintervals.

Sequential sampling – The collection of samples for discrete time intervals repeated several times over a longer period of time (e.g., for 5 min out of every hour over 24 h). Such sampling is usually used to obtain estimates of average concentrations over the full period of sampling. It is usually automated and coupled with sequential analysis of the collected sample.

Sign test – Test designed to determine whether the true theoretical median equals a specified value.

Signed Wilcoxon test – Test for the median in which the sample is taken to contain independent, identically distributed observations from a reference population having a symmetric probability density.

Sill – Upper bound of the rise in variance on a semivariogram. Its magnitude is the variance of the differences of random points.

Solvent blank – Blank that consists only of the solvent used to dilute the sample.

Sorbent – A solid or liquid medium in or upon which materials (usually gaseous) are collected by absorption, adsorption, or chemisorption.

Sorbent sampling – The removal of chemicals from a gaseous sample by passing the gas through or allowing it to come in contact with a sorptive medium. The chemicals are subsequently desorbed for analysis. Sorbent sampling can be either integrative or sequential.

Sorption – A process by which one material (the sorbent) takes up and retains another material (the sorbate) by the processes of absorption, adsorption, or chemisorption. Chemical reactions may accompany or follow sorption.

Spatial error – As described in Gy's theory of sampling heterogeneous media, it is the within-sample spatial variation.

Spatial gradient technique – Technique especially appropriate for sampling rivers for chemical constituents. By applying this technique, the distance between points on a transect or grid in a grid pattern can be determined.

Spatial outlier – Sample value that does not agree in magnitude with the values of its neighboring samples, especially the samples within a range of correlation.

Spatial variable – A random variable that is autocorrelated in space. Field samples closer together than the range of correlation are spatial variables. Spatial variables are called regional variables in the mining literature.

Specificity – A protocol with specificity produces responses unaffected by properties of the unit other than the property of interest. When applied to measurements, specificity is synonymous with selectivity.

Spiked sample – A sample prepared by adding a known mass of target analyte to a specified amount of matrix sample for which an independent estimate of target analyte concentration is available.

Split sample – Two or more representative portions taken from a sample or subsample and analyzed by different analysts or laboratories.

Standard deviation – Square root of the variance.

Standard error – Standard deviation of the average.

Stationarity – Decision to proceed with averaging and inference over a predetermined population or area in space.

Statistically based plan – Sampling plan that provides the basis for making probabilistic conclusions that are independent of personal judgment.

Statistically true value – Value that the mean of a population approaches as the number of samples increases.

Stratified random sampling – Sampling technique in which estimates of strata means are combined to yield estimates of the population mean.

System blank (instrument blank) – Measure of the instrument background, or baseline, response in the absence of a sample.

Systematic error – Error that usually results in a consistent deviation (bias) in a final result and cannot be estimated statistically.

Systematic sampling – Use of random numbers to lay a Cartesian grid over a region so that the coordinate origin and axis orientation are random. Samples are then taken at each grid intersection.

Time resolution – For an in situ measurement, it is the time required for the real-time sensor to reach 90% of the final response to a step change in concentration of the measured species.

Tolerable error – A subjective amount of acceptable error in data specified by the user of the data (e.g., ±10% of the true or assigned value of an analyte concentration).

Total sample variance – An estimate of the variability of an observation after consideration of all factors contributing to measurement error.

Toxicity characteristic leaching procedure (TCLP) – Test proposed by the EPA to determine if waste is hazardous by the characteristic of toxicity.

Training set – Set of exemplary situations used by an expert to "teach" a computer.

Trip blank (sampling media blank) – A clean sample of matrix that is carried to the sampling site and transported to the laboratory for analysis without being exposed to sampling procedures.

Two-sample Wilcoxon test – Test designed to determine whether the mean of one population differs from the mean of another population.

Type I error – *See false positive.*

Type II error – *See false negative.*

Unit – A subdivision that is homogeneous enough for the purposes of the study so that it can be described by a single measurement.

Vapor pressure – The pressure of a vapor in equilibrium with its liquid or solid form at a given temperature. It is usually determined by measuring the saturation concentration of the vapor over the condensed phase. Also called *saturation vapor pressure.*

Variance – Measure of the variability in a population.

Volatile organic chemicals – Organic compounds with saturation vapor pressures greater than 10^{-2} kPa at 25 °C.

Volatilization – Physical process in which volatile species can be lost to the atmosphere.

Whole-air sampling – The collection of an air sample into a sealable container such as a canister, bottle, or bag for subsequent analysis of its contents. Whole-air sampling can be instantaneous, integrative, or sequential.

Wind rose – Diagram that summarizes statistical information about the wind.

Windows – Directional classes used for grouping sample pairs.

Worst case – Situation that results in the most conservative results.

Indexes

Affiliation Index

Subject Index

Copy editing and indexing by Zeki Erim, Jr.
Production by Paula M. Bérard
Jacket design by Amy O'Donnell

Typeset by Betsy Kulamer, Washington, DC
Printed and bound by Maple Press Company, York, PA

Mail this card to get the *DQO-PRO* software program for use with this book free of charge. You can also download the software from the Instant Reference Sources home page on the Internet. The Internet address is http://www.instantref.com/inst-ref.htm. You may write to Lawrence H. Keith at Instant Reference Sources, Inc., 7605 Rockpoint Drive, Austin, TX 78731, or fax him at (512) 345-2386 for technical support.

Please send a free copy of *DQO-PRO* to:

Name

Mailing Address

Mailing Address

City State ZIP

Please send a free copy of *DQO-PRO* to:

Name

Mailing Address

Mailing Address

City State ZIP

Place
Stamp
Here

Instant Reference Sources, Inc.
7605 Rockpoint Drive
Austin, TX 78731

Place
Stamp
Here

Instant Reference Sources, Inc.
7605 Rockpoint Drive
Austin, TX 78731